Recent Advances in Reservoir Stimulation and EOR Technology in Unconventional Reservoirs

Recent Advances in Reservoir Stimulation and EOR Technology in Unconventional Reservoirs

Editors

Lufeng Zhang
Linhua Pan
Yushi Zou
Jie Wang
Minghui Li
Wei Feng

MDPI

Basel • Beijing • Wuhan • Barcelona • Belgrade • Novi Sad • Cluj • Manchester

Editors

Lufeng Zhang
Oil Production Department
Sinopec Petroleum
Exploration and Production
Development Research
Institute
Beijing
China

Linhua Pan
Oil Production Department
Sinopec Petroleum
Exploration and Production
Development Research
Institute
Beijing
China

Yushi Zou
Unconventional Oil and Gas
Science and Technology
Institute
China University of
Petroleum (Beijing)
Beijing
China

Jie Wang
School of Petroleum
Engineering
Yangtze University
Wuhan
China

Minghui Li
Oil & Gas and New
Energy Branch
PetroChina Corporation
Beijing
China

Wei Feng
Department of Geosciences
University of Padova
Padova
Italy

Editorial Office
MDPI
St. Alban-Anlage 66
4052 Basel, Switzerland

This is a reprint of articles from the Special Issue published online in the open access journal *Processes* (ISSN 2227-9717) (available at: https://www.mdpi.com/journal/processes/special_issues/4I558BUPX7).

For citation purposes, cite each article independently as indicated on the article page online and as indicated below:

Lastname, A.A.; Lastname, B.B. Article Title. *Journal Name* **Year**, *Volume Number*, Page Range.

ISBN 978-3-7258-0185-5 (Hbk)
ISBN 978-3-7258-0186-2 (PDF)
doi.org/10.3390/books978-3-7258-0186-2

Contents

processes

Editorial

Recent Advances in Reservoir Stimulation and Enhanced Oil Recovery Technology in Unconventional Reservoirs

Lufeng Zhang [1,*], Linhua Pan [1], Yushi Zou [2], Jie Wang [3], Minghui Li [2] and Wei Feng [4]

1 State Key Laboratory of Shale Oil and Gas Enrichment Mechanisms and Effective Development, Sinopec Petroleum Exploration and Production Development Research Institute, Beijing 100083, China; plh_cup@163.com
2 Unconventional Oil and Gas Science and Technology Institute, China University of Petroleum (Beijing), Beijing 102249, China; zouyushi@126.com (Y.Z.); lmhcupb@163.com (M.L.)
3 School of Petroleum Engineering, Yangtze University, Wuhan 430100, China; wangjie@yangtzeu.edu.cn
4 Department of Geosciences, University of Padova, 35131 Padova, Italy; wei.fengunipd@gmail.com
* Correspondence: zlfcupb@163.com

In the past decade, significant advances in reservoir stimulation and enhanced oil recovery technologies have resulted in rapid production growth in unconventional reservoirs. To further increase and stabilize the production of unconventional reservoirs, researchers continue to develop new technologies and apply them to this field. This Special Issue combines unconventional reservoir stimulation and enhanced oil recovery technologies and the latest research on geology, reservoir, drilling, and completion.

The rapid increase in the production of fossil energy has been made possible by effective reservoir stimulation and enhanced oil recovery (EOR) technologies for unconventional oil and gas reservoirs. As one of the most important reservoir stimulation technologies, hydraulic fracturing usually injects high-pressure fluid to create enough fractures in the target reservoir, which aims to improve the seepage conditions and increase the contact area between the target formation and production well. Such stimulation technologies usually involve complex fluid–solid coupling processes, including fracture initiation, propagation, conductivity, etc.

Enhanced oil recovery has been used to solve the problem of sharply declining oil rates after a production period. In this process, some special chemicals (e.g., surfactants and nano-emulsions) are injected into the reservoir to increase the recovery effectiveness of the residual oil. EOR processes often involve complex physical–chemical processes, including liquid emulsification, water–rock reactions, etc. [1–4]. Therefore, the progress of reservoir stimulation and EOR technology will contribute to the rapid development of unconventional oil and gas resources. Meanwhile, these technologies are also used to develop geothermal and coal resources [5–8].

This collection, which accompanies the Special Issue of Processes, emphasizes theory, technology, and application innovation and compiles 35 current publications on original applications of new ideas and methodologies in unconventional oil and gas reservoirs.

Citation: Zhang, L.; Pan, L.; Zou, Y.; Wang, J.; Li, M.; Feng, W. Recent Advances in Reservoir Stimulation and Enhanced Oil Recovery Technology in Unconventional Reservoirs. *Processes* **2024**, *12*, 234. https://doi.org/10.3390/pr12010234

Received: 9 January 2024
Accepted: 11 January 2024
Published: 22 January 2024

Review of Research Presented in This Special Issue

The papers published in this Special Issue describe recent advances in reservoir stimulation and EOR technology in unconventional reservoirs. These studies are divided into four categories.

The first type involves numerical and experimental simulation studies on fracture propagation patterns in unconventional reservoir stimulation techniques. For numerical simulation, Ran et al. systematically investigated the expansion patterns of multi-branch hydraulic fractures using finite element and extended finite element methods [9]. The characteristics of the target reservoir and the expansion patterns of hydraulic fractures under conditions such as formation dip angle, wellbore orientation, fracturing fluid backflow rate, low temperature, stress effects, artificial plugging, interlayer heterogeneity, and the presence of multiple branched horizontal wells were systematically investigated. For experimental studies, the influence of various factors on drag reducer performance was studied. Moreover, utilizing microscopic methods, the drag reduction mechanism of the drag reducer was also investigated. Shi et al. and their colleagues investigated the crack propagation patterns and influencing factors during hydraulic fracturing in four different reservoir types: offshore unconsolidated sandstone, high-salinity reservoirs, gravel formations, and deep shale formations [10].

The second type is acidizing and acid fracturing technology in unconventional reservoir stimulation technology. The acidizing potential for enhanced oil recovery in gravel formations was evaluated by comparing changes in rock porosity, permeability, and rock mechanics before and after acidizing. A novel high-temperature cross-linked acid's acid corrosion capability and acid-induced fracture conductivity were assessed. The temperature field of acid-rock reactions and the incremental production effects of multistage acidizing processes during acidizing were studied based on theoretical analysis and mathematical model establishment.

The third type involves unconventional oil and gas reservoir-enhanced oil recovery (EOR) technologies. These EOR technologies include the use of 3D-printed artificial cores for experimental research, the development of salt-resistant displacement polymers, studies on fluid flow behavior in shale oil and gas reservoirs, optimization of gas injection methods in low-permeability heterogeneous gas reservoirs, exploration of post-pressurization water injection development techniques, and technologies aimed at improving recovery rates in heavy oil reservoirs and salt cavern gas storage facilities.

The fourth type involves reservoir, drilling, and completion technologies related to reservoir modification and enhanced oil recovery (EOR). In terms of reservoirs, the sedimentary history of the Qiongdongnan Basin in the northern South China Sea and the Yinggehai Basin was investigated. A model for calculating the axial force on the drill string was proposed in drilling, and a simulation study on the rheological properties of lost circulation materials during drilling was performed. Regarding completion, this includes casing safety assessment and case studies on the field application of a novel gas lift valve.

Many academics from various fields, from the natural sciences to engineering, have been researching reservoir stimulation and EOR technology in unconventional reservoirs. New theories and technologies are proposed in this Special Issue, including experimental methods, numerical simulation technology, and pilot cases that can help readers and researchers better understand and be inspired by the cutting-edge technologies in reservoir stimulation and EOR technology in unconventional reservoirs.

Author Contributions: Investigation, L.Z.; writing—original draft preparation, J.W.; writing—review and editing, L.Z., J.W., W.F., M.L., Y.Z. and L.P. All authors have read and agreed to the published version of the manuscript.

Conflicts of Interest: The authors declare no conflicts of interest.

List of Contributions

1. Ran, Q.; Zhou, X.; Dong, J.; Xu, M.; Ren, D.; Li, R. Numerical Simulation of Multi-Fracture Propagation Based on the Extended Finite Element Method. *Processes* **2023**, *11*, 2032.
2. Cai, J.; Li, F. Estimation of Fracture Height in Tight Reserviors via a Finite Element Approach. *Processes* **2023**, *11*, 1566.
3. Yang, K.; Wang, L.; Ge, J.; He, J.; Sun, T.; Wang, X.; Zhao, Y. Impact of formation dip angle and wellbore azimuth on fracture propagation for shale reservoir. *Processes* **2023**, *11*, 2419.
4. Ran, Q.; Zhou, X.; Dong, J.; Xu, M.; Ren, D.; Li, R. Study on the Fracture Propagation in Multi-Horizontal Well Hydraulic Fracturing. *Processes* **2023**, *11*, 1995.
5. Li, J.; Liu, S.; Li, J.; Liu, Z.; Chen, X.; Li, J.; Liang, T. Evaluation of Fracture Volume and Complexity of Tight Oil Wells Based on Flowback Data. *Processes* **2023**, *11*, 2436.

6. Cha, M.; Alqahtani, N.B.; Wang, L. Cryogenic Fracture Proliferation from Boreholes under Stresses. *Processes* **2023**, *11*, 2028.
7. Zou, J.; Zhang, Y.; Zhang, L.; Jing, J.; Fu, Y.; Wang, Y.; Zhang, G.; Zhou, F. Numerical Simulation Research on the Effect of Artificial Barrier Properties on Fracture Height. *Processes* **2023**, *11*, 310.
8. Wan, B.; Liu, Y.; Zhang, B.; Luo, S.; Wei, L.; Li, L.; He, J. Investigation of the Vertical Propagation Pattern of the 3D Hydraulic Fracture under the Influence of Interlayer Heterogeneity. *Processes* **2022**, *10*, 2449.
9. Zhao, Y.; Wang, L.; Ma, K.; Zhang, F. Numerical Simulation of Hydraulic Fracturing and Penetration Law in Continental Shale Reservoirs. *Processes* **2022**, *10*, 2364.
10. Zheng, M.; Sheng, L.; Ren, H.; Yiming, A.; Yao, E.; Zhang, K.; Zhao, L. Development and Performance Evaluation of Scale-Inhibiting Fracturing Fluid System. *Processes* **2022**, *10*, 2135.
11. Liu, Z.; Tian, Z.; Yuan, H.; Li, Y.; Ge, H.; Zhou, F. Effect of Shear Flow on Drag Reducer Performance and Its Microscopic Working Mechanism. *Processes* **2022**, *10*, 1485.
12. Shi, S.; Zhuo, R.; Cheng, L.; Xiang, Y.; Ma, X.; Wang, T. Fracture Characteristics and Distribution in Slant Core from Conglomerate Hydraulic Fracturing Test Site (CHFTS) in Junggar Basin, Northwest China. *Processes* **2022**, *10*, 1646.
13. Liu, Y.; Xu, P.; Zhang, L.; Zou, J.; Lan, X.; Sheng, M. Lab Experiments for Abrasive Waterjet Perforation and Fracturing in Offshore Unconsolidated Sandstones. *Processes* **2023**, *11*, 3137.
14. Hu, Z.; Chen, P.; Jiang, W.; Yang, Y.; Li, Y.; Zou, L.; Wang, H.; Sun, Y.; Peng, Y. Physical Simulation Experiments of Hydraulic Fracture Initiation and Propagation under the Influence of Deep Shale Natural Fractures. *Processes* **2023**, *11*, 1934.
15. Pan, L.; Wang, L.; Zheng, W.; Han, F.; Zibibula, A.; Zhu, Z.; Li, S. Study on Salt Dissolution Law of High Salinity Reservoir and Its Influence on Fracturing. *Processes* **2023**, *11*, 304.
16. Wang, L.; Jia, W.; Xu, Y.; Mou, J.; Liao, Z.; Zhang, S. Case Study on the Effect of Acidizing on the Rock Properties of the Mahu Conglomerate Reservoir. *Processes* **2023**, *11*, 626.
17. Lin, H.; Hou, T.; Wang, F.; Yue, L.; Liu, S.; Yuan, G.; Wang, G.; Liu, Y.; Wang, Q.; Zhou, F. Experimental Study of Acid Etching and Conductivity of High-Temperature-Resistant Cross-Linked Acid. *Processes* **2023**, *11*, 722.
18. Mou, J.; He, J.; Zheng, H.; Zhang, R.; Zhang, L.; Gao, B. A New Model of Temperature Field Accounting for Acid–Rock Reaction in Acid Fracturing in Shunbei Oilfield. *Processes* **2023**, *11*, 294.
19. Zhu, D.; Wang, Y.; Cui, M.; Zhou, F.; Wang, Y.; Liang, C.; Zou, H.; Yao, F. Acid system and stimulation efficiency of multistage acid fracturing in porous carbonate reservoirs. *Processes* **2022**, *10*, 1883.
20. Cruz-Maya, J.A.; Mendoza-de la Cruz, J.L.; Martínez-Mendoza, L.C.; Sánchez-Silva, F.; Rosas-Flores, J.A.; Jan-Roblero, J. Three-Dimensional Printing of Synthetic Core Plugs as an Alternative to Natural Core Plugs: Experimental and Numerical Study. *Processes* **2023**, *11*, 2530.
21. Li, X.; Xu, A.; Ma, M.; Liu, S.; Ni, J.; Zhao, L. Preparation of Polymer Solution for Profile Control and Displacement Using Wastewater with High Ca2+/Mg2+ and Fe2+ Concentrations. *Processes* **2023**, *11*, 325.
22. Xu, F.; Jiang, H.; Liu, M.; Jiang, S.; Wang, Y.; Li, J. NMR-Based Analysis of Fluid Occurrence Space and Imbibition Oil Recovery in Gulong Shale. *Processes* **2023**, *11*, 1678.
23. Lei, Y.; Wu, Z.; Wang, W.; Wu, J.; Ma, B. Study on the Flow Pattern and Transition Criterion of Gas-Liquid Two-Phase Flow in the Annular of Shale Gas Fractured Horizontal Wells. *Processes* **2022**, *10*, 2630.
24. Luo, J.; Wang, L. Research on Gas Channeling Identification Method for Gas Injection Development in High-Pressure Heterogeneous Reservoir. *Processes* **2022**, *10*, 2366.

25. Dong, L.; Li, L.; Dong, W.; Wang, M.; Chen, X. Investigation on the Injection Pattern of Intermittent Natural Gas Flooding in Ultra-Low Permeability Reservoirs. *Processes* **2022**, *10*, 2198.
26. Huang, T.; Peng, K.; Song, W.; Hu, C.; Guo, X. Change Characteristics of Heavy Oil Composition and Rock Properties after Steam Flooding in Heavy Oil Reservoirs. *Processes* **2023**, *11*, 315.
27. Liao, K.; Zhu, J.; Sun, X.; Zhang, S.; Ren, G. Numerical Investigation on Injected-Fluid Recovery and Production Performance following Hydraulic Fracturing in Shale Oil Wells. *Processes* **2022**, *10*, 1749.
28. Ren, G.; Ma, X.; Zhang, S.; Zou, Y.; Duan, G.; Xiong, Q. Optimization of Water Injection Strategy before Re-Stimulation Considering Fractures Propagation. *Processes* **2022**, *10*, 1538.
29. Wang, L.; Song, Z.; Huang, X.; Xu, W.; Chen, Z. Study on the Influence of Pressure Reduction and Chemical Injection on Hydrate Decomposition. *Processes* **2022**, *10*, 2543.
30. Wu, B.; Zhang, M.; Deng, W.; Que, J.; Liu, W.; Zhou, F.; Wang, Q.; Liang, T. Study and Mechanism Analysis on Dynamic Shrinkage of Bottom Sediments in Salt Cavern Gas Storage. *Processes* **2022**, *10*, 1511.
31. Ma, M.; Qi, J.; Ma, J.; Peng, H.; Lei, L.; Song, Q.; Zhang, Q.; Bai, M. Cenozoic Subsidence History of the Northern South China Sea: Examples from the Qiongdongnan and Yinggehai Basins. *Processes* **2023**, *11*, 956.
32. Nie, Z.; Shi, S.; Wu, B.; Huang, X. Axial Force Calculation Model for Completion String with Multiple Point Resistances in Horizontal Well. *Processes* **2023**, *11*, 2621.
33. Pu, L.; Xu, P.; Xu, M.; Zhou, J.; Liu, Q.; Song, J. Numerical Simulation Study on the Flow Properties of Materials for Plugging While Drilling in MWD. *Processes* **2022**, *10*, 1955.
34. Wang, H.; Li, M.; Zhao, Q.; Hao, W.; Zhang, H.; Li, Y.; Huang, P.; Zou, Y. Study on Casing Safety Evaluation in High-Temperature Wells with Annular Pressure Buildup. *Processes* **2023**, *11*, 1915.
35. Liu, Q.; Tang, J.; Ke, W.; Wang, H.; Orivri, U.D. Case Study: Successful Application of a Novel Gas Lift Valve in Low Pressure Wells in Fuling Shale Gas Field. *Processes* **2022**, *11*, 19.

References

1. Zhang, L.; Zhou, F.; Zhang, S.; Li, Z.; Wang, J.; Wang, Y. Evaluation of permeability damage caused by drilling and fracturing fluids in tight low permeability sandstone reservoirs. *J. Pet. Sci. Eng.* **2019**, *175*, 1122–1135.
2. Su, H.; Zhou, F.; Wang, Q.; Yu, F.; Dong, R.; Xiong, C.; Li, J.; Liang, T. Flow Physics of Polymer Nanospheres and Diluted Microemulsion in Fractured Carbonate Reservoirs: An Investigation into Enhanced Oil Recovery Mechanisms. *SPE J.* **2021**, *26*, 2231–2244. [CrossRef]
3. Wang, B.; Zhou, F.; Zou, Y.; Liang, T.; Wang, D.; Hu, J.; Gao, L. Effects of previously created fracture on the initiation and growth of subsequent fracture during TPMSF. *Eng. Fract. Mech.* **2018**, *200*, 312–326. [CrossRef]
4. Wang, B.; Zhou, F.; Yang, C.; Xu, C.; Liu, J.; Han, S.; Wang, D.; Ren, Z.; Liang, T. A Novel Experimental Method to Investigate the Plugging Characteristics of Diversion Agents within Hydro-fracture. *J. Pet. Sci. Eng.* **2019**, *183*, 106354. [CrossRef]
5. Liang, T.; Xu, K.; Lu, J.; Nguyen, Q.; DiCarlo, D. Evaluating the Performance of Surfactants in Enhancing Flowback and Permeability after Hydraulic Fracturing through a Microfluidic Model. *SPE J.* **2020**, *25*, 268–287. [CrossRef]
6. Liang, T.; Zhao, X.; Yuan, S.; Zhu, J.; Liang, X.; Li, X.; Zhou, F. Surfactant-EOR in tight oil reservoirs: Current status and a systematic surfactant screening method with field experiments. *J. Pet. Sci. Eng.* **2021**, *196*, 108097. [CrossRef]
7. Wang, D.; Zhou, F.; Ding, W.; Ge, H.; Jia, X.; Shi, Y.; Wang, X.; Yan, X. A numerical simulation study of fracture reorientation with a degradable fiber-diverting agent. *J. Nat. Gas Sci. Eng.* **2015**, *25*, 215–225. [CrossRef]
8. Bai, H.; Zhou, F.; Zhang, M.; Gao, X.; Xu, H.; Yao, E.; Wang, J.; Li, Y. Optimization and friction reduction study of a new type of viscoelastic slickwater system. *J. Mol. Liq.* **2021**, *344*, 117876. [CrossRef]
9. Ran, Q.; Zhou, X.; Dong, J.; Xu, M.; Ren, D.; Li, R. Numerical Simulation of Multi-Fracture Propagation Based on the Extended Finite Element Method. *Processes* **2023**, *11*, 2032. [CrossRef]
10. Shi, S.; Zhuo, R.; Cheng, L.; Xiang, Y.; Ma, X.; Wang, T. Fracture Characteristics and Distribution in Slant Core from Conglomerate Hydraulic Fracturing Test Site (CHFTS) in Junggar Basin, Northwest China. *Processes* **2022**, *10*, 1646. [CrossRef]

processes

MDPI

Article

Lab Experiments for Abrasive Waterjet Perforation and Fracturing in Offshore Unconsolidated Sandstones

Yigang Liu [1], Peng Xu [2], Liping Zhang [1], Jian Zou [1], Xitang Lan [1] and Mao Sheng [2,*]

1 Bohai Petroleum Research Institute, Tianjin Branch, CNOOC (China) Co., Ltd., Tianjin 300452, China; liuyg@cnooc.com.cn (Y.L.); zhanglp17@cnooc.com.cn (L.Z.); zoujian@cnooc.com.cn (J.Z.); lanxt@cnooc.com.cn (X.L.)
2 National Key Laboratory of Petroleum Resources and Engineering, China University of Petroleum (Beijing), Beijing 102249, China; 13126690311@163.com
* Correspondence: shengmao@cup.edu.cn

Abstract: Multistage hydraulic fracturing has been proven to be an effective stimulation method to extract more oil from the depleted unconsolidated sandstone reservoirs in Bohai Bay, China. The offshore wellbores in this area were completed with a gravel pack screen that is much too difficult to be mechanically isolated in several stages. Hydra-jet fracturing technology has the advantages of multistage fracturing by one trip, waterjet perforation, and hydraulic isolation. The challenges of hydraulic-jet fracturing in offshore unconsolidated sandstone reservoir can be summarized as follows: the long jet distance, high filtration loss, and large pumping rate. This paper proposes full-scale experiments on the waterjet perforation of unconsolidated sandstone, waterjet penetration of screen liners and casing, and pumping pressure prediction. The results verified that multistage hydra-jet fracturing is a robust technology that can create multiple fractures in offshore unconsolidated sandstone. Lab experiments indicate that the abrasive water jet is capable to perforate the screen-casing in less than one minute with an over 10 mm diameter hole. The water jet perforates a deep and slim hole in unconsolidated sandstone by using less than 20 MPa pumping pressure. Recommended perforating parameters: maintain 7% sand concentration and perforate for 3.0 min. Reduce sand ratio to 5%, maintain 3.0 m^3/min flow rate, and continue perforating for 7.0 min. The injection drop of the nozzle accounts for more than 62% of the tubing pump pressure. The recommended nozzle combinations for different fracturing flow rates are 8 × ø6 mm or 6 × ø7 mm for 2.5 m^3/min and 3.0 m^3/min, and 8 × ø7 mm for 3.5 m^3/min and 4.0 m^3/min. A one-trip-multistage hydra-jet fracturing process is recommended to be used for horizontal wells in offshore unconsolidated sandstone reservoirs.

Keywords: offshore; unconsolidated sandstone; hydra-jet fracturing; perforation experiment

Citation: Liu, Y.; Xu, P.; Zhang, L.; Zou, J.; Lan, X.; Sheng, M. Lab Experiments for Abrasive Waterjet Perforation and Fracturing in Offshore Unconsolidated Sandstones. *Processes* **2023**, *11*, 3137. https://doi.org/10.3390/pr11113137

Academic Editors: Carlos Sierra Fernández, Jie Wang, Lufeng Zhang, Linhua Pan, Minghui Li, Wei Feng and Yushi Zou

Received: 22 September 2023
Revised: 18 October 2023
Accepted: 30 October 2023
Published: 2 November 2023

1. Introduction

Hydraulic fracturing of unconsolidated sandstones has become an important technique used to enhance oil recovery for the offshore reservoirs [1]. The first hydraulic fracturing of offshore unconsolidated sandstone appeared in the Gulf of Mexico [2], following successful cases in Brazil, Nigeria, and Bohai Bay. However, gravel pack screen completion is popular in unconsolidated sandstone oil wells and it is much too difficult to deploy multistage hydraulic fracturing using mechanical isolation [3]. Flexible multistage hydraulic fracturing technology for use in unconsolidated sandstones is required.

Abrasive waterjet (AWJ) fracturing stimulation, also called hydra-jet fracturing, has been accepted as an effective and efficient stimulation technique for multistage well completion used with casing, slotted liners, and even open hole [4]. Major technical advantages include the integration of AWJ perforation and fracturing, hydraulic isolation capacity, pinpoint fracture initiation, unlimited stages, and high efficiency [5]. It has become a flexible technology used to achieve multistage fracturing in offshore reservoir stimulation [6].

Hydra-jet fracturing was first used in horizontal wells with uncemented and pre perforated liners, off the shore of Brazil in 2004, which proved its effectiveness in offshore multistage stimulation [7]. Following the first implementation of hydra-jet fracture, acidizing was used with great success in deep water fields off the shore Brazil in 2005 [8]. Hydra-jet propped fracturing was tested in mature offshore oil fields in Congo with low to moderate permeability of sandstone in 2008 and 2010 [9]. The first completed hydra-jet fracturing multizone application was performed in an offshore high-permeability oil well located in the Bozhong oil field in Bohai Bay, China, and has proven successful since 2020 [10]. Although several successful cases are present in offshore reservoir stimulation [11], engineering challenges still exist to be solved, as illustrated in Figure 1. First, abrasive waterjet perforation becomes more challenging since several penetration layers are present, including the slotted linear layer, the gravel pack layer, the casing, and the cement shield [12]. The waterjet standoff distance becomes larger in contrast to the onshore cemented casing [13], so the waterjet energy reduces too much to penetrate a large and deep perforating hole in unconsolidated sandstone. It is significant to evaluate the capacity of waterjet perforation for offshore unconsolidated sandstone [14,15]. The jet rate, sand ratio, and injection time need to be verified. Second, unconsolidated sandstone is a type of medium with a low strength, less than 10 MPa, and high permeability, over 200 mD [16–18]. The perforation morphology of unconsolidated sandstone is different from that of conventional rock, which will affect the jet gun design and injection parameter design.

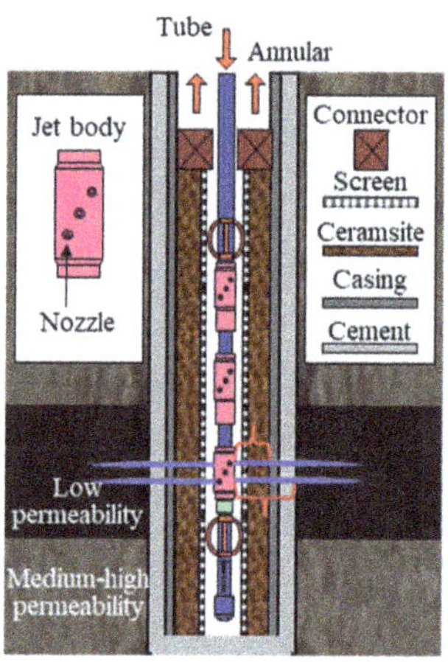

Figure 1. Illustration of abrasive waterjet perforation and multistage fracturing in unconsolidated sandstones.

This study proposes lab experiment schemes to validate abrasive waterjet perforation and multistage fracturing in unconsolidated sandstone, including full-scale abrasive waterjet penetration through the screen liner and waterjet rock perforation. In Section 2, the recommended jet rate is obtained via lab experiments. In Section 3, the nozzle combination is optimized using the recommended jet rate. A bottom-hole tool string and procedure design for offshore Hydra-jet fracturing are recommended. In Section 4, the feasibility of hydra-jet fracturing in an offshore unconsolidated sandstone reservoir is verified using a well in Bohai Bay.

2. Abrasive Waterjet Perforation in Gravel Pack Completion

Abrasive waterjet perforation in an offshore unconsolidated sandstone reservoir is shown in Figure 2. The perforating fluid enters the tubing and is accelerated through the nozzle [19–21]. Several layers, including the screen liner, the gravel pack, casing, cement, and formation rock, are penetrated by waterjet. In order to avoid serious damage to the unconsolidated sandstone formation, we plan to first penetrate the screen liner, gravel, casing, and cement, and then perforate the unconsolidated sandstone. A full-scale experiment was proposed to obtain the relationship between the abrasive waterjet rate and the perforating time. The feasibility of the waterjet perforating the unconsolidated sandstone was verified by comparing the waterjet's impact on unconsolidated sandstone and red sandstone. The characteristics of the perforating shape in unconsolidated sandstone were obtained.

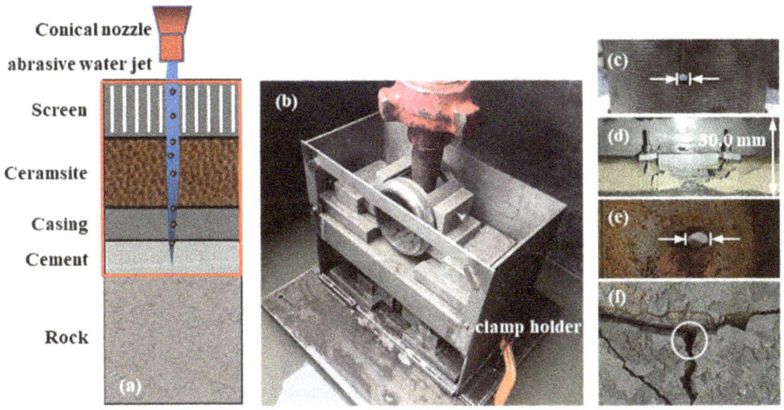

Figure 2. Abrasive waterjet impact on screener-casing. (**a**) Schematic diagram of the experimental method. (**b**) Experimental facility. (**c**) Screener after experiment. (**d**) Gravel pack after experiment. (**e**) Casing after experiment. (**f**) Cement after experiment.

2.1. Capacity of Abrasive Waterjet Penetrating Screen Liner and Casing

The full-scale lab simulator was developed to simulate the physical behaviors of abrasive waterjet penetration through several layers including the screen liner, the gravel pack, the casing, and the cement shield, as illustrated in Figure 2. The specific parameters of the experiment are designed according to typical well parameters, and this information is shown in Table 1. An STP 600 plunger pump (Sinopec Oilfield Equipment Corporation, China) with a maximum flow rate of 1000 L/min was used to generate the abrasive waterjet. The distance from the nozzle outlet to the screen was set to 5 mm. The experiment was carried out under submerged conditions. Concentric holes appeared in the screen liner and casing (Figure 2c,e). The gravel pack and cement were easily penetrated by the abrasive waterjet (Figure 2d,f). It took only a moment for the abrasive waterjet to destroy the gravel pack and cement. In addition, the gravel layer and cement had little effect on screen and casing damage. Thus, we only studied the perforation of the screen liner and casing.

We obtained the relationship between the abrasive waterjet rate and jet time by looking at the penetration time and decreasing the injection velocity step by step (Figure 3). When the nozzle pressure approaches the limit 35 MPa, the jet velocity is 220 m/s, which is taken as the upper limit. We observed the screen and casing penetration every minute until the casing was penetrated. For example, when the jet rate was 220 m/s, after one minute the screen was penetrated and the casing was slightly abraded. After another minute, the casing was penetrated. Therefore, the critical jet time at 220 m/s is 2 min. As the pumping rate decreases, the ability of the abrasive waterjet to penetrate the screen-casing decreases. The critical jet times for 190 m/s, 160 m/s, and 150 m/s are 4 min, 6 min, and 9 min, respectively. When the jet velocity is 150 m/s, the jet time approaches the limit of 10 min, thus 150 m/s is taken as the lower limit. The penetration diameter is comprehensively affected by jet distance, jet rate, and jet time. Under the experimental conditions used in this study, the penetration diameter of the screen is 1.36 to 1.96 times the diameter of the nozzle. The penetration diameter of the casing is 1.40 to 2.56 times the diameter of the nozzle.

Table 1. Comparison of field and experimental materials.

Materials	Field Parameters	Experimental Parameters
Nozzle	conical, outlet diameter 5–7 mm	conical, outlet diameter 5 mm
Fluid	fracturing fluid	water
Abrasive	quartz sand, garnet, ceramisite, 20/40 mesh	ceramisite, 20/40 mesh, volume density of 1620 kg/m^3, apparent density of 2950 kg/m^3, compressive strength of 69 MPa
Sand concentration	6–8% Volume ratio	5% Volume ratio
Screener	139.7 mm wire-wound screener, base pipe of 25.3 kg/m and N80 rank	139.7 mm wire-wound screener nipple, base pipe of 25.3 kg/m and N80 rank, side window for nozzle
Gravel pack	ceramist, 20/40 mesh, thickness 37 mm	ceramist, 20/40 mesh, thickness 30 mm
Casing	244.5 mm casing, 86.9 kg/m and N80 rank	244.5 mm casing, 86.9 kg/m and N80 rank
Cement	Portland cement, Water-cement ratio 0.44, thickness 33 mm	Portland cement, Water-cement ratio 0.44, thickness 30 mm

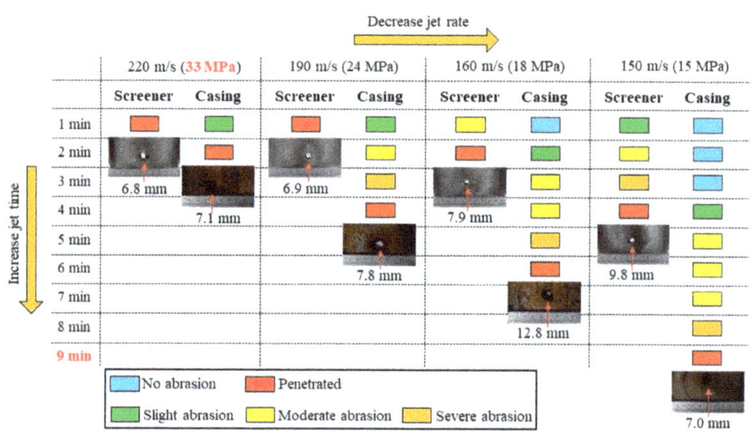

Figure 3. The process of abrasive water jet penetrating screen-casing at different jet rates.

2.2. Capacity of Waterjet Perforation in Unconsolidated Sandstone

The unconsolidated sandstone is our target material and these samples were collected from a drilling core from the Bohai Bay formation at a depth of 1600 m. In order to make the blank group, the red sandstone was selected from natural outcrops in Sichuan. Table 2 compares their physical and mechanical properties.

Table 2. Comparison of physical parameters between unconsolidated sandstone and red sandstone.

Physical Parameters	Unconsolidated Sandstone	Red Sandstone
Density, kg/m^3	1990	2230
Porosity, %	21.9	17.8
Permeability, mD	56.3	32.6
Elasticity modulus, GPa	0.29	8.71
Poisson ratio	0.34	0.38
uniaxial compressive strength, MPa	2.7	39.5

Considering the coring size of ø25 × 50 mm and the large jet distance, a nozzle with 1 mm outlet diameter is used in this experiment (Figure 4a). The injection time is 10 s.

For these two types of rocks, jet rock breaking was carried out six times, including five experimental conditions (Figure 4b,c). The unconsolidated sandstone is perforated under waterjet as result #1 (Figure 4b). For results #2 to #6, the red sandstone is jetted, and the jet conditions are changed until the perforating depth of the red sandstone is close to that of the unconsolidated sandstone. The feasibility of the waterjet perforating the unconsolidated sandstone is verified via comparison with the red sandstone, and the perforating characteristics of unconsolidated sandstone are obtained via CT scan.

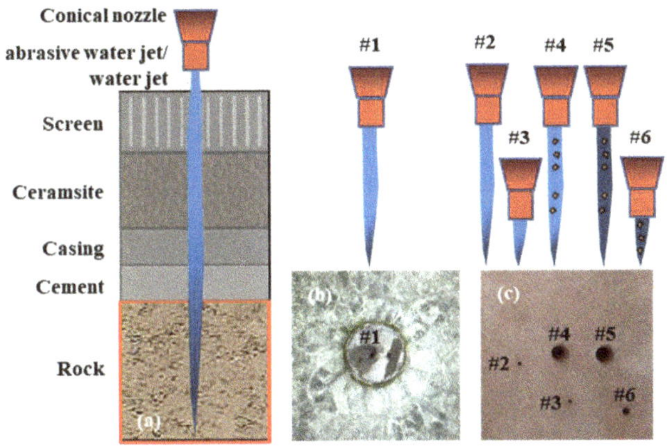

Figure 4. Waterjet perforation experiments. (**a**) Schematic diagram of the experimental method. (**b**) Unconsolidated sandstone after experiment. (**c**) Red sandstone after experiment.

In Figure 5a, result #1 and result #2 show that the rock breaking depth of unconsolidated sandstone is 7.7 times that of red sandstone under the same jet conditions. By changing the waterjet (WJ) into an abrasive waterjet (AWJ), increasing the jet rate (JR) to 270 m/s, and reducing the jet distance (JD) to 5 mm, the rock-breaking depth of #6 of the red sandstone is approximated to the rock-breaking depth of #1 of the unconsolidated sandstone. In addition, the jet parameters of #6 are similar to those of onshore construction, which proves that the waterjet has the ability to perforate unconsolidated sandstone under a large jet distance. Figure 5b,c shows that the rock breaking diameter of unconsolidated sandstone is generally larger than that of red sandstone. Through the rock breaking diameter of the red sandstone under different jet conditions, it is found that the higher the rock-breaking efficiency and the larger the jet distance, the larger the rock-breaking diameter. Unconsolidated sandstone is easy to break, and the offshore perforation is mostly at a large jet distance. Therefore, tools and processes need to be optimized to avoid large perforation diameters in offshore unconsolidated sandstone reservoirs.

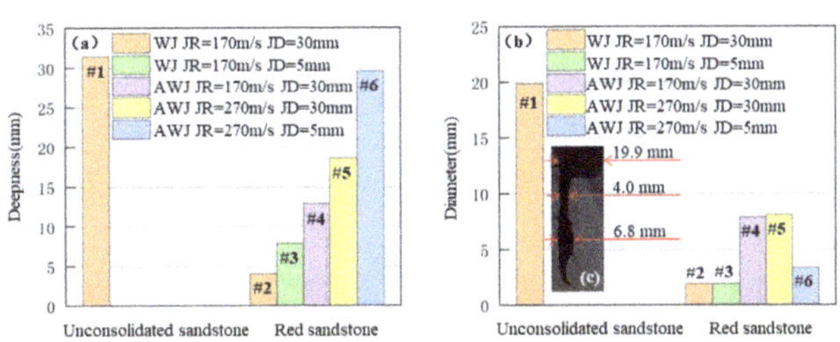

Figure 5. Comparison of jet impact on unconsolidated sandstone and red sandstone. (**a**) Rock breaking depth. (**b**) Rock breaking diameter. (**c**) CT scanning of #1 unconsolidated sandstone perforated by water jet.

3. Optimization of Bottom-Hole Tool and Procedures of Offshore Hydra-Jet Fracturing

3.1. Optimization of Key Parameters of Waterjet Nozzles

The waterjet nozzles are critical parts in the transfer of high-pressure energy to kinetic energy with a high velocity impact. The nozzle diameter and its number are two key parameters for the hydra-jet fracturing tool. Two aspects of nozzle design should be taken into account. The first point is to reach the minimum waterjet velocity to reserve enough energy for the perforation. The second aspect is to satisfy the requirement of pumping rate. The formula for nozzle pressure drop is as follows:

$$P_b = \frac{513.559V^2\rho}{C^2}$$

where P_b is nozzle pressure drop, MPa; $V = \frac{Q}{A}$ is jet rate, m/s; Q is flow rate, L/s; $A = 0.25\pi D^2$ is outlet area of all nozzles, mm^2; D is nozzle diameter, mm; ρ is fluid density, g/cm^3; and C is discharge coefficient of nozzle, generally 0.9.

Figure 6 illustrates the workflow to optimize the nozzles parameters. The waterjet perforation experiments indicate that the critical waterjet velocity required is up to 190 m/s to make a deep and large perforating hole. Figure 6b indicates the correlation between pumping rate and waterjet velocity. If the required pumping rate is above 3.0 m^3/min, the corresponding nozzle diameter and number can be optimized as 8 × ø6 mm or 6 × ø7 mm. If the pumping rate is over 3.5 m^3/min, the corresponding nozzle diameter and number can be 8 × ø7 mm. Except for the nozzle diameter and numbers, the nozzle distance between the two layers, as shown in Figure 6c, is another key parameter to be considered. According to the perforation experiment on unconsolidated sandstone, the hole diameter is 20 times that of of the nozzle diameter. Therefore, the recommended nozzle distance between the two layers is 200 mm to avoid the connection of multiple perforating holes.

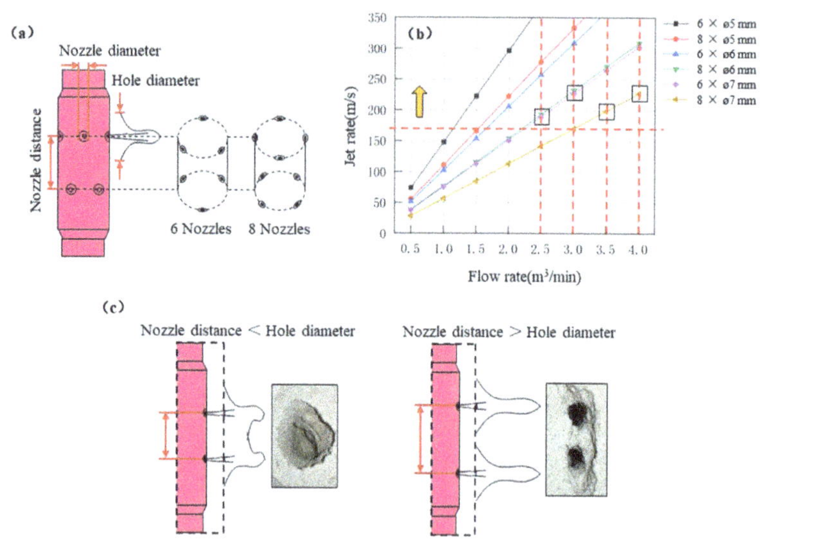

Figure 6. Illustration of bottom-hole tool and the optimization of waterjet nozzles. (**a**) Key parameters of waterjet nozzles. (**b**) Correlations between pumping rate and waterjet velocity (**c**) Nozzle distance between two layers and its effects on perforating holes.

3.2. Design of Bottom-Hole Tool String for Offshore Hydra-Jet Fracturing

The key point of bottom-hole tool design is to avoid the sand sticking issue while trialing the hydra-jet tool. Thus, we selected an elastic, deformable centralizer and a spherical guide shoe. The trailing-tool was recommended for horizontal multistage hydraulic-jet fracturing in an offshore unconsolidated sandstone reservoir. Figure 7 illustrates the tool string, including guide shoe, multi-hole pipe, one-way valve, Hydra-jet body with nozzles, and the elastic, deformable centralizers. The spherical guide shoe is used to ensure the tool is capable of passing the inner steps of wellbore. The multi-hole pipe and one-way valve allows pre-washing job and reverse circulation washing. The elastic deformable centralizer makes the hydra-jet body centralized and reduces the risk of sand sticking issue.

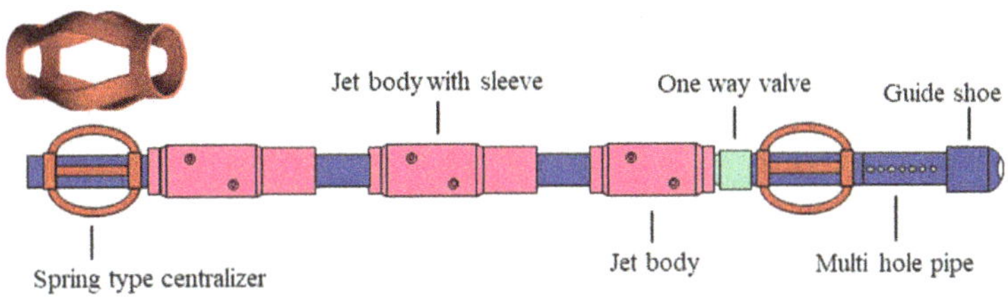

Figure 7. Bottom-hole tool string for offshore hydra-jet fracturing.

3.3. Hydra-Jet Fracturing Process

There are two typical types of hydra-jet fracturing, involving the trailing frac-string process and non-tripping frac-string process [22–24]. In order to reduce the risk of the sand sticking issue, the trailing frac-string process was recommended for offshore sand packing well completion. The detailed steps include the hydra-jet tool trip-in, waterjet perforation, and hydraulic fracturing.

(1) Hydra-jet tool trip-in: The hydra-jet tool is trip-in to the target depth. A wellbore cleanout was required using fluid circulation from tubing and hydra-jet tool to casing annular. Then, the one-way ball was pumped through the tubing, with a low pumping rate less than 1.0 m^3/min. Once the tubing pressure increases sharply, it indicates the one-way ball is effectively seated on the valve.

(2) Waterjet perforation: The abrasive particles were mixed with perforation fluid. The recommended parameters can be listed as: waterjet velocity of 190 m/s, 20–40 mesh ceramist, 6–8% volume ratio of sand concentration, perforation time of 10–15 min.

(3) Hydraulic fracturing: Reduce tubing flow and slowly close the plug valve of the annular choke line. Increase the flow rate of the tubing to the designed fracturing rate and continue jetting. Then use the annular pumps gel or water, which can keep enough net pressure to propagate fractures and complement fluid leakage in fractures. All of the gel and chemical additive is injected through the tubing to avoid eroding the casing. Finally, the overflow rate is calculated.

4. Case Study

A hydraulic fracturing design for one candidate well has been carried out and the case study has been analyzed to indicate the workflow.

4.1. Reservoir Characteristics

Well SZ36-X is located in the southern Bohai Sea. Many fault blocks and fault anticlinal traps are formed due to the complex fault system. In the field, the sedimentary microfacies types mainly include an underwater distributary channel, an estuarine bar, and a remote sand bar deposit. The main reservoirs are relatively concentrated vertically. The thickness of the single sand layer is generally not more than 10.0 m. Reservoir interlayers are relatively developed. The reservoir is shallow buried. Compaction and diagenesis are weak. The reservoir is relatively unconsolidated. The reservoir space is dominated by primary intergranular pores. The average porosity of the reservoir is 30.5%. Reservoir permeability is more than 50 mD. The reservoir in this area has the characteristics of thin thickness, poor physical properties, and strong heterogeneity. The crude oil in this field is a heavy oil with a high density, high viscosity, high content of colloidal asphalt, low sulfur content, low wax content, and low freezing point. The viscosity of the surface crude oil is between 23.4 mPa·s and 11,355.0 mPa·s. The viscosity of the underground crude oil is between 24.1 mPa·s and 452.0 mPa·s. The saturation pressure is between 5.0 MPa and 13.7 MPa. The pressure coefficient is about 1.03. The original formation pressure is 14.3 MPa (corresponding to the altitude −1450.0 m), and the temperature gradient is 3.22 °C/100 m, which belongs to the normal temperature system.

4.2. Pump Pressure Checking

Checking the pump pressure of the tubing and casing is the key to verifying the feasibility of jet fracturing [25,26]. Tubing pump pressure is used to generate jet fracturing power, counter flow friction, and balance casing pump pressure. Jet fracturing power accelerates the jets through nozzles to aid in hydraulic perforation and hydraulic isolation. The flow friction includes the tubing part and the annulus part. Casing pump pressure is used to replenish formation energy. Tubing and annulus are pressure-connected, so part of the tubing pump pressure needs to balance the casing pump pressure. During the perforating stage, the tubing enters the fluid, the annulus returns the fluid, and the casing pressure is 0 MPa [27,28]. During the fracturing stage, the fluid is replenished in the annulus, and the casing pressure is related to the fracture generation and the hydraulic isolation. Table 3 lists the parameters of the case.

Table 3. The calculation parameters of cases.

Classification	Name	Parameter
Geology	Vertical depth, m	2000
	Fracture initiation pressure gradient, MPa/m	0.0185
	Fracture extension gradient, MPa/m	0.0150
Well	Oblique depth, m	2500
	Inside diameter of casing, mm	224.4
Tool	Nozzle combination	6 × ø7 mm/8 × ø7 mm
	Inside diameter of tubing, mm	76
	Outside diameter of tubing, mm	88.9
Process	Pumping rate, m^3/min	2.5, 3.0 3.5, 4.0
	Fracturing fluid density, kg/m^3	1050
	Fracturing fluid viscosity, mPa·s	1.12
	Flow coefficient	0.22

The calculation formula of fluid friction loss in tubing and annulus:

$$\mathrm{Re}=\begin{cases}\dfrac{\rho d^{n}v^{(2-n)}}{8^{(n-1)}\mu\left(\frac{3n+1}{4n}\right)^{n}}, & \text{tube}\\[2ex] \dfrac{\rho(D_1-D_2)^{n}v^{(2-n)}}{12^{(n-1)}\mu\left(\frac{2n+1}{3n}\right)^{n}}, & \text{annular}\end{cases}$$

$$f=\begin{cases}\dfrac{16}{\mathrm{Re}}, & \mathrm{Re}\le 2100\\[2ex] \dfrac{\lg(n)+3.93}{50\mathrm{Re}^{\frac{1.75-\lg(n)}{7}}}, & \mathrm{Re}>2100\end{cases}$$

$$P_f=\begin{cases}\dfrac{2f_1\rho L v^2}{d}\cdot 10^{-6}, & \text{tubing}\\[2ex] \dfrac{2f_2\rho L v^2}{D_1-D_2}\cdot 10^{-6}, & \text{annulus}\end{cases}$$

where Re is the Reynolds number; d is the inside diameter of the tubing, m; D_1 is the inside diameter of the casing, m; D_2 is outside diameter of the tubing, m; n is the flow coefficient; v is the average flow rate in tubing or annulus, m/s; μ is the viscosity, mPa·s; f is the fluid friction coefficient; and P_f is fluid friction loss, MPa.

The calculation formula of pumping pressure in tubing and annulus:

$$P_{tubing} = P_b + P_{ftubing} + 0.4P_{annulus}$$

$$P_{annulus} = \begin{cases} 0 & \text{, perforation} \\ P_{frac\_i} - P_h - P_{boost} & \text{, fracture initiation} \\ P_{frac\_e} - P_h & \text{, fracture extension} \end{cases}$$

where P_{tubing} is the tubing pump pressure, MPa; $P_{annulus}$ is the casing pump pressure, MPa; P_b is the injection drop, MPa; $P_{ftubing}$ is the flow friction of the tubing, MPa; P_h is the head of liquid, MPa; P_{frac_i} is the fracture initiation pressure, MPa; P_{frac_e} is the fracture extension pressure; and P_{boost} is the injection boost, 8.0 MPa.

Figure 8a shows a comparison of the three tubing pump pressure components affected by pumping rate. When the pumping rate is 4.0 m^3/min, the flow friction of the annular is 0.4 MPa. Compared with the onshore 5–1/2 inch casing, the offshore 9–5/8 inch casing has a much larger annular flow area, so the flow friction of the annular is negligible. At these four flow rates, the injection drop is at least 2.2 times the flow friction of the tubing. Injection drop accounts for more than 62% of the tubing pump pressure, and optimizing the nozzle combination can significantly reduce the tubing pump pressure. Figure 8b shows that all cases of hydra-jet fracturing satisfy the tubing pressure limit, which is below 56 MPa.

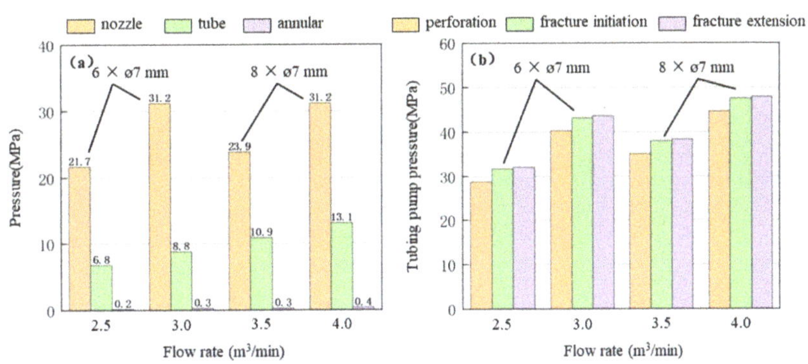

Figure 8. Pump pressure distribution and change throughout the jet fracturing process under different fracturing flow rates. (**a**) Comparison of pump pressure components affected by flow rate. (**b**) Comparison of pump pressure at different stages of jet fracturing.

4.3. Operation and Requirements

Close the BOP and the four-way annular injection wing valve. Pump the low flow rate. Set the tubing flow rate to 0.5–1.0 m^3/min. Fill the tubing with base fluid. Drop the valve ball; low feed the ball to block the check valve. After the base fluid is injected into the tubing at 11.0 m^3, the design flow rate is increased to 3.0 m^3/min. If the tubing pressure reaches 40.0–42.0 MPa, it indicates that the valve ball is in place and the following steps are carried out. If this pressure is not reached, continue to lower it to 3.0 m^3, increase the design flow rate to 3.0 m^3/min, and again judge whether the valve ball is in place. Increase tubing flow rate to 3.0 m^3/min. Begin sand mixing with 20/40 mesh ceramic particles/sand ratio of 7%. Ensure the flow rate and sand ratio are stable. Maintain the

7% sand ratio and perforate for 3.0 min. Reduce sand ratio to 5%, maintain 3.0 m^3/min flow rate, and continue perforating for 7.0 min. Stop adding sand, maintain the flow rate of the tubing, and pump gel to replace the ceramic in the tubing. Reduce the tubing flow rate to 1.0 m^3/min, slowly close the annular return valve, and open the annular injection valve. Increase tubing flow rate to 3.0 m^3/min. Start the annulus injection at a pressure not greater than the maximum design annulus pressure. In the first stage, the calculated value is 7 MPa. In the second and third stages, the maximum annulus pressure shall be determined according to the pump stop pressure in the first stage. Set annulus flow rate to 0.5~1.0 m^3/min. Proppant is then pumped in. When the first stage of the fracturing pump injection is completed, stop pumping. When the pressure is reduced to 0.0 MPa, the well is washed. Rotate the string, and drag the string to the next injection point when there is no abnormality. If the string becomes stuck in the sand, reverse circulation should be used.

During the construction process, the construction flow rate and sand ratio should be adjusted according to the construction pressure. The fluid volume is calculated according to the actual running fracturing string. The fluid volume should not exceed 1.5 times the calculated column volume. Annulus pressure should be monitored throughout the fracturing stage. Within the allowable range of casing pressure, the annular flow rate can be appropriately increased. After fracturing is complete, wellbore losses and spills should be observed before the string is drawn up. If the jet gun fails during the fracturing process, the pump should be stopped, and the ball should be thrown to open the slide sleeve of the standby gun for hydraulic jet fracturing.

5. Conclusions

This paper proposed full-scale experiments for the waterjet perforation of unconsolidated sandstone, the waterjet penetration of screen liners and casing, and pumping pressure prediction. The results verified that multistage hydra-jet fracturing is a robust technology that can be to create multiple fractures in offshore unconsolidated sandstone. The study can be concluded as follows:

(1) The abrasive water jet is capable of perforating the screen-casing in less than one minute with an over 10 mm diameter hole. The water jet perforates a deep and slim hole in unconsolidated sandstone by using less than 20 MPa pumping pressure. Recommended perforating parameters include: maintain 7% sand ratio and perforate for 3.0 min, reduce sand ratio to 5%, maintain 3.0 m^3/min flow rate, and continue perforating for 7.0 min.

(2) Nozzle pressure drop accounts for more than 62% of the tubing pump pressure. Optimizing the nozzle combination can significantly reduce the pump pressure. The recommended nozzle combinations for different fracturing flow rates are 8 × 6 mm or 6 × 7 mm for 2.5 m^3/min and 3.0 m^3/min, and 8 × 7 mm for 3.5 m^3/min and 4.0 m^3/min.

(3) To avoid the sand sticking issue, a one-trip-multistage jet fracturing process is recommended for use in horizontal wells in offshore unconsolidated sandstone reservoirs.

Author Contributions: Writing—original draft, P.X.; Writing—review and editing, M.S.; Project administration, Y.L., L.Z., J.Z. and X.L. All authors have read and agreed to the published version of the manuscript.

Funding: This work was financially supported by the CNOOC Ltd. Scientific Research Project (No. YXKY-2021-TJ-02).

Data Availability Statement: Data are contained within the article.

Conflicts of Interest: The authors declare no conflict of interest.

References

1. Yan, C.; Chen, Y.; Chen, T.; Cheng, Y.; Yan, X. Experimental Study of Hydraulic Fracturing for Unconsolidated Reservoirs. *Rock Mech. Rock Eng.* **2022**, *55*, 3399–3424. [CrossRef]
2. Monus, F.L.; Broussard, F.W.; Ayoub, J.A.; Norman, W.D. Fracturing unconsolidated sand formations offshore Gulf of Mexico. In Proceedings of the SPE Annual Technical Conference and Exhibition, Washington, DC, USA, 4–7 October 1992.
3. Molenaar, M.M.; Cox, B.E. Field cases of hydraulic fracture stimulation diagnostics using fiber optic distributed acoustic sensing (DAS) measurements and Analyses. In Proceedings of the SPE Unconventional Gas Conference and Exhibition, Muscat, Oman, 28–30 January 2013.
4. Sheng, M.; Huang, Z.-W.; Tian, S.-C.; Zhang, Y.; Gao, S.-W.; Jia, Y.-P. CFD analysis and field observation of tool erosion caused by abrasive waterjet fracturing. *Pet. Sci.* **2020**, *17*, 701–711. [CrossRef]
5. Surjaatmadja, J.B.; Bezanson, J.; Lindsay, S.; Ventosilla, P.; Rispler, K. New Hydrajet Tool Demonstrates Improved Life for Perforating and Fracturing Applications. In Proceedings of the SPE/ICoTA Coiled Tubing and Well Intervention Conference and Exhibition, The Woodlands, TX, USA, 1–2 April 2008.
6. Huang, Z.; Li, G.; Tian, S.; Song, X.; Sheng, M.; Shah, S. *Abrasive Water Jet Perforation and Multi-Stage Fracturing;* Gulf Professional Publishing: Oxford, UK, 2017.
7. Surjaatmadja, J.B.; Willett, R.; McDaniel, B.; Rosolen, M.A.; Franco, M.L.; dos Santos, F.C.R.; Fernandes, P.D.; Carneiro, F.A.; De Lima, B.B.; Cortes, M. Hydrajet-fracturing stimulation process proves effective for offshore brazil horizontal wells. In Proceedings of the SPE Asia Pacific Oil and Gas Conference and Exhibition, Richardson, TX, USA, 18–20 October 2004.
8. Rodrigues, V.F.; Neumann, L.F.; Rosolen, M.A.; Fernandes, P.D.; Lima, C.; Surjaatmadja, J.B.; Gonzalez, C.M.; Carneiro, F. First Implementation of Hydrajet Fracture Acidizing in Deepwater Offshore Brazil Fields. In Proceedings of the SPE European Formation Damage Conference, Sheveningen, The Netherlands, 25–27 May 2005.
9. McDaniel, B.W.; Surjaatmadja, J.B.; East, L.E. Use of hydrajet perforating to improve fracturing success sees global expansion. In Proceedings of the CIPC/SPE Gas Technology Symposium 2008 Joint Conference, Calgary, AB, Canada, 16–19 June 2008.
10. Fu, J.; Ma, Y.; Zhang, M.; Ma, C.; Liu, Z.; Zhou, S.; He, L.; Feng, L. First Offshore Hydraulic-Jet-Fracturing Multizone Completion Application Case Study: Challenges and Lessons Learned in the Bozhong Oil Field, Bohai Bay, China. In Proceedings of the SPE Asia Pacific Oil & Gas Conference and Exhibition, Virtual, 17–19 November 2020.
11. Li, S.; Yi, X.; Kuang, M.; Xie, Y.; Zhang, Z.; Wang, X.; Gao, Y. Integrated Technologies in the Floating Hydraulic Fracturing for Unconventional Offshore Reservoir. In Proceedings of the Offshore Technology Conference, Houston, TX, USA, 4–7 May 2020.
12. Diaz, O.G.; Luna, G.G.; Liao, Z.; Axinte, D. The new challenges of machining Ceramic Matrix Composites (CMCs): Review of surface integrity. *Int. J. Mach. Tools Manuf.* **2019**, *139*, 24–36. [CrossRef]
13. Kamel, A.H. RJD: A cost effective frackless solution for production enhancement in marginal fields. In Proceedings of the SPE Eastern Regional Meeting, Canton, OH, USA, 13–15 September 2016.
14. Huang, M.; Wu, L.; Ning, F.; Wang, J.; Dou, X.; Zhang, L.; Liu, T.; Jiang, G. Research progress in natural gas hydrate reservoir stimulation. *Nat. Gas Ind. B* **2023**, *10*, 114–129. [CrossRef]
15. Bi, G.; Wang, X.; Han, F.; Wu, J.; Yuan, P.; Fu, S.; Ma, Y. Study on the Mechanism of High-Efficiency Rock Breaking by Hydraulic Jet Based on Explicit Dynamics. *Processes* **2023**, *11*, 2165. [CrossRef]
16. Meese, C.; Mullen, M.; Barree, R. Offshore Hydraulic Fracturing Technique. *J. Pet. Technol.* **1994**, *46*, 226–229. [CrossRef]
17. Pandey, V.J.; Burton, R.C.; Nozaki, M. Evolution of Frac-Pack Design and Completion Procedures for High Permeability Gas Wells in Subsea Service. In Proceedings of the SPE Hydraulic Fracturing Technology Conference, Richardson, TX, USA, 4–6 February 2014.
18. Shi, X.; Zhang, W.; Xu, H.; Xiao, C.; Jiang, S. Experimental study of hydraulic fracture initiation and propagation in unconsolidated sand with the injection of temporary plugging agent. *J. Pet. Sci. Eng.* **2019**, *190*, 106813. [CrossRef]
19. Di, J.; Zhao, X.; Lv, J.; Rao, L. Application of Abrasive Waterjet Technology for Large and Multi-Casing Cutting. In Proceedings of the SPE/IADC Middle East Drilling Technology Conference and Exhibition, Abu Dhabi, United Arab Emirates, 23–25 May 2023.
20. Surjaatmadja, J.B.; Bailey, A.; Sierra, S. HydraJet Testing Under Deep Well Conditions Defines New Requirements for Hard-Rock Perforating. In Proceedings of the SPE Rocky Mountain Petroleum Technology Conference, Denver, CO, USA, 14–16 April 2009.
21. Pan, Y.; Zhai, S.; Meng, X.; Pei, K.; Huo, F. Study on the Fracturing of Rock by High-Speed Water Jet Impact. *Processes* **2022**, *11*, 114. [CrossRef]
22. Shaykamalov, R.; Gaponov, M.; Mukhametshin, M.; Bildanov, V.; Katermin, A.; Bashirov, I. Multistage horizontal wells refracturing by means of abrasive jet perforation+ frac technology. In Proceedings of the SPE Symposium: Hydraulic Fracturing in Russia: Experience and Prospects, Richardson, TX, USA, 22–24 September 2020.
23. Lee, K.K.; Chen, J.L.; Liu, Y.J.; Li, Y.K.; Zeng, X.Z.; Zhou, B.; Hu, G.; Wang, S.L. Application of Pin-Point Sand Jet Perforating Annular Fracturing in Multiple Coal Seams Unleashing Clean Energy in Southwest China. In Proceedings of the SPE Annual Technical Conference and Exhibition, Richardson, TX, USA, 9–11 October 2017.
24. Zhu, D.; Wang, Y.; Cui, M.; Zhou, F.; Wang, Y.; Liang, C.; Zou, H.; Yao, F. Acid System and Stimulation Efficiency of Multistage Acid Fracturing in Porous Carbonate Reservoirs. *Processes* **2022**, *10*, 1883. [CrossRef]
25. Li, G.; Sheng, M.; Tian, S.; Huang, Z.; Li, Y.; Yuan, X. Multistage hydraulic jet acid fracturing technique for horizontal wells. *Pet. Explor. Dev.* **2012**, *39*, 107–112. [CrossRef]

26. Li, G.; Sheng, M.; Tian, S.; Huang, Z.; Li, Y.; Yuan, X. New technique: Hydra-jet fracturing for effectiveness of multi-zone acid fracturing on an ultra deep horizontal well and case study. In Proceedings of the IADC/SPE Asia Pacific Drilling Technology Conference and Exhibition, Tianjin, China, 9–11 July 2012.
27. Sheng, M.; Li, G.; Huang, Z.; Tian, S.; Qu, H. Experimental study on hydraulic isolation mechanism during hydra-jet fracturing. *Exp. Therm. Fluid Sci.* **2012**, *44*, 722–726. [CrossRef]
28. Huang, Z.; Li, G.; Tian, S.; Shen, Z.; Luo, H. Mechanism and numerical simulation of pressure stagnation during water jetting perforation. *Pet. Sci.* **2008**, *5*, 52–55. [CrossRef]

processes

MDPI

Article

Axial Force Calculation Model for Completion String with Multiple Point Resistances in Horizontal Well

Zhen Nie, Shuzhe Shi *, Bohong Wu and Xueqin Huang

Research Institute of Petroleum Exploration and Development, Beijing 100083, China; niezhen@petrochina.com.cn (Z.N.); wubohong@petrochina.com.cn (B.W.); huangxueqin@petrochina.com.cn (X.H.)
* Correspondence: shishuzhe@petrochina.com.cn

Abstract: Frequent accidents may happen during the string run-down and pull process due to the lack of accuracy in the prediction of string force analysis. In order to precisely predict the completion string axial force in horizontal wells, a new model is established, and an in-house software has been developed. The model aims to predict the multiple local resistances that occur at different points on the completion string, which makes up for the technical defects of the commonly used software. It can calculate resistance at different points of the string, which will lead to varying hook load amplification. This method can also predict the axial force of the completion string. By changing the hook load, location, and direction, the resistance can be determined more accurately. Based on the calculation and analysis, the relationship between local resistance, the blocking point, and the amplification factor is also obtained. Furthermore, this model is used to analyze the local resistance of a horizontal well with multiple external packers in the low-permeability Sadi Reservoir of Halfaya Oilfield, Iraq. The recorded data from in-site operations are compared with the predicted results from this model. The results show that the relative errors between the recorded data and model calculation are within the range of 10%, which indicates that the calculated values are reliable. Meanwhile, the results indicate the success of the subsequent completion design and the construction of the oilfield.

Keywords: horizontal well; completion string; local resistance; axial force; hook load

Citation: Nie, Z.; Shi, S.; Wu, B.; Huang, X. Axial Force Calculation Model for Completion String with Multiple Point Resistances in Horizontal Well. *Processes* **2023**, *11*, 2621. https://doi.org/10.3390/pr11092621

Academic Editors: Linhua Pan, Yushi Zou, Jie Wang, Minghui Li, Wei Feng and Lufeng Zhang

Received: 14 July 2023
Revised: 12 August 2023
Accepted: 16 August 2023
Published: 2 September 2023

1. Introduction

Horizontal well drilling technology can be traced back to 1927, and it was first put into practice in the 1940s [1,2]. In the 1950s and 1960s, it experienced a small upsurge of development and then stopped. In the 1970s and 1980s, with the rise in oil prices, an improvement in drilling matching technology, horizontal well development technology, and horizontal well technology was widely seen in the United States, Canada, and France. Nowadays, almost all types of reservoirs, such as depleted reservoirs, tight gas reservoirs, low-permeability reservoirs, marginal reservoirs, and high-permeability reservoirs, can be developed using horizontal well technology. In tight low-permeability reservoirs, natural fractures act as the only oil flow channel. The advantages of horizontal wells are particularly obvious. More than 70% of the horizontal wells drilled in the mid-1990s were completed in such reservoirs [3–5]. The study of horizontal well reservoir fracturing stimulation technology began in the 1980s. Since the long horizontal section of horizontal wells often crosses the reservoir, it is difficult to achieve effective reservoir reconstruction by acidizing and fracturing the whole well section. Therefore, it is necessary to conduct fracturings in multiple sections to form ideal independent artificial fractures and maximize the technical advantages of horizontal wells in developing low-permeability oilfields. Compared to general fracturing of vertical wells and horizontal wells, staged fracturing can effectively induce reservoir fractures, increase the discharge area, and greatly improve the production of oil and gas wells [6–8]. In recent years, with the continuous expansion of shale gas, tight oil/gas, and coalbed methane, horizontal well technology has been developed on an unprecedented scale, and new horizontal well fracturing technologies have been emerging [9–16].

One-trip multi-stage fracturing completion is a comprehensive method performed in open-hole horizontal wells. It combines the completion string and fracturing string as a one-trip string and runs down together. Tools run down the well, including bidirectional anchoring hangers, expandable open-hole packers, dropping ball promotion sliding sleeves, and differential pressure opening sleeves. Open-hole packers seal the horizontal section to achieve isolation from the fracturing operation part. In this way, the whole horizontal section can be completely fractured. By 2013, 21 fractured sections in one well had been achieved in the Changqing Oilfield in China, in which tight gas accounts for 70~80% of the total gas [17]. Multi-stage acidizing technology and fracturing technology are commonly used in low-permeability reservoirs [18].

The open-hole staged fracturing completion of horizontal wells involves a long horizontal open-hole section and a number of large-diameter tools. During the construction process, it is difficult to ensure that the completion string can be smoothly run to the specific position, and this is the key to successful fracturing completion [19]. For example, one staged fracturing string stuck to the wellbore while being run down the well, and eight packers were set locally [20]. This forced the researchers to change the fracturing stage from six to three. The actual length of the horizontal section was 1169 m, while the string only entered at 439 m, with a 730 m loss.

The running resistance of the string comes from two aspects: one is the axial resistance or rotation moment caused by friction after the string comes into contact with the wellbore, and the other is the local resistance caused by the string itself or wellbore conditions. In 2020, Guo and Rashid proposed an analytical model for axial force transfer in extended-reach drilling [21]. Generally, friction refers to prior circumstances, so the model is established mainly based on the previous one developed in this work. The local blocking force is considered empirically according to the carrying capacity of rigid tools.

The investigation shows that although there is some theoretical analysis of the friction resistance of complex completion strings in horizontal wells, and the field has rich operation experience, accidents occur from time to time when the string cannot be run down to the bottom of the well. Such accidents are fatal, often leading to the abandonment of wells and huge economic losses. Running a completion string is a systematic operation, and even commercial software, such as Landmark, cannot predict the probability of resistance. Therefore, according to the field construction process, developing a new calculation method for pipe string axial force is necessary.

2. Materials and Methods

2.1. Characteristics of the String Resistance

The lower multi-stage integrated fracturing string is sent into the horizontal open hole by the landing string. The overall structure of pipe string from bottom to top is as follows:

Guide shoe + ball seat + casing + differential pressure sleeve + casing + open-hole packer + casing + ball sliding sleeve + casing + · · · + open-hole packer + casing + hanging packer + plug back sealing barrel + release sub + landing string + ground-driven equipment.

As shown in Figure 1, during the running process, the fracturing string must pass through the vertical section, the curved section, and the horizontal section. After entering the curved section, the elastic bending deformation of the string occurs, and the friction resistance increases rapidly. After entering the horizontal open-hole section, the packer is affected by debris, shoulder, grooves, and other factors, resulting in various additional resistances. These resistances are transmitted to the wellhead, which is reflected by the change in hook load.

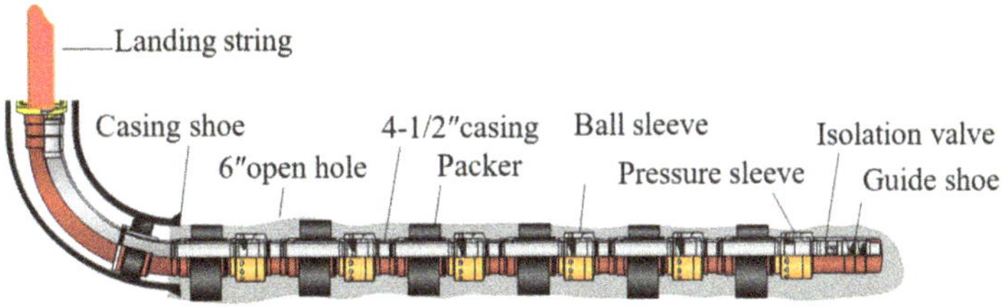

Figure 1. Wellbore and string structure of the lower section.

2.2. Basic Model of Mechanical Analysis

2.2.1. One-Dimensional Borehole

Assuming the string is subjected to axial tension in a one-dimensional borehole, the bending effect of the string and packer is ignored due to the simple load condition. If the string unit length buoyant weight is q, the string length is L, the string outer diameter is r_c, the well inclination angle is β, and the friction coefficient is μ, then

The normal contact force between the centralizer and wellbore is $N = qLsin\beta$;

The axial friction force is $F_a = \mu N$;

The circumferential friction force is $F_a = \mu N r_c$;

Therefore, in this case, the string friction is independent of the axial force.

2.2.2. Two-Dimensional Borehole

For the horizontal well, the two-dimensional borehole focuses on friction calculation. Previous scholars have conducted several studies on the axial force and friction of the string in horizontal wells [22–29], and there is not much difference in these models.

Based on the theory of plane elastic beam, Lubinski established the elastic line differential equation of drilling strings with axial tension and axial compression [30]. However, in practical analyses, it often uses the simplified method. First of all, it is considered that the string curvature is equal to that of the borehole, a micro-segment force analysis is shown in Figure 2. Given the initial angle of the element φ, the length of the element is $dS = Rd\varphi$, the string is subjected to the axial tension P, and the string contacts with the upper side of the wellbore. In the string running process, the friction force is opposite to the moving direction. In Figure 2, R is the curvature radius of the well section, w is the weight of string line, dN is the normal contact force between the element and the wellbore, μ is the friction factor between the string and the wellbore, and M is the bending moment of string.

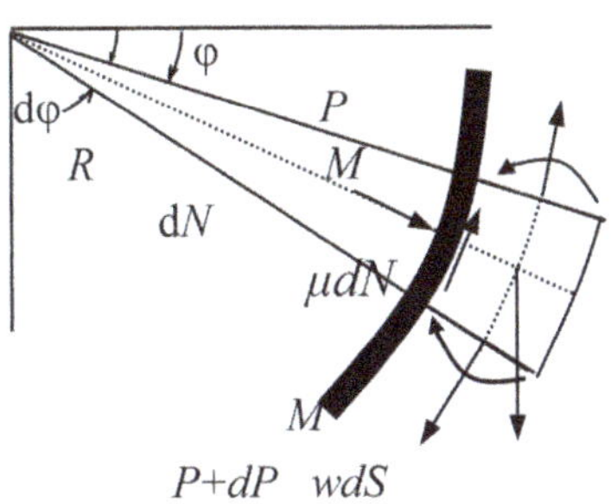

Figure 2. Mechanical analysis of smooth string and wellbore contact.

Assuming that the element curvature is constant, the element shear force is 0, and the string bending moment is $M = EI/R$.

The normal contact force between the unit length string and the wellbore is $N = P/R$.

For the pipe string under axial pressure, the string makes contacts with the lower side of the wellbore, and the calculation of the bending moment and normal contact force per unit length are the same as above.

2.2.3. Three-Dimensional Borehole

Choose a three-dimensional discrete string section. When the axial force T_2 at the lower side of the string and the lateral force per unit length F_n are known, the axial force at the upper side of the string element can be calculated by the following equation:

$$T_1 = T_2 + \frac{L_s}{cos(\theta/2)}[qcos\overline{\alpha}] \pm \mu F_n \tag{1}$$

where L_s is the length of the string section; q is the string effective weight per unit length; μ is the friction factor between wellbore and string, which is taken as "+" when the string moves upward and "−" when the string moves downward; $\overline{\alpha} = (\alpha_1 + \alpha_2)/2$ is the average deviation angle; $\theta = arccos[cos\alpha_1 cos\alpha_2 + sin\alpha_1 sin\alpha_2 cos(\varphi_2 - \varphi_1)]$ is the full angle change rate of the string element; α_1 and φ_1 are the deviation angle and azimuth angle of the upper endpoint of the string element; α_2 and φ_2 are the deviation angle and azimuth angle of the lower end point of the string element; and F_n is contact force caused by the axial force of string and gravity.

For the friction torque calculation model in a three-dimensional borehole, the contact force is calculated as follow:

$$N_i = \sqrt{(T_i \Delta\theta sin\alpha)^2 + (|T_i \Delta\alpha| \pm W_m dl sin\alpha)^2} \tag{2}$$

where N_i is the contact force between the micro-element section of pipe string and the wellbore, T_i is the axial force at the micro-element bottom side; $\Delta\theta$ is the micro-segment azimuth increment, α is the deviation angle, $\Delta\alpha$ is the deviation angle increment of micro-element, W_m is the string unit's buoyant weight and $W_m = W_a \times \left(1 - \frac{\rho_m}{\rho_s}\right)$, W_a is the string unit weight in the air, ρ_m is the liquid density, ρ_s is the string density, and dl is the micro-segment length.

In Equation (2), when $\Delta\theta = 0$, this equation can be used to calculate the contact force of a two-dimensional well, and when $\Delta\alpha = 0$, it can be considered as one-dimension.

The axial force:

$$T_{i+1} = T_i + W_m dl cos\alpha \pm \mu N \tag{3}$$

where T_{i+1} is the axial force at the bottom side of the micro-element.

Since the curve section of the horizontal well is very long and contains special well parts, the buckling problem should be considered when calculating the friction torque. After the pipe string buckles, the contact between the pipe string and the wellbore will inevitably occur. The contact force caused by buckling should also be considered when calculating friction resistance. The pipe string buckle condition can be judged by calculating and comparing whether the equivalent axial force exceeds the critical buckling force of the pipe string.

The critical string buckling forces in the inclined section and the horizontal section are calculated by the following equations:

$$F_{cr} = 2\sqrt{\frac{W_m EI sin\alpha}{r}} \tag{4}$$

$$F_{hel} = 2\sqrt{\frac{2W_m EI sin\alpha}{r}} \tag{5}$$

where F_{cr} is the string sinusoidal critical buckling force, F_{hel} is the string helical buckling critical force, and EI is the string bending rigidity.

The critical buckling force of the string in the vertical section is calculated according to the following equations:

$$F_{cr} = 2.55\sqrt[3]{EIq^2} \tag{6}$$

$$F_{hel} = 5.55\sqrt[3]{EIq^2} \tag{7}$$

The micro-element contact force between the string and the wellbore caused by buckling is calculated as follows:

$$N_{ei} = \frac{rT^2}{4EI} \tag{8}$$

The total contact force between the string and the wellbore is as follows:

$$N = N_i + N_{ei} \tag{9}$$

2.3. *Multi-Point Resistance Calculation Method*

2.3.1. Disadvantages of Traditional Resistance Calculation Method

The traditional models are simple and convenient for programming, but they have the following disadvantages:

(1) The additional resistance caused by special tools, such as packers, cannot be accurately described;
(2) The local resistance caused by wellbore grooves and debris accumulation cannot be reflected in the model;
(3) It is impossible to distinguish the influence of resistance at different positions from the hook load.

At present, commercial software does not have such analytical functions. In order to overcome these shortcomings, this work establishes a precise extrapolation method based on the micro-element analysis method.

2.3.2. Establishment of the Precise Calculation Method

The axial force of the whole string is obtained by superposition calculation from the bottom section of the string. When the pipe string is blocked at some point, the local resistance is calculated separately and added to the axial force.

The specific methods and processes are as follows:

1. Calculation method of overall axial force of strings

For horizontal wells, to obtain the axial force at the upper section of the string, the axial force during the run-down and pull-up processes can be calculated from the bottom of the string to the wellhead and finally verified by the hook load. The equivalent axial force at the bottom of the string is set to 0. By using the basic model, the deformation, contact force and friction resistance of each pipe section are calculated, and then the axial force of the whole string is obtained.

2. Local resistance

Whether it is caused by the string or the wellbore, once the string is subjected to local resistance, the resistance will be applied to the calculation node of the string, and its influence on the hook load can be transmitted. According to the characteristics of resistance, it can be divided into upward resistance, downward resistance, and bidirectional resistance. The upward resistance only exists when the string is pulling up, but does not exist when the string is running down; the downward resistance only exists when the string is running down, but does not exist when the string is pulling up; and the bidirectional resistance will exist whether the string is pulling up or running down.

3. Process of multi-point resistance calculation and analysis

The process of multi-point resistance calculation and analysis is shown in Figure 3. The corresponding calculation software is developed based on this process.

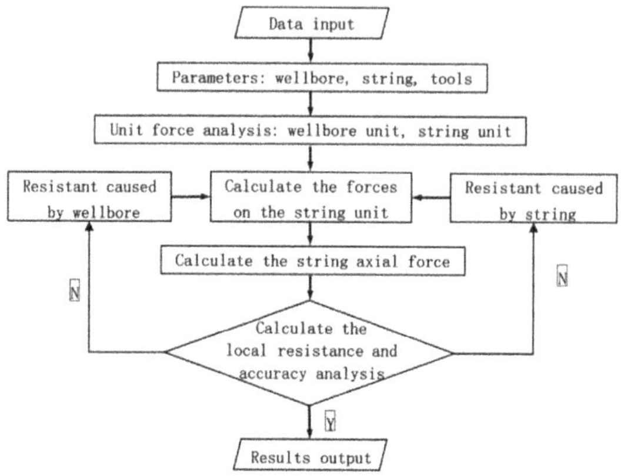

Figure 3. Flow chart of multi-point resistance calculation.

The software used for the proposed model is based on the Python development environment, combining the optimized equations and methods with string mechanics, wellbore conditions, string structure, and construction process to calculate the axial force distribution and friction resistance when running down the string. The torsional force calculation is also added in this software, which can analyze the characteristics of hook load when some tools, such as external packers, are stuck, to provide reference values to more accurately determine the position of the stuck point and the stuck character.

4. Advantages and innovation of this method
    (1) The ability to accurately calculate the axial force of the whole string from the bottom to the wellhead by calculating the deformation, contact force, and friction force;
    (2) The division of the resistance into pull-up resistance, run-down resistance, and bi-directional resistance can describe the tool effect more accurately;
    (3) In the calculation of the axial force, the numerical method is used to calculate the relationship between the contact force and the axial force;
    (4) The description of the force and deformation of the string with the packer in the curved wellbore with the third-order differential equation, allows for the resistance of the string near the centralizer and packer to be obtained by numerical simulation;
    (5) The ability to quickly judge the key position and key cause in case of jamming.

3. Results

3.1. Basic Data

Based on the data of well S from the Halfaya Oilfield in Iraq, the mechanical analysis is carried out.

Wellbore parameters: the well depth is 3737 m; the kick-off point depth is 2234 m; the production casing outer diameter is 177.8 mm, the wall thickness is 10.36 mm, the running depth is 2926 m, and the well deviation is 90°. the open hole horizontal section is from 2926 m to 3737 m with a diameter of 152.4 mm.

A drill pipe and heavy-weight (HW) drill pipe are used to send and release the completion casing string. The string has eight open-hole packers, and the distance from the packers to the string toe section is 90 m, 150 m, 235 m, 340 m, 475 m, 580 m, 684 m, and 760 m, respectively.

The top of the casing is the hanger packer, and above the hanger packer is the drill pipe and heavy-weight drill pipe. The main parameters of the overall string are shown in Table 1. The friction factor is 0.25 in the casing and 0.3 in the open hole.

Table 1. Main structural parameters of string.

Number	Name	External Diameter (mm)	Internal Diameter (mm)	Length (m)
1	Drill pipe	101.60	82.55	2130
2	Hw drill pipe	101.60	65.09	228
3	Drill pipe	101.60	82.55	202
4	Casing	114.30	101.60	1137

3.2. String Axial Force

From previous models and parameters, the hook load is calculated along the depth, which is shown in Figure 4. It can be seen that the hook load has the following characteristics:

(1) The hook load difference between running down and pulling up gradually becomes larger when the string passes the kick-off point;
(2) During the pulling-up process, the hook load increases with the string length;
(3) During the running-down process, the hook load increases monotonously before the string enters the horizontal section, and decreases slowly after entering the horizontal section.

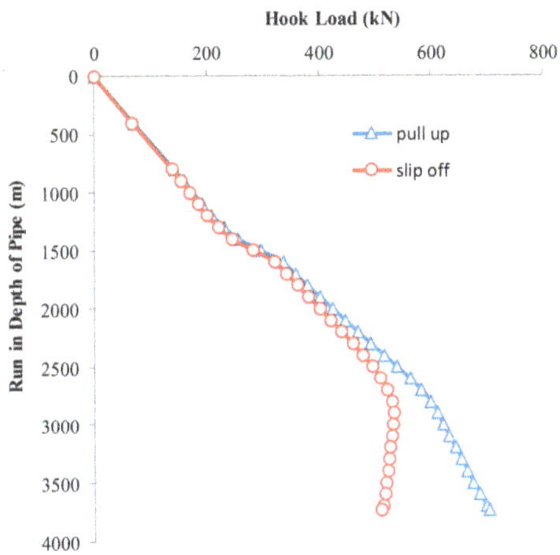

Figure 4. Hook load at different depths.

When the string reaches the bottom, the axial force distribution is shown in Figure 5. It can be seen that the maximum axial compressive force occurs around 2700–2800 m, about 50 kN. However, this section is curved, and the string will not be subject to spiral buckling. In the horizontal section, the axial compressive force decreases to about 40 kN. For the 4-1/2″ casing, the critical value of spiral buckling is about 220 kN in horizontal wells. Therefore, the string will not buckle.

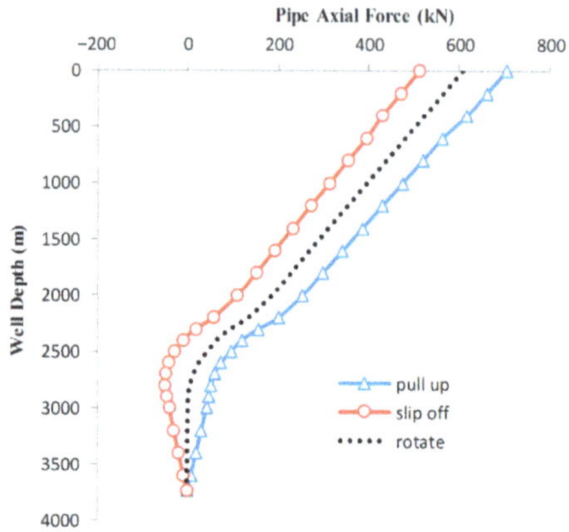

Figure 5. String axial force distribution.

3.3. *Effect of Local Resistance Amplification*

It is assumed that the completion string of well S has been run to the bottom. In order to investigate the relationship between the local resistance and the hook load, it is assumed that the string encounters axial resistance of 0 kN, 50 kN, and 100 kN near its bottom. The corresponding hook load variation is shown in Table 2.

Table 2. Influence of string axial resistance at the bottom on hook load.

Local Resistance Fs (kN)	Pull-Up Hook Load (kN)			Run-Down Hook Load (kN)		
	Value	Increase	Amplification Factor	Value	Increase	Amplification Factor
1	705.33	-	-	512.56	-	-
2	783.07	77.74	1.555	443.94	68.62	1.372
3	863.91	158.58	1.586	371.40	141.16	1.412

In this table, the hook load increases during the pulling-up process, and the hook load decreases during the running-down process when compared to the case of no local resistance. The amplification factor is the ratio of the increase and decrease value to the local resistance value.

The following main features can be obtained:

(1) The resistance at the string bottom is amplified during the upward transmission process;
(2) The amplification factor of the pulling-up process is greater than that of the running-down process;
(3) The larger the local resistance, the larger the amplification factor.

Figure 6 shows the hook load differences between the pull-up process and run-down process under different *Fs* conditions. As can be seen, *Fs* = 0 kN means that there is no blocking in the well, and that the calculated hook load is 705.33 kN. If the blocking point resistance is 50 kN, the hook load will be increased by 77.74 kN instead of 50 kN. And if the blocking point resistance is 100 kN, the hook load will be increased by 158.58 kN instead of 100 kN. According to the relationship between the blocking point resistance and the hook load, the amplification factor can be obtained.

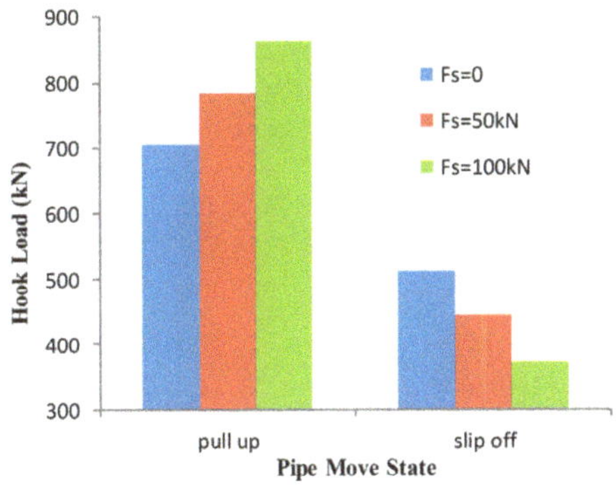

Figure 6. Hook load comparison when completion string meets axial resistance at the bottom hole.

To study the influence of blocking position on hook load, several blocking point locations were assumed. For Case 1, in which assumes that the blocking point is located in the middle of the kick-off segment and with a depth of 2580 m (1156 m away from the bottom), the corresponding hook load variation is shown in Table 3.

Table 3. Hook load versus string blocking point at the middle kick-off segment.

Fs (kN)	Pull-Up Hook Load (kN)			Run-Down Hook Load (kN)		
	Value	Increase	Amplification Factor	Value	Increase	Amplification Factor
1	705.33	-	-	512.56	-	-
2	768.65	63.32	1.266	459.67	52.89	1.058
3	832.07	126.74	1.267	403.95	108.61	1.086

For Case 2, in which assumed that the blocking point is located at the beginning of the kick-off segment and with a depth of 2230 m (1506 m away from the bottom), the corresponding hook load variation is shown in Table 4.

Table 4. Hook load versus string blocking point at the beginning of the kick-off segment.

Fs (kN)	Pull-Up Hook Load (kN)			Run-Down Hook Load (kN)		
	Value	Increase	Amplification Factor	Value	Increase	Amplification Factor
1	705.33	-	-	512.56	-	-
2	758.70	53.37	1.067	465.65	46.91	0.938
3	812.07	106.74	1.067	418.52	94.04	0.940

Table 5 shows the amplification factor changes under different blocking situations. Also, the amplification factor with two blocking point situations is calculated.

Table 5. The amplification factors under different blocking situations.

Fs (kN)	Amplification Factor (Middle Section)		Amplification Factor (Beginning Section)		Amplification Factor (Both Blocking)	
	Pull-Up	Run-Down	Pull-Up	Run-Down	Pull-Up	Run-Down
50	1.266	1.058	1.555	1.372	1.412	1.248
100	1.267	1.086	1.586	1.412	1.429	1.287

The calculation results showed that:

(1) The amplification factor increases as the distance of blocking point from the bottom hole increases;
(2) The amplification factor is affected by the resistance value, position, and other factors;
(3) The axial resistances at different positions have mutual amplification and accumulation effects;
(4) The variation of the hook load caused by two resistances is greater than the sum of hook load variations caused by a single resistance.

4. Discussion

4.1. Hook Load Comparison

The open-hole completion string of this well is equipped with eight external packers, and the running process is smooth. After the string runs into the hole, the hook load of the pull-up and run-down processes are recorded, as shown in Table 6. At the same time, the predicted hook load before the operation is given in the table.

Table 6. Comparison of recorded and calculated hook load.

Operation	Recorded (kN)	Predicted (kN)	Relative Difference (%)
Pull-up	770.00	719.70	−6.53
Run-down	510.00	523.00	2.55

When the string is running down, the hook load relative difference is −2.55%, which means that the running operation of the completion string is smooth, and thtt the quality of the well is good. There are no obvious shoulders, no diameter shrinkage, no collapse, and the hole is clean. Furthermore, the design and use of completion mud, wiper trip technology, and completion string are reasonable.

When the string is pulling up, the hook load difference is 6.53%, which indicates the completion string encounters additional resistance; compared to the run-down relative difference, it is shown that the pull-up process will encounter a larger resistance.

During the in-site process of string pull-up and run-down, some of the parameters cannot be accurately obtained, such as the fineness of the wellbore description and the controllability of the construction equipment. Therefore, from the industrial analysis perspective, there will be errors between the recorded data and theoretical analysis results.

4.2. Error Analysis and Resistance Source Judgment

Based on the in-site pigging data (4″ drill pipe and 4″ heavy weight drill pipe, with 1.23 g/cm^3 mud density). The well string load limit conditions are 770 kN during pull-up, and 510 kN during run-down, and the calculated friction factors are shown in Table 7.

Table 7. Comparison of recorded data and calculated results.

Friction Factor		Axial Force (kN)					
Casing	Openhole	Pull-Up			Run-Down		
		Recorded	Calculated	Error	Recorded	Calculated	Error
0.3	0.3	77	76.69	−0.40%	51	51.14	−1.98%
0.25	0.3	77	75.37	−2.11%	51	52.39	0.05%
0.25	0.25	77	74.20	−3.63%	51	53.62	2.10%
0.2	0.25	77	72.99	−5.21%	51	54.76	2.04%

The maximum error and average error of the axial force during the pull-up process are −5.21% and −0.40%, respectively. The maximum error and average error of axial force during the run-down process are −2.10% and 0.05%, respectively. All the relative errors between calculated results and in-site limited values are within the range of 10%, which indicates that the calculated values are reliable.

The calculated friction factor is applied to the drilling operation of the double-pigging process and the operation of the completion string. The error between the calculated value and the in-site recorded value is small, indicating that the stability of the mud system is good, which will ensure the smooth running of the completion strings with multiple external packers in the Sadi horizontal wells of Halfaya Oilfield.

The work results were successfully applied and verified, providing a successful experience for the future completion design and construction. Combined with calculation results and analysis, the following experience and guidance are provided:

(1) Use the parameters recorded during drilling to obtain the friction factor of pipe string in the wellbore. Verify it through the pigging string to ensure that the value is reasonable, and then apply it to the friction analysis of the string running operation.
(2) There is an error between the recorded value and the calculated value of the completion string load. Since the completion string cannot be rotated, the specific source of resistance cannot be confirmed, which highlights the importance of the early mud system and well drilling plan. At the same time, it also shows that after the external packer is installed, there is still room for improvement in the calculation of the pulling-up and run-down resistance.

In general, error sources also include the following parts:

(1) In the kick-off section, the contact force between the string and the wellbore is relatively high, which makes it more difficult to conduct the pipe string operation. At the same time, it is difficult to guarantee the smoothness of the curved well section, and the fine description of the whole process is even more difficult.
(2) In the target well interval, the wellbore's geometry may be down-dipped, horizontal, up-dipped, or even wavy shape. When the target section is down-dipped, it is more difficult to pull the pipe string up than it is to run the string down. Meanwhile, various factors lead to differences in the wellbore cleanliness, which will affect the calculation accuracy.
(3) The pipe string structure, such as the completion pipe strings, is equipped with multiple packers, centralizers, and other tools. On the one hand, it changes the bending stiffness of pipe string, and on the other hand, it is also extremely sensitive to micro-steps and cuttings beds on the wellbore.
(4) Rotary run-down casing, floating run-down casing, and other processes can reduce friction while increasing the difficulty of calculating accuracy.

5. Conclusions

The proposed model focuses on dealing with multiple resistances of different points of the completion string. This model makes up for the technical defect of some common software, such as Landmark. It is suitable for the simulation of the casing running process with multiple external packers. Through the change in hook load, blocking position, and resistance character, the resistance value can be estimated more accurately.

Based on the case study of a horizontal open hole completion string from the low permeability Sadi reservoir of Halfaya Oilfield, Iraq, the following conclusions are obtained:

1. The string resistance at the horizontal section will be amplified during the pulling-up process;
2. The amplification factor is larger during the pulling-up process than during the running-down process;
3. The larger the local resistance, the larger the amplification factor;
4. The amplification factor is affected by resistance value, position, and other factors;
5. The axial resistances at different positions have mutual amplification and accumulation effects;

6. The change in hook load caused by both resistances is greater than the sum of hook load changes caused by a single resistance.

Author Contributions: Conceptualization, Z.N. and B.W.; methodology, Z.N. and S.S.; formal analysis, Z.N. and S.S.; resources, X.H. and B.W.; data curation, X.H.; writing—original draft preparation, Z.N. and S.S.; writing—review and editing, X.H. All authors have read and agreed to the published version of the manuscript.

Funding: This work was supported by the Major Science and Technology Project of PetroChina (2023ZZ19-06).

Data Availability Statement: Not applicable.

Conflicts of Interest: The authors declare no conflict of interest. The data presented in this study are available on request from the corresponding author. Some of the data are not publicly available due confidentiality issues relating to the oilfield policy.

References

1. Sheppard, M.C.; Wick, C.; Burgess, T. Designing Well Paths to Reduce Drag and Torque. *SPE Drill. Eng.* **1987**, *2*, 344–350. [CrossRef]
2. Helmy, M.W.; Khalaf, F.; Darwish, T. Well design using computer model. *SPE Drill. Complet.* **1998**, *13*, 42–46. [CrossRef]
3. Lietard, O.; Ayoub, J.; Pearson, A. Hydraulic Fracturing of Horizontal Wells: An Update of Design and Execution Guidelines. In Proceedings of the International Conference on Horizontal Well Technology, Calgary, AB, Canada, 18–20 November 1996.
4. Thomson, D.W.; Nazroo, M.F. Design and Installation of a Cost-Effective Stimulation Completion System for Horizontal Chalk Wells Where Multiple Zones Require Acid. *SPE Drill. Complet.* **1998**, *13*, 151–156. [CrossRef]
5. Mcdaniel, B.W. Review of Current Fracture Stimulation Techniques for Best Economics in Multilayer, Low-Permeability Reservoirs. In Proceedings of the SPE Eastern Regional Meeting, Morgantown, WV, USA, 14–16 September 2005.
6. Dees, J.M.; Freet, T.G.; Hollabaugh, G.S. Horizontal Well Stimulation Results in the Austin Chalk Formation, Pearsall Field, Texas. In Proceedings of the SPE Annual Technical Conference and Exhibition, New Orleans, LA, USA, 23–26 September 1990.
7. Thomas, R.L.; Milne, A. The Use of Coiled Tubing during Matrix Acidizing of Carbonate Reservoirs. In Proceedings of the SPE Asia Pacific Oil and Gas Conference, Kuala Lumpur, Malaysia, 20–22 March 1995.
8. Norris, M.R.; Berntsen, B.A.; Skartveit, L.; Teesdale, C. Multiple Proppant Fracturing of Horizontal Wellbores in a Chalk Formation: Evolving the Process in the Valhall Field. In Proceedings of the European Petroleum Conference, The Hague, The Netherlands, 20–22 October 1998.
9. Mc Daniel, B.W.; Willett, R.M. Stimulation Techniques for Low-Permeability Reservoirs with Horizontal Completions That Do not Have Comented Casing. In Proceedings of the SPE Gas Technology Symposium, Calgary, AB, Canada, 30 April–2 May 2002.
10. East, L.; Willett, R.; Surjaatmadja, J.; McDaniel, B.W. Application of New Fracturing Technique Improves Stimulation Success for Openhole Horizontal Completions. In Proceedings of the SPE International Symposium and Exhibition on Formation Damage Control, Lafayette, LA, USA, 18–20 February 2004.
11. Granger, M.J.; Stickling, R.W.; E1-Rabaa, A.M.; Ortez, Y.P. Horizontal Well Applications in the Guymon-Hugoton Field: A Case Study. In Proceedings of the SPE Gas Technology Symposium, Calgary, AB, Canada, 28 April–1 May 1996.
12. Wang, Y. *Analysis and Research on Downhole Ability of Completion String of Horizontal Well;* China University of Petroleum (East China): Dongying, China, 2008.
13. Su, H.; Xie, Q. Study on completion technology of horizontal wells in oil fields. *China Pet. Chem. Stand. Qual.* **2019**, *14*, 215–216.
14. Jia, J. Study on completion technology of horizontal well in new buried hill reservoir. *Tech. Superv. Pet. Ind.* **2019**, *35*, 10–13.
15. Pershin, I.M.; Papush, E.G.; Kukharova, T.V.; Utkin, V.A. Modeling of Distributed Control System for Network of Mineral Water Wells. *Water* **2023**, *15*, 2289. [CrossRef]
16. Kazanin, O.I.; Sidorenko, A.A.; Sidorenko, S.A.; Ivanov, V.V.; Mischo, H. High productive longwall mining of multiple gassy seams: Best practice and recommendations. *Acta Montan. Slovaca* **2022**, *27*, 152–162.
17. Yuan, D.; Sun, H.; Li, Y.; Zhang, M.; Chi, X.; He, X. Multi-Stage Fractured Tight Gas Horizontal Well Test Data Interpretation Study. In Proceedings of the International Petroleum Technology Conference, Beijing, China, 26–28 March 2013.
18. Wang, D.; Sun, J.; Li, Y.; Peng, H. An efficient hybrid model for nonlinear two-phase flow in fractured low-permeability reservoir. *Energies* **2019**, *12*, 2850. [CrossRef]
19. Zhang, X.; Jiang, T.; Ye, C. Study on Optimization of Running Technology of Open Hole Fracturing String in Horizontal Well. *Inn. Mong. Petrochem. Ind.* **2014**, *10*, 123–124.
20. Franco, C.; Solares, R.; Marri, H.; Hussain, H. The Use of SatgeFrac New Technology to Complete and Stimulate Horizontal Wells: Field Case. In Proceedings of the SPE Saudi Arabia Section Technical Symposium, Al-Khobar, Saudi Arabia, 10–12 May 2008.
21. Boyun, G.; Rashid, S. An Analytical Model for Axial Force Transfer and the Maximum Compression Point of Work Strings in Extend Reach Drilling. *Insights Min. Sci. Technol.* **2020**, *2*, 555592.
22. Mitchell, R.F. Effects of Well Deviation on Helical Buckling. *SPE Drill. Complet.* **1997**, *12*, 1–24. [CrossRef]

23. Wu, J. Drill-Pipe Bending and Fatigue in Rotary Drilling of Horizontal Wells. In Proceedings of the SPE Eastern Regional Meeting, Columbus, OH, USA, 23–25 October 1996.
24. Zhao, J. Calculation of drill string friction in horizontal well. *Pet. Drill. Prod. Technol.* **1995**, *17*, 6–11.
25. Wang, J. Analytical analysis of contact friction resistance of drill string in horizontal well. *Pet. Mach.* **1995**, *23*, 44–50.
26. Ma, S. Calculation of friction and friction moment of drill string in horizontal well. *J. Pet. Univ. Nat. Sci. Ed.* **1996**, *20*, 24–28.
27. Han, Z. Calculation of axial force and strength check of bending drill string. *Pet. Drill. Technol.* **1996**, *24*, 7–12.
28. Zu, F. Prediction and application of drill string friction in directional wells. *Pet. Drill. Prod. Technol.* **2001**, *23*, 9–11.
29. Chen, H. Analysis of lateral force between casing string and drill string. *Nat. Gas Ind.* **2001**, *21*, 63–65.
30. Lubinski, A. *Advances in Drilling Engineering*; Petroleum Industry Press: Beijing, China, 1994; Volume 1, pp. 248–249.

processes

Article

Three-Dimensional Printing of Synthetic Core Plugs as an Alternative to Natural Core Plugs: Experimental and Numerical Study

Juan Antonio Cruz-Maya [1,*], José Luis Mendoza-de la Cruz [2], Luis Carlos Martínez-Mendoza [3], Florencio Sánchez-Silva [3], José Alfredo Rosas-Flores [1] and Janet Jan-Roblero [4,*]

1 Unidad Profesional Interdisciplinaria en Ingeniería y Tecnologías Avanzadas, Instituto Politécnico Nacional, Mexico City 07340, Mexico
2 Laboratorio de Termodinámica y Síntesis de Productos Químicos, Instituto Mexicano del Petróleo, Mexico City 07730, Mexico
3 Laboratorio de Ingeniería Térmica e Hidráulica Aplicada, Escuela Superior de Ingeniería Mecánica y Eléctrica, Unidad Profesional Adolfo López Mateos, Instituto Politécnico Nacional, Mexico City 07738, Mexico
4 Laboratorio de Biotecnología Ambiental, Departamento de Microbiología, Escuela Nacional de Ciencias Biológicas, Instituto Politécnico Nacional, Mexico City 11350, Mexico
* Correspondence: jacruz@ipn.mx (J.A.C.-M.); jjanr@ipn.mx (J.J.-R.)

Abstract: This paper proposes three-dimensional (3D) additive fabrication of synthetic core plugs for core flooding experiments from spheres and grains of Berea Sandstone using a digital particle packing approach. Samples were generated by systematically combining the main textural parameters of the rock reservoir to design synthetic core plugs Numerical flow simulation was per-formed using the lattice Boltzmann method (LBM) to verify the flow distribution and permeability for comparison with the experimentally measured permeability and to that obtained from correlations in the literature. The digital porosity of the sample was compared to the porosity measured using an HEP-P helium porosimeter. The numerical and experimental results for permeability and porosity differed by a maximum of 18%.

Keywords: three-dimensional (3D) additive manufacturing; digital rock physics; core flooding test

Citation: Cruz-Maya, J.A.; Mendoza-de la Cruz, J.L.; Martínez-Mendoza, L.C.; Sánchez-Silva, F.; Rosas-Flores, J.A.; Jan-Roblero, J. Three-Dimensional Printing of Synthetic Core Plugs as an Alternative to Natural Core Plugs: Experimental and Numerical Study. *Processes* **2023**, *11*, 2530. https://doi.org/10.3390/pr11092530

Academic Editors: Jie Wang, Lufeng Zhang, Linhua Pan, Minghui Li, Wei Feng and Yushi Zou

Received: 16 July 2023
Revised: 12 August 2023
Accepted: 16 August 2023
Published: 23 August 2023

1. Introduction

Many areas of applied science and engineering involve the study of fluid flow in porous media. Fields as diverse as geology, hydrogeology, and petroleum engineering rely heavily on it to solve problems. In the oil field, understanding multiphase flow in porous media is the basis for developing recovery scenarios and oil/gas production strategies. The oil and gas industry has traditionally used core flooding tests to experimentally measure permeability, relative permeability, saturation change, porosity, and fluid–rock interactions in reservoir and outcrop rock. These tests involve forcing fluids at high confinement pressure through a core plug within a pressure vessel (core holder) [1]. The fluids used may include nitrogen, reservoir brine, crude oil, drilling fluids, and drilling mud filtrate. Other fluids specifically designed to improve or enhance oil recovery (EOR) are also used in core flooding tests to evaluate the effect of treatment and characterize fluid mobility for injection test design.

EOR is a technique for recovering crude oil from a reservoir after primary production has been exhausted due to natural reservoir pressure. EOR processes consist of injecting fluids into the porous medium of the reservoir to create a physical or chemical stimulation effect that increases oil recovery. Various fluids, such as gas, steam, foams, polymers, surfactants, liquid CO_2, alkalis, microbial products, or combinations of these fluids, are injected into the reservoir [2]. These fluids are selected according to the desired recovery effect to be induced in the reservoir. The effectiveness of a potential EOR process can be tested experimentally via core flooding tests prior to expensive field implementation.

However, there are several problems associated with core flooding tests. One of the main problems is obtaining intact cuttings from the reservoir rock to obtain reliable experimental results. Reservoir rock is cut into cylinders to produce core plugs for experimental testing. This is difficult because of the destruction of rock samples and contamination from drilling mud particles that occur during the oil well drilling process, which make it difficult to obtain sufficient material in sufficiently good condition to produce reliable core plug samples. Each core has a specific mineralogy and a unique pore network configuration, even for cores from the same well, formation, and orientation. As a result, researchers must use cores of different morphologies and geometries to extend or repeat their experiments, and the inherent uncertainties involved make it difficult to interpret experimental results. In addition, many of the core flooding tests are destructive (e.g., reactive flow). Therefore, it is impossible to repeat and design experiments because rock samples can only be used once for a single test. In addition, most tests require sample decontamination to remove residual oil, accumulated salts, drilling mud, and other contaminants that can alter the original morphology of the rock. Because of the difficulty of obtaining natural rock cuttings, the oil industry has used outcrop rock or sandstone cuttings as substitutes for natural rock cuttings [3]. However, even in sandstones, which have a very consistent morphology, it is impossible to obtain samples with identical pore networks for repeated testing. Efforts have been made for decades to represent the geometry of porous rock reliably. Starting from very simplified arrangements using clusters of spheres or bundles of capillary tubes of the same diameter [4].

In recent years, a branch of study has focused on reproducing the real microstructure of porous rocks to generate digital rock samples, using techniques such as microtomography and spatial microscopy and disciplines such as statistical physics [5,6]. The combination of 3D microtomographic image reconstruction of natural rocks and numerical simulation of pore-level flow complements, and in some cases replaces, traditional laboratory core flood testing [7]. Another approach to the reconstruction technique involves the development of porous media packages with different geometric, morphological, and anisotropic configurations using physics-based algorithms to simulate the sedimentation, compaction, and cementation processes that lead to rock formation [8–10]. This approach is very useful in developing of morphologically manipulated porous samples for research purposes, where it is necessary to control specific morphological parameters to tune or validate analytical and numerical models. This approach is enhanced by combining it with additive manufacturing, which offers the possibility of 3D printing digital representations of porous structures obtained from CT images and physics-based algorithms [9].

The purpose of study was to investigate whether 3D additive manufacturing of porous samples could reproduce the morphological characteristics of real rocks and serve as viable samples for core flooding experiments. The design of the digital samples was based on the manipulation of typical morphological parameters of the rock, such as grain size and shape, grain size variation (sorting), and cementation. The morphological manipulation of the porous rock provides the opportunity to produce many digital and 3D-printed specimens on an ad hoc basis for research and academic studies, and for tuning numerical and analytical models of flow in porous media.

We used lattice Boltzmann simulations as a complementary method to calculate the flow and predict the permeability of the digitally designed sample prior to 3D printing. The LBM provides an accurate, high-throughput method for solving fluid flow problems in porous media with complex geometries, such as those generated via digital rock physics. LBM is widely used to model pore-scale flow in porous structures [11–20]. For a general introduction to the application of lattice Boltzmann theory in porous media, see [11].

This paper describes (i) the digital construction of core plugs for core flooding experiments by packing samples with spheres of different sizes and realistic grains with different shapes, obtained via micro-CT of Berea sandstone; (ii) numerical simulation of flow in a digital core plug by means of the LBM method to determine the permeability and verify the connectivity of the pore network; and experimental determination of the permeability and porosity of 3D-printed core plugs via core flooding experiments and a helium porosimeter (HEP-P), respectively.

2. Methods

2.1. Digital Core Plug Design

The digital construction of the samples was based on the main morphological parameters of the rock. The systematic combination of these parameters made it possible to control the two main petrophysical properties of the porous rock medium: permeability and porosity. These are the main macroscopic properties used to characterize and classify porous media and describe fluid flow. Permeability and porosity were determined numerically and experimentally for the digital and 3D-printed samples, respectively, to permit a comparison of the values obtained. The results were used to evaluate the potential of 3D-printed core plugs for use in core flooding tests. Some advantages of producing 3D-printed replicas of digitally generated samples while controlling the morphological parameters of the core samples are as follows.

- The 3D-printed replicas can be designed ad hoc according to the experimental requirements by manipulating the morphological parameters to produce specific samples to experimentally calibrate numerical and analytical models of fluid flow in a porous medium.
- The availability of 3D-printed samples with the same pore network allows the design of experiments (DoE) where multiple input factors are manipulated to determine their effect on a desired outcome (response). The DoE can also include intentional changes in the morphological structure of the sample to allow for multilevel factorial experiments.
- The 3D replication of the samples allows for the repeatability of experiments. In contrast, the pore network of natural rock is regularly destroyed during experimental tests [6,7,21,22].
- 3D-printed replicas offer a low-cost alternative to expensive original samples [23].
- It is possible to directly compare numerical simulation results and experimental results for samples with the same digital and 3D-printed pore networks, respectively.
- Morphological manipulation at the digital level allows the development of samples with predicted future morphologies derived from natural changes in the rock structure of reservoirs.
- During core flooding experiments, unconsolidated rock samples can be 3D-printed to create replicas that can withstand extreme pressures and temperatures [23].
- The development of digitals specimens, combined with 3D printing, allows for the resizing of specimens for enhanced oil recovery testing, where fluid mobility characterization requires a minimum characteristic length.

Sample packing is performed using a digital particle packing approach based on the physical processes of sedimentation, compaction, and cementation that rocks undergo during formation [10]. In this approach, the particles are a collection of voxels that move in a grid with six orthogonal and 20 diagonal movements. For the digital construction of samples with the morphology of real rocks, digital grains of natural sandstone from Berea were used as particles for the packing of the samples. Digital grains were extracted from X-ray computed tomography of a cube of Berea sandstone with a resolution of 4.87 μm/voxel [9].

In nature, a porous rock medium of the rock is affected by a process of diagenesis, characterized by the consolidation of material between the grains or particles of the porous medium, caused by a dynamic process of maturation of a sediment as it passes through the rock (lithification). This process is characterized by the formation of mineral sediments in the spaces between the rock grains, which causes, among other things, the cementation of the porous medium.

Cementation causes a reduction in pore space, which reduces the porosity and permeability of the porous medium. A voxel-based geometric cementation process was used to simulate this phenomenon during the construction of the specimens. This process consisted of digitally reconstructing bridges between grains using particles with different geometries to simulate the geometric configuration of diagenesis in real rocks [24]. Digital voxel-based geometric cementation allows control of the porosity of the samples by adding or reducing bridges between the interstitial spaces of the grains. The steps for developing core plug samples are as follows:

- The center of mass and voxel coordinates for each particle were stored in a database. This information was used to create a set of connected 3D voxels for each particle. Particles were extracted from this database to form the packed samples. The geometries ranged from spheres to Berea sandstone grains. The texture of the grains could be modified by changing the voxelization to a convex hull (rounded grains) [9].
- The sizes and shapes of the particles used to build the sample were selected according to the morphological characteristics desired for the sample. We started with the digital development of typical porous media samples, such as those formed by homogeneous spheres, to represent the porous medium in a simplified way and establish a reference according to Darcy's law. The samples evolve in their digital construction through the systematic combination of morphological parameters until samples with realistic rock morphology are obtained.
- Sample packing was performed using the DigiPac digital particle packing tool [10], which uses stochastic and deterministic approaches to particle packing. Digital voxel-based geometric cementation was then added using the digital tool. During this process, the porosity of the sample was determined from the voxelization of the sample.
- Prior to 3D printing, the permeability of the sample was determined by numerical simulation of fluid flow using LBM.
- Once the morphological characteristics of the sample and the required petrophysical properties of permeability and porosity were determined, 3D printing was performed. The samples can be rescaled depending on the resolution of the printer and the needs of the experiment. Figure 1 shows the most complex digital packages for the study cases presented in this work, corresponding to study cases 4, 6, 7, and 8 described in Section 3.3. Digital packaging e (Figure 1) was used to calibrate the parameters of the sedimentation models used to develop the digital packages.

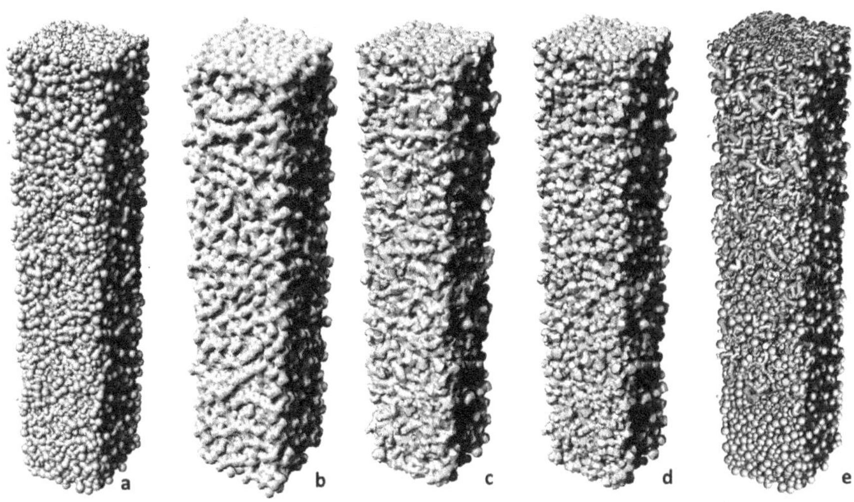

Figure 1. Digital core plug packaging: (**a**) mixed spheres with cementation (case 4); (**b**) mono-sized grains with ellipsoids as geometric cementation (case 6); (**c**) multi sized grains with bridges between grains (case 7); (**d**) multi-sized grains with hyperboloids and ellipsoids as geometric cementation (case 8); (**e**) random packing of different geometries used to calibrate numerical sedimentation models.

2.2. Lattice Boltzmann Method

The lattice Boltzmann method is used to calculate the flow and predict the permeability of a digital design before 3D printing. The fluid is represented by a set of fictitious particles moving independently (flow) at different velocities (impulse discretization) in each grid domain (space discretization) along time steps (time discretization) [13]. LBM consists of two main steps, propagation, and collision of particles. The propagation step represents the movement of particles by which their densities are shifted from one node to another according to the values of the distribution function for each of the possible directions. The collision step represents the interaction between the particles, which change their directional velocities when they reach the next node. The lattice Boltzmann equation (Equation (1)) describes the evolution of a discretized particle distribution function f_i $(x, \vec{e_i}, t)$, which represents the probability of finding a particle moving and colliding in a lattice, with a given discrete velocity $\vec{e_i}$ at time t at a given location in the domain x.

$$f_i(\vec{x} + \vec{e_i}\Delta t, t + \Delta t) - f_i(\vec{x}, t) = -\frac{\Delta t}{\tau}[f_i(\vec{x}, t) - f_i^{eq}(\vec{x}, t)] \quad (1)$$

where f_i $(\vec{x}$ $t)$ is the ith direction (i = 0, 1, 2... 18 in our case) density distribution function at the lattice site x and time t with a discrete velocity $\vec{e_i}$, f_i^{eq} $(\vec{x}$ $t)$ is the equilibrium distribution function, τ is the dimensionless relaxation time related to the viscosity, and Δt is the lattice time step, which is in the lattice units Δt = 1.

The LBM algorithm is implemented in two steps: first, particle collision controls relaxation to equilibrium, and second, particle propagation moves the distribution functions to adjacent lattice cells.

$$\text{collision}: f_i(\vec{x}, t) = f_i(\vec{x}, t) - \frac{1}{\tau}[f_i(\vec{x}, t) - f_i^{eq}(\vec{x}, t)] + F_i(\vec{x}, t)\Delta_t \quad (2)$$

$$\text{propagation}: f_i(\vec{x} + \vec{e_i}\Delta t, t + \Delta t) = f_i(\vec{x}, t) \quad (3)$$

constant external force F_i, defined as a constant pressure gradient to drive the fluid flow, is added to the RHS of Equation (2), which is defined as $F_i = -\nabla p/\rho_o$ and the single relaxation time (SRT) τ is related to the kinematic lattice viscosity, ν.

$$\nu = (\tau - 0.5)c_s^2 \quad (4)$$

where c_s is the nondimensional sound speed (c_s = $1/\sqrt{3}$). In this study, we consider the Bhatnagar–Gross–Krook (BGK) approximation for the collision implementation (right side of Equation (1) and defined in Equation (2)), which models a collision as a linear relaxation of the distribution function toward equilibrium in the Lattice Boltzmann equation. The local equilibrium distribution function f_i^{eq} is given as follows:

$$f_i^{eq} = w_i\rho\left(1 + \frac{\vec{e_i} \cdot u}{c_s^2} + \frac{(\vec{e_i} \cdot u)^2}{2c_s^4} - \frac{|u|^2}{2c_s^2}\right) \quad (5)$$

where w_i is the weight associated with the velocity, $\vec{e_i}$; and ρ and u are the density and macroscopic velocity, respectively, that must satisfy the requirement for a low Mach number, i.e., u/c_s = Ma << 1. These macroscopic quantities (density and velocity) can be computed in terms of the moments of the velocity distribution functions as follows:

$$\rho = \sum_i f_i \quad (6)$$

and

$$u = \sum_i \vec{e_i} f_i + \tau\rho F_i' \quad (7)$$

where, $F_i' = F_i/\rho_0$, the body force is restricted to be along the X axis. D3Q19 model (D is for dimension and Q for the number of discrete directions) is used in the present work, which has the velocity vectors $\vec{e}_i = [\vec{e}_{ix}, \vec{e}_{iy}, \vec{e}_{iz}]$, with w_i, a lattice set of weighting coefficients, which are: $w_0 = 1/3$; $w_{1-6} = 1/18$ and $w_{7-18} = 1/36$.

Packed samples are incorporated directly into the LBM code. The particles (grains or spheres) are fixed during the simulation procedure. The boundary conditions applied were bounce back (for the internal solid no-slip boundary) and periodical (for the external boundary). Numerical simulation of fluid flow in digital rocks using the LBM is described in detail in the references [11–20].

The input parameter (body force) was varied by trial and error until surface velocities below 10^{-4} were obtained to calibrate the simulation. The numerical simulation of the flow for each case study presented in this work is performed in a computational domain larger than the Representative Element Volume (REV) of the specimen, which is the smallest computational domain, to ensure the reliability of the numerical results [15,16]. One of the advantages of numerical simulation is the visualization of the flow distribution in the pore network of the specimen, which allows the identification of errors in the digital design through pore connectivity and flow channeling. In addition, the LBM method can be directly integrated with X-ray micro-tomography or digitally generated microstructures.

2.3. Numerical Permeability

The permeability of the sample can be determined using the Darcy equation, expressed in lattice units (LU) and time step (TS), as shown in Equation (8) [9].

$$k = \frac{U\nu}{f} \tag{8}$$

where k is the permeability of the core plug (LU^2); $\nu = (2\tau - 1)/6$ is the kinematic viscosity ($LU^2\ TS^{-1}$); f is a body force ($LU\ TS^{-2}$); and U is the calculated velocity averaged over the weighted pore area obtained from the numerical simulation over the entire flow domain, including the solids ($LU\ TS^{-1}$). The input parameters for the LBM numerical simulation are the relaxation parameter (τ) and the body force (F_i), which must be greater than 0.5 and less than 0.015, respectively, for numerical stability. Various external forces, including gravity, Lorentz, and Coriolis, can be applied to the fluid. The external pressure gradient applied in the x-direction can also be considered an external force field. In this study, we used the body force F_i, as $(dp/dx)/\rho$ (where ρ is the mass density of the fluid). Figure 2 shows the velocity profiles obtained for case studies 5, 6, and 7 in LU.

To interpret the simulation results independent of the physical units, it is common to write the permeability equation in dimensionless form by scaling with the characteristic length. Therefore, the permeability k obtained from Equation (8) is normalized using the squared mean diameter D^2, as shown in Equation (9).

$$k' = \frac{k}{D^2} \tag{9}$$

where the mean particle (grain or sphere) diameter D serves as the characteristic length in LU or physical units, which is calculated as the harmonic average of the individual particle sizes [8], for each grain packing configuration defined in the case studies in Section 3.3.

$$D = \frac{\sum_i^n f_i V_i}{\sum_i^n \frac{f_i V_i}{D_i}} \tag{10}$$

where f_i is the number fraction of the ith particle with diameter D_i and volume V_i.

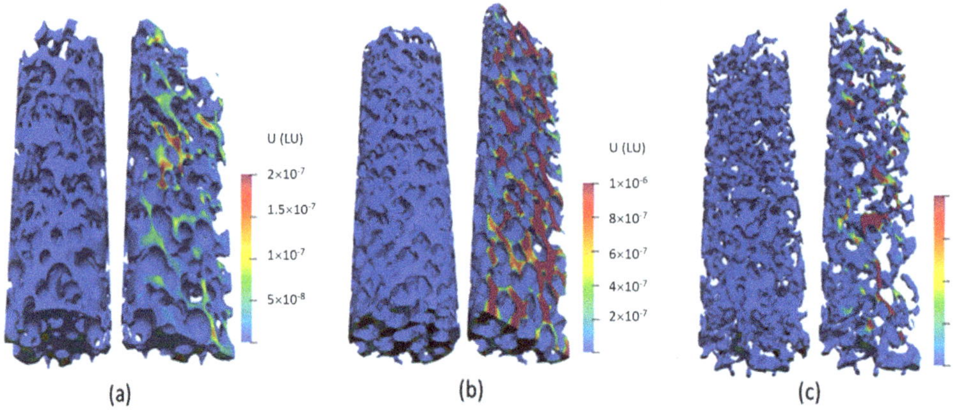

Figure 2. Contour velocity magnitude (LU): (**a**) mono-sized grains (case 5); (**b**) mono-sized grains with ellipsoids as geometric cementation (case 6); (**c**) multi sized grains with bridges between grains (case 7).

The volume-equivalent diameter, defined as the diameter of the sphere with the same volume as the grain (Equation (10)), was used to determine the individual diameter of each grain ($\boldsymbol{D_i} = \sqrt[3]{((6\boldsymbol{V_i})/\pi)}$), with LU or physical units. For each case study, the numerical permeability was compared with the dimensionless permeability obtained from correlations reported in the literature. The correlations used have the structure of Equation (11), where the coefficients $\boldsymbol{a}$ and $\boldsymbol{b}$ depend on the characteristic morphological parameters of the granular medium, such as the shape and distribution of the grains and the cementation in the pore spaces [8].

$$\frac{k}{D^2} = a\varnothing^b \quad (11)$$

Torskaya et al. [8] used sedimentation algorithms to construct grain packings of various geometries. They constructed a base case consisting of sandstone grain packing with a realistic shape and distribution. The packing grains were replaced by spheres, spherical grains, and ellipsoidal grains. They used fluid flow modeling via finite difference approximation and tomographic imaging to determine the permeability of the packings and obtain the values of the parameters $\boldsymbol{a}$ and $\boldsymbol{b}$ in Equation (11). Torskaya's work was used as a reference to compare the dimensionless permeability results obtained in this study.

2.4. Three-Dimensional Additive Manufacturing of Core Plug

MultiJet Printing (MJP) technology, which uses an inkjet printing process with piezoelectric print heads to deposit photo-curable plastic resin and molten wax layer by layer, was used in this study to additively manufacture artificial core plugs. The MJP 3600 Max (3D Systems, Rock Hill, SC, USA) industrial printer was used with a resolution of 16 μm/layer. This wax-assisted technology keeps the pores free of resin material to improve the representation of voids as rock pore networks. The core plug was digitally embedded in a housing (core holder) that confines the core plug particle packing, preventing particle deformation and disintegration during core flooding tests. The core holder is a pressurized chamber that directs fluids through the sample. The array core plug and core holder were digitally assembled and saved as an STL file for 3D printing. This array can be scaled to other dimensions while maintaining the morphological structure of the porous sample. This is useful for adapting to the resolution of the printer used or to the needs of the experimental design.

The procedure for printing the core plug consists of four steps. The first step is to design a target 3DP pattern using particle stacking and densification tools such as Digipac [10], AutoCAD® V 2023, or proprietary digital tools. The second step is to convert the digital structure of the 3D rock model into an STL file that most 3D printers can recognize. In this step, the core plug and core housing assembly can be rescaled to the requirements of the experimental design (DoE) or the resolution of the printer. The third step is to print a target solid rock model. The fourth step is to remove support material from the physical rock models (core plugs). Figure 3 shows the selection of grains for the core plug with different levels of voxelization, from the digitized grains of the Berea Sandstone microtomography to smoothed and rounded grains (convex hull). The grains are packed to form the digital core. The figure al-so shows a natural Berea sandstone core for comparison to the digitally constructed core.

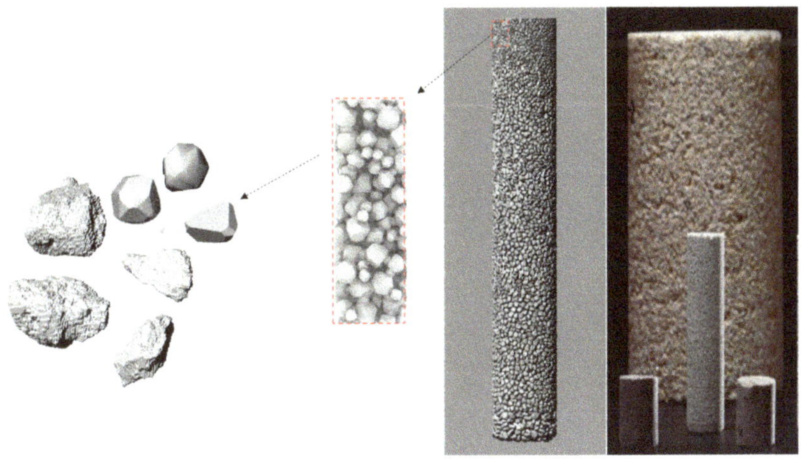

Figure 3. Synthetic core plug developed with sedimentation algorithms using Berea sandstone grains as packing particles vs. Berea sandstone core plug used for core flooding tests.

One of the greatest challenges in 3D printing core plugs using the FDM printing technique is the removal of the support material (wax) from the pore cavities. First, the sample was pre-treated with an ultrasonic cleaning system at a temperature of 90.0 °C (363.15 K) to remove as much wax as possible. The core block was then placed in the core flooding system to circulate a commercial cleaning solution (RESINAWAY®; Monocure 3D, Chicago, IL, USA) at temperatures ranging from 70.0 °C (343.15 K) to 90.0 °C (363.15 K) using a computerized positive displacement pump (PDP). The core flooding system is thermoregulated using an air bath oven (maximum operating temperature of 363.15 K). The cleaning times are long and depend on sample porosity, temperature, and materials used. The characterization of the removal of support material is beyond the scope of this work. Each sample was allowed to drain for days prior to experimental testing to evaporate any remaining cleaning solution prior to the core flooding test. Figure 4 shows a 3D-printed core plug, Figure 4a shows only the core plug, and Figure 4c show the core plug is embedded in a core holder.

Figure 4. Three-dimensionally printed core plugs from the case studies in Section 3.3: (**a**) grain core plug, case 7; (**b**) mix spheres with cementation, case 4; (**c**) multi-sized grains with cementation, case 8.

2.5. Core Flooding Experimental Setup

The experimental setup of the core flooding system used in this study for the determination of the experimental permeability of the manufactured core plugs is shown in Figure 5. The setup consisted of the following components: (i) a BFSP-1000-15 Vinci syringe computerized positive displacement pump (PDP) (Vinci Technologies, Nanterre, France), (ii) a stainless-steel high-pressure cylinder (500 mL capacity) equipped with a floating piston, (iii) a 3D-printed core block, (iv) a differential pressure transmitter (DPT- EJX115A; Yokogawa, Mexico City, Mexico), and (v) a back-pressure regulator (BPR). The DPT was connected to the inlet and outlet ports of the core block, which was calibrated against a deadweight balance with an uncertainty of ±0.01% at full scale. A graphical user interface (GUI) and data acquisition system was used to control and monitor the pressure, volume, and temperature in the core flooding system.

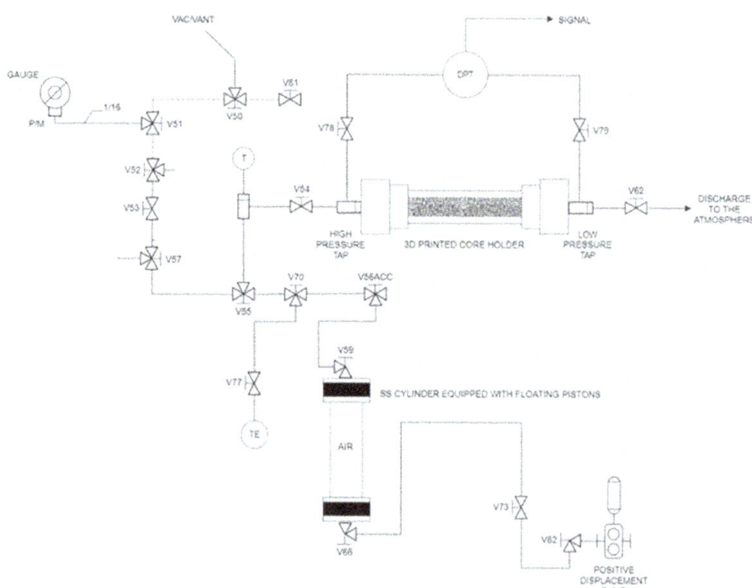

Figure 5. Diagram of the experimental setup.

The key element in this approach is the 3D-printed core block, which consists of the core holder and the core plug (porous medium). The advantage of 3D printing the core block is that the core plug particle assembly is contained within the core holder, preventing the deformation or disaggregation of the particles during experimental testing. The core block caps are removable to facilitate the removal of the support material (wax) from the porous media sample during the cleaning process (Figure 6). The caps have a built-in plug and distribution plate that can be designed in various configurations. The core block is a "plug and play" element, designed ad hoc for the needs of the experimental design, which can be inserted and replaced directly into the experimental setup and replaced as needed, and the same core block can be printed as many times as necessary to validate and repeat experiments. A constant overburden pressure may be required on the core holder to withstand the confined pressures required during the execution of experimental tests, and various commercially available 3D printing materials can be used to achieve the required pressures and temperatures. Today, 3D printing material technology has advanced rapidly, and the market offers a wide range of available materials, such as ULTEM 9085 for pressures up to 69 MPa and temperatures up to 153.0 °C (426.15 K).

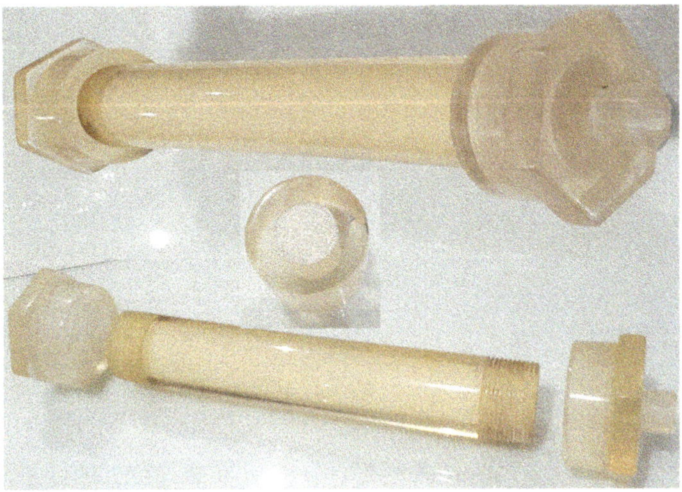

Figure 6. Three-dimensionally printed array core plug and core holder for core flooding tests.

3. Results

3.1. Experimental Permeability

Eight core flood experiments were conducted using the morphological configurations described in Section 3.3. The samples used in the experimental tests were 20 mm (0.788 in) in diameter and up to 200 mm (7.874 in) long. Figure 6 shows the 3D-printed array core plug array and core holder for the core flooding tests.

The core block was installed horizontally in the experimental setup (Figure 5) and the PDP was operated in constant-flow (steady-state) mode for all cases studied. Pressure ports are located at both ends of the core block to measure the differential pressure at the inlet-outlet sample. The data was transferred to the computer and plotted over time. When the differential pressure has stabilized and showed gradual variation on the graph, it was used with the volumetric flow rate data to calculate the average gas permeability (***k***) according to Darcy's modified equation (Equation (12)) for compressible gas flow [25,26]. This equation is valid for laminar flow when there is a linear relationship between the pressure gradient and gas volume flow.

$$\frac{P_2^2 - P_1^2}{2P_1 L} = \frac{\mu}{k}\frac{Q}{A} \tag{12}$$

where k is the permeability of the sample, μ is the gas viscosity (Pa-s), L is the length of the core plug (m), Q is the volumetric flow rate (m^3/s), and P_1 and P_2 are the pressures at the low- and high-pressure sides of the 3D-printed core holder (Pa), respectively. In cases of turbulent flow (Reynolds numbers greater than 10) and very low sample permeabilities (less than 10 md), inertial effects and gas slip effects (the so-called Klinkenberg effect), respectively, must be considered in Equation (12) [27,28]. The experimental permeability obtained using Equation (12) is divided by the square of the harmonic mean diameter of each sample's individual particle sizes (Equation (10)) to obtain the dimensionless experimental permeability (Equation (9)). The core flooding experiments were performed over a volume flow range of 600 to 1800 mL/h to obtain Reynolds numbers less than 1, and the permeabilities of the samples were greater than 10 mD, so that the Klinkenberg effects were not included in the Equation (12). The dimensionless experimental permeability obtained for each of the study cases is presented in Section 3.3.

3.2. Experimental Porosity

In natural rocks, the original (primary) porosity is modified via post-depositional processes to produce secondary porosity. These processes result from rock compaction and chemical and biological processes that lead to the formation of mineral deposits or cementation in the pore space. Cementation causes a reduction in pore space, which reduces the porosity and permeability of the rock. To simulate cementation in the constructed samples, the original porosity of the sample was modified by developing voxel-based geometric cementation by inserting particles of different geometries between the grains of the sample (hyperboloids, ellipsoids, and cylinders). This was achieved by building digital bridges in the pore space of the sample using a proprietary voxel-based geometric algorithm developed in FreeCAD software (Version 0.18.3-2019) [29]. Cementation allows porosity to be created in the sample according to the needs of the experimental design. Digital voxel-based geometric cementation allows control of the porosity of the samples by adding or reducing the bridges between the interstitial spaces of the grains.

In this study, a Vinci HEP-P helium porosimeter was used to determine the effective porosity ($\varnothing_{eff}$) of the samples, which is a measure of the fluid storage capacity of a rock and is defined as the fraction of the rock's total volume corresponding to spaces that can store fluids. The limits of its values for any porous medium range from zero to one. Most porosimeters measure the effective porosity (volume of interconnected pores), which is of interest for estimating oil and gas in place and is mathematically defined as shown in Equation (13)

$$\varnothing_{eff} = \frac{V_{iv}}{V_b} \tag{13}$$

where $\varnothing_{eff}$ is the effective porosity and V_{iv} and V_b are the volume of the interconnected voids and the bulk volume, respectively.

The result of the experimental process is the measurement of the volume of the solids placed in the expansion chamber of the porosimeter. For 3D-printed samples, the total volume (bulk volume) of the solid consists of the volume of the particles, spheres or grains, and the volume of the cementation, plus the volume of the sample holder. The expansion porosimeter is based on Boyle's ideal gas law, which describes isothermal expansion for gases with ideal behavior. The pressure range typically used is 90 to 100 psi, which allows the gas to be modeled as ideal. The system was calibrated using steel discs to obtain a reference volume. Each measurement was repeated three times and averaged to an accuracy of 0.1%. Helium gas has a low molecular weight, which allows the molecules to penetrate the pore space due to its high diffusivity. In addition, helium does not adhere to the walls of the sample and core holder. The samples used for porosity measurements had a core diameter of 1.5 in and a length of up to 2 in. Figure 7 shows the digital porosity profile obtained via voxelization of the core plug and the 3D-printed sample used to find the porosity of case study 4. Details of the helium porosimeter porosity measurement procedure can be found in References [30,31]. The digital porosity results and the experimental helium porosity results are shown in the next section.

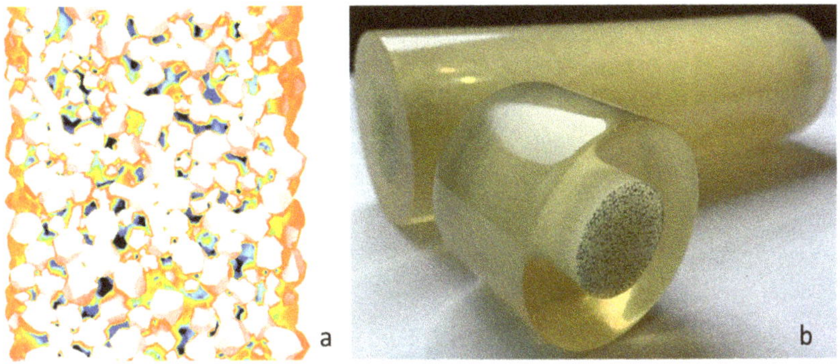

Figure 7. (**a**) Digital porosity profile, and (**b**) 3D-printed core plug used to determine the porosity via HEP-P.

3.3. Study Cases

The first case study corresponds to a sample formed f-homogeneous sphere, which has been widely used to represent porous media and was taken as a reference case. In case 2, the spheres are increased in size to embed each other and achieve a reduction in porosity. Case 3 is a mixture of spheres of different sizes to manipulate the permeability of the sample. The morphological complexity of the subsequent case studies was progressively increased through mixing particles of different sizes and shapes. In addition, smaller particles were incorporated between the pore spaces to control the porosity of the sample, simulating the natural cementation of the rock. For cases 5 and 6, grains obtained from microtomography of Berea sandstone were used as packing particles. In case 5, the core plugs were formed with the same grain (size and shape), while in case 6, the porosity was reduced by using ellipsoidal bridges between the grains of case 5. Finally, for cases 7 and 8, samples were formed with grains of different sizes and shapes, with geometric diagenesis simulated by placing ellipsoidal bridges between grains to generate realistic samples.

The numerical and experimental permeabilities were made dimensionless by dividing them by the square of the harmonic mean diameter of each sample (Equation (9)) for the purpose of comparing the numerical and experimental results. The numerical dimensionless permeability was also compared to the dimensionless permeability obtained via the correlation shown in Equation (11). The results are presented in Table 1. The coefficients $\boldsymbol{a}$ and $\boldsymbol{b}$ in Equation (11) for cases 1–4 are 0.021 and 3.52, respectively. These coefficients correspond to packed spheres with the same grain size distribution of sandstone. For cases 5–8, the coefficients $\boldsymbol{a}$ and $\boldsymbol{b}$ are 0.020 and 3.49, respectively, for packed samples with spherical grains. Table 1 also shows the porosity values obtained digitally and experimentally with the Vinci HEP-P helium porosimeter (HeP) (Vinci Technologies, Nanterre, France).

The experimental dimensionless permeability values are lower than the dimensionless permeability values obtained through numerical simulation and those obtained through the correlation (Equation (11)). For the non-cemented samples (cases 1, 3, 5, and 7), the percentage differences between the experimental and numerical dimensionless permeabilities are close to 2%. For the cemented samples (cases 2, 4, 6, and 8), the percentage differences reach values close to 18%. The experimental results for permeability show the same trend as those for porosity, with values lower than the numerical results. Although porosity does not appear in Equation (12), permeability is dependent on porosity. A factor to be considered in reducing the porosity of the samples and, consequently, the permeability is the presence of residues of the support material in the pore space of the sample. In particular, the percentage reductions in permeability and porosity are higher for the cases with cementation between the grains (cases 2, 4, 6, and 8) than for those without cementation (cases 1, 3, 5, and 7) and even higher than for the results obtained from the numerical simulation, in which, of course, support material in the pore space is not considered.

Table 1. Comparison of results obtained for dimensionless permeability and porosity of 3D-printed cores.

Case	Sample Morphology	Mean Part. Diameter µm	Porosity		Differ. (%)	Dimensionless Permeability (k/D²) × 10⁴			Differ. Numer–Analyt. (%)	Differ. Numer–Exptl. (%)
			Digit.	HeP		Num.	Analyt.	Exptl.		
1	Mono-sized spheres	30	0.41	0.38	7	9.6120	9.0959 †	8.8564	5.36	7.86
2	Mono-sized embedded spheres	12.63	0.277	0.26	6	2.4980	2.2924 †	2.1473	8.23	14.03
3	Mixed spheres	26.58	0.407	0.29	7	9.7700	8.8713 †	8.5691	9.19	12.29
4	Mixed spheres with cementation	31.4	0.331	0.30	10	4.7846	4.3115 †	3.9369	9.88	17.71
5	Mono-sized grains	22.85	0.419	0.40	5	10.674	9.6544 ‡	9.500	9.55	10.99
6	Mono-sized grains with ellipsoids	54.61	0.350	0.33	6	5.6915	5.1265 ‡	4.710	9.92	17.24
7	Multi-sized grains	38.95	0.412	0.40	3	9.8362	9.1038 ‡	8.718	7.44	11.35
8	Multi-sized grains with cementation	48.98	0.300	0.27	10	2.8944	2.9935 ‡	2.378	3.42	17.84

Coefficients a and b in Equation (11) for cases 1–4 † are 0.021 and 3.52, respectively. For cases 5–8 ‡, coefficients a and b are 0.020 and 3.49, respectively.

4. Discussion

The core flooding experiments were conducted with the overall objective of determining whether 3D-printed core plugs have potential as a substitute for real rock core plugs. To perform this evaluation, we proposed the digital design and 3D additive manufacturing of eight samples (core plugs), varying the main morphological parameters for each case study, such as the grain size and shape, sorting, and cementation. These parameters dominate the sample porosity and permeability. These petrophysical properties are the most representative of the porous medium. Therefore, we determined the porosity and permeability value of each of the 3D-printed samples by three means—numerical flow simulation, an analytical model reported in the literature, and experimental core flooding tests—to permit direct comparisons between the results obtained and assess whether porosity and permeability are reproducible in 3D-printed cores designed ad hoc for experimental research purposes.

We believe that 3D printing is a good way to produce cores with morphologies that can be adapted to research needs. In EOR processes, in particular, adequate characterization of the mobility of injection fluids in the reservoir is crucial for the design and control of the injection test. Having samples with the same pore network and morphological configuration susceptibility for which the porosity and permeability can be manipulated represents an opportunity to extend the experimental design (DoE). The numerical and experimental results for dimensionless permeability and porosity show differences of 18% and 10%, respectively. We consider these acceptable degrees of difference that can be reduced with technological advances in 3D printing and the development of new materials.

5. Conclusions

The strength of this approach is the digital construction of the sample by manipulating morphological parameters such as the shape and size of the particles (grains or spheres) and the degree of cementation between them to achieve the desired permeability and porosity. Digital rock physics, combined with particle packing algorithms and 3D printing technology, allows the physical construction of porous samples on an ad hoc basis for experimental designs required for purely experimental research as well as for the validation of analytical/numerical formulations of fluid flow in hydrocarbon production systems. In particular, the samples presented in the present work are oriented towards fluid mobility studies in laboratory-scale EOR tests (Figure 6). The disadvantage of the MJP 3D printing technique used in this work is the difficulty in removing the support material, especially for samples with interparticle bridges (cementation cases). However, the development of new 3D printing and removal materials is increasing the reliability of 3D printing as an alternative replacement for reservoir rock samples.

In nature, it is virtually impossible to find porous media with the same morphological characteristics, even in sandstones such as Berea, which has a slightly heterogeneous morphology. Historically, the inability to find porous media with the same morphological characteristics has severely limited experimental work. The generation of synthetic and 3D-printed rocks offers the possibility of having replicas to perform a series of experiments under different process conditions in the same pore network. This fact will change how flow is studied in porous media created ad hoc according to the morphology required by the target research, since it will be possible to compare numerical and experimental results in the same pore network. Having 3D-printed replicas will allow the design of factorial experiments (DoE). In most experiments in porous media related to hydrocarbon recovery, the porous sample is permanently damaged. The approach presented in this work allows the development of samples with more complex morphologies, with fractures and vugular structures, for the digital and physical reproduction of naturally fractured rock reservoirs.

Author Contributions: Conceptualization, J.A.C.-M. and L.C.M.-M.; methodology, F.S.-S. and J.J.-R.; software, J.A.R.-F.; validation, J.L.M.-d.l.C.; formal analysis, J.A.C.-M. and J.J.-R.; investigation, J.A.R.-F. and F.S.-S.; writing—original draft preparation, J.A.C.-M.; writing—review and editing, J.A.R.-F. All authors have read and agreed to the published version of the manuscript.

Funding: This work was supported the by Instituto Politécnico Nacional (IPN), Secretaría de Investigación, and Posgrado grants SIP20221939, SIP20220781, and SIP20230286.

Data Availability Statement: Not applicable.

Acknowledgments: The authors (J.A.C.-M., J.J.-R, and F.S.-S.) would like to acknowledge the support of IPN through the EDI and COFAA sponsorships and for the SNI-CONACYT. In addition, the scholarships CONACYT (462386) and SIP-BEIFI awarded to L.C.M.-M.

Conflicts of Interest: The authors declare no conflict of interest.

References

1. Baldygin, A.; Nobes, D.; Mitra, S. New Laboratory Core Flooding Experimental System. *Ind. Eng. Chem. Res.* **2014**, *53*, 13497–13505. [CrossRef]
2. Lake, L. *Enhanced Oil Recovery*; Prentice-Hall Inc.: Upper Saddle River, NJ, USA, 1989; 550p.
3. Bjørlykke, K.; Jahren, J. Sandstones and Sandstone Reservoirs. In *Petroleum Geoscience*; Springer: Berlin/Heidelberg, Germany, 2010; pp. 113–140. [CrossRef]
4. Martys, N.; Torquato, S.; Bentz, D. Universal scaling of fluid permeability for sphere packings. *Phys. Rev. A* **1994**, *50*, 403–408. [CrossRef] [PubMed]
5. Hasan Al, M. Digital Rock Physics: Using CT Scans to Compute Rock Properties. *IEEE Signal Process. Mag.* **2018**, *35*, 121–131.
6. Andrä, H.; Combaret, N.; Dvorkin, J.; Glatt, E.; Han, J.; Kabel, M.; Keehm, Y.; Krzikalla, F.; Lee, M.; Madonna, C.; et al. Digital rock physics benchmarks part II: Computing effective properties. *Comput. Geosci.* **2013**, *50*, 33–43. [CrossRef]
7. Dvorkin, J.; Armbruster, M.; Baldwin, C.; Fang, Q.; Derzhi, N.; Gomez, C.; Nur, B.; Nur, A.; Mu, Y. The future of rock physics: Computational methods vs. lab testing. *First Break* **2008**, *26*, 63–68. [CrossRef]
8. Torskaya, T.; Shabro, V.; Torres-Verdín, C.; Salazar-Tio, R.; Revil, A. Grain Shape Effects on Permeability, Formation Factor, and Capillary Pressure from Pore-Scale Modeling. *Transp. Porous Media* **2014**, *102*, 71–90. [CrossRef]
9. Cruz, J.; Martínez, L.; Sánchez, F.; Rosas, J.; Jan, J. Three-Dimensional AdditiveManufacturing of Artificial Oil Reservoir Rock Core Plugs for Core Flooding Experimental Tests. *3D Print. Addit. Manuf.* **2022**, *9*, 233–244. [CrossRef]
10. Jia, X.; Willians, R.A. A packing algorithm for particles of arbitrary shapes. *Powder Technol.* **2001**, *120*, 175–186. [CrossRef]
11. Succi, S.; Benzi, R.; Massaioli, F. A review of the Lattice Boltzmann Method. *Int. J. Mod. Phys. C* **1993**, *04*, 409–415. [CrossRef]
12. Liu, H.; Kang, Q.; Leonardi, C.; Schmieschek, S.; Narváes, A.; Jones, B.; Williams, J.; Valocchi, A.; Harting, J. Multiphase lattice Boltzmann simulations for porous media applications. *Comput. Geosci.* **2016**, *20*, 777–805. [CrossRef]
13. Zhang, J. Lattice Boltzmann method for microfluidics: Models and applications. *Microfluid. Nanofluid.* **2011**, *10*, 1–28. [CrossRef]
14. Boek, E.S.; Venturoli, M. Lattice-Boltzmann studies of fluid flow in porous media with realistic rock geometries. *Comput. Math. Appl.* **2010**, *59*, 2305–2314. [CrossRef]
15. Martínez, L.; Sánchez, F.; Martínez, E.; Cruz, J. Numerical study of fluid flow at pore scale in packed bed of spheres and grains to obtain the REV. *Comptes Rendus Méc.* **2020**, *348*, 769–779. [CrossRef]
16. Wang, P. Lattice Boltzmann simulation of permeability and tortuosity for flow through dense porous media. *Math. Probl. Eng.* **2014**, *2014*, 694350. [CrossRef]
17. Fattahi, E.; Waluga, C. Lattice Boltzmann methods in porous media simulations: From laminar to turbulent flow. *Comput. Fluids* **2016**, *140*, 247–259. [CrossRef]

18. Farahani, M.; Nezhad, M. On the effect of flow regime and pore structure on the flow signatures in porous media. *Phys. Fluids* **2022**, *34*, 115139. [CrossRef]
19. Soleimani, R.; Norouzi, S.; Reza, M. Investigation of gas condensate drop-out effect on gas relative permeability by Lattice Boltzmann modelling. *Can. J. Chem. Eng.* **2019**, *97*, 1921–1930. [CrossRef]
20. Norouzi, S.; Soleimani, R.; Farahani, M. Pore-scale simulation of capillary force effect in water-oil immiscible displacement process in porous media. In Proceedings of the 81st EAGE Conference and Exhibition 2019, London, UK, 3–6 June 2019.
21. Gao, Y.; Wu, T.; Zhou, Y. Application and prospective of 3D printing in rock mechanics: A review. *Int. J. Miner. Metall. Mater.* **2021**, *28*, 1–17. [CrossRef]
22. Ishutov, S.; Hasiuk, F.J. 3D Printing Berea Sandstone: Testing a New Tool for Petrophysical Analysis of Reservoirs. *Petrophysics* **2017**, *58*, 592–602.
23. Almetwally, A.G.; Jabbari, H. 3D-Printing replication of porous media for lab-scale characterization research. *ACS Omega* **2021**, *6*, 2655–2664. [CrossRef]
24. Theocharis, A.; Roux, J.N.; Langlois, V. Elasticity of model weakly cemented granular materials: A numerical study. *Int. J. Solids Struct.* **2020**, *193–194*, 13–27. [CrossRef]
25. Takeuchi, S.; Nakashima, S.; Tomiya, A. Permeability measurements of natural and experimental volcanic materials with a simple permeameter: Toward an understanding of magmatic degassing processes. *J. Volcanol. Geotherm. Res.* **2008**, *177*, 329–339. [CrossRef]
26. Jeevan, J.; Kuntikana, G.; Singh, D. Investigations on gas permeability in porous media. *J. Nat. Gas. Sci. Eng.* **2019**, *64*, 81–92.
27. Wu, Y.; Pruess, K.; Persoff, P. Gas flow in porous media with Klinkenberg effects. *Transp. Porous Media* **1998**, *32*, 117–137. [CrossRef]
28. Bear, J. *Dynamics of Fluids in Porous Media*; Dover Publications: New York, NY, USA, 1988.
29. Riegel, J.; Mayer, W.; Havre, Y. FreeCAD (Version 0.18.3). 2019. Available online: https://www.freecad.org (accessed on 12 July 2019).
30. Liu, L.; Li, H.; Zhou, H.; Lin, S.; Li, S. Design and Application of a Rock Porosity Measurement Apparatus under High Isostatic Pressure. *Minerals* **2022**, *12*, 127. [CrossRef]
31. Washburn, E.; Bunting, E. Determination of porosity by the method of gas expansion. *J. Am. Ceram. Soc.* **1922**, *5*, 112–129. [CrossRef]

processes

Article

Evaluation of Fracture Volume and Complexity of Tight Oil Wells Based on Flowback Data

Jie Li [1], Sen Liu [2], Jianmin Li [1], Zhigang Liu [2], Xi Chen [1], Jiayan Li [1] and Tianbo Liang [1,*]

1 CNPC Engineering Technology Research Institute of Xinjiang Oilfield Company, Karamay 834000, China; lj688@petrochina.com.cn (J.L.); zy-ljm@petrochina.com.cn (J.L.); chenxi@petrochina.com.cn (X.C.); fcljyan@petrochina.com (J.L.)

2 National Key Laboratory of Petroleum Resources and Engineering, China University of Petroleum at Beijing, Beijing 102249, China; 15501123700@163.com (S.L.); lzhigang134@gmail.com (Z.L.)

* Correspondence: liangtianboo@163.com

Abstract: For tight reservoirs, horizontal wells and multi-stage fracturing can generate a complex fracture network that realizes economic and effective development. The volume and complexity of the fracture network are of great significance to accurately predicting the productivity of tight oil wells. In this work, a mathematical model of a multiphase flow is proposed to evaluate the stimulation effect based on the early flowback data. The model showing the early slope of the material balance time (*MBT*) and production balance pressure (*RNP*) can help estimate the effective stimulated volume of the horizontal well. The linear flow region can be determined from the slope of the log–log plot of the *MBT* versus *RNP* curve, which equals 1. The method is verified by commercial simulation software, and the calculated stimulated volume is consistent with the statistical results of simulation results. Results also show that the flow pattern of the fracture–matrix system can be judged by the slope of the flowback characteristic curve in the early stage of flowback, and then the complexity of the fracture network can also be obtained. The proposed method can provide an avenue to evaluate the fracturing work using the flowback data quickly.

Keywords: tight oil; hydraulic fracturing; flowback analysis; stimulation volume

Citation: Li, J.; Liu, S.; Li, J.; Liu, Z.; Chen, X.; Li, J.; Liang, T. Evaluation of Fracture Volume and Complexity of Tight Oil Wells Based on Flowback Data. *Processes* **2023**, *11*, 2436. https://doi.org/10.3390/pr11082436

Academic Editor: Li Xi

Received: 7 July 2023
Revised: 4 August 2023
Accepted: 5 August 2023
Published: 13 August 2023

1. Introduction

After the hydraulic fracturing of horizontal wells, the fracturing fluid flows back before production starts. It is generally believed that the retention of fracturing fluid in the formation will cause damage that reduces the formation's permeability. However, if the flowback is too rapid after fracturing, proppants can flow back and this causes the closure of the generated fracture network, which in turn reduces the well's productivity.

Currently, the total flowback rate of fracturing fluid in a tight oil reservoir in the Mahu reservoirs is around 10–80% in the first year. It is generally believed that there are two directions for the imbibed fracturing fluid; it can be trapped in smaller pores due to the capillary force or in closed natural or hydraulic fractures. Considering that the flowback fluid fills the main and secondary fractures during hydraulic fracturing, information on the generated fracture network is likely to be contained in the flowback data.

Numerical simulation methods for hydraulic fracturing have been continuously studied with the development of unconventional reservoirs. The calculation models have developed from the initial two-dimensional KGD model and PKN model to a quasi-three-dimensional model that combines KGD and PKN models while including the fracture toughness and leak-off of the fracturing fluid [1]. Models have further developed into full three-dimensional models that consider more realistic formation conditions [2]. The calculation models for hydraulic fractures have been continuously optimized, and the accuracy of the models has been continuously improved.

Several methods are commonly used for simulating hydraulic fractures, including the extended finite element method (XFEM), boundary element method (BEM), and discrete element method (DEM). Each of these calculation methods has its own advantages and disadvantages. Wang et al. investigated the numerical method and equivalent continuum approach (ECA) of fluid flow in the fractured porous media. The commonly used discrete fracture model (DFM) without upscaling needs full discretization of all fractures. It can capture each fracture accurately but will get in trouble with mesh partition and low computational efficiency, especially when a complex geometry is involved [3]. In addition, Wang et al. presented a numerical investigation of fluid flow in the heterogeneous porous media with a consideration of the flux connection of the fracture–cavity network. A hybrid-dimensional modeling approach combined with the dual fracture-pore model is presented [4]. The UFM simulation based on the boundary element method uses a fully coupled numerical method to solve the problem, which fully considers the heterogeneity of the reservoir, the anisotropy of stress, and the stress shadow effect, and can simulate complex fracture network shapes [5]. Not only can it simulate the mechanism of fracture propagation and proppant transport, but it can also effectively simulate the interaction between hydraulic fractures and natural fractures [6].

Scholars have conducted extensive studies on the issue of trapped hydraulic fracturing fluid. The fracturing fluid is distributed in the three-dimensional space surrounding the fractures, including the upper and lower reservoirs of the fractures and the reservoir spaces parallel to the fractures; due to the influence of gravity, the amount of fracturing fluid distributed below the fractures is greater than that above [7]. These can affect the material balance when estimating the volume of the generated fracture network. Furthermore, the reservoir may contain many natural fractures. These geological weak planes have significant differences in mechanical properties and matrix, and during hydraulic fracturing, the hydraulic fractures intersect with the natural fractures in a complex manner. Pagels et al. developed the retention area of the fracturing fluid and found that fracturing fluid would be retained in the natural fracture space that is not connected to the main fractures, as well as in the micro-pores of the formation that are penetrated by imbibition [8]. Cheng et al. found that as the capillary force increases, the amount of imbibition in the rock matrix and natural fractures increases [9]. The wider the natural fractures are, the more fracturing fluid is trapped inside; McClure et al. used numerical simulation to show that the stress field near the intersection of hydraulic fractures and natural fractures can cause fracture closure during fluid flowback, which trapped the fracturing fluid therein [10]. In addition to these disconnected spaces, the reservoir below the fractures affected by gravity may be another main space for fracturing fluid retention; Agrawal et al. used numerical simulation to prove that gravity has a significant impact on the flowback of fracturing fluid [7]; Parmar et al. established the effects of surface tension, gravity, and wettability on the flowback rate of fracturing fluid and found that gravity is the primary influencing factor for the retention of fracturing fluid [11].

Besides the influence of natural fractures and formation properties, researchers have found that the properties of hydraulic fractures themselves can also affect the flowback pattern of the fracturing fluid. Liu et al. found that the tortuosity of complex fractures can affect fluid flowback, mainly because the higher the tortuosity of the fracture is, the weaker is the effect of gravity and the stronger is the capillary imbibition into the rock matrix [12]. Water can invade the matrix adjacent to the complex fracture network, which is unable to flow back [13]. Song et al. found that the diversion ability of hydraulic fractures is the main factor affecting the flowback rate of the fracturing fluid; the poorer the fracture conductivity ability, the lower the flowback rate [14]. Warpinski et al. found that increasing the choke size can improve the flowback rate and enhance production [15]; they also found that the more complex the network of hydraulic fractures is, the higher the corresponding flowback rate is, and thus the complexity of the network of hydraulic fractures can be estimated based on the flowback rate of the fracturing fluid. Modeland et al. found that relatively soft reservoirs may cause proppant embedment, leading to a decrease in fracture conductivity ability and consequently affecting the flowback pattern of the fracturing fluid flowback [16]. Mayerhofer et al. analyzed microseismic data and found that complex fracture networks tend to form in shale reservoirs after hydraulic fracturing [17]. McClure et al. found that the connectivity of complex fracture networks can be reduced due to fracture closure, and therefore, the more complex the fracture network is, the lower the flowback rate is [18]. In addition, Yang et al. believed that the structure and connectivity of complex fracture networks also affect the flowback of fracturing fluid [19].

Together with the rapid development of unconventional reservoirs, researchers have proposed different analysis methods to evaluate the productivity of the horizontal well after hydraulic fracturing. Lee et al. established a trilinear flow model from the conventional bilinear flow model from the matrix through the hydraulic fracture and to the wellbore, and further used this model to study the influence of different fracture parameters on flowback and production curves [20]. Daviau et al. believed that pressure interference exists between fractures due to different fracturing times of multistage fractured horizontal wells, and that it is necessary to calculate multistage fractures separately and then add the inter-fracture interference [21]. Larsen et al. analyzed the flow stages of horizontal wells after hydraulic fracturing and divided them into four stages as follows; linear flow in the fracture, radial flow in the fracture, linear flow in the formation, and pseudo-radial flow [22]. Ozkan et al. optimized the theoretical model based on Lee's trilinear flow model and analyzed its dynamic characteristics of pressure and productivity based on the solution results [23]. E. Stagorova et al. considered the significant difference in permeability between the formation near the fracture and the distant area and drew typical well-testing curves to analyze the production decline of a well [24].

In general, the flow process of fracturing fluid during flowback can be roughly divided into three stages. In the first stage, the flowing fluid in the fracture is a single-phase fracturing fluid, the pressure in the fracture is high, and the permeability of the fracture is much higher than that of the matrix. After production starts, the matrix flow can be ignored, and the flow in the fracture is considered to be linear. This stage lasts for a short time. In the second stage, the fracture system gradually depressurizes, and due to differences in fracture length and permeability, some fracturing fluids in certain fractures may still be in the linear flow stage, while other fractures may have completely back-flowed. This stage is a transitional period. In the third stage, the pressure in all fractures has propagated to the effective fracture boundary, and the entire fracture system appears to be in the linear flow stage from the matrix to the fracture surface. At this point, there is no supply in the formation for the single-phase fracturing fluid, and the flow enters the boundary-controlled flow stage. Because different-sized fractures exhibit a unified boundary response, the flow in this stage is regular and easy to identify.

In this work, a mathematical model of a multiphase flow is proposed to evaluate the stimulation effect based on the early flowback data. The model showing the early slope of the material balance time (*MBT*) and production balance pressure (RNP) can help distinguish different flow stages of the well and further estimate its effective stimulated volume.

2. Flowback Model

For the tight conglomerate reservoir in the Mahu area, the matrix permeability ranges from 0.05 to 42.40 mD with an average permeability of 3.45 mD, while the permeability of hydraulic fractures is usually several orders of magnitude larger, typically in the tens of thousands of millidarcies. Due to the long-term retention of fracturing fluid in the reservoir, it can cause reservoir damage. Therefore, after hydraulic fracturing, the well is opened for flowback as soon as possible to reduce the imbibition of the fracturing fluid into the deep reservoir. Most of the fracturing fluid is distributed in the fracture and around the fracture, resulting in higher pore pressure in the hydraulic fracture.

2.1. Model Assumptions

In the early stages of flowback, based on the difference in permeability and the distribution characteristics of the fracturing fluid, it can be assumed that all the fluids produced in the initial stage come from the fractures, and the fluid supply from the matrix can be ignored. Based on this assumption and for the sake of simplification in the calculation, it is further assumed that the average density of fluids in the wellbore, matrix, and fracture is the same; the pressures in the fracture, wellbore, and matrix are interconnected, and have similar pressure change rates. The compressibility of the fluid in the matrix is equal to that of the fluid in the fracture. The model also assumes that the reservoir has no bottom water and ignores the compressibility of the wellbore, and that the driving energy is only elastic energy. Schematic diagram of the model is shown in Figure 1.

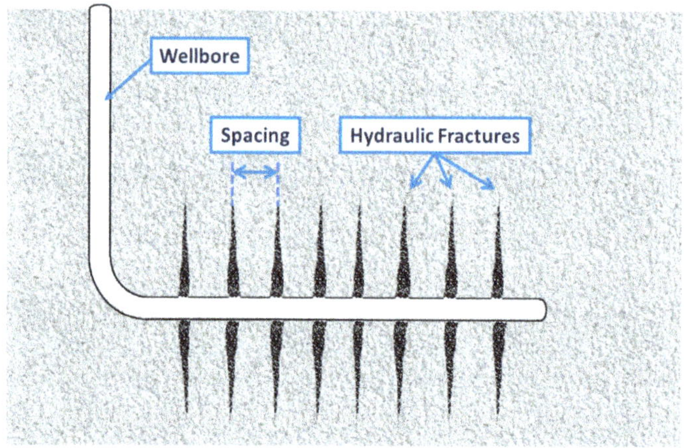

Figure 1. Schematic diagram of reservoir after hydraulic fracturing.

2.2. Mass Balance Equation

Because the mobile formation water is ignored in the reservoir, the flowback of fracturing fluid relies on the elastic energy released by fluid volume expansion caused by the decline in formation pressure in the fractures and pore spaces. Assuming that the effective fracture and matrix are two independent flow systems, the total volume of liquid discharged by the elastic energy relying on rock and liquid within the entire effective fracture and pore volume is equal to the difference between the liquid volume flowing into the matrix by permeation and the liquid volume flowing out to the wellbore, as shown below:

$$q_m \rho_m B_m - q_s \rho_s B_s = \frac{d}{d_t}\left(\rho_f V_f + \rho_{wb} V_{wb}\right) \tag{1}$$

$$C_f = \frac{1}{\rho_f}\frac{d\rho_f}{dP_f} \quad C_{wb} = \frac{1}{\rho_{wb}}\frac{d\rho_{wb}}{dP_{wb}} \tag{2}$$

Substituting (2) for (1) and simplifying the equation can obtain the material balance as shown in (3):

$$q_m \rho_m B_m - q_s \rho_s B_s = \rho_f \frac{dV_f}{dP_f}\frac{dP_f}{dt} + V_f \rho_f C_f \frac{dP_f}{dt} + V_{wb}\rho_{wb}C_{wb}\frac{dP_{wb}}{dt} \tag{3}$$

Consider the model assumptions as follows:

$$\begin{cases} \rho_s \approx \rho_{wb} \approx \rho_f \\ \frac{dP_{wb}}{dt} \approx \frac{dP_f}{dt} \approx \frac{d\overline{P}}{dt} \\ B_m = B_s = B \end{cases} \tag{4}$$

The material balance equation can be further simplified to (5):

$$(q_m - q_s)B = \left(\frac{dV_f}{dP_f} + V_f C_f + V_{wb} C_{wb}\right)\frac{d\overline{P}}{dt} \tag{5}$$

From this equation, it can be derived that

$$\frac{d\overline{P}}{dt} = \frac{(q_m - q_s)B}{C_{st}} \tag{6}$$

where

$$C_{st} = \frac{dV_f}{dP_f} + V_f C_f + V_{wb} C_{wb} \tag{7}$$

Therein, q_m represents the liquid provided by the matrix; ρ_m represents the density of the liquid in the matrix; B_m represents the volume factor of fluid in the matrix; q_s represents the wellhead flow rate; ρ_s represents the wellhead liquid density; B_s represents the wellhead liquid volume factor; ρ_f represents the fluid density in the fracture; V_f represents the fracture volume; ρ_{wb} represents the fluid density in the wellbore; V_{wb} represents the wellbore volume; C_f represents the fracture compression factor; P_f represents the pressure in the fracture; C_{wb} represents the compression factor in the wellbore; P_{wb} represents the pressure in the wellbore; C_{st} represents the total compression factor. Detailed nomenclatures are in Appendix A.

2.3. Linear Flow Equation for Fractures

After hydraulic fracturing, a considerable amount of fracturing fluid not only remains in the fractures but also imbibes the rock matrix adjacent to the created fractures, where the imbibition rate is related to the rock porosity and permeability. Considering that during the flowback, the pressure of the fracturing fluid in fractures decreases rapidly, this drives the imbibed fracturing fluid in the matrix back into the fractures. Assuming that the permeability in the fracture is significantly larger than the rock, the pressure at different locations of fractures can be treated as equal during the flowback; furthermore, the flow of fracturing fluids in fractures can be treated as linear, as shown in Figure 2 below.

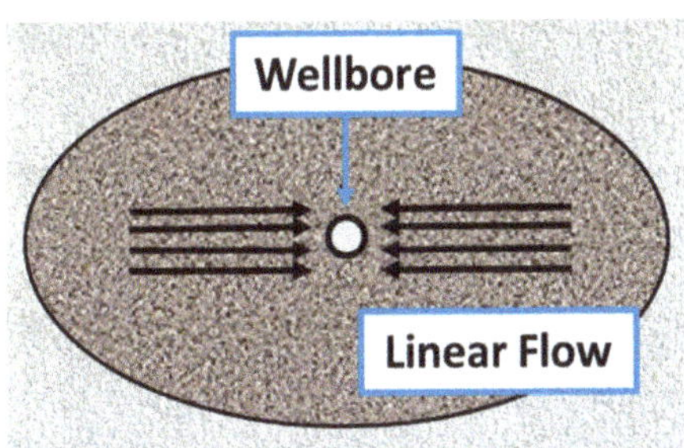

Figure 2. Linear flow of fracturing fluid in fracture to wellbore.

In one wing of the fracture, the pressure distribution during the boundary-controlled flow stage can be approximated as follows:

$$\begin{cases} \dfrac{\partial^2 P_f}{\partial x^2} = \dfrac{\phi_f C_t \mu}{K_f}\dfrac{\partial P_f}{\partial t} \\ \left.\dfrac{\partial P_f}{\partial x}\right|_{x=x_f} = 0 \\ \left.P_f\right|_{x=R_w} = P_{wf} \end{cases} \tag{8}$$

where x_f represents the half-length of the fracture and K_f represents the fracture permeability.

2.4. Model Solution

By substituting the material balance equation into the linear flow equation, the following equation can be obtained:

$$\frac{\partial^2 P_f}{\partial x^2} = -\frac{\phi_f C_t \mu}{K_f}\frac{q_s B}{C_{st}} \tag{9}$$

Combined with the boundary conditions, it can be seen that the pressure distribution at each location of the fracture is as follows:

$$P_f(x,t) = P_{wf} - \frac{\phi_f C_t \mu}{K_f}\frac{q_s B}{C_{st}}\left(\frac{x^2}{2} - x_f x\right) \tag{10}$$

Using the area-weighted average method, the average fluid pressure in the fracture can be expressed as follows:

$$\overline{P}(t) = \frac{\int_0^{x_f} P_f \cdot dV_f}{\int_0^{x_f} dV_f} \tag{11}$$

$$\overline{P} = P_{wf} + \frac{\phi_f C_t \mu}{K_f}\frac{q_s B}{3C_{st}}{x_f}^2 \tag{12}$$

Based on the assumption that the volume of fracturing fluid flowing from the matrix into the fracture is significantly smaller than the volume of fracturing fluid flowing out of the fracture (13), the initial fluid production can be determined using the total compressibility coefficient as shown in (14):

$$q_m \ll q_s \tag{13}$$

$$N_P B = -C_{st}\left(\overline{P} - P_i\right) \tag{14}$$

If

$$RNP = \frac{p_i - p_{wf}}{q_s} \quad MBT = \frac{N_p}{q_s}, \tag{15}$$

the average fluid pressure in the fracture can be simplified as follows:

$$RNP = \frac{B}{C_{st}}MBT + \frac{\phi_f C_t \mu B}{3C_{st}K_f}{x_f}^2 \tag{16}$$

From the above equation, it can be found that after the linear flow stage which is controlled by the boundary, the *MBT–RNP* curve is a straight line. Therefore, the slope "m" of the *MBT–RNP* curve can show the geometry information of the fracture:

$$\mathrm{m} = \frac{B}{C_{st}} = \frac{B}{\frac{dV_f}{dP_f} + V_f C_f + V_{wb} C_{wb}} = \frac{B}{V_f\left(C_w + C_f\right) + V_{wb} C_{wb}} \tag{17}$$

If we ignore the volume change in the wellbore, the volume of the fracture can be obtained using the slope of the *MBT–RNP* curve and the fracture compressibility coefficient as follows:

$$V_f = \frac{B}{m\left(C_w + C_f\right)} \tag{18}$$

We take the logarithm of both sides of Equation (18) and obtain the following equation:

$$log_{10}RNP = log_{10}\left(\frac{B}{C_{st}}MBT + \frac{\phi_f C_t \mu B}{3C_{st}K_f}{x_f}^2\right) \tag{19}$$

$$log_{10}RNP = log_{10}MBT + log_{10}\left(\frac{B}{C_{st}} + \frac{1}{MBT}\frac{\phi_f C_t \mu B}{3C_{st}K_f}{x_f}^2\right) \quad (20)$$

Let:

$$Y = log_{10}RNP \quad X = log_{10}MBT \quad a = \tfrac{B}{C_{st}} \quad b = \tfrac{\phi_f C_t \mu B}{3C_{st}K_f}{x_f}^2 \quad (21)$$

Then:

$$MBT = 10^X \quad (22)$$

$$Y = X + log_{10}\left(a + \frac{1}{10^X}b\right) \quad (23)$$

We take the derivative:

$$Y' = 1 - \frac{1}{1 + \frac{a}{b}10^X} \quad (24)$$

Therein, μ represents the liquid viscosity; ϕ_f represents the fracture porosity; R_w represents the wellbore radius; C_t represents the total compression factor of matrix; P_{wf} represents the flowing bottomhole pressure; $\overline{P}$ represents the average pressure; N_P represents the cumulative volume of liquid; P_i represents the initial pressure; RNP represents the rate-normalized pressure; MBT represents the material–balance time; C_w represents the compression factor of water; V_m represents the matrix volume; L represents the cluster spacing; P_m represents the matrix pressure.

From the above equation, it can be seen that as time increases, MBT increases, and the slope of the $log_{10}MBT$-$log_{10}RNP$ curve increases and eventually approaches 1. Therefore, the linear flow stage controlled by the boundary can be identified by the slope of the $log_{10}MBT$-$log_{10}RNP$ curve.

3. Validation of the Flowback Model

To validate the mathematical model proposed above, hydraulic fractures were uniformly generated using the Kinetix simulator. Fractures and the matrix were described using unstructured grids of various sizes. An INTERSECT simulator was then utilized to obtain the flowback rate and production rate of this fractured well. By injecting the same amount of fracturing fluid as during the fracturing, fluid invasion into the rock matrix adjacent to hydraulic fractures can be obtained. By further simulating the flowback and production process using the field operation scheme, flowback data can be obtained, from which the geometry of the created fracture network can be calculated using the proposed model, and compared with the model in the simulator.

3.1. Numerical Simulation Process

Numerical simulation on hydraulic fracturing, flowback, and production is conducted in Petrel, whose schedule is shown below.

(1) The geological model of the reservoir was established based on the well-logging interpretation data. Rock mechanical properties are calculated from well-loggings of Mahu-18 wells, which provide an important basis for horizontal well fracture and obtain geomechanical parameters such as Young's modulus and Poisson's ratio. The geological model of the reservoir is established by these data. Considering that only one well is fractured in the case, the structural and stress changes of the formation can be ignored, and thus a multi-layer geological model was established as shown in Table 1.

(2) Once the reservoir was established, the horizonal well was put into the model according to the hydraulic fracturing design. Key parameters used in this case are shown in Table 2 below.

Table 1. Reservoir Parameters in Numerical Simulation.

Layers	Thickness of Layers (m)	Layer Type	Pore Pressure (MPa)	Minimal Horizontal Stress (MPa)	Permeability (mD)	Porosity	Young's Modulus (GPa)	Poisson's Ratio	Oil Saturation
1	30	Barrier	56.34	75	0.012	2.10%	26.69	0.3	0%
2	20	Reservoir	56.75	65	0.675	3.46%	35	0.25	10%
3	10	Reservoir	57.01	65	0.675	3.46%	35	0.25	65%
4	20	Reservoir	57.27	65	0.675	3.46%	35	0.25	10%
5	30	Barrier	57.70	75	0.012	2.10%	26.69	0.3	0%

Table 2. Hydraulic Fracturing Parameters.

Well Length (m)	Clusters per Stage	Cluster Spacing (m)	Mesh of Proppant	Liquid-Proppant Ratio	Pumping Rate (m^3/min)	Liquid Volume (m^3)	Proppant Volume (m^3)	Proppant Intensity (m^3/m)
700	4	17.5	40/70	16	10	16,000	700	1.0

(3) After hydraulic fracturing whose pumping schedule is shown in Appendix B, the dimensions of each fracture can be obtained as shown in Figure 3. Fractures do not uniformly propagate due to the stress shadow effect among fractures, and the average fracture length is 277.83 m in this case, as shown in Table 3.

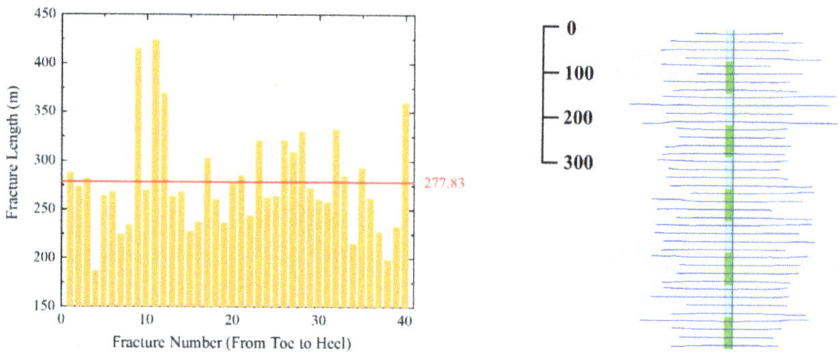

Figure 3. Distribution of Fracture Lengths Obtained from Numerical Simulation.

Table 3. Average Fracture Parameter.

Fracture Volume (m^3)	Average Fracture Length (m)	Average Fracture Width (mm)	Average Fracture Height (m)	Average Fracture Conductivity (mD·m)
3066	277.83	9.37	30.00	660.66

(4) After hydraulic fractures are generated, as shown above, unstructured grids are used for meshing all fractures in order to conduct the production simulation. After the fracturing simulation, the 3D fracture model is coarsened using unstructured grids, where grids around fractures are locally refined. Grids around each fracture are small enough to ensure the convergence of calculation, and they become larger towards the rock matrix to increase the calculation efficiency.

Different relative permeability curves were assigned to the fracture and rock matrix. Relative permeability curves of the rock matrix were measured using the reservoir core samples (Figure 4), while a pair of straight lines was used for hydraulic fractures (Figure 5). The measured relative permeability curves show a limited multiphase flow region for such a low-permeability rock, where the average residual water saturation is around 38%, and the average residual oil saturation is around 27%.

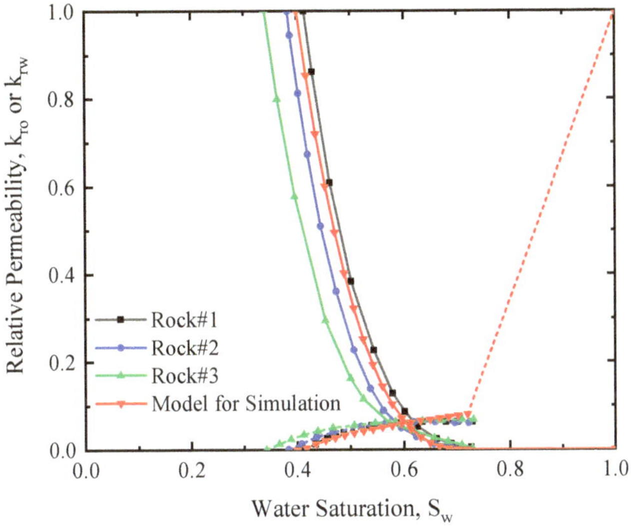

Figure 4. Oil–water relative permeability curves measured from reservoir rock samples.

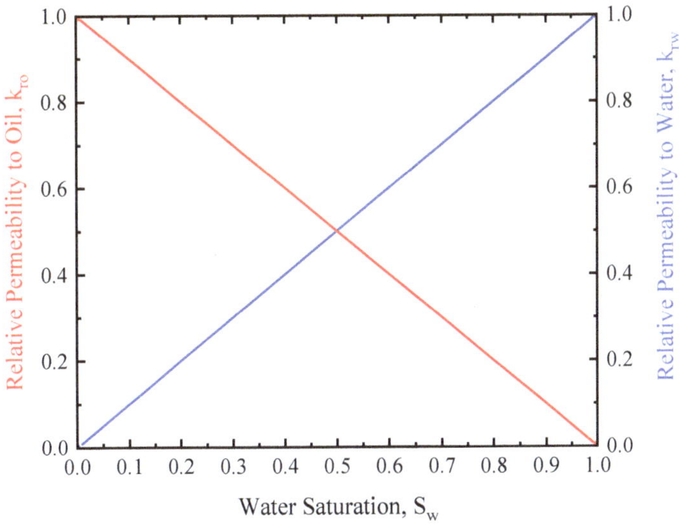

Figure 5. Oil–water relative permeability curves of hydraulic fractures used in the simulation.

(5) During the production simulation, the production rate was kept at 50 m^3/d until the bottom-hole pressure decayed to 35 MPa, after which the production was continued at a constant pressure mode. The calculated production data are shown in Figure 6. As shown in the Figure 6, the daily production of water (i.e., the fracturing fluid) is large in the first 15 days, which represents the main flowback region; in this region, flowback water is mainly from the propped fractures. As production continued, the daily water production decreased and daily oil production increased; after 10 days of production, the oil–water ratio increased beyond 1 in this simulation case.

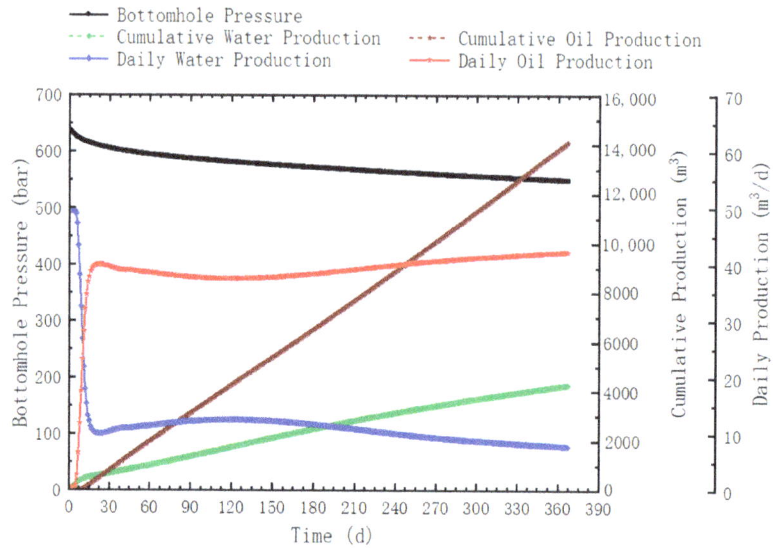

Figure 6. Changes in water production and bottom-hole pressure obtained from numerical simulation.

3.2. Flowback Pattern of the Fracturing Fluid

Along with the production, the fracturing fluid flows back with the oil. Typically, the early flowback has a large water saturation and gradually decreases with time as shown above. Numerical simulation reveals two stages in the production; in the first stage, the flowback water mainly comes from hydraulic fractures, while in the second stage the flowback water mainly comes from the rock matrix. Flowback in the first stage shows the geometry and permeability of the fracture network, while flowback in the second stage shows the geometry and permeability of both the fracture network and the rock matrix.

(1) Flowback Pattern of the Fracturing Fluid from Hydraulic Fractures

A simple double-wing fracture model is firstly used to understand water flowback after hydraulic fracturing. From the change in water saturation in different grids, it can be seen that after approximately 15 days, water saturation in the fracture drops quickly to around 0 (i.e., residual water saturation as shown in Figure 7), indicating that fracturing fluid in the fracture has almost completely flowed back. However, there is still a large amount of fracturing fluid waiting to flow back in the invaded region adjacent to the fracture.

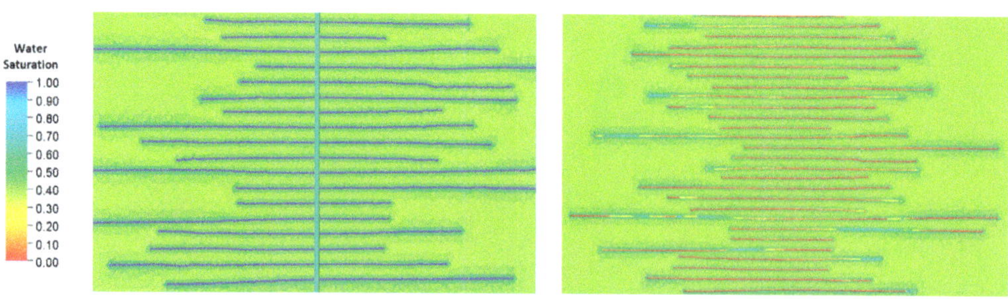

Figure 7. Change in water saturation in and around hydraulic fractures during flowback.

Using the porosity and water saturation of the fracture grids, the reduction in the fracturing fluid in fractures during the flowback can be determined and compared with the cumulative produced water at the wellhead. During the early stage of flowback, the reduction in hydraulic fracturing fluid within the fracture is slightly less than the cumulative water production at the wellhead; this is because the pore pressure of fractures is larger than the matrix after fracturing, which drives water to imbibe the rock matrix adjacent to fractures. After 17 days of flowback, the cumulative water production at the wellhead exceeded the reduction in hydraulic fracturing fluid within the fracture, indicating that all of the hydraulic fracturing fluid within the fracture had flowed back. At this point, the volume of hydraulic fracturing fluid within the fracture remained relatively stable, and the hydraulic fracturing fluid at the wellhead was supplied by the surrounding invaded zone. As shown in Figure 8, the time point of significant difference between the two curves occurred between 10 and 11 days of flowback, indicating the transition period during which the surrounding invaded zone began to contribute to the flowback of the fracturing fluid.

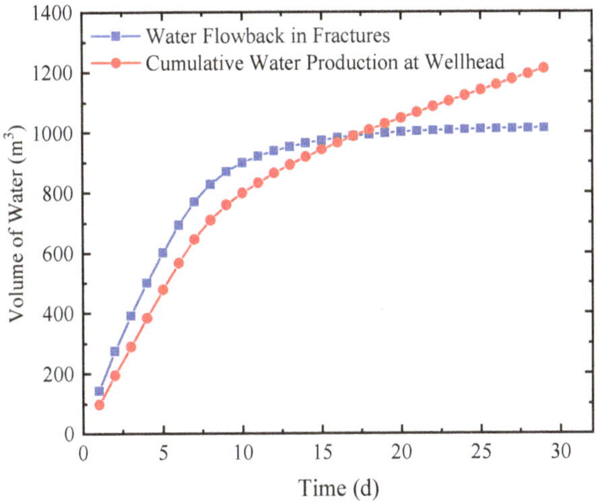

Figure 8. Comparison of water flowback in fractures and the cumulative water production at the wellhead.

When we plot the change in the flowback ratio in fractures (i.e., the daily reduction volume of fracturing fluid in fractures to the initial volume of fracturing fluid in fractures), it can be seen that the flowback ratio decreases almost linearly at the early stage but logarithmically at the later stage. These two stages are more obvious in the log–log plot as shown in Figure 9, where the first stage represents the linear flow of the flowback from fractures, and the later stage represents the flowback from the low-permeability rock matrix. The transition between the two stages lasts for several days.

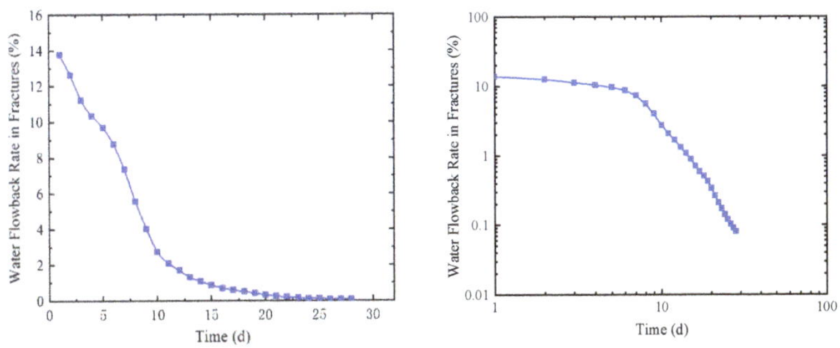

Figure 9. Changes in flowback ratio with time.

(2) Flowback Pattern of the Fracturing Fluid from the Invaded Rock Matrix

After 17 days of production, the amount of fracturing fluid in the fracture remains relatively stable. At this point, the flowback rate of the fracturing fluid from the invading region can be calculated, which is approximately equal to the daily water production rate observed at the wellhead. This stage of flowback continues for a long time. After another 100 days of production, the average daily water production at the wellhead is 18.25 m^3/d, which can be considered as the return rate of the fracturing fluid from the invaded zone under the current production regime.

Vertically, the invaded zone where the fracturing fluid invades can be divided into four regions as shown in Figure 10: the reservoir above the fracture, the reservoir surrounding the upper half of the fracture, the reservoir surrounding the lower half of the fracture, and the reservoir below the fracture. Contributions of the fracturing fluid flowing back from these regions change with the production as shown in Figure 11. At the beginning of flowback and production, the fracturing fluid imbibes the reservoir rock due to the high pressure in the fractures, which leads to negative initial flowback rates in the top and bottom reservoir regions; due to gravity, the amount of fracturing fluid imbibing the bottom region of the reservoir is larger than that imbibing the top region of the reservoir. The specific difference is influenced by the reservoir's permeability, the total amount of fracturing fluid, and pore pressure changes in fractures after hydraulic fractures. In this numerical simulation model, the amount of fracturing fluid that enters the reservoir below the fracture is tens of times larger than the amount that enters the reservoir above the fracture.

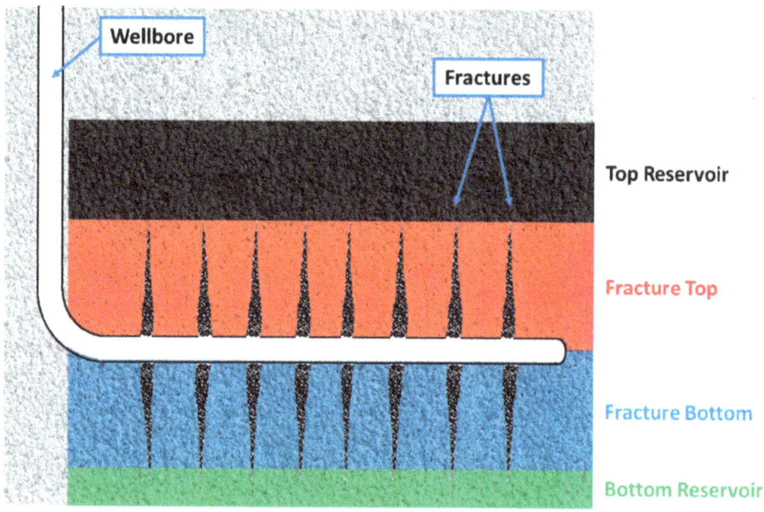

Figure 10. Schematic of water-invaded regions.

When the second flowback stage starts, the fracturing fluid mainly comes from the rock matrix, and imbibition of the fracturing fluid continues, especially in the lower reservoir region due to gravity. In the early period of this stage, the pressure in fractures and their adjacent regions is still high, and the capillary force also plays a role; thus, the fracturing fluid overcomes the gravity and stays in the reservoir above the fracture. However, in the later period of this stage, after the pressure and water saturation around the fracture decrease significantly, the fracturing fluid in the reservoir above the fractures gradually flows back (after 10 months).

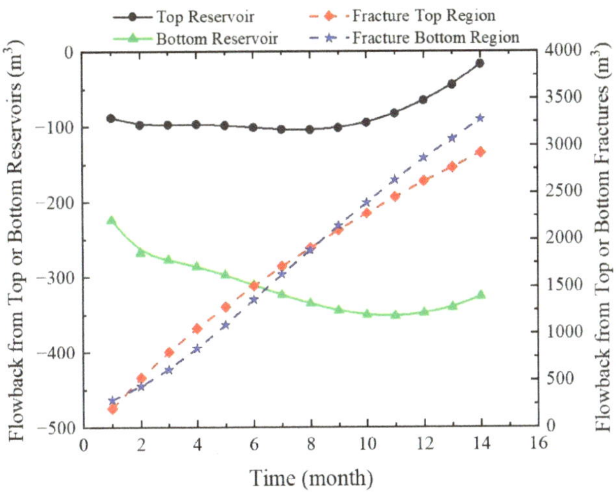

Figure 11. Contributions of flowback from different water–invaded regions.

3.3. Validation of Flowback Model for Single Cluster

To minimize the interference among fractures, a single-cluster fracturing simulation is conducted to validate the proposed flowback model. Figure 12 shows the changes in production and bottom-hole pressure of this single-cluster case.

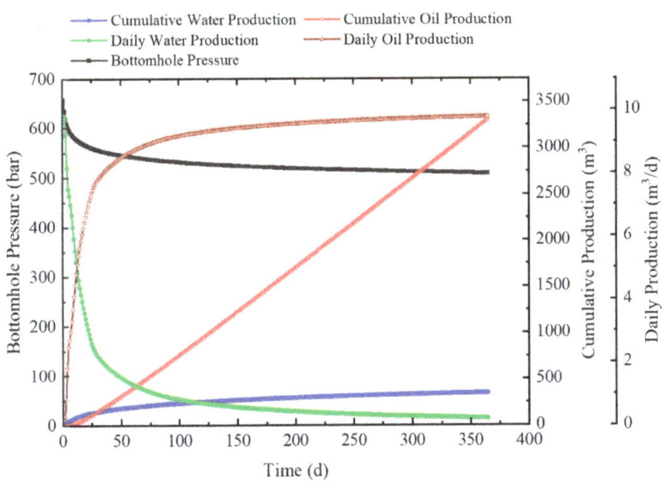

Figure 12. Changes in production and bottom-hole pressure of single-cluster simulation case.

The production data are normalized to obtain the rate-normalized pressure (RNP) and material-balance-time (*MBT*) as shown below:

$$RNP = \frac{p_i - p_{wf}}{q_s} \; MBT = \frac{N_p}{q_s}$$

After converting the data into *MBT* and *RNP*, a plot of the *MBT*–*RNP* curve (also known as the flowback characteristic curve) can be generated as shown in Figure 13. The *MBT*–*RNP* curve for a cluster shows a good linear trend in the later stages, as expected.

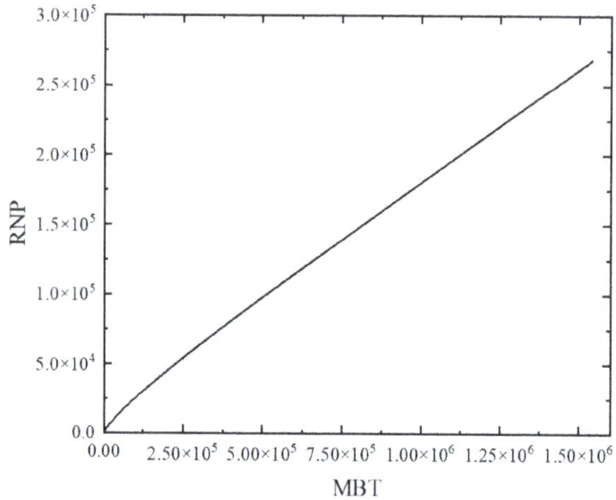

Figure 13. Flowback characteristic curve of the single-cluster case.

Taking the logarithm of both x and y axes, Figure 13 is converted to a plot of *logMBT* versus *logRNP*, as shown in Figure 14. The red curve in Figure 14 shows the change in the slope, which is slightly below 1 before 4.5, and plateaus at 1 thereafter. The slope below 1 is likely attributed to the closure of the fracture at the early stage; once the dimension of the fracture is stabilized, the flowback is a standard linear flow and the slope equals 1.

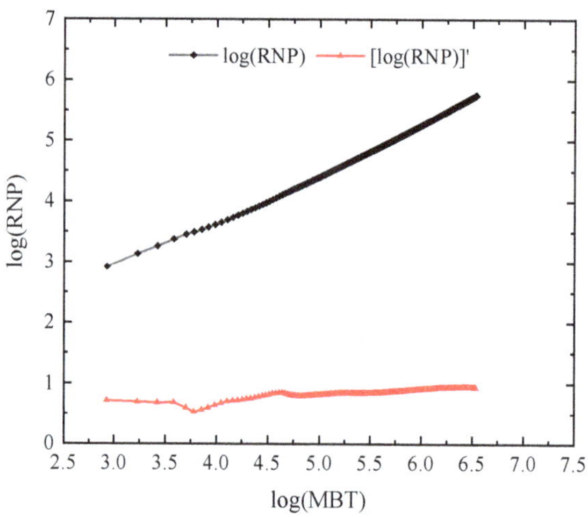

Figure 14. Log–log plot of the flowback characteristic curve of the single-cluster case.

3.4. Validation of Flowback Model for the Entire Well

Figure 15 shows the flowback characteristic curve of the entire-well case that is calculated from Figure 6. Similar to the single-cluster case, the *RNP* increases almost linearly with the *MBT* while the slope changes only slightly at the early stage.

After converting this flowback characteristic curve in the log–log plot as shown in Figure 16, it can be seen that the trend of the *logRNP-logMBT* of the entire-well case is similar to that of the single-cluster case. By carefully comparing the detailed characteristics of the flowback curve and the fracture network generated by the simulator, it can be found that the flowback data of the single-cluster case are close to those predicted by the ideal mathematical model; however, for the entire-well case, the characteristics of the flowback curve change due to the heterogeneity of the created fractures and their interactions. As shown in Figure 16, the minimal slope of the *logRNP-logMBT* curve appears after 4.5, which is clearly later than the single-cluster case; this indicates a slower transition to the second stage of the flowback that is mainly determined by the invaded water in the rock matrix.

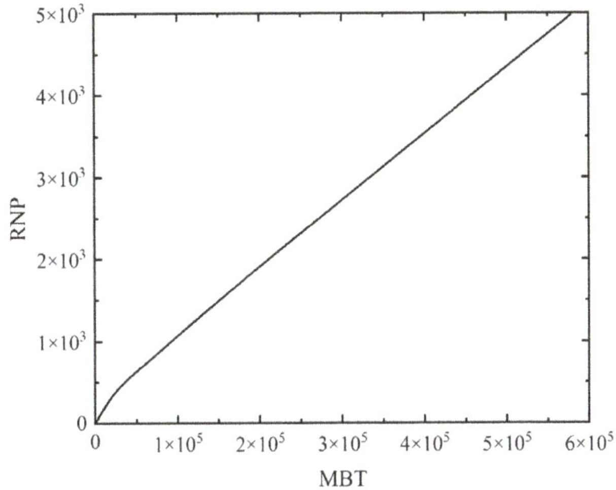

Figure 15. Flowback characteristic curve of the entire-well case.

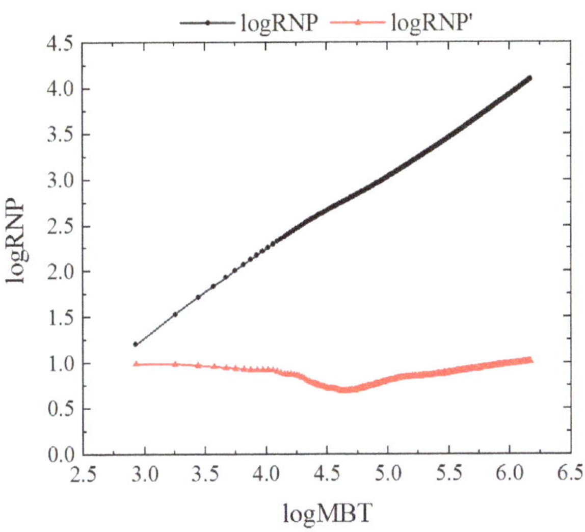

Figure 16. Log–log plot of the flowback characteristic curve of the entire-well case.

Furthermore, the total volume of the created fracture network can be calculated based on the slope of the *RNP* curve in the early stages of the flowback, whose detailed steps are as follows.

(1) Obtain the slope of the *RNP* curve and the corresponding bottom-hole pressure.

(2) Calculate the compressibility coefficient of the fracture system based on the bottom-hole pressure drop.

(3) Calculate the volume of the fracture network based on the slope and compressibility coefficient.

As shown in Figure 17, when calculating the slope, it is best to select the early flowback data. However, due to the limitation of data acquisition frequency in the field, hourly data are unachievable; instead, fitting data are used which are obtained from the average daily data in the first 10 days of flowback.

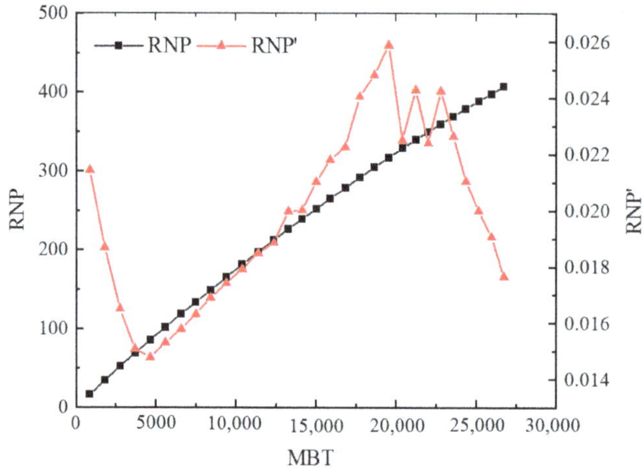

Figure 17. RNP curve and its slope in 30 days of flowback.

In the first few days, the slope (m) of the RNP curve is about 0.0214. An early-proposed empirical formula is used to calculate the compressibility coefficient of the fracture system, which is about 0.00201 MPa^{-1} [25]. Assuming that the fluid compressibility (C_w) is a constant of 2.59×10^{-6} MPa^{-1} and the volume factor B is a constant of 1.02, the total volume of the fracture system can be calculated by (25). which is about 3761.9 m^3.

$$V_m = \frac{B}{m\left(C_w + C_f\right)} \tag{25}$$

In numerical simulation, the total fracture volume is 3066.0 m^3, which gives an error of about 22%. Because the empirical formula is established for tight sandstones that are different from the target reservoir, the compressibility coefficient of the fracture system is corrected using the fracture volume from the simulation, which gives 0.00246 MPa^{-1} for future usage.

4. The Impact of Fracture Complexity

In the horizontal well-fracturing of tight conglomerate reservoirs in the Mahu well area, complex fracture networks are likely to be generated due to the heterogeneous nature of the conglomerate. In fracturing designs, complex fracture networks are not considered, and instead, fractures are assumed to propagate symmetrically with bi-wing fractures. Therefore, we start from such a fracturing design and use natural fractures to generate the complex fracture network formed in tight conglomerate reservoirs. In this section, the density of natural fractures is changed to enable us to understand how it affects the flowback curve.

The natural fracture density comparison is shown in Figure 18, with reservoir properties and fracturing parameters being consistent with those used in previous sections. The key parameters of hydraulic fractures are shown in Table 4.

Table 4. Hydraulic fracture parameters under different natural fracture densities.

Natural Fracture Densities	Fracture Volume (m^3)	Total Fracture Area (m^2)	Average Fracture Length (m)	Average Fracture Width (mm)	Average Fracture Conductivity (mD·m)
None	3066.00	334,861.60	277.83	9.37	660.66
Sparse	2649.00	324,490.60	193.61	8.80	618.30
Dense	2378.00	328,308.80	182.64	8.07	567.21

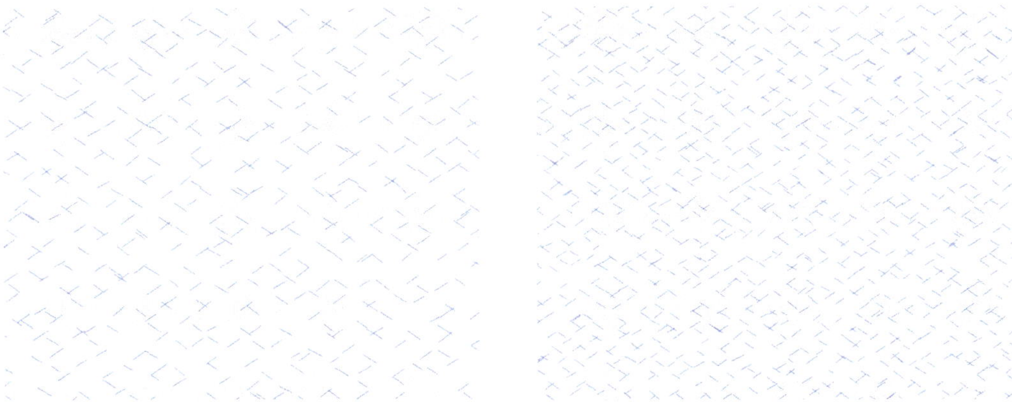

Figure 18. Different densities of natural fractures in the reservoir for simulation comparison.

As shown in Figure 19, in the absence of natural fractures, the hydraulic fracturing creates symmetrical bi-wing fractures. The denser the natural fractures are, the more complex the created fracture with shorter fracture lengths, narrower fracture widths, lower fracture conductivities, smaller total fracture volumes, and greater fluid loss will be. However, the total fracture surface area remains approximately equal in all three cases.

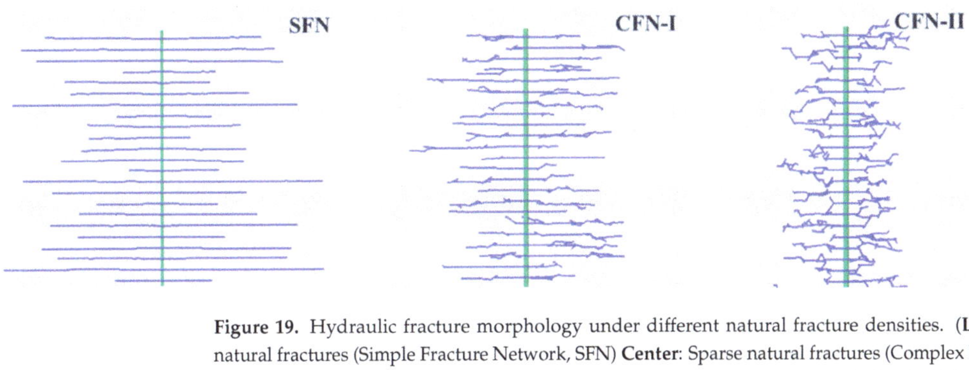

Figure 19. Hydraulic fracture morphology under different natural fracture densities. (**Left**: No natural fractures (Simple Fracture Network, SFN) **Center**: Sparse natural fractures (Complex Fracture Network I, CFN-I) **Right**: Dense natural fractures (Complex Fracture Network II, CFN-II)).

Using the same numerical simulation procedure as in Section 2.1, a post-fracturing production simulation is conducted, and the production curves are shown in Figure 20.

The degree of fracturing complexity is also reflected in the *MBT–RNP* plot, as well as the log–log plot shown in Figures 21 and 22. The degree of complexity of fractures does not significantly affect the overall trend of the *MBT–RNP* curve. As shown in Figure 23, as the complexity of the fractures increases, the *MBT–RNP* curve shifts upwards with a smaller initial slope. Despite this, the initial slope can still be used to accurately calculate the volume of fractures.

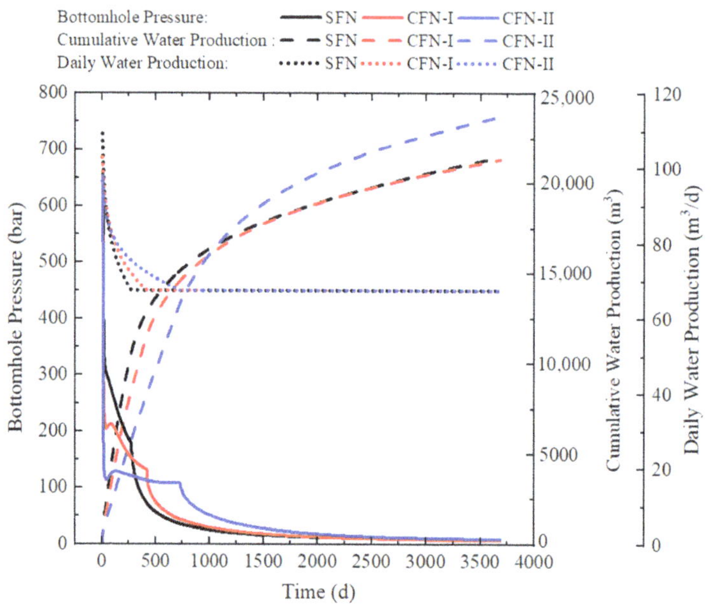

Figure 20. Production curves under different fracture complexities.

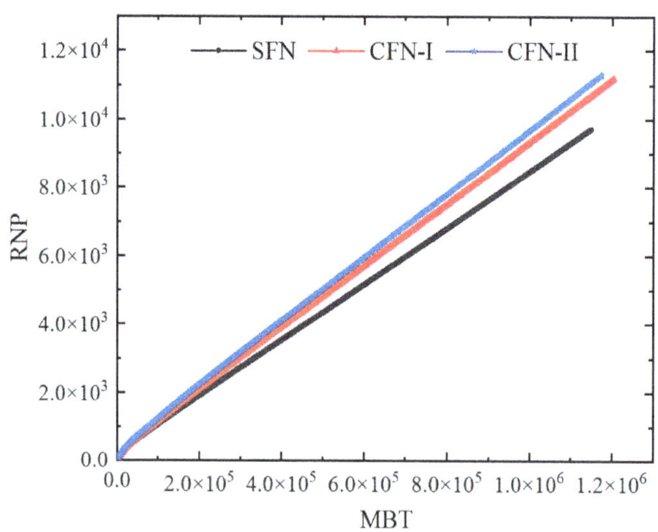

Figure 21. RNP difference under different fracture complexity.

The effect of fracture complexity on the *MBT–RNP* double-logarithmic curve trend is negligible, as the overall trend remains constant. However, the position of the *MBT–RNP* double-logarithmic curve shifts upward with the increasing fracture complexity. This upward shift is attributed to the increased resistance to fluid flow due to the tortuous path of the fracture. Furthermore, the initial slope of the derivative curve for complex fractures is less than 1, indicating that the linear flow is not present during the initial flowback phase.

This non-linear flow behavior is due to the flow resistance caused by the complex fracture path. Water imbibes the matrix adjacent to the fracture network due to the capillary force, which decreases water saturation in hydraulic fractures. The more complex the fracture network is, the bigger the fracture volume is, the longer the time water remains in the formation, the harder the flowback and the lower the flowback ratio will be.

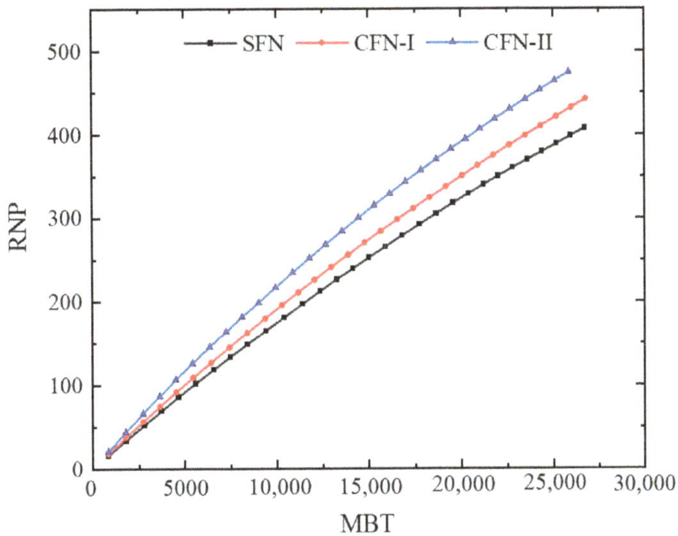

Figure 22. RNP difference under different fracture complexity (flowback for 30 days).

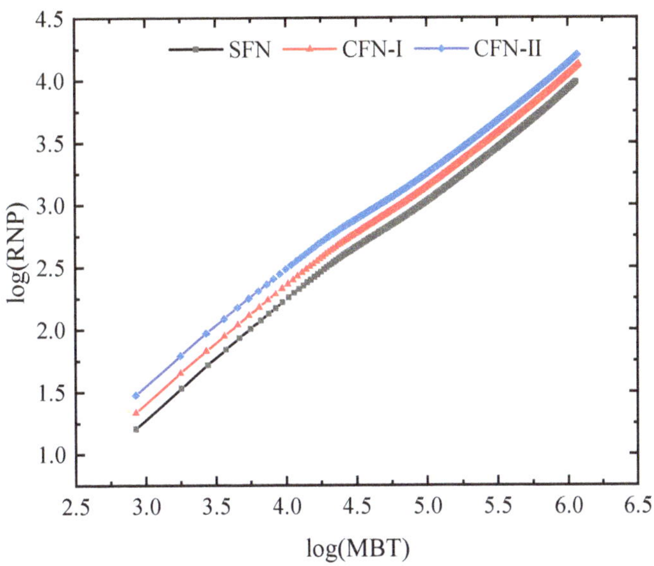

Figure 23. LogRNP difference under different fracture complexity.

As shown in Figure 24, comparing the *logMBT-logRNP′* slope for fractures with different complexities, it is observed that the slope decreases as the fracture complexity increases. This is because the flow path of the fracturing fluid becomes more complex as the fracture complexity increases, resulting in a greater deviation from linear flow in the characteristic curve. Accurate measurement of the *MBT–RNP* double-logarithmic curve and *logMBT-logRNP′* slope for complex fractures is essential for modeling the flowback of the fracturing fluid, which in turn is necessary for calculating the total volume of the fracture network.

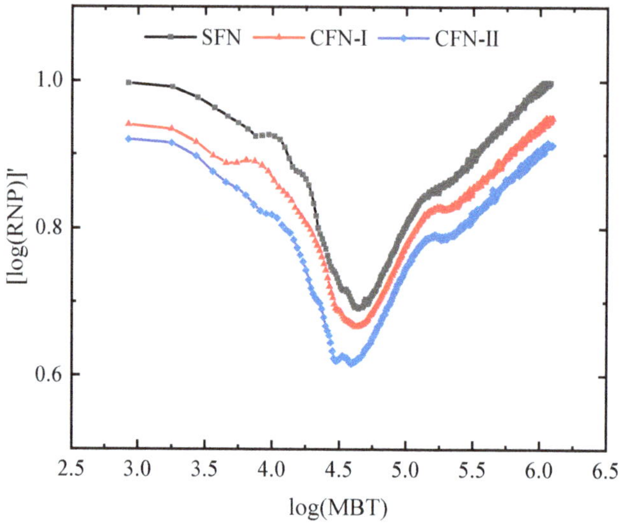

Figure 24. LogRNP′ difference under different fracture complexity.

5. Conclusions

During hydraulic fracturing, the pressure differential between the fracture and the reservoir can result in a significant loss of fracturing fluid into the formation, particularly adjacent to the fracture faces. During the shut-in, the fluid imbibes the rock matrix from hydraulic fractures, while during the flowback, the fluid quickly returns to fractures and the wellhead. To accurately model the flowback of the fracturing fluid, early and late-time flowback models were developed based on the principles of linear flow within the fracture and from the matrix into the fracture. Notably, the determination of the linear flow is based on the slope of the log–log plot of the MBT versus the RNP curve, which is equal to 1. To achieve greater precision in calculating the total volume of the fracture network, it is essential to obtain accurate pressure and production data in the first few days of the flowback, where the fracturing fluid mainly comes from the hydraulic fractures. The complexity of the flow path of the fracturing fluid during the initial flowback stage determines the deviation of the characteristic flowback curve from the linear flow region, which enables people to know the complexity of the fracture network from the flowback data. Ultimately, with a precise characterization of the fracture compressibility and the early flowback characteristic curve, the total volume of the fracture network can be calculated using the proposed model in this study. In the future, this model can refer to the grid strategy of the discretization complex domain to establish a three-dimensional heterogeneous porous media multiphase flow simulation method.

Author Contributions: Conceptualization, J.L. (Jianmin Li); Data curation, X.C.; Formal analysis, J.L. (Jie Li), S.L., J.L. (Jianmin Li) and J.L. (Jiayan Li); Investigation, Z.L.; Methodology, J.L. (Jie Li), S.L., Z.L. and T.L.; Resources, J.L. (Jie Li), J.L. (Jianmin Li), X.C. and J.L. (Jiayan Li); Software, S.L.; Validation, J.L. (Jie Li), S.L. and Z.L.; Writing—original draft, S.L., Z.L. and T.L.; Writing—review & editing, T.L. All authors have read and agreed to the published version of the manuscript.

Funding: This work was financially supported by the General Program Grant from the National Natural Science Foundation of China (52274051 and 52174045), the Strategic Cooperation Technology Projects of CNPC and CUPB (ZLZX2020-01), and the Foundation for Innovative Research Groups of the National Natural Science Foundation of China (51521063).

Data Availability Statement: Not applicable.

Acknowledgments: This work was financially supported by the General Program Grant from the National Natural Science Foundation of China (52274051 and 52174045), the Strategic Cooperation Technology Projects of CNPC and CUPB (ZLZX2020-01), and the Science and Technology Project of PetroChina (2022ZS0607). We sincerely thank Schlumberger for donating Petrel subsurface software to China University of Petroleum in Beijing.

Conflicts of Interest: The authors declare no conflict of interest.

Appendix A. Nomenclature

Symbol Names	Units	Meaning
q_m	m^3/min	Liquid provided by the matrix
ρ_m	kg/m^3	Density of liquid in the matrix
B_m	-	Volume factor of fluid in the matrix
q_s	m^3/min	Wellhead flow rate
ρ_s	kg/m^3	Wellhead liquid density
B_s	-	Wellhead liquid volume factor
ρ_f	kg/m^3	Fluid density in the fracture
V_f	m^3	Fracture volume
ρ_{wb}	kg/m^3	Fluid density in the wellbore
V_{wb}	m^3	Wellbore volume
C_f	MPa^{-1}	Fracture compression factor
P_f	MPa	Pressure in the fracture
C_{wb}	MPa^{-1}	Compression factor in the wellbore
P_{wb}	MPa	Pressure in the wellbore
C_{st}	MPa^{-1}	Total compression factor
x_f	m	Half-length of the fracture
K_f	mD	Fracture permeability
μ	mPa·s	Liquid viscosity
ϕ_f	-	Fracture porosity
R_w	m	Wellbore radius
C_t	MPa^{-1}	Total compression factor of matrix
P_{wf}	MPa	Flowing bottomhole pressure
$\overline{P}$	MPa	Average pressure
N_P	m^3	Cumulative volume of liquid
P_i	MPa	Initial pressure
RNP	$MPa{\cdot}min/m^3$	Rate-normalized pressure
MBT	min	Material-balance time
C_w	MPa^{-1}	Compression factor of water
V_m	m^3	Matrix volume
L	m	Cluster spacing
P_m	MPa	Matrix pressure

Appendix B. Pumping Schedules

Steps	Pump Rate (m^3/min)	Fluid Name	Fluid Volume (m^3)	Slurry Volume (m^3)	Prop. Conc (kg/m^3)	Proppant	Pump Time (min)
1	10	Slickwater	180	180	0	None	2
2	10	Slickwater	55.33	75.33	30	Sand 40/70	5.6
3	10	Slickwater	83.05	158.38	60	Sand 40/70	8.49
4	10	Slickwater	92.29	250.67	90	Sand 40/70	9.54
5	10	Slickwater	110.77	361.44	120	Sand 40/70	11.58
6	10	Slickwater	132.88	494.32	150	Sand 40/70	14.04
7	10	Slickwater	156.86	651.18	180	Sand 40/70	16.76
8	10	Slickwater	174.02	825.2	210	Sand 40/70	18.79
9	10	Slickwater	186.89	1012.09	240	Sand 40/70	20.39
10	10	Slickwater	215.27	1227.36	270	Sand 40/70	23.73
11	10	Slickwater	110.77	1338.13	300	Sand 40/70	12.34
12	10	Slickwater	50.38	1388.51	330	Sand 40/70	5.67

References

1. Chen, B.; Barboza, B.R.; Sun, Y.; Bai, J.; Thomas, H.R.; Dutko, M.; Cottrell, M.; Li, C. A Review of Hydraulic Fracturing Simulation. *Arch. Comput. Methods Eng.* **2022**, *29*, 1–58. [CrossRef]
2. Wang, J.; Mao, X.; Peng, C.; Chen, J.; Deng, H.; Liu, Z.; Wang, W.; Fu, Z.; Wang, C. Three-Dimensional Refined Modelling of Deep Structures by Using the Level Set Method: Application to the Zhaoping Detachment Fault, Jiaodong Peninsula, China. *Math. Geosci.* **2023**, *55*, 229–262. [CrossRef]
3. Wang, L.; Golfier, F.; Tinet, A.; Chen, W.; Vuik, C. An efficient adaptive implicit scheme with equivalent continuum approach for two-phase flow in fractured vuggy porous media. *Adv. Water Resour.* **2022**, *163*, 104186. [CrossRef]
4. Wang, L.; Chen, W.; Vuik, C. Hybrid-dimensional modeling for fluid flow in heterogeneous porous media using dual fracture-pore model with flux interaction of fracture–cavity network. *J. Nat. Gas. Sci. Eng.* **2022**, *100*, 104450. [CrossRef]
5. Weng, X.; Kresse, O.; Cohen, C.E.; Wu, R.; Gu, H. Modeling of Hydraulic-Fracture-Network Propagation in a Naturally Fractured Formation. *SPE Prod. Oper.* **2011**, *4*, 368–380.
6. Kresse, O.; Weng, X.; Gu, H.; Wu, R. Numerical Modeling of Hydraulic Fractures Interaction in Complex Naturally Fractured Formations. *Rock Mech. Rock Eng.* **2013**, *46*, 555–568. [CrossRef]
7. Agrawal, S.; Sharma, M.M. Liquid Loading within Hydraulic Fractures and Its Impact on Unconventional Reservoir Productivity. In Proceedings of the SPE/AAPG/SEG Unconventional Resources Technology Conference, Denver, CO, USA, 12–14 August 2013; Volume 2013, pp. 1314–1318.
8. Pagels, M.; Willberg, D.M.; Edelman, E.; Zagorski, W.; Frantz, J. Quantifying Fracturing Fluid Damage on Reservoir Rock to Optimize Production. In Proceedings of the SPE/AAPG/SEG Unconventional Resources Technology Conference, Denver, CO, USA, 12–14 August 2013; Volume 2013, pp. 1766–1774.
9. Cheng, Y. Impact of Water Dynamics in Fractures on the Performance of Hydraulically Fractured Wells in Gas Shale Reservoirs. *J. Can. Pet. Technol.* **2012**, *51*, 143–151. [CrossRef]
10. McClure, M. The Potential Effect of Network Complexity on Recovery of Injected Fluid Following Hydraulic Fracturing. In Proceedings of the SPE Unconventional Resources Conference, The Woodlands, TX, USA, 1–3 April 2014.
11. Parmar, J.; Dehghanpour, H.; Kuru, E. Displacement of water by gas in propped fractures: Combined effects of gravity, surface tension, and wettability. *J. Unconv. Oil Gas Resour.* **2014**, *5*, 10–21. [CrossRef]
12. Liu, Y. Modeling of Recovery and In-Situ Distribution of Fracturing Fluid in Shale Gas Reservoirs Due to Fracture Closure, Proppant Distribution and Gravity Segregation. Master's Thesis, University of Alberta, Edmonton, AB, Canada, 2017.
13. Liu, N.; Liu, M.; Zhang, S. Flowback patterns of fractured shale gas wells. *Nat. Gas Ind.* **2015**, *35*, 50–54. [CrossRef]
14. Song, B. Model for Fracturing Fluid Flowback and Characterization of Flowback Mechanisms. Ph.D. Thesis, Texas A & M University, College Station, TX, USA, 2014.
15. Warpinski, N.R.; Mayerhofer, M.J.; Vincent, M.C.; Cipolla, C.L.; Lolon, E. Stimulating Unconventional Reservoirs: Maximizing Network Growth While Optimizing Fracture Conductivity. *J. Can. Pet. Technol.* **2009**, *10*, 39–51. [CrossRef]
16. Modeland, N.; Buller, D.; Chong, K.K. Statistical Analysis of the Effect of Completion Methodology on Production in the Haynesville Shale. In Proceedings of the North American Unconventional Gas Conference and Exhibition, The Woodlands, TX, USA, 14–16 June 2011.
17. Mayerhofer, M.J.; Lolon, E.P.; Youngblood, J.E.; Heinze, J.R. Integration of Microseismic Fracture Mapping Results With Numerical Fracture Network Production Modeling in the Barnett Shale. In Proceedings of the SPE Annual Technical Conference and Exhibition, San Antonio, TX, USA, 24–27 September 2006.
18. McClure, M.W.; Horne, R.N. Correlations between formation properties and induced seismicity during high pressure injection into granitic rock. *Eng. Geol.* **2014**, *175*, 74–80. [CrossRef]
19. Yang, R.; Huang, Z.; Li, G.; Yu, W.; Sepehrnoori, K.; Lashgari, H.R.; Tian, S.; Song, X.; Sheng, M. A Semianalytical Approach To Model Two-Phase Flowback of Shale-Gas Wells with Complex-Fracture-Network Geometries. *SPE J.* **2017**, *22*, 1808–1833. [CrossRef]
20. Lee, S.T.; Brockenbrough, J.R. A New Approximate Analytic Solution for Finite-Conductivity Vertical Fractures. *SPE Form. Eval.* **1986**, *1*, 75–88. [CrossRef]
21. Daviau, F.; Mouronval, G.; Bourdarot, G.; Curutchet, P. Pressure Analysis for Horizontal Wells. *SPE Form. Eval.* **1988**, *3*, 716–724. [CrossRef]
22. Al-Rbeawi, S. Transient and Pseudo-Steady-State Inflow Performance Relationships for Multiphase Flow in Fractured Unconventional Reservoirs. *Transp. Porous Media* **2019**, *126*, 743–777. [CrossRef]
23. Denney, D. Practical solutions for pressure transient responses of fractured horizontal wells in unconventional reservoirs. *J. Pet. Technol.* **2010**, *62*, 63–64. [CrossRef]
24. Stalgorova, E.; Mattar, L. Practical Analytical Model to Simulate Pro duction of Horizontal Wells with Branch Fractures. In Proceedings of the SPE Canadian Unconventional Resources Conference, Calgary, AB, Canada, 30 October–1 November 2012.
25. Aguilera, R. Recovery Factors and Reserves in Naturally Fractured Reservoirs. *J. Can. Pet. Technol.* **1999**, *38*. [CrossRef]

processes

Article

Impact of Formation Dip Angle and Wellbore Azimuth on Fracture Propagation for Shale Reservoir

Kefeng Yang [1,*], Lei Wang [2], Jingnan Ge [3], Jiayuan He [4], Ting Sun [5], Xinliang Wang [5] and Yanxin Zhao [2]

[1] China Petrochemical Corporation, Beijing 100728, China
[2] School of Petroleum Engineering, Yangtze University, Wuhan 434000, China; wang-lei@yangtzeu.edu.cn (L.W.); wxl094315@outlook.com (Y.Z.)
[3] PetroChina Zhejiang Oilfield Company, Hangzhou 311100, China; gejn85@petrochina.com.cn
[4] Sinopec Petroleum Exploration and Production Development Research Institute, Beijing 102206, China; hejy.syky@sinopec.com
[5] College of Safety and Ocean Engineering, China University of Petroleum, Beijing 100083, China; ting.sun@cup.edu.cn (T.S.); wxlcup@foxmail.com (X.W.)
* Correspondence: yangkefeng@sinopec.com

Abstract: The significant vertical heterogeneity, variations in ground stress directions, and irregular bedding interfaces make it extremely challenging to predict fracture propagation in continental shale reservoirs. In this article, we conducted a series of triaxial laboratory experiments on continental shale outcrop rocks to investigate the effects of formation dip angle and wellbore orientation on crack propagation under horizontal well conditions. Our study revealed that fracture propagation features can be categorized into four distinct types: (1) hydraulic fractures pass through the bedding interface without activating it; (2) fractures pass through and activate the bedding interface; (3) hydraulic fractures open and penetrate the bedding interface while also generating secondary fractures; and (4) hydraulic fractures open but do not penetrate the bedding interface. We found that as the dip angle decreases, the likelihood of fractures penetrating through the bedding interface increases. Conversely, as the dip angle increases, fractures are more likely to simply open the interface without penetrating it. Moreover, we observed that the well azimuth significantly affects the degree of fracture distortion. Specifically, higher azimuth angles corresponded to a higher degree of fracture distortion.

Keywords: continental shale; hydraulic fractures; formation dip angle; borehole azimuth

Citation: Yang, K.; Wang, L.; Ge, J.; He, J.; Sun, T.; Wang, X.; Zhao, Y. Impact of Formation Dip Angle and Wellbore Azimuth on Fracture Propagation for Shale Reservoir. *Processes* **2023**, *11*, 2419. https://doi.org/10.3390/pr11082419

Academic Editor: Qingbang Meng

Received: 18 July 2023
Revised: 6 August 2023
Accepted: 9 August 2023
Published: 11 August 2023

1. Introduction

China possesses abundant shale gas resources, estimated to be approximately 1.3 times greater than traditional gas reserves [1]. However, due to significant vertical heterogeneity, variations in formation stress directions, and irregular bedding interfaces, controlling fracture propagation in shale gas reservoirs proves to be extremely challenging [2]. Consequently, this results in low efficiency in enhancing the recovery factor. In order to gain a better understanding of the influence of formation dip angle and wellbore azimuth on fracture propagation, a series of true triaxial laboratory experiments were conducted in this study.

To date, numerous researchers have performed true triaxial experiments on outcrops to investigate the behavior of hydraulic fracture propagation. Some studies [3–10] utilized outcrops consisting of sandstone and sand-coal inter-beds. The findings indicate that when the fracture propagation reaches the interface, the fracture either ceases growth, changes its propagation direction, bifurcates into multiple directions, or continues to propagate along the original path and penetrates the bedding interface. The presence of weak cementation interfaces hinders the vertical propagation of hydraulic fractures, while high vertical stress differences and strong interfacial strength between sand layers facilitate vertical fracture propagation. Additionally, when the horizontal stress difference reaches 3 MPa, hydraulic fracture propagation changes its direction, connecting to natural fractures in proximity to wells or weak cementation interfaces. Li Zhi et al. [11] conducted a study on the influence of bedding interfaces on fracture propagation. They highlighted the presence of wide openings and shear zones after the main fracture reached the bedding interface. These shear zones are typically longer than the opening zones and serve as the main channels for fluid flow. Altammar [12] and colleagues investigated the impact of vertical stress and interlayer properties on fracture height using cement samples measuring 30 cm × 30 cm × 10 cm. Their research demonstrated that the manner in which fractures penetrate through bedding interfaces depends on formation properties/conditions, and operational parameters. Sun Keming et al. [13,14] analyzed the influence of bedding dip angles and strength on fracture growth. Their findings indicated that when the main fractures, which experience the minimum vertical stress, reach the bedding interface, the smaller the angle between the bedding interface and fracture direction, the more likely the fracture is to change its original direction and extend along the bedding interface. Conversely, the larger the angle between the bedding interface and the original fracture direction, the greater the likelihood that the main fracture will penetrate the bedding interface and continue growing in its original direction. Zhou Tong [15], Liu Liming [16], Pang Tao [17], and others examined the impact of formation dip angles on fracture propagation through laboratory experiments and numerical simulations. They discovered that an increasing formation dip angle can restrict fracture growth in the vertical direction while enhancing the opening of the bedding interface. As the dip angle becomes larger, it becomes more challenging for the fracture to penetrate through the bedding interface. Jia Changgui, Hou Bing, and colleagues [18–22] conducted a series of true triaxial experiments on wells with high dip angles to study the effects of perforation phase angle, well dip angle, and well azimuth on initiating hydraulic fractures. Their results revealed that the well dip angle primarily influences the degree of fracture distortion, while the perforation phase angle mainly affects the number of hydraulic fractures. Furthermore, an increase in the azimuth angle leads to an increase in the degree of distortion for hydraulic fractures.

The existing literature primarily focuses on the impact of geological and engineering factors on hydraulic fracture propagation, particularly in relation to parallel bedding interfaces. However, there are few studies that have thoroughly examined the influence of bedding dip angles on fracture propagation. Furthermore, the majority of the existing literature concentrates on deviated wells rather than horizontal wells. As a result, this study aims to investigate the effects of formation dip angle and wellbore azimuth angle on fracture expansion behavior in continental shale. To accomplish this, we will conduct true triaxial physical simulation experiments on outcrop rock samples and subsequently clarify the fracture propagation law of continental shale. By evaluating fracture expansion behavior in relation to varying formation dip angles and wellbore azimuth angles, the study aims to provide comprehensive insights into the influence of these factors on fracture propagation. This research will contribute to a better understanding of hydraulic fracturing in continental shale formations and aid in optimizing drilling and fracturing operations.

2. Design of the Experiments

2.1. Experiment Protocol and Procedure

In this study, we utilized a large-scale true triaxial physical simulation system (refer to Figure 1) to investigate fracture propagation in continental shale outcrops. The selected outcrop samples were obtained from the Jiashengqi area in Anwen Town, Qijiang District, Chongqing City. These outcrop samples are representative of continental shale and exhibit favorable conditions, characterized by a high concentration of dark organic matter. To prepare the samples for experimentation, they were cut into 300 mm × 300 mm × 300 mm test specimens, as depicted in Figure 2. The preparation procedure for these samples is outlined in Figure 3. In the figure, α denotes the formation dip angle, which represents the angle between the bedding interface and the horizontal plane. For a more comprehensive understanding of the parameters of the experimental rock samples, please refer to Figure 3.

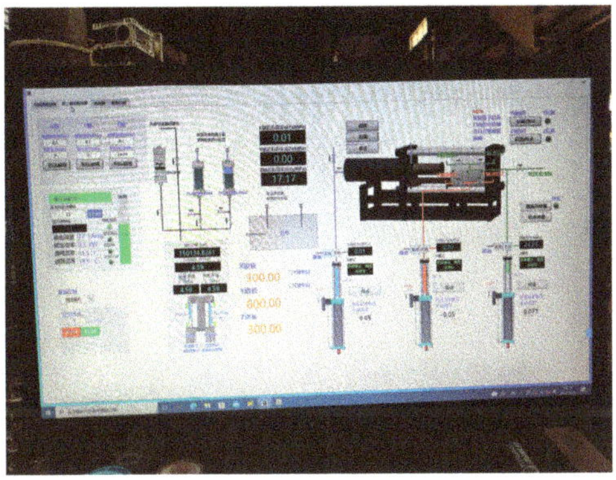

Figure 1. Large scale true triaxial physical simulation system.

Figure 2. Process of cutting a shale outcrop.

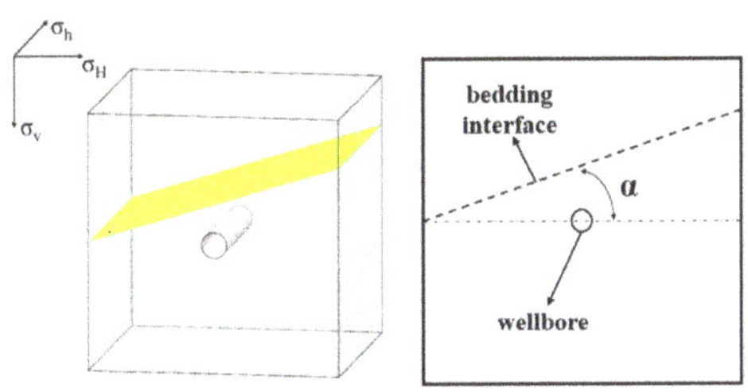

Figure 3. Schematic for preparing rock samples.

Figure 4 illustrates the four different drilling schemes employed in the study, each having distinct wellbore azimuths. Wellbore azimuth refers to the angle between the wellbore axis and the direction of the maximum horizontal principal stress. To simulate the actual drilling process, a drill bit with a diameter of 30 mm is used to bore the hole in accordance with the scheme depicted in Figure 4. Following the drilling process, the simulated wellbore is sealed using epoxy resin anchorage glue, as shown in Figure 5. In order to prevent the anchorage glue from obstructing the perforation holes, insulating tape is wrapped around the upper portion of the wellbore prior to sealing. The simulated wellbore possesses the following specifications: an outer diameter of 22 mm, an inner diameter of 10 mm, and a sealed bottom. The length of the wellbore can be adjusted based on the desired wellbore azimuth. Additionally, four 2 mm-diameter holes are drilled 10 mm above the bottom of the wellbore to simulate the presence of perforation holes.

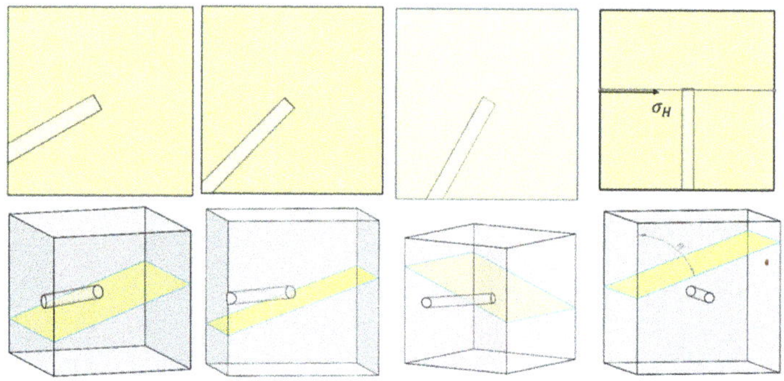

Figure 4. Drilling schemes for different wellbore azimuths.

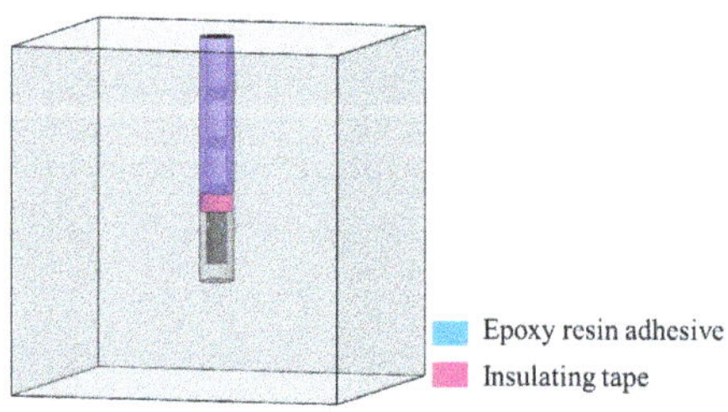

Figure 5. Sealing and "well cementation" for rock samples.

2.2. Methodology of the Experiments

In this study, a total of nine experiments were conducted based on the similarity criterion to investigate the influence of the dip angle and wellbore azimuth angle on fracture propagation [23]. The objective of the first five sets of experiments was to explore the impact of different bed dip angles on fracture propagation. The remaining four sets of experiments specifically assessed the impact of altering the wellbore azimuth on fracture propagation. For more precise information on the experimental parameters, including the specific bed dip angles, wellbore azimuths, and their effects on fracture propagation, please refer to Table 1 in the paper.

Table 1. Experiment parameters.

Rock Sample No.	Drainage m^3/min	Viscosity mPa·s	σ_h MPa	σ_H MPa	σ_v MPa	Formation Dip Angle	Wellbore Azimuth
Q1			17	30		0	
Q2			17	30		10	
Q3	0.0001	5	17	30	25	20	90
Q4			17	30		30	
Q5			17	30		40	
F1			17	30			30
F2	0.0001	5	17	30	25	20	45
F3			17	30			60
F4			17	30			90

2.3. Experimental Results

The experiment involved the use of fluorochrome to track the pattern of fracture propagation in samples. After the experiment, the samples were cut along the fracture propagation plane to observe the pattern, which was marked by fluorochrome. Table 2 and Figure 6 provide a summary of the laboratory results. It is evident from Figure 6 that the main fracture in all samples initiated in a direction perpendicular to the wellbore. As it propagated vertically, it activated and opened at least two bedding interfaces and fully penetrated at least one bedding interface. In Figure 6, the left side represents the post-compression rock sample anatomy diagram, while the right side shows the recomposition of the fracture morphology. To clarify the color representation in the fracture morphology recomposition: the red color represents the main hydraulic fracture plane, which is the primary fracture that propagated through the rock sample. The blue color represents the activated interlayer interface, indicating the bedding interfaces that have been affected or opened by the hydraulic fracture. The green color represents the secondary hydraulic fracture plane generated by the activated interlayer interface. These secondary fractures occur along the activated bedding interfaces. The yellow color indicates the direction of the bedding interface within the rock sample. Based on Figure 6, the relationship between the main fracture and the bedding interface can be categorized into three distinct patterns: (1) Fracture Net with '十' Pattern: The hydraulic fracture opened and passed through the bedding interface, resulting in a fracture network resembling the shape of the Chinese character '十'. This pattern is depicted in Figure 6d,i.; (2) Fracture Net with 'T' Pattern: The hydraulic fracture opened the bedding interface but did not penetrate through it, forming a fracture network that resembles the shape of a letter 'T'. This pattern is illustrated in Figure 6g,h; and (3) Fracture Net with 'Y' Pattern: The hydraulic fracture opened and penetrated through the bedding interface, generating secondary hydraulic fractures along the bedding interface. This pattern forms a fracture network that resembles the shape of the letter 'Y'. Additionally, it should be noted that not all bedding interfaces in the rock samples were accurately identified. The observed bedding interfaces are those that were activated and opened during the experiment. It is speculated that there may be another pattern of fracture propagation net with an 'I' shape, where the hydraulic fracture passes through the bedding interface directly without activating it.

Table 2. Experiment results summary.

Rock Sample No.	Summary of Fracture Propagation for All the Samples
Q1	Having activated, opened, and penetrated 2 bedding interfaces and generated 1 secondary fracture.
Q2	Having activated, opened, and penetrated 2 bedding interfaces and generated 2 secondary fractures.
Q3	Having activated, opened, and penetrated 2 bedding interfaces and generated 1 secondary fracture.
Q4	Having activated, opened, and penetrated 2 bedding interfaces
Q5	Having activated, opened, and penetrated 2 bedding interfaces
F1	Having activated, opened, and penetrated 2 bedding interfaces and generated 1 secondary fracture.
F2	Having activated and opened, 2 bedding interfaces and penetrated only 1 bedding interface.
F3	Having activated and opened, 2 bedding interfaces and penetrated only 1 bedding interface.
F4	Having activated, opened, and penetrated 3 bedding interfaces.

The use of fluorochrome in the experiment allowed for precise tracking and visualization of fracture propagation patterns. By cutting the samples along the fracture propagation plane and observing the marked patterns, researchers were able to gain deeper insights into the behavior of fractures in response to variations in bed dip angle and wellbore azimuth. Table 2 provides a comprehensive summary of the laboratory results, offering valuable information regarding the experiments conducted. The results presented in Figure 6 serve as graphical representations, displaying the anatomical and morphological characteristics of the fracture patterns. Analyzing Figure 6, it becomes evident that the main fracture consistently initiated perpendicular to the wellbore direction in all samples. The observation results show that the state of ground stress has a decisive influence on the initiation path and extension pattern of hydraulic fractures. As the fracture propagated vertically, it exhibited a remarkable ability to activate and open multiple bedding interfaces within the rock samples. It was also noted that the fracture fully penetrated at least one bedding interface, emphasizing the force and extent of its propagation. To aid in the interpretation of the fracture morphology recomposition in Figure 6, color-coded visualizations were employed. The red color represents the main hydraulic fracture plane, which denotes the primary fracture path that propagated through the rock sample. The blue color represents the activated interlayer interface, signifying the bedding interfaces affected or opened by the hydraulic fracture. The green color represents the secondary hydraulic fracture plane generated along the activated bedding interface, indicating the occurrence of additional fractures parallel to the bedding interfaces. Finally, the yellow color represents the orientation of the bedding interface within the rock sample. Examining the relationship between the main fracture and the bedding interface, three distinct patterns emerged. The first pattern, known as the "Fracture Net with '十' Pattern," was characterized by the hydraulic fracture passing through the bedding interface, resulting in a fracture network resembling the shape of the Chinese character '十'. This pattern is exemplified in Figure 6d,i. The second pattern, referred to as the "Fracture Net with 'T' Pattern," was marked by the hydraulic fracture opening the bedding interface without complete penetration, forming a fracture network resembling the shape of the letter 'T'. This pattern is illustrated in Figure 6g,h. Finally, the third pattern, known as the "Fracture Net with 'Y' Pattern", featured the hydraulic fracture opening and penetrating the bedding interface, generating secondary hydraulic fractures along the bedding interface. This pattern forms a fractured network resembling the shape of the letter 'Y'. It is important to note that while the observed bedding interfaces were accurately identified in the rock samples, there may be additional patterns of fracture propagation that have not been thoroughly investigated. One potential pattern of interest is the "Fracture Net with 'I' Pattern", where the hydraulic fracture directly passes through the bedding interface without activating it.

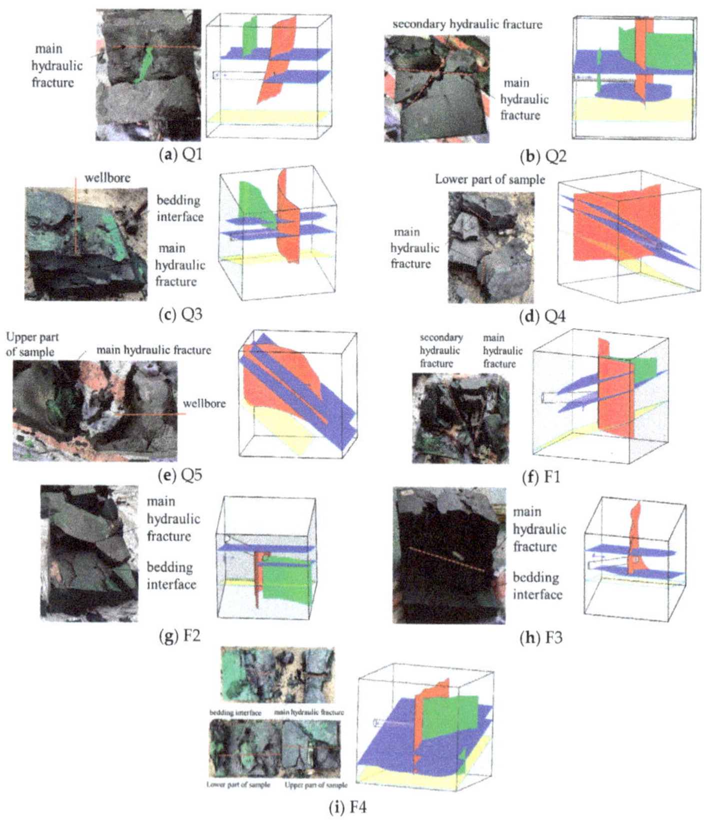

Figure 6. Experimental results under different formation dip angles.

Figure 7 provides a summary of the injection pressure curves observed in the experiments. Based on the characteristics of these pressure curves, we can identify at least two fracturing modes: (1) Peak Fracturing Pressure: Some injection pressure curves exhibit a distinct peak, indicating a clear fracturing event. This phenomenon is observed in rock samples Q4 and F3; and (2) No Peak Fracturing Pressure: On the other hand, other injection pressure curves do not show a clear peak, suggesting the presence of weak bedding interfaces around the perforated holes. In these cases, the hydraulic fractures initially propagate along these weak bedding interfaces, resulting in a lower fracturing pressure. This behavior can be observed in rock samples Q1 and F4. Additionally, by considering the extension pressure characteristics in combination with the observations of fracture propagation, we can identify two fracture extension modes: (1) Slowly Rising Trend: Some injection pressure curves exhibit a gradual and slow rise, indicating limited vertical fracture propagation. This trend is evident in samples Q3 and F1; and (2) Slowly Declining Trend: Conversely, other injection pressure curves demonstrate a gradual and slow decline, indicating unrestricted vertical fracture propagation. This trend can be observed in samples Q2 and Q5. Furthermore, sample Q3 exhibits a complex fracture pattern, which could be attributed to the presence of intercrossing fractures within the sample.

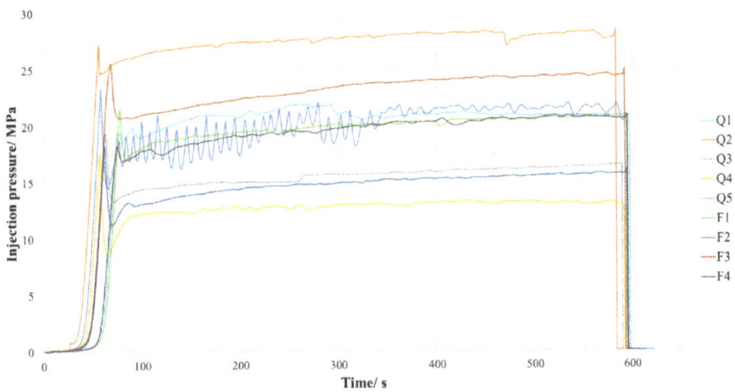

Figure 7. Comparison of the water-injection pressure curves.

3. Analysis of Impact Factors

3.1. Impact of Formation Dip Angles

The main objective of this section is to examine how changes in the dip angle of the formation impact the propagation of fractures. To achieve this, five experimental sets were conducted, with the formation dip angle being the only variable parameter while keeping all other factors constant. The formation dip angles tested were 0°, 10°, 20°, 30°, and 40°, as clearly stated in Table 1. The laboratory results obtained from the experiments are presented in Figure 6a–e, and a comprehensive comparison of these results can be found in Table 3.

Table 3. Comparison of the experimental results under different formation dip angles.

Rock Sample No.	Depth of Fracture Extension/cm	Number of Activated Bedding Interfaces	Degree of Opened Bedding Interfaces	Number of Secondary Fractures
Q1	23.1	2	1.5	1
Q2	22.6	2	1.5	2
Q3	21.8	2	1.7	1
Q4	20.5	2	1.7	1
Q5	19.8	2	2	/

Based on the observations depicted in Figure 6 and summarized in Table 3, it can be inferred that the main fractures in all five samples were oriented perpendicular to the wellbore, with each fracture successfully penetrating two bedding interfaces. Analysis of the results revealed that as the formation dip angle increased, the opening of the bedding interfaces also increased, while the vertical length of the fractures decreased. Consequently, the study's conclusion states that small formation dip angles facilitate vertical extension of fractures and reduce the likelihood of activating the bedding interfaces. This phenomenon can be attributed to several factors. Firstly, when the vertical ground stress remains constant, higher dip angles result in lower stress on the bedding interfaces [24]. Consequently, the bedding interfaces become more susceptible to shear stress-induced destruction and are more easily activated through shear failure. This implies that fractures propagate along the weak bedding interfaces, reducing the vertical length of the fractures. Additionally, the higher the dip angle, the less likely it is for secondary fractures to occur on the activated bedding interface. This can be explained by the larger opening of the bedding interfaces, which leads to greater filtration loss and a decrease in water injection accumulation. A higher dip angle creates a larger opening, limiting the formation of secondary fractures and resulting in more focused fracture propagation.

In summary, when the formation dip angle is small (such as 0°), hydraulic fractures can smoothly extend vertically and penetrate the bedding interfaces with a lower likelihood of activation. Conversely, larger dip angles (greater than 40°) significantly increase the likelihood of activating the bedding interfaces, making it challenging for fractures to grow vertically. When the dip angle ranges from 10° to 30°, vertical fracture growth is comparatively easier, while the likelihood of activating the bedding interfaces is relatively higher, resulting in a more complex fracture propagation pattern.

The presence of natural fractures introduces various challenges for accurately predicting and controlling the formation and distribution of hydraulic fractures. These pre-existing fractures can create complex fracture networks as the injected fracturing fluid can propagate along these pathways, causing the hydraulic fractures to deviate from their intended trajectory. This can complicate the management of fracture geometry and the optimization of reservoir stimulation. Moreover, natural fractures contribute to fracturing fluid loss. Areas of poor fracture connectivity or high porosity can lead to fluid leakage, resulting in pressure loss and inefficient utilization of fracturing fluid. Apart from operational difficulties and increased costs, this fluid loss can also have adverse environmental impacts. However, it is important to note that natural fractures can also impede the vertical expansion of hydraulic fractures, limiting their ability to adequately communicate with high-quality reservoirs. If a natural fracture intersects vertically with a hydraulic fracture or is shielded within the target zone, the hydraulic fracture may not be able to extend effectively into the reservoir, reducing the overall efficiency and productivity of the fracturing operation.

Therefore, it becomes crucial to thoroughly evaluate the geological characteristics of the formation, understand the nature of the fracture network, and consider the operational requirements when designing fracturing strategies. By analyzing field geological data and incorporating the geological model, it becomes possible to optimize the vertical expansion of hydraulic fractures, improve the connectivity of natural fractures, and shape a complex fracture network. This optimization can be achieved through careful adjustment of injection pressures, construction parameters, and the sequence of fracturing fluid injection.

3.2. *Impact of Wellbore Azimuth Angles*

In order to thoroughly investigate the influence of wellbore azimuth on fracture propagation, a series of meticulously designed laboratory experiments were conducted. These experiments aimed to vary the wellbore azimuth while maintaining all other parameters constant, thereby isolating the effect of azimuth on fracture behavior. The tested wellbore azimuths were 30°, 45°, 60°, and 90°, as specified. The results obtained from these experiments are illustrated in Figure 6f–i, and a comparison of the experimental outcomes can be found in Table 4.

Table 4. Comparison of the experimental results under different wellbore azimuth angles.

Rock Sample No.	Amplitude/MPa	Distortion Length/cm	Depth of Fracture Extension/cm	The Number of Penetrated Layers/Layer
F1	4.43	20.6	21.5	2
F2	3.56	9.6	17.8	1
F3	3.65	8.2	20.3	1
F4	3.93	0	20.8	3

Based on the observations depicted in Figure 6 and summarized in Table 4, it can be concluded that the fractures consistently initiated perpendicular to the wellbore, regardless of the tested wellbore azimuths. However, variations in wellbore azimuth did exhibit significant differences in fracture behavior, specifically in terms of distortion level and initiation direction. The length of the fractures did not show a significant variation with different wellbore azimuths. However, the distortion level of the fractures was found to be notably different. When the wellbore azimuth was smaller, the distortion level of the fracture increased. This can be attributed to the fracture initially propagating perpendicular to the wellbore. However, due to existing ground stress, the fracture changed its propagation direction to align along the maximum horizontal formation stress. As a result, the fracture experienced higher distortion when the wellbore azimuth was smaller. In contrast, when the wellbore azimuth was higher, fracture initiation occurred along the maximum horizontal formation stress, and the fracture continued to propagate in that direction without changing its trajectory. This resulted in less distortion of the fracture compared to smaller wellbore azimuths. Furthermore, it is worth noting that smaller wellbore azimuths required higher formation fracture pressures. This is because when the fracture initiates perpendicular to the wellbore, it needs to overcome higher stress to propagate and crack the formation.

In summary, the wellbore azimuth has a significant impact on the distortion level and initiation direction of fractures. Smaller wellbore azimuths result in higher distortion levels as the fracture changes its trajectory, while larger wellbore azimuths lead to fractures initiating and propagating parallel to the maximum horizontal formation stress. Additionally, smaller wellbore azimuths require higher formation fracture pressures due to the increased stress the fracture needs to overcome to propagate.

Therefore, it is crucial to integrate geological data in order to optimize the location of drilling before commencing the drilling process. By taking into account various factors such as geological characteristics, lithology, fault distribution, and other relevant data, it becomes possible to identify the optimal position for the borehole, ensuring a matching trajectory with the stress distribution within the formation. Through judicious selection of the relationship between the borehole trajectory and ground stress, fluid flow can be enhanced, the challenges associated with fracturing construction can be reduced, and the effectiveness of hydraulic fractures can be improved.

4. Conclusions

The paper utilized outcrop rock samples to conduct laboratory experiments, focusing on the influence of formation dip angle and wellbore azimuth on fracture propagation under horizontal well conditions. The key conclusions drawn from the study are as follows:

(1) Natural Fractures: The presence of natural fractures poses challenges in predicting and controlling hydraulic fracturing. It can result in multiple propagation characteristics of hydraulic fractures, which can be categorized into four types. These include fractures passing through bedding interfaces without activating them, fractures acti-

vating and opening bedding interfaces, fractures opening and passing through bedding interfaces while generating secondary fractures within them, and fractures opening bedding interfaces without penetrating them;

(2) Formation Dip Angle: A smaller dip angle of the formation leads to a greater vertical extension of the hydraulic fracture, making it easier to penetrate through interfaces. Additionally, a smaller dip angle reduces the likelihood of activating and opening bedding interfaces. On the other hand, a larger wellbore azimuth results in smoother fracture patterns and easier vertical propagation. However, when the formation dip angle falls within the range of 10–30°, the fracture behavior becomes more complex due to the balanced probability of the hydraulic fracture both opening and penetrating the bedding interfaces;

(3) Design Considerations: Prior to drilling, considering factors such as geological characteristics, lithology, and fault distribution can help determine the optimal drilling location. This aids in reducing the difficulties encountered during hydraulic fracturing operations. When designing hydraulic fracturing construction plans, it is essential to optimize construction parameters and pumping schedules based on a comprehensive evaluation of various factors. This approach facilitates achieving the optimal expansion of hydraulic fractures and shaping the fracture networks.

Author Contributions: Conceptualization, K.Y.; Methodology, K.Y. and L.W.; Investigation, K.Y. and L.W.; Resources, K.Y.; Writing—original draft, K.Y., L.W., J.H. and T.S.; Writing—review & editing, K.Y., L.W., J.G., J.H., T.S., X.W. and Y.Z. All authors have read and agreed to the published version of the manuscript.

Funding: This research was funded by Research on drilling and production enhancement technology of deep coalbed methane (Grant number P23207).

Data Availability Statement: No new data were created.

Conflicts of Interest: The authors declare no conflict of interest.

References

1. Zhang, J.; Zhou, Z.; Song, T. Comparison of exploration and development history, geological characteristics and exploitation conditions of shale gas in China and the United States and its enlightenment. *Acta Pet. Sin.* **2022**, *43*, 1687–1701.
2. Ma, K.; Wang, L.; Xu, W. Physical simulation of fracture propagation in lacustrine shale during hydraulic fracturing. *Chin. Sci. Technol. Pap.* **2022**, *17*, 539–545.
3. Daneshy, A.A. Hydraulic fracture propagation in layered formations. *Soc. Pet. Eng. J.* **1978**, *18*, 33–41. [CrossRef]
4. Mahjour, S.K.; Faroughi, S.A. Selecting representative geological realizations to model subsurface CO_2 storage under uncertainty. *Int. J. Greenh. Gas Control.* **2023**, *127*, 103920. [CrossRef]
5. Ahmadi, S.; Motie, M.; Soltanmohammadi, R. Proposing a modified mechanism for determination of hydrocarbons dynamic viscosity, using artificial neural network. *Pet. Sci. Technol.* **2020**, *38*, 699–705. [CrossRef]
6. Warpinski, N.R.; Schmidt, R.A.; Northrop, D.A. In-Situ stresses: The predominant influence on hydraulic fracture containment. *J. Pet. Technol.* **1982**, *34*, 653–664. [CrossRef]
7. Yang, J.; Wang, Y.; Li, Q. Experimental study on propagation mechanism of complex hydraulic fracture in coal-bed. *J. Coal Sci.* **2012**, *37*, 73–77.
8. Fu, H.; Cai, B.; Xiu, N. The study of hydraulic fracture vertical propagation in unconventional reservoir with beddings and field monitoring. *Nat. Gas Geosci.* **2021**, *32*, 1610–1621.
9. Li, D.Q.; Zhang, S.C.; Zhang, S.A. Experimental and numerical simulation study on fracturing through interlayer to coal seam. *J. Nat. Gas Sci. Eng.* **2014**, *21*, 386–396. [CrossRef]
10. Zou, Y.; Zhang, S.; Ma, X.; Zhou, T.; Zeng, B. Numerical investigation of hydraulic fracture network propagation in naturally fractured shale formations. *J. Struct. Geol.* **2016**, *84*, 1–13. [CrossRef]
11. Zhi, L.; Changgui, J.; Chunhe, Y. Propagation of hydraulic fissures and bedding planes in hydraulic fracturing of shale. *J. Rock Mech. Eng.* **2015**, *34*, 12–20.
12. Altammar, M.J.; Agrawal, S.; Sharma, M.M. Effect of geological layer properties on hydraulic-fracture initiation and propagation: An experimental study. *SPE J.* **2019**, *24*, 757–794. [CrossRef]
13. Sun, K.; Zhang, S.; Xin, L. Impacts of bedding directions of shale gas reservoirs on hydraulically induced crack propagation. *Nat. Gas Ind.* **2016**, *36*, 45–51. [CrossRef]

14. Sun, K.; Ji, H.; Zhang, S.-C. Influence of bedding azimuth and strength on hydraulic fracturing in shale. *Exp. Mech.* **2020**, *35*, 343–348.
15. Zhou, T. Investigation of Hydraulic Fracture Propagation Mechanism in Laminated Shale Gas Reservoirs. Ph.D. Thesis, China University of Petroleum, Beijing, China, 2017.
16. Liu, L. Study on the Propagation Law of Hydraulic Fracture at the Formation Interface. Ph.D. Thesis, Southwest Petroleum University, Chengdu, China, 2014.
17. Pang, T.; Jiang, Z.; Li, H. Influence of the spatial position of horizontal well in roof strata of crushed soft coal seam on the propagation of fracturing fractures. *J. Coal Sci.* **2022**, *47* (Suppl. S1), 196–203. [CrossRef]
18. Jia, C.; Li, Z.; Deng, J. Large-scale three-dimensional simulation test for directional perforation and fracturing in deflected well. *J. Southwest Pet. Univ.* **2007**, *29*, 135–137+193.
19. Hou, B.; Zhang, R.; Diao, C. Experimental study on hydraulic fracture propagation in highly deviated wells. *China Offshore Oil Gas* **2016**, *28*, 85–91.
20. Li, N. Investigation into the Fracturing Treatment for Shale Formation with Large Dip Angle in Northeast of Chongqing. Ph.D. Thesis, China University of Petroleum, Beijing, China, 2016.
21. Chen, M.; Chen, Z.; Huang, R. Study on initiation of water pressure fracture in high inclination well. *J. Pet. Univ. (Nat. Sci. Ed.)* **1995**, *19*, 30–35.
22. Li, M.; Zhou, F.; Hu, X.-D. Numerical simulation of multi-cluster hydraulic fracture propagation in highly deviated wells. *Sci. Technol. Eng.* **2020**, *20*, 11555–11561.
23. Liu, G.; Pang, F.; Chen, Z. Development of scaling laws for hydraulic fracture simulation tests. *J. Pet. Univ. (Nat. Sci. Ed.)* **2000**, *24*, 45–48. [CrossRef]
24. Renshaw, C.E.; Pollard, D.D. An experimentally verified criterion for propagation across unbounded frictional interfaces in brittle, linear elastic materials. *Int. J. Rock Mech. Min. Sci.* **1995**, *32*, 237–249. [CrossRef]

processes

MDPI

Article

Numerical Simulation of Multi-Fracture Propagation Based on the Extended Finite Element Method

Qiquan Ran, Xin Zhou *, Jiaxin Dong, Mengya Xu, Dianxing Ren and Ruibo Li

Research Institute of Petroleum Exploration and Development, No. 20 Xueyuan Road, Haidian District, Beijing 100083, China; ranqq@petrochina.com.cn (Q.R.); djx1021@petrochina.com.cn (J.D.); xumengya@petrochina.com.cn (M.X.); rendianx@petrochina.com.cn (D.R.); liruibo01@petrochina.com.cn (R.L.)
* Correspondence: zhouxin510@petrochina.com.cn

Abstract: Multi-stage, multi-cluster fracturing in horizontal wells is widely used as one of the most effective methods for unconventional reservoir transformation. This study is based on the extended finite element method and establishes a multi-hydraulic fracturing propagation model that couples rock damage, stress, and fluid flow, and the influence of horizontal stress difference and cluster spacing on fracture propagation is quantitatively analyzed. The simulation results show that changes in horizontal stress differences and inter-cluster spacing have a significant impact on the final propagation morphology of hydraulic fractures, and the change of the fracture initiation sequence forms different stress shadow areas, which in turn affects the propagation morphology of the fractures. When two fractures simultaneously propagate, they will eventually form a "repulsive" deviation, and a smaller stress difference and a decrease in inter-cluster spacing will lead to a more significant deviation of the fracture. Specifically, when the horizontal stress difference is 4 MPa and the cluster spacing is 6 m, the offset of the fracture tip along the direction of minimum horizontal principal stress is about 1.6 m, compared to the initial perforation position. When two fractures propagate sequentially, the fractures do not significantly deviate and propagate along the direction of maximum horizontal principal stress. When fractures propagate sequentially, the stress difference has little effect on the morphology of the fracture, but changes in inter-cluster spacing will significantly affect the length of the fracture. This study quantifies the effect of inter-fracture interference on fracture propagation morphology, providing guidance for optimizing the construction parameters of multi-stage hydraulic fracturing.

Keywords: hydraulic fracturing; fracture interference; stress shadow; XFEM

Citation: Ran, Q.; Zhou, X.; Dong, J.; Xu, M.; Ren, D.; Li, R. Numerical Simulation of Multi-Fracture Propagation Based on the Extended Finite Element Method. *Processes* **2023**, *11*, 2032. https://doi.org/10.3390/pr11072032

Academic Editors: Weizhong Dai, Jie Wang, Lufeng Zhang, Linhua Pan, Minghui Li, Wei Feng and Yushi Zou

Received: 12 May 2023
Revised: 24 June 2023
Accepted: 5 July 2023
Published: 7 July 2023

1. Introduction

Hydraulic fracturing is an important stimulation measure for the exploration and development of some unconventional reservoirs, such as shale reservoirs and tight sandstone reservoirs [1,2]. A single-plane fracture is not enough to realize the economic exploitation of unconventional oil and gas resources, and the complex fracture network is a necessary condition to improve the oil and gas production in low-permeability reservoirs, which is usually realized by multi-stage hydraulic fracturing of horizontal wells. In order to improve the operation efficiency, the usual construction method is to form several perforation clusters and create multiple hydraulic fractures in the same section of the reservoir through multi-stage hydraulic fracturing of horizontal wells, thereby increasing the volume of reservoir reconstruction [3–5]. However, according to some scholars' statistical analysis of production data, less than half of the perforation clusters contribute to the oil and gas production within a section, which means that most perforation clusters do not achieve the expected effect of increasing production [6,7]. The real increase in oil and gas production comes from the combination of unconventional well structures and large-scale hydraulic fracturing. Single-stage fracturing has evolved into multi-stage production enhancement, and single-well-fracturing has evolved into simultaneous fracturing of branch wells to increase reservoir control and improve well productivity. This has become a commonly used measure to increase production in the industry today [8–10].

Therefore, there is an urgent need to improve the efficiency of production and reduce extraction costs; however, there are many factors that affect the propagation of hydraulic fractures: (1) The physical properties of the reservoir near the wellbore are heterogeneous, including natural faults, in situ stress, and rock yield strength. (2) Interactions between different hydraulic fractures within the same horizontal wellbore may inhibit further propagation of some fractures, especially if a fracture is subjected to additional stress from adjacent fractures. This phenomenon is known as the stress shadow effect, which can occur even in homogeneous reservoirs [11–13]. (3) The dynamic fluid distribution between hydraulic fractures is related to wellbore hydraulics and perforation characteristics, and is influenced by wellbore friction, perforation friction, and hydraulic fracture propagation control.

Regarding the required number of fractures or spacing between fractures, it is important to note that the optimal production should be achieved through a cost-effective and efficient fracture network. Roussel and Sharma have studied this issue from the perspectives of production and geomechanics [14]. A study was conducted on the distance between hydraulic fractures from a production perspective. Yu and Sepehrnoori [15] observed that fractures too close to each other did not significantly increase production. According to their research results, fracture spacing, length, and well spacing can be optimized under certain porosity, permeability, and fracture conductivity conditions. However, this study did not consider the impact of geological factors of the reservoir on determining the optimal distance between fractures. Many studies have shown that geological and geo-mechanical information about rocks and their variations in the reservoir is also important, as stress primarily controls the initiation and propagation of fractures [16,17]. Stress shadow mainly refers to the phenomenon of local high stress in the direction perpendicular to the fracture surface near the fracture center, which will lead to the reorientation of the direction of the maximum stress in the stress-affected area, and then make the subsequent fracture propagation deviate or even parallel to the wellbore axis. Therefore, it is necessary to optimize the fracture spacing to obtain the maximum number of fractures perpendicular to the wellbore [18].

Currently, there are many numerical methods used for hydraulic fracturing analysis, among which the finite element method (FEM) is the most widely used. The FEM is essentially a numerical method for the mechanics of continuous media. In order to use the finite element method for analyzing hydraulic fracturing and other discontinuous problems, improvements need to be made to the traditional finite element method. Methods for improving the finite element method for non-continuous problems can be divided into two categories: unfixed-mesh methods and fixed-mesh methods [19,20]. Physics-based hydraulic fracturing models usually include different coupled components [21,22]: rock deformation and fracture opening and closing, fluid flow within the fracture, and fracture initiation and propagation. Many scholars have already carried out comprehensive research on the propagation of fractures through the use of numerical simulation methods. Sendon [23] established an analytical model of induced stress around a single planar fracture, which became the cornerstone for subsequent research on fracture interference. Bunger and Peirce conducted research on the propagation of winged fractures and the interaction between them in isotropic media, proposing simple measures to promote synchronized fracture propagation [24]. Zhang et al. studied the deflection and propagation of hydraulic fractures encountering planar bedding planes using boundary element and finite difference methods. The results showed that the deflection of fractures and fluid invasion into weak interfaces mainly depend on the local stress and deformation state at the intersection [25]. Most of the existing studies on fracture interactions have focused on static descriptions, with less attention paid to changes in the magnitude and orientation of the inter-fracture stress during dynamic fracture propagation. Furthermore, there has been little research on the impact of the order of fracture initiation on competitive fracture propagation, and a lack of quantitative exploration of the stress shadow effect on fracture propagation.

In this study, we employ the extended finite element method (XFEM) to simulate hydraulic fracture propagation, which is a type of fixed-grid method. The key idea of XFEM is to enrich the finite element interpolation space with additional discontinuous functions, so that the displacement jumps associated with the fracture can be modeled within the element [26,27]. Different from the existing research, based on the bilinear T-S criterion, this study uses stiffness degradation to describe the damage evolution process of fractures, and realizes the coupling process of fluid flow and solid damage deformation to simulate the expansion of two clusters of different fractures in the same section of horizontal wells. Using the advantages of finite element software simulation, the dynamic change process of the stress shadow size and influence area during fracture propagation is intuitively characterized. The fracture length and width data are extracted by the plug-in program, and the influence of cluster spacing, the horizontal stress difference, and the fracture initiation sequence on fracture propagation morphology is emphatically explored.

2. Materials and Methods

2.1. Fluid–Solid Coupling Simulation

During hydraulic fracturing in saturated porous media, the fluid acting on the fracture surface causes deformation of the solid phase in the porous media. The change in rock pore pressure caused by the injection of fracturing fluid leads to changes in fracture morphology and permeability. Therefore, the process of fracture propagation in geological formations is a dynamic coupling process of viscous fluid flow and rock deformation [4,28,29]. In this study, we investigated the extension of hydraulic fractures and the deformation of the rock matrix during hydraulic fracturing. A homogeneous and isotropic 2D elastic medium was chosen as the region for hydraulic fracture propagation. The rock deformation was described by linear elasticity theory, and the equilibrium equation was satisfied by:

$$\begin{cases} \sigma_{ij,j} + f_i = 0 \\ \varepsilon_{ij} = \left(u_{j,i} + u_{i,j}\right)/2 \\ \sigma_{ij} = C_{ijkl}\varepsilon_{kt} \end{cases} \tag{1}$$

where, σ_{ij} is the stress tensor, f_i is the volume force on the rock mass, ε_{ij} is the strain tensor, u is the medium displacement, and C is the elastic tensor.

To simulate fluid flow in hydraulic fractures, the porous media continuity equation was applied in weak form by imposing pore pressure at each node to simulate the flow of fracturing fluid within the porous medium:

$$\frac{\mathrm{d}}{\mathrm{d}t}\left(\int_V \rho_w \varphi_w \mathrm{d}V\right) + \int_S \rho_w \varphi_w n q_w \mathrm{d}S = 0 \tag{2}$$

where ρ_w is the density of the fluid, $\mathrm{kg/m^3}$, φ_w is the matrix porosity, q_w is the average flow velocity of the fluid in the matrix, $\mathrm{m^3/s}$, and n is the outer normal direction of the surface S.

The flow equation in the above equation satisfies the Darcy equation:

$$q_w = -\frac{1}{n_w g \rho_w} k' \cdot (\nabla p_w - \rho_w g) \tag{3}$$

In this study, the concept of the permeability coefficient was used to characterize the fluid permeability, which is expressed as follows:

$$k' = \frac{k \rho g}{\mu} \tag{4}$$

where, k' is the permeability coefficient, m/s, k is the permeability, m^2, μ is the fluid viscosity, Pa·s, n_w is the ratio of fluid volume to total volume, ∇P_w is the pressure gradient in the direction of fracturing fluid flow, and g is the free-fall acceleration, m/s^2.

2.2. The Mathematical Model of Extended Finite Element

2.2.1. Extended Finite Element Method

In the traditional finite element method, the problem domain is discretized into a series of small elements, and the solution is approximated using polynomial functions within each element. This method performs well in dealing with simple geometries and continuous media. However, when there are local discontinuities such as fractures, it requires increasing the mesh refinement to capture these discontinuities, leading to increased computational costs. The extended finite element method is based on the classical finite element method and incorporates additional generalized degrees of freedom to represent local discontinuities. This allows simulations to be performed on relatively coarse grids without the need for mesh refinement around fractures [26,30].

The extended finite element method improves the interpolation shape function within the element. The method incorporates the use of the Heaviside step function, $H_{(x)}$, and the asymptotic fracture tip function, $F_\alpha(x)$, to account for the discontinuity of the fracture surfaces, where $H_{(x)}$ is used to characterize the intermittent displacement field:

$$H(x) = \text{sgn}(\varphi(x)) = \begin{cases} 1, \varphi(x) > 0 \\ -1, \varphi(x) < 0 \end{cases} \tag{5}$$

$F_\alpha(x)$ is used to characterize the singular displacement field and describe the singularity of the fracture tip stress. The expression in polar coordinates is:

$$F_\alpha(x) = \left[\sqrt{r}\sin\frac{\theta}{2}, \sqrt{r}\cos\frac{\theta}{2}, \sqrt{r}\sin\theta\sin\frac{\theta}{2}\ \sqrt{r}\sin\frac{\theta}{2}\cos\frac{\theta}{2}\right], \alpha = 1,2,3,4 \tag{6}$$

Therefore, its displacement vector function, u, which characterizes the overall division characteristics, is:

$$u = \sum_{I=1}^{N} N_I(x)\left[u_I + H(x)a_I + \sum_{\alpha=1}^{4} F_\alpha(x)b_I^\alpha\right] \tag{7}$$

where $N_1(x)$ is the ordinary nodal displacement form function, u_I is the continuous part of the displacement solution, a_I and b_I^α are the nodal extended degree of freedom vectors, $H(x)$ is the intermittent jump function of the fracture surface, and $F_\alpha(x)$ is the fracture tip stress asymptotic function [31,32]. The first part of the equation applies to all nodes within the model, the second part applies to nodes whose functional shape is formed by internal fractures, and the third part is used only for nodes whose shape function is separated by the fracture tip [33]. As shown in Figure 1, when an element is undamaged, each virtual node lies entirely on the primary node; when the element is cracked, the fracture element is divided into two parts, each of which consists of several virtual nodes and primary nodes dependent on the fracture direction. Each virtual node is unbound from its origin and can be moved [34].

The expressions for the interpolation of the pressure field and the displacement field have the same form:

$$p = \sum_{I=1}^{N} N_I(x)\left[p_I + \varphi(x)p_I + \sum_{\alpha=1}^{4} P_\alpha(x)p_I^\alpha\right] \tag{8}$$

where, p_I, p_I^α are the node pressure expansion degrees of freedom, and $\varphi(x)$ and $P_\alpha(x)$ are the enhancement functions of the pressure nodes.

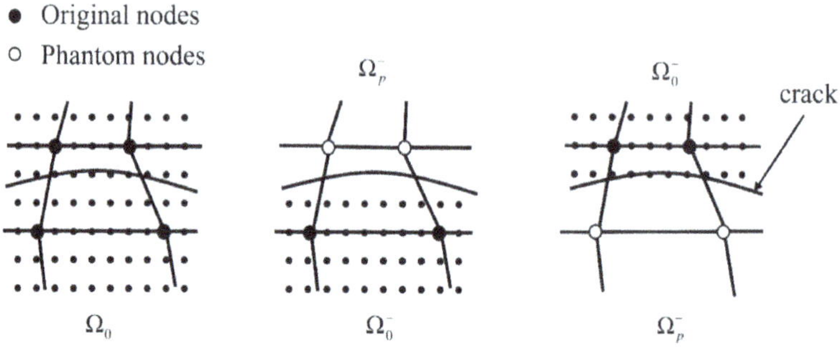

Figure 1. The XFEM using phantom nodes.

2.2.2. Fracture Propagation and Extension Criteria Based on Extended Finite Element

In this study, the extended finite element method was used to simulate the expansion of hydraulic fractures. The tensile-separation criterion based on the damage mechanics of cohesive units was used for the initiation and extension of hydraulic fractures [35]. As shown in Figure 2, the first part involves determining the effective displacement at complete damage, δ_m^f, relative to the effective displacement at the onset of damage, δ_m^0, or the energy dissipation due to failure, G_C. The second part mainly defines the evolution of the damage parameter, D, between initial failure and complete failure. This parameter can be specified by a linear or exponential softening law, which is directly provided in a table that shows the relationship between effective displacement and damage. When using the linear displacement-based extension criterion, it can be expressed as:

$$D = \frac{\delta_{\mathrm{m}}^{\mathrm{f}}\left(\delta_{\mathrm{m}}^{\max} - \delta_{\mathrm{m}}^{0}\right)}{\delta_{\mathrm{m}}^{\max}\left(\delta_{\mathrm{m}}^{\mathrm{f}} - \delta_{\mathrm{m}}^{0}\right)} \tag{9}$$

where δ_m^{max} is the maximum displacement of the unit, δ_m^f is the displacement at which the unit is opened, and δ_m^0 is the displacement at which the unit starts to be damaged.

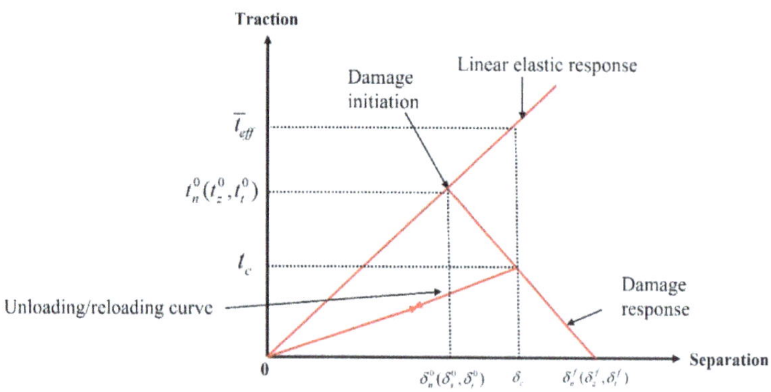

Figure 2. Tensile—separation curve.

The constitutive model of rock before damage is linear elasticity, and the element that degrades to complete failure after damage. In this study, the maximum principal stress criterion was used as the fracture initiation criterion, that is:

$$f = \frac{\sigma_{\max}}{\sigma_{\max c}} \tag{10}$$

where, σ_{maxc} is the maximum critical stress to which the rock is subjected, MPa, and σ_{max} is the maximum principal stress to which the rock is subjected, MPa.

In this study, the type of damage evolution for hydraulic fracture propagation was selected as displacement, and mode-independent was used as the mixed-mode behavior. When the initial damage reached 0.001 mm, the element was considered completely broken. During the process of hydraulic fracture propagation, the formation's permeability and porosity will change with the change of the effective stress of the porous medium in the formation, so the coupling relationship between the formation stress field and the seepage flow field must be considered. According to the principle of virtual work, the stress balance equation can be obtained as follows:

$$\int_V \sigma \delta \varepsilon dV = \int_S t \delta v dS + \int_V \hat{f} \delta v dV \tag{11}$$

where δv is the virtual velocity, m/s, $\delta \varepsilon$ is the virtual variation rate, s^{-1}, $\hat{f}$ is the volume force per unit volume, N/m^3, t is the external surface force per unit area, N/m^3, and σ is the total stress in the porous medium of the formation, Pa.

Based on the principle of mass conservation, the continuity equation of fluid media can be obtained as follows:

$$\int_V \delta v \frac{1}{J} \frac{\mathrm{d}}{\mathrm{d}l} (J \rho_{\mathrm{w}} n_{\mathrm{w}}) \mathrm{d}V + \int_V \delta v \frac{\partial}{\partial x} (\rho_{\mathrm{w}} n_{\mathrm{w}} v_{\mathrm{w}}) \mathrm{d}V = 0 \tag{12}$$

where, J is the rate of change of the formation pore volume, n_{w} is the ratio of formation liquid volume to total volume, ρ_{w} is the density of liquid in the formation pore, kg/m^3, and v_{w} is the flow velocity of the formation pore liquid, m/s.

2.3. *Stress Field Superimposed on Fracture Propagation of Multiple Clusters in Horizontally Oriented Well*

During the fracturing process of the horizontal well, the formation of horizontal wellbore will significantly affect the original in situ stress field. Therefore, the injection of fracturing fluid will cause the pore pressure around the wellbore to increase and generate additional stress due to the difference between the pore pressure of the wellbore and that of the formation. In situ stress is the natural stress that exists in a rock mass, and it is regularly distributed in a three-dimensional state. There are many factors that affect the change of in situ stress, among which the obvious ones are wellbore pressure, the original tectonic stress component, and the change caused by fracturing fluid seepage [36]. For the process of propagation of multi-cluster fractures in the horizontal well, besides considering the stress disturbance caused by the propagation of different fractures, the significant temperature difference of the reservoir at different well depths and the thermal effect of the drilling fluid and fracturing fluid injected on the wellbore and fractures also bring additional stress effects. Figure 3 shows the multi-fracture stress superposition model, which regards the rock as an infinitesimal deformation, porous elastic body, according to the principle of stress superposition, the disturbed stress field of horizontal well-fracturing is the sum of the initial in situ stress, wellbore pressure, fracturing fluid seepage, fracture-induced stress, and additional stress caused by thermal effects, that is:

$$\sigma_{\mathrm{sum}} = \sigma_{\mathrm{ini}} + \sigma_{\mathrm{w}} + \sigma_{1} + \sigma_{\mathrm{f}} + \sigma_{T} \tag{13}$$

where, σ_{sum} is the total fracture disturbance stress, σ_{ini} is the initial ground stress, σ_{w} is the stress around the wellbore after drilling, σ_{1} is the fracture fluid percolation stress, σ_{f} is the fracture-induced stress, and σ_{T} is the thermal stress.

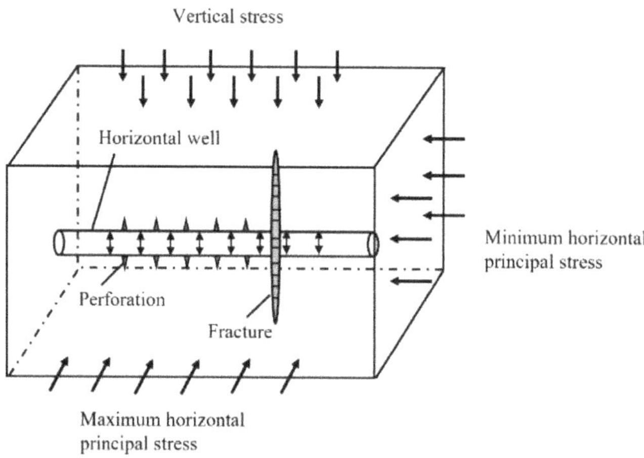

Figure 3. Multi-fracture stress superposition model.

3. Model Construction and Validation

3.1. Model Construction

In the simulation of simultaneous expansion of multiple clusters of fractures in horizontal wells, a total of three fractures with equal spacing were set in the horizontal wells, where the direction of the horizontal wellbore was the direction of the minimum horizontal principal stress to ensure that the formed hydraulic fractures were transverse fractures. The size of the numerical simulation model was 100 m $\times$ 100 m, and the related geo-mechanical parameters and fracturing construction parameters are shown in Table 1.

Table 1. Mechanical parameters of stratigraphic rocks.

Variables	Unit	Value of Reservoir Rock
Young's modulus	GPa	15
Poisson's ratio	/	0.25
Permeability coefficient	m/s	1×10^{-7}
Initial pore ratio	/	0.1
Filtration loss factor	m/(Pa·s)	1×10^{-14}
Tensile strength	MPa	6

The simulation adopted the super-hydrostatic pressure system, with a fracturing fluid flow rate and viscosity of 0.12 m^3/min and 1 mPa·s, respectively. The perforation length was 1 m, and the pumping time of each hydraulic fracture was maintained for 100 s. To simplify the calculation, the horizontal wellbore was located at the bottom of the model, and the perforation direction was aligned with the maximum horizontal principal stress direction. The established geometric model is shown in Figure 4. A constant-pressure water fracturing system was applied, and the boundary pore pressure was set to 0 MPa and kept constant during the fracturing process. The reservoir rock matrix was represented by a structured, standardized, linear quadrilateral plane strain element type known as CPE4P, and the hydraulic fracture propagation was simulated with two-node linear truss elements (T2D2).

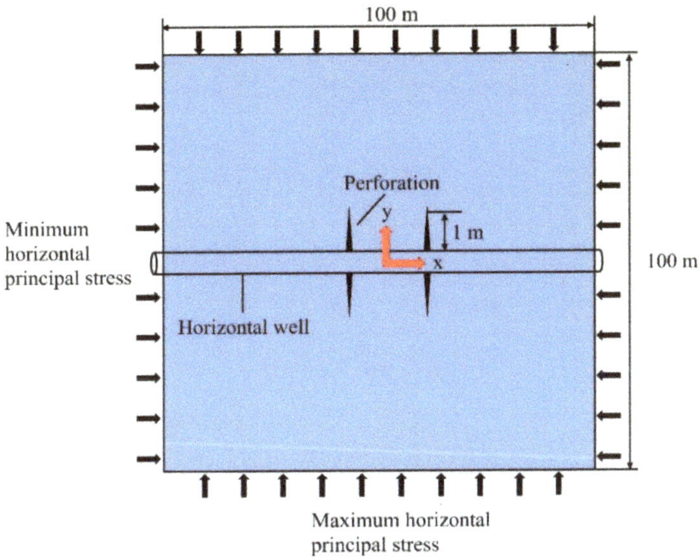

Figure 4. Geometric model of horizontal well multi-cluster fracturing.

In the XFEM, fracture is modeled independent of mesh configuration and element type. This means that no remeshing is required, and discontinuity, including fractures, need not be aligned with element boundaries [37]. However, different mesh configurations have an impact on the simulation convergence and effectiveness. Therefore, we conducted an investigation on various mesh sizes and selected four different mesh sizes for evaluation: 0.3 m, 0.5 m, 0.8 m, and 1 m, respectively. The relevant mechanical parameters are referenced from Table 1. The cluster spacing was 6 m, the horizontal differential stress was 4 MPa, and the fracturing fluid displacement and viscosity were 0.12 m^3/min and 1 mPa·s, respectively. The length of the perforation was 1 m, and the injection time was set to 100 s.

Figure 5 displays the morphology of fractures after simultaneous initiation and propagation of dual-cluster fractures extracted using the plugin under different mesh sizes. As observed from Figure 5, it can be inferred that changing the mesh size had minimal impact on the final morphology of the fractures after propagation. Two cases with the largest difference in half-fracture height were selected for comparison. When the mesh size was 0.3 m, the half-fracture height on the right side was 15.37 m, while with a mesh size of 0.8 m, the half-fracture height was 14.91 m. The difference between the two cases was only 0.46 m. Considering that smaller mesh sizes are suitable for smaller initial perforation lengths, it is important to balance the computational efficiency and simulation accuracy. Finer mesh configurations tend to significantly increase the computation time and can lead to convergence issues [38]. Taking into account both the computational time and the simulation effectiveness, a mesh size of 0.5 m was selected as the optimal choice for the fracture propagation numerical model in this study. This mesh size strikes a balance between capturing essential details and maintaining an acceptable computational efficiency, ensuring a reasonable compromise between accuracy and computational resources.

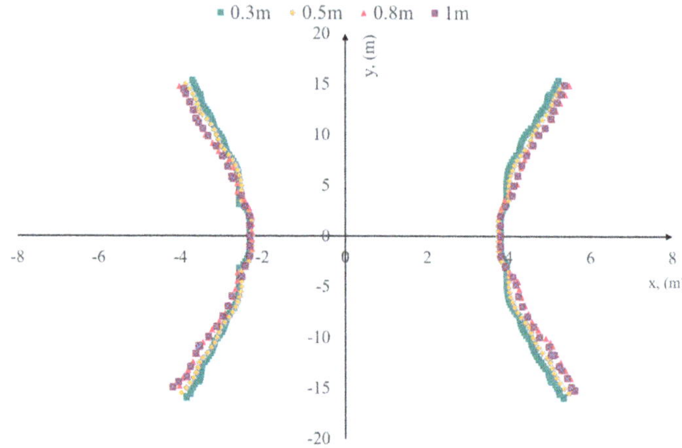

Figure 5. Fracture morphology of different mesh sizes.

3.2. Model Verification

The interaction among multiple fractures plays a crucial role in determining the geometric shape of the fractures during their propagation. In this study, we conducted a comparative analysis of the dual-cluster fracture propagation model for horizontal wells proposed by Wu et al. [39]. The wellbore and initial perforation locations in their model were similar to those depicted in Figure 4. The horizontal wellbore was aligned parallel to the minimum horizontal principal stress direction, while the initial perforation locations were positioned along the maximum horizontal principal stress direction. The model input parameters are presented in Table 2, and the simulation results are shown in Figure 6. Figure 6a illustrates the fracture opening after expansion in the current model, while Figure 6b displays a comparison of the fracture propagation paths between the two models. In a horizontal well, parallel fractures may deviate from each other due to the stress shadow effect. Once the fractures open, they exert additional stresses on the surrounding rock and neighboring fractures. This can lead to local variations in the direction of the horizontal principal stress and deviations in the fracture paths from the planar geometry.

Table 2. Input parameters (after Wu et al., 2012 [39]).

Variables	Unit	Value
Young's modulus	psi	4.35×10^6
Poisson's ratio	/	0.35
Maximum horizontal stress	psi	6903
Minimum horizontal stress	psi	6773
Injection rate	bbl/min	40
Fluid viscosity	cp	1
Distance between initiation points	ft	33

From Figure 6, it can be observed that the fracture propagation paths of the two models exhibited a high degree of consistency, with only slight deviations at the fracture tips. The analysis suggests that these deviations might be attributed to the different boundary effects caused by the variations in model sizes. As the fractures propagated, they encountered different resistances at the fracture tips, leading to slight differences in the final morphology of the fractures. Based on the above analysis, it can be concluded that the simulation results of the two models mentioned earlier exhibited good agreement. This confirmed the effectiveness and accuracy of the model used in this study.

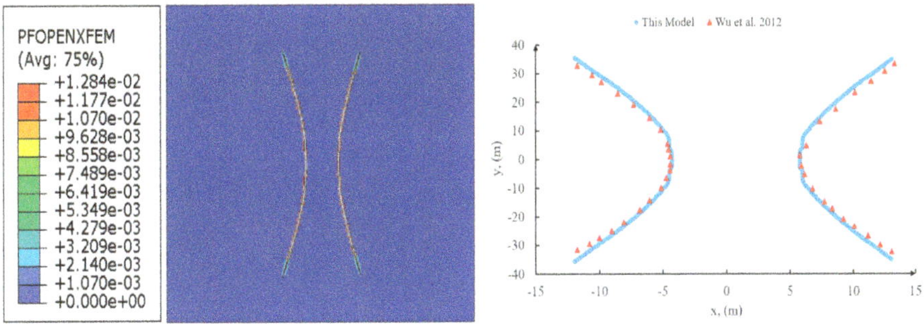

(a) (b)

Figure 6. (**a**) Fracture propagation geometry. (**b**) The comparison of fracture propagation paths between this model and Wu et al. [39]'s model.

4. Numerical Simulation Results

To investigate the competition of fracture propagation under simultaneous initiation of two fractures, two perforated intervals with a perforation depth of 1 m were placed horizontally along the direction of the minimum horizontal principal stress in the study area as the initial fractures. The injection time was set to 100 s and the injection rate was chosen as 0.002 m/s. The two perforated intervals concurrently injected fluid. This paper also explores the interference effect of the stress disturbance caused by the propagation of the first fracture on the subsequent initiation fracture under the condition of two fractures initiating at different times. In order to simulate the stress shadow effect induced by fracture propagation under reservoir conditions, two fractures were initiated at different times with a constant injection rate of 0.002 m/s and a duration of 100 s each, followed by a 600 s pressure depletion after the first fracture was completed. The simulation of simultaneous and sequential fracture initiation revealed how stress field and fracture spacing influenced the complexity and diversity of fracture propagation patterns. To investigate the effects of stress difference and fracture spacing on fracture propagation, we performed simulations with different values of these parameters. We varied the stress difference from 2 to 8 MPa and the fracture spacing from 6 to 12 m. Here, we present the results and analysis of how these factors influenced fracture propagation.

4.1. Horizontal Differential Principal Stress

In the simulations under different reservoir stress conditions, we set different values of minimum and minimum horizontal principal stress, such that the horizontal principal stress difference was 2, 4, 6, or 8 MPa. The overburden stress was set to 15 MPa. Table 3 shows the reservoir rock mechanical parameters for this area. The other basic rock mechanical parameters are presented in Table 1. The perforation cluster spacing was kept at 10 m. As shown in Figure 7, the distribution of reservoir pore pressure underwent changes as the stress difference increased. Furthermore, the deviation angles of the fracture morphology gradually decreased as a result, and stress concentration phenomena appeared at the tips of the fractures. By programming and extracting the nodal information of the two-dimensional fractures, the post-propagation coordinate information of the fractures can be obtained. Based on Figure 8, the morphology of the fractures exhibited consistency under different stress difference conditions when the two fractures simultaneously initiated and propagated. This consistency is attributed to the initial perforation cluster being symmetrically located in the model and having consistent boundary conditions. Under a stress difference of 2 MPa, the deviation of the fracture tip was significant, deflecting approximately 1.02 m along the direction of the minimum horizontal principal stress, as compared to the initial initiation point. As the stress difference increased, the variation in the fracture morphology notably decreased. Under the stress difference conditions of 6 MPa and 8 MPa, the fracture morphology was approximately consistent.

Table 3. Reservoir ground stress conditions and cluster spacing.

Case	Vertical Stress/MPa	Minimum Horizontal Principal Stress/MPa	Maximum Horizontal Principal Stress/MPa	Horizontal Principal Stress Difference/MPa	Cluster Spacing/m
Case 1	15	6	8	2	10
Case 2	15	6	10	4	10
Case 3	15	6	12	6	10
Case 4	15	6	14	8	10

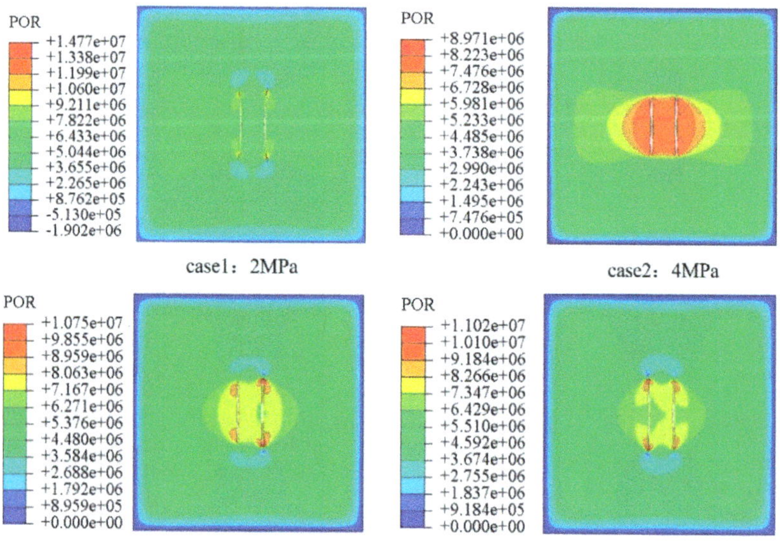

Figure 7. Pore pressure distribution after simultaneous fracture initiation under different stress difference conditions.

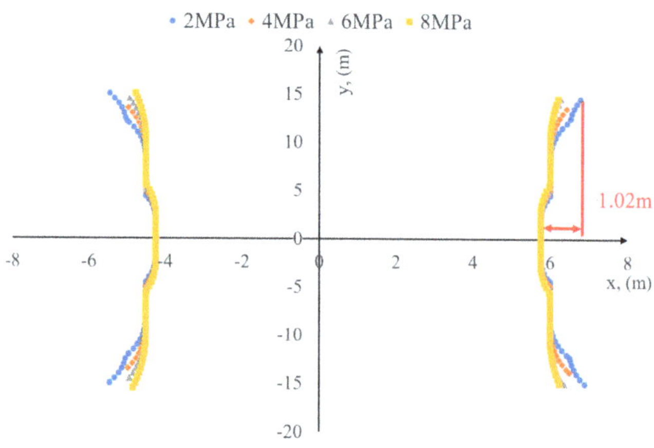

Figure 8. The morphology of the simultaneously initiated fractures under different stress difference conditions.

In contrast, for simulation calculations involving the sequential initiation of two fractures, the interference from the stress shadow effect generated by the first fracture after expansion may affect the propagation of the second fracture. Thus, the morphology of the fractures will exhibit significant differences compared to the case where both fractures simultaneously initiated. As illustrated in Figure 9, the change in the horizontal principal stress difference had little influence on the final morphology of the fracture. However, due to the stress interference from the previously initiated fracture on the subsequently initiated fracture, the average half-length of the later-initiated fracture decreased by approximately 6 m under the four different stress difference conditions. In contrast, the length of the previously initiated fracture did not decrease compared to the simultaneous initiation condition. These findings suggest that stress interference between fractures can have a detrimental impact on reservoir fracturing and enhanced oil recovery by reducing the hydraulic fracturing effective area.

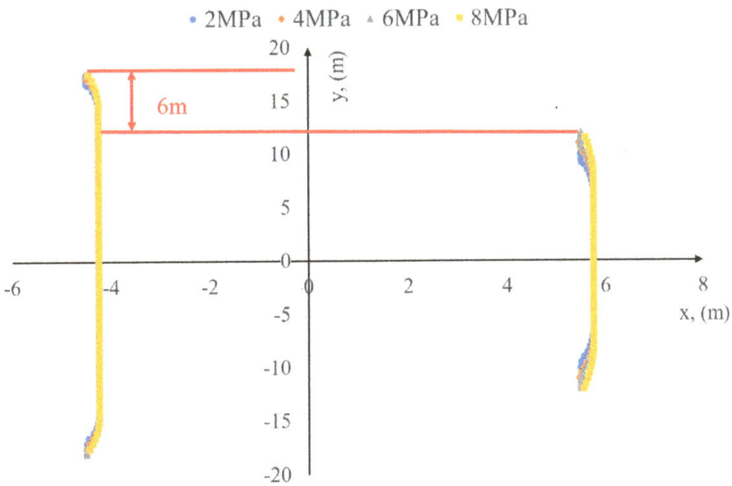

Figure 9. The morphology of the sequentially initiated fractures under different stress difference conditions.

Figure 10 shows the variation of the minimum horizontal principal stress vector under the condition of a stress difference of 2 MPa. Due to the default setting of the simulator, the vector represents compressive stress as a negative value. From the figure, it can be observed that after the first hydraulic fracture extended, the surrounding stress field was disturbed, resulting in an elliptical-shaped stress shadow area that restricted the propagation of the second fracture. Figure 11 shows the distribution of the minimum horizontal principal stress at the ends of the two sequentially propagating fractures under the four different stress conditions. It can be observed from the figure that the minimum horizontal principal stress at the ends of the initial fracture reached its maximum value, resulting in a high-stress area at the tip of the fracture. When the second fracture propagated, it eventually stopped due to the fact that the pore pressure inside the fracture was lower than the fracture pressure of the element.

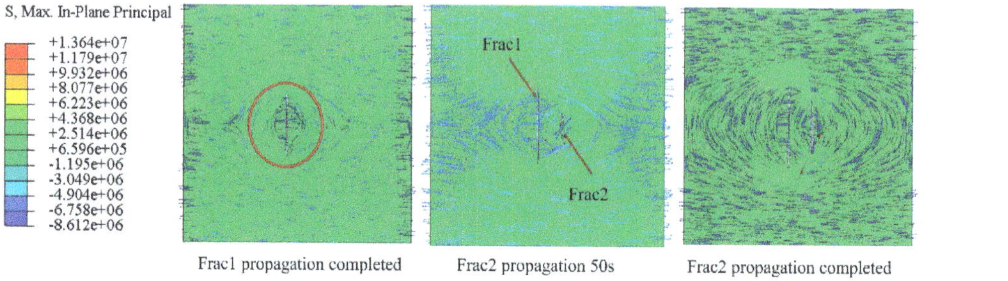

Figure 10. Minimum horizontal principal stress vector at different moments under the condition of a 2 MPa stress difference.

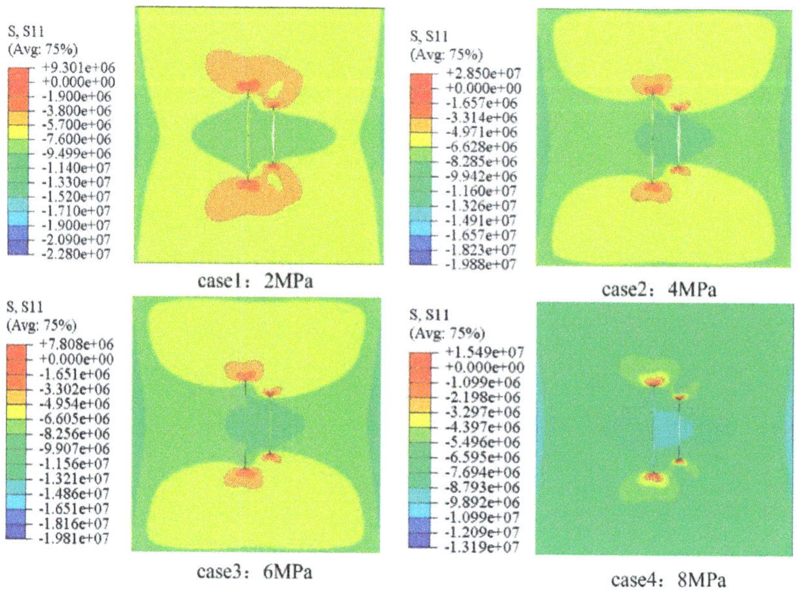

Figure 11. The distribution of minimum horizontal principal stress after sequential fracturing under different stress difference conditions.

The subsequent fracture initiation pressure was also affected by stress interference. Under the condition of staggered fracture initiation, the difference in the pore pressure of the fracture elements was not significant among the four stress difference conditions. Taking the stress difference of 2 MPa as an example, Figure 12 shows the pore pressure changes with time at the perforation locations of the two fractures under the stress difference of 2 MPa. It can be seen from the figure that as time increased, the pore pressure of the two fractures first increased, and then decreased to a certain level and remained stable. The first fracture started to propagate in 0–100 s, followed by the pressure-relief stage of the first fracture from 100 to 700 s, and then the second fracture started to propagate. The initiation pressure of the first fracture was around 36.31 MPa, while the subsequent fracture initiation pressure increased to 43.08 MPa. This suggests that the propagation of the first fracture hindered the propagation of the second fracture, and ultimately, the pore pressure in both fractures tended to be consistent.

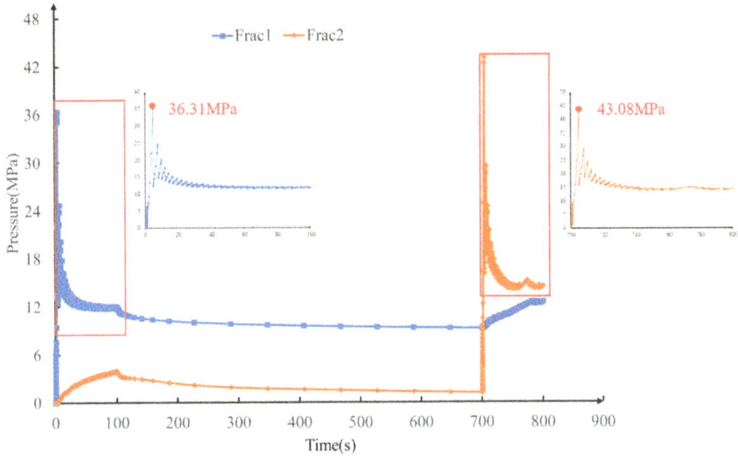

Figure 12. The evolution of pore pressure at the location of the perforation over time.

4.2. Perforation Cluster Spacing

The variation of cluster spacing also had a significant impact on the morphology of the fracture propagation. In this study, the effect of different perforation cluster spacing on fracture propagation was investigated. Cluster spacings of 6 m, 8 m, 10 m, and 12 m were set, respectively (Table 4). The mechanical parameters of the reservoir rock are still based on Table 1, with the minimum horizontal principal stress set to 6 MPa, the maximum horizontal principal stress set to 10 MPa, and the overlying stress set to 15 MPa.

Table 4. Mechanical parameters of reservoir rocks and cluster spacing.

Case	Vertical Stress /MPa	Minimum Horizontal Principal Stress /MPa	Maximum Horizontal Principal Stress /MPa	Cluster Spacing /m
Case 5	15	6	10	6
Case 6	15	6	10	8
Case 7	15	6	10	10
Case 8	15	6	10	12

Figure 13 shows the distribution of fracture pore pressure after simultaneous initiation and propagation of fractures under different cluster-spacing conditions. It can be seen from the figure that the maximum pore pressure was generated at the fracture tip. Combined with the fracture morphology in Figure 14, it can be seen that as the cluster spacing decreased, the inter-fracture deflection angle gradually increased, and the end of the hydraulic fracture with a cluster spacing of 6 m deviated from the initial perforation cluster position by about 1.6 m. It should be noted that, in contrast to the simulation results under the condition of the changing stress difference (Figure 8), the two simultaneously initiated fractures under the conditions of 8 m and 12 m cluster spacing showed obvious asymmetry. The reason for this is that, under the distance condition of the shot-hole cluster, the initial shot-hole positions of the two fractures were not in the middle of the model, and the influence of the boundary effect led to greater resistance to the propagation of the second fracture.

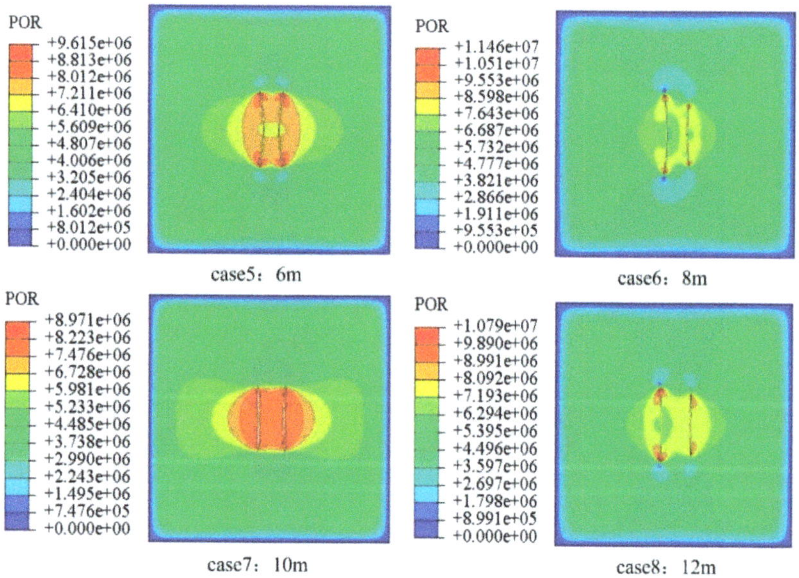

Figure 13. Pore pressure distribution after simultaneous fracture initiation under different cluster-spacing conditions.

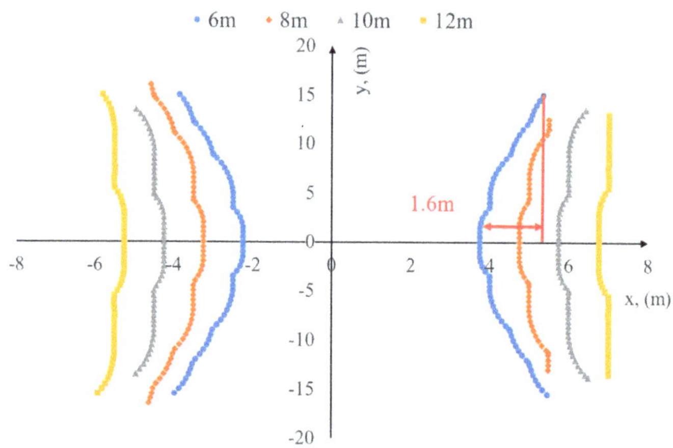

Figure 14. The morphology of the simultaneously initiated fractures under different cluster-spacing conditions.

Similarly, simulations were conducted to study the sequential initiation and propagation of fractures under different cluster-spacing conditions. The obtained fracture patterns are shown in Figure 15, which exhibited significant differences in the propagation of the two fractures. The later-initiated fracture was subjected to interference from the first fracture, resulting in a significant decrease in its half-length, but with a smaller deflection. This indicates that fracture deflection is more susceptible to interference when multiple fractures are competing for propagation at the same time. Moreover, the length of the fractures significantly varied during sequential propagation, and the interference became more pronounced as the cluster spacing decreased.

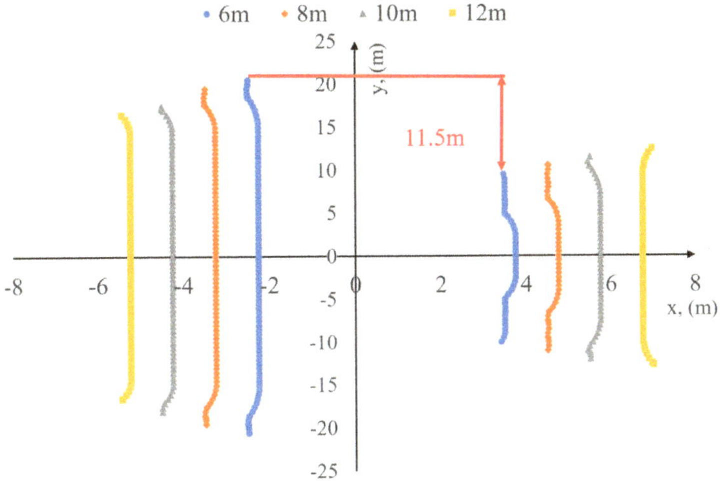

Figure 15. The morphology of the sequentially initiated fractures under different cluster-spacing conditions.

When the cluster spacing was 6 m, the half-length of the second fracture was reduced by about 11.5 m. By referring to the minimum horizontal principal stress vector map of the fracture propagation process under the cluster spacing of 6 m, as shown in Figure 16, it can be inferred that the propagation of the first fracture resulted in significant stress interference, which changed the magnitude and direction of the minimum horizontal principal stress and created stress shadow effects around the fracture, where the minimum horizontal principal stress around the fracture was higher than the original minimum horizontal principal stress, resulting in the stress shadow effect around the fracture. When the second fracture propagated, the same pumping displacement generated the same pore pressure at the fracture tip. However, due to the presence of high-stress zones, the fracture propagation was hindered, and the half-length of the fracture significantly decreased. In addition, for the case of sequential initiation, in the simulation examples of cases 5–8, the average initiation pressure of the second fracture was lower than that of the first fracture, with a difference of about 8 MPa. The trend of the relationship between the initial perforation pore pressure and time was similar to that shown in Figure 12.

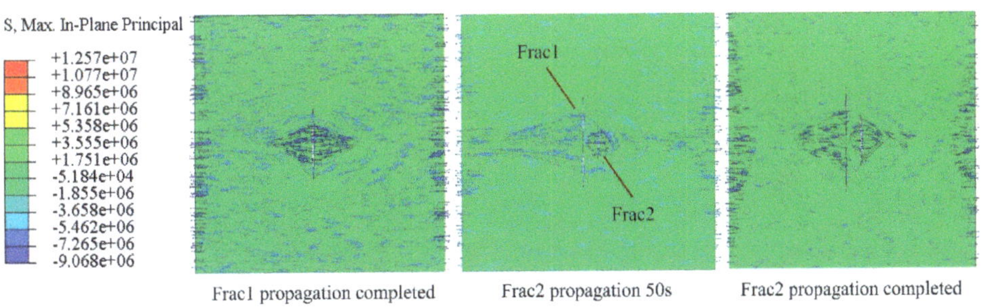

Figure 16. Minimum horizontal principal stress vector at different moments when the cluster spacing was 6 m.

5. Conclusions

In this study, the bilinear T-S criterion and stiffness degradation were used to describe the damage evolution of fractures, and a coupled model of fluid flow and solid damage deformation was constructed. Based on the extended finite element model, the propagation simulation of multiple cluster fractures during the horizontal well-fracturing transformation process was realized, and the following main conclusions were obtained.

(1) During the process of multi-cluster fracture propagation and extension in horizontal wells, the stress difference and the size change of inter-cluster clusters had a greater impact on the final morphology of fractures and the area of reservoir transformation. The fracture initiation sequence also directly affected the final fracture morphology. When the two fractures simultaneously initiated, a more obvious fracture deflection occurred; when the fractures initiated at different times, the fracture length of the later-initiated fracture was significantly reduced due to the stress interference caused by the earlier-propagated fracture.

(2) When simultaneously initiating, changing the stress difference and inter-cluster spacing formed two relatively symmetrical deflected fractures. The smaller the minimum horizontal principal stress difference and the inter-cluster spacing, the more obvious the fracture deflection. When the horizontal stress difference was 4 MPa and the inter-cluster spacing was 6 m, the fracture tip deviated from the initial perforation position by about 1.6 m along the direction of the minimum horizontal principal stress.

(3) When the two fractures separately initiated, the fractures basically extended along the direction of the maximum horizontal principal stress. Due to the propagation of the first fracture, the size and direction of the surrounding stress field were changed, and a high-stress zone was induced around the fracture, forming an elliptical stress shadow area, which hindered the propagation of subsequent fractures. The variation in the stress difference had a minimal impact on the fracture length. Under four different stress difference conditions, the length of the subsequent fracture decreased by approximately 6 m. On the other hand, the change in cluster spacing had a significant influence on the fracture length. When the cluster spacing was 6 m, the half-length of the fracture decreased the most. The subsequently expanding fracture had a reduction of approximately 11.5 m compared to the initially expanding fracture.

(4) The pore pressure at the perforation location also experienced significant changes with variations in the fracturing sequence. When the two fractures were simultaneously initiated, the fracture initiation pressure for both fractures was around 36 MPa. However, in the case of sequential fracturing, the initiation pressure for the subsequent fracture increased by 7 MPa, reaching 43.08 MPa. This is because the expansion of the first fracture altered the original stress field, resulting in an increased resistance for subsequent fracture initiation. Eventually, the pore pressure at the perforation locations of both fractures stabilized at approximately 13 MPa.

Author Contributions: Writing—original manuscript, Q.R.; supervision, X.Z.; investigation, J.D.; data curation, M.X.; formal analysis, D.R. and R.L. All authors have read and agreed to the published version of the manuscript.

Funding: This research was funded by the Key Core Technology Research Projects of PetroChina Company Limited (No. 2020B-4911) and the APC was funded by the Research Institute of Petroleum Exploration and Development.

Data Availability Statement: No new data were created or analyzed in this study. Data sharing is not applicable to this article.

Conflicts of Interest: The authors declare no conflict of interest.

References

1. Zou, J.; Zhang, Y.; Zhang, L.; Jing, J.; Fu, Y.; Wang, Y.; Zhang, G.; Zhou, F. Numerical Simulation Research on the Effect of Artificial Barrier Properties on Fracture Height. *Processes* **2023**, *11*, 310. [CrossRef]
2. Dong, R.; Wang, Q.; Wheeler, M.F. *Prediction of Mechanical Stability of Acidizing-Induced Wormholes through Coupled Hydro-Chemo-Mechanical Simulation*; OnePetro: Richardson, TX, USA, 2019.
3. Dong, R.; Alpak, F.O.; Wheeler, M.F. Accurate Two-Phase Flow Simulation in Faulted Reservoirs by Combining Two-Point Flux Approximation and Mimetic Finite Difference Methods. *SPE J.* **2023**, *28*, 111–129. [CrossRef]
4. Zhao, Y.; Wang, L.; Ma, K.; Zhang, F. Numerical Simulation of Hydraulic Fracturing and Penetration Law in Continental Shale Reservoirs. *Processes* **2022**, *10*, 2364. [CrossRef]
5. Jia, S.; Dai, Z.; Zhou, Z.; Ling, H.; Yang, Z.; Qi, L.; Wang, Z.; Zhang, X.; Thanh, H.V.; Soltanian, M.R. Upscaling Dispersivity for Conservative Solute Transport in Naturally Fractured Media. *Water Res.* **2023**, *235*, 119844. [CrossRef] [PubMed]
6. Rybak, Y.; Khayrutdinov, M.; Kongar-Syuryun, C.; Tyulyayeva, Y. Resource-Saving Technologies for Development of Mineral Deposits. *Sustain. Dev. Mt. Territ.* **2021**, *13*, 405–415.
7. Bosikov, I.I.; Klyuev, R.V.; Mayer, A.V. Comprehensive Assessment of Hydraulic Fracturing Technology Efficiency for Well Construction during Hydrocarbon Production. *Записки Горного Института* **2022**, *258*, 1006–1013. [CrossRef]
8. Wang, Y.; Zhou, F.; Zhang, Y.; Wang, Y.; Su, H.; Dong, R.; Wang, Q.; Bai, H. Numerical Studies and Analysis on Reaction Characteristics of Limestone and Dolomite in Carbonate Matrix Acidizing. *Geoenergy Sci. Eng.* **2023**, *222*, 211452. [CrossRef]
9. Liu, C.; Cui, J.; Zhang, Z.; Liu, H.; Huang, X.; Zhang, C. The Role of TBM Asymmetric Tail-Grouting on Surface Settlement in Coarse-Grained Soils of Urban Area: Field Tests and FEA Modelling. *Tunn. Undergr. Space Technol.* **2021**, *111*, 103857. [CrossRef]
10. Zhang, X.; Wang, Z.; Reimus, P.; Ma, F.; Soltanian, M.R.; Xing, B.; Zang, J.; Wang, Y.; Dai, Z. Plutonium Reactive Transport in Fractured Granite: Multi-Species Experiments and Simulations. *Water Res.* **2022**, *224*, 119068. [CrossRef] [PubMed]
11. Bunger, A.P.; Peirce, A.P. Numerical Simulation of Simultaneous Growth of Multiple Interacting Hydraulic Fractures from Horizontal Wells. In Proceedings of the Shale Energy Engineering 2014, Pittsburgh, PA, USA, 21–23 July 2014; American Society of Civil Engineers: Pittsburgh, PA, USA, 2014; pp. 201–210.
12. Singh, A.; Xu, S.; Zoback, M.; McClure, M. *Integrated Analysis of the Coupling between Geomechanics and Operational Parameters to Optimize Hydraulic Fracture Propagation and Proppant Distribution*; OnePetro: Richardson, TX, USA, 2019.
13. Zhu, T.; Ding, H.; Wang, C.; Liu, Y.; Xiao, S.; Yang, G.; Yang, B. Parameters Calibration of the GISSMO Failure Model for SUS301L-MT. Chin. *J. Mech. Eng.* **2023**, *36*, 20. [CrossRef]
14. Roussel, N.P.; Sharma, M.M. *Strategies to Minimize Frac Spacing and Stimulate Natural Fractures in Horizontal Completions*; OnePetro: Richardson, TX, USA, 2011.
15. Yu, W.; Sepehrnoori, K. Optimization of Multiple Hydraulically Fractured Horizontal Wells in Unconventional Gas Reservoirs. *J. Pet. Eng.* **2013**, *2013*, 151898. [CrossRef]
16. Xu, W.; Calvez, J.L.; Thiercelin, M. *Characterization of Hydraulically-Induced Fracture Network Using Treatment and Microseismic Data in a Tight-Gas Formation: A Geomechanical Approach*; OnePetro: Richardson, TX, USA, 2009.
17. Abousleiman, Y.; Tran, M.; Hoang, S.; Bobko, C.; Ortega, A.; Ulm, F.-J. *Geomechanics Field and Laboratory Characterization of Woodford Shale: The Next Gas Play*; OnePetro: Richardson, TX, USA, 2007.
18. Morrill, J.C.; Miskimins, J.L. *Optimizing Hydraulic Fracture Spacing in Unconventional Shales*; OnePetro: Richardson, TX, USA, 2012.
19. Ren, Q.; Dong, Y.; Yu, T. Numerical Modeling of Concrete Hydraulic Fracturing with Extended Finite Element Method. *Sci. China Ser. E-Technol. Sci.* **2009**, *52*, 559–565. [CrossRef]
20. Zhang, X.; Ma, F.; Dai, Z.; Wang, J.; Chen, L.; Ling, H.; Soltanian, M.R. Radionuclide Transport in Multi-Scale Fractured Rocks: A Review. *J. Hazard. Mater.* **2022**, *424*, 127550. [CrossRef] [PubMed]
21. Chen, B.; Barron, A.R.; Owen, D.R.J.; Li, C.-F. Propagation of a Plane Strain Hydraulic Fracture with a Fluid Lag in Permeable Rock. *J. Appl. Mech.* **2018**, *85*, 091003. [CrossRef]
22. Dong, R.; Wheeler, M.F.; Ma, K.; Su, H. *A 3D Acid Transport Model for Acid Fracturing Treatments with Viscous Fingering*; OnePetro: Richardson, TX, USA, 2020.
23. Sneddon, I.N.; Mott, N.F. The Distribution of Stress in the Neighbourhood of a Fracture in an Elastic Solid. Proceedings of the Royal Society of London. *Ser. A Math. Phys. Sci.* **1997**, *187*, 229–260. [CrossRef]
24. Meehan, C.L.; VanBriesen, J.M.; Vahedifard, F.; Yu, X.; Quiroga, C. *Shale Energy Engineering 2014*; American Society of Civil Engineers: Reston, VA, USA, 2014; ISBN 978-0-7844-1365-4.
25. Zhang, X.; Jeffrey, R.G. Role of Overpressurized Fluid and Fluid-Driven Fractures in Forming Fracture Networks. *J. Geochem. Explor.* **2014**, *144*, 194–207. [CrossRef]
26. Moës, N.; Dolbow, J.; Belytschko, T. A Finite Element Method for Fracture Growth without Remeshing. *Int. J. Numer. Methods Eng.* **1999**, *46*, 131–150. [CrossRef]
27. Peng, J.; Xu, C.; Dai, B.; Sun, L.; Feng, J.; Huang, Q. Numerical Investigation of Brittleness Effect on Strength and Microcracking Behavior of Crystalline Rock. *Int. J. Geomech.* **2022**, *22*, 04022178. [CrossRef]
28. Wan, B.; Liu, Y.; Zhang, B.; Luo, S.; Wei, L.; Li, L.; He, J. Investigation of the Vertical Propagation Pattern of the 3D Hydraulic Fracture under the Influence of Interlayer Heterogeneity. *Processes* **2022**, *10*, 2449. [CrossRef]
29. Zhu, H.; Deng, J.; Jin, X.; Hu, L.; Luo, B. Hydraulic Fracture Initiation and Propagation from Wellbore with Oriented Perforation. *Rock Mech. Rock Eng.* **2015**, *48*, 585–601. [CrossRef]

30. Daux, C.; Moës, N.; Dolbow, J.; Sukumar, N.; Belytschko, T. Arbitrary Branched and Intersecting Fractures with the Extended Finite Element Method. *Int. J. Numer. Methods Eng.* **2000**, *48*, 1741–1760. [CrossRef]
31. Liu, C.; Liu, H.; Zhang, Y.; Deng, D.; Wu, H. Optimal Spacing of Staged Fracturing in Horizontal Shale-Gas Well. *J. Pet. Sci. Eng.* **2015**, *132*, 86–93. [CrossRef]
32. Haddad, M.; Sepehrnoori, K. XFEM-Based CZM for the Simulation of 3D Multiple-Cluster Hydraulic Fracturing in Quasi-Brittle Shale Formations. *Rock Mech. Rock Eng.* **2016**, *49*, 4731–4748. [CrossRef]
33. Abdullah, E.; Ferrero, J.-F.; Barrau, J.-J.; Mouillet, J.-B. Development of a New Finite Element for Composite Delamination Analysis. *Compos. Sci. Technol.* **2007**, *67*, 2208–2218. [CrossRef]
34. Heidari-Rarani, M.; Sayedain, M. Finite Element Modeling Strategies for 2D and 3D Delamination Propagation in Composite DCB Specimens Using VCCT. CZM and XFEM Approaches. *Theor. Appl. Fract. Mech.* **2019**, *103*, 102246. [CrossRef]
35. Salehi, S.; Nygaard, R. Full Fluid-Solid Cohesive Finite-Element Model to Simulate Near Wellbore Fractures. *J. Energy Resour. Technol.* **2014**, *137*, 012903. [CrossRef]
36. Liu, F.; Gordon, P.A.; Valiveti, D.M. Modeling Competing Hydraulic Fracture Propagation with the Extended Finite Element Method. *Acta Geotech.* **2018**, *13*, 243–265. [CrossRef]
37. Sepehri, J.; Soliman, M.Y.; Morse, S.M. *Application of Extended Finite Element Method (XFEM) to Simulate Hydraulic Fracture Propagation from Oriented Perforations*; OnePetro: Richardson, TX, USA, 2015.
38. Wu, K.; Olson, J.E. Simultaneous Multifracture Treatments: Fully Coupled Fluid Flow and Fracture Mechanics for Horizontal Wells. *SPE J.* **2015**, *20*, 337–346. [CrossRef]
39. Wu, R.; Kresse, O.; Weng, X.; Cohen, C.; Gu, H. *Modeling of Interaction of Hydraulic Fractures in Complex Fracture Networks*; OnePetro: Richardson, TX, USA, 2012.

processes

MDPI

Article

Cryogenic Fracture Proliferation from Boreholes under Stresses

Minsu Cha [1,*], Naif B. Alqahtani [2] and Lei Wang [3]

1 Department of Civil Engineering, Jeju National University, Jeju-si 63243, Republic of Korea
2 Carbon Management Technologies Institute, King Abdulaziz City for Science and Technology (KACST), Riyadh 11442, Saudi Arabia; nqahtani@kacst.edu.sa
3 Petroleum Engineering Department, Chengdu University of Technology, Chengdu 610059, China; leiwang.cup@gmail.com
* Correspondence: mcha@jejunu.ac.kr

Abstract: Cryogenic fracturing has been explored in recent years as a waterless fracturing method for well stimulation to avoid issues encountered in water-based hydraulic fracturing. Cryogenic stimulation using liquid nitrogen applies large thermal gradients on reservoir rocks to induce fractures. This study investigates the initiation and proliferation of cryogenic fractures from boreholes under external stress on specimens. We flowed liquid nitrogen through boreholes drilled through the center of transparent PMMA cylinders under uniaxial stress and monitored fracture proliferation, temperatures, and borehole pressures. Our results show that the effect of stress resembles that of hydraulic fractures such that fractures propagate more in the direction of the stress. Under loading perpendicular to the borehole axis, a cloud of annular and longitudinal fractures extends more in the direction of loading. Under loading parallel to the borehole axis, longitudinal fractures dominate, and annular fractures become more suppressed and more sparsely distributed than those of unconfined specimens. Even if fractures are driven to initiate against the influence of stress, such as those from a boundary edge of a high stress concentration, they gradually deflect in the direction of stress, similar to hydraulic fractures from perforation holes that curve toward a direction perpendicular to the minimum stress direction.

Keywords: cryogenic fracturing; thermal fracture; fracture propagation; liquid nitrogen; thermal shock; thermal stress; hydraulic fracturing; well stimulation; unconventional reservoir; hot dry rock; enhanced geothermal system

Citation: Cha, M.; Alqahtani, N.B.; Wang, L. Cryogenic Fracture Proliferation from Boreholes under Stresses. *Processes* **2023**, *11*, 2028. https://doi.org/10.3390/pr11072028

Academic Editor: Qingbang Meng

Received: 7 June 2023
Revised: 23 June 2023
Accepted: 25 June 2023
Published: 6 July 2023

1. Introduction

To increase hydrocarbon mobility in low-permeability reservoirs, most wells are completed with well stimulations. Multistage hydraulic fracturing in horizontally drilled wells has been the most popular method for well stimulation. The large amount of water-based fracturing fluid in hydraulic fracturing, however, creates water-related problems such as potential impacts on the environment and groundwater, formation damage resulting from capillary retention and clay swelling, and the high level of consumption of freshwater, placing demands on local water supplies. Consequently, to avoid or minimize these water-related issues, the research and development into waterless or reduced-water fracturing methods have been active in recent years [1–5]. Cryogenic fracturing has been researched as a potentially effective waterless fracturing method.

In cryogenic fracturing, a cryogen, typically liquid nitrogen (LN), is applied on a much warmer rock under downhole conditions, exerting a very large thermal gradient and thermal stresses on reservoir formation to initiate and propagate fractures [6]. Laboratory and numerical studies have been conducted to understand cryogenic fracturing and its implications for the field [7]. Cryogenic treatments increase the pore size in saturated sandstone and expand micro-fissures resulting from thermal stress and frost force [8]. The shale pore structure undergoes significant alteration after freeze–thaw cycles, and some micro-pores are aggregated to form micro- and macro-cracks [9]. LN cooling increases permeability [10–12], although it is less effective on sandstone than on other sedimentary rocks [13,14]. Conductive cracks, the size and number of which increase with the initial temperature, are also generated on the surface and near the wellbore of shale [10,15]. After LN treatment, the strength and brittleness of shale decline, which reduces the initiation and propagation pressure of reservoir stimulation [9,16,17].

Nevertheless, we still do not understand how cryogenic fractures initiate and proliferate from boreholes over time. Consequently, Cha et al. [18] studied the process of cryogenic fracture proliferation from wellbores of transparent PMMA samples under no external confining stress. They revealed that annular fracture, longitudinal fracture, and exclusion distance are unique patterns of cryogenic fractures in the early stage. In addition, while cryogenic stimulation produces tortuous fractures, higher borehole pressure promotes straighter growth. Cha et al. [18], however, did not apply external stress to the specimens.

Therefore, in this follow-up study, we investigated the proliferation of cryogenic fractures from boreholes under external stress. For comparison, we applied two uniaxial loading schemes: (1) loading perpendicular to the borehole axis and (2) loading parallel to the borehole axis. Like Cha et al. [18], this study focused on pure thermal shock with negligible borehole hydraulic loading during the coolant flow. We flowed LN through boreholes drilled through the center of transparent PMMA blocks to visualize fracture initiation and propagation. Fracture propagation, temperature, and borehole pressure were recorded during the flow to assess cryogenic fracturing processes and behaviors.

2. Laboratory Study

2.1. Devices and Procedure

LN was supplied from a Dewar and transported through a vacuum-jacketed hose to the specimen and injected into the borehole and then directed to an outlet (Figure 1). This configuration allowed liquid nitrogen (LN) to be continuously flushed, which maximized thermal shock and the thermal gradient on borehole walls and thus cooled the rock around the borehole as rapidly as possible. For the flow of LN through the borehole, we applied a coaxial flow design in which LN entered the borehole through the central smaller-diameter inlet tubing (blue tubing and arrow—Figure 1a), which passed through a larger-size cross-shaped fitting. Warmed nitrogen exited through the annulus between the inlet tubing and the casing, and then through the space in the cross-shaped fitting (orange arrows—Figure 1a). Cryogenic fracturing was carried out via pure thermal shock; the pressure inside the boreholes was less than 50 kPa during the LN flow. A small diameter (30 AWG) of T-type thermocouple wires (TT-T-30 from Omega Engineering, Norwalk, CT, USA) was selected for fast thermal responses (<0.1 s) and access to the borehole. More information about the equipment and measurements appears in Cha et al. [19] and Lu and Cha [20].

Uniaxial loading schemes include (1) loading perpendicular to the borehole axis and (2) loading parallel to the borehole axis (Figure 1 and Table 1). Loading perpendicular to the borehole axis was attained via vertical stress on a horizontal well, in which a loading was applied along a strip due to the cylindrical shape of the specimens used in this study (Figure 1b). The loading parallel to the borehole axis was achieved using vertical stress on a vertical well, in which case, uniform stress was applied over the whole end faces (Figure 1c). Unlike triaxial loading, uniaxial loading enabled real-time visual monitoring because the sides perpendicular to the other two axes were exposed. Teflon sheets placed between the specimen and loading platens uniformly distributed the loading and minimized friction.

2.2. Specimens

We conducted the experiments in poly(methyl methacrylate) (PMMA), a transparent, thermoplastic polymer, with the advantage being that one could visually observe fracture proliferation. Researchers have used PMMA for hydraulic fracturing tests because of its known properties and proximity to shale in terms of certain mechanical properties, including fracture toughness and brittleness [21–23]. Table 2 summarizes the mechanical and thermal properties of PMMA at both room temperature and cryogenic temperature, for it exhibits temperature-dependent properties. We used three cylindrical PMMA specimens, the outer sizes of which were 10.2 cm in diameter, and embedded the casing in the borehole to a depth of 4.45 cm in all three cases. The full dimensions of each specimen and its borehole are provided in detail in Figure 2. The stress conditions applied to each specimen and roughness conditions are listed in Table 1.

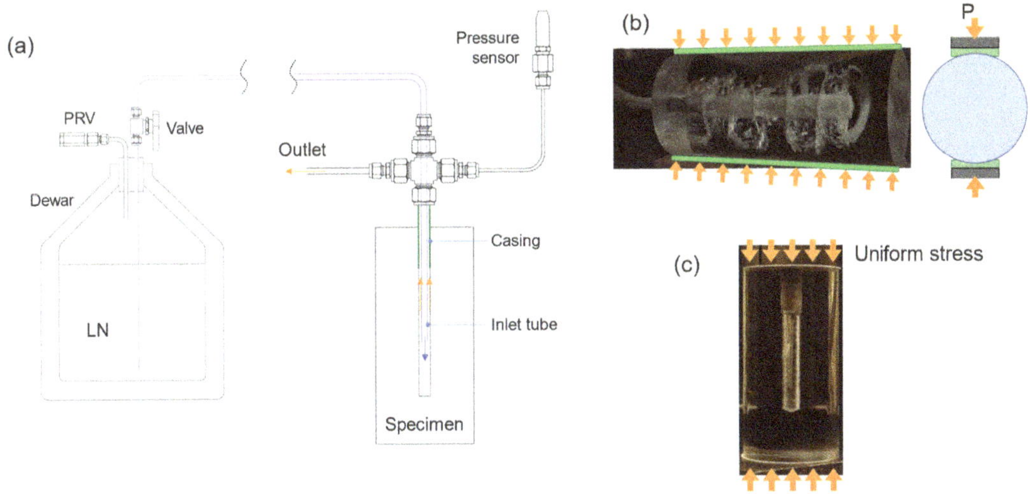

Figure 1. Experimental setup for the cryogenic stimulation of specimens under stress. (**a**) Overall setup without the loading device. (**b**) Vertical stress on the horizontal well (i.e., stress perpendicular to the borehole axis). (**c**) Vertical stress on the vertical well (i.e., stress parallel to the borehole axis).

Table 1. Conditions of externally applied stresses and roughness of borehole surfaces.

	Stress Direction	Stress Magnitude	Borehole Surface
Specimen A	Perpendicular to the borehole axis	101 kPa (average) 191 kPa (peak) (P = 2.35 kN)	Rough
Specimen B	Parallel to the borehole axis	6.9 MPa	Rough
Specimen C	Parallel to the borehole axis	3.45 MPa	Rough (upper part) Smooth (lower part)

Table 2. Properties of liquid nitrogen and PMMA (Updated from Cha et al. [18]). Properties of shale, gas nitrogen, and water are shown for comparison.

Properties	PMMA [(a)]	PMMA (Low-T) [(b)]	Shale [(c)]
Density (g/cm^3)	1.18–1.19	-	2.39
Unconfined compressive strength (MPa)	90–120	250 (−40 °C)	54.6
Tensile strength (MPa)	55–76	100–110 (−40 °C)	8.48 (Splitting)
Static Young's modulus (GPa)	2.4–3.3	5.1 (−173 °C)	41.4
Surface energy (dyn/cm)	41	-	46.6
Fracture toughness K_{IC} ($MPa \cdot m^{0.5}$)	~1.5	0.59–0.96 (−80 °C)	1.5
Poisson's ratio	0.35–0.4	-	0.27
Specific heat capacity (J/(kg·K))	1450	450 (−196 °C)	990
Thermal conductivity (W/(m·K))	0.18–0.19	0.14 (−196 °C)	
Linear thermal expansion coeff. (K^{-1})	$(60–80) \times 10^{-6}$	26×10^{-6} (−196 °C)	11×10^{-6}

Table 2. *Cont.*

	Liquid nitrogen [(d)]	Gas nitrogen [(e)]	Water [(e)]
Viscosity (cP)	0.158	1.76×10^{-2}	1.002
Density (g/mL)	0.807	0.0012	0.998
Surface tension (dyn/cm) (against air)	8.85	-	72.8
Specific heat (kJ/(kg·K))	2.04	1.04	4.18
Thermal conductivity (W/(m·K))	0.140	0.025	0.591

[(a)] At 20 °C and 1 atm. From manufacturers except the surface energy [24] and fracture toughness [25]. [(b)] At low temperatures [26–30]. [(c)] At 20 °C and 1 atm. Niobrara shale data from Cha et al., 2018 [17], except the surface energy [31] and fracture toughness [32] for Barnett shale. [(d)] At −196 °C (77 K) and 1 atm. [(e)] At 20 °C and 1 atm.

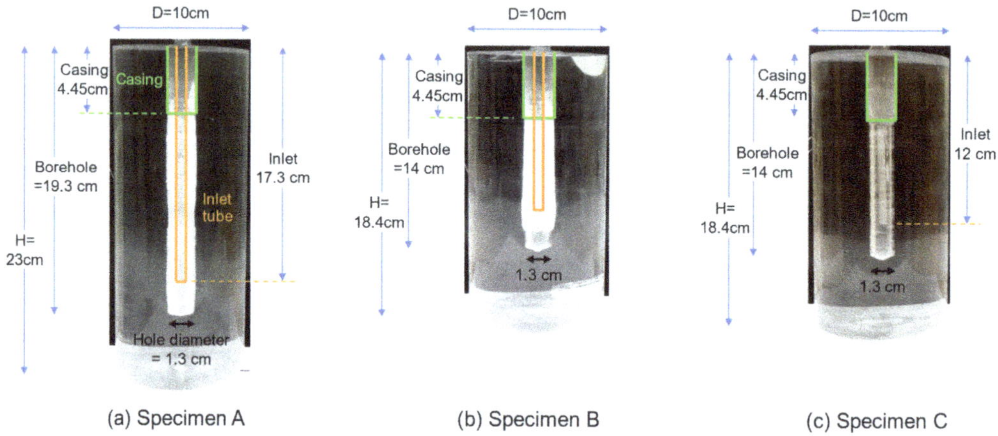

Figure 2. Dimensions of PMMA specimens and placement of the stainless-steel casings and inlet tubes.

3. Results and Interpretations

3.1. Temperature and Pressure

With the introduction of LN, the borehole temperatures of the specimens dropped rapidly and reached the boiling point of nitrogen in 3.14 min for Specimen A, 3.32 min for Specimen B, and 2.4 s for Specimen C (Location 1 and 2 of Figure 3). The temperature at the outer surfaces, however, remained high, above 0 °C, and also depended on their proximity to the cracks, resulting from the entrance of LN into the crack apertures (Location 3, 4, and 5 of Figure 3). As Location 5 was the farthest from the cracks, the temperature remained the highest. When the LN flow stopped, the temperature at Location 4 recovered because the LN evaporated from the apertures of adjacent cracks, so the area began to warm up. At Location 3, however, the temperature continued to decrease because of its proximity to the region of higher crack density and the steel loading block, the temperature of which was low due to contact with the flow tubing and leaked LN. This temperature distribution demonstrates the effect of the presence of cracks on temperature propagation. During the flows, the ΔT between the borehole surface and the outer surface remained above 200 K, creating an average thermal gradient between the borehole and the outer surfaces of about 46.0 K/cm.

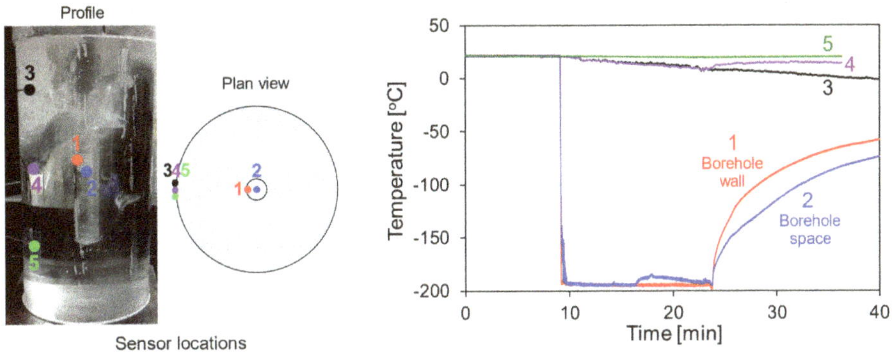

Figure 3. Typical data of the temperature evolution at various locations during the flow of LN (Specimen C). The colored dots in the left figure show the locations of temperature sensors.

As LN was directly transferred from the Dewar to the borehole, the pressure inside the borehole remained closely associated with the pressure inside the Dewar. The pressure measured at the borehole fell in the range of 90–100 kPa for Specimen A, 20–80 kPa for Specimen B, and 30–60 kPa for Specimen C (Figure 4a). Unlike Specimen A, Specimen B and C had leakage, which reduced the borehole pressure. In fact, because of excessive leaking, we terminated the experiment for Specimen C.

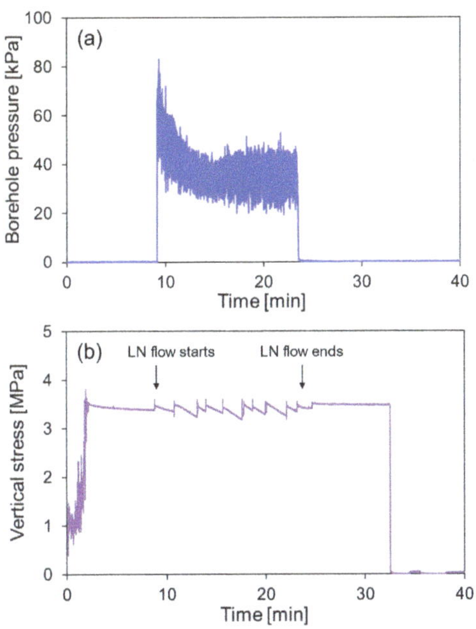

Figure 4. (**a**) Typical borehole pressure and (**b**) vertical stress responses controlled by the oil pressure of the hydraulic actuator, indicating the shrinkage of the specimen during the flow of LN (Specimen C).

Global, longitudinal contractions of specimens were indicated by pressure responses of the hydraulic system (Figure 4b). That is, when the specimen was cooled down, the specimen shrank and the pistons in contact with the specimen decompressed, which decreased the hydraulic pressure. In Figure 4b, each jump represents a manual adjustment that made up for pressure decay in the hydraulic lines to maintain the vertical stress at around 3.45 MPa.

3.2. Fracture Propagation

Behaviors observed in unconfined specimens still held under uniaxial stresses. Fractures started orthogonally from the borehole surfaces, and initial cracks developed primarily in two patterns—fractures perpendicular to the borehole axis (annular fractures) and fractures parallel to the borehole axis (longitudinal fractures)—which were governed by simple pre-crack stress components: circumferential thermal contraction and longitudinal thermal contraction. While the fractures initiated from the borehole perpendicular to the borehole surface, they soon curved one way or another. Fracture growth was characterized by abrupt starts and stops, and as the fracture propagated outward, the growth appeared to be more continuous.

Specimens A, B, and C took 1.83 min, 2.17 min, and 0.5 min, respectively, for the initial cracks to emerge after the initial flow. The average speeds of radial fracture propagation were quantified from time-lapse photos using the longest cracks. The average speeds of fracture propagation in the radial direction were 6.4 mm/min for Specimen A, 7.9 mm/min for Specimen B, and 5.7 mm/min for Specimen C. As Specimen C had a major fracture with consistent curvature, its lengthwise crack speed was 7.1 mm/min. These crack speeds were slightly faster compared to those in the unconfined test [18]. As fractures propagate radially, the fracture propagation rate tends to decrease.

3.3. Effect of Stress

3.3.1. Loading Perpendicular to the Borehole Axis (Specimen A)

Annular fractures emerged slightly earlier than longitudinal fractures and at a quasi-equal distance (final distance between the fractures ≈ 2.5 cm) (Figures 5 and 6). Longitudinal fractures followed about 15 s after and tended to initiate mainly from the top (ceiling) and bottom parts of the borehole, oriented vertically, which was the effect of externally applied stresses.

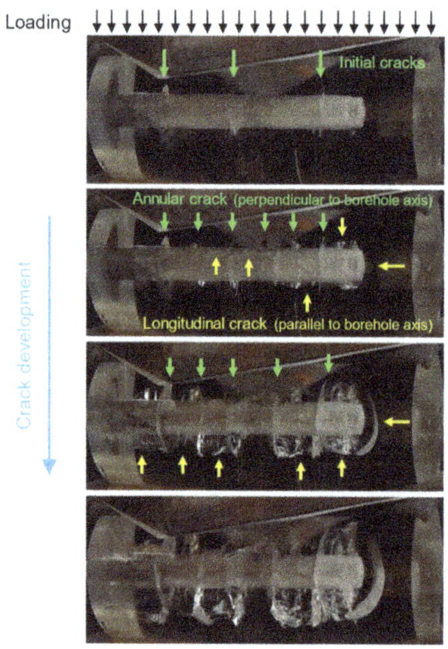

Figure 5. Selected steps of crack development in Specimen A.

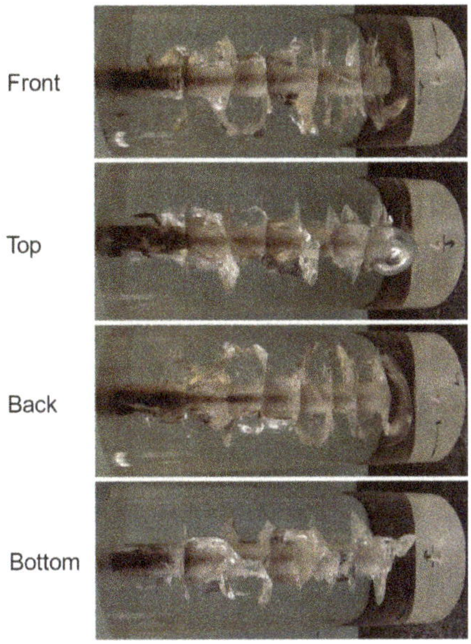

Figure 6. Crack view from the four sides after completing the experiment on Specimen A.

Although silicone pads were placed to distribute the load a bit, the configuration, not including the borehole, resembled Brazilian tests (Figures 1c and 7a). In Brazilian test loading, compressive stress occurs along the loading direction within the specimen, while tensile stress occurred perpendicular to the loading direction, which is a favorable stress distribution for vertical crack initiation and propagation. Figure 7b shows that the axial loading created the compressive stress of 191 kPa on the horizontal plane at the center of the disk and the tensile stress of −64 kPa on the vertical plane. Although the tensile stress was far below the tensile strength 65 MPa of PMMA (Table 2), it still contributed to preferential fracture propagation along the direction of the loading axis by creating a relative difference between the stresses of the two directions.

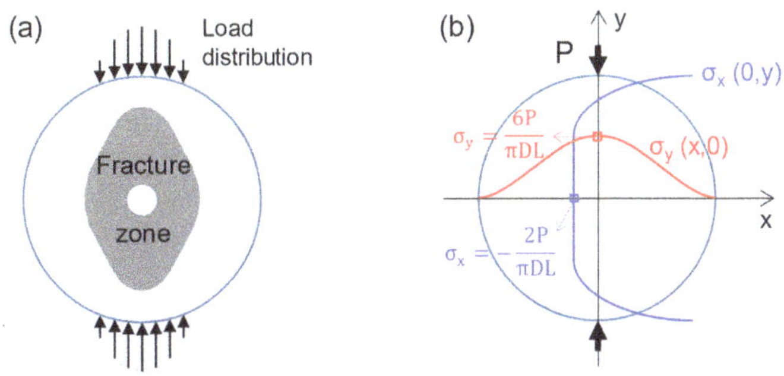

Figure 7. (**a**) Schematic of the approximate radial extent of fractures observed in Specimen A (cross-sectional view). (**b**) Stress distribution within a cylindrical specimen under Brazilian-test-type loading.

Overall, both annular and longitudinal fractures extended more in the direction of the loading (Figures 6 and 7a). The likely explanation for why the fractures did not propagate all the way to the end was the sharply increasing compressive stress near both ends (Figures 6 and 7).

Following the initiation of the annular fractures, longitudinal cracks initiated from and/or formed between the annular fractures, as pre-existing defects (i.e., annular fractures) act as stress concentrators (Figure 5).

3.3.2. Loading Parallel to the Borehole Axis (Specimen B)

For Specimen B, vertical stress of 6.9 MPa was applied on the vertical borehole (Table 1). Longitudinal fractures appeared first, and then, after about 50 s (which is significant), annular fractures followed. Longitudinal fractures remained dominant throughout the stimulation, and the annular fractures in Specimen B were more suppressed and more sparsely distributed (final distance between the fractures $\approx$ 3.5 cm) than those in Specimen A (Figure 8). After all, unlike in an unconfined borehole, in which cooling immediately creates tension longitudinally and circumferentially, the wall of the vertical borehole was initially under compressive stress longitudinally due to the external loading. With continued LN flow, cooling reduced the compressive stress and then created tensile stress. Unlike Specimen A, in which fractures propagated preferentially in the direction of loading (Figure 7a), the radial propagation of the fractures in Specimen B were roughly equally distributed without a preferential direction (Figure 9).

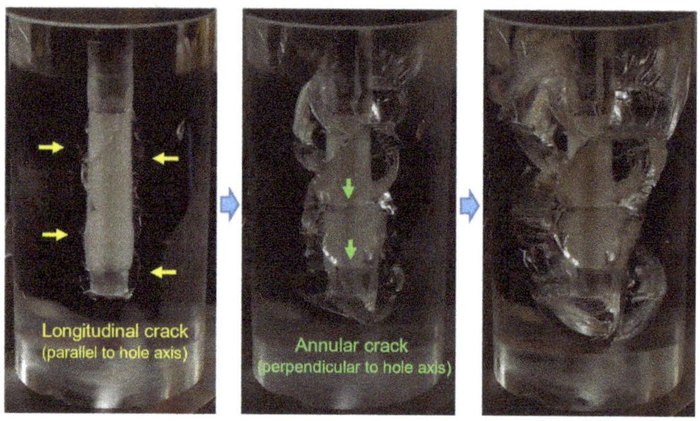

Figure 8. Selected steps of crack development in Specimen B.

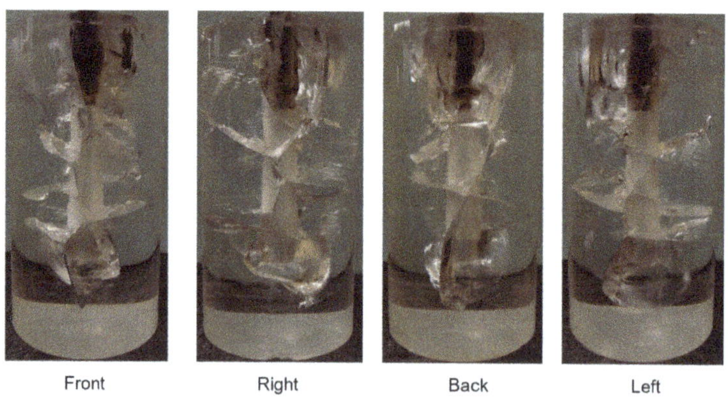

Figure 9. Crack view from the four sides after completing the experiment in Specimen B.

3.4. Effect of the Distribution of Surface Defects on Crack Initiation

It was observed that during drilling, due to the elevated temperature, borehole surfaces underwent plastic deformation and developed a rough texture. In addition, residual stresses may have existed near the borehole walls because of the cooling of the hot surface during drilling, although this was not verified using such tools as photoelasticity. On the other hand, the smooth or "flawless" surface was created when the temperature of the materials was kept low by slow drilling and a sufficient supply of cooling fluid. Therefore, different levels of surface roughness existed (Figure 2).

The upper part of Specimen C, where the casing was embedded, had a rough surface but a smooth surface below (Figure 2c). All of the fractures in this specimen initiated from the rough surface, but no fractures initiated from the smooth/polished surface (Figures 10 and 11). A major fracture initiated from the circumference (green arrow—Figure 10) where the hole diameter changes abruptly (the diameter of the hole for the embedded casing was slightly larger). As fractures only occurred on the rough surface in the upper part, they reached the top boundary, causing leakage (Figure 10).

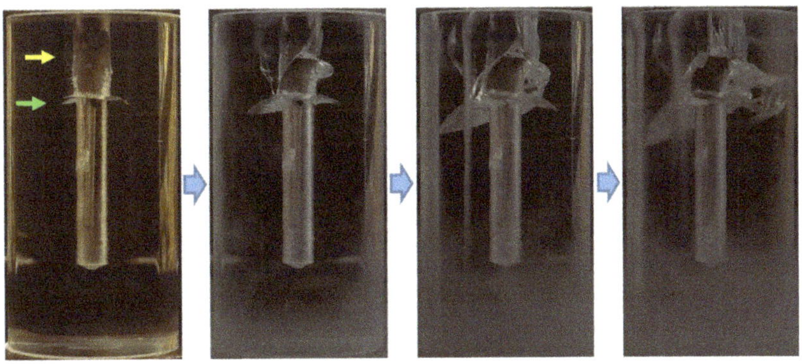

Figure 10. Selected steps of crack development in Specimen C.

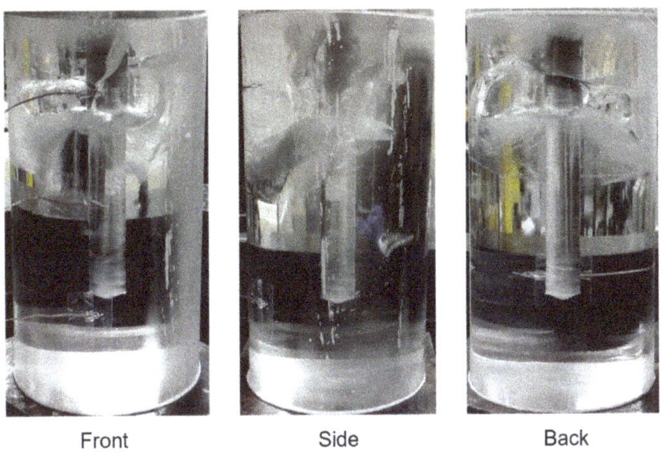

Figure 11. View of cracks from the sides after completing the experiment in Specimen C.

Hence, it follows that the uniformly rough surface provided a random distribution of defects, which acted as stress concentrators when tensile stress was applied, and thus served as spots for the nucleation of cracks. Conversely, polished/flawless surfaces do not have defects (or have defects on a much smaller scale) and thus do not facilitate crack nucleation as readily as rough surfaces. Thus, surface roughness lowers threshold stress for crack initiation, as predicted using the critical stress at the tip of a defect/crack required for propagation, given by Griffith's criterion [33]

$$\sigma_f = \sqrt{\frac{2E\gamma}{\pi a}} \text{ for plane stress, and} \tag{1a}$$

$$\sigma_f = \sqrt{\frac{2E\gamma}{\pi a(1-\nu^2)}} \text{ for plane strain,} \tag{1b}$$

where a is the depth of a micro crack on the surface, γ is the surface energy per unit area, and ν is Poisson's ratio. With crack length a = 0.1 mm and other properties in Table 2, we obtain σ_f = 0.86 MPa for the plane stress condition and σ_f = 0.93 MPa for the plane strain condition. Also, Irwin's criterion for crack propagation in terms of fracture toughness is [34]

$$\sigma_f = \frac{K_{Ic}}{Y\sqrt{\pi a}} \tag{2}$$

where K_{Ic} is the fracture toughness of the material, and Y is a dimensionless geometric factor that depends on the shape and size of the crack. Applying a value of Y = 1.3 for a semi-elliptical surface crack [35] and the fracture toughness of 0.78 at −80 °C (Table 2), we obtained σ_f = 33.6 MPa.

While Griffith's criterion yields a much lower value than Irwin's criterion, both predict lower threshold stresses for initiating a crack than the tensile strength of 55–76 MPa of PMMA. Incidentally, the change in the tangential tensile stress induced by temperature changes on the wellbore wall estimated using $\sigma_{max}^{\Delta T} = \frac{E}{1-\nu}\alpha \cdot \Delta T$ was calculated to be 67 MPa with the specimen properties in Table 2 (E = 2.85 GPa, $\alpha = 70 \times 10^{-6}$/K, and ν = 0.38) and ΔT = 210 °C. Overall, the formulas indicate the role of defects in lowering the stress required to initiate fractures. The difference between the depth of roughness of the rough and smooth surfaces was at least one order of magnitude, which effectively prevented crack initiation on the smooth surface.

Therefore, it follows from the observations that if the borehole is rough throughout, then the distributed fractures with exclusion distance occur (Specimen A and B). However, if a polished surface exists, cracks occur outside the polished area such as rough surfaces or weak boundaries (Specimen C).

These observations suggest that the initiation and pattern of cryogenic fractures from a wellbore in the field would be significantly influenced by surface characteristics such as drilling-induced textures or naturally existing cracks.

3.5. *Cryogenic Fracture vs. Hydraulic Fracture: Effect of Stress*

The orientation of a hydraulically induced fracture is dependent on the in situ stress state (magnitude and orientation). Hydraulic fractures form perpendicular to the minimum stress direction (Figure 12a). Under the same stress states as those applied in this study, Figure 12 shows the schematics of the expected orientation of hydraulically induced fractures from an open borehole. As no stresses were present in either horizontal direction (albeit not a realistic condition), this particular situation would have created a vertical fracture in any lateral direction (Figure 12b,c). Hydraulic fracturing exhibits one major fracture plane (per stage in staged fracturing) (Figure 12), although complex features can arise locally if a reservoir formation contains natural fractures. In addition, boundary effects may exist in laboratory specimens.

On the other hand, cryogenic thermal shock with a negligible influence of borehole hydraulic loading induces multiple initiations (vs. the bi-wing initiation of hydraulic fracturing) and curved propagation (vs. the straight propagation of hydraulic fracturing). If stress is applied uniaxially, in addition to manifesting these two features, the effect of stress resembles that of hydraulic fractures such that fractures propagate more in the direction of maximum stress (Figures 5–9).

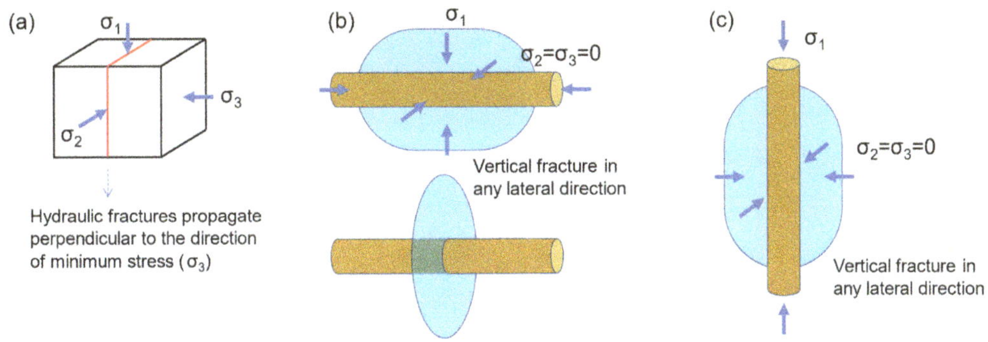

Figure 12. Expected orientation of hydraulically induced fractures. (**a**) General principle. (**b**) The case of horizontal wells with $\sigma_1 \neq 0$ and $\sigma_2 = \sigma_3 = 0$ (Specimen A). (**c**) The case of vertical wells with $\sigma_1 \neq 0$ and $\sigma_2 = \sigma_3 = 0$ (Specimen B and C).

Unlike in the cases that present a regular fracture pattern nucleated from randomly, homogeneously distributed defects (Specimen A and B), in Specimen C, a major annular fracture was driven to initiate perpendicular to the surface of the ringed edge because of a large stress concentration there even though the specimen was under loading parallel to the borehole axis. However, the annular fracture gradually became vertical (the direction of stress), which was clearly the influence of the applied stress (Figures 10 and 11).

Similar behaviors occur in hydraulic fracturing in perforated holes. After hydraulic fractures are forced to initiate at the tip of perforation holes, they curve toward the direction perpendicular to the minimum stress direction [36–39]. Figure 13 displays an idealized diagram of the influence of the stress ratio σ_H/σ_h and the injection rate.

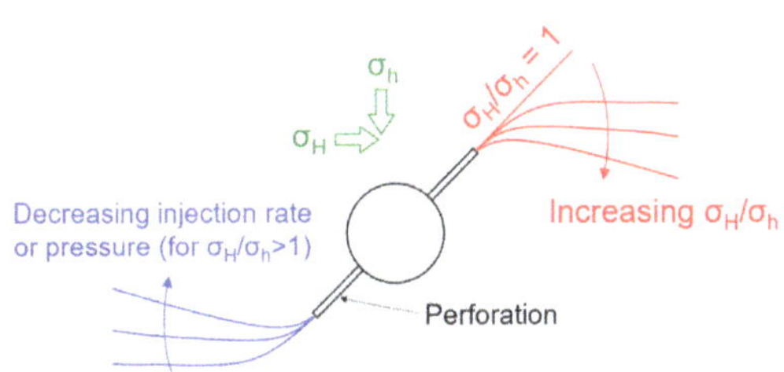

Figure 13. Idealized diagram of the effect of the stress ratio σ_H/σ_h and the injection rate on hydraulic fracturing from perforated holes.

3.6. Cooling Rate of the Borehole Surface

The cooling rates of the borehole surfaces from the initial introduction of LN to about −190 °C, where the temperature quickly stabilizes, were 1.09 °C/s for Specimen A, 1.03 °C/s for Specimen B, and 90.8 °C/s for Specimen C (Figure 3). Thus, the thermal transfer was much more facilitated in Specimen C.

According to the results of previous studies on the cooling of a concrete borehole [17,19], the cooling rates of Specimen A and B were comparable to cases in which the film boiling regime is relevant, where an insulating vapor film is created between the liquid and a hot surface, which slows heat transfer (i.e., the Leidenfrost effect). On the other hand, the magnitude of the cooling rate in Specimen C was comparable to that of the contact boiling regime, which allows much more rapid cooling. In Cha et al.'s study [17], rapid cooling with contact boiling was achieved in the high-pressure LN flow through a borehole, probably due to suppressed vapor film under high pressure.

In pursuit of an explanation for the significantly different cooling rates in our study, although the roughness level of the specimens is a factor affecting boiling regimes, a variety of other surface properties also affect boiling regimes. Many studies have reported that roughness destabilizes the vapor film or elevates the Leidenfrost point, thus enhancing heat transfer [40–42]. On the other hand, some studies have reported that certain texture patterns reduce the Leidenfrost point [43,44]. Other surface properties such as wettability and porosity affect the boiling regime [45,46]. A further complexity in our study was the rapid flow condition, which has not been reported in the literature.

A high initial cooling rate is favorable for inducing thermal shock because it creates high local thermal gradients transiently. However, the distribution of pre-existing surface defects appears to be a more dominant factor in fracture initiation.

4. Discussions

4.1. Crack Spacing: Exclusion Distance

A tendency toward regularly spaced annular fractures was displayed in Specimen A and B (Figures 5 and 8). Regular crack spacing or an exclusion distance is present because major cracks cannot form closer than a certain length due to the limited amount of thermal contraction. The crack spacing pattern or exclusion distance has also been reported and modeled in thermal cracking in simple two-dimensional media [47–49].

For the geometry of a slab subjected to cooling on one face, the spacing and penetration of the cracks in a regular array depends on the following dimensionless quantity:

$$\frac{L}{L_c} = \frac{(\alpha \Delta T)^2 EL}{G_F} \tag{3}$$

where L is the slab thickness, L_c is the characteristic length (or Griffith crack length) of a shrinking solid, and G_F is the fracture energy of crack propagation (N/m). L_c is based on thermal tensile stress (σ_{th}), Young's modulus (E), and fracture energy:

$$L_c = \frac{G_F}{(\alpha \Delta T)^2 E} = \frac{G_F E}{\sigma_{th}^2} \tag{4}$$

The significance of L_c is that cracking will occur when the diffusion of heat has penetrated the slab to a depth of roughly L_c. As L/L_c increases, cracking starts earlier, the final crack spacing decreases, and penetration increases [50,51]. From the equation, increasing ΔT accelerates the trend. In the 3D borehole environment, where fractures at later stages begin to interact in a complex manner, the exclusion distance of annular fractures is more obvious in the early stages (Figures 5 and 8).

4.2. Fracture Tortuosity

A non-dimensional fracture toughness (κ) was introduced to categorize the pattern of tortuosity or waviness of thermally induced crack propagation in brittle solid [52]:

$$\kappa = \frac{K_{Ic}}{\alpha E \Delta T \sqrt{\pi b}}, \tag{5}$$

where α is the linear thermal expansion coefficient, and b is crack spacing. Cracks propagate straight when κ is relatively large, while they may form wavy paths for relatively small κ values. In our experimental results, κ was 0.141, categorized as "wavy". Although the classification based on simple glass plates may not allow quantitative comparison with our results from borehole geometry, the equation predicts that smaller fracture toughness and larger ΔT induce more wavy fractures, supported by simulation results. High temperature differences induce high T-stress, which may lead to the turning of the crack. When its local propagating direction becomes normal to its macroscopic propagation direction, it is likely to be arrested because further development of the cooled layer does not increase the stress intensity factor (K_I) [52].

The pattern and degree of tortuosity and fracture merging as observed in the borehole geometry of this study were more complex than those observed in the simple geometries of previous thermal shock studies [53–55]. Possible causes for fracture tortuosity and the merging observed in this study were the geometry of the borehole and the flow of cryogenic liquid nitrogen in fracture apertures, driving complex, dynamic thermo-mechanical conditions and crack propagation. In fact, we observed that LN flowed almost to the tip of a crack in its liquid state assisted by its low viscosity and surface tension (see Table 2) even though no permeation into the materials occurred.

The curvatures in this study exhibited similarities to "caged fractures" around underbalanced wellbores, where fractures curl around wellbores within a fracture cage [56,57], within which, the principal deviatoric compressional stress follows concentric rings, not the far-field stress. From the principle of a crack following the path of least resistance, the tendency of the fractures curved to be contained within a cylindrical volume rather than propagated outward was induced by the rotation of the local principal stress caused by changes in the internal stress field.

4.3. Effect of Rock Types

PMMA was selected for this study because of its transparency for observations of real-time fracture processes, and PMMA has been used in the literature as a surrogate for rocks in borehole fracturing experiments owing to its similarity to shale regarding certain mechanical properties [21–23]. Among rock formations, the closest counterpart of PMMA, which is glassy and impermeable, is shale. In Cha et al.'s study [17], breakdown tests after cryogenic thermal shock on fine-grained shale specimens with low permeability revealed distinct curvatures in fractures even though penetration was limited due to large triaxial stresses.

On the other hand, fractures created via cold-water thermal shock on hot concrete specimens under negligible borehole pressure exhibited a rough fracture surface without consistent curvature [20]. Porous sandstone also exhibited a fracture surface similar to that of concrete, but penetration was significantly less than that of concrete even in an unconfined condition in our unpublished work.

Two differences between fine-grained rock and coarse-grained/porous rock related to thermal fracture patterns can be pointed out. Coarse-grained or porous rocks such as concrete specimens and sandstone are characterized by significant porosity and large grain sizes. Fluid permeates into the porous formation, affecting the temperature field differently from that in a non-permeation condition. Large grains can divert or bifurcate cracks in grain scales, possibly affecting subsequent crack propagation as well. For the latter aspect of large grain sizes, laboratory specimens may not be large enough to reduce the relative effect of grain size and manifest curvature patterns in a greater scale.

4.4. Limitations

Although the degree of the locality of the thermal process was sufficient to create the level of fractures observed in this study, thermal contractions around boreholes were not fully local, since we observed some global contraction in Figure 4b. In a large-scale test or field-scale test where the volume influenced by the thermal process is significantly smaller than the surrounding volume, the locations of measurement or the boundary would be sufficiently far from the thermal influence; thus, measurement would take place under fixed displacement, and global contraction would be negligible. The findings are applicable to open holes.

5. Conclusions

In this study, we performed cryogenic fracturing tests in the boreholes of transparent PMMA specimens under external stresses and directly observed dynamic cryogenic fracturing processes and morphology with time in the borehole geometry. Salient, unprecedented observations of cryogenic fracture proliferation from boreholes under stresses are summarized below.

The major differences between cryogenic fracturing and hydraulic fracturing are that cryogenic thermal shock induces multiple initiations (vs. the bi-wing initiation of hydraulic fracturing) and curved propagation (vs. the straight propagation of hydraulic fracturing). If external stress is applied uniaxially, while still manifesting these two features, the effect of stress resembles that of hydraulic fractures such that fractures propagate more in the direction of the stress.

Specifically, in the case of loading perpendicular to the borehole axis, the cloud of annular and longitudinal fractures extends more in the direction of loading than it does in the case of an unconfined specimen. In the case of loading parallel to the borehole axis, longitudinal fractures are dominant, and annular fractures are more suppressed and more sparsely distributed than they are in the cases of loading perpendicular to the borehole axis or the unconfined specimen.

Even if an annular fracture is driven to initiate from a boundary under the loading parallel to the axis because of high stress concentration, it gradually deflects in the direction of stress, much as hydraulic fractures initiated at the tip of perforation holes curve toward the direction perpendicular to the minimum stress direction.

The roughness of a borehole surface lowers the threshold stress for the initiation of cryogenic fractures. On a borehole surface with uniform roughness, distributed fractures with an exclusion distance will occur. If a surface presents various roughness levels, cracking is focused on rougher surfaces or boundaries. Therefore, in the field, the initiation of cryogenic fractures would be greatly influenced by surface characteristics such as drilling-induced texture or naturally existing cracks.

Author Contributions: Conceptualization, M.C.; methodology, M.C.; formal analysis, M.C.; investigation, M.C., N.B.A. and L.W.; writing—original draft preparation, M.C.; writing—review and editing, N.B.A. and L.W.; funding acquisition, M.C. All authors have read and agreed to the published version of the manuscript.

Funding: This work was supported by a research grant from Jeju National University in 2021.

Institutional Review Board Statement: Not applicable.

Informed Consent Statement: Not applicable.

Data Availability Statement: The data presented in this study are available on request from the corresponding author.

Conflicts of Interest: The authors declare no conflict of interest.

References

1. Wang, L.; Yao, B.; Cha, M.; Alqahtani, N.B.; Patterson, T.W.; Kneafsey, T.J.; Miskimins, J.L.; Yin, X.; Wu, Y.-S. Waterless fracturing technologies for unconventional reservoirs-opportunities for liquid nitrogen. *J. Nat. Gas Sci. Eng.* **2016**, *35 Pt A*, 160–174. [CrossRef]
2. Liew, M.; Danyaro, K.U.; Zawawi, N.A.W.A. A comprehensive guide to different fracturing technologies: A review. *Energies* **2020**, *13*, 3326. [CrossRef]
3. Nianyin, L.; Chao, W.; Suiwang, Z.; Jiajie, Y.; Yinhong, D. Recent advances in waterless fracturing technology for the petroleum industry: An overview. *J. Nat. Gas Sci. Eng.* **2021**, *92*, 103999. [CrossRef]
4. Fu, C.; Liu, N. Waterless fluids in hydraulic fracturing–A review. *J. Nat. Gas Sci. Eng.* **2019**, *67*, 214–224. [CrossRef]
5. Liu, J.; Xie, J.; Yang, B.; Li, F.; Deng, H.; Yang, Z.; Gao, M. Experimental Study on the Damage Characteristics and Acoustic Properties of Red Sandstone with Different Water Contents under Microwave Radiation. *Materials* **2023**, *16*, 979. [CrossRef]
6. Tang, S.; Wang, J.; Chen, P. Theoretical and numerical studies of cryogenic fracturing induced by thermal shock for reservoir stimulation. *Int. J. Rock Mech. Min. Sci.* **2020**, *125*, 104160. [CrossRef]
7. Huang, Z.; Zhang, S.; Yang, R.; Wu, X.; Li, R.; Zhang, H.; Hung, P. A review of liquid nitrogen fracturing technology. *Fuel* **2020**, *266*, 117040. [CrossRef]
8. Cai, C.; Li, G.; Huang, Z.; Shen, Z.; Tian, S.; Wei, J. Experimental study of the effect of liquid nitrogen cooling on rock pore structure. *J. Nat. Gas Sci. Eng.* **2014**, *21*, 507–517. [CrossRef]

9. Han, S.; Cheng, Y.; Gao, Q.; Yan, C.; Han, Z. Experimental study of the effect of liquid nitrogen pretreatment on shale fracability. *J. Nat. Gas Sci. Eng.* **2018**, *60*, 11–23. [CrossRef]
10. Wu, X.; Huang, Z.; Li, R.; Zhang, S.; Wen, H.; Huang, P.; Dai, X.; Zhang, C. Investigation on the damage of high-temperature shale subjected to liquid nitrogen cooling. *J. Nat. Gas Sci. Eng.* **2018**, *57*, 284–294. [CrossRef]
11. Alqatahni, N.B.; Cha, M.; Yao, B.; Yin, X.; Kneafsey, T.J.; Wang, L.; Wu, Y.-S.; Miskimins, J.L. Experimental Investigation of Cryogenic Fracturing of Rock Specimens Under True Triaxial Confining Stresses. In Proceedings of the SPE Europec featured at 78th EAGE Conference and Exhibition, Vienna, Austria, 30 May 2016; p. 24.
12. Elwegaa, K.; Emadi, H. The effect of thermal shocking with nitrogen gas on the porosities, permeabilities, and rock mechanical properties of unconventional reservoirs. *Energies* **2018**, *11*, 2131. [CrossRef]
13. Du, M.; Gao, F.; Cai, C.; Su, S.; Wang, Z. Study on the surface crack propagation mechanism of coal and sandstone subjected to cryogenic cooling with liquid nitrogen. *J. Nat. Gas Sci. Eng.* **2020**, *81*, 103436. [CrossRef]
14. Cha, M.; Yin, X.; Kneafsey, T.; Johanson, B.; Alqahtani, N.; Miskimins, J.; Patterson, T.; Wu, Y.-S. Cryogenic fracturing for reservoir stimulation—Laboratory studies. *J. Pet. Sci. Eng.* **2014**, *124*, 436–450. [CrossRef]
15. Yang, R.; Chunyang, H.; Huang, Z.; Wen, H.; Li, X.; Huang, P.; Liu, W.; Chen, J. Liquid Nitrogen Fracturing in Boreholes under True Triaxial Stresses: Laboratory Investigation on Fractures Initiation and Morphology. *SPE J.* **2021**, *26*, 135–154. [CrossRef]
16. Jiang, L.; Cheng, Y.; Han, Z.; Gao, Q.; Yan, C.; Wang, H.; Fu, L. Effect of liquid nitrogen cooling on the permeability and mechanical characteristics of anisotropic shale. *J. Pet. Explor. Prod. Technol.* **2019**, *9*, 111–124. [CrossRef]
17. Cha, M.; Alqahtani, N.B.; Yao, B.; Yin, X.; Kneafsey, T.J.; Wang, L.; Wu, Y.-S.; Miskimins, J.L. Cryogenic Fracturing of Wellbores Under True Triaxial-Confining Stresses: Experimental Investigation. *SPE J.* **2018**, *23*, 1271–1289. [CrossRef]
18. Cha, M.; Alqahtani, N.B.; Yin, X.; Wang, L.; Yao, B.; Kneafsey, T.J.; Miskimins, J.L.; Wu, Y.-S. Propagation of Cryogenic Thermal Fractures from Unconfined PMMA Boreholes. *Energies* **2021**, *14*, 5433. [CrossRef]
19. Cha, M.; Alqahtani, N.B.; Yin, X.; Kneafsey, T.J.; Yao, B.; Wu, Y.-S. Laboratory system for studying cryogenic thermal rock fracturing for well stimulation. *J. Pet. Sci. Eng.* **2017**, *156*, 780–789. [CrossRef]
20. Lu, Y.; Cha, M. Thermally induced fracturing in hot dry rock environments—Laboratory studies. *Geothermics* **2022**, *106*, 102569. [CrossRef]
21. Gan, Q.; Elsworth, D.; Alpern, J.S.; Marone, C.; Connolly, P. Breakdown pressures due to infiltration and exclusion in finite length boreholes. *J. Pet. Sci. Eng.* **2015**, *127*, 329–337. [CrossRef]
22. Khadraoui, S.; Hachemi, M.; Allal, A.; Rabiei, M.; Arabi, A.; Khodja, M.; Lebouachera, S.E.I.; Drouiche, N. Numerical and experimental investigation of hydraulic fracture using the synthesized PMMA. *Polym. Bull.* **2020**, *78*, 3803–3820. [CrossRef]
23. Alpern, J.; Marone, C.; Elsworth, D.; Belmonte, A.; Connelly, P. Exploring the Physicochemical Processes That Govern Hydraulic Fracture Through Laboratory Experiments. In Proceedings of the 46th U.S. Rock Mechanics/Geomechanics Symposium, Chicago, IL, USA, 24–27 June 2012; p. 6.
24. Zhang, C.; Zhou, Y.; Shao, T.; Xie, Q.; Xu, J.; Yang, W. Hydrophobic treatment on polymethylmethacrylate surface by nanosecond-pulse DBDs in CF4 at atmospheric pressure. *Appl. Surf. Sci.* **2014**, *311*, 468–477. [CrossRef]
25. Kinloch, A.J. *Fracture Behaviour of Polymers*; Springer Science & Business Media: Berlin/Heidelberg, Germany, 2013.
26. Yu, P.; Yao, X.; Tan, S.; Han, Q. A Macro-Damaged Viscoelastoplastic Model for Thermomechanical and Rate-Dependent Behavior of Glassy Polymers. *Macromol. Mater. Eng.* **2016**, *301*, 469–485. [CrossRef]
27. Richeton, J.; Ahzi, S.; Vecchio, K.S.; Jiang, F.C.; Adharapurapu, R.R. Influence of temperature and strain rate on the mechanical behavior of three amorphous polymers: Characterization and modeling of the compressive yield stress. *Int. J. Solids Struct.* **2006**, *43*, 2318–2335. [CrossRef]
28. Zhang, X.; Sun, Z.; Hu, X. Low temperature fracture toughness of PMMA and crack-tip conditions under flat-tipped cylindrical indenter. *Polym. Test.* **2014**, *38*, 57–63. [CrossRef]
29. Bilotti, E.; Fenwick, O.; Schroeder, B.C.; Baxendale, M.; Taroni-Junior, P.; Degousée, T.; Liu, Z. 6.14 Organic Thermoelectric Composites Materials. In *Comprehensive Composite Materials II*; Beaumont, P.W.R., Zweben, C.H., Eds.; Elsevier: Oxford, UK, 2018; pp. 408–430. Available online: https://www.sciencedirect.com/science/article/abs/pii/B9780128035818100244 (accessed on 6 June 2023).
30. Esposito, M.; Buontempo, S.; Petriccione, A.; Zarrelli, M.; Breglio, G.; Saccomanno, A.; Szillasi, Z.; Makovec, A.; Cusano, A.; Chiuchiolo, A.; et al. Fiber Bragg Grating sensors to measure the coefficient of thermal expansion of polymers at cryogenic temperatures. *Sens. Actuators A Phys.* **2013**, *189*, 195–203. [CrossRef]
31. Nguyen, D.; Phan, T.; Hsu, T.-P.; Phan, J. Adhesion and surface energy of shale rocks. *Colloids Surf. A Physicochem. Eng. Asp.* **2017**, *520*, 712–721. [CrossRef]
32. Taghichian, A.; Hashemalhoseini, H.; Zaman, M.; Yang, Z.-Y. Geomechanical optimization of hydraulic fracturing in unconventional reservoirs: A semi-analytical approach. *Int. J. Fract.* **2018**, *213*, 107–138. [CrossRef]
33. Griffith, A.A., VI. The phenomena of rupture and flow in solids. *Philos. Trans. R. Soc. Lond. Ser. A Contain. Pap. A Math. Phys. Character* **1921**, *221*, 163–198.
34. Irwin, G.R. Analysis of stresses and strains near the end of a crack traversing a plate. *J. Appl. Mech.* **1957**, *24*, 361–364. [CrossRef]
35. Anderson, T.L. *Fracture Mechanics: Fundamentals and Applications*; CRC Press: Boca Raton, FL, USA, 2017.
36. Liu, L.; Li, L.; Elsworth, D.; Zhi, S.; Yu, Y. The impact of oriented perforations on fracture propagation and complexity in hydraulic fracturing. *Processes* **2018**, *6*, 213. [CrossRef]

37. Dong, Z.; Tang, S. Numerical study of near-wellbore hydraulic fracture propagation. *Theor. Appl. Fract. Mech.* **2019**, *103*, 102274. [CrossRef]
38. Feng, Y.; Gray, K. Modeling of curving hydraulic fracture propagation from a wellbore in a poroelastic medium. *J. Nat. Gas Sci. Eng.* **2018**, *53*, 83–93. [CrossRef]
39. Xi, X.; Yang, S.; Shipton, Z.; Cai, M. Modelling the near-wellbore rock fracture tortuosity: Role of casing-cement-rock well system, perforation and in-situ stress. *Int. J. Rock Mech. Min. Sci.* **2022**, *157*, 105182. [CrossRef]
40. Kruse, C.; Anderson, T.; Wilson, C.; Zuhlke, C.; Alexander, D.; Gogos, G.; Ndao, S. Extraordinary shifts of the Leidenfrost temperature from multiscale micro/nanostructured surfaces. *Langmuir* **2013**, *29*, 9798–9806. [CrossRef] [PubMed]
41. Kwon, H.-m.; Bird, J.C.; Varanasi, K.K. Increasing Leidenfrost point using micro-nano hierarchical surface structures. *Appl. Phys. Lett.* **2013**, *103*, 201601. [CrossRef]
42. Weickgenannt, C.M.; Zhang, Y.; Sinha-Ray, S.; Roisman, I.V.; Gambaryan-Roisman, T.; Tropea, C.; Yarin, A.L. Inverse-Leidenfrost phenomenon on nanofiber mats on hot surfaces. *Phys. Rev. E* **2011**, *84*, 036310. [CrossRef]
43. Vakarelski, I.U.; Patankar, N.A.; Marston, J.O.; Chan, D.Y.; Thoroddsen, S.T. Stabilization of Leidenfrost vapour layer by textured superhydrophobic surfaces. *Nature* **2012**, *489*, 274–277. [CrossRef]
44. Arnaldo del Cerro, D.; Marin, A.G.; Römer, G.R.; Pathiraj, B.; Lohse, D.; Huis in't Veld, A.J. Leidenfrost point reduction on micropatterned metallic surfaces. *Langmuir* **2012**, *28*, 15106–15110. [CrossRef]
45. Zhong, L.; Guo, Z. Effect of surface topography and wettability on the Leidenfrost effect. *Nanoscale* **2017**, *9*, 6219–6236. [CrossRef]
46. Kim, H.; Truong, B.; Buongiorno, J.; Hu, L.-W. On the effect of surface roughness height, wettability, and nanoporosity on Leidenfrost phenomena. *Appl. Phys. Lett.* **2011**, *98*, 083121. [CrossRef]
47. Bourdin, B.; Marigo, J.-J.; Maurini, C.; Sicsic, P. Morphogenesis and Propagation of Complex Cracks Induced by Thermal Shocks. *Phys. Rev. Lett.* **2014**, *112*, 014301. [CrossRef] [PubMed]
48. Jenkins, D.R. Determination of crack spacing and penetration due to shrinkage of a solidifying layer. *Int. J. Solids Struct.* **2009**, *46*, 1078–1084. [CrossRef]
49. Jiang, C.P.; Wu, X.F.; Li, J.; Song, F.; Shao, Y.F.; Xu, X.H.; Yan, P. A study of the mechanism of formation and numerical simulations of crack patterns in ceramics subjected to thermal shock. *Acta Mater.* **2012**, *60*, 4540–4550. [CrossRef]
50. Jenkins, D.R. Optimal spacing and penetration of cracks in a shrinking slab. *Phys. Rev. E* **2005**, *71*, 056117. [CrossRef]
51. Amarasiri, A.L.; Kodikara, J.K. Effect of characteristic lengths of fracture on thermal crack patterns. *Int. J. Geomech.* **2015**, *15*, 04014071. [CrossRef]
52. Sumi, Y.; Mu, Y. Thermally induced quasi-static wavy crack propagation in a brittle solid. *Mech. Mater.* **2000**, *32*, 531–542. [CrossRef]
53. Wang, Y.; Zhou, X.; Kou, M. An improved coupled thermo-mechanic bond-based peridynamic model for cracking behaviors in brittle solids subjected to thermal shocks. *Eur. J. Mech.-A/Solids* **2019**, *73*, 282–305. [CrossRef]
54. Shao, Y.; Zhang, Y.; Xu, X.; Zhou, Z.; Li, W.; Liu, B. Effect of crack pattern on the residual strength of ceramics after quenching. *J. Am. Ceram. Soc.* **2011**, *94*, 2804–2807. [CrossRef]
55. Xu, X.; Lin, Z.; Sheng, S.; Yuan, W. Evolution mechanisms of thermal shock cracks in ceramic sheet. *J. Appl. Mech.* **2016**, *83*, 071001. [CrossRef]
56. Weijermars, R.; Wang, J. Stress Reversals near Hydraulically Fractured Wells Explained with Linear Superposition Method (LSM). *Energies* **2021**, *14*, 3256. [CrossRef]
57. Weijermars, R. Stress cages and fracture cages in stress trajectory models of wellbores: Implications for pressure management during drilling and hydraulic fracturing. *J. Nat. Gas Sci. Eng.* **2016**, *36*, 986–1003. [CrossRef]

processes

Article

Study on the Fracture Propagation in Multi-Horizontal Well Hydraulic Fracturing

Qiquan Ran, Xin Zhou *, Jiaxin Dong, Mengya Xu, Dianxing Ren and Ruibo Li

Research Institute of Petroleum Exploration and Development, No. 20 Xueyuan Road, Haidian District, Beijing 100083, China; ranqq@petrochina.com.cn (Q.R.); djx1021@petrochina.com.cn (J.D.); xumengya@petrochina.com.cn (M.X.); rendianx@petrochina.com.cn (D.R.); liruibo01@petrochina.com.cn (R.L.)
* Correspondence: zhouxin510@petrochina.com.cn

Abstract: Multi-horizontal well hydraulic fracturing is a widely employed and highly effective method for stimulating tight and shale reservoirs. However, most existing studies primarily focus on investigating the impact of intra-well interference on fracture propagation while neglecting the influence of inter-well interference. Here, a multi-well hydraulic-fracture-propagation model is established to examine the effects of inter-well interference on fracture propagation within a multi-well system. In this study, based on the bilinear T-S criterion, the stiffness degradation is used to describe the damage and evolution process of fracture, the coupling process of fluid flow and solid damage and deformation is realized, and the dynamic distribution of inter-fracture flow is realized by using Kirchhoff function on the basis of the cohesive zone method (CZM) finite element model. Finally, the fracture-propagation model of multiple horizontal wells is established. Based on this model, the mechanism of inter-well interference on fracture propagation is studied, and the influence law of Young's modulus and fracture displacement on fracture propagation in multi-wells is investigated. The results show that the reservoir can be divided into self-influence area, tension area and compression area according to the stress distribution state in the hydraulic fracture propagation of multi-wells. The propagation rate of hydraulic fractures in horizontal wells is significantly accelerated when they propagate to the local tension area generated by the fracture tip of neighboring wells, and rapidly decreases as the hydraulic fractures continue to propagate to the compression area of neighboring wells. Rocks with a lower Young's modulus tend to be more plastic, forming hydraulic fractures with usually lower fracture lengths and usually larger fracture widths. The hydraulic fracture has an inhibitory effect on the propagation of fractures closer to each other in neighboring wells, and this inhibitory effect gradually increases as the distance decreases. The dominance of the dominant fracture to propagate in the self-influence area gradually decreases under inter-well and intra-well interference. As the dominant fracture propagates into the tension and compression areas of the neighboring well fractures, the feed fluid will show a brief rise and then eventually stabilize. This study quantifies the effect of inter-well interference on fracture propagation and lays the foundation for treatment optimization of small well spacing hydraulic fracturing.

Keywords: fracture propagation; inter-well interference; cohesive; hydraulic fracturing; numerical simulation

Citation: Ran, Q.; Zhou, X.; Dong, J.; Xu, M.; Ren, D.; Li, R. Study on the Fracture Propagation in Multi-Horizontal Well Hydraulic Fracturing. *Processes* **2023**, *11*, 1995. https://doi.org/10.3390/pr11071995

Academic Editor: Qingbang Meng

Received: 18 May 2023
Revised: 25 June 2023
Accepted: 27 June 2023
Published: 2 July 2023

1. Introduction

Tight reservoirs and shale reservoirs have gradually become the focus of oil and gas exploitation, and infill horizontal wells and small well spacing hydraulic fracturing technology can fully stimulate the reservoir and make underground crude oil efficiently used, which has now become the most important stimulation method for these reservoirs [1,2]. Previous studies have focused on the effect of intra-well interference on fracture propagation [3,4], but the effect of inter-well interference on fracture propagation has become particularly important in small spacing stimulation [5,6]. Since the small well spacing exploitation method is still in its infancy, the mechanism and law of inter-well interference on fracture propagation are still unclear.

During hydraulic fracture propagation, the rock is damaged and fractures are formed under the injected high-pressure fluid. Currently, the majority of mathematical models employed for simulating fracture propagation rely on linear elasticity theory [7]. The drawbacks of those models are that they cannot accurately describe the fracture propagation process and their simulation results are severely controlled by the mesh shape [8]. In addition, these methods only discretize the body boundary, and this phenomenon can cause serious problems during iterative calculations. To address this phenomenon, a finite element method incorporating the CZM has been employed to simulate the propagation process of hydraulic fractures, taking into account the effects of propagation and softening at the fracture tip [9–11]. The finite element model based on CZM simulates the coupling process of incompressible viscous fluid and rock deformation, and the results show that the model is more accurate for describing the extension process of hydraulic fracturing [12]. In a related study, Guo et al. [13] utilized a novel CZM model that couples the flow-damage field to simulate the interaction between hydraulic fractures and natural fractures. Additionally, Li et al. [14] proposed an innovative cohesive unit for pore pressure, incorporating the Coulomb friction contact model to depict fracture contact behavior. Overall, the CZM-based finite element method exhibits considerable advantages over alternative approaches, as it encompasses fracture tip damage and evolution, rendering it good for modeling the fracture propagation.

To explore the impact of stress shadows and complex stress fields on fracture propagation, various methodologies, such as XFEM, BEM, FEM, UFM, and DEM, have been developed [15,16]. Notably, during the propagation of hydraulic fractures, pull-up damage arises at the fracture tip, leading to a reduction in stress within that particular region [17]. Lecampion et al. [18] conducted a study on fracture propagation in impermeable media using the extended finite element method (XFEM) and proposed the implementation of a specialized enrichment function at the fracture tip to capture pore size and pressure. However, this investigation did not account for the longitudinal fluid flow. In a different study, Chen et al. [19] examined fracture propagation in impermeable media by considering longitudinal flow and utilizing a cohesion law-controlled boundary element method (BEM). The obtained results displayed favorable agreement with field monitoring outcomes. Moreover, the opening of hydraulic fractures can cause compression of the surrounding reservoir, leading to alterations in pore pressure and stress fields adjacent to the fracture [20,21]. Sun et al. [22] employed the finite element method (FEM) to investigate hydraulic fracturing parameters in porous media, revealing a gradual decrease in fluid pressure within the fracture as the fracture length increases. Importantly, field diagnostic techniques have indicated that multiple fractures often propagate simultaneously, leading to uneven propagation due to the influence of stress shadows [23]. Subsequent researchers have highlighted that, in addition to stress shadowing, fracture propagation is also influenced by stress shadows, and perforation friction plays a crucial role in the dynamic distribution of flow [3,24]. However, the majority of existing studies have primarily focused on investigating intra-well interference of hydraulic fractures within a single well, with minimal consideration given to the dynamic distribution of fracturing fluid between fractures. In comparison to the aforementioned methods, the CZM offers notable advantages, including its robust node compilation capability, convenient coupling calculations, and fast computational speed. Consequently, CZM proves advantageous in terms of coupling dynamic flow distribution in this aspect.

To elucidate the behavior of fracture propagation in multi-horizontal well hydraulic fracturing, this study employs stiffness degradation to characterize the damage and evolution process of fractures based on the bilinear T-S criterion. Furthermore, it achieves the coupling of fluid flow and solid damage and deformation. By integrating the CZM finite element model, the dynamic distribution of flow in multiple fractures is realized through the application of Kirchhoff's law. Ultimately, a coupled model for fracture propagation in multi-horizontal wells is established. Based on this model, the study investigates the mechanism of inter-well interference on fracture propagation, examining the impact of Young's modulus and fracture location on fracture propagation in multi-horizontal wells.

2. Mathematical Model

The core problem of the study on the interference of multi-fractures between wells on fracture propagation is to accurately describe the fracture propagation and the change of the stress field between wells. The CZM finite element method can accurately describe the damage and evolution of fractures, the tangential and normal flow of fracturing fluid in the fractures, the flow of fluid in porous media and the deformation of solids and has been widely used to describe the fracture initiation and propagation process during hydraulic fracturing.

2.1. Fluid Flow in the Fracture

The fracturing fluid opens the reservoir and forms a fracture at a pressure higher than the breakdown pressure. As shown in Figure 1, part of the fluid in the fracture flows tangentially along the heel to toe of the fracture, and the other part of the fluid leaks off along the wall of the reservoir under the pressure difference between the fracture and the matrix. For the tangential flow of fluid in the fracture, assuming that the fracturing fluid is an incompressible Newtown fluid, the overall volume balance equation for tangential flow established by Poiseuille is used in this study as follows [25].

$$q = -\frac{w^3}{12\mu}\nabla p \tag{1}$$

where q is the flow velocity of the flow along the fracture propagation direction, ∇p is the pressure difference of the flow along the fracture propagation direction, w is the fracture width in this cohesive unit and μ is the fracturing fluid viscosity.

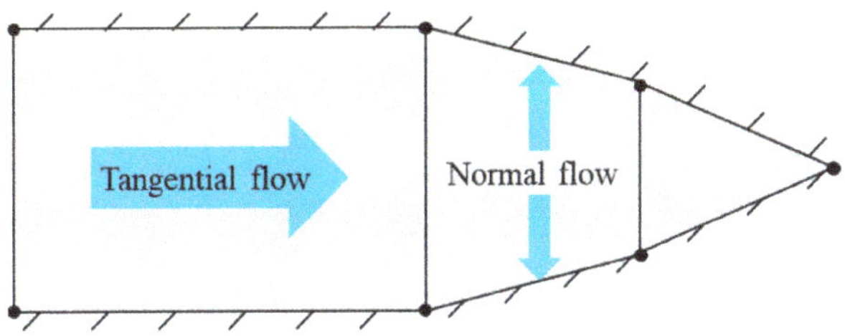

Figure 1. Flow patterns of fluids in the cohesive element.

To account for the leakage of fracturing fluid along the fracture wall, two permeable cohesive layers are introduced. These layers facilitate fluid leak-off along the fracture wall, as illustrated in Figure 2. The normal leak-off at the fracture surface can be characterized as follows:

$$\begin{cases} q_t = c_t\left(p_f - p_t\right) \\ q_b = c_b\left(p_f - p_b\right) \end{cases} \tag{2}$$

where p_t and p_b are the pore pressure in the top and bottom layer unit, respectively, p_f is the pressure of the cohesive unit, c_t and c_b are the leak-off coefficients at the top and bottom layers, respectively, and q_t and q_b are the normal volume leak-off rates at the top and bottom layers, respectively.

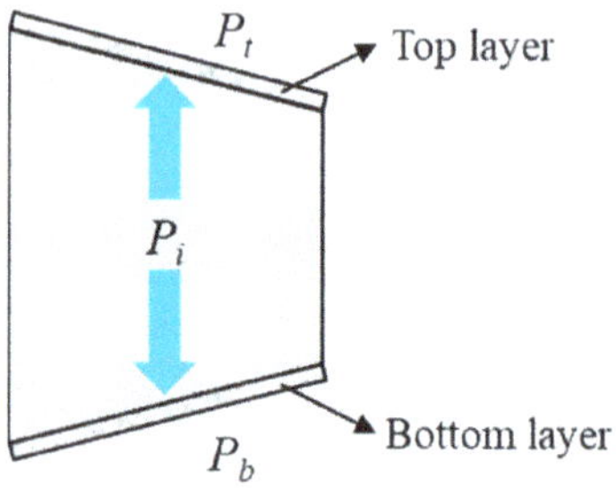

Figure 2. Normal flow for cohesive surface.

Considering the flow along the fracture surface in the fracture, the normal flow of the fluid perpendicular the fracture surface and the variation of the fracture width during the hydraulic fracturing, the continuity equation for fluid in the fracture can be expressed as:

$$\frac{\partial w}{\partial t} + \nabla \cdot q + (q_t + q_b) = Q(t)\delta(x, y) \quad (3)$$

where *Q(t)* is the fluid source for this unit.

2.2. Damage and Evolution of Fractures

The flow of fluid in the fracture at a macroscopic level, the flow of fluid in the porous medium at a microscopic level and the elastic deformation and damage of the rock are all interconnected. The CZM employs the traction-separation law to model the behavior of interfaces. In the model, Biot's effective stress principle is employed to describe the effective stress [26]. This principle considers the interaction between fluid flow and solid deformation, accounting for the impact of fluid pressure on the effective stress within the rock matrix. This principle considers the interaction between fluid flow and solid deformation, accounting for the impact of fluid pressure on the effective stress within the rock matrix.

$$\sigma_{ij} = \sigma'_{ij} + \alpha p \delta_{ij} \quad (4)$$

where $\sigma_{i,j}$ and $\delta'_{i,j}$ is the total stress and the effective stress, respectively, α is the Biot coefficient and $\delta_{i,j}$ is the tensor of Kronecker.

The damage model of the cohesive unit establishes the relationship between tensile stresses and displacements between adjacent meshes. V. Tomar et al. [27] proposed a bilinear T-S criterion, depicted in Figure 3, which comprises two components for damage evolution. The first component involves determining the relationship between the effective displacement, denoted as δ_m^f, at complete failure, relative to the effective displacement, δ_m^0 at the onset of damage. This component also considers the energy dissipation, G_c, attributed to the damage. The second component characterizes the stiffness degradation index, D, between the initial stage of damage and complete damage. The stiffness degradation index, D, is directly defined in terms of effective displacement and the relationship between effective displacement and damage. When adopting the linear displacement expansion criterion, the expression is

$$D = \frac{\delta_m^f \left(\delta_m^{max} - \delta_m^0\right)}{\delta_m^{max}\left(\delta_m^f - \delta_m^0\right)} \quad (5)$$

where δ_m^{max} is the maximum displacement of the cohesive unit undergoing damage, δ_m^f is the displacement in the cohesive unit during the fracture propagation and δ_m^0 is the displacement when the cohesive unit is transformed from the elastic phase to the destructive phase.

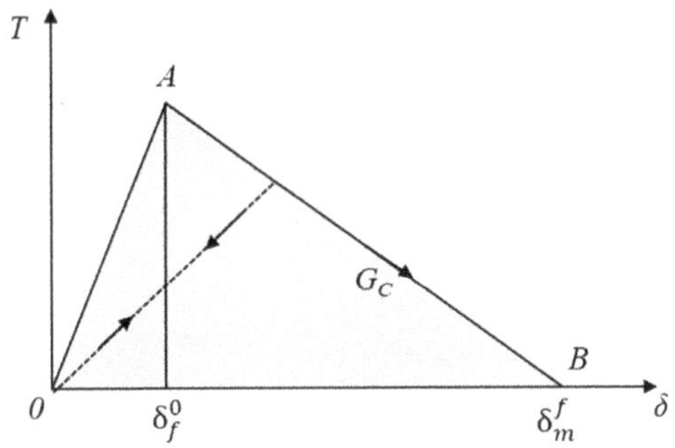

Figure 3. Cohesive linear-softening for the criterion.

The damage incurred by the rock can be quantified using the D. It is important to note that this condition is also valid when the rock sample is subjected to compressive stress. Therefore, the stiffness degradation criterion employed to characterize the fracture unit can be expressed as follows:

$$t_n = \begin{cases} (1-D)T_n, & T_n \geq 0 \\ T_n, & T_n < 0 \end{cases} \quad (6)$$

$$t_s = (1-D)T_s \quad (7)$$

$$t_t = (1-D)T_t \quad (8)$$

where T_n, T_s and T_t represent the stresses obtained from 3 directions.

The D depends on the softening nature of the rock in the reservoir.

$$\delta_m = \sqrt{\langle \delta_n \rangle^2 + \delta_s^2 + \delta_t^2} \quad (9)$$

where δ_m is the effective displacement obtained during the fracture propagation.

In this study, the B-K criterion is introduced to describe the fracture propagation.

$$G_n^c + (G_s^c - G_n^c)\left\{\frac{G_s + G_t}{G_s + G_t + G_n}\right\}^\eta = G^c \quad (10)$$

where G represents the energy release rate, n, s and t correspond to the 3 directions, and η represents a constant associated with the reservoir.

2.3. Fluid-Solid Coupling Equilibrium Equation

The upward normal leak-off of fluid along the fracture wall leads to an increase in pore pressure within the rock matrix surrounding the fracture. The effective stress in this context is influenced by both the rock skeleton and the pore pressure, p. The rise in pore pressure results in alterations in parameters such as reservoir porosity and fluid flow rate, subsequently impacting the normal fluid leak-off along the fracture wall [28].

In accordance with the principle of virtual work, the equilibrium equation can be expressed as follows:

$$\int_V (\overline{\sigma} - p_w I)\delta\varepsilon dV = \int_S \boldsymbol{t} \cdot \delta \boldsymbol{v} dS + \int_V f \cdot \delta v dV \quad (11)$$

where $\overline{\sigma}$ and p_w are the Boit effective stress and pore pressure, respectively, δ_ε and δ_v are the virtual strain rate and virtual velocity, respectively, and $\boldsymbol{t}$ and f are the surface displacement per unit area and body force per unit volume, respectively.

The change in fluid mass is equal to the mass of fluid flowing into and out of the cohesive surface per unit time.

$$\frac{1}{J}\frac{\partial}{\partial t}(J\rho_w n_w) + \frac{\partial}{\partial X}(\rho_w n_w v_w) = 0 \tag{12}$$

where J is the body variation ratio of reservoir rock, ρ_w is the fracturing fluid density in the fracture and n_w is the porosity.

Utilizing Darcy's law to describe the flow of fluid within the reservoir, the continuity equation for fluid leak-off can be expressed as follows:

$$v_w = -\frac{1}{n_w g \rho_w}\boldsymbol{k}\left(\frac{\partial p_w}{\partial x} - \rho_w g\right) \tag{13}$$

where $\boldsymbol{k}$ is the permeability of the reservoir.

2.4. Dynamic Distribution of Fluids

In hydraulic fracturing treatment, multiple fractures are usually formed when fracturing one or more sections of a well. As the formed hydraulic fractures squeeze the surrounding reservoir, stress shadows are formed, which can further influence the opening and propagation of the surrounding fractures. The flow pattern of the fracturing fluid in the wellbore and perforation cluster is shown in Figure 4. Due to the effect of wellbore friction, shot hole friction and resistance to fracture propagation, the amount of fluid intake during propagation of different hydraulic fractures in the same well is a dynamic equilibrium process. Similar to the shunt current in an electric circuit, the flow of fluid in each fracture needs to satisfy Kirchhoff's law [29].

$$Q = \sum_{1}^{N} Q_I \tag{14}$$

$$p_o = p_{pf,i} + p_{cf,i} + p_{w,i} \tag{15}$$

where Q is the total flow velocity in the wellbore, Q_i is the flow velocity into the ith fracture, p_o is the total pressure, p_{wi} is the pressure in the first unit in the ith fracture, $p_{cf,i}$ is the pressure loss in the horizontal wellbore and $p_{pf,i}$ is the pressure loss from shot hole.

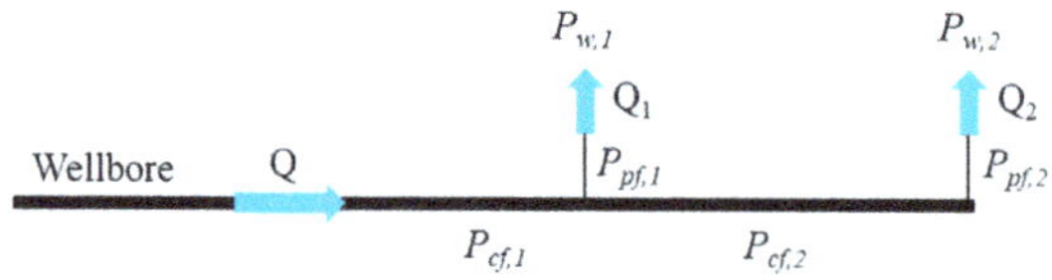

Figure 4. Dynamic distribution mode for fluid.

To address the issue of shot hole friction, a large number of researchers have shown that it is related to fluid properties, shot hole parameters, flow rate and other parameters. Among them, Crump [30] has established the relationship between shot hole resistance and fluid and shot hole parameters and flow rate based on a large number of experiments. This method is simple, efficient and highly integrated with Bernoulli's equation, and has been widely used by the industry.

$$\Delta p^I_{fric} = 0.807249\frac{\rho}{{n_p}^2{D_p}^4C^2}Q_I^2 \tag{16}$$

where Δp^I_{fric} is the pressure difference of perforation cluster I, Q_I is the flow velocity of perforation cluster I, n_p is the perforation number, D_p is the perforation diameter and C is the dimensionless coefficient characterizing the shape of perforation hole.

According to the calculation characteristics of perforation friction, the wellbore unit type is set to FP2D2, and the perforation unit type is set to FPC2D2 in the model. Because Churchill has the advantages of good convergence and wide application, this study uses it to calculate the perforation friction of perforation unit.

$$\Delta P - \rho g \Delta Z = K\frac{\rho Q^2}{2A^2} \tag{17}$$

where ΔP is the pressure difference, ΔZ is the elevation difference and K is the loss coefficient.

The values of K and A can be determined by combining Equations (16) and (17).

$$K = 2 \times 0.807249 \tag{18}$$

$$A = n_p \times C \times D_p^2 \tag{19}$$

2.5. Model Description

The schematic diagram of the model in this study is shown in Figure 5. Considering the size of fractures in hydraulic fracturing, the size of the model in this work is set to 160 m × 160 m, and one horizontal well is arranged at the upper and lower boundaries, respectively. There are one and two perforation clusters in wellbore 1 and wellbore 2 in the study area, respectively. Perforation cluster 1 (Perf 1) and perforation cluster 2 (Perf 2) satisfy the dynamic distribution of flow rate during the fracturing process. The injection rates of fracturing fluid in wellbore 1 and wellbore 2 are u_1 and u_2, respectively. The boundaries of the model are considered constant pressure boundaries. The initial pore pressure in the reservoir is denoted as P_b, and the minimum horizontal principal stress, maximum horizontal principal stress, and vertical stress are represented by S_h, S_H and S_v, respectively. The model employs the parameters listed in Table 1. The model was simulated using the software Abaqus, which has the CZM method embedded in it. Previous studies have demonstrated its good applicability for describing the hydraulic fracturing process [18,19]. It is worth noting that the stresses in all simulation results are effective stresses.

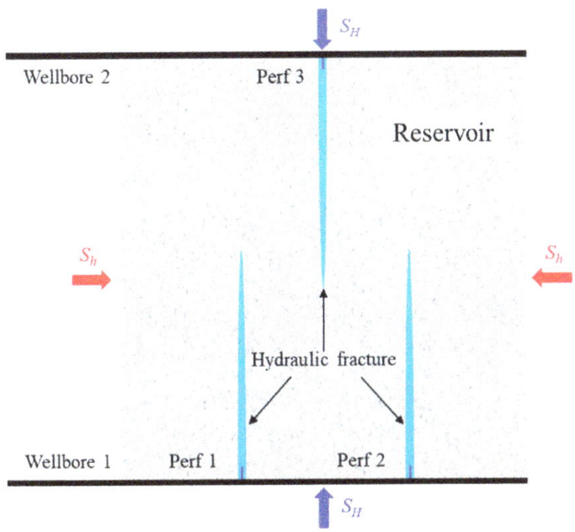

Figure 5. Schematic diagram of fracture propagation in the multi-horizontal well.

Table 1. The parameters used in this study.

Parameter	Value	Parameter	Value
S_h	65 MPa	Young's modulus	50 GPa
S_H	70 MPa	Poisson's ratio	0.25
S_v	75 MPa	Permeability	0.1 mD
P_b	48 MPa	Porosity	8.9%
Fluid viscosity	100 mPa·s	C	0.6
n_p	16	D_p	12 mm
Fluid density	1000 kg/m^3		

3. Result and Discussion

3.1. Mechanism of Inter-Well Interference

Fracture propagation can change the pore pressure and in situ stress in the reservoir, and clarifying the mechanism of inter-well fracture interference during fracturing is essential to optimize the effect of multi-well fracturing treatment. This section investigates fracture propagation and stress changes in two horizontal wells during the fracturing process. The spacing of the injection clusters in wellbore 1 is 60 m. The Young's modulus of the rock in the reservoir is 50 GPa. u_1 and u_2, the injection rate of the two wells, are 0.016 m/s and 0.008 m/s, respectively. The simulation results of the fracture propagation in the two wells during hydraulic fracturing are shown in Figure 6a. As can be seen from the figure, the fracture width of well 1 is smaller than that of well 2, but the fracture length of well 1 is larger than that of well 2. This is due to the fact that the fluid in the fracture needs to keep the fracture open during the fracture opening and propagation. The hydraulic fracture opening causes an increase in pressure on the rock around the fracture, resulting in an increase in the minimum principal stress in a certain area around the fracture (Figure 6b), which means the stress shadow is formed. The degree of this stress decreases gradually along the fracture propagation direction. Under the influence of the stress shadow, the maximum horizontal principal stress also shows the same trend (Figure 6c). Under the influence of the stress shadow generated by the adjacent fractures, the fractures in well 1 require more fluid pressure to maintain the fracture opening during the actual propagation (Figure 6d), and therefore the fracture width of the fractures formed in well 1 is smaller than that of the fractures formed in well 2. The smaller fracture width and larger fluid pressure allowed the fractures to propagate more rapidly because the fluid intake for each fracture was essentially the same in both wells. In summary, proper stress shade can increase the distance of fracture propagation.

The minimum horizontal principal stress is the key parameter to determine the fracture propagation, and the smaller the minimum horizontal principal stress is, the smaller the resistance to fracture opening and propagation. Figure 7 shows the distribution of the minimum horizontal principal stress at different stages of the fracture propagation process. As can be seen from the figure, the opening of the fracture in the reservoir will squeeze the rocks in the area around the fracture wall, thus increasing the minimum horizontal principal stress in the area. In the process of fracture propagation, the fracture tip will be tension damaged by the fracture fluid, and the rocks in the fracture tip area will be subjected to a certain degree of tension, thus making the minimum horizontal principal stress in the area decrease to different degrees. The fracture length in horizontal well 1 is shown in Figure 8. The fracture in horizontal well 1 is mainly influenced by its own intra-well interference at the early stage of propagation. After the fracture in well 1 propagates to the tension area formed by the fracture in horizontal well 2, the propagation rate of the hydraulic fracture in well 1 increases rapidly in this area because the minimum horizontal principal stress in this area decreases. With the continuous propagation of the fracture, when the hydraulic fracture of well 1 propagates to the compression area of well 2, the propagation rate of the fracture decreases rapidly at this time. As the fracture continues to propagate, the degree of decline is greater. There are two main reasons for this. One is that the closer to the root of the well 2 fracture, the greater the increase in the minimum horizontal principal stress in the compression zone. The second is that as the fracture continues to propagate, the leak-off of fluid increases, and the propagation rate of the fracture itself gradually decreases.

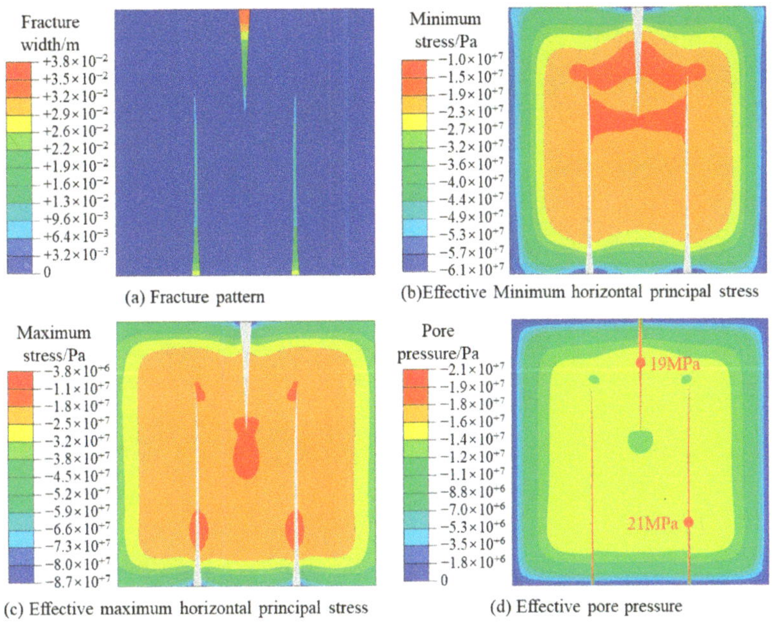

Figure 6. Fracture parameters and stress distribution during fracture propagation (150 s).

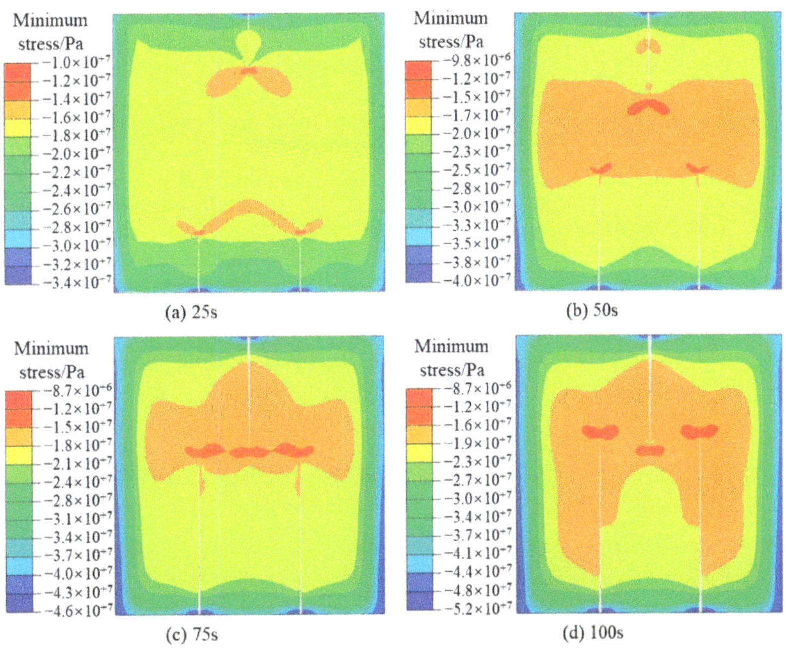

Figure 7. Distribution of the effective minimum horizontal principal stress during propagating.

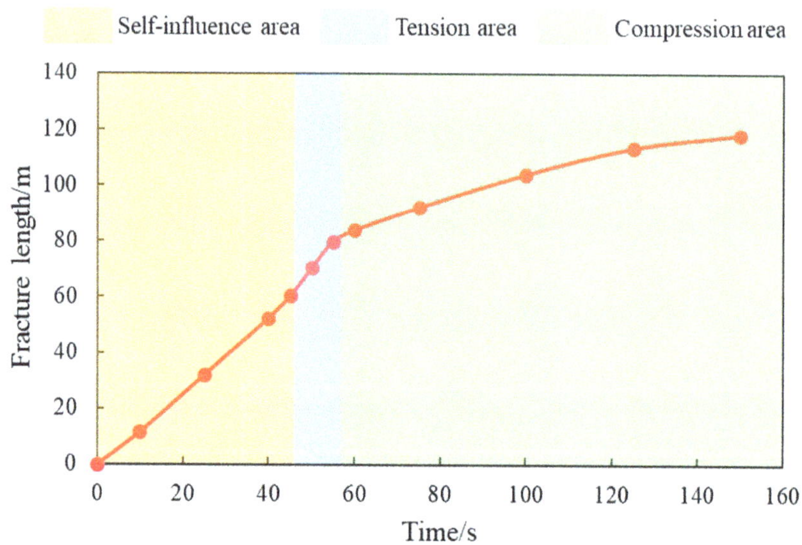

Figure 8. Hydraulic fracture length versus time in horizontal well 1.

3.2. *Effect of Young's Modulus*

The fracturing fluid enters the reservoir along the wellbore from the perforation cluster, and the high-pressure fluid opens the reservoir to form a fracture. The fluid in the fracture squeezes the rock around the fracture wall and causes deformation. Young's modulus is an important parameter for quantitatively evaluating the stress–strain relationship of rocks. In this section, the propagation pattern of the fracture is studied for Young's modulus of the reservoir rock at 30 GPa, 40 GPa, 50 GPa and 60 GPa, respectively, and the simulation results are shown in Figures 9 and 10. From the figures, it can be seen that with the increasing Young's modulus, the fracture width shows a decreasing trend, but the fracture length shows an increasing trend. Among them, the maximum fracture widths of Perf 1 are 3.12 cm, 2.70 cm, 2.49 cm and 2.34 cm, respectively. The maximum fracture widths of Perf 3 are 3.82 cm, 3.58 cm, 3.55 cm and 3.54 cm, respectively. The fracture lengths of Perf 1 are 80.54 m, 103.24 m, 113.76 m and 124.02 m, respectively. The maximum fracture lengths of Perf 3 are 60 m, 63.51 m, 63.89 m and 65.29 m. It is worth noting that proper intra-well interference can increase the stimulation pressure and reduce the fracture width, which can enhance the fracture propagation distance under a constant pumping rate in the field [16]. Rocks with lower Young's modulus tend to be more plastic and form hydraulic fractures with typically shorter fracture lengths and larger fracture widths [25]. The fracture width in well 1 is further reduced compared to that in well 2 due to stress shade, resulting in a hydraulic fracture more sensitive to Young's modulus in well 1. In addition, combined with the relative positions of the fractures between wells in Figure 10, it can be seen that with the increase in Young's modulus, the fractures will propagate to the tension area more quickly, and the existence of the tension area will further increase the distance of fracture propagation. When the fracture continues to propagate to the compression area, the propagation rate of the fracture will decrease. In summary, the fracture length and width of multi-horizontal wells are controlled by both Young's modulus and inter-well interference.

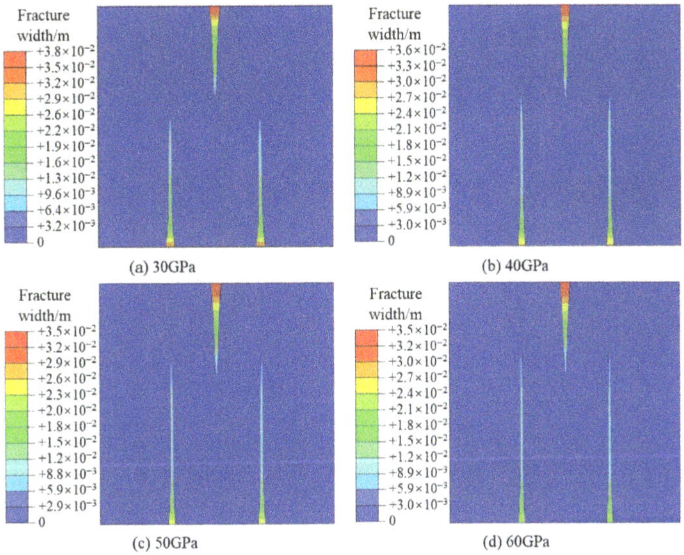

Figure 9. Fracture width of hydraulic fractures at different Young's modulus (140 s).

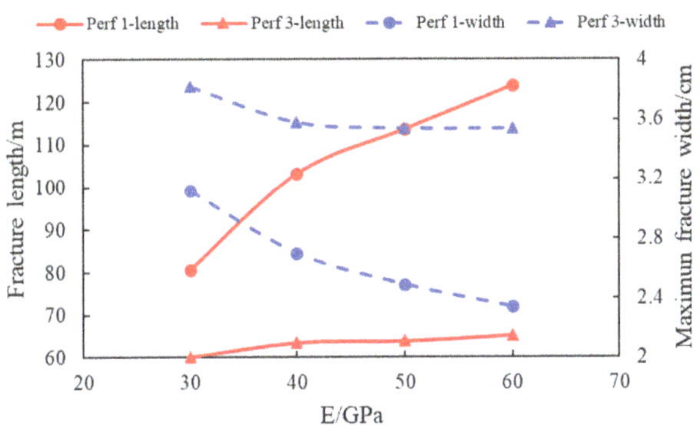

Figure 10. Fracture length and maximum fracture width at different Young's modulus.

3.3. *Effect of Fracture Placement Location*

In hydraulic fracturing stimulation, the relative position of the perforation clusters in the wellbore is often not uniformly distributed. Under the combined influence of geological and engineering factors, the distance between perforation clusters usually varies. Due to fracturing interference between multi-horizontal wells, the relative position of the perforation clusters between wells is an important factor affecting fracture propagation. This section investigates the fracture propagation results for the lateral distances of Perf 3 to Perf 1 and Perf 2 at 30 m and 30 m (30 m–30 m), 20 m–40 m and 10 m–50 m, respectively. In order to more intuitively reflect the influence of the location of the fractures, we chose the fracture propagation results after 100 s of fluid injection (Figure 11) as the object of comparison. From the fracture propagation results, it can be seen that the formed hydraulic fracture has an inhibitory effect on the propagation of the closer fracture in the neighboring wells, and this inhibitory effect gradually increases as the distance decreases. In the case of 10 m–50 m, the lengths of Perf 1 and Perf 2 are 128.7 m and 36.9 m, respectively. In the case of 30 m–30 m, 20 m–40 m and 10 m–50 m fracture placement patterns, the ratio of the fracture lengths of Perf 2 and Perf 1 have fracture length ratios of 1, 1.76 and 3.49, respectively.

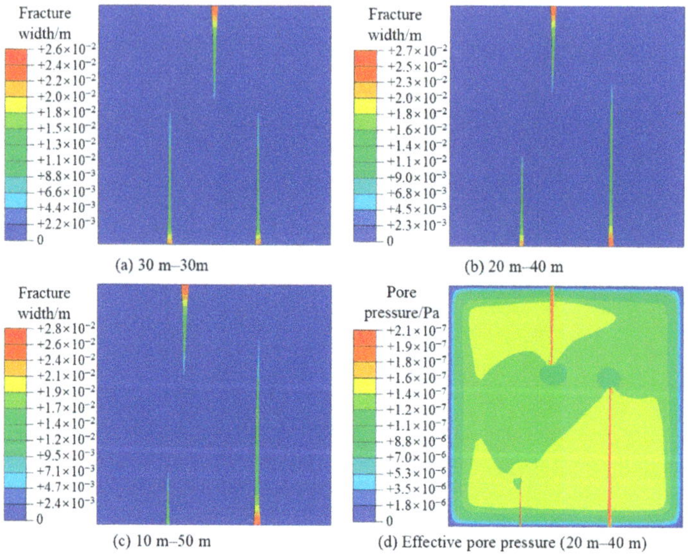

Figure 11. The propagation state of hydraulic fractures at different placement positions (100 s).

It is clear from the distribution of pore pressure that the formation of the dominant fracture (Perf 2) due to the difference in interference caused by the placement pattern will in turn force the flow of fluids from the neighboring well fracture (Perf 3) to the other side, thus further inhibiting the propagation process of the inferior fracture (Perf 1). Due to the difference in propagation resistance, the fluid intake of the perforation clusters in a well is a dynamic distribution process, and the relationship between the flow rates of Perf 1 and Perf 2 during fracture propagation is shown in Figure 12. As can be seen from the figure, at the early stage of propagation, the interference between wells is small, and at this time the perforation clusters have the same flow distribution ratio among them. As the fracture continues to propagate, the gradually increasing inter-well interference makes Perf 1 become the inferior fracture and Perf 2, which becomes the dominant fracture, receives more fluid. As Perf 1 continues to feed fluid, the interference from Perf 1 to Perf 2 gradually increases, causing the amount of fluid fed to Perf 2 to decrease slightly. When Perf 2 propagates into the tension area of the neighboring well fracture, the fluid feed at this point shows a brief increase. After finally entering the compression area, the fluid volume of Perf 1 and Perf 2 gradually stabilizes.

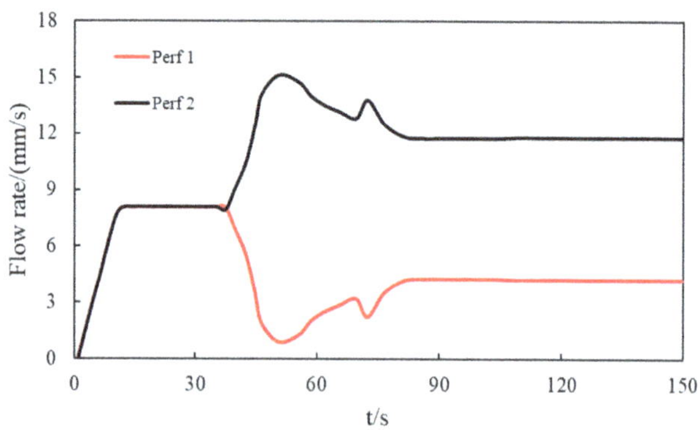

Figure 12. Dynamic distribution of flow rates in Perf 1 and Perf 2 for the case of 10 m–50 m.

4. Conclusions

In this study, the coupled model of fluid flow and solid damage and deformation is constructed by using the bilinear T-S criterion and the stiffness degradation to describe the damage and evolution of fractures, and the dynamic distribution of fluid due to uneven fracture propagation in the hydraulic fracturing is realized based on the CZM finite element model. The main conclusions are obtained as follows.

(1) Stress shade of multiple fractures can reduce the fracture width while increasing the fracture length. According to the stress distribution state in the hydraulic fracture propagation in multi-horizontal wells, the reservoir can be divided into the self-influence area, tension area and compression area. The propagation rate of the hydraulic fractures in horizontal wells accelerates significantly when they propagate to the local tension area generated by the fracture tip of neighboring wells and decreases rapidly as the hydraulic fractures continue to propagate to the compression area of neighboring wells.

(2) Proper intra-well interference can increase the fracture propagation distance. Reservoirs with higher Young's modulus are usually more brittle, and the fracture width and fracture length formed during hydraulic fracturing are smaller and longer, respectively.

(3) Hydraulic fractures have an inhibitory effect on the propagation of closer fractures in neighboring wells, and this inhibitory effect gradually increases as the distance decreases. The formation of the dominant fracture will further influence the pore pressure field, thus inhibiting the propagation of the inferior fracture in the same well. The dominance of the dominant fracture to propagate in the self-influence area gradually decreases under inter-well and intra-well interference. As the dominant fracture propagates into the tension and compression areas of the neighboring well fractures, the feed fluid will show a brief rise and then eventually stabilize.

In this study, the effect of the stress shadow on fracture propagation during the fracturing process of multi-horizontal wells is studied more systematically. However, the reservoir permeability of 0.1 mD in the model does not consider the effect of pore elasticity on fracture propagation during hydraulic fracturing. The coupling of fluid and pore elasticity deformation on fracture propagation in multi-horizontal wells hydraulic fracturing still needs to be studied continuously.

Author Contributions: Writing-Original manuscript, Q.R.; supervision, X.Z.; investigation, J.D.; data curation, M.X.; formal analysis, D.R. and R.L. All authors have read and agreed to the published version of the manuscript.

Funding: This research was funded by the Key Core Technology Research Projects of PetroChina Company Limited (No. 2020B-4911) and the APC was funded by the Research Institute of Petroleum Exploration and Development.

Data Availability Statement: No new data were created or analyzed in this study. Data sharing is not applicable to this article.

Conflicts of Interest: The authors declare no conflict of interest.

References

1. Wang, Y.; Zhou, F.; Zhang, Y.; Wang, Y.; Su, H.; Dong, R.; Wang, Q.; Bai, H. Numerical studies and analysis on reaction characteristics of limestone and dolomite in carbonate matrix acidizing. *Geoenergy Sci. Eng.* **2023**, *222*, 211452. [CrossRef]
2. Li, M.; Zhou, F. Multi-fracture initiation sequence and breakdown pressure in horizontal wells during TDPF: A visualization experimental investigation based on PMMA. *J. Pet. Sci. Eng.* **2022**, *215*, 110645. [CrossRef]
3. Wu, K.; Olson, J.E. Mechanisms of Simultaneous Hydraulic-Fracture Propagation from Multiple Perforation Clusters in Horizontal Wells. *SPE J.* **2016**, *21*, 1000–1008. [CrossRef]
4. Huang, W.; Kamenski, L. A geometric discretization and a simple implementation for variational mesh generation and adaptation. *J. Comput. Phys.* **2015**, *301*, 322–337. [CrossRef]
5. Chu, H.; Chen, Z.; Liao, X.; Lee, W.J. Transient behavior modeling of a multi-well horizontal pad in a reservoir with irregular boundary using boundary element method. *J. Pet. Sci. Eng.* **2022**, *209*, 109852. [CrossRef]

6. Seth, P.; Manchanda, R.; Kumar, A.; Sharma, M.M. Analyzing Pressure Interference between Horizontal Wells during Fracturing. *J. Pet. Sci. Eng.* **2021**, *204*, 108696. [CrossRef]
7. Dong, R.; Wheeler, M.F.; Ma, K.; Su, H. A 3D Acid Transport Model for Acid Fracturing Treatments with Viscous Fingering. Presented at the SPE Annual Technical Conference and Exhibition, Virtual, 26–29 October 2020. [CrossRef]
8. Dong, R.; Wheeler, M.F.; Su, H.; Ma, K. Modeling Multistage Acid Fracturing Treatments in Carbonate Reservoirs. Presented at the SPE Hydraulic Fracturing Technology Conference and Exhibition, Virtual, 4–6 May 2021. [CrossRef]
9. Carrier, B.; Granet, S. Numerical modeling of hydraulic fracture problem in permeable medium using cohesive zone model. *Eng. Fract. Mech.* **2012**, *79*, 312–328. [CrossRef]
10. Sarris, E.; Papanastasiou, P. The influence of the cohesive process zone in hydraulic fracturing modelling. *Int. J. Fract.* **2011**, *167*, 33–45. [CrossRef]
11. Dong, R.; Alpak, F.O.; Wheeler, M.F. Accurate Two-Phase Flow Simulation in Faulted Reservoirs by Combining Two-Point Flux Approximation and Mimetic Finite Difference Methods. *SPE J.* **2023**, *28*, 111–129. [CrossRef]
12. Lin, H.; Deng, J.; Liu, W.; Xie, T.; Xu, J.; Liu, H. Numerical simulation of hydraulic fracture propagation in weakly consolidated sandstone reservoirs. *J. Cent. South Univ.* **2018**, *25*, 2944–2952. [CrossRef]
13. Guo, J.; Zhao, X.; Zhu, H.; Zhang, X.; Pan, R. Numerical simulation of interaction of hydraulic fracture and natural fracture based on the cohesive zone finite element method. *J. Nat. Gas Sci. Eng.* **2015**, *25*, 180–188. [CrossRef]
14. Li, Y.; Deng, J.G.; Liu, W.; Feng, Y. Modeling hydraulic fracture propagation using cohesive zone model equipped with frictional contact capability. *Comput. Geotech.* **2017**, *91*, 58–70. [CrossRef]
15. Zou, J.; Zhang, Y.; Zhang, L.; Jing, J.; Fu, Y.; Wang, Y.; Zhang, G.; Zhou, F. Numerical Simulation Research on the Effect of Artificial Barrier Properties on Fracture Height. *Processes* **2023**, *11*, 310. [CrossRef]
16. Zhu, D.; Wang, Y.; Cui, M.; Zhou, F.; Wang, Y.; Liang, C.; Zou, H.; Yao, F. Acid System and Stimulation Efficiency of Multistage Acid Fracturing in Porous Carbonate Reservoirs. *Processes* **2022**, *10*, 1883. [CrossRef]
17. Li, J.; Dong, S.; Hua, W.; Li, X.; Pan, X. Numerical Investigation of Hydraulic Fracture Propagation Based on Cohesive Zone Model in Naturally Fractured Formations. *Processes* **2019**, *7*, 28. [CrossRef]
18. Lecampion, B. An extended finite element method for hydraulic fracture problems. *Commun. Numer. Methods Eng.* **2009**, *25*, 121–133. [CrossRef]
19. Chen, Z.; Bunger, A.P.; Zhang, X.; Jeffrey, R.G. Cohesive Zone Finite Element-Based Modeling of Hydraulic Fractures. *Acta Mech. Solida Sin.* **2009**, *22*, 443–452. [CrossRef]
20. Wu, B.; Zhang, M.; Deng, W.; Que, J.; Liu, W.; Zhou, F.; Wang, Q.; Li, Y.; Liang, T. Study and Mechanism Analysis on Dynamic Shrinkage of Bottom Sediments in Salt Cavern Gas Storage. *Processes* **2022**, *10*, 1511. [CrossRef]
21. Yuan, K.; Huang, W.; Chen, X.; Cao, Q.; Fang, X.; Lin, T.; Jin, C.; Li, S.; Wang, C.; Wang, T. The Whole-Aperture Pore Structure Characteristics and Their Controlling Factors of the Dawuba Formation Shale in Western Guizhou. *Processes* **2022**, *10*, 622. [CrossRef]
22. Sun, S.; Zhou, M.; Lu, W.; Davarpanah, A. Application of Symmetry Law in Numerical Modeling of Hydraulic Fracturing by Finite Element Method. *Symmetry* **2020**, *12*, 1122. [CrossRef]
23. Zhu, D.; Wang, Y.; Cui, M.; Zhou, F.; Zhang, Y.; Liang, C.; Zou, H.; Yao, F. Effects of spent viscoelastic-surfactant acid flow on wormholes propagation and diverting performance in heterogeneous carbonate reservoir. *Energy Rep.* **2022**, *8*, 8321–8332. [CrossRef]
24. Zhu, H.; Wang, H.; Tang, X.; Li, Y. Hydraulic Fracture Propagation in Sand-Mudstone Interbedded Reservoir Integrated with Different Fluid Flow of Multi-Perforated Fractures. Presented at the ARMA-CUPB Geothermal International Conference, Beijing, China, 5–8 August 2019. Available online: https://onepetro.org/ARMACUPB/proceedings-abstract/CUPB19/All-CUPB19/125225 (accessed on 17 March 2023).
25. Peng, Y.; Zhao, J.; Sepehrnoori, K.; Li, Z. Fractional model for simulating the viscoelastic behavior of artificial fracture in shale gas. *Eng. Fract. Mech.* **2020**, *228*, 106892. [CrossRef]
26. Biot, M.A. General Theory of Three-Dimensional Consolidation. *J. Appl. Phys.* **1941**, *12*, 155–164. [CrossRef]
27. Tomar, V.; Zhai, J.; Zhou, M. Bounds for element size in a variable stiffness cohesive finite element. *Int. J. Numer. Methods Eng.* **2004**, *61*, 1894–1920. [CrossRef]
28. Peng, Y.; Zhao, J.; Sepehrnoori, K.; Li, Z.; Xu, F. Study of delayed creep fracture initiation and propagation based on semi-analytical fractional model. *Appl. Math. Model.* **2019**, *72*, 700–715. [CrossRef]
29. Elbel, J.L.; Piggott, A.R.; Mack, M.G. Numerical Modeling of Multilayer Fracture Treatments. Presented at the Permian Basin Oil and Gas Recovery Conference, Midland, TX, USA, 18–20 March 1992. [CrossRef]
30. Crump, J.B.; Conway, M.W. Effects of Perforation-Entry Friction on Bottomhole Treating Analysis. *J. Pet. Technol.* **1988**, *40*, 1041–1048. [CrossRef]

processes MDPI

Article

Physical Simulation Experiments of Hydraulic Fracture Initiation and Propagation under the Influence of Deep Shale Natural Fractures

Zhou Hu [1], Pengfei Chen [2], Wei Jiang [1], Yadong Yang [1], Yizhen Li [1], Longqing Zou [1], Huaming Wang [1], Yuping Sun [3,*] and Yu Peng [4,*]

1 Shale Gas Exploration and Development Department, CNPC Chuanqing Drilling Engineering Co., Ltd., Chengdu 610051, China
2 Research Institute of Natural Gas Technology, PetroChina Southwest Oil and Gas Field Company, Chengdu 610213, China
3 PetroChina Research Institute of Petroleum Exploration & Development, Beijing 100083, China
4 State Key Laboratory of Oil and Gas Reservoir Geology and Exploitation, Southwest Petroleum University, Chengdu 610500, China
* Correspondence: sunyuping01@petrochina.com.cn (Y.S.); pengyu_frac@foxmail.com (Y.P.)

Abstract: Horizontal wells' multi-section and multi-cluster hydraulic fracturing plays an important role in the efficient development of shale gas. However, the influence of the perforating hole and natural fracture dip angle on the process of hydraulic fracture initiation and propagation has been ignored in the current researches. This paper presents the results related to a tri-axial large-scale hydraulic fracturing experiment under different natural fracture parameters. We discuss the experimental results relating to the near-wellbore tortuosity propagation of hydraulic fractures. Experimental results showed that the triaxial principal stress of the experimental sample was deflected by the natural fracture, which caused significant near-wellbore tortuosity propagation of the hydraulic fractures. The fractures in most rock samples were not perpendicular to the minimum horizontal principal stress after the experiment. As well, the deflection degree of triaxial principal stress direction and the probability of hydraulic fractures near-wellbore tortuosity propagation decreased with the increase of the natural fracture dip angle. After hydraulic fractures' tortuous propagation, the hydraulic fractures will propagate in the direction controlled by the triaxial stress in the far-wellbore area. For reservoirs with natural fractures, proppant in hydraulic fracturing should be added after the fractures are fully expanded to prevent sand plugging in tortuous fractures. When the permeability of natural fractures is low, the volume of fracturing fluid entering natural fractures is small, and hydraulic fractures are easy to pass through the natural fractures.

Keywords: shale gas; hydraulic fracturing; fracture dip angle; near wellbore distortion; fracture permeability

Citation: Hu, Z.; Chen, P.; Jiang, W.; Yang, Y.; Li, Y.; Zou, L.; Wang, H.; Sun, Y.; Peng, Y. Physical Simulation Experiments of Hydraulic Fracture Initiation and Propagation under the Influence of Deep Shale Natural Fractures. *Processes* **2023**, *11*, 1934. https://doi.org/10.3390/pr11071934

Academic Editor: Carlos Sierra Fernández

Received: 6 June 2023
Revised: 25 June 2023
Accepted: 26 June 2023
Published: 27 June 2023

1. Introduction

Shale reservoirs have extremely low porosity and permeability, which can be obviously impacted by mineral deposition [1]. In the development of shale gas, hydraulic fracturing plays an important role in connecting the wellbore and reservoir matrix, establishing the gas flow path, and reducing gas seepage resistance [2]. The hydraulic fracture influences the whole life of the shale gas well, including the fracturing period, shut-in period, and production [3]. Numerical simulation is the main means of studying the impact of hydraulic fractures on reservoir fluid flow and production [4], especially in the production of shale gas reservoirs [5]. However, the propagation law of hydraulic fractures is difficult to obtain through numerical simulation and needs to be studied through fracturing experiments [6]. The interaction between hydraulic fractures and natural fractures has been widely studied, which directly determines the complexity of the fracture network, the invasion range of fracturing fluid, and the stimulation effect of shale gas wells [7]. The complexity of the fracture network deeply impacts the production of a shale gas well [8]. As an important fracture parameter, the influence of natural fracture dip angle on the propagation of hydraulic fractures is rarely discussed [9]. As well, the fracture initiation and fracture propagation patterns near the horizontal wellbore, which seriously impact the near-wellbore friction during fracturing and the tortuosity and resistance of shale gas flow during production, have always been regarded as the research focus of hydraulic fracturing. Therefore, the analysis of fracture propagation patterns and influencing factors in the near wellbore area is of great significance for improving the effectiveness of fracturing construction.

Natural fractures are widely distributed in shale reservoirs. Due to the low porosity and permeability of the shale matrix, effective activation or connection of natural fractures has become an important way to improve the production of shale gas wells. Building a fracture network has become the main goal of current fracturing construction. Therefore, the study of the influence of natural fractures on the propagation of hydraulic fractures is of great significance. Many researchers discussed the intersection mechanism of hydraulic fracture and natural fracture through laboratory experiments [10]. Lamont et al. [11] and Daneshy et al. [12] used outcrop rock samples as experimental samples for the first time to conduct fracturing experiments and discussed the influence of natural fractures on hydraulic fracture propagation. Their research shows that the scale of natural fractures has a significant impact on the expansion of hydraulic fractures. Real triaxial large-scale physical simulation experiments were carried out using natural shale outcrop rock samples and artificial gypsum rock samples. As well, the impact of the intersection angle and horizontal principal stress difference on the hydraulic fracture passing through the naturally weak surface was discussed. [13–15]. As well, the influence of natural fracture shear strength on fracture intersection criterion is discussed by fracturing experiment [16,17]. Natural fractures and formation stress control the geometry and propagation behavior of hydraulic fractures [18]. Gu et al. verified the intersection criterion between hydraulic fractures and natural fractures through experiments [19]. This shows that natural fractures are of great significance to the construction of complex fracture networks. As well, complex fracture network is conducive to the efficient development of Shale gas [20]. The influence of the angle between natural fractures and hydraulic fractures on fracture propagation and the intersection criterion between hydraulic fractures and natural fractures is mainly discussed. However, hydraulic fractures in the horizontal well of the shale reservoir may have the same angle as natural fractures with different dip angles. Therefore, the influence of natural fracture inclination on hydraulic fracture propagation is worth studying. Fatahi et al. [21] discussed the interaction between hydraulic fractures and natural fractures through hydraulic fracturing experimental research and numerical simulation methods. The research shows that the smaller the angle between the natural fracture and the minimum horizontal principal stress, the easier the hydraulic fracture will pass through the natural fracture. The influence of bedding on fracture propagation in coal was also discussed through experiments and numerical simulation [22]. Based on these studies, the optimization scheme of hydraulic fracturing operation is also proposed [23]. Zou et al. [24] use CT scanning technology to analyze the hydraulic fracture geometry of shale rock samples with natural fractures after a fracturing experiment. The shear slip characteristics of natural fractures during hydraulic fracturing were analyzed through experiments [25]. The hydraulic fracture approach angle will affect the shear slip degree of natural fractures [26]. The vertical propagation mechanism of hydraulic fracture is analyzed by fracturing experiments with samples containing interlayers and bedding [27]. The influence of bedding and natural fractures on hydraulic crack propagation has also been confirmed through numerical simulation [20]. Above all, the impact of shale bedding on fracture height and the interaction between hydraulic fractures and natural fractures are considered in a large-scale hydraulic fracturing physical simulation experiment. A large number of studies show that natural fractures significantly affect the expansion of hydraulic fractures [28,29]. As well, many intersection mechanism models between hydraulic fractures and natural fractures have been proposed [30]. However, few studies have discussed the influence of fracture dip angle on hydraulic fracture propagation with complex natural fracture networks and hydraulic fractures with near-wellbore tortuosity propagation. The migration of proppant in fractures is affected by the near-wellbore tortuosity propagation of hydraulic fractures. This is because curved fractures increase the migration resistance of proppant in the fracturing fluid, which can easily cause proppant to accumulate at the fracture opening and block the flow channel. This situation will seriously impact fracture propagation and hinder the construction of the fracture network. Therefore, the study of hydraulic fractures' near-wellbore tortuosity propagation is of great significance.

In this paper, cement samples with natural fractures were used to study the influence of natural fracture dip angle on hydraulic fracture initiation and propagation through a tri-axial large-scale hydraulic fracturing experiment. The influence of natural fracture dip on the near-wellbore triaxial principal stress and hydraulic fractures' near-wellbore tortuosity propagation was discussed. As well, the influence of horizontal principal stress differences on the propagation of hydraulic fractures and the intersection of hydraulic fractures and natural fractures is analyzed.

2. Materials and Methods

2.1. Experimental Equipment and Sample

The real tri-axial large-scale hydraulic fracturing initiation and propagation physical simulation system (Figure 1) is composed of a triaxial stress loading chamber, a liquid pumping system, a pneumatic control system, and a digital control system. There is a hydraulic jack in the X-Y-Z direction in the triaxial stress loading chamber, which can provide compressive stress to simulate the triaxial principal stress in the formation. One of the hydraulic jacks is installed at the bottom of the loading chamber to apply vertical stress to the experimental sample. The other two are installed in a horizontal direction to apply two mutually perpendicular horizontal stresses to the experimental sample. Set stress values in three directions in the experimental system, and the experimental equipment will load corresponding stresses in the three directions to meet the experimental conditions.

Figure 1. The real tri-axial large-scale hydraulic fracturing initiation and propagation physical simulation system.

A liquid pumping system, which is connected to an artificial wellbore in the cement sample, can provide constant pressure pumping and constant displacement pumping. The pump rate range is from 1 mL/min to 120 mL/min. The pump pressure range is from 0 MPa to 100 MPa. The maximum volume of the liquid injection pump is 500 mL. The pipeline and wellbore should have strong pressure-bearing capacity to meet the requirements of the fracturing experiment. During the experiment, the liquid is pumped into the intermediate container, and the dyed liquid, which is used to simulate fracturing fluid, is injected into the artificial wellbore by the intermediate container. Throughout the entire experiment, the pump pressure of the liquid will be recorded until the end of the experiment.

Experimental samples are formed by cement and sand in the mold (Figure 2b). An artificial wellbore, which is a steel pipe with an external thread for enhancing the friction between the wellbore and sample, is inserted into the mold during the cement setting. It should be noted that the steel pipe did not fully fill the wellbore of the experimental sample. In the experiment described in this paper, the wellbore of the rock sample is 5 mm longer than the steel pipe, which allows the hydraulic fractures to propagate from the bottom of the well after the fracturing experiment, and the effect is comparable to the perforation process in actual construction. The experimental samples were cut by a water jet with reference to different natural fracture dip angles (Table 1) and recombined with cement. For each rock sample, we used the same material for bonding to ensure that the natural fracture permeability in each rock sample tends to be consistent. This can effectively avoid the impact of large differences in natural fracture permeability on the fracturing experimental results.

(a) casing pipe (b) Cement sample preparation

(c) Cutting sample to build natural fractures

Figure 2. Experimental sample preparation.

Table 1. Experimental scheme.

Sample No	Sizes of Rock Sample (mm)	Depth of the Wellbore (mm)	Length of Steel Pipe (mm)	Inner Diameter of Steel Pipe (mm)	Triaxial Principal Stress X-Y-Z (MPa)	Number of Natural Fracture	Natural Fracture Dip Angle (°)
P1-1	300 × 300 × 300	150	145	25	50-60-60	1	80
P1-2					50-65-65	1	80
P1-3					50-70-70	1	80
P2-1					50-60-60	1	60
P2-2					50-65-65	1	60
P2-3					50-70-70	1	60
P3-1					50-60-60	1	40
P3-2					50-65-65	1	40
P3-3					50-70-70	1	40
P4					50-65-65	2	80/80
P5					50-65-65	2	80/−80
P6					50-65-65	2	40/40
P7					50-65-65	2	40/−40
P8					50-65-65	2	80/0

2.2. Experimental Method and Procedure

Experimental research is the main means to explore the propagation of hydraulic fractures and the interaction criteria between hydraulic fractures and natural fractures. The experimental methods and steps in this paper are as follows.

(1) Polish the irregular residue on the sample surface to make the surface of the samples regular and flat. The installation process of the rock sample can be performed smoothly.
(2) Put the experimental sample into the predetermined position of the triaxial stress loading chamber through the lifting machine. Then install metal cushion blocks between the hydraulic pump and rock sample to make it fully fit with the surface of the sample and ensure uniform stress on the sample.
(3) Add the pre-configured fracturing fluid, which is prepared with glycerin and dye, into the fracturing intermediate container. Then connect the fracturing fluid pipeline outlet with the pressure sensor, and then connect the pipeline with the upper part of the wellbore.
(4) Apply triaxial stress to the rock sample by using the hydraulic pump. In this process, it is necessary to maintain the slow and synchronous loading of three-dimensional stress.
(5) After the triaxial stress of the rock sample reaches the predetermined value, pump fluid into the sample at a small pumping rate to fill the fracturing fluid injection pipeline and wellbore space. When the pipeline pressure has an upward trend, inject fracturing fluid into the wellbore with a pre-designed pumping rate, and record the changes in pumping pressure and pumping rate during fracturing through the data acquisition system. When the pressure curve of the computer acquisition system window shows a sudden drop in pressure, it indicates that the sample has successfully fractured. Then continue to record the change rule of injection pressure with injection displacement.
(6) Stop the pump to complete the test when the predetermined pumping volume is reached. Then remove the sample from the triaxial stress loading chamber with a lifting machine. Use a large cutting machine to cut the sample along the fracture surface, observe and record the space coordinate position of the residual tracer trace from the cutting surface, and determine the shape and extension direction of the crack initiation.

To discuss the influence of natural fracture dip and horizontal stress difference on fracture propagation, each group of experiments was set with the same rock sample size, wellbore depth, and steel pipe size, only changing the three-dimensional principal stress, the number of fractures, and natural fracture dip. The experimental design scheme was shown in Table 1. The natural fracture of each experimental sample was shown in Figure 3. The pump rate is set to 30 mL/min in the fracturing experiment. Although temperature changes can impact the structure of shale [31], the fracturing construction time is relatively short, so the influence of temperature on crack propagation was ignored in the experimental design of our work.

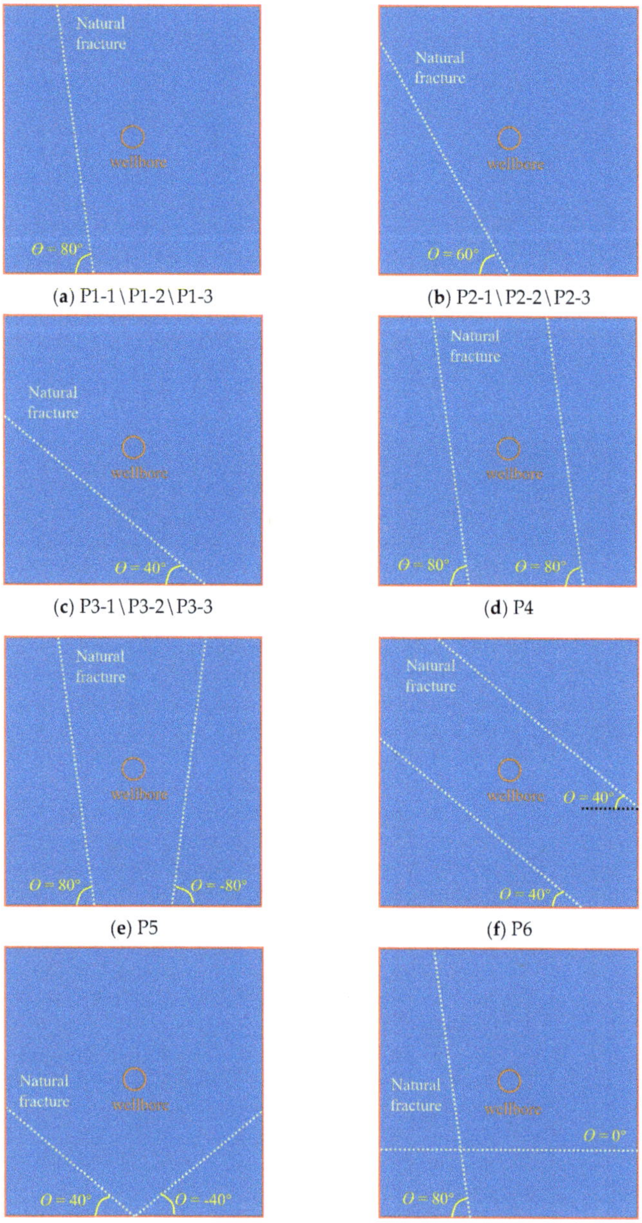

(**a**) P1-1\P1-2\P1-3 (**b**) P2-1\P2-2\P2-3

(**c**) P3-1\P3-2\P3-3 (**d**) P4

(**e**) P5 (**f**) P6

(**g**) P7 (**h**) P8

Figure 3. Schematic diagram of experimental sample and natural fracture (the view is wellbore direction).

3. Results and Discussion

The sample was broken by physical tools after the hydraulic experiment. The morphology of hydraulic fractures in rock samples is recorded (Figure 4). Due to the staining of the fracturing fluid, the hydraulic fracture morphology in the rock sample after the fracturing experiment and the intrusion of the fracturing fluid into the natural fractures can be clearly distinguished. As well, the fractures in the sample were redrawn with the drawing software (Figure 5). Based on the redrawn fracture morphology, the propagation mechanism of hydraulic fractures can be analyzed. As well, the influence of natural fracture dip on fracture propagation is discussed.

(**a**) Before fracturing

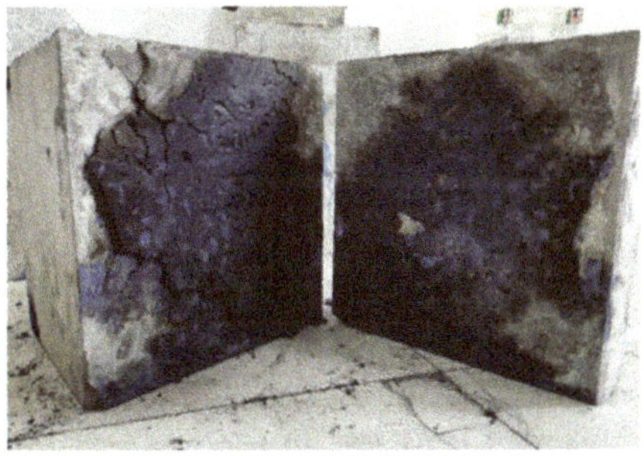

(**b**) Fracturing fluid in the natural fracture

(**c**) Hydraulic fracture penetrating sample

(**d**) Hydraulic fracture in sample

Figure 4. Experimental sample after fracturing.

From the hydraulic fracture morphology of the above 14 sets of experiments, it is obvious that hydraulic fractures do not pass through natural fractures in most cases. When the inclination angle of natural fractures is 80°, hydraulic fractures all pass through natural fractures, such as P1-1, P1-2, P1-3, P4, and P5. When the inclination angle of the natural fracture is 60°, only the hydraulic fractures in the case of P2-1 pass through the natural fracture. This result is in line with many current studies on the intersection of natural and hydraulic fractures, where the larger the inclination angle of natural fractures, the easier it is for them to pass through natural fractures [19]. However, there were many hydraulic fractures in the experimental results that were not perpendicular to the minimum horizontal principal stress set in the experiment. Tortuosity propagation of hydraulic fractures occurs in the near-wellbore area. This indicates that there is a difference between the horizontal principal stress and the pre-set principal stress during the experimental process. This is caused by natural crack-induced stress and perforation hole stress, and the smaller the inclination angle of the natural crack, the more the main stress is affected.

It is generally believed that the expansion of hydraulic fractures will be affected by the maximum and minimum horizontal principal stress. The initiation and propagation of hydraulic fractures are in the direction perpendicular to the minimum horizontal principal stress at the borehole. This phenomenon has also been confirmed in most large physical simulation experiments without prefabricated fractures [32]. Under the influence of perforation and induced stress of natural fracture near-wellbore in this experiment, the initiation and propagation mode of hydraulic fractures becomes very complex in our experiment. It shows a near-wellbore distortion effect that is not completely controlled by the maximum and minimum horizontal principal stress.

The hydraulic fracture morphology of shale is strongly influenced by natural fractures [33]. The fractures perpendicular to the wellbore in P1–2 do not intersect with the bottom of the well. First, the fractures parallel to the wellbore communicate with the natural fractures, forming a large leak-off, and then the fracture initiates and propagates from the natural fracture in the direction perpendicular to the wellbore. This phenomenon shows that the fracture distortion effect influenced by complex natural fractures is only effective in the near-wellbore position. After the fracture propagates far away from the stress concentration area, the fracture propagation is controlled by triaxial principal stress and conforms to the stress control theory. In the samples with only distorted fractures, such as P2-2, P2-3, P3-1, P3-3, P8, etc., the fractures may propagate according to the mode of crustal stress control after pumping for a period of time. When the natural fracture density of the reservoir is high, there will be a large number of fractures with various dip angles near the well. In this case, sand plugging is easy to form, and the amount of pre-fluid needs to be increased. After the fracture perpendicular to the direction of the minimum horizontal principal stress is fully expanded, a sand-adding operation can be carried out.

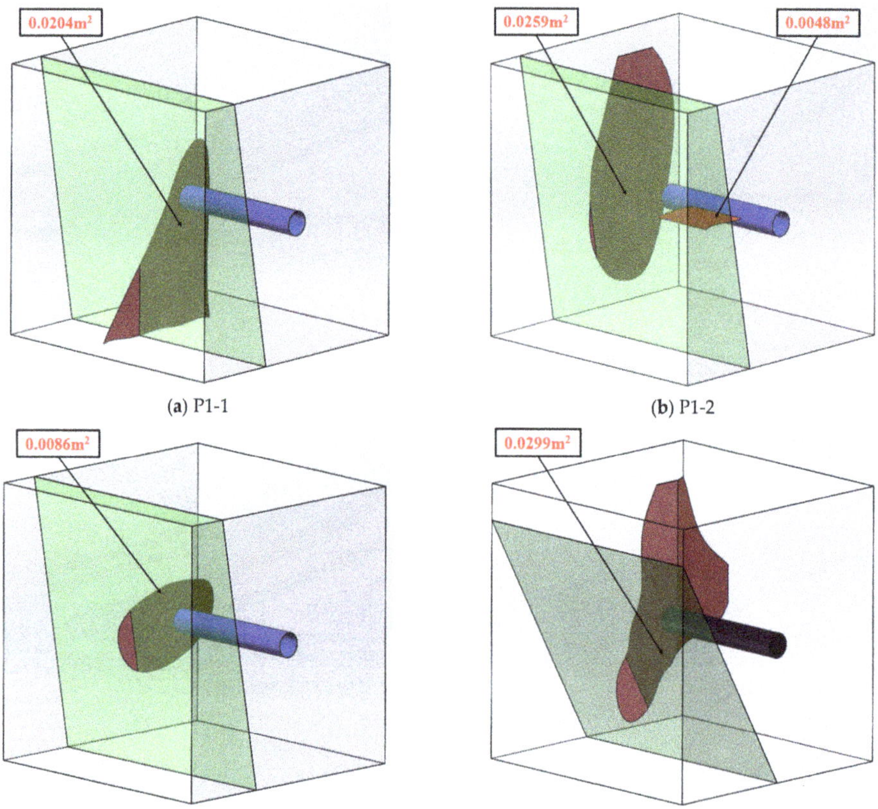

(**a**) P1-1 (**b**) P1-2 (**c**) P1-3 (**d**) P2-1

Figure 5. *Cont.*

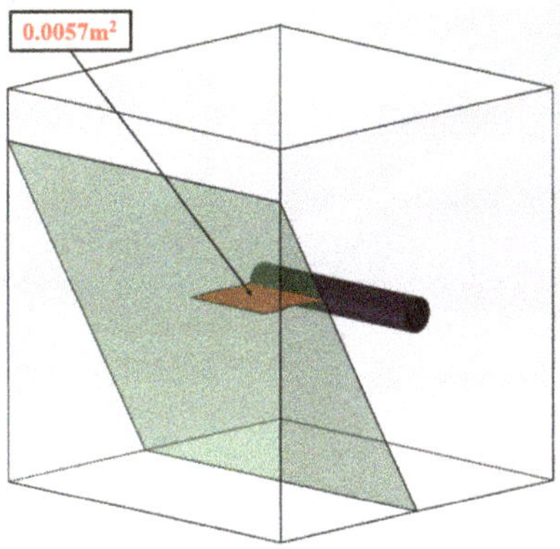

(**e**) P2-2

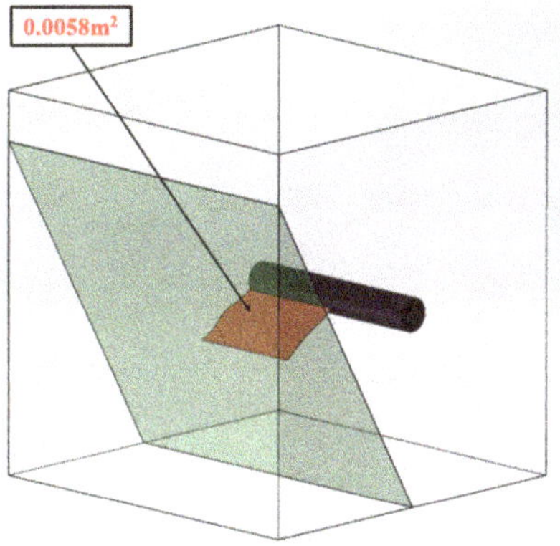

(**f**) P2-3

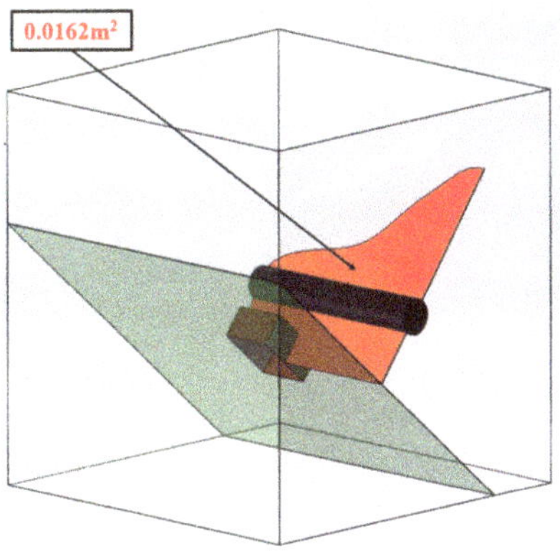

(**g**) P3-1

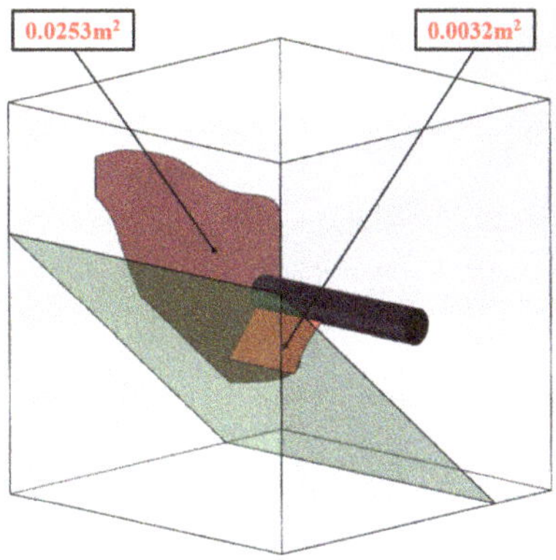

(**h**) P3-2

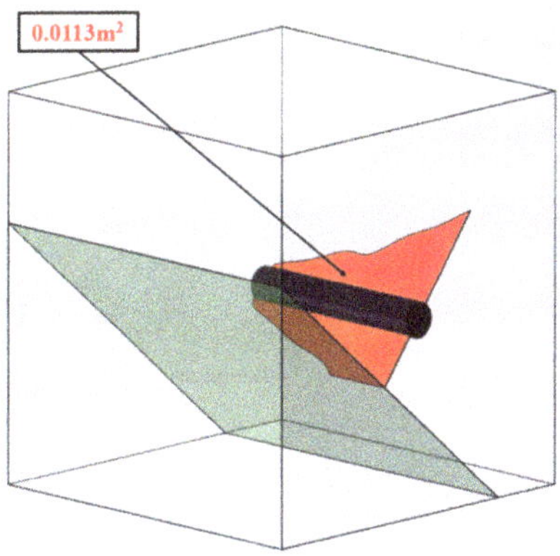

(**i**) P3-3

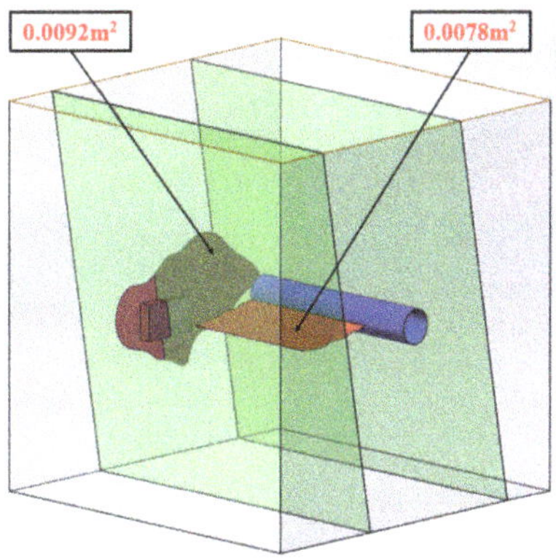

(**j**) P4

Figure 5. *Cont.*

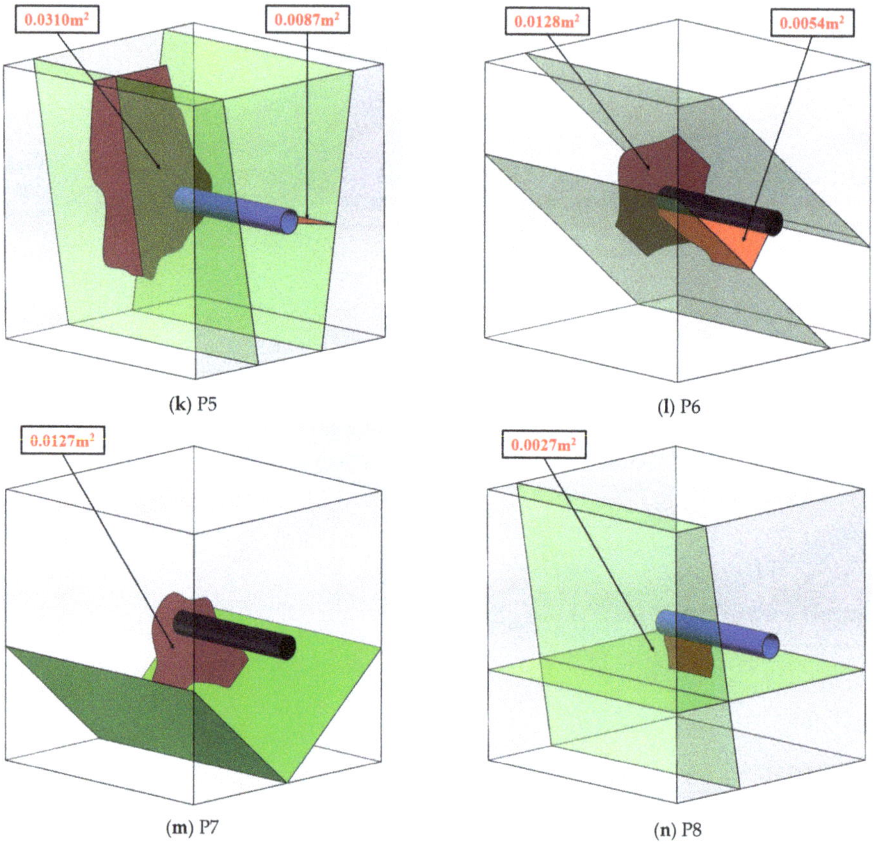

Figure 5. Fracture morphology after the experiment.

The permeability of hydraulic fractures has an extremely important impact on the expansion of fractures. According to previous experimental results, under the maximum and minimum horizontal principal stress differences (10, 15, and 20 MPa) set in this experiment, the artificial fractures should completely pass through the natural fractures. Under the experimental conditions in this paper, natural fractures are cemented and filled again by cement, but the leak-off volume of fracturing fluid will increase significantly after hydraulic fractures intersect with natural fractures. In the experimental results, only a few of the natural fractures in the sample are not completely colored. The massive fracturing fluid leak-off in natural fractures causes the artificial fractures to turn along the natural fractures instead of passing through the natural fractures. It has greatly increased our confidence in building artificial fracture networks in deep shale gas reservoirs.

When the natural fracture dip angle is greater than 80°, hydraulic fractures easily penetrate natural fractures, and the hydraulic fracture propagation direction is mainly in the direction of the vertical minimum horizontal principal stress [34,35], such as P1-1, P1-2, and P1-3. With the prefabricated natural fracture dip angle decreasing (from 80° to 60° and then to 40°), the deflection degree of the maximum and minimum horizontal principal stress directions increases. As well, the shape of the hydraulic fracture gradually deviates from the control of maximum and minimum horizontal principal stresses. When the natural fracture dip angle is 60°, the fractures parallel to the wellbore direction appear in experiments P2-2 and P2-3. When the natural fracture dip angle is 40°, the fractures parallel to the wellbore direction appear in the experiments of P3-1, P3-2, and P3-3.

The stress-sensitive effect of natural fractures impacts the permeability of natural fractures and the fracturing fluid leak-off in natural fractures. From the three groups of experiments (P1-1P1-3, P2-1P2-3, and P3-1P3-3), it can be analyzed that the leak-off volume of fracturing fluid in natural fractures gradually decreases with the increase of the difference between the maximum and minimum horizontal principal stresses, as shown in Figure 6. Especially when the difference between the horizontal principal stresses is more than 20 MPa, the spread of fracturing fluid in the natural fracture will be strongly inhibited, and only a part of the natural fracture area in the sample is stained. This is because it is difficult to open natural fractures, making it difficult for fracturing fluid to enter natural fractures under high stress. In the experiments of P4P8, the difference between the maximum and minimum horizontal principal stress is 15 MPa. As well, leak-off inhibition only occurs in the P6 sample. Therefore, in the fracturing construction of deep shale gas wells, the fracturing fluid leak-off in natural fractures may be inhibited when the maximum and minimum horizontal principal stress differences are greater than 20 MPa. This will impact the hydration fracture network and reduce the fracturing effect. The permeability of natural fractures and matrix can be enhanced by corrosion filler [36], which can promote the invasion of fracturing fluid in natural fractures and enhance the complexity of the fracture network. Therefore, the effect of stimulation can be improved by the corrosion of fracture-filling materials in fracturing construction.

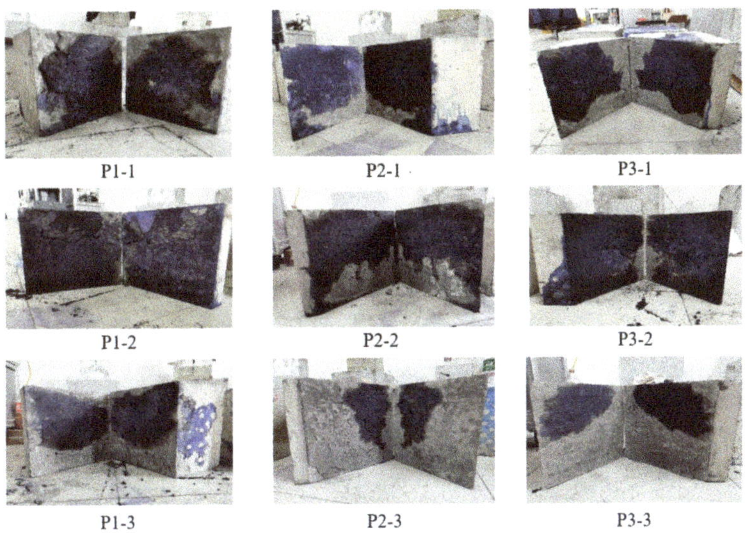

Figure 6. Fluid intrusion in natural fractures.

The experiments of P4–P8 show hydraulic fracture initiation and propagation under the influence of multiple groups of natural fractures. Comparing P4, P5, P6, and P7, it is shown that hydraulic fractures are easier to pass through high-dip-angle fractures near the wellbore. We assumed that the influence of a high dip angle fracture on horizontal principal stress is not obvious, and hydraulic fractures are easier to crack and expand in the non-dip mode. The experiments of P8 show that the maximum and minimum horizontal principal stresses are deflected by horizontal natural fractures and high-angle fractures. After the fracture initiated from the perforation hole, the horizontal fracture was directly communicated along the minimum horizontal principal stress. After the natural fracture was communicated, a large amount of leakage occurred, and the fracture almost no longer propagated to the rest.

4. Conclusions

In this work, 14 sets of fracturing experiments were completed by the real tri-axial large-scale hydraulic fracturing initiation and propagation physical simulation system. By observing the morphology of hydraulic and natural fractures in experimental rock samples, we redrew the fracture morphology through software. Through the analysis of experimental results, the influence of natural fracture dip on hydraulic fracture propagation is discussed. The influence of natural fracture permeability on the propagation of hydraulic fractures under different stress conditions is analyzed. This work also conducted an analysis and discussion on the influencing factors of near-wellbore distortion propagation.

(1) The direction of triaxial principal stress will be deflected by the near-wellbore natural fracture, which causes significant near-wellbore tortuosity propagation of hydraulic fractures. As well, the deflection degree of the triaxial principal stress direction and the probability of hydraulic fractures near-wellbore tortuosity propagation is negatively correlated with the natural fracture dip angle.
(2) The influence of high-dip angle fracture on the maximum and minimum horizontal principal stresses is not obvious. In this case, the propagation of hydraulic fracture is also controlled by the three-dimensional principal stress. With the prefabricated natural fracture dip angle decreasing (from 80° to 60° and then to 40°), the deflection degree of the maximum and minimum horizontal principal stress directions increases. As well, the shape of the hydraulic fracture gradually deviates from the control of maximum and minimum horizontal principal stresses.
(3) The hydraulic fractures will propagate in the direction controlled by triaxial stress in the far-wellbore area after tortuously propagating. For reservoirs with natural fractures, proppant in hydraulic fracturing should be added after the fractures are fully expanded, and the amount of pre-fluid should be increased to prevent sand from plugging in tortuous fractures.
(4) The stress-sensitive effect of natural fractures impacts the permeability of natural fractures and the fracturing fluid leak-off in natural fractures. Natural fractures are easily crossed by hydraulic fractures when their permeability is small. When the difference between the horizontal principal stresses is more than 20 MPa, the spread of fracturing fluid in the natural fracture will be strongly inhibited, and only a part of the natural fracture area in the sample is stained. Therefore, enhancing the permeability of natural fractures with corrosion fracture filler can increase the complexity of the fracture network.

Author Contributions: Conceptualization, Z.H. and P.C.; methodology, W.J.; validation, Y.Y.; formal analysis, Y.L.; investigation, L.Z.; resources, H.W.; writing—original draft preparation, Y.S. writing—review and editing, Y.P. All authors have read and agreed to the published version of the manuscript.

Funding: This research was funded by the National Natural Science Foundation of China (No. 52204051 and U21B2071) and the Natural Science Foundation of Sichuan Province (No. 2022NSFSC0971).

Data Availability Statement: Not applicable.

Conflicts of Interest: The authors declare no conflict of interest.

References

1. Khormali, A.; Petrakov, D.G.; Farmanzade, A.R. Prediction and Inhibition of Inorganic Salt Formation under Static and Dynamic Conditions—Effect of Pressure, Temperature, and Mixing Ratio. *Int. J. Technol.* **2016**, *7*, 943–951. [CrossRef]
2. Peng, Y.; Zhao, J.; Sepehrnoori, K.; Li, Z.; Xu, F. Study of delayed creep fracture initiation and propagation based on semi-analytical fractional model. *Appl. Math. Model.* **2019**, *72*, 700–715. [CrossRef]
3. Li, Y.; Luo, A.; Chen, X.; Cheng, L.; Chang, C.; Xie, W. A Novel model for simulating the integration process of hydraulic fracturing, shut-in period, and well production. *Front. Energy Res.* **2022**, *10*, 979526. [CrossRef]
4. Zhang, X.; Zhang, J.; Xiao, C. Experimental Investigation of Seepage Mechanism on Oil-Water Two-Phase Displacement in Fractured Tight Reservoir. *Geofluids* **2022**, *2022*, 6792154. [CrossRef]

5. Luo, A.; Li, Y.; Wu, L.; Peng, Y.; Tang, W. Fractured horizontal well productivity model for shale gas considering stress sensitivity, hydraulic fracture azimuth, and interference between fractures. *Nat. Gas Ind. B* **2021**, *8*, 278–286. [CrossRef]
6. Zhai, M.; Wang, D.; Li, L.; Zhang, Z.; Zhang, L.; Huang, B.; Li, A.; Zhang, Q. Investigation on the mechanism of hydraulic fracture propagation and fracture geometry control in tight heterogeneous glutenites. *Energy Explor. Exploit.* **2021**, *40*, 246–278. [CrossRef]
7. Xu, W.; Zhao, J.; Rahman, S.S.; Li, Y.; Yuan, Y. A Comprehensive Model of a Hydraulic Fracture Interacting with a Natural Fracture: Analytical and Numerical Solution. *Rock Mech. Rock Eng.* **2018**, *52*, 1095–1113. [CrossRef]
8. Chen, Z.; Liao, X.; Yu, W. A semianalytical well-testing model of fracture-network horizontal wells in unconventional reservoirs with multiple discretely natural fractures. *Nat. Gas Ind. B* **2020**, *7*, 567–582. [CrossRef]
9. Zhao, J.; Li, Y.; Wang, S.; Jiang, Y.; Zhang, L. Simulation of complex fracture networks influenced by natural fractures in shale gas reservoir. *Nat. Gas Ind. B* **2014**, *1*, 89–95.
10. Guo, J.; Lu, Q.; He, Y. Key issues and explorations in shale gas fracturing. *Nat. Gas Ind. B* **2023**, *10*, 183–197. [CrossRef]
11. Lamont, N.; Jessen, F.W. The effects of existing fractures in rocks on the extension of hydraulic fractures. *J. Pet. Technol.* **1963**, *15*, 203–209. [CrossRef]
12. Daneshy, A.A. Hydraulic fracture propagation in the presence of planes of weakness. In *SPE European Spring Meeting*; OnePetro: Richardson, TX, USA, 1974.
13. Warpinski, N.R.; Teufel, L.W. Influence of Geologic Discontinuities on Hydraulic Fracture Propagation. *J. Pet. Technol.* **1987**, *39*, 209–220. [CrossRef]
14. Blanton, T.L. Propagation of hydraulically and dynamically induced fractures in naturally fractured reservoirs. In *SPE Unconventional Gas Technology Symposium*; OnePetro: Richardson, TX, USA, 1986.
15. Blanton, T.L. An experimental study of interaction between hydraulically induced and pre-existing fractures. In *SPE Unconventional Gas Recovery Symposium*; OnePetro: Richardson, TX, USA, 1982.
16. Zhou, J.; Chen, M.; Jin, Y.; Zhang, G.-Q. Analysis of fracture propagation behavior and fracture geometry using a tri-axial fracturing system in naturally fractured reservoirs. *Int. J. Rock Mech. Min. Sci.* **2008**, *45*, 1143–1152. [CrossRef]
17. Renshaw, C.; Pollard, D. An experimentally verified criterion for propagation across unbounded frictional interfaces in brittle, linear elastic materials. *Int. J. Rock Mech. Min. Sci. Geomech. Abstr.* **1995**, *32*, 237–249. [CrossRef]
18. Zhou, J.; Jin, Y.; Chen, M. Experimental investigation of hydraulic fracturing in random naturally fractured blocks. *Int. J. Rock Mech. Min. Sci.* **2010**, *47*, 1193–1199. [CrossRef]
19. Gu, H.; Weng, X.; Lund, J.; Mack, M.; Ganguly, U.; Suarez-Rivera, R. Hydraulic Fracture Crossing Natural Fracture at Nonorthogonal Angles: A Criterion and Its Validation. *SPE Prod. Oper.* **2012**, *27*, 20–26. [CrossRef]
20. Gong, T.; Xia, B.; Liu, L.; Peng, Z.; Gao, Y. Propagation of hydraulic fracture under the joint impact of bedding planes and natural fractures in shale reservoirs. *Energy Sci. Eng.* **2019**, *7*, 2690–2702. [CrossRef]
21. Fatahi, H.; Hossain, M.; Sarmadivaleh, M. Numerical and experimental investigation of the interaction of natural and propagated hydraulic fracture. *J. Nat. Gas Sci. Eng.* **2017**, *37*, 409–424. [CrossRef]
22. Ai, C.; Li, X.-X.; Zhang, J.; Jia, D.; Tan, W.-J. Experimental investigation of propagation mechanisms and fracture morphology for coalbed methane reservoirs. *Pet. Sci.* **2018**, *15*, 815–829. [CrossRef]
23. Jiang, T.; Zhang, J.; Wu, H. Experimental and numerical study on hydraulic fracture propagation in coalbed methane reservoir. *J. Nat. Gas Sci. Eng.* **2016**, *35*, 455–467. [CrossRef]
24. Zou, Y.; Zhang, S.; Zhou, T.; Zhou, X.; Guo, T. Experimental Investigation into Hydraulic Fracture Network Propagation in Gas Shales Using CT Scanning Technology. *Rock Mech. Rock Eng.* **2016**, *49*, 33–45.
25. Hu, L.; Ghassemi, A. Laboratory-Scale Investigation of the Slippage of a Natural Fracture Resulting from an Approaching Hydraulic Fracture. *Rock Mech. Rock Eng.* **2021**, *54*, 2547–2558. [CrossRef]
26. Shi, Z.; Chen, J.; Gong, D.; He, S.; Jiang, X.; Sun, C.; Fan, L. Experimental study on the shear-slip characteristics of natural fractures in shale reservoirs. *Energy Sci. Eng.* **2022**, *10*, 1240–1250. [CrossRef]
27. Zhang, J.; Yu, Q.; Li, Y.; Pan, Z.; Liu, B. Hydraulic Fracture Vertical Propagation Mechanism in Interlayered Brittle Shale Formations: An Experimental Investigation. *Rock Mech. Rock Eng.* **2023**, *56*, 199–220. [CrossRef]
28. Dehghan, A.N. An experimental investigation into the influence of pre-existing natural fracture on the behavior and length of propagating hydraulic fracture. *Eng. Fract. Mech.* **2020**, *240*, 107330. [CrossRef]
29. Zhang, X.; Yin, M. Investigation of the Hydraulic Fracture Propagation Law of Layered Rock Strata Using the Discrete-Particle model. *Geofluids* **2022**, *2022*, 8038085. [CrossRef]
30. Kolawole, O.; Ispas, I. Interaction between hydraulic fractures and natural fractures: Current status and prospective directions. *J. Pet. Explor. Prod. Technol.* **2020**, *10*, 1613–1634. [CrossRef]
31. Kang, Y.; Hao, Z.; Chen, Q.; You, L.; Chen, M.; Tian, J.; Li, B. Heating-Induced Enhancement of Shale Gas Transport and Its Application for Improving Hydraulic Fracturing Performance. *Energy Fuels* **2022**, *36*, 14682–14695. [CrossRef]
32. Xia, Y.; Tan, P.; Wang, X.; Ren, L. Differences of fracture propagation behavior for two typical fractured formations. *Nat. Gas Ind. B* **2022**, *9*, 264–270. [CrossRef]
33. Chang, X. Laboratory analysis of liquid injection method on hydraulic fracturing initiation and propagation in deep shale formation. *Nat. Gas Ind. B* **2019**, *6*, 652–658. [CrossRef]
34. Daneshy, A. Three-Dimensional Analysis of Interactions Between Hydraulic and Natural Fractures. In *The SPE Hydraulic Fracturing Technology Conference and Exhibition*; OnePetro: Richardson, TX, USA, 2019.

35. He, Q.; Suorineni, F.; Oh, J. Modeling Interaction between Natural Fractures and Hydraulic Fractures in Block Cave Mining. In *The 49th U.S. Rock Mechanics/Geomechanics Symposium*; OnePetro: Richardson, TX, USA, 2015.
36. Chen, Q.; Hu, H.; Kang, Y.; You, L.; Zhou, Y.; Cheng, Q.; Liang, Y. Mitigating water blockage in shale matrix by acidizing or oxidizing treatment: Evidence from transport pathway change and diffusivity enhancement. *J. Pet. Sci. Eng.* **2022**, *219*, 111132. [CrossRef]

processes MDPI

Article

Study on Casing Safety Evaluation in High-Temperature Wells with Annular Pressure Buildup

Hao Wang [1,2], Mu Li [1,2], Qing Zhao [1], Weiwei Hao [1], Hui Zhang [2,*], Yafei Li [1], Pengpeng Huang [1] and Yi Zou [1]

[1] CNPC Engineering Technology R&D Company Limited, Beijing 102206, China; wanghaodr@cnpc.com.cn (H.W.)
[2] College of Petroleum Engineering, China University of Petroleum, Beijing 102249, China
* Correspondence: zhanghuicup2018@163.com

Abstract: In high-temperature wells, annular pressure buildup (APB) caused by temperature increase is a widespread phenomenon in production, especially in offshore thermal recovery wells. It increases the load on the tubing and casing and consequently threatens the wellbore integrity. Hence, research on casing safety evaluation and APB management has great significance for field production. In this paper, the tubing and casing safety evaluation and APB limit determination methods are presented considering the effect of thermal stress and APB. Based on the case study of an offshore thermal recovery well, an APB-management chart and the recommended optimal range of APB are provided. Finally, an analysis of three commonly used mitigation methods is presented. The effect and the recommended parameters of these mitigation methods are further discussed. The research results show that the thermal stress and APB phenomena affect the stress distribution of the casing and may bring great danger to the wellbore integrity. Maintaining the APB in the safety range is necessary for field production. It is recommended that the annular pressure be kept below the critical value given in this paper. Injecting nitrogen in annulus A and installing rupture disks are both effective methods to improve casing safety. In the case study, the APB decrease percentage is more than 75% when nitrogen is injected in annulus A. However, the nitrogen pressure, the rupture pressure and the installation depth of the rupture disk need to be determined via casing safety evaluation. The effect of optimizing the steel grade and thickness of the tubing and casing is not significant. They can be used as assistance methods when other mitigation methods are adopted.

Keywords: high-temperature well; casing safety; annular pressure buildup; offshore thermal recovery well; pressure management

Citation: Wang, H.; Li, M.; Zhao, Q.; Hao, W.; Zhang, H.; Li, Y.; Huang, P.; Zou, Y. Study on Casing Safety Evaluation in High-Temperature Wells with Annular Pressure Buildup. *Processes* **2023**, *11*, 1915. https://doi.org/10.3390/pr11071915

Academic Editor: Zhiwei Gao

Received: 19 May 2023
Revised: 20 June 2023
Accepted: 20 June 2023
Published: 26 June 2023

1. Introduction

With the increase in drilling depth and the exploitation of offshore oil and gas, the problem of high-temperature well production is becoming more and more prominent. As an unconventional fossil resource produced by high-temperature wells, heavy oil is considered to be a significant source of world oil and gas growth [1]. In Bohai Bay, China, 50.4% of total offshore oil reserves are heavy oil [2]. Most of the offshore heavy oil reservoirs are exploited via the thermal recovery method [3] (mostly is steam or thermal fluid injection). In the later process of steam injection, wellbore temperature reaches the maximum, leading to the riskiest situation [4]. The temperature increase of strings and annulus fluid is caused by heat transfer and it may increase the annular pressure, which is known as the annular pressure buildup (APB) phenomenon [5]. The annuli between the casings are defined as "A", "B" and "C" from the inside out. The wellbore structure is shown in Figure 1. APB is a serious problem in high-temperature well production and has been widespread internationally. It may cause casing failure or destroy wellbore integrity, and further lead to significant safety accidents and economic loss [4,5]. The casing collapsed due to APB in the well Pompano A-31 in the Gulf of Mexico [6]. APB also resulted in a casing failure in the Marlin A-2 well [7,8]. APB in the A annulus can be released through the surface equipment in most onshore and offshore wells. However, the release may not be possible in outer annuli in subsea wells because of the wellhead limits [9]. Therefore, research on casing safety evaluation and APB limit determination has great engineering significance and is favorable for maintaining wellbore integrity and ensuring the efficient development of the heavy oil [10].

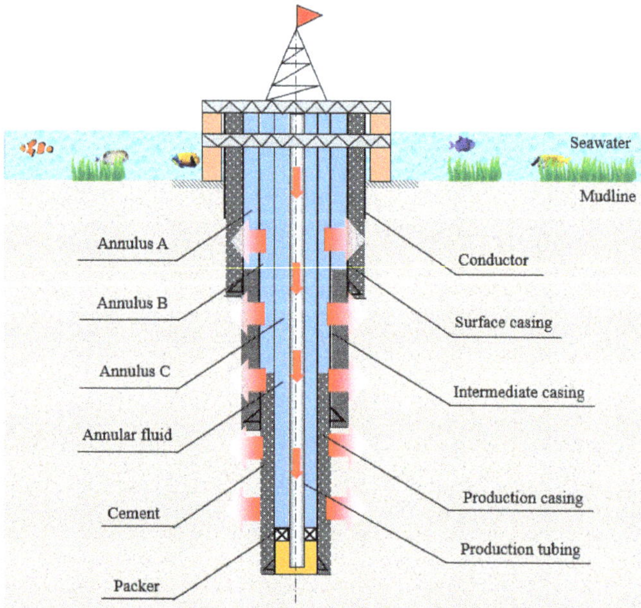

Figure 1. The conventional wellbore structure of offshore wells.

Many scholars have researched APB prediction and casing failure. Considering the coupling effect of pressure and volume, a mathematical model for calculating APB was derived by Oudeman et al. [11,12]. Liu et al. [13] developed an APB prediction model that could be used for offshore wells. Yin et al. [14] further optimized the model by taking the variance of the physical parameters into account. Hasan et al. proposed a semi-steady state temperature prediction method of the wellbore and annulus fluid [15]. A higher production rate will cause higher APB [16]. Wang et al. proposed a novel prediction model to predict the APB in an annulus with gas and liquid. They also conducted physical experiments to further analyze the effect of the dissolved gas and cement on APB [17,18]. Xu et al. analyzed the causes of the casing damage that occurred in Du 84. The effect of several prevention measures was also discussed in their work [19]. Liu et al. analyzed the casing fatigue in thermal recovery wells through numerical simulations. Then four fatigue life models were adopted to analyze the casing fatigue life based on a field case [20]. Gao considered the effect of temperature on casing properties and the triaxial thermal stresses in the casing were calculated [21]. Liang analyzed the casing thermal stress and wellhead growth phenomenon. The cement level was further optimized to reduce the casing thermal stress [22]. Ferreira et al. evaluated the application of vacuum insulated tubing (VIT) and its effect on APB. The optimal installation length and position of the VIT are also given in their paper [23].

Scholars have also researched the mitigation of APB [24], including the following:

- Well structure: make the top of cement (TOC) below the previous casing shoe; full cementing; liner cementing [25];

- String materials: casing covered with compressible material [26]; optimization of casing size and steel grade [27]; vacuum insulated tubing [23];
- Equipment: rupture disk [28]; unidirectional pressure control tool;
- Technological measures: injecting N_2 or compressible liquid into annuli [29]; adopting foam spacer.

However, each of the above methods has its limits and application range. Designing appropriate operating parameters for these methods is an important job. Safety and economy must both be considered. The casing safety evaluation and the APB limit determination are prerequisites and essential for the design of mitigation methods. However, there is a lack of research on tubing and casing safety evaluation under high-temperature conditions. APB limit determination and management are also important topics that need to be studied urgently.

In this work, considering the thermal stress, the safety of the casing under APB is evaluated based on APB prediction. Then a method of APB limit determination is proposed through a case study of an actual offshore well and APB-management charts and the recommended range are provided. Furthermore, three commonly used mitigation methods are analyzed and operation parameters are also recommended in this work. It is hoped that this work could provide help for the engineering design of APB management and mitigation.

2. APB-Prediction Model

According to the equation of state, fluid pressure is the function of volume, temperature and mass. Therefore, the relationship of the pressure change, the temperature change and the volume change of the annular fluid is established to achieve the APB value. When the annulus is sealed, the equation is expressed as follows [17,18]:

$$\int_{T_{ini}}^{T_{fin}} a_{isob} dT - \int_{P_{ini}}^{P_{fin}} k_{isot} dP = \ln\left(\frac{V_{fin}}{V_{ini}}\right) \tag{1}$$

where

$$\begin{cases} V_{fin} = V_{ini} + \Delta V_f \\ \Delta V_f = \Delta V_{isob} + \Delta V_{isot} \\ \Delta V_{ann} = \Delta V_f \end{cases} \tag{2}$$

The isobaric thermal expansion coefficient and isothermal compression coefficient vary with the temperature and pressure of the fluid. Hence, they can be rewritten as:

$$\begin{aligned} a_{isob} &= f(P, T) \\ k_{isot} &= g(P, T) \end{aligned} \tag{3}$$

The function can be fitted with experimental data or calculated with the equation of state [23,28]. In this work, the function is obtained from experimental data fitting [30]. The equations are shown in Equations (4) and (5).

$$\alpha_{isob} = \frac{p_{11} + p_{12}T + p_{13}T^2 + p_{14}T^3 + p_{15}P + p_{16}P^2}{1 + p_{17}T + p_{18}P} \tag{4}$$

$$k_{isot} = \frac{p_{21} + p_{22}T + p_{23}T^2 + p_{24}P}{1 + p_{25}T + p_{26}P + p_{27}P^2} \tag{5}$$

From Equation (2), the volume change of the annular fluid includes two parts. The first part is the volume change caused by isobaric thermal expansion. The second is the volume change caused by isothermal compression. When the annulus is sealed, the volume change of the annular fluid is equal to the annular volume change. The annular volume change can be obtained as follows [29].

(1) Radial displacement of the casing caused by thermal expansion [31]:

$$u_t = (1 + 2\mu)\alpha r \Delta T_r \tag{6}$$

(2) Radial displacement of the casing caused by internal and external pressures [31]:

$$u_p = \frac{1+\mu}{E\left(r_o^2 - r_i^2\right)}\left[-\frac{r_i^2 r_o^2 (P_o - P_i)}{r} + (1-2\mu)\left(P_i r_i^2 - P_o r_o^2\right) r\right] \tag{7}$$

The volume change of the annulus is:

$$\Delta V_{ann} = \pi \int_0^L \left[(r_o + u_t + u_p)^2 - (r_i + u_t + u_p)^2 - \left(r_o^2 - r_i^2\right)\right] dz \tag{8}$$

The fluid mostly consists of liquid and gas. The volume change of the liquid can be obtained from Equation (1) and the volume change of the gas can be obtained from the equation of state as follows [32]:

$$P = \frac{RT}{\tilde{v} - b} - \frac{a_c a}{\tilde{v}^2 + 2b\tilde{v} - b^2} \tag{9}$$

where

$$a = \left[1 + m\left(1 - T_r^{0.5}\right)\right]^2 \quad a_c = 0.457235 R^2 T_c^2 / P_c$$
$$m = 0.374640 + 1.54226\omega - 0.26992\omega^2 \quad b = 0.077796 R T_c / P_c$$

Hence, the annular pressure can be found through the process shown in Figure 2.

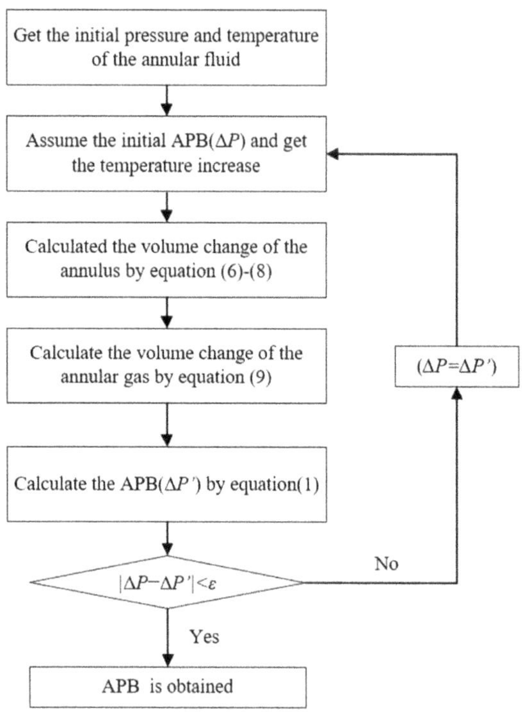

Figure 2. The APB-calculation process.

3. Tubing and Casing Safety Evaluation

3.1. Tubing and Casing Safety Evaluation

To determine the limit of APB, it is necessary to first conduct a safety evaluation of the tubing and casing. Due to the high temperature and the constraint of the wellhead, the thermal stress caused by temperature has a great effect on the casing stress distribution. Hence, the casing stress consists of the thermal stress and the stress caused by internal and external pressure. The casing stress is shown in Figure 3.

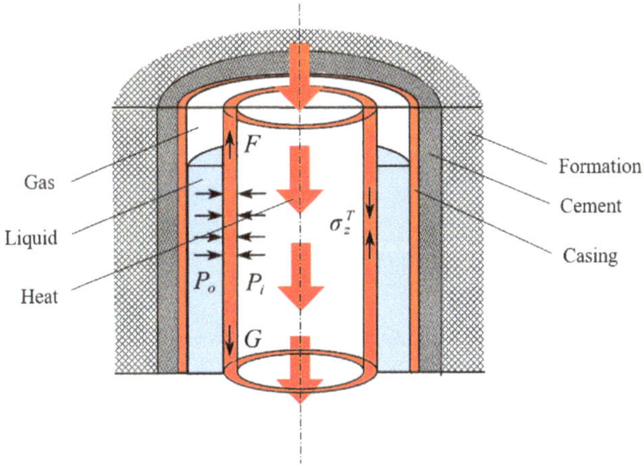

Figure 3. The casing stress.

According to the elastic mechanics, the stress caused by pressure and temperature can be obtained as follows [33]:

$$\begin{cases} \sigma_r = \frac{P_i r_i^2 - P_o r_o^2}{r_o^2 - r_i^2} + \frac{r_i^2 r_o^2 (P_o - P_i)}{r^2 \left(r_o^2 - r_i^2\right)} \\ \sigma_\theta = \frac{P_i r_i^2 - P_o r_o^2}{r_o^2 - r_i^2} - \frac{r_i^2 r_o^2 (P_o - P_i)}{r^2 \left(r_o^2 - r_i^2\right)} \\ \sigma_z = \frac{F - G}{\pi \left(r_o^2 - r_i^2\right)} + \mu(\sigma_\theta + \sigma_r) \end{cases} \tag{10}$$

$$\begin{cases} \sigma_\theta^t = \frac{E\alpha\Delta T_w}{2(1-\mu)} \left(\frac{1 - \ln K_r}{\ln K} - \frac{K_r^2 + 1}{K^2 - 1}\right) \\ \sigma_r^t = \frac{E\alpha\Delta T_w}{2(1-\mu)} \left(-\frac{\ln K_r}{\ln K} + \frac{K_r^2 - 1}{K^2 - 1}\right) \\ \sigma_z^t = \frac{E\alpha\Delta T_w}{2(1-\mu)} \left(\frac{1 - 2\ln K_r}{\ln K} - \frac{2}{K^2 - 1}\right) - aE\Delta T_r \end{cases} \tag{11}$$

The pressure on the casing can be calculated with the following equations.

$$\begin{cases} P_i = \Delta P_i + g \int_0^H \rho_i \cos\theta dz \\ P_o = \Delta P_o + g \int_0^H \rho_o \cos\theta dz \end{cases} \tag{12}$$

Hence, the total casing stress can be written as:

$$\begin{cases} \sigma_r = \sigma_r^p + \sigma_r^T \\ \sigma_\theta = \sigma_\theta^p + \sigma_\theta^T \\ \sigma_z = \sigma_z^p + \sigma_z^T \end{cases} \tag{13}$$

The von Mises criterion is used in this work to conduct the casing stress check and evaluate casing safety [31]. The von Mises stress equation is shown in Equation (14). The safety factor is shown in Equation (15). When the total stress on the casing exceeds the yield strength, casing failure may occur.

$$\sigma_{\text{Mises}} = \frac{\sqrt{2}}{2}\sqrt{(\sigma_r - \sigma_\theta)^2 + (\sigma_\theta - \sigma_z)^2 + (\sigma_z - \sigma_r)^2} \le Y_p \tag{14}$$

$$S = \frac{\sigma_{\text{Mises}}}{Y_p} \tag{15}$$

Once the temperature distribution of the wellbore and the APB in annuli are obtained, the casing stress can be calculated via Equations (10) and (11). Then the safety factor of the casing at different depths can be found with Equation (15).

3.2. APB Limit Determination

The APB limit is the maximum or minimum allowable APB that does not cause casing damage. Therefore, the APB limit can be determined by combining the casing safety-evaluation method and the APB-prediction model. The key is to determine the minimum pressure difference when casing collapse or burst occurs. The process of the determination of the APB limit is shown in Figure 4.

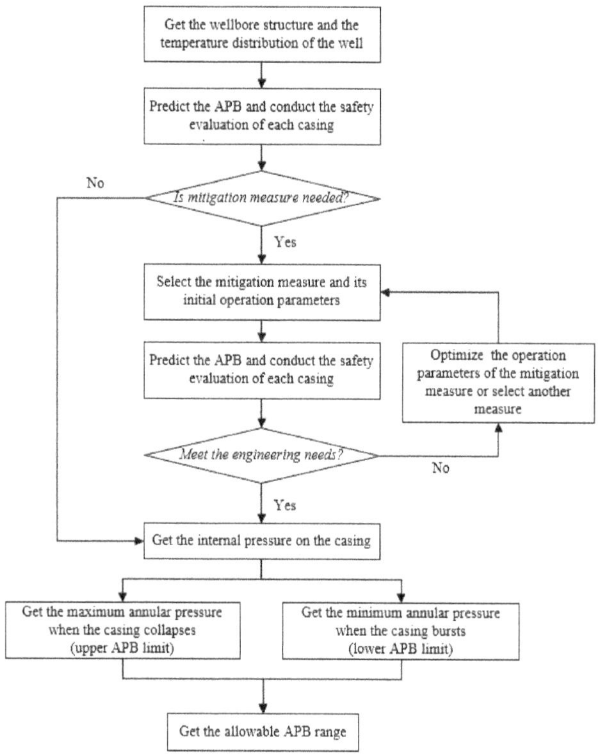

Figure 4. The process of the determination of the APB limit.

Through the process in Figure 4, the APB upper and lower limit values could be determined. Keeping the annular pressure in the allowable range could ensure the safe production of high-temperature wells with APB and avoid unnecessary economical loss.

4. Application and Discussion

4.1. Case Study

The case study is based on the parameters of a field offshore vertical well. Wellbore structure is demonstrated in Figure 1. The mudline depth is 148 m and the well depth is 1850 m. The depth of the packer is 1712 m. To reduce heat loss, vacuum insulated tubing (VIT) is used. The casing program parameters are shown in Table 1. The input data used for calculation are shown in Table 2.

Table 1. Casing program parameters.

Casing Program	Outer Diameter (mm)	Thickness (mm)	Depth (m)	TOC (m)
Conductor	914.4	38.1	218	-
Surface casing	508.0	12.7	867	0
Intermediate casing	339.7	12.3	1263	651
Production casing	244.5	11.9	1850	1050
Production tubing	88.9	6.5	1850	-

Table 2. List of input data for modeling.

Parameters (Units)	Values
Geothermal gradient (°C/m)	0.03
Steam temperature at wellhead (°C)	250
Mudline temperature (°C)	4
Steam injection rate (t/d)	110
Steam injection time (d)	10
Tubing thermal conductivity (W/(m·°C))	0.1
Casing thermal conductivity (W/(m·°C))	40
Elasticity modulus of casing (GPa)	210
Poisson's ratio of tubing and casing	0.3
Poisson's ratio of cement	0.15
Isobaric expansion coefficient of tubing and casing ($°C^{-1}$)	0.000012
Isobaric expansion coefficient of cement ($°C^{-1}$)	0.00001

In this case, the wellbore temperature distribution was calculated with the semi-steady state temperature prediction model [15], which is commonly used in wellbore temperature prediction. The temperature distribution of this well is shown in Figure 5.

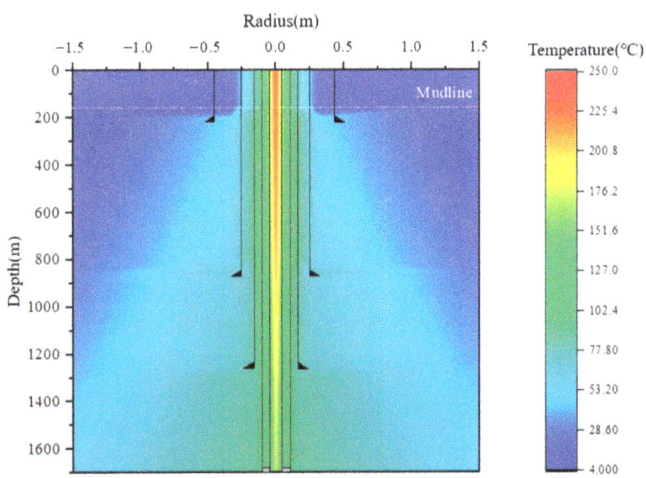

Figure 5. The wellbore temperature distribution.

From Figure 5, the heat transfers from the tubing to the outside and it increases the temperature around the wellbore. This is the essential cause of the APB problem and the thermal stress of the casing. The temperature of the steam is also decreased by heat dissipation when it arrives at the bottom of the well. In this well, the steam temperature decreases from 250 °C to 176 °C, while 29.6% of the heat is lost. Hence, control of the heat dissipation in the thermal recovery well is of great importance. Using high-quality vacuum insulated tubing (VIT) or heat-insulated spacer fluid can not only ensure the efficiency of heavy oil recovery but also reduce the risk of APB.

The APB in three annuli is calculated under the temperature distribution. The casing safety evaluation is conducted and the APB limit is also determined accordingly. The APB results are shown in Table 3.

Table 3. APB calculation results.

Annulus	Average Temperature Increment (°C)	APB (MPa)
A	105.12	140.88
B	77.42	63.23
C	61.65	28.37

The casing safety evaluation results are shown in Figure 6.

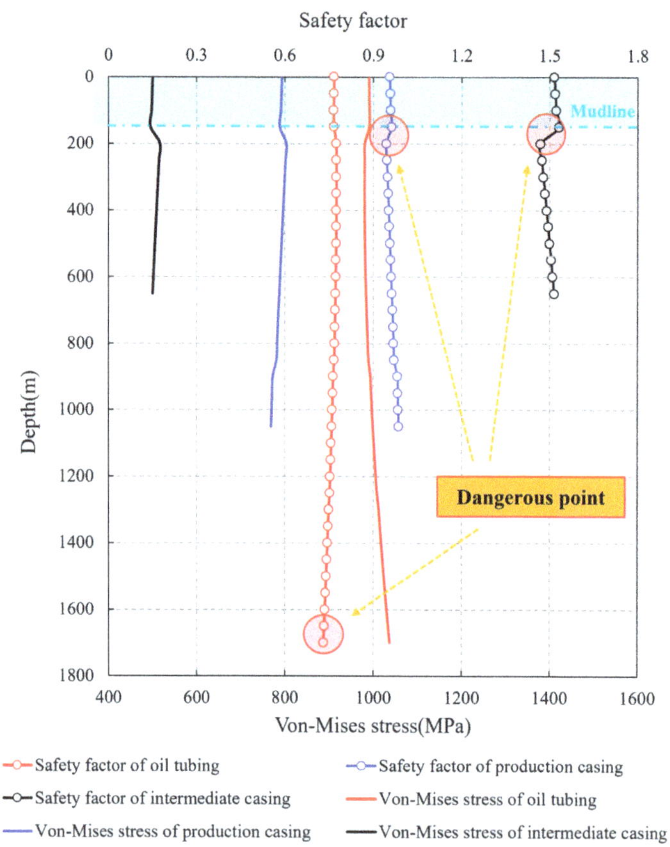

Figure 6. The safety evaluation results of tubing and casing.

Figure 6 presents the casing stress and safety factors. The safety factors of tubing and production casing are less than 1.0, which means damage risk. The dangerous point of the tubing is at the bottom. This is because the external pressure increases with the increasing depth. However, the dangerous point of the production and intermediate casing is around the mudline. This is due to the great temperature difference between the heated fluid and the surrounding environment.

In accordance with the safety-evaluation method of tubing and casing, the APB limits can be obtained as shown in Figure 7.

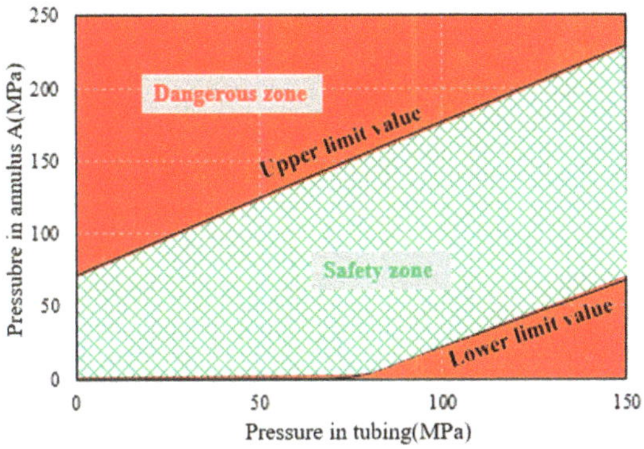

(**a**) pressure in tubing—pressure in annulus A.

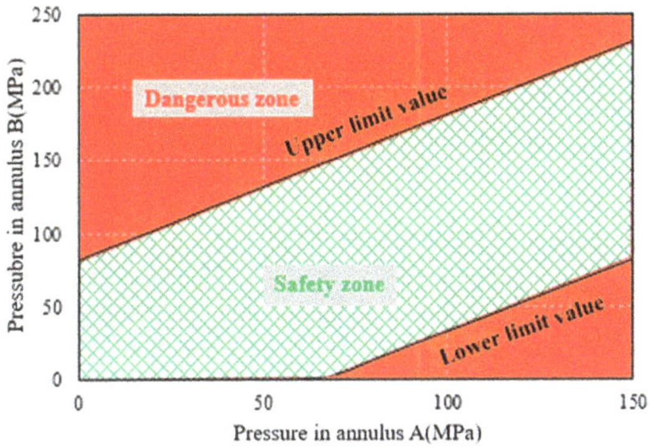

(**b**) pressure in annulus A—pressure in annulus B

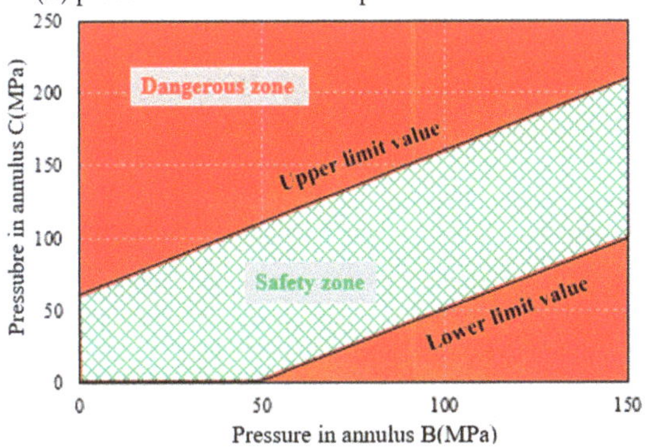

(**c**) pressure in annulus B—pressure in annulus C

Figure 7. The APB limit-management chart.

In Figure 7, the green shaded area is the safety zone; it can be seen that keeping the annular pressure in this range could ensure the safety of the casing and the wellbore integrity. The upper and lower limit value is the maximum and minimum allowable annular pressure. It may bring a great risk of damage to the casing if the annular pressure is in the dangerous zone. Hence, maintaining the pressure in the safety zone is necessary and the pressure beyond the upper limit or below the lower limit value is strictly prohibited in field production.

To provide better guidance for APB management, the recommended optimal range of annular pressure is given in Figure 8 in accordance with the APB limit analysis. The lower limit value is set as 0 MPa, which makes the critical value lower than the upper limit value. This is more appropriate for field pressure management and the casing safety is more fully guaranteed. When the pressure is beyond the critical value, it may bring damage to the casing on one side but this is not inevitable. The more the pressure exceeds the critical value, the greater the probability of casing failure.

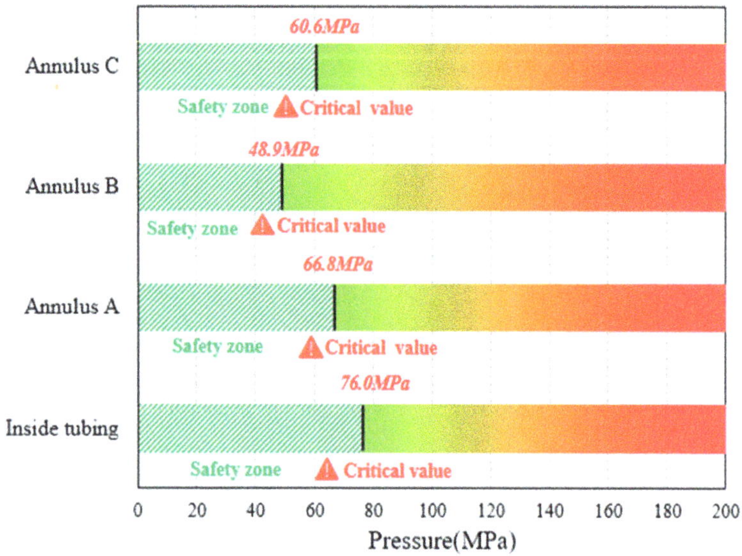

Figure 8. The recommended optimal range of annular pressure.

It is important to note that the critical value of APB in annulus B is lower compared to other annuli. Hence, the management of APB in annulus B is more important considering that pressure release in annulus B is more difficult.

In the field operation of this well, the annular pressure should be kept below the critical value in Figure 8. If the pressure of an annulus exceeds the critical value in some cases, the adjacent annular pressure should be controlled strictly according to the management chart in Figure 7.

4.2. *Analysis of the Mitigation Methods*

4.2.1. Nitrogen or Foam Injection

In some thermal recovery wells, nitrogen is injected into annulus A for thermal insulation and to reduce heat loss [34]. In the well in the case study, the fluid in annulus A is fully replaced by nitrogen with 20 MPa. The temperature distribution when the annulus A is filled with nitrogen is shown in Figure 9.

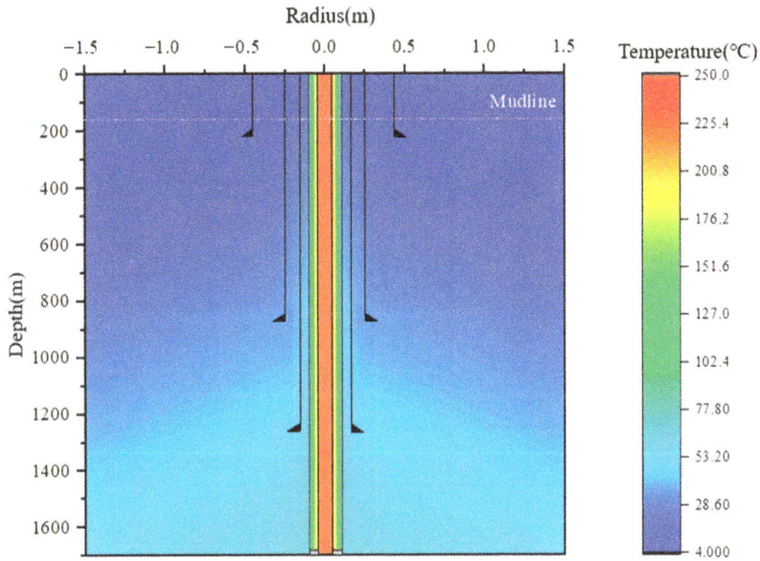

Figure 9. The temperature distribution of the wellbore with nitrogen in annulus A.

The average temperature increment and APB in each annulus are presented in Table 4.

Table 4. APB calculation results when annulus A is filled with nitrogen.

Annulus	Average Temperature Increment (°C)	APB (MPa)	APB Decrease Percentage
A	111.54	32.71	76.78%
B	16.95	5.97	90.56%
C	13.06	2.59	90.87%

When annulus A is filled with nitrogen, the temperature of the steam at the bottom increases from 176 °C to 234 °C. This means 78.38% of the heat loss is avoided. The average temperature increment in annulus A increases from 105.12 °C to 111.54 °C, while that in annulus B and C decreases to 16.95 °C and 13.06 °C. The heat insulation property of nitrogen ensures the high temperature of the steam. Meanwhile, the APB in each annulus undergoes a large decrease and the decrease percentages are all more than 75% because of the good compression property of the nitrogen.

The casing safety evaluation results are shown in Figure 10.

In Figure 10, the safety factors of tubing and casing increase significantly with the nitrogen in annulus A. The dangerous point of production casing changes to the bottom and that of tubing changes to the wellhead. This is because the external pressure on the tubing and the thermal stress of the casing decrease. According to the management chart in Figures 7 and 8, the APB in each annulus is all in the safety zone and the wellbore integrity is guaranteed. It is concluded that injecting nitrogen into annulus A is an effective method to improve casing safety.

To maintain tubing and casing safety, the nitrogen injection pressure is optimized. The optimization results are shown in Figure 11.

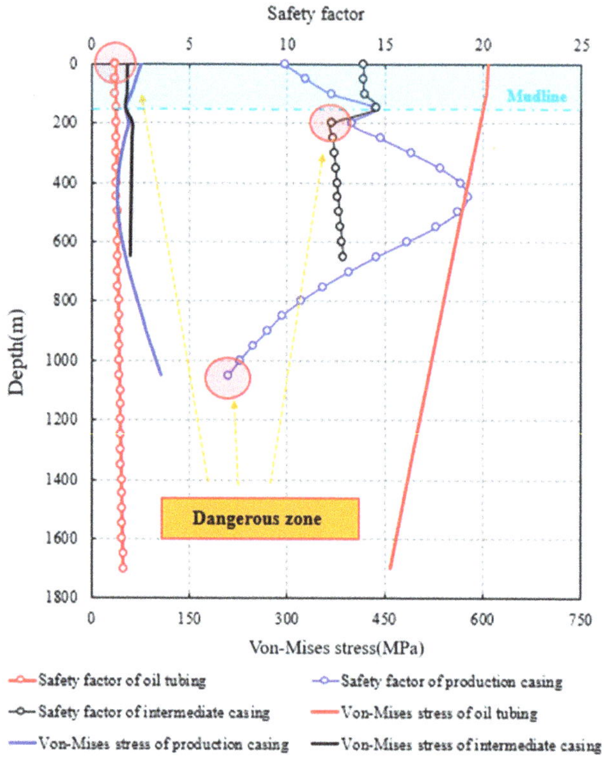

Figure 10. The safety evaluation results of tubing and casing with nitrogen in annulus A.

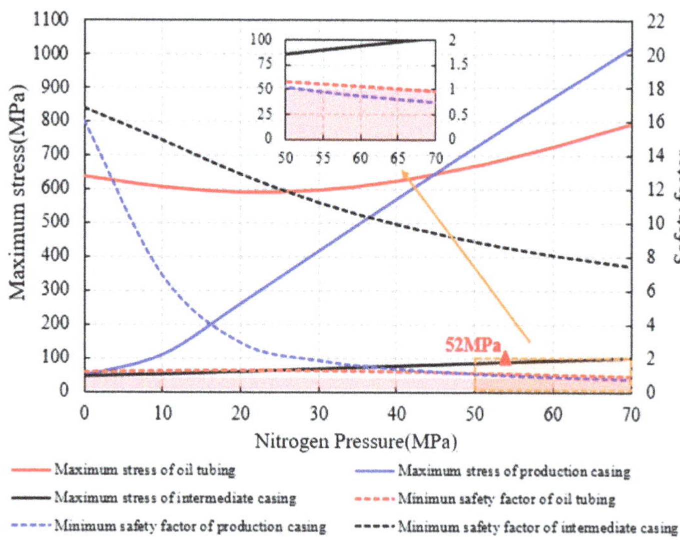

Figure 11. The maximum stress and minimum safety factor of tubing and casing under different nitrogen pressures.

With the increase in nitrogen pressure, the tubing and casing stress increases and the safety factor decreases. Failure may occur on tubing and production casing when the nitrogen pressure is more than 52 MPa and 65 MPa, respectively. Hence, the nitrogen pressure should be kept below 52 MPa and as low as possible.

4.2.2. Selection of the Rupture Disk

The rupture disk is a device that ruptures under a certain pressure difference. It is often installed on the casing string. When it ruptures, the adjacent annuli are connected and fluid mass exchange occurs. The pressure difference on the casing is eliminated and the external and internal pressure become equal. In this work, the selection of rupture pressure of the rupture disk and its maximum installation depth are both analyzed. The rupture pressure of the disk is as follows:

Casing failure mainly involves collapse and burst. To avoid both accidents, the safe pressure in Equation (16) is the minimum pressure difference when the casing collapses and bursts. The minimum allowable pressure difference of the casing is shown in Figure 12.

$$P_{work} \leq P_{rup} \leq P_{safe} \quad (16)$$

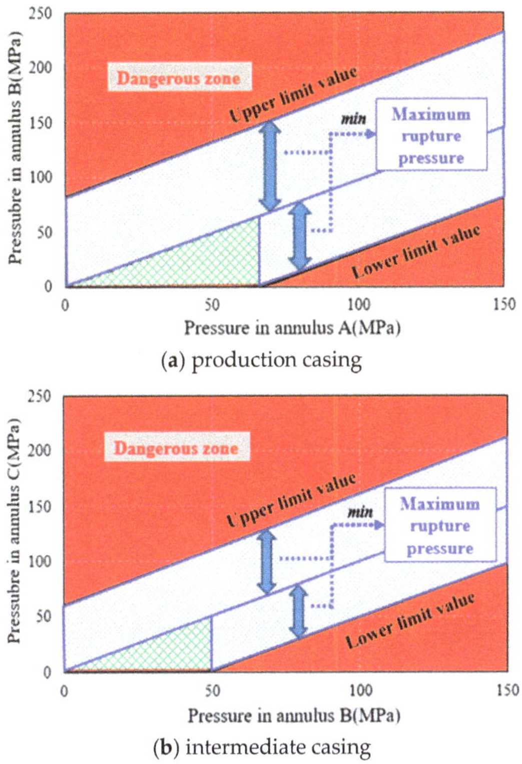

Figure 12. The maximum rupture pressure.

To avoid casing failure, the rupture pressure is set at 80% of the minimum allowable pressure difference of the casing. The maximum rupture pressure of the disk for the different casing is shown in Table 5.

Table 5. Maximum rupture pressure of the disk.

Casing	Maximum Allowable Pressure Difference of the Casing (MPa)	Maximum Rupture Pressure of the Disk (MPa)
Production casing	66.8	53.4
Intermediate casing	48.9	39.1

In the production of the well in the case study, the maximum work pressure is 20 MPa and the design safety coefficient of the minimum rupture pressure is 1.2. The rupture pressure of the disk should be 24 MPa–53.4 MPa for the production casing and 24 MPa-39.1 MPa for the intermediate casing. For this well, a rupture disk with 27.6 MPa (4000 psi) or 34.5 MPa (5000 psi) rupture pressure is appropriate.

The maximum installation depth of the rupture disk can be calculated as follows:

$$H_{rup} = \frac{P_{rup} - C_s P_{work}}{|\Delta\rho| \cdot g} \tag{17}$$

The maximum installation depth of the rupture disk is shown in Figure 13.

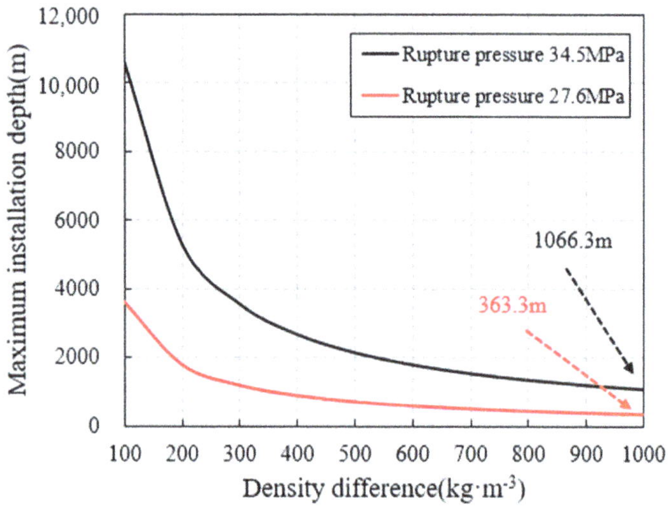

Figure 13. The maximum installation depth of rupture disk.

With an increase in the density difference between the fluid in adjacent annuli, the maximum installation depth of the rupture disk decreases. When the density of the annular fluid is 1000 kg/m^3 and one of the annuli is fully empty, the maximum installation depth is only 363.3 m for 27.6 MPa rupture disk and 1066.3 m for 34.5 MPa rupture disk. Hence, the 34.5 MPa rupture disk is recommended in this well.

To conclude, the rupture disk should be selected according to the casing safety evaluation and Equation (16). It is also necessary to install the rupture disk above the maximum installation depth and as close to the wellhead as possible.

4.2.3. Optimization of the Casing Grade and Thickness

Adopting appropriate casing is also a convenient and economical method to protect wellbore integrity. In this work, a safety evaluation of tubing and casing with different steel grades and thicknesses is conducted.

The steel grade determines the yield strength of the tubing and casing and accordingly affects the wellbore integrity. Thickness affects the stress distribution of the casing and its safety. Based on the well parameters in the case study, the safety evaluation results of the tubing and casing of different steel grades and thicknesses are shown in Figure 14.

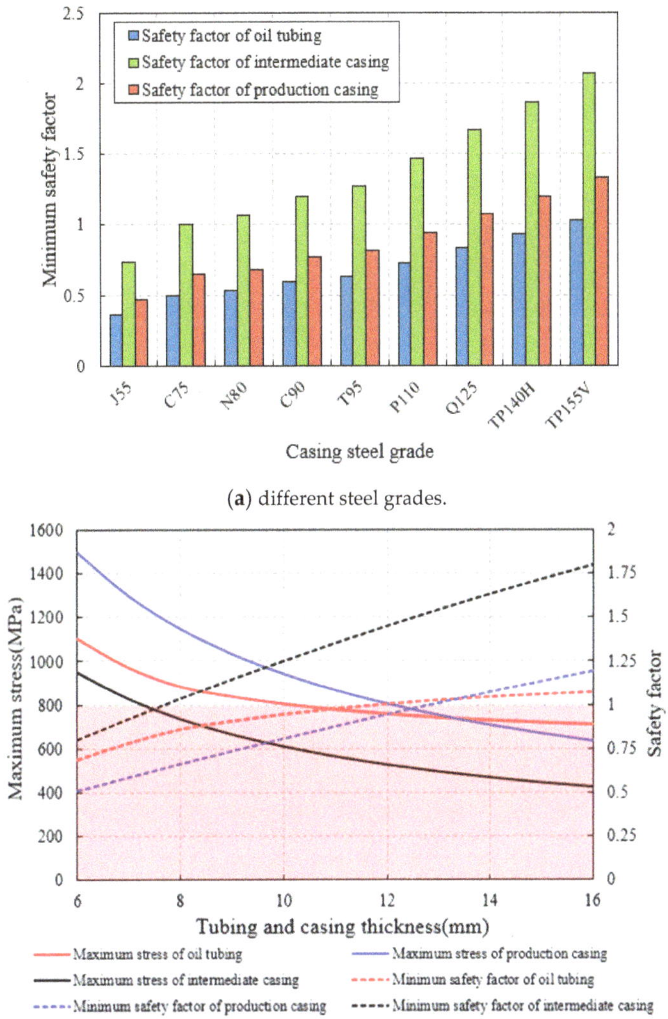

(**a**) different steel grades.

(**b**) different thicknesses

Figure 14. The safety evaluation results of the tubing and casing of different steel grades and thicknesses.

As shown in Figure 14a, the casing steel grades with high yield strength have higher safety factors. The safety factors of tubing and casing are all greater than 1 if the grade TP155V is adopted. However, the safety factor of oil tubing is only 1.03, which means a very small margin of safety is left. Hence, just optimizing the grade of tubing and casing is not enough to ensure wellbore safety. It is necessary to adopt other mitigation methods.

From Figure 14b, the safety factor increases with the increase of the thickness. The minimum thickness that could ensure the tubing safety is 12 mm. For production and intermediate casing, it is 13 mm and 8 mm. This is because the pressure on the intermediate casing is less than that on production casing and oil tubing, which means a thinner thickness is enough to bear the load. Meanwhile, with the increase in the thickness, the increase in the safety factor becomes slower. This means the effect of increasing thickness on improving tubing and casing safety decreases gradually. Hence, the thickness should be more than 13 mm if other mitigation methods are not implemented.

To conclude, optimizing the steel grade and thickness of the tubing and casing has a weaker effect compared to other APB-mitigation methods. Selecting a thicker casing with high steel grade could contribute to ensuring the safety of the wellbore. The priority of the selection of these two parameters depends on the economic cost.

5. Conclusions and Suggestions

(1) Based on the APB-prediction model proposed, the casing safety evaluation and APB limit determination methods of the high-temperature wells are presented in this work. Research shows that the APB phenomena and the thermal stress caused by high temperature affect the stress distribution of the casing and may bring great danger to the wellbore integrity.
(2) The establishment method of the APB-management chart and the recommended optimal range are given in the case study. Maintaining the annular pressure in the safety zone is necessary in field production. The annular pressure should be kept below the critical value recommended in this work. If the pressure in an annulus exceeds the critical value, the adjacent annular pressure should be controlled strictly according to the APB-management chart.
(3) Nitrogen injection in annulus A is an effective method to improve casing safety. The heat insulation and compression properties of nitrogen ensure the high temperature of the steam and reduce the APB in each annulus. The APB decrease percentage is more than 75% in the case study. With the increase in the nitrogen pressure, the safety factors of the tubing and casing decrease. The nitrogen pressure should be controlled below the maximum allowable pressure obtained from casing safety evaluation.
(4) When the rupture disk is installed on the casing, its rupture pressure should be between the maximum operating pressure and the minimum casing safety pressure, and the safety margin is recommended because of the pressure surge. Its maximum installation depth also needs to be determined according to the density of the annular fluid. In the case study, the maximum installation depth of 27.6 MPa rupture disk is only 363.3 m, so the 34.5 MPa rupture disk is recommended.
(5) The effect of optimizing the steel grade and thickness of the tubing and casing is not significant. They can be used as assistance methods when other mitigation methods are adopted. Selecting a thicker casing with high steel grade could contribute to ensuring the safety of the wellbore. The priority of the selection of these two parameters depends on the economic cost.

Author Contributions: All the authors conceived and designed the study. Conceptualization, H.W.; Writing—original draft, H.W. and M.L.; Supervision, Q.Z. and H.Z.; Funding acquisition, W.H. and Y.L.; Writing—review & editing, P.H. and Y.Z. All authors have read and agreed to the published version of the manuscript.

Funding: This research was funded by CNPC research project grant number [2022ZG06] and the subject grant number [2021DJ4105] under the CNPC research project grant number [2021DJ4100].

Data Availability Statement: Not applicable.

Acknowledgments: This work is supported by CNPC research project "Research on Key Technologies and Equipment for Ultra-Deep Oil and Gas Resource Drilling over 10,000 Meters" (No. 2022ZG06) and the subject "Research on Cementing Seal Integrity technology in Complex Ultra-Deep Wells" (No. 2021DJ4105) under the CNPC research project "Research on optimal and fast drilling technology for deep and ultra-deep Wells" (No. 2021DJ4100).

Conflicts of Interest: The authors declare no conflict of interest.

Nomenclature

T_{ini}	initial annular temperature (K)
T_{fin}	final annular temperature (K)
α_{isob}	isobaric thermal expansion coefficient of the annular fluid (1/K)
T	annular fluid temperature (K)
P_{ini}	initial annular pressure (MPa)
P_{fin}	final annular pressure (MPa)
K_{isot}	isothermal compressibility of the annular fluid (1/MPa)
P	annular fluid pressure (MPa)
V_{fin}	final volume of the annular fluid (m^3)
V_{ini}	initial volume of the annular fluid (m^3)
V_{fin}	volume change of the annular fluid (m^3)
ΔV_{isob}	volume change of the annular fluid caused by isobaric thermal expansion (m^3)
ΔV_{isot}	volume change of the annular fluid caused by isothermal compression (m^3)
ΔV_{ann}	volume change of the annulus (m^3)
p_{11}	fitting coefficient, $p_{11} = -2.4475 \times 10^{-5}$
p_{12}	fitting coefficient, $p_{12} = 8.3417 \times 10^{-6}$
p_{13}	fitting coefficient, $p_{13} = -3.9802 \times 10^{-8}$
p_{14}	fitting coefficient, $p_{14} = 5.4896 \times 10^{-11}$
p_{15}	fitting coefficient, $p_{15} = 1.1241 \times 10^{-5}$
p_{16}	fitting coefficient, $p_{16} = -2.4564 \times 10^{-8}$
p_{17}	fitting coefficient, $p_{17} = -3.1943 \times 10^{-3}$
p_{18}	fitting coefficient, $p_{18} = 9.1821 \times 10^{-3}$
p_{21}	fitting coefficient, $p_{21} = 6.1393 \times 10^{-4}$
p_{22}	fitting coefficient, $p_{22} = -4.6437 \times 10^{-6}$
p_{23}	fitting coefficient, $p_{23} = 1.4114 \times 10^{-8}$
p_{24}	fitting coefficient, $p_{24} = 3.1346 \times 10^{-6}$
p_{25}	fitting coefficient, $p_{25} = -2.6949 \times 10^{-3}$
p_{26}	fitting coefficient, $p_{26} = -3.1586 \times 10^{-3}$
p_{27}	fitting coefficient, $p_{27} = 1.9188 \times 10^{-4}$
u_t	casing deformation caused by thermal expansion (m)
μ	Poisson's ratio of the casing
α	linear expansion coefficient of the casing (1/K)
r	radius of calculation position (m)
ΔT_r	temperature change at the calculation position (°C)
u_p	casing deformation caused by internal and external pressure (m)
E	elastic modulus of the casing (MPa)
r_i	inner radius of the casing (m)
r_o	outer radius of the casing (m)
P_i	inner pressure of the casing (MPa)
P_o	external pressure of the casing (MPa)
L	length of the annulus (m)
z	well depth (m)
R	gas constant ($J \cdot mol^{-1} \cdot K^{-1}$), $R = 8.314\ J \cdot mol^{-1} \cdot K^{-1}$
$\tilde{v}$	gas molar volume (m^3)
P_r	reduced pressure, $P_r = P/P_c$
T_r	reduced temperature, $T_r = T/T_c$
P_c	critical pressure (MPa), $P_c = 3.394$ MPa for nitrogen
T_c	critical temperature (K), $T_c = 126.15$ K for nitrogen
ω	Pitzer's acentric factor, $\omega = 0.045$ for nitrogen
σ_r^P	radial stress caused by pressure (MPa)
σ_θ^P	circumferential stress caused by pressure (MPa)
σ_z^P	axial stress caused by pressure (MPa)
F	hanging force (10^{-6} N)
G	gravitational force (10^{-6} N)
σ_r^T	radial thermal stress (MPa)

σ_θ^T	circumferential thermal stress (MPa)
σ_z^T	axial thermal stress (MPa)
K	the ratio of the outer radius to the inner radius
K_r	the ratio of the outer radius to the radius of the calculation position
ΔT_w	the temperature difference between the inside and outside walls of the casing (°C)
ΔP_i	annular pressure buildup in the inner annulus (MPa)
ΔP_o	annular pressure buildup in the outer annulus (MPa)
ρ_i	density of inner annular fluid (kg/m^3)
ρ_o	density of outer annular fluid (kg/m^3)
θ	wellbore inclination angle (kg/m^3)
σ_r	total radial stress (MPa)
σ_θ	total circumferential stress (MPa)
σ_z	total axial stress (MPa)
σ_{Mises}	von-Mises stress (MPa)
Y_p	yield strength of the casing (MPa)
S	safety factor
P_{work}	operation pressure in the production (MPa)
P_{rup}	rupture pressure (MPa)
P_{safe}	the minimum casing safety pressure (MPa)
C_s	design safety coefficient
$\Delta\rho$	density difference between the fluid in the inner and outer annuli (kg/m^3)
g	gravitational acceleration (m/s^2)

References

1. Fan, H.J.; Fan, T.E.; Deng, J.H.; Zhang, L.J.; Zheng, W.; Chen, L.F.; Ge, Z.Z.; Xie, H.J.; Liang, X. The influence of interlayer on the development of steam chamber in steam stimulation during heavy oil recovery. *Processes* **2023**, *11*, 1742. [CrossRef]
2. Pan, G.; Chen, J.; Zhang, C.; Liu, D.; Wu, J.; Li, H.; Fang, Z.; Qu, J.; Zhang, J. Combined technology of weak gel flooding assisting thermal huff and puff enhances oil recovery for offshore heavy oil field. In Proceedings of the SPE Annual Technical Conference and Exhibition, Dubai, United Arab Emirates, 26–28 September 2016. SPE-181626-MS. [CrossRef]
3. Ji, Y.M.; Li, B.L.; Han, Z.Y.; Wang, J.; Li, Z.M.; Li, B.F. Study on flow characteristics of flue gas and steam co-injection for heavy oil recovery. *Processes* **2023**, *11*, 1406. [CrossRef]
4. Zhang, Z.; Wang, H. A calculation method for thermal expansion mechanics of sealed annulus between multiple packers and its application. *Nat. Gas Ind.* **2016**, *36*, 65–72.
5. Yang, J.; Tang, H.; Liu, Z.; Yang, L.; Huang, X.; Yan, D.; Tian, R. Prediction model of casing annulus pressure for deepwater well drilling and completion operation. *Pet. Explor. Dev.* **2013**, *40*, 661–664. [CrossRef]
6. Pattillo, P.D.; Cocales, B.W.; Morey, S.C. Analysis of an annular pressure buildup failure during drill ahead. *SPE Drill Compl.* **2006**, *21*, 242–247. [CrossRef]
7. Bradford, D.W.; Fritchie, D.G.; Gibson, D.H.; Gosch, S.W.; Pattillo, P.D.; Sharp, J.W.; Taylor, C.E. Marlin failure analysis and redesign; part 1, description of failure. In Proceedings of the IADC/SPE Drilling Conference, Dallas, TX, USA, 26–28 February 2002; SPE-88814-PA. [CrossRef]
8. Ellis, R.C.; Fritchie, D.G.; Gibson, D.H.; Gosch, S.W.; Pattillo, P.D. Marlin failure analysis and redesign; part 2, redesign. In Proceedings of the IADC/SPE Drilling Conference, Dallas, TX, USA, 26–28 February 2002; SPE-88838-PA. [CrossRef]
9. Williamson, R.; Sanders, W.; Jakabosky, T.; Serio, M.; Griffith, J.E. Control of contained-annulus fluid pressure buildup. In Proceedings of the SPE/IADC Drilling Conference, Amsterdam, The Netherlands, 19–21 February 2003. SPE-79875-MS. [CrossRef]
10. Wang, X.Y.; Jiang, T.X.; Zhang, Y.Y.; Zhou, J.; Xiao, H.C.; Li, W.D. A three-dimensional analytical solution of stress field in casing-cement-stratum system considering initial stress state. *Processes* **2023**, *11*, 1164. [CrossRef]
11. Oudeman, P.; Bacarreza, L.J. Field trial results of annular pressure behavior in a high-pressure/high-temperature well. *SPE Drill Compl.* **1995**, *10*, 84–88. [CrossRef]
12. Oudeman, P.; Kerem, M. Transient behavior of annular pressure build-up in HP/HT wells. *SPE Drill. Complet.* **2006**, *21*, 234–241. [CrossRef]
13. Liu, B.; Yang, J.; Zhou, B.; Yan, D.; Tian, R.; Liu, Z.; Luo, J.; Huang, X. Study of casing annulus pressure for deepwater drilling and completions. In Proceedings of the SPE Deepwater Drilling and Completions Conference, Galveston, TX, USA, 10–11 September 2014. SPE-170318-MS. [CrossRef]
14. Yin, F.; Gao, D. Improved calculation of multiple annuli pressure buildup in subsea HPHT wells. In Proceedings of the IADC/SPE Asia Pacific Drilling Technology Conference, Bangkok, Thailand, 25–27 August 2014. [CrossRef]
15. Hasan, A.R.; Kabir, C.S. Wellbore heat-transfer modeling and applications. *J. Petrol. Sci. Eng.* **2012**, *86*, 127–136. [CrossRef]
16. Hasan, A.R.; Izgec, B.; Kabir, C.S. Sustaining production by managing annular pressure buildup. *SPE Drill. Complet.* **2010**, *25*, 195–203. [CrossRef]

17. Wang, H.; Zhang, H.; Li, J.; Chen, A.; Liu, J.; Sun, T.; Lin, C. Study on annular pressure buildup in offshore heavy oil thermal recovery wells considering dissolved gas contained in annuli. *Energies* **2021**, *14*, 3213. [CrossRef]
18. Wang, H.; Zhang, H.; Li, J.; Sun, T. Study on annular pressure buildup phenomenon in subsea wells considering the effect of cement. *Energy Sci. Eng.* **2021**, *10*, 81–95. [CrossRef]
19. Xu, J.; Liu, H. Casing damage prevention and control technology of super heavy oil thermal production well. In Proceedings of the 2nd 2016 International Conference on Sustainable Development, Xi'an, China, 2–4 December 2016.
20. Liu, Z.; Samuel, R.; Gonzales, A.; Kang, Y. Casing integrity: Strain-based fatigue life estimation coupled with numerical thermal-flow simulation and multistring stress analysis. In Proceedings of the SPE Thermal Well Integrity and Design Symposium, Banff, AB, Canada, 23–25 November 2015. SPE-178448-MS. [CrossRef]
21. Gao, B. Casing stress analysis with effects of temperature on material properties. In Proceedings of the Fifth International Conference on Nonlinear Mechanics, Shanghai, China, 11–14 June 2007.
22. Liang, Q.J. Casing thermal stress and wellhead growth behavior analysis. In Proceedings of the SPE Asia Pacific Oil and Gas Conference and Exhibition, Perth, Australia, 22–24 October 2012. SPE-157977-MS. [CrossRef]
23. Ferreira, M.V.; Santos, A.R.; Vanzan, V. Thermally insulated tubing application to prevent annular pressure buildup in Brazil offshore fields. In Proceedings of the SPE Deepwater Drilling and Completions Conference, Galveston, TX, USA, 20–21 June 2012; SPE-151044-MS. [CrossRef]
24. Zhang, B.; Guan, Z.; Lu, N.; Hasan, A.R.; Wang, Q.; Xu, B. Trapped annular pressure caused by thermal expansion in oil and gas wells: A review of prediction approaches, risk assessment and mitigation strategies. *J. Petrol. Sci. Eng.* **2019**, *172*, 70–82. [CrossRef]
25. Sathuvalli, U.B.; Payne, M.L.; Pattillo, P.D.; Rahman, S.; Suryanarayana, P.V. Development of a screening system to identify deepwater wells at risk for annular pressure build-up. In Proceedings of the SPE/IADC Drilling Conference, Amsterdam, The Netherlands, 23–25 February 2005; SPE-92594-MS. [CrossRef]
26. Sathuvalli, U.B.; Pilko, R.M.; Gonzalez, R.A.; Pai, R.M.; Sachdeva, P.; Suryanarayana, P.V. Design and performance of annular pressure build-up APB mitigation techniques. In Proceedings of the IADC/SPE Drilling Conference and Exhibition, Fort Worth, TX, USA, 1–3 March 2016; SPE-178886-MS. [CrossRef]
27. Zeng, B.; Zhou, X.J.; Cao, J.; Zhou, F.; Wang, Y.; Wang, Y.Z.; Song, Y.; Hu, J.J.; Du, Y.R. A casing deformation prediction model considering the properties of cement. *Processes* **2023**, *11*, 695. [CrossRef]
28. Liu, Z.; Samuel, R.; Gonzales, A.; Kang, Y. Modeling and simulation of annular pressure buildup APB management using syntactic foam in HP/HT deepwater wells. In Proceedings of the SPE Deepwater Drilling and Completions Conference, Galveston, TX, USA, 14–15 September 2016; SPE-180308-MS. [CrossRef]
29. Zhang, B.; Guan, Z.; Wang, Q.; Xuan, L.; Liu, Y.; Sheng, Y. Appropriate completion to prevent potential damage of annular pressure buildup in deepwater wells. In Proceedings of the IADC/SPE Asia Pacific Drilling Technology Conference, Singapore, 22–24 August 2016; SPE-180542-MS. [CrossRef]
30. Liu, G.; Ma, L.; Liu, J. *Chemical and Physical Properties Manual*; Chemical Industry Press: Beijing, China, 2002.
31. Yang, G.T. *Introduction to Elasticity and Plasticity*; Tsinghua University Press: Beijing, China, 2004.
32. Danesh, A. *PVT and Phase Behaviour of Petroleum Reservoir Fluids*; Elsevier Science B.V.: Amsterdam, The Netherlands, 1998.
33. Ping, X.; Guo, T.; Li, A. *Thermal Stress and Thermal Fatigue*; National Defense Industry Press: Beijing, China, 1984.
34. Sun, T.; Zhang, X.; Liu, S.; Cao, Y.; Xie, R. Annular pressure buildup calculation when annulus contains gas. *Chem. Tech. Fuels Oils* **2018**, *54*, 484–492. [CrossRef]

processes

Article

NMR-Based Analysis of Fluid Occurrence Space and Imbibition Oil Recovery in Gulong Shale

Fei Xu, Hanqiao Jiang *, Ming Liu, Shuai Jiang, Yong Wang and Junjian Li

State Key Laboratory of Petroleum Resources and Prospecting, China University of Petroleum (Beijing), Beijing 102249, China; 2020310133@student.cup.edu.cn (F.X.); 2021210279@student.cup.edu.cn (M.L.); shuaijiang614@163.com (S.J.); wycupb@163.com (Y.W.); lijunjian@foxmail.com (J.L.)
* Correspondence: jhqlf@163.com

Abstract: The Gulong shale oil reservoir is situated in freshwater to slightly saline lacustrine basins mainly consisting of a pure shale geological structure, which is quite different from other shale reservoirs around the world. Currently, the development of Gulong shale oil mainly relies on hydraulic fracturing, while the subsequent shut-in period for imbibition has been proven to be an effective method for enhancing shale oil recovery. To clarify the characteristics of the fluid occurrence space and the variation in the fluid occurrence during saltwater imbibition in Gulong shale, this paper carried out porosity and permeability tests on Gulong shale cores and analyzed the fluid occurrence space characteristics and imbibition oil recovery based on nuclear magnetic resonance (NMR). In the porosity and permeability tests, T_2 distributions were used to correct the porosity measured by the saturation method to obtain the NMR porosity. Combined with the identification of fractures in shale cores using micro-CT and the analysis of porosity and permeability parameters, it was found that the permeability of the shale cores was related to the development of fractures in the shale cores. Through the testing and analysis of T_1-T_2 maps of the shale cores before and after saturation with oil, it was found that the shale mainly contained heavy oil, light oil, and clay-bound water, and they were distributed in different regions in the T_1-T_2 maps. Finally, the T_1-T_2 maps of the shale cores at different imbibition stages were analyzed, and it was found that saltwater mainly entered the minuscule inorganic pores of clay minerals during the imbibition process and squeezed the larger-sized inorganic pores containing light oil through the hydration expansion effect, thus expelling the light oil from the shale core and achieving the purpose of enhanced oil recovery.

Keywords: nuclear magnetic resonance; shale oil; occurrence space characteristics; imbibition

Citation: Xu, F.; Jiang, H.; Liu, M.; Jiang, S.; Wang, Y.; Li, J. NMR-Based Analysis of Fluid Occurrence Space and Imbibition Oil Recovery in Gulong Shale. *Processes* **2023**, *11*, 1678. https://doi.org/10.3390/pr11061678

Academic Editor: Qingbang Meng

Received: 29 April 2023
Revised: 25 May 2023
Accepted: 29 May 2023
Published: 31 May 2023

1. Introduction

With the increasingly growing demand for oil and gas resources worldwide, unconventional oil and gas resources are becoming increasingly important in the global energy supply against the backdrop of a sustained consumption of oil and gas resources and a a gradual decrease in newly discovered conventional oil and gas reserves [1]. In recent years, with the sustained exploration of the Songliao Basin by the Daqing Oilfield, the huge development potential of Gulong shale oil has been gradually tapped [2,3]. Gulong shale oil is located in the Songliao Basin in Daqing, China, and is situated in sedimentary deposits of freshwater to slightly brackish water lake basins. Its main body is a pure shale geological structure [4–6]. Gulong shale exhibits significant differences from other shale reservoirs around the world in terms of its rock composition, physical properties, oil content, and shale oil mobility [7,8]. The reservoir space of Gulong shale is mainly composed of matrix pores and bedding fractures, with the main types of pores being organic matter pores, dissolution pores, and interlayer pores of clay minerals. The nanoscale pore–throat system controlled by the horizontal bedding greatly improves the physical properties of the reservoir. Under the confining pressure conditions, the horizontal permeability ranges from 0.011 to 1.620 × 10^{-3} μm^2, with an average of 0.580 × 10^{-3} μm^2. Gulong shale is mainly composed of clay felsic shales, with an average clay mineral content of 35.6%. The predominant clay mineral in Gulong shale is illite accompanied by a minor amount of chlorite [9–11].

To evaluate the oil-bearing characteristics of Gulong shale, previous studies have conducted NMR experiments. NMR is a non-destructive testing method that can accurately characterize the pore size distribution characteristics within the entire pore scale of a core through T_2 spectra. Shale porosity can be measured quickly and simply by weighing or NMR testing [12,13]. NMR porosity measurement is conducted to measure the T_2 signal amplitude of a shale core saturated with a single-phase fluid, compare it with a standard rock sample, and finally calculate the NMR porosity. However, it should be noted that both the weighing method and NMR method require oil washing operations on the core, which will undoubtedly cause damage to the organic pores of the shale. Otherwise, pseudo-solid protons (bitumen, kerogen, and structural water) in shale cores also have NMR T_2 signals and can also affect porosity measurements [14]. Many scholars have carried out research on NMR porosity correction [15,16], and Kausik et al. [17] found that the T_2 signal of pseudo-solid protons is very short and usually less than 0.1 ms, which inspired us to measure signals with T_2 > 0.1 ms to avoid the influence of pseudo-solid protons. T_2-D maps or T_1-T_2 maps can distinguish the different types of fluids in a core [18–21]. Habina et al. [22] analyzed the T_1-T_2 maps and T_2 distributions of water and kerogen in different clay minerals and shale cores and found that hydroxyl groups from crystallographic lattices in clay minerals also showed up on the two spectra, which meant that the total NMR porosity was larger than the actual porosity due to the T_2 signal of the hydroxyl group from the crystallographic lattice. Liu et al. [23] used T_1-T_2 maps to analyze the signals of organic matter, water, light hydrocarbons, and hydroxyl compounds in Gulong shale, which clarified the oil saturation and movable oil saturation of the shale. However, the preservation process of the core before testing may inevitably result in the volatilization of light hydrocarbons, which has a certain impact on the results. Pang et al. [24] used T_2 distributions and T_1-T_2 maps to analyze the pore size distribution and fluid composition in Gulong shale. However, their focus was mainly on the impact of fractures on the oil saturation and fluid mobility, and their study did not provide a detailed classification of other fluids. In addition, Wei [25], Yan [26], and Li [27] have applied NMR technology to study the oil saturation, the sensitivity of the pore structure to external fluids, and other characteristics of Gulong shale, respectively, achieving new insights into Gulong shale and demonstrating the feasibility and effectiveness of NMR technology in shale analysis.

Currently, the main method of production for Gulong shale oil is hydraulic fracturing, which yields light, high-quality crude oil with a low density and viscosity [28]. During the shut-in period of the reservoir, a significant improvement in shale oil production could be achieved through the mechanism of imbibition [29]. There is currently limited research on the imbibition of Gulong shale. Wu et al. [30] studied the imbibition and enhanced oil recovery effects of an emulsion system in Gulong shale, and the washing oil efficiency of imbibition could reach 54.6%. Liu et al. [31] conducted a spontaneous imbibition experiment on Gulong shale cores and found that the oil recovery tended to stabilize after 100 h. The above studies indicate that imbibition is an important method of improving the recovery of shale oil. However, the oil–water displacement process during the imbibition of Gulong shale is not clear yet, and the variation in the fluid distribution in different pore sizes of shale under imbibition requires further study.

To analyze the occurrence space characteristics of different fluids, especially the oil phase, and the variation in the fluid distribution in cores during the imbibition process, this paper intended to conduct relevant research using NMR technology. First, conventional core experiments were carried out to analyze the porosity and permeability characteristics and fracture development of Gulong shale core samples. To avoid damaging the organic pores of the shale by washing oil, dry rock samples that had not been treated in any way were directly saturated with kerosene, and the porosity measured by the saturation method was corrected to obtain the NMR porosity based on the T_2 distribution values of the cores before and after oil saturation. Next, by measuring and analyzing the T_1-T_2 maps of the shale cores before and after oil saturation, the occurrence space characteristics of the fluids in the shale, especially the oil phase, were studied. Finally, a saltwater imbibition experiment was performed on the shale cores, and the variation in the fluid distribution during the imbibition process was studied and analyzed by T_1-T_2 maps at different imbibition times.

2. Experiment Apparatus and Materials

2.1. Experiment Apparatus

In this experiment, the gas permeability of Gulong shale cores was measured using PDP-200 pulse decay permeameters which was from Core Lab in Houston, TEX, USA. The permeability measurement range of the apparatus was 10^{-5}–10 mD, and the maximum confining pressure of the core holder could reach 70 MPa, which could simulate the overburden pressure of most oil and gas reservoirs and meet the requirements of this experiment. The dry shale samples form Gulong shale were saturated with kerosene using a conventional saturation device. By comparing the difference in the weight and T_2 spectrum signal value of the cores before and after saturation with oil, the accurate porosity of the shale was calculated. A SPEC-023 nuclear magnetic resonance flow experiment analyzer from SPEC Co. in Beijing, China was used to measure the T_2 distributions and T_1-T_2 maps of the Gulong shale cores in the dry and saturated oil states. The magnetic field frequency of the instrument was 9.38 MHz, and the experimental parameters were set as follows: sampling interval of 1 us, echo number of 1024, echo time of 120 us, waiting time of 2500 ms, TIMin of 100 us, and TIMax of 1.5 s. It should be noted that the echo time that was set to 120 us was carefully adjusted to ensure that the T_2 values of the measured fluid were all greater than 0.1 ms to avoid interference from pseudo-solid protons. A conventional imbibition bottle was used to conduct the shale imbibition experiments after saturation with oil.

2.2. Experiment Materials

Three pieces of dry Gulong shale cores are shown in Figure 1, with the top and bottom surfaces left exposed and the side of the cylindrical cores wrapped in thermoplastic tubing to protect the shale cores from breaking. Kerosene was used to saturate the dry rock cores. KCl solution was used for conducting the imbibition experiments on the Gulong shale cores saturated with oil. The salinity of the KCl solution was 6000 ppm, which was consistent with the salinity of the fluid in a shale reservoir in Daqing.

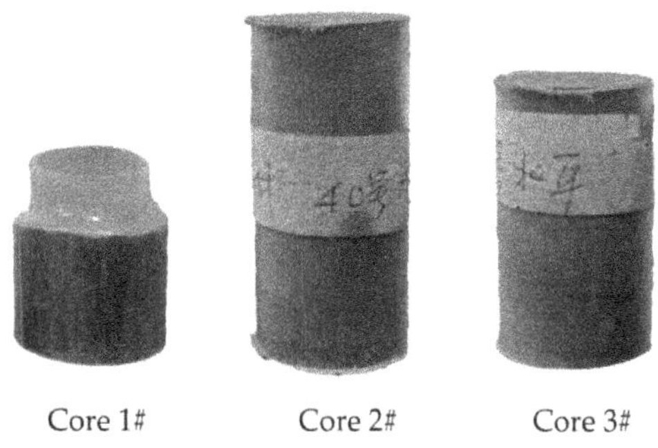

Figure 1. Photograph of the Gulong shale.

3. Experimental Principle and Method

3.1. Experimental Principle

The main principle of 1D NMR can be described by the following equation:

$$\frac{1}{T_{2a}} = \frac{1}{T_{2b}} + \rho\frac{s}{v} \tag{1}$$

where T_{2b} is the intrinsic relaxation time of the fluid, ms; T_{2a} is the measured apparent relaxation time in the shale core, ms; ρ is the surface relaxation rate of the fluid in the pore, m/s; and $\frac{s}{v}$ is the surface-area-to-volume ratio of the fluid in the pore, μm^{-1}, which is inversely proportional to the pore size.

For fluids with a long intrinsic relaxation time, T_{2b}, such as water and light oil, $\frac{1}{T_{2b}} \ll \rho\frac{s}{v}$, and T_{2a} is mainly determined by $\rho\frac{s}{v}$, so the 1D NMR spectrum could reflect the distribution of the fluids in the internal pores.

Conventional cores need to be cleaned and dried before measuring their permeability and porosity. However, shale cores differ from conventional cores. Organic matter in shale, such as kerogen, can dissolve in the organic solvent used for cleaning, which can cause irreversible damage to the organic pores in the shale if cleaned. However, without prior oil washing, the saturation method cannot accurately measure the porosity of shale pores that contain residual oil and bound water, particularly organic pores. This inevitably leads to an underestimation of porosity. Therefore, for an accurate evaluation of shale porosity, it is imperative to correct the porosity using NMR and obtain the NMR porosity.

The total signal intensity of the T_2 distributions, i.e., the maximum cumulative signal amplitude, was positively correlated with the fluid content, and the higher the fluid content in shale samples, the greater the total T_2 signal intensity. The total signal intensity of the T_2 distributions obtained from the dry scan, S_1, and that obtained from the saturated shale after being saturated with kerosene, S_2, were related to the volume of residual oil and bound water in the dry shale sample, V_1, and the volume of kerosene that saturated the shale, V_{kero}, as per Equation (2). It should be noted that, as stated in Ref. [22], T_2 signals detected from dry cores may contain a signal from hydroxyl groups from crystallographic lattice. This will be further elaborated on later.

$$\frac{S_1}{S_2} = \frac{V_1}{V_1 + V_{kero}} \tag{2}$$

Then, we have the following relationship:

$$V_1 = \frac{S_1}{S_2 - S_1}V_{kero} \tag{3}$$

The NMR porosity can be obtained as follows:

$$\phi_{NMR} = \frac{V_1 + V_{kero}}{V_{shale}} = \frac{\frac{S_1}{S_2 - S_1}V_{kero} + V_{kero}}{V_{shale}} = \frac{S_2}{S_2 - S_1}\frac{V_{kero}}{V_{shale}} = \frac{S_2(m_2 - m_1)}{(S_2 - S_1)\rho_o V_{shale}} \tag{4}$$

where m_1 is the dry core sample mass; m_2 is the core sample mass saturated with oil; S_1 is the total signal intensity of the T_2 spectrum of the dry core; S_2 is the total signal intensity of the T_2 distributions of the oil-saturated core; ρ_o is the density of kerosene; and V_{shale} is the volume of the shale sample.

Core NMR analysis can be conducted in two ways: T_2-D and T_1-T_2 maps. Studies on T_2-D map analyses of rock cores mainly focus on sandstones, where the diffusion characteristics of fluids are reflected by varying the echo time T_E over a wide range. The T_E value determines the accuracy of measuring the pore size, with smaller T_E values allowing for the measurement of smaller pores. In shale, where nanometer-scale pores dominate, a small T_E value is required to ensure measurement accuracy. However, the T_2-D maps cannot guarantee a consistently small T_E, making it difficult to meet the precision requirements of nanometer-scale pores in shale measurements. Therefore, T_2-D maps are only suitable for conventional reservoirs and cannot accurately evaluate shale. When measuring T_1-T_2 maps, it is necessary to ensure that the echo spacing T_E is a small value, thereby ignoring the influence of the diffusion coefficient D. This property just meets the measurement requirements of nanometer-scale pores in shale, so T_1-T_2 maps can more accurately evaluate shale.

3.2. Experimental Method

3.2.1. Measurement of Permeability in Shale Core

The gas permeability experiment used PDP-200 pulse decay permeameters to measure the ultra-low permeability. The experimental apparatus is shown in Figure 2. The experimental procedure was as follows: (1) Preheat the PDP equipment and check the airtightness of the apparatus. (2) Perform a test on a mock core to detect system errors. (3) Place the Gulong shale core into the apparatus, apply a confining pressure of 30 MPa, and conduct a nitrogen gas permeability test. (4) Analyze the data using software to obtain the permeability value. Applying a confining pressure of 30 MPa simulated the pressure environment of shale in actual formations. If this was not carried out, the low confinement pressure would result in a lower closure of the fractures in the shale, which would affect the accuracy of the shale permeability measurement.

Figure 2. PDP-200 pulse decay permeameters.

3.2.2. Saturation of Shale Core with Kerosene

To ensure that the shale core was fully saturated with kerosene, it was necessary to conduct long-term vacuum and high-pressure saturation experiments on the shale cores with kerosene. The experimental procedure was as follows: (1) Check the leakage of the instrument pipeline. (2) Place the three shale cores in the pressure vessel and evacuate the vessel for 15 days to achieve a vacuum degree of 0.1 Pa, ensuring that the shale cores are fully evacuated. (3) Saturate the shale cores with kerosene under a pressure of 30 MPa for 15 days, ensuring that the shale is fully saturated with kerosene.

3.2.3. Porosity Measurement and Correction

The steps of the porosity measurement and correction were as follows: (1) Measure the mass, volume, and T_2 distributions of the dry cores. (2) After the cores are saturated with kerosene, measure the mass and NMR T_2 distributions of the cores saturated with kerosene. (3) Calculate the core porosity based on the change in the core mass. (4) Calculate the porosity according to Equation (4).

3.2.4. Shale Imbibition Experiment

The steps of the shale imbibition experiment were as follows: (1) Before imbibition, measure the mass and T_1-T_2 map of the cores saturated with kerosene. (2) Place the cores into the imbibition bottle and soak the cores in a KCl solution with a salinity of 6000 ppm.

(3) After the core is soaked in the KCl solution for a period of time, take out the core and wipe it clean, and then measure the mass and the T_1-T_2 map. (4) Repeat steps 2 and 3 until the core quality changes are minimal.

3.2.5. NMR Testing

The dry cores, kerosene-saturated cores, and the cores taken out after each interval during the imbibition process were measured for their T_2 distributions and T_1-T_2 maps using a SPEC-023 nuclear magnetic resonance permeameter. The T_2 distribution data were recorded, and the T_1-T_2 maps were generated by using a particular software.

4. Result and Discussion

4.1. Analysis of Porosity and Permeability

The permeability of the three shale cores measured by the PDP-200 pulse decay permeameters and the porosity of the cores calculated from the changes in the shale core mass before and after saturation with oil are shown in Table 1. The porosity and permeability of Core 1# were both the highest among the three cores. Core 2# and Core 3# had a similar porosity, but their permeability differed significantly. The permeability of Core 2# was about three times that of Core 3#.

Table 1. Porosity and permeability parameters of Gulong shale core samples.

Core Number	Dry Weight /g	Weight after Saturation /g	Volume of Kerosene /cm^3	Core Volume /cm^3	Porosity /%	Permeability /mD
1#	27.22	27.895	0.84	10.47	8.06	0.422
2#	75.786	77.218	1.79	28.43	6.30	0.235
3#	61.794	62.951	1.45	23.44	6.17	0.080

Due to the presence of bound water or heavy oil that was difficult to evaporate in the dry core samples, the porosity measured above did not take into account the residual fluids in the core. Therefore, the T_2 distributions were used to calculate the NMR porosity, as shown below.

The T_2 distributions of the three shale cores before and after saturation with oil are shown in Figure 3:

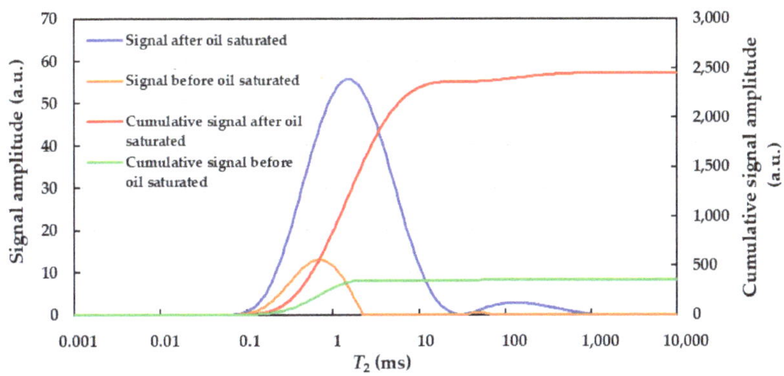

(a)

Figure 3. *Cont.*

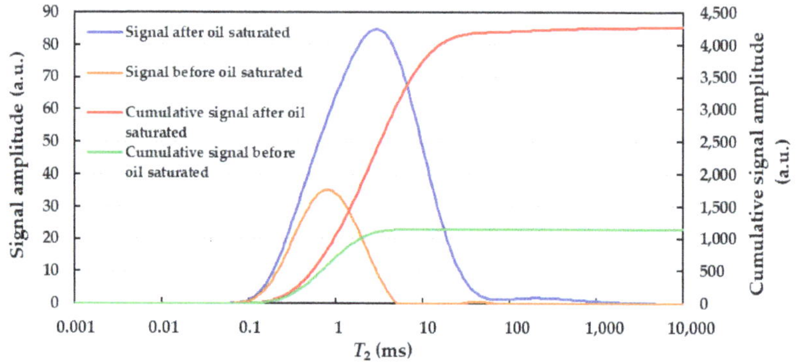

(b)

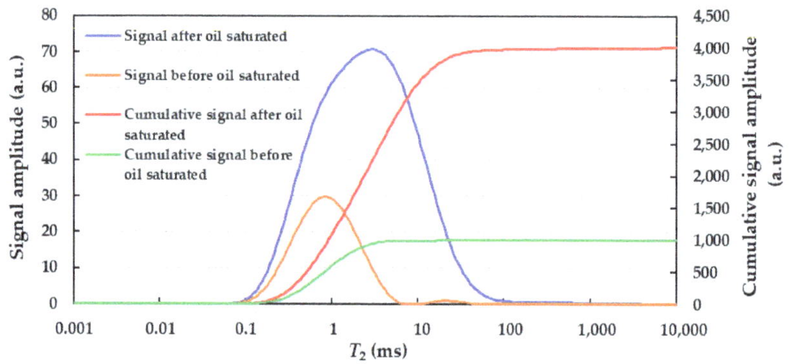

(c)

Figure 3. T_2 distributions of Gulong shale cores. (**a**) T_2 distribution curve for Core 1#; (**b**) T_2 distribution curve for Core 2#; (**c**) T_2 distribution curve for Core 3#.

The porosity corrected based on the T_2 distributions is shown in Table 2. In addition, as mentioned above, the T_2 signals detected from dry cores may have contained hydroxyl group signals from the crystallographic lattice. The predominant clay mineral in Gulong shale is illite and is accompanied by a minor amount of chlorite. Ref. [22] showed that the maximum T_2 value for illite was 0.13 ms, while that of chlorite was 0.1 ms. Therefore, a porosity of $T_2 < 0.13$ ms was also calculated, as shown in Table 2. The porosity of $T_2 < 0.13$ ms was observed to be extremely small, i.e., less than 0.2 ‰, rendering it entirely negligible. This also implied that the signal of the hydroxyl groups from the crystallographic lattice constituted a minute fraction of the T_2 signal in the dry cores.

Table 2. Corrected porosity of Gulong shale cores.

Core Number	Initial Porosity	NMR Porosity	Porosity of $T_2 < 0.13$ ms	Permeability /mD
1#	8.06%	9.40%	0.14‰	0.422
2#	6.30%	8.59%	0.19‰	0.235
3#	6.17%	8.23%	0.16‰	0.080

A comparison of the NMR porosity and permeability of the shale cores showed that the NMR porosity of Core 1# was still the highest, but it was only 0.81% higher than that of

Core 2# and 1.17% higher than that of Core 3#. However, the permeability of Core 1# was significantly higher than that of Core 2# and Core 3#. Moreover, while the NMR porosity of Core 2# was only 0.36% higher than that of Core 3#, the permeability of Core 2# was about three times that of Core 3#. To analyze the reasons for the differences in the NMR porosity and permeability mentioned above, micro-CT scanning was conducted on the three cores, and the imaging results are shown in Figure 4. Due to equipment limitations, the scanning height of each core was only about 2 cm, and the scan area was the middle of the core. The scanning width was the cross-sectional diameter of the core, which was 2.5 cm. Although none of the three cores scanned the whole core, the bedding fractures in the Gulong shale cores had a very good continuity, that is, once fractures were detected in the middle of the core, there were very likely to be fractures in other parts of the core; meanwhile, if no fractures were detected in the middle of the core, there were also very likely to be no fractures in other parts. Therefore, the results obtained by scanning the local part of the core could characterize the whole core. Otherwise, the results of the micro-CT analysis could be verified against the T_2 distributions results, which could also minimize the impact of scanning a part of the cores instead of the whole cores.

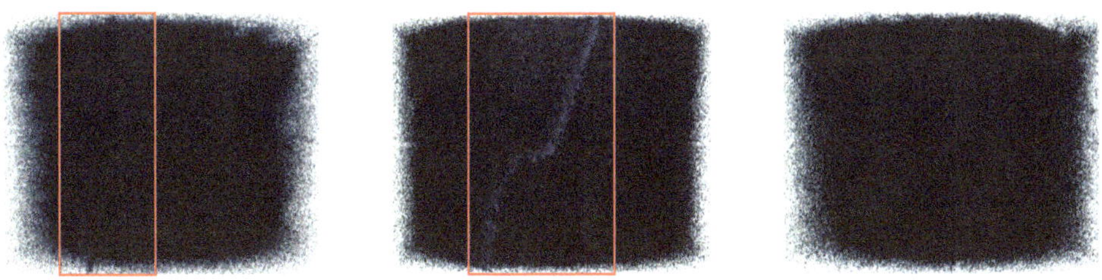

Figure 4. Micro-CT scan images of the Gulong shale cores. (**a**) Core 1#; (**b**) Core 2#; (**c**) Core 3#. The red box marked the area where the fracture was located.

The micro-CT analysis could identify fluids in the pores of the cores, but because the device had a resolution of only 14.5 microns, it could only identify fluids in micrometer-sized pores. Since the pores in the Gulong shale samples were mostly nanometer-sized, it was impossible to identify fluids in the pores of the shale cores using micro-CT. However, the fractures in the shale were on a micrometer scale, and the fluid in the fractures could be identified by micro-CT scanning. As can be seen from Figure 4, the left half of the micro-CT scanning image of Core 1# showed an obvious vertical fracture, the middle of the micro-CT scanning image of Core 2# showed an obvious meandering fracture, and there was no trace of fractures in the micro-CT scanning image of Core 3#. These micro-CT scan images explained why Core 3# had the lowest permeability and Core 1# and Core 2# had much higher permeabilities than Core 3#. Otherwise, the T_2 distributions of the shale cores saturated with oil, as shown in Figure 3, could also indicate if there were fractures in the cores. It can be seen that Core 1# had a clear peak with a T_2 value of 100 ms, which was the signal of the oil phase in the fracture. Core 2# also had a small peak with a T_2 value of 100–1000 ms and with a peak value that was significantly smaller than that of Core 1#, indicating that the fracture volume of Core 2# was smaller than that of Core 1#, and the fracture of Core 1# was more developed, meaning that the permeability of Core 1# was higher than that of Core 2#. Core 3# had almost no peak with a T_2 value of 100–1000 ms, indicating that there was no fracture development in Core 3#. Therefore, the permeability of the shale was significantly correlated with the development of fractures, i.e., the more developed the fracture was, the higher the permeability of the shale was.

4.2. Fluid Occurrence Space Analysis

According to previous studies [20–24], the classification of fluid types is mainly based on the value of T_1. The T_1 value of water is the lowest, the T_1 value of light oil is greater than that of water, and the T_1 value of heavy oil is the highest. In addition, the same fluid has different T_1 values at different T_2 values. For example, the T_1 value of light oil in organic pores with small T_2 values is less than that of inorganic pores with large T_2 values. Therefore, the value of T_1/T_2 should be used to distinguish fluids. Usually, water has a T_1/T_2 value of around 1, light oil has a T_1/T_2 value of around 10, and heavy oil has a T_1/T_2 value of around 100. In addition, the T_1-T_2 maps might also contain the signal of hydroxyl groups from the crystallographic lattice. However, based on the T_2 distribution analysis above, it could be observed that the T_2 signal of hydroxyl groups from the crystallographic lattice exhibited a significantly lower intensity compared to that of the residual fluid present in the dry core. Therefore, the signal of hydroxyl groups from the crystallographic lattice could also be ignored in the T_1-T_2 maps shown below.

The distribution of fluids within the shale cores was analyzed using the results of the T_1-T_2 map tests on the cores before and after saturation with oil. The T_1-T_2 maps of Core 1# before and after saturation with oil are shown in Figure 5.

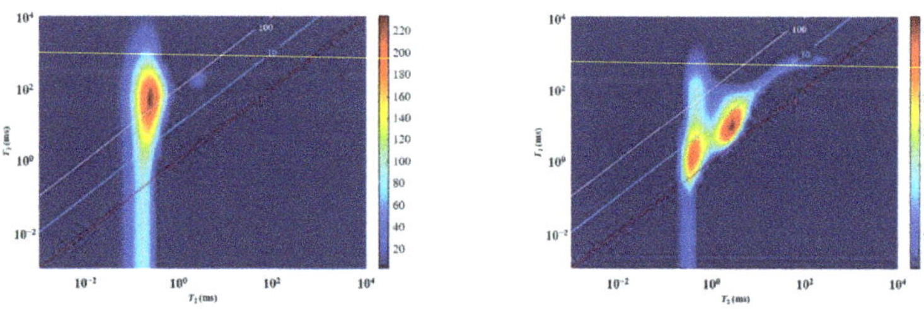

(**a**) (**b**)

Figure 5. The comparison of the T_1-T_2 maps of Core 1# before and after being saturated with oil. (**a**) Core 1# before saturation with oil; (**b**) Core 1# after saturation with oil.

As shown in Figure 5a, the T_1-T_2 map of Core 1# in its dry state had only one peak signal area shown in red. The signals originating from the fluids within this region primarily exhibited a distribution within the T_2 range of 0.1–1 ms and with $T_1/T_2 \approx 100$. This indicated that the fluids present in Core 1# before oil saturation were mainly heavy oil components such as asphaltene. As shown in Figure 5b, after saturation with oil, two red peak signal areas appeared in the T_1-T_2 map, with T_2 ranges of 0.1–1 ms and 1–10 ms, respectively, both of which were distributed in region with a T_1/T_2 range of 1–10. Since the pores in the shale that contained oil were mainly organic and inorganic pores and the pore sizes of organic pores are usually smaller than those of inorganic pores [14–18], it could be inferred that the left red peak signal area corresponded to the oil phase saturated in the organic pores, while the right red peak signal area corresponded to the oil phase saturated in the inorganic pores. In addition, in the T_1-T_2 map after oil saturation, there were light blue signals appearing in the region with a T_2 range of 100–1000 ms and with $T_1/T_2 \approx 10$, which mainly corresponded to the signals of the oil phase in the shale fractures. This indicated that the proportion of the oil phase in the fractures of the shale rock core was very low.

The T_1-T_2 maps of Core 2# before and after oil saturation are shown in Figure 6. As shown in Figure 6a, it can be seen that the T_1-T_2 map of Core 2# in the dry state had a red bar-shaped signal region, and the signals originating from the fluids within this region primarily exhibited a distribution with a T_2 range of 0.1–1 ms and a T_1/T_2 value below 10, indicating that the fluids in Core 2# before oil saturation were mainly composed of clay-bound water and light oil components in organic pores. According to Figure 6b, after saturation with oil, the two red peak signal areas appeared in the T_1-T_2 map, with T_2 values ranging from 0.1–1 ms and 1–10 ms and T_1/T_2 values ranging from 1–10 and near 10, respectively, which was similar to the trend observed in Core 1#.

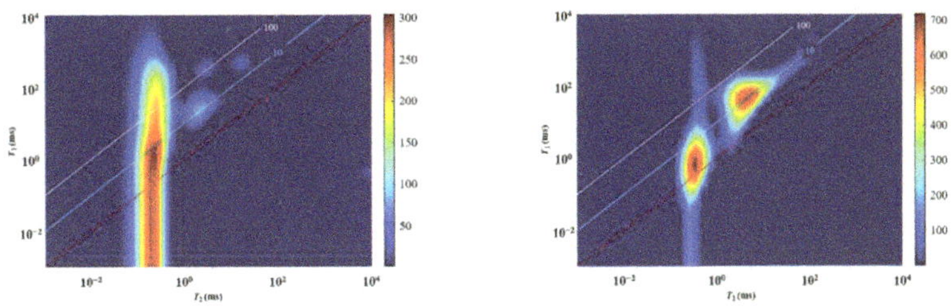

(a) (b)

Figure 6. The comparison of the T_1-T_2 maps of Core 2# before and after being saturated with oil. (**a**) Core 2# before saturation with oil; (**b**) Core 2# after saturation with oil.

The T_1-T_2 maps of Core 3# before and after saturation with oil are shown in Figure 7. From Figure 7a, it can be seen that the dry state T_1-T_2 map of Core 3# also had a red bar-shaped signal area, indicating that the fluids contained in Core 3# before oil saturation included heavy and light oil components in organic pores, as well as clay-bound water. According to Figure 7b, after being saturated with oil, two red peak signal areas appeared in the T_1-T_2 maps. The T_2 ranges of these areas were about 0.1–1 ms and 1–10 ms, respectively, and the T_1/T_2 values were between 1–10 and around 10, which was similar to the pattern of the first two cores. In addition, in the T_1-T_2 map after oil saturation, in the region where T_1/T_2 was about 10 and T_2 was around 100–1000 ms, the fluid signal of Core 3# was much weaker than that of Core 1# and slightly weaker than that of Core 2#, indicating that Core 3# was basically not developed with fractures, which was consistent with the micro-CT scanning results mentioned earlier.

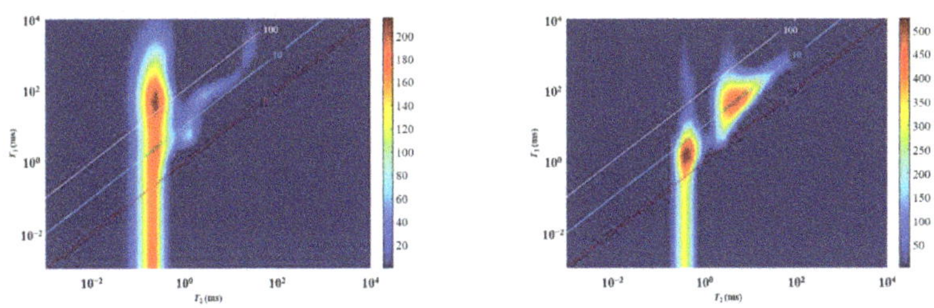

(a) (b)

Figure 7. The comparison of the T_1-T_2 spectra of Core 3# before and after being saturated with oil. (**a**) Core 3# before saturation with oil; (**b**) Core 3# after saturation with oil.

The analysis of the NMR T_1-T_2 maps of the three shale cores before and after oil saturation indicated that the fluid types in the Gulong shale cores in this experiment mainly included heavy oil, light oil, and clay-bound water. Heavy oil was mainly distributed in organic pores, while light oil was distributed in both organic pores, inorganic pores, and fractures. The organic pores were mainly distributed in the region with a T_2 range of 0.1–1 ms in the T_1-T_2 maps, and the heavy oil in the organic pores was distributed in the region with a T_1/T_2 value of 100, while the light oil in the organic pores was distributed in the region with a T_1/T_2 value range of 1–10. The light oil in the inorganic pores was distributed in the region with a T_2 range of 1–100 ms and a T_1/T_2 value of 10. The clay-bound water was mainly distributed in the region with a T_2 range of 0.1–1 ms and a T_1/T_2 value of 1.

4.3. Analysis of the Imbibition Effect

To investigate the effect of saltwater imbibition on the oil mobilization and fluid distribution in Gulong shale, a 120 h imbibition experiment with a 6000 ppm KCl solution was conducted on Core 3#. During the imbibition process, the core was removed multiple times for weighing and T_1-T_2 map measurement. During the imbibition process, it was observed that the core released oil, as shown in Figure 8.

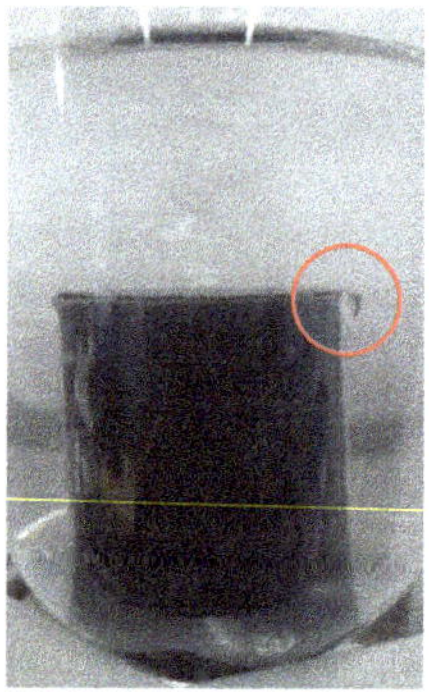

Figure 8. Expulsion of oil from the core during the imbibition process. The red circle marked kerosene that had been released from the shale core.

Using the mass difference of Core 3# obtained by weighing at each time, the approximate amount of oil extracted by imbibition could be calculated. The recovery rate curve of Core 3# during the imbibition process was then plotted, as shown in Figure 9. Due to the wiping of the core before weighing, the calculated results had some errors, leading to negative oil recovery values in the early stage of imbibition. As the duration of the imbibition increased, the water phase was gradually imbibed into the core and displaced the oil phase, resulting in a positive and gradual increase in the recovery. The imbibition rate decreased gradually as the imbibition time increased, with the fastest imbibition rate being at the beginning and the slowest being at the end. At the end of the 120 h imbibition experiment, the imbibition recovery rate reached 36.6%.

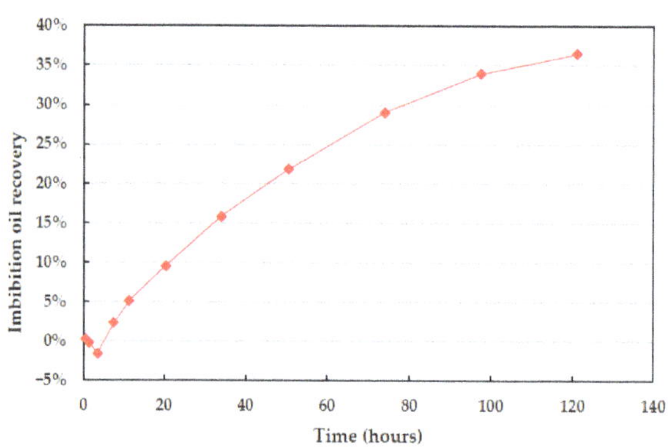

Figure 9. Imbibition oil recovery curve of Core 3#.

The results of the T_1-T_2 map tests of the shale core at different imbibition times are shown in Figure 10. With the increase in the imbibition time, the color of the signal peak in the region where the clay-bound water was present, i.e., the region with a T_2 range of 0.1–1 ms and T_1/T_2 values near 1, gradually changed from light yellow to dark red, indicating that the signal value continuously increases.

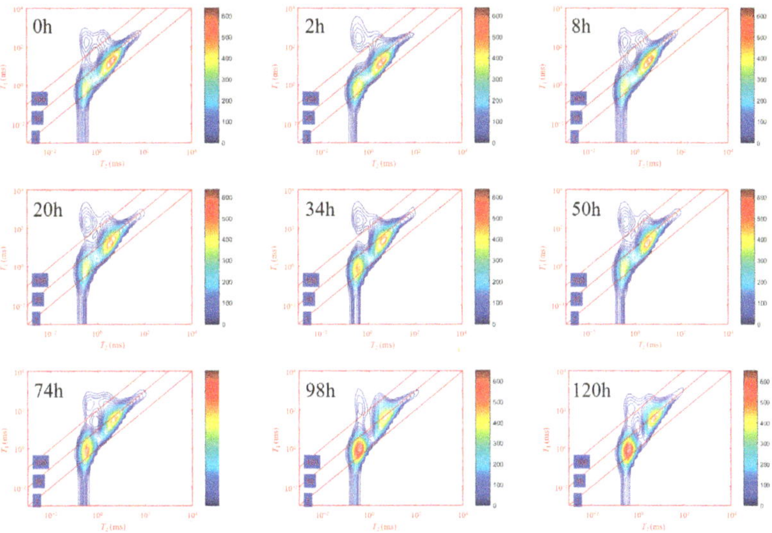

Figure 10. T_1-T_2 spectra of shale core under different imbibition times.

On the other hand, the color of the signal peak in the area where light oil in inorganic pores was present, i.e., the region with a T_2 value of 1–100 ms and T_1/T_2 values near 10, changed from red to yellow, indicating that the signal value in this area continuously decreased with the increase in the imbibition time. In addition, it is worth noting that the oil signals in the vicinity of a T_2 value of 1 ms also continuously weakened with the progress of the imbibition, and the signal values were basically reduced to zero at the end of the imbibition. The above phenomenon indicated that the saltwater imbibition into the core mainly entered the minuscule inorganic pores of clay minerals, thus reinforcing the clay-bound water signal. The light oil in the larger inorganic pores was expelled because the clay minerals underwent a certain degree of hydration expansion after the saltwater entered the minuscule clay mineral inorganic pores, thus squeezing the larger pores and forcing the oil phase in them to be expelled. The small-sized inorganic pores containing the oil phase were fully squeezed by the hydration expansion effect, and most of the oil phase was expelled, which caused the oil phase signal with a T_2 value of 1 ms to be reduced to almost zero at the end of the imbibition process. The signal of light oil in the organic pores overlapped with that of the water phase, and the signal value kept increasing, making it difficult to determine whether the light oil in the organic pores had been mobilized. However, considering that mobilizing the oil in the organic pores was the most difficult and that organic pores are oleophilic, it was inferred that the light oil in the organic pores was not mobilized or was only slightly mobilized. Therefore, the oil extracted by saltwater imbibition mainly came from the oil in the inorganic pores.

5. Conclusions

In this study, the pore and permeability characteristics, the fluid occurrence space, and the saltwater imbibition oil recovery characteristics of Gulong shale cores were studied using NMR. The conclusions are as follows:

(1) The porosity and permeability characteristics of the Gulong shale cores were analyzed using T_2 distributions. This method could measure the NMR porosity without oil washing, thus avoiding the destruction of organic matter in the shale cores. The difference in the porosity between the three shale cores was small (9.40%, 8.59%, and 8.23%, respectively), but the difference in the permeability was large (0.422 mD, 0.235 mD, and 0.080 mD, respectively). Combined with the analysis of the micro-CT and NMR T_2

spectra, it could be seen that the permeability of the Gulong shale cores was related to the degree of fracture development, with a greater fracture development resulting in a higher permeability.

(2) According to the T_1-T_2 maps, the fluid types in the Gulong shale cores measured in this experiment mainly included heavy oil, light oil, and clay-bound water, with light oil being further divided into organic pore light oil and inorganic pore light oil. The organic pores were mainly distributed in the region with a T_2 value range of 0.1–1 ms, and the inorganic pores mainly distributed in the region with a T_2 value larger than 1 ms. The heavy oil in the organic pores was distributed in the region with a T_1/T_2 value of 100, while the light oil in the organic pores was distributed in the region with a T_1/T_2 value range of 1–10. The light oil in the inorganic pores was distributed in the region with a T_2 value range of 1–100 ms and around a T_1/T_2 value of 10. The clay-bound water was mainly distributed in the region with a T_2 value range of 0.1–1 ms and a T_1/T_2 value of 1.

(3) Saltwater imbibition could remove a certain amount of oil from the shale core, and the imbibition recovery rate could reach 36.6%. Combined with the T_1-T_2 maps, it was determined that during the process of imbibition, saltwater mainly entered the minuscule inorganic pores of clay minerals under the action of capillary force, and after undergoing hydration expansion, it squeezed the inorganic pores, thereby removing the light oil in the inorganic pores. The oil phase in the organic pores was rarely affected.

Author Contributions: Conceptualization, F.X. and J.L.; methodology, H.J.; software, M.L. and S.J.; validation, J.L.; formal analysis, H.J.; investigation, F.X. and S.J.; resources, J.L.; data curation, Y.W.; writing—original draft preparation, F.X.; writing—review and editing, S.J.; visualization, M.L.; supervision, H.J.; project administration, Y.W. All authors have read and agreed to the published version of the manuscript.

Funding: This research was funded by the National Natural Science Foundation of China (No. 52174042).

Data Availability Statement: Not applicable.

Conflicts of Interest: The authors declare no conflict of interest.

References

1. Zou, C.; Dong, D.; Wang, Y.; Li, X.; Huang, J.; Wang, S.; Guan, Q.; Zhang, C.; Wang, H.; Liu, H. Shale Gas in China: Characteristics, Challenges and Prospects (II). *Petrol. Explor. Develop.* **2016**, *43*, 182–196. [CrossRef]
2. Li, Y.; Zhao, Q.; Qi, L.; Xue, Z.; Cao, X.; Liu, Z. Evaluation technology and practice of continental shale oil development in China. *Petrol. Explor. Develop.* **2022**, *49*, 1098–1109. [CrossRef]
3. Liu, B.; Shi, J.; Fu, X.; Lyu, Y.; Sun, X.; Gong, L.; Bai, Y. Petrological characteristics and shale oil enrichment of lacustrine fine-grained sedimentary system: A case study of organic-rich shale in first member of Cretaceous Qingshankou Formation in Gulong Sag, Songliao Basin, NE China. *Petrol. Explor. Develop.* **2018**, *45*, 884–897. [CrossRef]
4. Li, C.; Yan, W.; Wu, H.; Han, T.; Zheng, J.; Jun, Y.; Zhou, F.; Xu, H. Calculation of oil saturation in clay-rich shale reservoirs: A case study of Qing 1 Member of Cretaceous Qingshankou Formation in Gulong Sag, Songliao Basin, NE China. *Petrol. Explor. Develop.* **2023**, *49*, 1351–1363. [CrossRef]
5. Hua, G.; Wu, S.; Zhang, J.; Liu, R.; Guan, M.; Cai, Y.; Li, M.; Zhang, S. Laminar Structure and Reservoir Quality of Shales with High Clay Mineral Content in the Qingshankou Formation, Songliao Basin. *Energies* **2022**, *15*, 6132. [CrossRef]
6. Liu, B.; Wang, H.; Fu, X.; Bai, Y.; Bai, L.; Jia, M.; He, B. Lithofacies and depositional setting of a highly prospective lacustrine shale oil succession from the Upper Cretaceous Qingshankou Formation in the Gulong sag, northern Songliao Basin, northeast China. *AAPG Bull.* **2019**, *103*, 405–432. [CrossRef]
7. Niu, D.; Li, Y.; Zhang, Y.; Sun, P.; Wu, H.; Fu, H.; Wang, Z. Multi-scale classification and evaluation of shale reservoirs and 'sweet spot' prediction of the second and third members of the Qingshankou Formation in the Songliao Basin based on machine learning. *J. Pet. Sci. Eng.* **2022**, *216*, 110678. [CrossRef]
8. Li, S.; Li, S.; Guo, R.; Zhou, X.; Wang, Y.; Chen, J.; Zhang, J.; Hao, L.; Ma, X.; Qiu, J. Occurrence State of Soluble Organic Matter in Shale Oil Reservoirs from the Upper Triassic Yanchang Formation in the Ordos Basin, China: Insights from Multipolarity Sequential Extraction. *Nat. Resour. Res.* **2021**, *30*, 4379–4402. [CrossRef]
9. Sun, L.; Liu, H.; He, W.; Li, G.; Zhang, S.; Zhu, R.; Jin, X.; Meng, S.; Jiang, H. An analysis of major scientific problems and research paths of Gulong shale oil in daqing oilfield, NE China. *Petrol. Explor. Develop.* **2021**, *48*, 527–540. [CrossRef]
10. He, W.; Wang, M.; Wang, X.; Meng, Q.; Wu, Y.; Lin, T.; Li, J.; Zhang, J. Pore Structure Characteristics and Affecting Factors of Shale in the First Member of the Qingshankou Formation in the Gulong Sag, Songliao Basin. *ACS Omega* **2022**, *7*, 35755–35773. [CrossRef]

11. Huo, Z.; Hao, S.; Liu, B.; Zhang, J.; Ding, J.; Tang, X.; Li, C.; Yu, X. Geochemical characteristics and hydrocarbon expulsion of source rocks in the first member of the Qingshankou Formation in the Qijia-Gulong Sag, Songliao Basin, Northeast China: Evaluation of shale oil resource potential. *Energy Sci. Eng.* **2020**, *8*, 1450–1467. [CrossRef]
12. Kuila, U.; McCarty, D.K.; Derkowski, A.; Fischer, T.B.; Prasad, M. Total porosity measurement in gas shales by the water immersion porosimetry (WIP) method. *Fuel* **2014**, *117*, 1115–1129. [CrossRef]
13. Lewis, R.; Singer, P.; Jiang, T.; Rylander, E.; Sinclair, S.; Mclin, R.H. NMR T_2 distributions in the Eagle Ford shale: Reflections on pore size. In Proceedings of the SPE Unconventional Resources Conference, The Woodlands, TX, USA, 10–12 April 2013.
14. Birdwell, J.E.; Washburn, K.E. Multivariate analysis relating oil shale geochemical properties to NMR relaxometry. *Energy Fuels* **2015**, *29*, 2234–2243. [CrossRef]
15. Li, J.; Lu, S.; Chen, G.; Wang, M.; Tian, S.; Guo, Z. A new method for measuring shale porosity with low-field nuclear magnetic resonance considering non-fluid signals. *Mar. Pet. Geol.* **2019**, *102*, 535–543. [CrossRef]
16. Krzyzak, A.T.; Habina-Skrzyniarz, I.; Machowski, G.; Mazur, W. Overcoming the barriers to the exploration of nanoporous shales porosity. *Microporous Mesoporous Mater.* **2020**, *298*, 110003. [CrossRef]
17. Kausik, R.; Fellah, K.; Feng, L.; Freed, D.; Simpson, G. High-and low-field NMR relaxometry and diffusometry of the bakken petroleum system. In Proceedings of the SPWLA 57th Annual Logging Symposium, Reykjavík, Iceland, 25–29 June 2016.
18. Zhong, J.; Yan, R.; Zhang, H.; Feng, Y.; Liu, X. A decomposition method of nuclear magnetic resonance T_2 spectrum for identifying fluid properties. *Petrol. Explor. Develop.* **2020**, *47*, 740–752. [CrossRef]
19. Liu, Z.; Liu, D.; Cai, Y.; Yao, Y.; Pan, Z.; Zhou, Y. Application of nuclear magnetic resonance (NMR) in coalbed methane and shale reservoirs: A review. *Int. J. Coal Geol.* **2020**, *218*, 103261. [CrossRef]
20. Washburn, K.E.; Birdwell, J.E. Updated methodology for nuclear magnetic resonance characterization of shales. *J. Magn. Reson.* **2013**, *233*, 17–28. [CrossRef]
21. Kausik, R.; Fellah, K.; Rylander, E.; Singer, P.M.; Lewis, R.E.; Sinclair, S.M. NMR Relaxometry in Shale and Implications for Logging. *Petrophysics* **2017**, *57*, 339–350.
22. Habina, I.; Radzik, N.; Topor, T.; Krzyzak, A.T. Insight into oil and gas-shales compounds signatures in low field H-1 NMR and its application in porosity evaluation. *Microporous Mesoporous Mater.* **2017**, *252*, 37–49. [CrossRef]
23. Liu, B.; Bai, L.; Chi, Y.; Jia, R.; Fu, X.; Yang, L. Geochemical characterization and quantitative evaluation of shale oil reservoir by two-dimensional nuclear magnetic resonance and quantitative grain fluorescence on extract: A case study from the Qingshankou Formation in Southern Songliao Basin, northeast China. *Mar. Pet. Geol.* **2020**, *120*, 104481.
24. Pang, X.; Wang, G.; Kuang, L.; Zhao, F.; Li, C.; Wang, C.; Zhang, M.; Lai, J. Lamellation fractures in shale oil reservoirs: Recognition, prediction and their influence on oil enrichment. *Mar. Pet. Geol.* **2022**, *148*, 106032. [CrossRef]
25. Wei, J.; Fu, L.; Zhao, G.; Li, B.; Zhao, X.; Wang, A. Sensitivity of shale pore structure to external fluids for Gulong shaleoil reservoir in Songliao Basin. *Pet. Geol. Oilfield Dev. Daqing* **2022**, *41*, 120–129.
26. Yan, W.; Zhang, Z.; Chen, L.; Zhao, Z.; Wang, W. New evaluating method of oil saturation in Gulong shalebased on NMR technique. *Pet. Geol. Oilfield Dev. Daqing* **2021**, *40*, 78–86.
27. Li, B.; Deng, S.; Liu, Y.; Cao, S.; Jin, D.; Dong, D. Measurement method of movable fluid saturation of Gulong shale oil reservoir in Songliao Basin. *Pet. Geol. Oilfield Dev. Daqing* **2022**, *41*, 130–138.
28. He, W.; Meng, Q.; Feng, Z.; Zhang, J.; Wang, R. In-situ accumulation theory and exploration &development practice of Gulong shale oil in Songliao Basin. *Acta Pet. Sin.* **2022**, *43*, 1–14.
29. He, W.; Liu, B.; Zhang, J.; Bai, L.; Tian, S.; Chi, Y. Geological Characteristics and Key Scientific and Technological Problems of Gulong Shale Oil in Songliao Basin. *Earth Sci.* **2023**, *48*, 49–62.
30. Wu, X.; Hou, Z.; Yang, Y.; Chen, J.; Wang, H.; Nie, C. Development and performance evaluation for Gulong 1# emulsion system. *Pet. Geol. Oilfield Dev. Daqing* **2022**, *41*, 112–119.
31. Liu, G.; Yang, D.; Mei, X.; Yu, S.; Ma, W.; Fan, X. Method of well-soaking and controlled flowback after large-scalefracturing of Gulong shale oil reserviors in Songliao Basin. *Pet. Geol. Oilfield Dev. Daqing* **2020**, *39*, 147–154.

processes

MDPI

Article

Estimation of Fracture Height in Tight Reserviors via a Finite Element Approach

Jiujie Cai * and Fengxia Li

SINOPEC Petroleum Exploration and Production Research Institute, Beijing 102206, China
* Correspondence: caijj.syky@sinopec.com; Tel.: +86-(010)-5660-8085

Abstract: In tight reservoirs, the rock formations are typically less porous and permeable, which makes it more difficult for hydrocarbons to flow through them. In addition to length and conductivity, the height of a fracture is another critical parameter of the hydraulic fracturing treatments in unconventional tight/shale formations, which determines the stimulated reservoir volume. If the fracture height is too shallow, the volume of rock exposed to the fluid and proppant may not be sufficient to improve the reservoir's production significantly. Conversely, if the fracture height is too deep, the injected fluid may not be able to propagate high enough to reach the desired formation. However, after years of research, fracture height has often been simplified in traditional or recent studies of fracture simulation and estimation. The objective of this work is to propose an innovative way to simulate the hydraulic fracturing process in both horizontal and vertical directions in tight formations with a well-built finite element numerical model. Fracture toughness K_{IC} is calculated based on the Brazilian test. Vertical fracturing fluid was also considered, and the model was validated by fracture height monitoring data from a stimulated well in the Montney formation. The influence of rock and fluid properties on the fracture height propagation was studied thoroughly with sensitivity analysis. The results indicated the fracture height prediction model was in good accordance with the monitoring data collected from the field, with an error margin of 7.2%. Sensitivity analysis results showed that a high Young's modulus led to a larger stress intensity factor at the fracture tip, thus further advancing the fracture. Minimum horizontal stress also tends to facilitate the fracture to propagate. The influence of Poisson's ratio and fluid viscosity on fracture height propagation was also investigated.

Keywords: hydraulic fracturing; finite element method; fracture height; tight formation; fracture propagation

Citation: Cai, J.; Li, F. Estimation of Fracture Height in Tight Reserviors via a Finite Element Approach. *Processes* **2023**, *11*, 1566. https://doi.org/10.3390/pr11051566

Academic Editors: Linhua Pan, Yushi Zou, Jie Wang, Minghui Li, Wei Feng, Lufeng Zhang and Yidong Cai

Received: 11 April 2023
Revised: 9 May 2023
Accepted: 16 May 2023
Published: 21 May 2023

1. Introduction

Hydraulic fracturing is a stimulation process involving the injection of fluids under a high pressure into a formation to create and propagate induced fractures. In addition to the half-length and conductivity, fracture height is also important in the stimulation treatment. If the fracture height is too short, some areas of the productive zone may not be stimulated, limiting after-treatment productivity. Conversely, operators have strong economic incentives to ensure that fractures do not propagate beyond the formation and into the adjacent rock strata [1]. Fractures extending beyond the formation would not only be a waste of time, materials, and money but may also result in loss of the well and the associated oil/gas resources in some cases [2].

Extensive efforts have been devoted to developing accurate models for hydraulic fracture properties prediction since the 1950s [3–5], including two-dimensional (2D) models, pseudo-three-dimensional (P-3D) models, and three-dimensional (3D) models. The 2D models were proposed first, the most popular being the Perkins–Kern–Nordgren (PKN) model, the Khristianovic–Geertsma–de Klerk (KGD) model, and the radial model (penny shape model). In 1961, Perkins and Kern developed the PK model [6], which was later modified by Nordgren to consider fluid loss [7]. In this model, the height is fixed, and an elliptic intersection was assumed. The PKN model is applicable for long fractures with a limited height. The KGD model [8,9] is height-independent, with an elliptic horizontal intersection. This model is applicable for fractures with a small ratio of fracture length/height. The penny shape model is a radial model proposed by Sneddon and Green [10,11]. This model assumes the hydraulic fracturing fluid is injected from the center of the wellbore to the surrounding fractures and thus is applicable when the injection region is a point source. Pseudo-3D [12,13] and true 3D models were also developed to analyze fracture height growth and the fracture propagation process. Simonson et al. developed a pseudo-3D model to simulate height growth in a symmetric three-layer formation and investigated the effects of in situ stress and pressure gradients for the pay zone on the fracture height. Warpinski et al. [14,15] studied the fracturing process in a layered formation and suggested the in-situ stresses had a more dominant influence on the propagation of hydraulic fractures. Different fracture propagation criteria in impermeable rocks were adopted [16], and several theoretical and numerical models were conducted to analyze the propagation of fluid-driven fractures. Two dissipative processes were studied: the fracturing of the rock (toughness) [17] and the dissipation in the fracturing fluid (viscosity) [18].

Numerical methods have also been applied to simulate the fracture propagation process, which include the boundary element method (BEM), the finite difference method (FDM), the discrete element method (DEM), and the finite element method (FEM). In this model, we used the FEM method due to its ability to handle reservoir heterogeneity and rock properties in the layered reservoirs. Compared to FEM, BEM requires discretization and calculation only on the domain boundaries and cannot address the rock properties change in the matrix [19]; the FDM is limited to calculations of the meshes of the whole domain and to dynamic fracture propagation [20], while the discrete element method (DEM) is often used in discontinuous, separate domains and emphasizes the solution of contact and impact between multiple bodies [21,22]. Ma et al. [23–25] suggested that numerical simulation models based on a three-phase flow process can provide accurate predictions of water-silt inrush hazards in fault rocks and presented the viewpoint that numerical modeling is essential to avert potential disasters caused by water-silt inrush hazards.

However, fracture height is assumed to be a constant in traditional 2D analytical models such as the KGD and PKN models and is equal to fracture length in a radial model. In pseudo-3D models, fracture height is calculated explicitly based on fracturing fluid leak-off theory, which does not take rock mechanics and fracture propagation criterion into account. Moreover, facture height is often calculated in the numerical 3D propagation model of a homogeneous reservoir, which cannot consider the influence of complex bounding layers. In this study, a fully coupled finite element model was built to simulate fracture propagation in the vertical direction in order to analyze fracture height growth in an unconventional tight gas reservoir. Results of the model were compared with results from the analytical method and the fracture height monitoring data from a wellsite. The results indicated that the fracture height prediction model had a good correlation with the monitoring data collected from the field. A detailed analysis of the influence of fluid viscosity and element grid size on fracture height propagation is thus presented. Although the intention was to investigate fracture height, fracture length is also calculated in this study.

2. Model Description

The assumptions of the model are as follows: (1) The fracture geometry is plane-strain; (2) the fracturing fluid is incompressible, laminar, and Newtonian; (3) the lubrication theory can be applied for fluid flow calculation; (4) the linear elastic fracture mechanics theory is used as the criterion for fracture propagation; and (5) vertical stress is higher than the maximum horizontal stress. In addition, the M-integral method was used to calculate the stress intensity factor along the horizontal and vertical fracture propagation. Fracture tip stress intensity factors are first calculated as the fracture lengths propagate. Fracture height is then obtained for the in situ length when the stress intensity factor of the fracture tip in the vertical direction is equal to that in the horizontal direction.

2.1. Governing Equations

2.1.1. Mass Conservation

The fluid mass conservation law can be written as follows [26]:

$$\frac{\partial w}{\partial t} + \frac{\partial q}{\partial y} + C = 0 \tag{1}$$

where q is fluid flux, m^3/min; w is fracture width, m; t is time, s; y is vertical flow direction; and C is the leak-off term, $m/s^{0.5}$. Integrating Equation (1) twice yields the following:

$$2\int_0^l w dy + 2\int_0^t \int_0^{l(t')} C dy\, dt' = Q_0 t \tag{2}$$

2.1.2. Fluid Flow in the Fracture

Lubrication theory is used to describe fluid flow behavior within the fractures, which is described by the Poiseuille's law:

$$q = -\frac{w^3}{12\mu}\left(\frac{\partial p_f}{\partial y}\right) \tag{3}$$

where p_f is fluid pressure at the fracture surface, MPa; and μ is fluid viscosity, $Pa \cdot s$.

2.1.3. Fluid Leak-off

Carter's leak-off model is used to describe fluid leak-off to the surrounding matrix (Carter, 1957):

$$C(y,t) = \frac{2C_l}{\sqrt{t - t_0(y)}},\ t > t_0(y) \tag{4}$$

where C_l is leak-off coefficient, $m/s^{0.5}$; and $t_0(y)$ is the time the fracture tip arrives at position y, s.

2.1.4. Fracture Propagation Criterion

Tensile failure criterion is used for the reservoir rock to simulate fracture propagation, which is defined as follows:

$$K_I \geq K_{IC} \tag{5}$$

where K_I is the stress intensity factor, $MPa \cdot \sqrt{m}$; and K_{IC} is the critical stress intensity or fracture toughness, $MPa \cdot \sqrt{m}$.

2.1.5. Boundary Equations

Only a quarter of the fracture is modelled in this work, and the boundary conditions for fluid flow are as given below:

$$q(0) = {Q_0}\big/{4}, q(l) = 0 \tag{6}$$

where $q(0)$ is fluid flux at the start of the crack, m^3/min; and $q(l)$ is fluid flux at the fracture tip, m^3/min; Q_0 is the total injection flow rate, m^3/min.

2.1.6. Stress Intensity Factor Determination

The stress intensity factor K_I is calculated with the M-integral method [27,28]. The M-integral method is a path-independent line integral that is applied to a class of plane elastic fracture problems for a variety of elastic crack problems:

$$K_I = \frac{E}{2(1-v^2)}\left\{\int_D \left[\sigma_{ij}\frac{\partial u_i{}^a}{\partial x_1} + \sigma_{ij}^a\frac{\partial u_i}{\partial x_1} - \sigma_{mn}^a\varepsilon_{mn}\delta_{1j}\right]\frac{\partial \chi}{\partial x_j}dS - \int_{S_e} \chi p\frac{\partial u_i^a}{\partial x_1}dL\right\} \tag{7}$$

where domain D is a set of elements surrounding the fracture tip in the discretized finite element model. S_e is a set of edges of the finite elements in domain D, and these edges coincide with the fracture surface; σ_{ij} is the stress field; $x_j(j = 1,\ 2)$ is the local coordinate; $u_i(i = 1,\ 2)$ is the displacement field; δ is the Kronecker delta;χ is a scalar field; ε_{mn} is the strain field; σ_{ij}^a and $u_i{}^a$ are the auxiliary stress and displacement field, respectively.

The auxiliary stress and displacement fields can be expressed as follows:

$$\begin{Bmatrix} u_1^a \\ u_2^a \end{Bmatrix} = \frac{1}{2G}\sqrt{\frac{r}{2\pi}}\begin{Bmatrix} \cos\frac{\theta}{2}[\kappa - \cos\theta] \\ \sin\frac{\theta}{2}[\kappa - \cos\theta] \end{Bmatrix} \tag{8}$$

$$\begin{Bmatrix} \sigma_{11}^a \\ \sigma_{22}^a \\ \sigma_{12}^a \end{Bmatrix} = \frac{1}{\sqrt{2\pi r}}\cos\frac{\theta}{2}\begin{Bmatrix} 1-\sin\frac{\theta}{2}\sin\frac{3\theta}{2} \\ 1+\sin\frac{\theta}{2}\sin\frac{3\theta}{2} \\ \sin\frac{\theta}{2}\cos\frac{3\theta}{2} \end{Bmatrix} \tag{9}$$

where r is the distance from the fracture tip to the observation point x, and θ is the angle from the tangent to the fracture path. This angle equals $\frac{\pi}{2}$ as the fracture propagates in the vertical direction. In addition, G is the shear modulus; κ is the Kolosov constant, where $\kappa = (3 - \nu)/(1 + \nu)$ for the plane stress condition, and $\kappa = (3 - 4\nu)$ for the plane strain condition.

2.2. Numerical Implementation

The equations discussed in the previous sections are coupled together to obtain an implicit solution for all equations. The finite element method was used in the simulation model. Fracture propagation for each time step is solved with a three-step process, described as follows:

A proper time step is assumed first, and then, the finite element discretization of the relationship between the fracture width w and the pressure in the fracture can be determined:

$$w(y,t) = \frac{\pi(1-v^2)}{4E}\int_{l(t)} \ln\left|\frac{\sqrt{l(t)^2 - y^2} + \sqrt{l(t)^2 - y_1^2}}{\sqrt{l(t)^2 - y^2} - \sqrt{l(t)^2 - y_1^2}}\right| p(y_1,t)dy_1 \tag{10}$$

where $l(t)$ is the fracture half-height at time t.

The matrix form of equation of width and pressure is as follows:

$$\Delta w = M\Delta p \tag{11}$$

where Δw is width increment, m; and Δp is pressure increment, MPa; M is pressure influence coefficient matrix; and the physical meaning of Mij is the fracture width at certain point i induced by unit pressure at another point j within the fracture. According to the principle of variation, Equation (1) with any test function leads to its weak form as follows:

$$\int_L \left[-\nabla(\delta p) \cdot q + (\delta p)\frac{\partial w}{\partial t} + (\delta p)g\right]dl + \delta p(q \cdot n_1)|_s = 0 \quad (12)$$

where S is the collection of boundary conditions. Then, we can obtain the nonlinear finite equations and discretized equations.

An initial width w and pressure at time t are then assumed, and the Jacobian matrix generated in the process can be solved by a series of finite element equations. Backward Euler method is applied to solve the time difference in the integration. Newton–Raphson technique is conducted as the iteration method to solve the width and pressure increment in the next time step. K_I is calculated by Equation (7) and compared with the fracture propagation criterion K_{IC} (calculated based on the Brazilian test) in Equation (5). Such process is repeated until the solution converges.

Once the propagation criterion is met, the fracture propagates into the next grids, and fluid flows into the newly propagated fracture grid.

As the vertical stress is assumed to be the highest among all the stresses, fracturing fluid is assumed to flow in the vertical (y) direction in the early stage of the fracturing process. The model has 50 grids of 0.3 m in size in the vertical direction. In the horizontal (x) direction, grid size varies from 0.003 m at the fracture surface to several feet in the reservoir grids. Due to fracture symmetry, only a quarter of the fracture is modeled with half-length and half-width.

3. Fracture Height Calculation and Validation

3.1. Field Background

The Montney formation, located at the border of Alberta and British Columbia, Canada, is a sedimentary wedge that was deposited during the Early Triassic geologic period. Its lithology consists of interbedded fine-grained sandstones, siltstone, and dark gray shale. The reservoir depth ranges from 2800 m to 3500 m, reaching a maximum thickness of 280 m in the foothills of the northern Canadian Rockies and thinning out towards the north up to the Fort Nelson area and east to Peace River [29]. A large number of horizontal wells have been drilled in the heart of the Upper and Lower Montney gas plays [30]. The subject well is a horizontal well on the upper section of the Montney formation. A diagnostic fracture injection test (DFIT) was conducted before the well stimulation between the true vertical depth (TVD) of 2553 m and 2564 m. Table 1 shows the key parameters associated with the DFIT operations.

Table 1. Parameters in the simulated well.

Parameters	Unit	Value
Minimum horizontal stress σ	MPa	44.2
Young's modulus E	MPa	13,789.5
Poisson's ratio υ	/	0.2
Total injection flow rate Q	m^3/min	2.9×10^{-3}
Leak-off coefficient C_l	$m/s^{0.5}$	1.5×10^{-5}
Fluid viscosity μ	mPa·s	2.5

3.2. Fracture Toughness Determination

Fracture toughness K_{IC} is calculated based on the Brazilian test, which is a well-known indirect method used to measure rock tensile strength. The experiment is based on the observation that most rocks in biaxial stress fields fail due to tensile failure cracks along the loading diameter of the disc specimen [31].

The critical load in the experiment is shown as follows:

$$P_c = \frac{K_{IC}}{B\phi(c/R)} \tag{13}$$

where c is the fracture half-length, mm; P_c is critical load, MPa; R is radius of disc, mm; and B and $\phi(c/R)$ are defined as follows:

$$B = \frac{2}{\pi^{3/2}R^{1/2}t_1\alpha} \tag{14}$$

$$\phi\left(\frac{c}{R}\right) = \left(\frac{c}{R}\right)^{1/2} \int_0^{c/R} \left[\phi\left(\frac{r}{R}\right) / \left(\frac{c^2}{R} - \frac{r^2}{R}\right)^{1/2}\right] d\left(\frac{r}{R}\right) \tag{15}$$

where t_1 is the thickness of the disc (mm), and the value of the integral $\phi(c/R)$ for a given c/R can be estimated by a numerical integration method such as the Simpson's method.

The values of the parameters in Equations (13)–(15) can be found in Table 2, and K_{IC} is calculated to be 0.244 MPa $\cdot \sqrt{\text{m}}$.

Table 2. Parameters used for the K_{IC} calculation.

Parameters	Unit	Value
Specimen's angle α	degree	5
Radius of disc R	mm	38
Thickness of the disc t_1	mm	47
Critical load P_c	kN	4.85
Integral $\phi(\frac{c}{R})$	/	0.112

3.3. Numerical Model

A single fracture case was built via the finite element method, where only a quarter of fracture was simulated, as shown in Figure 1. The fracture grid was discretized into 50 cells with the size of 0.3 m in the vertical direction and varies from 0.003 m at the fracture surface to several feet in the horizontal direction to be compatible with the reservoir grids. The Young's modulus, Poisson's ratio, minimum horizontal stress, and fluid property values in the basic model are shown in the Table 1. In the basic model, a low-viscosity fluid (2.5 mPa·s) was injected. In this model, when stress intensity factor in the vertical tip equals its value in the horizontal direction, the model stops, and the corresponding fracture heights will be calculated, after which the fracture mainly propagates in the horizontal direction, and the height of fracture stops increasing.

3.4. Fracture Height Calculation

As the fracture length propagates, stress intensity factor at the fracture tip under static condition is calculated by the M-integral method with the parameters K_{IC} listed in Table 1. The relationship between the stress intensity factor and the fracture length is shown in Figure 2. The stress intensity factor decreases as the fracture half-length increases. When the fracture half-length reaches 16.8 m, K_I decreases to 0.2520 MPa $\cdot \sqrt{\text{m}}$, which is slightly above the fracture toughness (0.2440 MPa $\cdot \sqrt{\text{m}}$).

While calculating fracture heights, K_I under the different fracture half-lengths is used to replace K_{IC}. Specially, when the stress intensity factor in the vertical tip is equal to the critical stress intensity factor, the model stops, and the corresponding fracture heights will be outputted. We thus simulated the basic model with different half-fracture-lengths, and the calculated fracture heights and corresponding fracture lengths are shown in Figure 3.

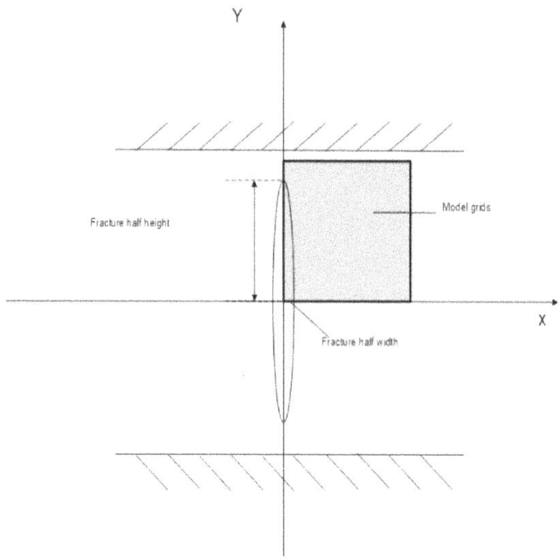

Figure 1. Schematic diagram for a hydraulic fracture in the tight formation.

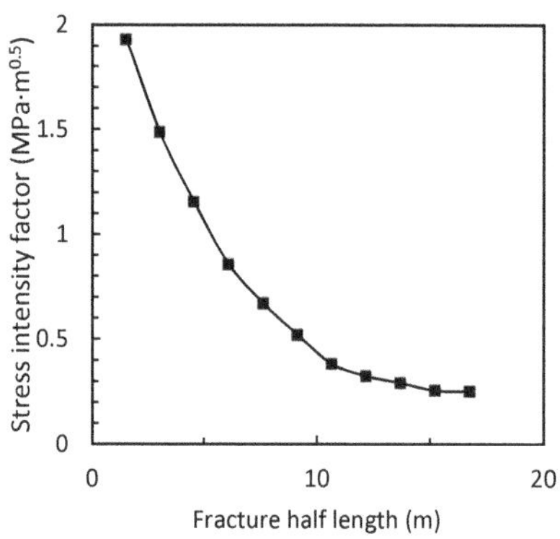

Figure 2. Change of the stress intensity factor as fracture half-length propagates.

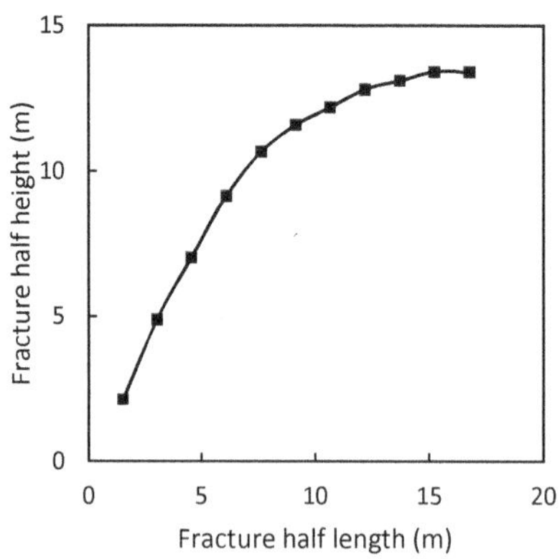

Figure 3. Propagation of the fracture half-height as the fracture half-length extends.

Figure 3 shows that fracture propagates first in the vertical and horizontal directions simultaneously at a similar rate. Once the height reaches a certain value, the horizontal growth dominates the propagation. In other words, the fracture will mainly propagate in the horizontal direction while keeping the same height. For example, while the fracture half-length grows from 13.7 m to 15.2 m, the half-height of the fracture increases only from 13.1 m to 13.4 m. When the fracture half-length reaches 15.2 m, the height of the fracture stays at 13.4 m, and vertical propagation stops.

3.5. Reference Case

To compare with the results from analytical method, a simulation was run with our numerical model for 9 minutes' fluid injection. The elastic modulus, Poisson's ratio, minimum horizontal stress, and fluid property values in the basic model are the same as parameters in Table 1. The fracture half-length reaches 15.2 m at the end of the early fracturing process, after which the fracture will mainly propagate in the horizontal direction, and the height of fracture will stop increasing. The result is shown in Figure 4.

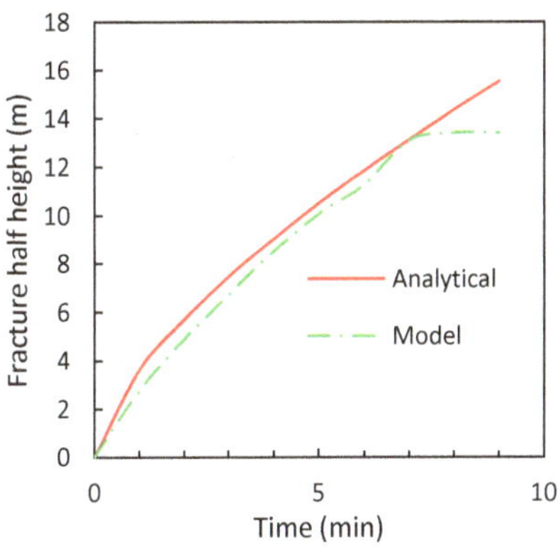

Figure 4. Fracture heights propagation with time.

3.6. Analytical Case

An analytical model has been proposed in the literature where the pressure drop within the fracture is ignored, and the fracturing fluid is assumed to fill the entire fracture. An assumed fracture length is also required in the analytical model, and the fracture half-height could be calculated via the equation below:

$$H = \left(\frac{QEt}{\sqrt{\pi}(1 - v^2)K_{IC}L} \right)^{2/3} \tag{16}$$

where Q is a quarter of the total injection rate at reservoir pressure, which is 0.00287 m^3/min in this study, and L is the fracture half-length. It can be seen from Equation (16) that the fracture height is very sensitive to the fracture toughness. The fracture half-height is calculated to be 14.36 m after 9 min of injection.

Figure 4 compares the calculated fracture heights derived from the analytical model and the finite element method applied in this study. The dotted line represents fracture height using the finite element method, while the solid curve depicts calculated fracture height using the analytical model. The two calculated fracture half-heights are similar within the first 7 min. However, once the fracture height reaches 13.4 m, the model used in this study stops growing, while that of the analytical model continues to increase at the same rate. This is because the fracture length is assumed to be a constant value in the analytical model, and the fracture does not propagate in the horizontal direction. When the fracture length is given, the fracture would continue to propagate in the vertical direction.

However, in fact, fracture propagation occurs in both vertical and horizontal direction, which is considered in this study. In the finite element model, the stress intensity factor is calculated for each time step and used as a key parameter to identify when the fracture propagation along the vertical direction stops. When the value of fracture tip toughness in the vertical direction is much larger than that of the horizontal direction, the fracture will stop extending in fracture height and focus on propagating in the horizontal direction. Fracture height will continue until the K_{IC} in both direction reaches equality in some certain time step.

3.7. Validation via Tracer Measurement

A radioactive tracer was run in the proppant stages of the treatment to measure the fracture height near the wellbore. The radioactive tracer was injected to measure the fracture height near the wellbore. The fracture treatments remain very restricted in the formation of interest with a fracture height of approximately 25 m (half-height is 12.5 m) [32]. The relative error of our model with real fracture height is 7.2%, indicating that the model is relatively accurate for fracture height calculation.

4. Sensitivity Analysis

This section discusses the characteristics of the response of the fracture height propagation and the effects of the main parameters on fracture height, including the grid size in the model, fluid viscosity, Young's modulus, Poisson's ratio, and minimum horizontal stress.

4.1. Fracturing Fluid Viscosity

The viscosity of the fracturing fluid affects pressure distribution along the fractures. Pressure drop within the fracture cannot be ignored, especially when the fracture length and/or height is long. Three fluid viscosities were studied in this work—1 mPa · s, 2.5 mPa · s, and 100 mPa · s. The results are shown in Figure 5. Fracture height decreases slightly as the fluid viscosity increases. This is because a higher fluid viscosity leads to a higher pressure drop along the fracture and a lower fluid pressure at the fracture tip.

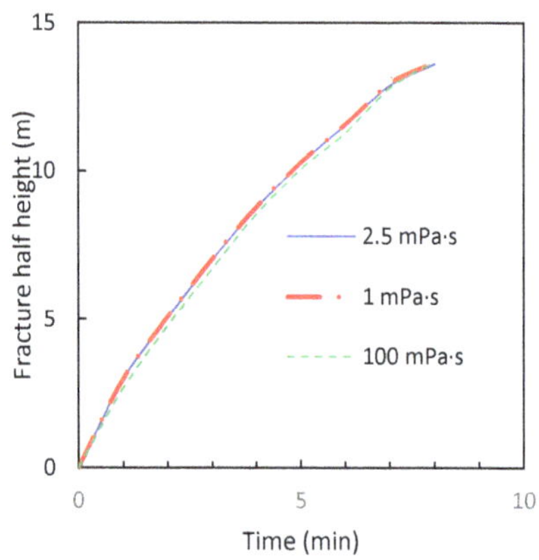

Figure 5. Fracture heights sensitivity to fluid viscosity.

4.2. Reservoir Rock Properties

4.2.1. Young's Modulus

Young's modulus indicates the stiffness of the reservoir rock, and it affects the stress intensity factor calculated at the fracture tip. Three Young's moduli were examined—13,789.5 MPa, 6894.8 MPa, and 34,473.8 MPa. Figure 6 depicts the fracture half-height calculated with the three Young's moduli, respectively. The Young's modulus can significantly affect the calculated half-height of the fracture. The higher the Young's modulus, the larger the calculated fracture height will be. The calculated fracture half-height increased by nearly 100% when the Young's modulus increased from 13,789.5 MPa to 34,473.8 MPa. This is because a high Young's modulus leads to a large stress intensity factor at the fracture tip at each time step in the simulation, and thus, the fracture tends to propagate further into the formation. In addition, a large modulus indicates that the reservoir matrix is stiffer. Thus, the fracture propagates further, while the width of fracture tends to be narrower.

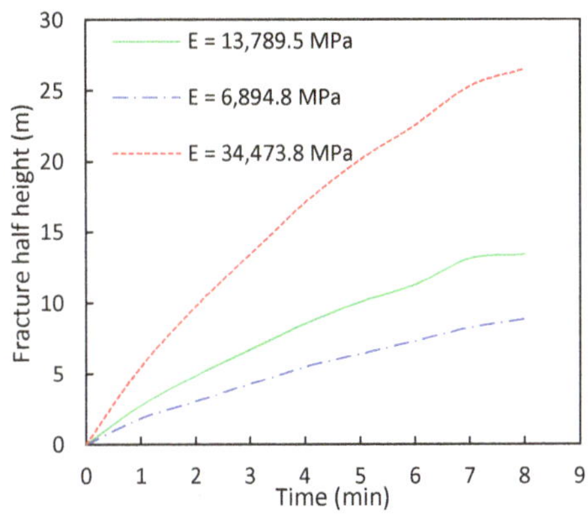

Figure 6. Fracture half-heights' sensitivity to Young's modulus.

4.2.2. Poisson's Ratio

The Poisson's ratio is another key parameter that affects the stress intensity factor at the fracture tip. Three Poisson's ratios—0.1, 0.2, and 0.5—were investigated in this study to evaluate its effect on fracture height. The results are shown in Figure 7, which demonstrates that a slight difference exists between scenarios when the Poisson's ratios are 0.1 and 0.2, where the fracture half-height increases from 13.1 m to 13.4 m at the end of simulation. A noticeable difference is shown when the Poisson's ratio increases to 0.5, where the calculated fracture half-height is increased to 14.6 m.

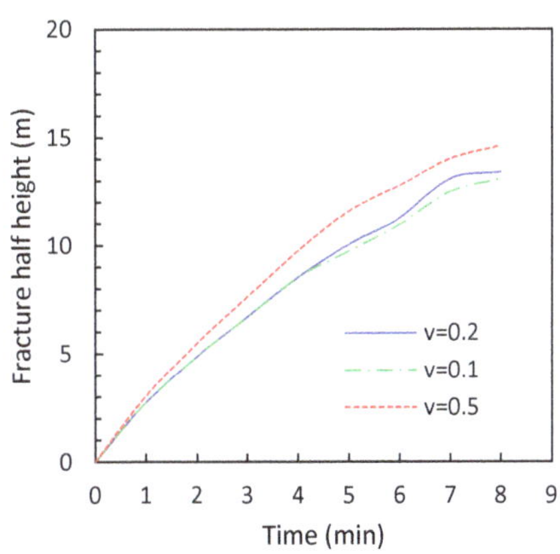

Figure 7. Fracture half-heights' sensitivity to Poisson's ratio.

4.3. *Minimum Horizontal Stress*

Minimum horizontal stress is an essential parameter to calculate the stress intensity factor at the fracture tip and the net pressure, which is the difference between fluid pressure in the fracture and rock stress on the fracture surface. Three minimum horizontal stresses of

34.5 MPa, 44.2 MPa, and 55.2 MPa were simulated in this study, and the results are shown in Figure 8. Calculated fracture height can be greatly affected by minimum horizontal stress. The fracture half-height reached 20.1 m for the scenario with a 55.2 MPa as the minimum horizontal stress compared to 13.4 m and 11.9 m for scenarios of 44.2 MPa and 34.5 MPa. The fracture propagates further under a smaller minimum horizontal stress. This is mainly because a lower minimum horizontal stress in the rock leads to a relatively large net pressure in the fracture and large stress intensity factor at the fracture tip.

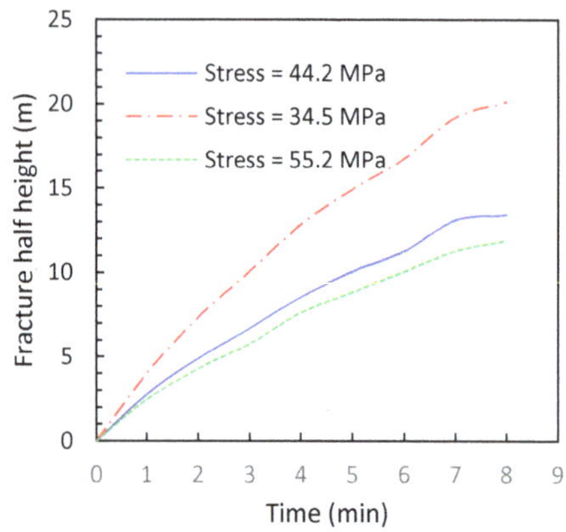

Figure 8. Fracture half-heights' sensitivity to minimum horizontal stress.

4.4. Grid Size

In this study, fracture half-height is calculated to be 13.4 m with the vertical grid size of 0.3 m. To examine the sensitivity of the grid size to calculated half-height results, the model was repeated twice with a grid size of 0.075 m and 0.038 m in the fracture height direction. The results are shown in the Figure 9. A slight difference was found for the three grid sizes, indicating that 0.3 m of the grid size is sufficient for the calculation. With grid size changing from 0.3048m to 0.0762 m and 0.0381 m, the fracture half-height at the end of injection increased from 13.41 m to 13.62 m and 13.71 m. When the model grid is smaller, the dotted line of fracture half-height becomes smoother, but the influence of grid size on the fracture height is not significant.

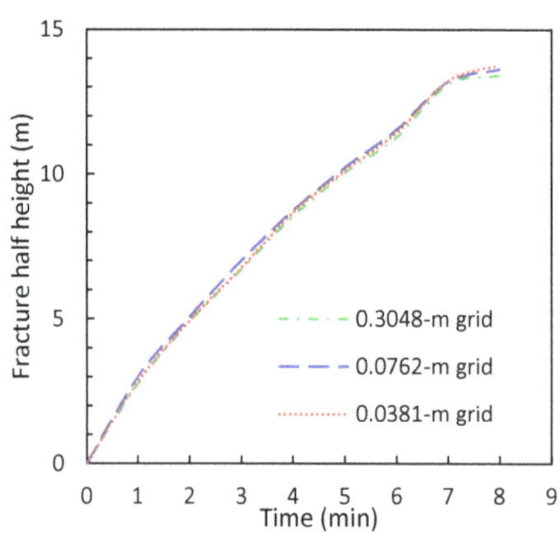

Figure 9. Fracture half-heights' sensitivity to grid size.

5. Conclusions

1. An innovative numerical model that fully couples the hydraulic fracture propagation, fluid flow in the fracture, and fluid leak-off into the reservoir matrix by finite element method is established to calculate fracture height in the tight formation using the proposed model;
2. The well-based numerical model is successfully used in the filed case of Montney, which indicates a relative error of 7.2% compared with the field tracer result;
3. The sensitivity analysis indicates that fracture height can be significantly affected by Young's modulus and minimal horizontal stress. A high Young's modulus leads to an increased stress intensity factor at the fracture tip for each time step, which prompts the fracture to advance further, while the width of fracture becomes smaller. The influence of grid size on height fracture is not significant. When the model grid is smaller, the trend of fracture height propagation becomes smoother.

Author Contributions: Conceptualization, J.C.; methodology, J.C.; software, J.C.; validation, J.C.; formal analysis, J.C.; investigation, J.C.; resources, J.C. and F.L.; data curation, J.C.; writing—original draft preparation, J.C.; writing—review and editing, J.C.; visualization, J.C.; supervision, F.L.; project administration, F.L.; funding acquisition, F.L. All authors have read and agreed to the published version of the manuscript.

Funding: This research was funded by YK-2022-34-3, project "Numerical simulation and software module development for hydraulic fracture propagation in a complex reservoir with the ultra-tight sandstone gas" from SINOPEC Petroleum Exploration and Production Research Institute.

Institutional Review Board Statement: Not applicable.

Informed Consent Statement: Not applicable.

Data Availability Statement: Not applicable.

Conflicts of Interest: The authors declare no conflict of interest.

Nomenclatures

w	Fracture width, m
Δw	Fracture width increment, m
q	Fluid flux, $m^3 \cdot s^{-1}$
t	Time, seconds
C	Leak-off term, $m \cdot s^{-1}$
p_f	Fluid pressure in the fracture, mpa
Δp	Pressure increment, mpa
μ	Fluid viscosity, mpa·s
M	Pressure influence coefficient matrix
C_l	Leak-off coefficient, $m \cdot s^{-0.5}$
$t_0(y)$	The time of fracture tip arrives at y, seconds
K_I	Stress intensity factor, $mpa \cdot m^{0.5}$
K_{IC}	Critical stress intensity factor, $mpa \cdot m^{0.5}$
E	Young's modulus, mpa
v	Poisson's ratio
u	Displacement, m
σ	Stress, mpa
ε	Strain
σ^a	Auxiliary stress, mpa
u^a	Auxiliary displacement, m
δ	The Kronecker delta
χ	Scalar field
r	Distance from the fracture tip, m
θ	The angle from the tangent to the fracture path, radians

κ	The Kolosov constant
G	Shear modulus, mpa
$l(t)$	The fracture half-height at time t, m
p_0	Initial net fluid pressure, mpa
P_c	Critical load, mpa
c	Fracture half-length in the disc, mm
R	Radius of disc, mm
t_1	Thickness of the disc, mm

References

1. Parshall, J. Barnett Shale showcases tight-gas development. *J. Pet. Technol.* **2008**, *60*, 48–55. [CrossRef]
2. Spellman, F.R. *Hydraulic Fracturing Wastewater: Treatment, Reuse, and Disposal*; CRC Press: Boca Raton, FL, USA, 2017.
3. Harrison, E.; Kieschnick, W.F., Jr.; McGuire, W.J. The mechanics of fracture induction and extension. *Trans. AIME* **1954**, *201*, 252–263. [CrossRef]
4. Howard, G.C.; Fast, C.R. Optimum fluid characteristics for fracture extension. In *Drilling and Production Practice*; American Petroleum Institute: Washington, DC, USA, 1957.
5. Crittendon, B.C. The mechanics of design and interpretation of hydraulic fracture treatments. *J. Pet. Technol.* **1959**, *11*, 21–29. [CrossRef]
6. Perkins, T.K.; Kern, L.R. Widths of hydraulic fractures. *J. Pet. Technol.* **1961**, *13*, 937–949. [CrossRef]
7. Nordgren, R.P. Propagation of a vertical hydraulic fracture. *Soc. Pet. Eng. J.* **1972**, *12*, 306–314. [CrossRef]
8. Khristianovic, S.; Zheltov, Y. Formation of vertical fractures by means of highly viscous fluids. In Proceedings of the 4th World Petroleum Congress, Rome, Italy, 6–15 June 1955; Volume 2, pp. 579–586.
9. Geertsma, J.; De Klerk, F. A rapid method of predicting width and extent of hydraulically induced fractures. *J. Pet. Technol.* **1969**, *21*, 1571–1581. [CrossRef]
10. Sneddon, I.N.; Elliot, H.A. The opening of a Griffith crack under internal pressure. *Q. Appl. Math.* **1946**, *4*, 262–267. [CrossRef]
11. Green, A.E.; Sneddon, I.N. The distribution of stress in the neighborhood of a flat elliptical crack in an elastic solid. In *Mathematical Proceedings of the Cambridge Philosophical Society*; Cambridge University Press: Cambridge, UK, 1950; Volume 46, pp. 159–163.
12. Simonson, E.R.; Abou-Sayed, A.S.; Clifton, R.J. Containment of massive hydraulic fractures. *Soc. Pet. Eng. J.* **1978**, *18*, 27–32. [CrossRef]
13. Fung, R.L.; Vilayakumar, S.; Cormack, D.E. Calculation of vertical fracture containment in layered formations. *SPE Form. Eval.* **1987**, *2*, 518–522. [CrossRef]
14. Warpinski, N.R.; Schmidt, R.A.; Northrop, D.A. In-situ stresses: The predominant influence on hydraulic fracture containment. *J. Pet. Technol.* **1982**, *34*, 653–664. [CrossRef]
15. Warpinski, N.R.; Teufel, L.W. Influence of geologic discontinuities on hydraulic fracture propagation (includes associated papers 17011 and 17074). *J. Pet. Technol.* **1987**, *39*, 209–220. [CrossRef]
16. Renshaw, C.E.; Pollard, D.D. An experimentally verified criterion for propagation across unbounded frictional interfaces in brittle, linear elastic materials. *Int. J. Rock Mech. Min. Sci. Geomech. Abstr.* **1995**, *32*, 237–249. [CrossRef]
17. Paris, P.C.; Sih, G.C. Stress analysis of cracks. In *Fracture Toughness Testing and Its Applications*; ASTM International: West Conshohocken, PA, USA, 1965.
18. Detournay, E. Propagation regimes of fluid-driven fractures in impermeable rocks. *Int. J. Geomech.* **2004**, *4*, 35–45. [CrossRef]
19. Sheibani, F.; Olson, J. Stress intensity factor determination for three-dimensional crack using the displacement discontinuity method with applications to hydraulic fracture height growth and non-planar propagation paths. In *Effective and Sustainable Hydraulic Fracturing*; InTech: Vienna, Austria, 2013.
20. Chen, Y.M. Numerical computation of dynamic stress intensity factors by a Lagrangian finite-difference method (the HEMP code). *Eng. Fract. Mech.* **1975**, *7*, 653–660. [CrossRef]
21. Pande, G.; Beer, G.; Williams, J. Numerical methods in rock mechanics. *Int. J. Rock Mech. Min. Sci.* **2002**, *39*, 409–427.
22. Olson, J.E.; Taleghani, A.D. Modeling simultaneous growth of multiple hydraulic fractures and their interaction with natural fractures. In Proceedings of the SPE Hydraulic Fracturing Technology Conference, The Woodlands, TX, USA, 19–21 January 2009.
23. Ma, D.; Duan, H.; Zhang, J.; Liu, X.; Li, Z. Numerical Simulation of Water–Silt Inrush Hazard of Fault Rock: A Three-Phase Flow Model. *Rock Mech. Rock Eng.* **2022**, *55*, 5163–5182. [CrossRef]
24. Ma, D.; Duan, H.; Zhang, J.; Bai, H. A state-of-the-art review on rock seepage mechanism of water inrush disaster in coal mines. *Int. J. Coal Sci. Technol.* **2022**, *9*, 50. [CrossRef]
25. Ma, D.; Duan, H.; Zhang, J. Solid grain migration on hydraulic properties of fault rocks in underground mining tunnel: Radial seepage experiments and verification of permeability prediction. *Tunn. Undergr. Space Technol.* **2022**, *126*, 104525. [CrossRef]
26. Boone, T.J.; Ingraffea, A.R. A numerical procedure for simulation of hydraulically-driven fracture propagation in poroelastic media. *Int. J. Numer. Anal. Methods Geomech.* **1990**, *14*, 27–47. [CrossRef]
27. Freund, L.B. Stress intensity factor calculations based on a conservation integral. *Int. J. Solids Struct.* **1978**, *14*, 241–250. [CrossRef]
28. Yau, J.F.; Wang, S.S.; Corten, H.T. A mixed-mode crack analysis of isotropic solids using conservation laws of elasticity. *J. Appl. Mech.* **1980**, *47*, 335–341. [CrossRef]

29. Dembicki, M.; Nevokshonoff, G.; Johnsen, J.; Spence, M. The super pad—A multi-year integrated approach to resource development in the montney. In Proceedings of the Unconventional Resources Technology Conference, San Antonio, Texas, 20–22 July 2015; Society of Exploration Geophysicists: Tulsa, OK, USA; American Association of Petroleum Geologists: Tulsa, OK, USA; Society of Petroleum Engineers: Houston, TX, USA; pp. 2684–2695.
30. Stevens, S.; Ruuskraa, V. Special report: Gas shale-1: Seven plays dominate North America activity. *Oil Gas J.* **2009**, *29*, 36–41.
31. Aziz, N.I.; Schmidt, L.C. Rock fracture-toughness determination by the Brazilian test. *Eng. Geol.* **1993**, *33*, 177–188.
32. Popp, M. Completion and Stimulation Optimization of Montney Wells in the Karr Field. Ph.D. Thesis, University of Calgary, Calgary, AB, Canada, 2015.

processes

MDPI

Article

Cenozoic Subsidence History of the Northern South China Sea: Examples from the Qiongdongnan and Yinggehai Basins

Ming Ma [1], Jiafu Qi [2,*], Jinshan Ma [3], Heng Peng [4], Linlin Lei [5], Qian Song [1], Qing Zhang [1] and Mengen Bai [1]

1 Department of Geological Engineering, Qinghai University, Xining 810016, China; maming1002@126.com (M.M.)
2 State Key Laboratory of Petroleum Resource and Prospecting, College of Geosciences, China University of Petroleum, Beijing 102249, China
3 Xining Land Surveying and Planning Research Institute, Xining 810008, China
4 Department of Geology, Northwest University, Xi'an 710077, China
5 Research Institute of Logging Application, China Petroleum Logging Co., Ltd., Xi'an 710077, China
* Correspondence: qijiafu@cup.edu.cn

Abstract: The Qiongdongnan and Yinggehai Basins are important petroliferous basins. To study the Cenozoic subsidence characteristics of these two basins, their controlling factors, and their implications, we studied the basins' subsidence characteristics via one-dimensional, two-dimensional, and holistic subsidence. Then, we compared the basins' subsidence characteristics based on the evolution of several particular geological processes that occurred in the South China Sea (SCS) and adjacent areas. The results indicated that the change in the holistic subsidence of both basins occurred episodically. In addition, the subsidence in these two basins differed, including their subsidence rates, the migration of the depocenters, and the changes in the holistic subsidence. The dynamic differences between the two basins were the main factors controlling the differences in the subsidence in the two basins. In the Qiongdongnan Basin, the subsidence characteristics were primarily controlled by the mantle material flowing under the South China Block in the Eocene and the spreading of the SCS from the Oligocene to the Miocene. In the Yinggehai Basin, the subsidence characteristics were primarily controlled by the coupling between the uplift of the Tibetan Plateau and the strike-slip motion of the Red River Fault before the Early Miocene and by only the effect of the strike-slip motion of the Red River Fault from the Middle Miocene to the Late Miocene. Since the Pliocene, the subsidence characteristics of both basins have been principally controlled by the dextral strike-slip motion of the Red River Fault. The major faults contributed to the spaciotemporal variations in the subsidence within each basin.

Keywords: Qiongdongnan basin; Yinggehai basin; subsidence characteristics; variations; controlling factors

Citation: Ma, M.; Qi, J.; Ma, J.; Peng, H.; Lei, L.; Song, Q.; Zhang, Q.; Bai, M. Cenozoic Subsidence History of the Northern South China Sea: Examples from the Qiongdongnan and Yinggehai Basins. *Processes* **2023**, *11*, 956. https://doi.org/10.3390/pr11030956

Academic Editors: Linhua Pan, Yushi Zou, Jie Wang, Minghui Li, Wei Feng and Lufeng Zhang

Received: 31 December 2022
Revised: 13 March 2023
Accepted: 16 March 2023
Published: 21 March 2023

1. Introduction

The South China Sea and the adjacent areas are characterized by complicated and diverse geological phenomena and are surrounded by the Indian, Eurasian, and Pacific plates. Thus, these areas are also referred to as a natural geological laboratory [1]. Regarding the formation mechanism and the dynamic processes of the evolution of the SCS and the adjacent areas, various dynamic models have been proposed by researchers around the world based regarding different methods, including the back-arc-spreading model [2], the collision-extrusion model [3,4], the mantle-upwelling model [5,6], and the expanding-continental-margin model [7]. However, there are still some conflicts among these models, and each model has limitations [8] Even for the same model, discrepancies exist between different studies. The Qiongdongnan Basin (QDNB) and the Yinggehai Basin (YGHB), located in the northern continental margin of the SCS, are two important petroliferous basins. However, the tectonic locations of these two basins are particularly specific, and the sedimentary stratigraphy and the structures of these two basins are significantly different. In addition, several particular geological processes, such as the uplift of the Tibetan Plateau, the strike-slip motion of the Red River Fault, and the expansion of the SCS during the Cenozoic period, have dramatically affected the subsidence, filling, and evolution of these two basins. However, the role of each geological process in the basins' evolutions is still not clear. In addition, the geodynamics of the basins are poorly understood and controversial.

A basin's subsidence history is the reestablishment of the basin's tectonic and sedimentary evolution, and it records the significant geological events and reflects the basin's evolution [9]. Conversely, the characteristics of a basin's subsidence are fundamentally controlled by the basin's dynamics [10]. In view of this, in this study, the similarities and the differences in the Cenozoic subsidence characteristics of the two basins were investigated via comparative analysis, and then, the subsidence features of the two basins were, as compared to the geological processes that occurred in the SCS and its adjacent areas to ascertain the main factors controlling the mechanisms of the subsidence found in these two basins. The results of this study provide important evidence for the reconstruction of the geodynamic processes as well as important theoretical guidance for the deployments of oil-and-gas exploration in these two basins [11–13].

2. Geological Setting

The QDNB and the YGHB are located in the northern continental margin of the SCS and are surrounded by the Indian, Eurasian, and Pacific Plates. They are adjacent basins but have significantly different formation mechanism [14]. The QDNB, located at the intersection of the northern SCS margin and the Red River Fault Zone, is NE trending and consists of three major tectonic units from northwest to southeast [15]: the northern depression, the central uplift, and the central depression. The structural framework of the QDNB is characterized by alternating depressions and uplifts (Figure 1). The YGHB is diamond-shaped, trends NW, and consists of three major tectonic units: the Yingdong slope; a central depression composed of the Yinggehai sag, Henei sag, and Lingao rise; and the Yingxi slope (Figure 1). The evolution of the QDNB and the YGHB can be divided into two stages as a whole: a rifting stage during the Paleogene period and a post-rifting stage from the Neogene to the Quaternary period, which can be further divided into depression and thermal subsidence stages [16,17] (Figure 2).

The strata in the QDNB and the YGHB have been studied in detail, except for the Paleocene and Eocene strata, which have not been sampled or assigned specific formation names (Figure 2). The Paleocene strata primarily consist of volcanic rocks in these two basins, but the difference is that a small amount of limestone was deposited in the YGHB. The Eocene strata primarily consist of lacustrine and alluvial-facies conglomerate, sandstone, and mudstone. The sediments of the Lower Oligocene Yacheng Formation, including the fluvial, lacustrine, swamp, and fan-delta facies, are finer grained than those of the previous period and consist of mudstone, sandstone, and siltstone. The difference between these two basins is that a coal seam developed in the QDNB, while shale developed in the YGHB. The upper Oligocene Lingshui Formation consists of more fine-grained sediments, including interbedded siltstone and mudstone, because the water depth increased, and the sedimentary environment changed into delta and littoral facies. From the Early Miocene to the end of the Middle Miocene, when the Sanya and Meishan Formations were deposited, the sedimentary characteristics were different in the two basins. The sediments in the QDNB mostly consisted of interbedded siltstone and mudstone under the shallow platform and littoral facies. However, most sediments in the YGHB are mudstone, which developed under the littoral-shallow facies. From the Late Miocene until the present, i.e., the deposition time of the Huangliu–Ledong Formations, thick mudstone and thin siltstone formed in the bathyal facies in the QDNB, whereas in the YGHB, the proportion of siltstone is higher [17].

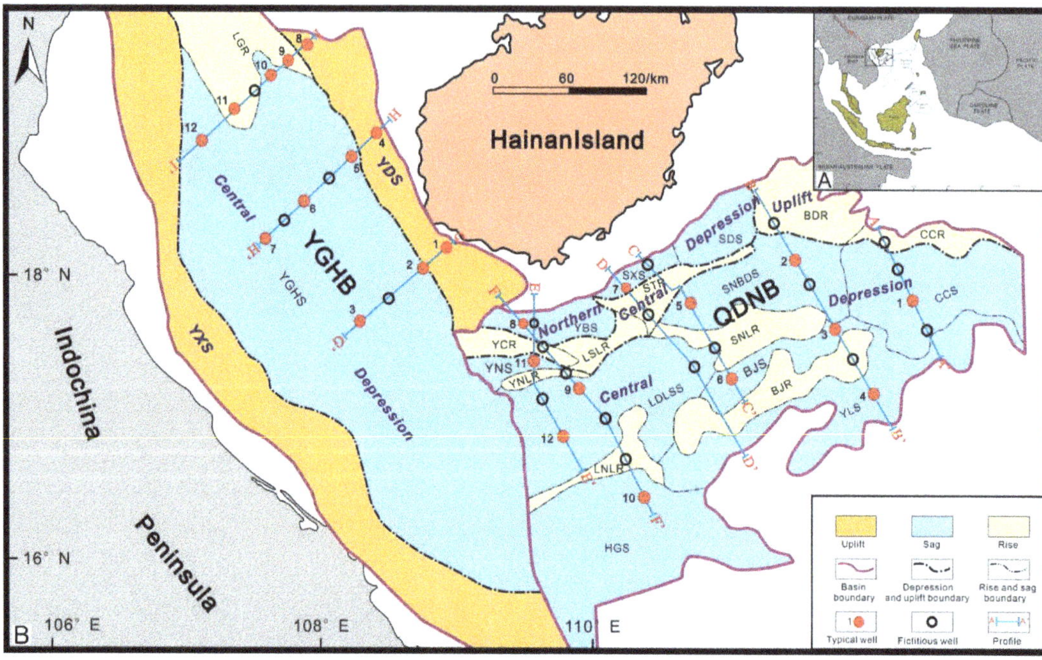

Figure 1. Map showing the tectonic units in the Qiongdongnan and Yinggehai Basins. YGHB, Yinggehai Basin; QDNB, Qiongdongnan Basin; YBS, Yabei sag; SXS, Songxi sag; SDS, Songdong sag; YNS, Yanan sag; SNBDS, Songnanbaodao sag; LDLSS, Ledonglingshui sag; BJS, Beijiao sag; CCS, Changchang sag; YLS, Yongle sag; HGS, Huangguang sag; YGHS, Yinggehai sag; YCR, Yacheng rise; LSLR, Lingshui low rise; YNLR, Yanan low rise; STR, Songtao rise; BDR, Baodao rise; CCR, Changchang rise; SNLR, Songnan low rise; LNLR, Lingnan low rise; BJR, Beijiao rise; LGR, Linggao rise. The locations of the wells and seismic profiles are also shown (modified after Shi et al. [18], Zhu et al. [19]).

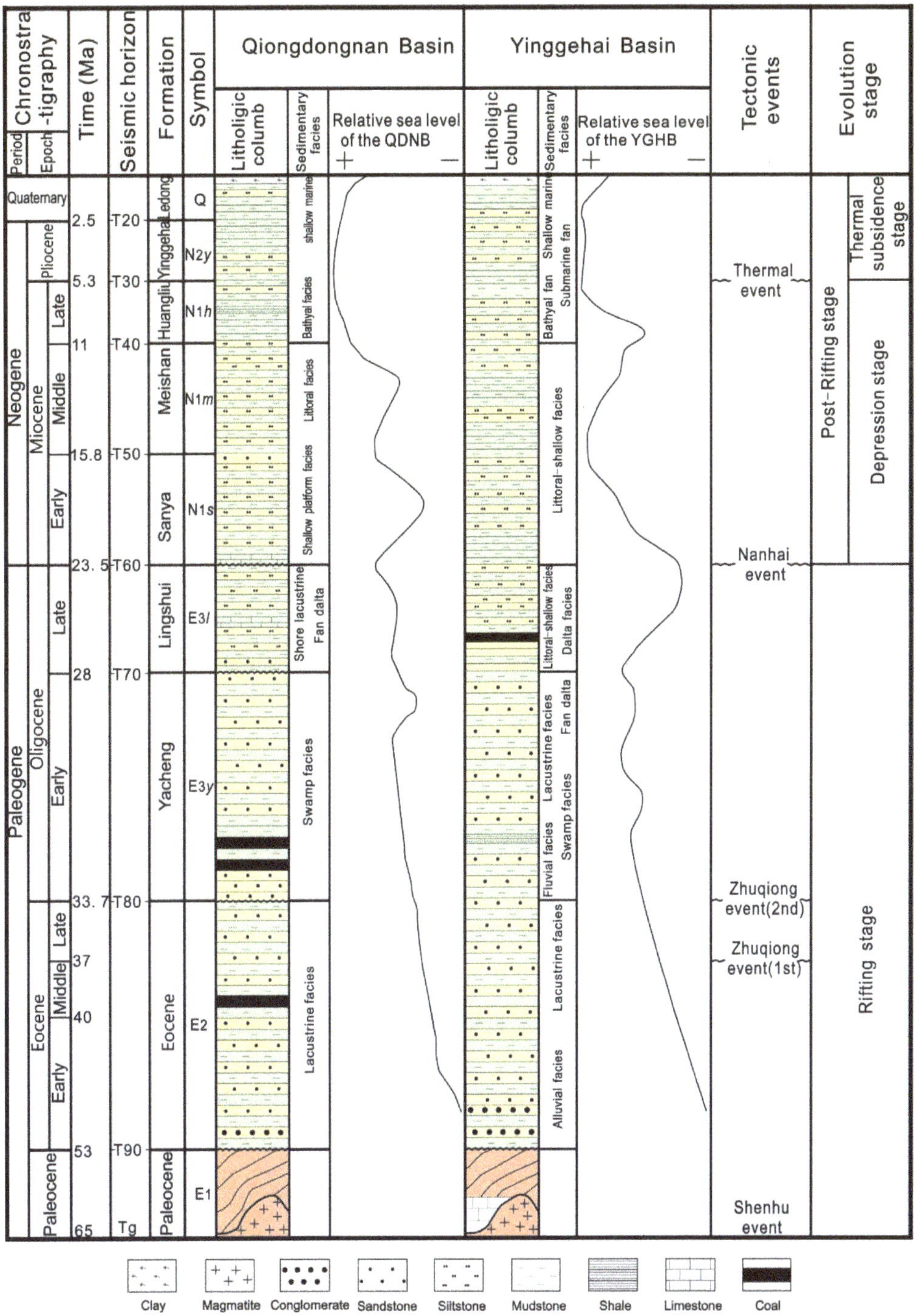

Figure 2. Simplified stratigraphic columns for the QDNB and the YGHB (modified after Zhu et al. [19], All China Commission of Stratigraphy [20], Wang et al. [21], Chen et al. [22], Liu et al. [23]).

3. Data and Methods

3.1. Data

Due to high acquisition costs, the well and seismic data in the QDNB and the YGHB are limited, and most of the early data are fragmentary in terms of the data needed to study subsidence, such as the lithology data. In this study, we chose three and six interpreted and most recently acquired seismic profiles for the analyses of the YGHB and QDNB, respectively. This effectively reduced the uncertainty and ensured the results were more reliable. The number of drilling wells was relatively small in the study area, as compared to the area of the two basins and the requirements of such a study. Therefore, several fictitious wells were created. To create this kind of well, the drilling wells were correlated with the seismic data first. Next, we chose several locations for the fictitious wells near the drilling wells on the seismic profiles for the calculations and analysis. Twenty-eight wells in QDNB and sixteen wells in the YGHB were selected to calculate and analyze their subsidence history (Figure 1). Most of the units were distributed by fictitious wells as soon as possible.

3.2. Method and Parameters

3.2.1. Methods

We analyzed the basin subsidence at three levels, i.e., the subsidence of single wells (1D), the subsidence along the profile (2D), and the subsidence of the entire basin. As for the subsidence calculation process, we synthesized and improved the traditional back-stripping method. The details are provided below.

The total subsidence (STT) of the basin could be divided into four components, namely, the tectonic subsidence induced by tectonism (ST), the loading subsidence caused by the sediments (D), the loading subsidence induced by sea-level changes (D(ΔL)), and the retroaction induced by the lithosphere (DT), which could be ignored in extensional basins [18,24]. Thus, the abovementioned relationships could be expressed mathematically as follows:

$$\mathrm{STT} = \mathrm{ST} + \mathrm{D} + \mathrm{D}(\Delta \mathrm{L}). \quad (1)$$

The total subsidence (STT) could be expressed as follows:

$$\mathrm{STT} = \mathrm{Hs} + \mathrm{Hw} + \Delta \mathrm{L}, \quad (2)$$

where Hs is the de-compacted stratum thickness, which could be obtained using the Authy equation [25]; Hw is the paleo-water depth; and ΔL is the value of the eustatic sea-level change. Hw and ΔL could be acquired in several ways, which is introduced in detail in a subsequent section.

The loading subsidence caused by the sediments (D) and the loading subsidence induced by sea-level change (D(ΔL)) could be acquired based on the Airy isostasy [26], as follows:

$$\mathrm{D} = \mathrm{Hs} \times (\varrho\mathrm{s} - \varrho\mathrm{w})/(\varrho\mathrm{m} - \varrho\mathrm{w}), \quad (3)$$

$$\mathrm{D}(\Delta \mathrm{L}) = \Delta \mathrm{L} \times \varrho\mathrm{w}/(\varrho\mathrm{m} - \varrho\mathrm{w}), \quad (4)$$

where ϱs, ϱm, and ϱw are the densities of the mean sediments, mantle, and water, respectively. Finally, we could easily acquire the tectonic subsidence (ST) using Equation (1).

The total subsidence and tectonic subsidence could be acquired simultaneously by using this technique. In addition, several controversial problems could be avoided by using this method, for example, the water-loaded basin subsidence and the uncertainty of the parameters in the tectonic subsidence calculation [27].

The calculations were conducted using MATLAB (R2014a) with a precision set at 10^{-5}; Excel software for further data processing and mapping; and finally, CorelDRAW software for map beautification.

3.2.2. Parameters

Many parameters were involved in the subsidence calculations and analysis, and every parameter had to be analyzed and processed scientifically and reasonably.

- Dates of strata boundaries: First, we identified the different degrees of the chronostratigraphic units and rock stratigraphic units in the seismic and well profiles [19]. Then, we determined the age of every chronostratigraphic unit's boundary after chronostratigraphic correlation and constructed a chronostratigraphic framework [20,21]. We also determined the ages of each rock stratigraphic unit's boundaries after the rock stratigraphic correlation and analysis of the biological fossils [22,23]. Finally, the stratigraphic columns of the basins were constructed to study the sedimentology and tectonics of the basin (Figure 2).
- Porosity, density, and compaction coefficient: According to the basin's stratigraphy, we analyzed the lithology and proportions of each component of each formation. Based on the results, the density of each formation was calculated by averaging their weighted components, which ensured the credibility of the results [17]. The densities of the mantle and water were 3330 kg/m^3 and 1000 kg/m^3, respectively [28]. For the surface porosity and compaction coefficient, previously published results on the subsidence in the QDNB and the YGHB were directly adopted, and these only considered the sandstone and mudstone while ignoring the other components of the strata in the North Sea Basins [29]. Furthermore, the sedimentary and diagenetic environments of the QDNB and the YGHB are also different from their counterparts in the North Sea Basins. The differences between the two study areas inevitably induced errors. To reduce these errors as much as possible, we acquired the surface porosity and compaction coefficient in the same way as the density. The surface porosity and compaction coefficient of each lithology are listed in Table 1.

Table 1. Surface coefficient of compaction, surface porosity, and density of different lithologies in a normal environment (according to Gao et al. [24]).

Lithology	Surface Coefficient of Compaction (km^{-1})	Surface Porosity (%)	Density (kg/m^3)
Mudstone	0.51	0.63	2720
Sandy mudstone	0.39	0.56	2680
Sandstone	0.27	0.49	2650
Conglomerate	0.22	0.46	2640

- Paleo-water depth and eustatic sea-level change: There are many ways to estimate the paleo-water depth, including paleontology, sedimentary facies analysis, geochemical indexes, and geomorphology back-stripping [30]. We obtained the paleo-water depth using the following steps: Firstly, we studied the types of sedimentary facies developed during each period in the basin, which were then analyzed based on updated data, including cuttings, cores, seismic data, and well logs [16,17,19]. Then, we obtained the paleo-water depth during each period according to the relationship between the water depth and sedimentary facies. Finally, to increase the accuracy, we also compared the results of the relative sea levels in the two basins, which were obtained from previous studies (Figure 2).

But it should be noted that the determination of the facies was based on an integrated analysis of sedimentary and fossil characteristics from various sources, which inevitably has led to mistakes in estimating the paleobathymetry and the subsidence [31]. A number of researchers have conducted studies on eustatic sea-level change [32–34]. In this study, we adopted the results of studies based on borehole data for the continental margin of eastern North America because the tectonic setting and evolution of the northern continental margin of the SCS are very similar to those of the eastern margin of North America [32].

4. Results

4.1. Subsidence Characteristics of Single Wells (1D)

4.1.1. 1D Subsidence Characteristics in the QDNB

In the QDNB, 28 wells were selected to calculate the 1D subsidence (Figure 1), and 12 typical wells were chosen to analyze the spatiotemporal variations in the Cenozoic subsidence of the basin (Figure 3).

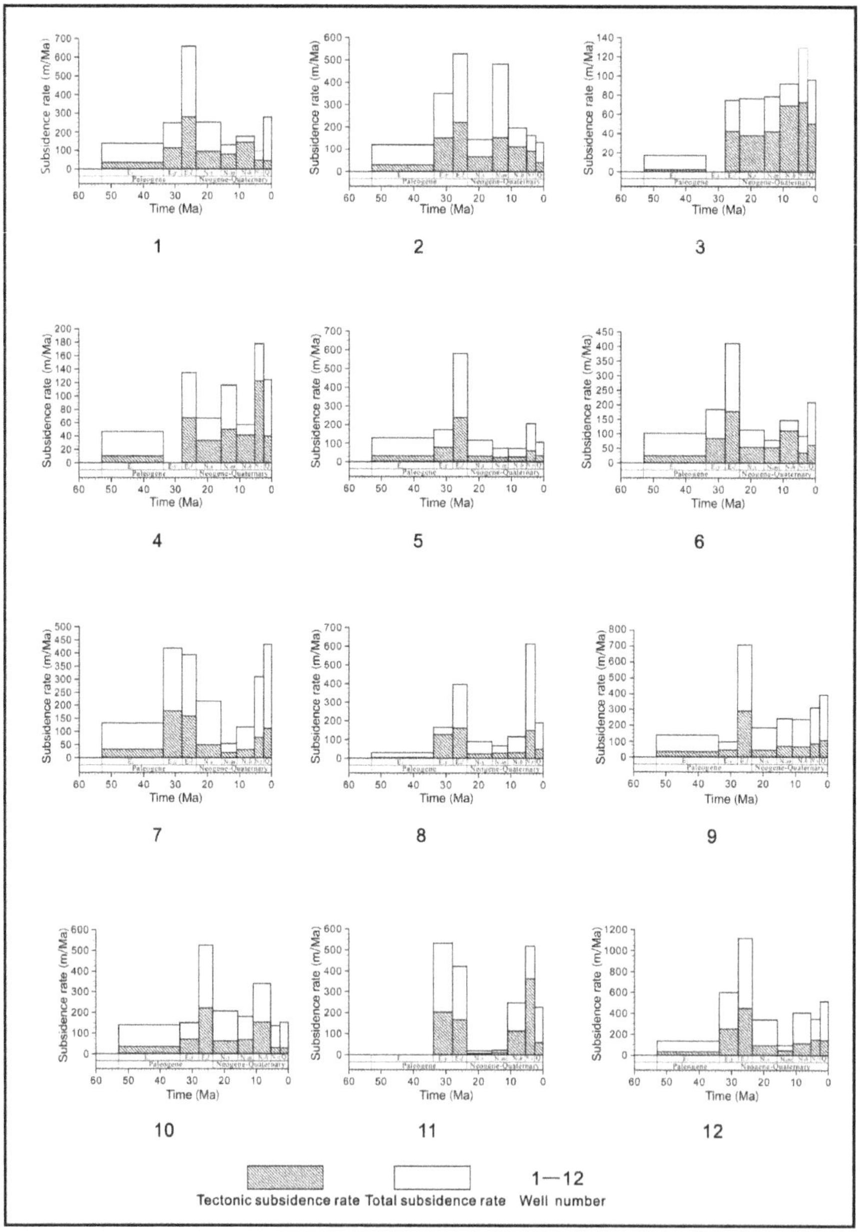

Figure 3. Subsidence rates of typical wells in the QDNB.

In the Eocene, the mean overall subsidence rate (Ra) was 83 m/myr and the mean tectonic subsidence rate (Rt) was 20 m/myr. The mean subsidence rate of one period in this paper was acquired by averaging the subsidence rates of all the wells for that period, which is described in a later section of this paper and was the same situation, so it is not introduced again. Based on the differences between the subsidence rates of the typical wells and the mean subsidence rate, the depocenters were divided into three types: (1) the main depocenters with subsidence rates greater than the mean subsidence rate, (2) the sub-depocenters with subsidence rates close to the mean subsidence rate; and (3) the non-depocenters with subsidence rates lower than the mean subsidence rate. The three types of depocenters are described in a later part of this paper, and the situation was the same, so they are not described here. The Changchang, Songnanbaodao, Songxi, western Ledonglingshui, and northern Huaguang sags were identified as the main depocenters (wells 1, 2, 5, 7, 9, 10, and 12 in Figures 1 and 3). The sub-depocenters were only distributed in the Beijiao sag (wells 3 and 6 in Figures 1 and 3). The Yongle and Yabei sags were the non-depocenters (wells 4 and 8 in Figures 1 and 3). The depocenters with the highest subsidence rates were distributed in the western Ledonglingshui and Yacheng sags. Apart from the Yabei and Yongle sags, the subsidence rates in the other areas were relatively high. Furthermore, the distribution of the main depocenters in the plane was an arc bending toward the north from east to west. In the Early Oligocene, the mean Ra and was 207 m/myr, and the Rt was 89 m/myr. The main depocenters were distributed in the eastern Songnanbaodao, Songxi, Yanan, and western Ledonglingshui sags (wells 2, 7, 11, and 12 in Figures 1 and 3). The Changchang, western Songnanbaodao, and western Beijiao sags were the sub-depocenters (wells 1, 5, and 6 in Figures 1 and 3). The non-depocenters included the eastern Beijiao, Yabei, northwestern Ledonglingshui, Huaguang, and Yongle sags (wells 3, 4, 8, 9, and 10 in Figures 1 and 3). The depocenters with the highest subsidence rates were still distributed in the western Ledonglingshui sag. The total subsidence features were similar to those in the previous period, except for the addition of subsidence areas in the south-eastern and northwestern parts of the basin. In the Late Oligocene, the mean Ra was 391 m/myr, and the mean Rt was 163 m/myr. The Changchang, Songnanbaodao, Ledonglingshui, and Huaguang sags were the main depocenters due to their high subsidence rates (wells 1, 2, 5, 6, 12, and 10 in Figures 1 and 3). The sub-depocenters included the western Beijiao, Songxi, and Yabei sags (wells 6, 7, 8, and 11 in Figures 1 and 3). The non-depocenters included the eastern Yabei and Yongle sags (wells 3 and 4 in Figures 1 and 3). The western Ledonglingshui sag was still the depocenter with the highest subsidence rate. Moreover, the planar distribution of the main depocenters was still a northward arc, and the sub-depocenters were distributed along the two sides of the arc. The southeastern part of the basin became the non-depocenter area, and the northwestern part of the basin became the sub-depocenters, as compared to the previous period. In the Early Miocene, the mean Ra was 129 m/myr, and the Rt was 41 m/myr. The main depocenters were the Changchang, Songxi, Huaguang, and western Ledonglingshui sags (wells 1, 7, 10, and 12 in Figures 1 and 3). The sub-depocenters included the Songnanbaodao sag, the western part of the Beijiao sag, and the northwestern Ledonglingshui and Yanan sags (wells 2, 5, 6, 9, and 11 in Figures 1 and 3). The eastern Beijiao, Yongle, and Yabei sags were the non-depocenters (wells 3, 4, and 8 in Figures 1 and 3). Overall, the subsidence features of this period were the same as those of the previous period, with only slight changes. The western part of the Ledonglingshui sag still had the highest subsidence rate among all units, and the planar distribution of the main depocenters and the sub-depocenters was still an arc bending toward the north, similar to that in the previous periods. However, the northwestern part of the basin became a non-depocenter area. In the Middle Miocene, the average Ra was 93 m/myr, and the mean Rt were and 41 m/myr. The main depocenters were primarily distributed in the Changchang, eastern Songnanbaodao, Yongle, northern Ledonglingshui, and Huaguang sags (wells 1, 2, 4, 9, and 10 in Figures 1 and 3). The only sub-depocenters were the eastern Beijiao and western Ledonglingshui sags (wells 3 and 12 in Figures 1 and 3). The western Songnanbaodao, western Beijiao, Songxi, Yabei, and Yanan sags were the non-depocenters (wells 5, 6, 7, 8, and 11 in Figures 1 and 3). The subsidence features of this period differed from those of the previous period. The depocenters with the largest subsidence rates were now in the northwestern Ledonglingshui sag. The range of the non-depocenters, which were distributed as an NW trending zone, was consistent with the trend of the Yacheng rise and the Lingshui and Songnan low rises and extended toward the inner part of the basin. Nevertheless, the depocenters, including the main and the sub-depocenters, were primarily distributed in the northeastern and southwestern parts of the non-depocenter zones. In the Late Miocene, the mean Ra was145 m/myr, and the mean Rt was 67 m/myr. The main depocenters included the Changchang, eastern Songnanbaodao, Ledonglingshui, and Huaguang sags (wells 1, 2, 9, 10, and 12 in Figures 1 and 3). The only sub-depocenter was the Beijiao sag (wells 3 and 6 in Figures 1 and 3). The Yongle, western Songnanbaodao, Songxi, Yabei, and Yanan sags were the non-depocenters (wells 4, 5, 7, 8, and 11 in Figures 1 and 3). The overall subsidence characteristics were similar to those during the previous period, but once again, the western Ledonglingshui sag became the depocenter with the highest subsidence rate. In addition, the non-depocenter zone, which had the same trend as the Lingshui low rise and the Yacheng and Songtao rises, shifted northward. In the Pliocene, the average Ra was 227 m/myr, and the average Rt was 79 m/myr. The Songxi, Yabei, Ledonglingshui, and Yanan sags were the main depocenters (wells 7, 8, 9, 12, and 11 in Figures 1 and 3). However, the sub-depocenters were primarily distributed in the Songnanbaodao, Yongle, and eastern Beijiao sags (wells 2, 5, 3, and 4 in Figures 1 and 3). The non-depocenter areas included the Changchang, western Beijiao, and Huaguang sags (wells 1, 6, and 10 in Figures 1 and 3). The subsidence features during this period were completely different from those of the previous periods. The depocenter with the largest subsidence rate was the Yabei sag, and the scope of the non-depocenters expanded toward the north, with a NE trend that was consistent with the trend of the Lingnan low rise and the Beijiao rise. The depocenters, including the main and the sub-depocenters, were distributed along the two sides of the non-depocenter zone, and the subsidence rate increased with increasing distance from the non-depocenter zone. In the Quaternary, the mean Ra was158 m/myr, and the mean Rt was 108 m/myr. The subsidence characteristics changed dramatically because the controlling factors varied from one area to another, and an anomalous subsidence occurred during this period. The northern part of the basin, e.g., the western Songnanbaodao, Songxi, Yabei, western Ledonglingshui, and Yanan sags (wells 5, 7, 8, 9, 11, and 12 in Figures 1 and 3), was characterized by a high Ra and a low Rt. In contrast, the southern part of the basin, e.g., the Changchang, eastern Songnanbaodao, Beijiao, Yongle, and Huaguang sags (wells 1, 2, 3, 4, 6, and 10 in Figures 1 and 3), was characterized by a low Ra and a high Rt.

The above analysis of the subsidence evolution indicated that from the Eocene to the Early Miocene, the subsidence features had remained similar overall. The northwestern and southeastern parts of the basin, especially the eastern Beijiao and Yongle sags, were the non-depocenter areas. The other areas of the basin were the depocenters, which were distributed in an arc shape, bending toward the north on a plane from east to west. The subsidence characteristics in the Middle and Late Oligocene were completely different from those during the previous periods, and the subsidence was characterized by the extension of the non-depocenter areas. The distribution of the non-depocenter areas during the Middle Miocene remained consistent with the trend of the Yacheng rise, Lingshui low rise, Songtao rise, and Songnan low rise, but it shifted to follow the trend of the Yacheng rise, Lingshui low rise, and Songtao rise in the Late Miocene. Generally, the depocenter areas were primarily distributed in the northeastern and southwestern parts of the basin, which had been separated by the non-depocenters during the Middle and Late Miocene periods. The subsidence features completely changed during the Pliocene. Most of the southern parts of the basin became the non-depocenters, the trend of which was consistently distributed with the Lingnan low rise and Beijiao rise, due to the northward migration of the depocenter areas. In the Quaternary, an anomalous subsidence occurred in the southern part of the basin.

4.1.2. 1D Subsidence Characteristics in the YGHB

In the YGHB, 16 wells were selected to calculate the 1D subsidence history, from which 12 typical wells were chosen to analyze the spatiotemporal variations in the subsidence of the basin (Figure 4). Because of the lack of exploration, only limited seismic and well data, along with limited information about the thick strata in the YGHB, were available. The strata systems of the basin that have been comprehensively studied include the Upper Oligocene and the Quaternary strata. Correspondingly, the time span of the subsidence history of the YGHB investigated in this paper was from the Late Oligocene to the Quaternary.

In the Late Oligocene, the mean Ra and Rt were 603 m/myr and 212 m/myr, respectively. The subsidence area was limited, and the main depocenters were distributed in the northwestern Yinggehai sag and the southeastern Linggao rise (wells 10, 11, and 12 in Figures 1 and 4). The non-depocenter areas were distributed only in the contact area between the northwestern Yinggehai sag and the southeastern Linggao rise (well 9 in Figures 1 and 4). In the Early Miocene, the average Ra was 159 m/myr, and the average Rt was 48 m/myr. The scope of the depocenter areas had extended further south during this period. The middle part of the Yinggehai sag was the main depocenter (wells 6 and 7 in Figures 1 and 4). The sub-depocenters were primarily distributed in the northeastern Yinggehai sag, which was located next to the Yingdong slope (wells 5, 9, and 10 in Figures 1 and 4). However, the northwestern Yinggehai sag and southeastern Linggao rise became the non-depocenters during this period (wells 11 and 12 in Figures 1 and 4). In the Middle Miocene, the mean Ra was 446 m/myr, and the mean Rt was 129 m/myr. The range of the depocenters persistently expanded southerly, as compared to their range during the previous period, and the entirety of the Yingdong slope subsided synchronously with different subsidence rates, which varied by location. The main depocenters were distributed in the southeastern Yinggehai sag and the Yingdong slope (wells 1, 2, and 3 in Figures 1 and 4). The main depocenters during the previous period became the sub-depocenters during this period (wells 6 and 7 in Figures 1 and 4). The northwestern part of the basin became the non-depocenter area (wells 5, 8, 9, 10, 11, and 12 in Figures 1 and 4). In the Late Miocene, the mean Ra and Rt were 87 m/myr and 21 m/myr, respectively. The subsidence features of this period were completely different from those of the previous periods, and the depocenters migrated to the northwestern part of the basin. The main depocenters were primarily distributed in the central and northwestern parts of the Yinggehai sag, the northwestern Yingdong slope, and the southeastern Linggao rise (wells 3, 6, 8, 10, 11, and 12 in Figures 1 and 4). The sub-depocenters were distributed in the western Yingge sag and the southeastern Linggao rise (wells 7 and 9 in Figures 1 and 4). The central and southeastern parts of the Yingdong slope, as well as the adjacent area, became the non-depocenter areas during this period (wells 1, 2, 4, and 5 Figures 1 and 4). In the Pliocene, the mean Ra was 483 m/myr, and the mean Rt was 130 m/myr. The main depocenters were concentrated in the central Yinggehai sag (wells 2, 3, 5, 6, and 7 in Figures 1 and 4). However, the contact area between the Yinggehai sag and the Linggao rise became the sub-depocenter area (wells 9, 10, and 11 in Figures 1 and 4). In addition, the entire Yingdong slope and the northwestern Yinggehai sag became the non-depocenter areas (wells 1, 4, 8, and 12 in Figures 1 and 4). The subsidence characteristics changed, as compared to the previous period. For example, the depocenters shifted from the northwest to the central part of the basin, and most of the non-depocenters were distributed on the Yingdong slope. In the Quaternary, the mean Ra was 140 m/myr, and the Rt was 27 m/myr. The overall subsidence features only changed slightly and were similar to those of the previous period overall. The range of the depocenters shrank, and they were distributed in the southeastern part of the basin, which caused the northwestern part of the basin to become the non-depocenter area. The main depocenters were distributed in the southeastern Yinggehai sag (wells 2 and 3 in Figures 1 and 4). However, the central Yinggehai sag and the northwestern Yingdong slope became the sub-depocenters (wells 5, 6, 7, and 8 in Figures 1 and 4). The non-depocenters were primarily distributed in the northeastern and northwestern parts of the basin (wells 1, 4, 9, 10, 11, and 12 in Figures 1 and 4).

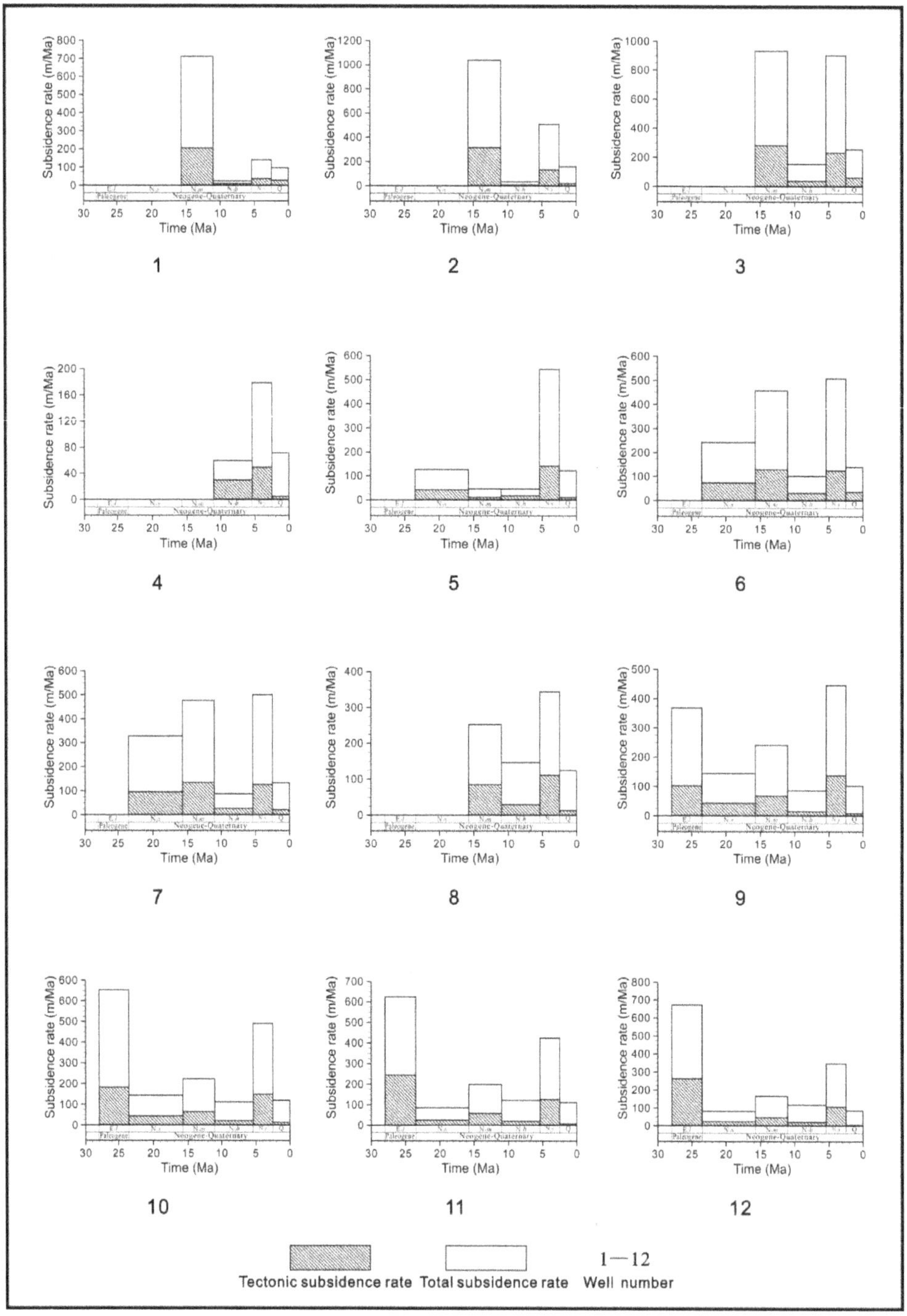

Figure 4. Subsidence rates of typical wells in the YGHB.

Based on the above analysis, we found that the spatiotemporal variations in the subsidence had changed regularly. From the Late Oligocene to the Middle Miocene, the depocenters had continuously shifted southeast, and the range of the depocenter area had expanded progressively. In the Late Miocene, the depocenters had migrated toward the northwest, and the range of the depocenter area was the largest, although the subsidence rate was relatively small. From the Pliocene to the present, the depocenters shifted to the east, and the scope of the depocenter area decreased. Consequently, the northwestern part of the basin became the non-depocenter area. The Late Miocene was an obvious and critical turning point, before and after which the subsidence features were dramatically different.

4.2. *Subsidence Characteristics of Profiles (2D)*

4.2.1. 2D Subsidence Characteristics in the QDNB

Six seismic profiles of the QDNB were studied, among which one profile in the eastern basin was selected to analyze the spatiotemporal variations in the 2D subsidence characteristics.

In the eastern part of the basin, the subsidence area was limited, and it only included the Songnanbaodao sag and the Baodao rise in the Eocene (Figure 5). The depocenters were primarily distributed in the Songnanbaodao sag, which experienced the maximum subsidence, reaching approximately 2420 m by the end of the Eocene. In the Early Oligocene, the scope of the subsidence, including several main depocenters, expanded to some extent, as compared to the previous period. By the end of the Early Oligocene, the maximum subsidence of the Yacheng Formation had reached 2900 m. In the Late Oligocene, the scope of the subsidence, which was characterized by nearly unchanged subsidence features, increased further. Moreover, the main depocenters were still distributed in the Songnanbaodao sag, and the maximum subsidence of the Lingshui Formation increased to 3080 m at the end of the Late Oligocene. In the Early Miocene, the subsidence features changed slightly, but the Songnanbaodao sag was still the area in which the maximum subsidence of the Sanya Formation had occurred, reaching approximately 1780 m by the end of the Early Miocene. For example, the stratum was distributed as a blanket in the plane, which led to a decrease in the subsidence differences between the depocenters. In addition, the subsidence in the Beijiao sag and the Beijiao rise decreased because these two tectonic units were uplifted during this period. During the Middle Miocene, the subsidence features varied significantly, except in the Songnanbaodao and Yongle sags. The subsidence of the northern Baodao rise, Beijiao sag, and Beijiao rise were small due to the remarkable amount of uplift. At the end of the Middle Miocene, the largest subsidence of the Meisha Formation was 1020 m. While in the Late Miocene, the subsidence characteristics of the Songnanbaodao sag, where the greatest subsidence of the Huangliu Formation reached 1050 m by the end of the Late Miocene, remained nearly unchanged. However, the subsidence features of the remaining areas changed dynamically. For example, the subsidence in the Beijiao sag was relatively large because it had subsided sharply during this period. The non-depocenters were distributed in the Beijiao rise and Yongle sag due to an incessant uplift. The subsidence features of all the areas changed in the Pliocene, during which the depocenters moved to the two sides of the locations of the depocenters during the previous period. The largest subsidence of the Yinggehai Formation increased to 1800 m by the end of the Pliocene and was located in the river channel. The uplift of the Beijiao had continued during this period, while the Yongle sag had subsided slightly. From the Quaternary to the present, the seawater intruded into the eastern QDNB with a large water depth because a transgression had begun during this period [14].

Based on the above analysis, we found that the differences in the subsidence of each unit were relatively high from the Eocene to the Oligocene. The reason for this was that this period was the rifting stage when the faults primarily developed, which had resulted in the development of the various grabens and half-grabens where the subsidence had occurred. However, after the Miocene, the differences in the subsidence of the tectonic units decreased. In the Pliocene, the southern part of the basin had been uplifted, and the northern part of the basin had subsided.

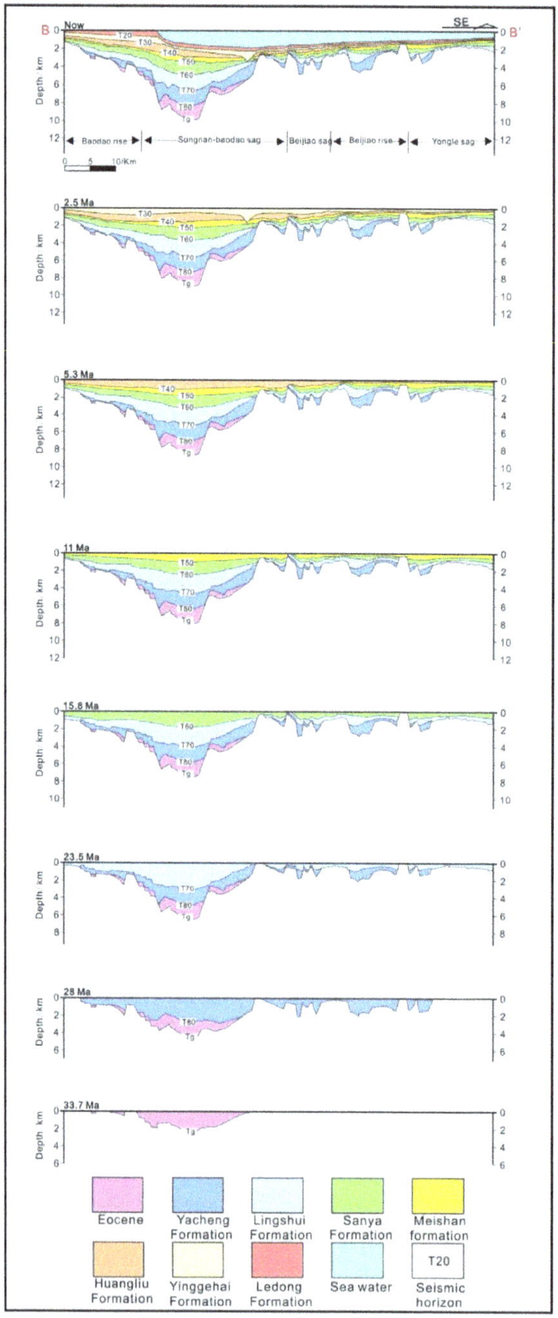

Figure 5. Regional subsidence evolution profiles in the QDNB (original profile from ref. [19], For location of the profile, see Figure 1).

4.2.2. 2D Subsidence Characteristics in the YGHB

In the YGHB, three seismic profiles distributed in the southern, central, and northern parts of the basin were analyzed, among which we selected the central profile to analyze the spatiotemporal variations of the subsidence in the basin. The YGHB was characterized by a low exploration level and a thick stratum in the central basin. In addition, due to the low quality of the seismic data, the stratum that could be used to analyze the subsidence history only included the Sanya Formation and the overlying formations (Figure 6).

In the Early Miocene, subsidence only occurred in the Yinggehai sag. The largest subsidence of the Sanya Formation was 2380 m by the end of the Early Miocene. In the Middle Miocene, the subsidence had still only occurred in the Yinggehai sag, but the amount of subsidence was significant, particularly in the central part of the sag. The shape of the profile of the Meishan Formation was a wedge because its thickness had increased gradually from east to west. The maximal subsidence of the Meishan Formation reached 3150 m at the end of the Middle Miocene. The subsidence characteristics had changed distinctly in the Late Miocene. For example, the subsidence range had expanded to the Yingdong slope. In addition, the thickness of the stratum varied significantly in the horizontal direction; however, the Yinggehai sag was still the area where the maximal subsidence of the Huangliu Formation occurred, reaching only 910 m by the end of the Late Miocene. Until the Pliocene, the subsidence features had only changed slightly. The scope of the subsidence remained unchanged, but the amount of subsidence varied significantly from the Yinggehai sag to the Yingdong slope because the former unit had experienced a large amount of subsidence and was the area where the depocenters had been distributed. In contrast, the latter unit was the area where the non-depocenters had been distributed due to its relatively small amount of subsidence. At the end of the Pliocene, the largest subsidence of the Yinggehai Formation was 2430 m. Since the Quaternary, the subsidence has increased gradually from west to east, but the largest subsidence of the Ledong Formation was still located in the Yinggehai sag, reaching 1090 m by the end of the Quaternary.

Based on the above analysis, we found that the subsidence of the YGHB during the Cenozoic was as follows. First, the Yinggehai sag, which had continuously subsided since the Miocene, was the main area where the depocenters had been distributed. However, the Yingdong slope had been uplifted on a large scale from the Early Miocene to the Middle Miocene, and on a small scale, during the Late Miocene, which had resulted in a lack of subsidence during the former two periods. Second, the YGHB was characterized by episodic subsidence during the Cenozoic based on the 2D subsidence characteristics, that is, the subsidence had changed in an orderly manner over time. For example, the subsidence had been significant in the Middle Miocene and Pliocene but minor during the Early Miocene, Late Miocene, and Quaternary.

4.3. Holistic Subsidence of the Entire Basin

To reveal the holistic subsidence of the entire basin, based on the subsidence differences between the two basins and the controlling factors of this phenomenon, we weighted all the well data to determine the holistic subsidence (Figures 7 and 8).

4.3.1. Holistic Subsidence of the QDNB

The holistic subsidence of the QDNB was episodic. According to the changes in the subsidence process, the holistic subsidence process of the QDNB consisted of three episodes. The first one occurred during the Eocene; the second episode occurred during the Oligocene and Middle Miocene; and the third episode occurred from the Late Miocene to the Quaternary (Figure 7).

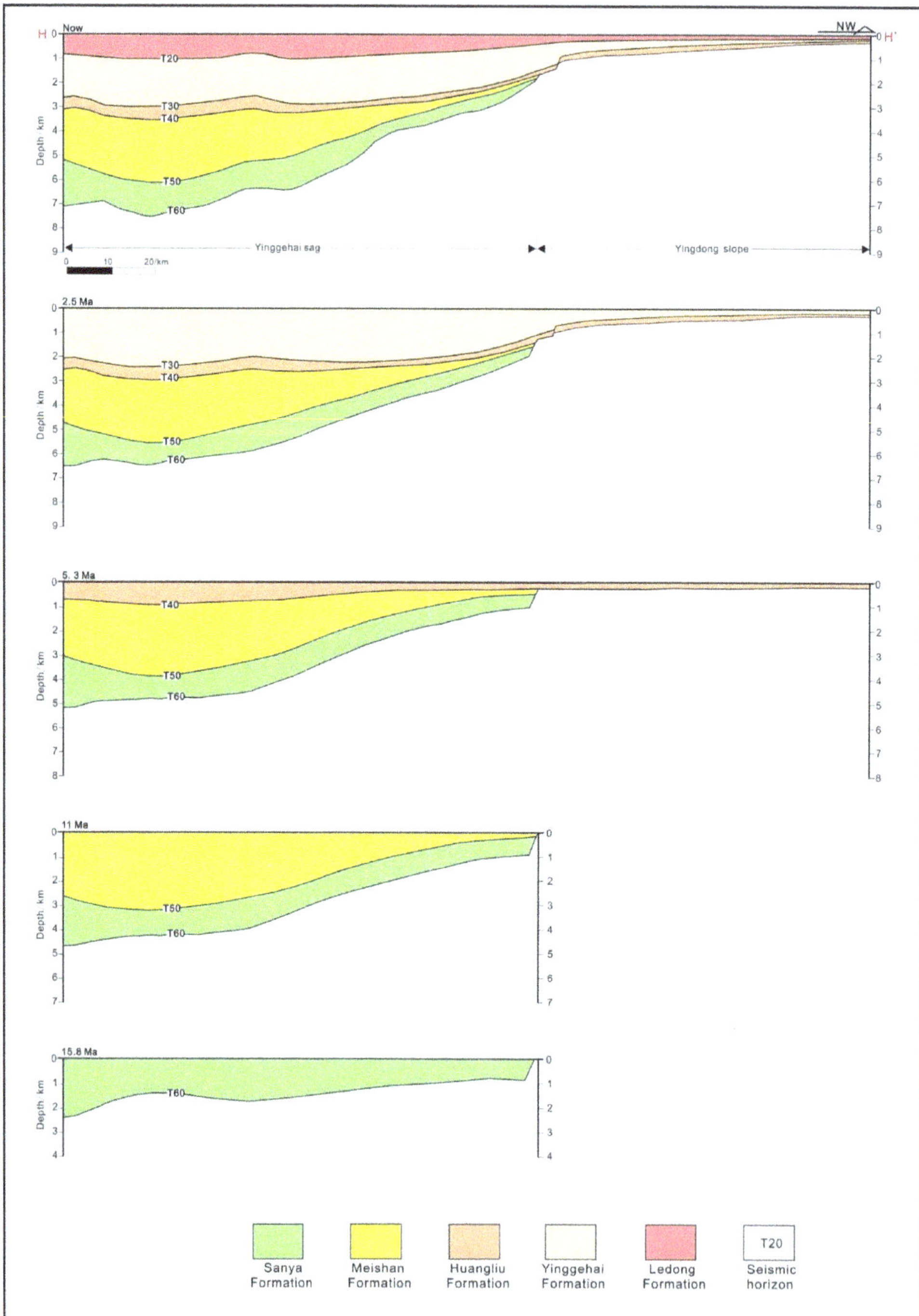

Figure 6. Regional subsidence evolution profile in the western YGHB (original profile from ref. [19]; for location of the profile, see Figure 1).

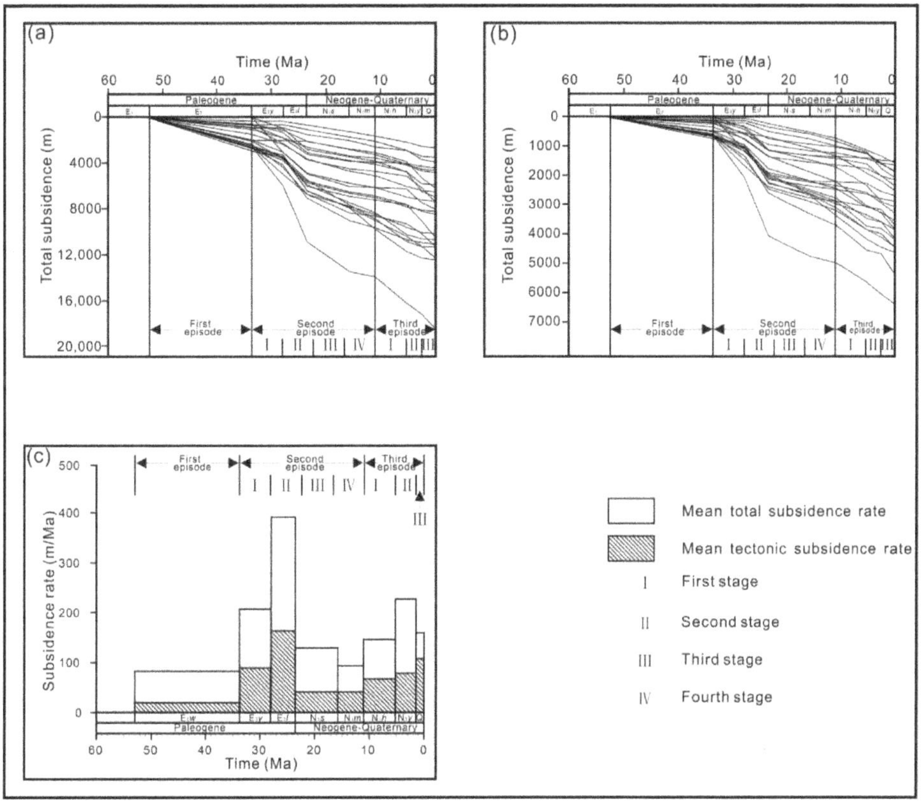

Figure 7. Holistic subsidence curves and subsidence rates in the QDNB: (**a**) total subsidence curve; (**b**) tectonic subsidence rate; (**c**) holistic total subsidence rate and holistic tectonic subsidence rate, which were acquired by averaging 28 wells. The lines in (**a**,**b**) are the subsidence curves of 28 fictitious wells.

During the first episode, the mean Ra was 80 m/myr, and the mean Rt was 20 m/myr, which only comprised 25% of the overall subsidence. During the second episode, because the subsidence rate increased, the Ra increased to 205 m/myr, and the Rt grown to 83.5 m/myr. In this study, the mean subsidence rate of a single episode was acquired by averaging the mean subsidence rates of all the periods in that episode (Figure 7). This is described in the latter part of the paper for the same situation and, thus, is not provided here. The proportion of the tectonic subsidence that occurred during the second episode comprised 40.7% of the Ra. Moreover, according to the changes in the subsidence rate, the second episode was divided into four stages. The first stage occurred in the Early Oligocene, during which the mean Ra was 207 m/myr, and the mean Rt was 89 m/myr. The tectonic subsidence during this stage comprised 42.9% of the overall subsidence. The second stage occurred in the Late Oligocene, during which the subsidence rate reached the maximum during the entire evolution of the basin. The mean Ra was 391 m/myr, the mean Rt was 163 m/myr, and the tectonic subsidence comprised 41.7% of the overall subsidence. The third stage occurred in the Early Miocene, during which the mean Ra and Rt were 129 m/myr and 41 m/myr, respectively. The tectonic subsidence during this stage comprised 31.8% of the overall subsidence. The fourth stage occurred in the Middle Miocene, during which the mean Ra was 93 m/myr and the mean Rt was still 41 m/myr (i.e., unchanged, as compared to the previous stage), accounting for 44.1% of the Ra. In the third episode, the mean Ra was 177 m/myr, and the mean Rt was 84 m/myr, comprising

47.5% of the Ra. Similar to the situation during the previous episode, the third episode was also divided into three stages. The first occurred during the Late Miocene. The mean Ra was 145 m/myr, and the mean Rt was 67 m/myr, so the tectonic subsidence comprised 46.2% of the overall subsidence. The second stage occurred during the Pliocene. The mean Ra was 227 m/myr, which was the highest rate during the entire third episode, and the mean Rt was 79 m/myr, accounting for 34.8% of the Ra. The third stage occurred during the Quaternary. The mean Ra was 159 m/myr, and the mean Rt was 108 m/myr. The tectonic subsidence comprised 67.9% of the overall subsidence and reached the highest rate during the entire evolution of the basin.

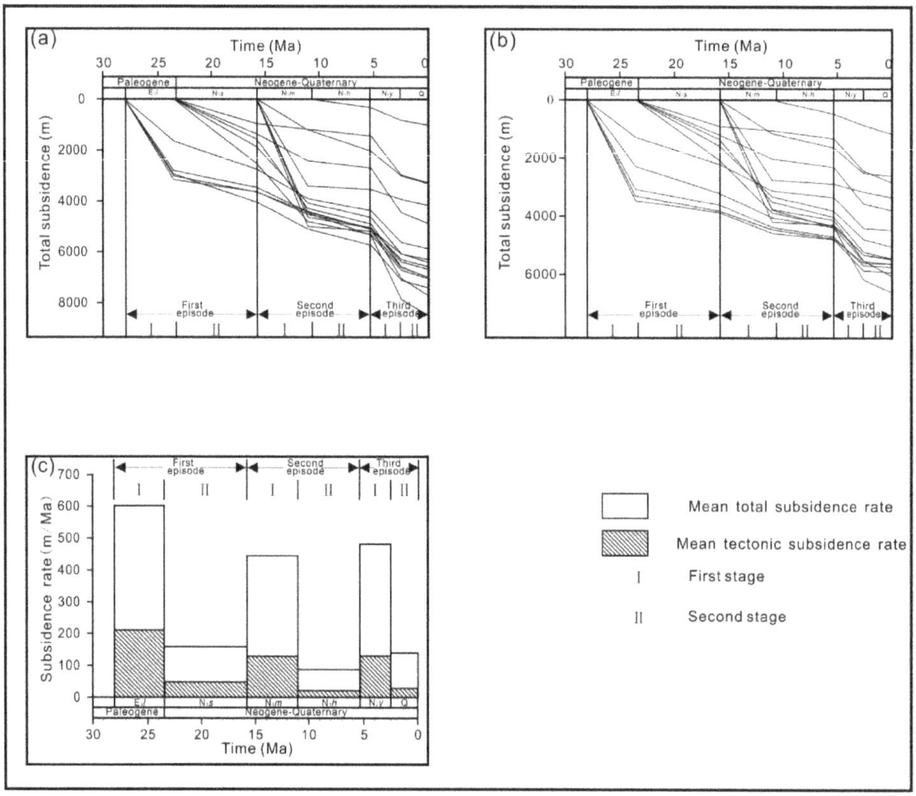

Figure 8. Holistic subsidence curves and subsidence rates in the YGHB: (**a**) total subsidence curve; (**b**) tectonic subsidence rate; (**c**) holistic total subsidence rate and holistic tectonic subsidence rate, which were acquired by averaging 16 wells. The lines in (**a**,**b**) are the subsidence curves of 16 fictitious wells.

Based on the above analysis, we concluded that the subsidence rate was low during the first episode, peaked during the second episode, and decreased in the third episode. However, the percentage of the overall subsidence determined by the tectonic subsidence increased continuously during the entire evolution of the basin, and it reached the high rate during the Quaternary.

4.3.2. Holistic Subsidence of the YGHB

The holistic subsidence of the YGHB was characterized by more distinct episodes, as compared to the QDNB. Based on the changes in the subsidence rate, since the Late Oligocene to the Quaternary, the basin subsidence process was divided into three episodes.

The first episode occurred from the Late Oligocene to the Early Miocene; the second episode occurred from the Middle Miocene to the Late Miocene; and the third episode occurred from the Pliocene to the Quaternary (Figure 8).

During the first episode, the mean Ra was 381 m/myr, the mean Rt was 130 m/myr, and the tectonic subsidence comprised 34.1% of the overall subsidence. This episode was further divided into two stages. The first stage occurred during the Oligocene, and the maximal subsidence rate in the entire basin subsidence process occurred during this period. During this stage, the mean Ra was 603 m/myr, and the mean Rt was 212 m/myr, comprising 35.1% of the Ra. The second stage, which was characterized by a sharp decrease in the subsidence rate, occurred in the Early Miocene. The mean Ra was 159 m/myr, and the mean Rt was 48 m/myr, accounting for 30.2% of the Ra. During the second episode, the subsidence rate lessened to some extent, and the mean Ra was 266.5 m/myr, and the mean Rt was 75 m/myr. The tectonic subsidence comprised 28.1% of the overall subsidence. Similarly, the second episode was separated into two stages. The first stage occurred during the Middle Miocene. The subsidence rate increased, the mean overall subsidence increased to 446 m/myr, and the mean Rt increased to 129 m/myr, accounting for 28.9% of the Ra. The second stage occurred during the Late Miocene. The minimum subsidence rate in the entire evolution of the basin occurred during this stage. The mean Ra was 87 m/myr, and the mean Rt was 21 m/myr. The tectonic subsidence comprised 24.1% of the overall subsidence. The third episode occurred in the Pliocene and Quaternary. The mean Ra was 311.5 m/myr, and the mean Rt was 78.5 m/myr, comprising 25.2% of the Ra. The third episode consisted of two stages. The first stage occurred in the Pliocene, during which the subsidence rate increased. The mean Ra was 483 m/myr, and the mean Rt was 130 m/myr, comprising 26.9% of the overall subsidence. The second stage occurred in the Quaternary, during which the subsidence rate was low. The mean Ra was 140 m/myr, and the mean Rt was only 27 m/myr, comprising 19.2% of the Ra.

The above analysis indicated that the most significant subsidence feature of the YGHB was the dramatically different episodes. Each subsidence episode consisted of a stage with a high subsidence rate and a stage with a low subsidence rate. In addition, the subsidence rate was high during the first episode and then decreased during the second and third episodes.

5. Discussion

5.1. Relationship between the Basin Subsidence and Tectonics

The QDNB and the YGHB were surrounded by the Indian, Eurasian, and Pacific plates in the Cenozoic. The interactions among these three plates have induced several particular geological processes, including the uplift of the Tibetan Plateau, the strike-slip motion of the Red River Fault, and the expansion of the SCS, all of which have influenced the basin evolution and subsidence. However, the effects of these three geological processes on the basin have varied for each basin due to the differences in their tectonic locations.

In the QDNB, the subsidence rate was low during the Eocene (Figure 9), and the northwestern and southeastern parts of the basin were the non-depocenters, according to our previous analysis (wells 4 and 8 in Figures 1 and 3). This was because, during the Eocene, the Indian Plate had moved northeastward [35], and the Pacific Plate had subducted under the Eurasian Plate [36]. Moreover, the rate of the movement of the Indian Plate toward the Eurasian Plate had been higher than the subduction rate of the Pacific Plate, relative to the Eurasian Plate, so there had been enough space to accommodate the mantle material under the South China Block, which had flowed from the northwest to the southeast [37]. Consequently, NE trending grabens and half-grabens developed under the comprehensive influence of these factors. The subsidence only occurred in these grabens and half-grabens. Therefore, the mean total subsidence of the entire basin was low. During the Early Oligocene, the influence of the flow of the mantle material had increased further, and the scale of the rifting increased significantly. Moreover, the SCS had also opened at 32 Ma [38], so the subsidence rate and the scale of the QDNB increased during this period.

The spreading rate of the SCS increased from the late Early Oligocene onwards, and then it peaked at the end of the Late Oligocene, which also caused the subsidence rate of the QDNB to reach its maximum during the Late Oligocene. The subsidence rate was also the highest during the entire evolution of the basin. The spreading of the SCS began to weaken in the Miocene, and it ceased at the end of the Early Miocene, which led to a rapid decrease in the subsidence rate and subsidence scale during the entire Miocene. In the Pliocene, a large-scale regional thermal event had occurred in the QDNB and the YGHB [39], and the dextral strike-slip motion of the Red River Fault had also commenced in the Pliocene. Thus, the subsidence rate of the QDNB increased again under the combined effects of these two geological processes. In the Quaternary, the subsidence rate of the QDNB decreased due to the smaller amount of dextral strike-slip of the Red River Fault. Similar to the other basins all over the world [40], one of the most obvious subsidence features of the QDNB was the anomalous tectonic subsidence that occurred in the Quaternary (wells 1, 2, 3, 4, 6, and 10 in Figures 1 and 3). These sags were characterized by a low total subsidence rate and a high tectonic subsidence rate in the Quaternary. Regarding the cause of this phenomenon, several researchers proposed the thermal cooling after the late magmatism [5]; however, other researchers have proposed a lower crustal flow as an explanation [41]. Based on the previous analysis, we found that the variations in the subsidence over time in the QDNB matched the changes in the flow of the mantle material in the Eocene and the variations in the spreading of the SCS from the Oligocene to the Miocene. However, since the Pliocene, the subsidence features have primarily been controlled by the comprehensive influence of the strike-slip motion of the Red River Fault and the thermal event.

In the YGHB, the subsidence rate had been high in the Late Oligocene (Figure 9), and the subsidence had primarily occurred in the western part of the basin (Figures 1 and 3). During this period, the uplift of the Tibetan Plateau ceased [42], which provided a sufficient sediment source. Accordingly, the scale of the strike-slip movement of the Red River Fault was significant during this period and provided large spatial accommodation because it had consumed the energy of the Indian Plate as it moved northeastward. Consequently, the subsidence rate reached its peak during the entire evolution. During the Early Miocene, the Tibetan Plateau began its second stage of uplift. In addition, in the Miocene, the strike-slip movement of the Red River Fault gradually became weaker because the Indian Plate had started to wedge into the Eurasian Plate [43], which resulted in the southeastward movement of the Eurasian Plate [44], but the velocity of the Eurasian Plate was slower than that of the Indochina Block. Thus, the subsidence rate had decreased in the Early Miocene. During the early Middle Miocene, the Tibetan Plateau remained static, but there was a certain amount of strike-trip movement of the Red River Fault that resulted in the subsidence rate being higher than during the previous period but lower than in the Late Oligocene. During the Late Miocene, the strike-slip movement of the Red River Fault nearly ceased, so the subsidence rate decreased even though the uplift of the Tibetan Plateau had stopped in the Late Miocene. In the Pliocene, the subsidence rate of the YGHB increased again and was higher than during the previous period. This was also the result of the comprehensive influence of the fast dextral strike-slip of the Red River Fault and the thermal event. In the Quaternary, the strength of the dextral strike-slip of the Red River Fault weakened, so the subsidence rate decreased. According to the previous analysis, we found that the subsidence characteristics of the YGHB were primarily controlled by the uplift of the Tibetan Plateau and the strike-slip motion of the Red River Fault before the Middle Miocene; however, the spreading of the SCS may have also contributed to the subsidence characteristics from the Late Oligocene to the Early Miocene. Since the Middle Miocene, the subsidence characteristics have primarily been controlled by the strike-slip movement of the Red River Fault, rather than by the uplift of the Tibetan Plateau. This was because during the first two stages of the uplift of the Tibetan Plateau, the strike-slip motion of the Red River Fault and the uplift of the Tibetan Plateau had been the two main consumers of the energy of the northeastward motion of the Indian Plate [42]. Therefore, the uplift of the Tibetan Plateau had been coupled with the strike-slip motion of the Red

River Fault [43,45], which then had controlled the subsidence characteristics of the YGHB. In the Middle Miocene, the Indian Plate had started to wedge into the Eurasian Plate [44], the Tibetan Plateau had uplifted through the large scale of the over-thrust tectonism, delamination, and melting of the lithosphere [42], which could not be coupled with the strike-slip movement of the Red River Fault and, thus, dramatically affected the subsidence characteristics of the YGHB.

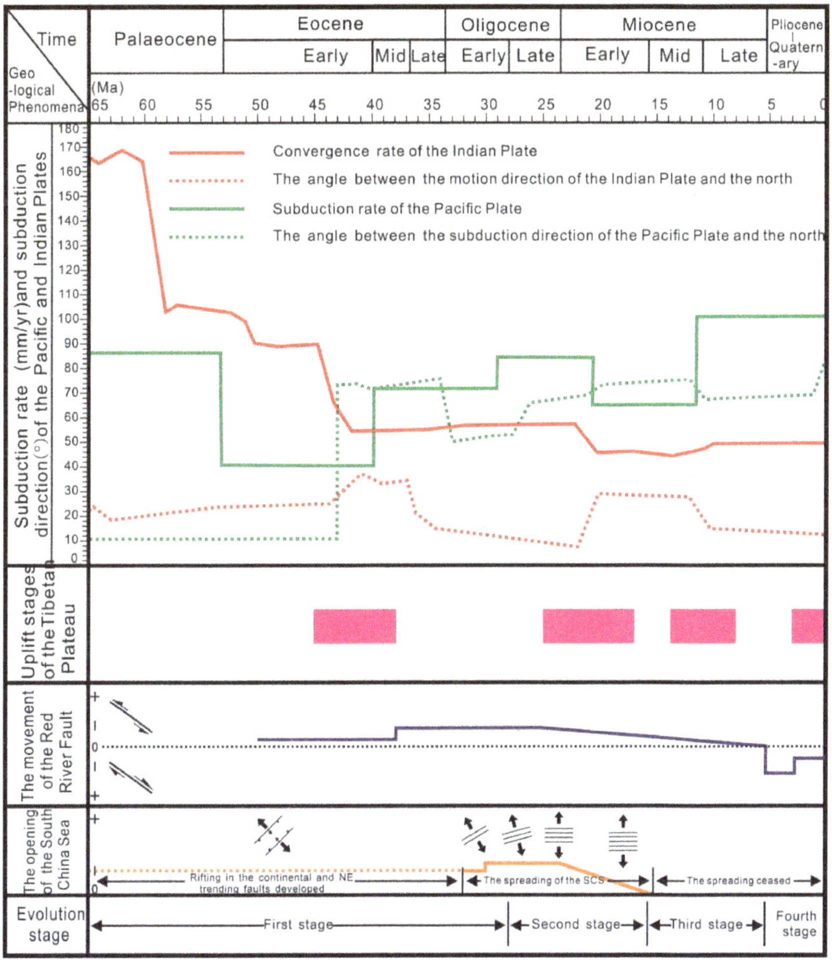

Figure 9. The evolutions of the geological phenomena surrounding the QDNB and the YGHB (i.e., the uplift stage of the Tibetan Plateau, according to Zhong et al. [42]). The movement of the Red River Fault was modified based on Sun et al. [43]. The spreading of the South China Sea was according to Yao [37] and Briais et al. [38].

5.2. Effect of the Faults on the Basin Subsidence

Major faults are always the boundaries of basins and tectonic units, and they determine the structural framework of a basin and play an important role in a basin's evolution, sedimentation, filling, and gas migration and release [46]. Abundant faults have developed in the SCS and its surrounding areas, which have been classified as lithospheric faults, crustal faults, and basement faults, according to their depths [47]. The QDNB and the YGHB are separated by a lithospheric fault (Figure 10). In the QDNB, there are three groups of faults: NW trending, NE trending, and nearly WE trending. The nearly WE trending faults distributed in the eastern part of the basin and the NW trending faults distributed in the western part of the basin are lithospheric faults. The NW trending and NE trending faults are distributed in the interior of the basin and are crustal and basement faults. Three NW trending faults have developed in the YGHB. The fault in the northeastern part of the basin is a lithospheric fault, and the other two faults are basement faults.

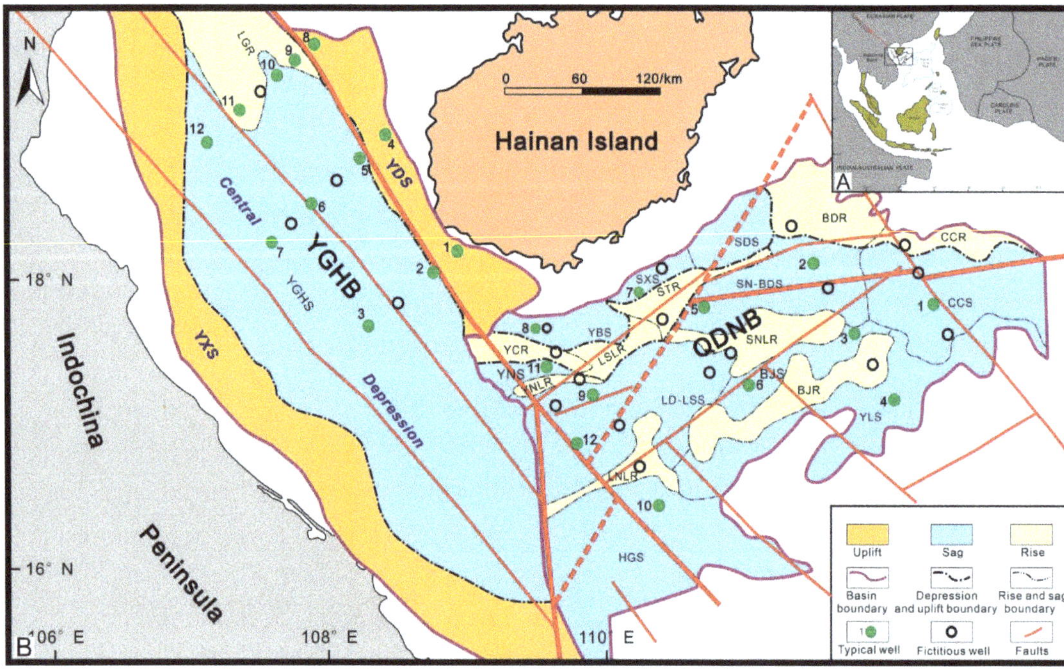

Figure 10. The relationship between the subsidence and basement faults. The abbreviations in this figure are the same as in Figure 1. (The map was modified according to Zhu et al. [19]. The faults are according to Lu et al. [47].)

The major faults significantly affected the subsidence in the two basins (Figure 10). For example, in the QDNB, the areas near the major faults were the subsidence areas in most of the periods of the basin's evolution, including the Changchang sag, Songnan Baodao sag, western Ledonglingshui sag, and Huaguang sag (wells 1, 2, 5, 9, 12, and 10 in Figures 3 and 10). In particular, the western Ledonglingshui sag was always a subsidence area, with a high subsidence rate because of the lithosphere fault that crosses it. In contrast, the areas farther away from the fault were the areas where the non-depocenters were distributed in most of the periods of the basin's evolution, such as the Yongle and Yabei sags (wells 4 and 8 in Figures 3 and 10). The subsidence in the YGHB was closely related to the strike-slip motion of the Red River Fault (Figures 9 and 10). The depocenters had been primarily distributed in the western part of the YGHB in the early period, which matched the left-lateral strike-slip movement of the Red River Fault. However, it shifted to the eastern part of the basin in the Pliocene due to the onset of the dextral strike-slip movement of the Red River Fault (Figures 4 and 10). Furthermore, the depocenters had continuously migrated from northwest to southeast with the southeastward propagation of the Red River Fault and the southeastward transportation of the sediment source. The faults had a significant influence on the subsidence in both basins, which resulted in spatiotemporal variations in the subsidence in the same basin.

6. Conclusions

The main conclusions of this study were as follows.

First, the differences in the subsidence characteristics of the QDNB and the YGHB were distinct, including the subsidence rates, the migrations of the depocenters, and the changes in the ranges of the subsidence.

Second, the changes in the holistic subsidence in both the QDNB and the YGHB occurred in episodes. In the QDNB, the holistic subsidence process could be divided into three episodes, according to the subsidence features. The first episode occurred in the Eocene; the second episode occurred from the Oligocene to the Middle Miocene; and the third episode occurred from the Late Miocene to the Quaternary. The variations in the holistic subsidence rate exhibited wavelike characteristics, with two peaks in the Late Oligocene and Pliocene. In the YGHB, the holistic subsidence process could also be divided into three episodes. The first episode occurred from the Late Miocene to the Early Miocene; the second episode occurred in the Middle and Late Miocene; and the third episode occurred from the Pliocene to the Quaternary. The changes in the holistic subsidence exhibited a distinct wavelike shape, with three crests and troughs, which was a typical subsidence feature of strike-slip basins.

Third, the dynamic variations were the primary factor controlling the differences in the subsidence of the QDNB and the YGHB. The variations in the subsidence characteristics matched the three main geological processes that occurred during this period. The variations in the subsidence in these two basins were obvious before the Pliocene. In the QDNB, the subsidence characteristics were primarily controlled by the variations in the flow of the mantle material under the South China Block in the Eocene and by the spreading of the SCS from the Oligocene to the Miocene. In the YGHB, the subsidence characteristics were chiefly controlled by the combined effects of the uplift of the Tibetan Plateau and the strike-slip motion of the Red River Fault before the Early Miocene and by the strike-slip motion of the Red River Fault from the Middle Miocene to the Late Miocene. Since the Pliocene, the changes in the subsidence in these two basins have been synchronous, and they have primarily been affected by the dextral strike-slip motion of the Red River Fault.

Finally, the major faults also affected the subsidence characteristics, and they induced the spatiotemporal variations in the subsidence within the same basin.

Author Contributions: Conceptualization, M.M. and J.Q.; methodology, H.P., M.M. and J.Q.; software, M.M. and J.M.; validation, H.P. and Q.S.; formal analysis, M.M.; investigation, Q.S.; resources, M.M. and Q.S.; data curation, L.L.; writing—original draft preparation, M.M.; writing—review and editing, M.M., J.Q., M.B., J.M. and L.L.; visualization, J.M.; supervision, Q.Z. and M.M.; project administration, M.M.; funding acquisition, J.Q. All authors have read and agreed to the published version of the manuscript.

Funding: This research was funded by the National Natural Science Foundation of China, grant number 41572202; and the Master's Degree Construction Project of Geological Resources and Geological Engineering of Qinghai University in 2023, grant number 41250103.

Data Availability Statement: The raw data supporting the conclusions of this article will be made available by the authors, without undue reservation.

Acknowledgments: The authors are thankful for the guidance of Qing Liu and Sen Feng in programming.

Conflicts of Interest: The authors declare no conflict of interest.

References

1. Hall, R.; Blundell, D.J. Tectonic evolution of SE Asia: Introduction. In *Tectonic Evolution of Southeast Asia*; Hall, R., Blundell, D.J., Eds.; Geological Society Special Publication: London, UK, 1996; Volume 106, pp. 353–431. [CrossRef]
2. Karig, D.E. Origin and Development of Marginal Basins in the Western Pacific. *J. Geophys. Res.* **1971**, *76*, 2542–2561. [CrossRef]
3. Tapponnier, P.; Molnar, P. Slip-line field theory and large-scale continental tectonics. *Nature* **1976**, *264*, 319–324. [CrossRef]
4. Leloup, P.H.; Amaud, N.; Lacassin, R.; Kienast, J.R.; Harrison, T.M.; Phan Trong, T.T.; Replumaz, A.; Tapponnier, P. New constraints on the structure, thermochronology, and timing of the Ailao Shan-Red River shear zone, SE Asia. *J. Geophys. Res.* **2001**, *106*, 6683–6732. [CrossRef]

5. Xie, X.; Müller, R.D.; Li, S.; Gong, Z.; Steinberger, B. Origin of anomalous subsidence along the Northern South China Sea margin and its relationship to dynamic topography. *Mar. Pet. Geol.* **2006**, *23*, 745–765. [CrossRef]
6. Hall, R.; Hattum, M.W.A.V.; Spakman, W. Impact of India–Asia collision on SE Asia. the record in Borneo. *Tectonophysics* **2008**, *451*, 366–389. [CrossRef]
7. Chen, G. The marginal extensional belt of East Asia Continent-investigating the origin of a discrete continental margin. *Geotecton. Metallog.* **1997**, *21*, 285–293. (In Chinese with English Abstract) [CrossRef]
8. Xia, B.; Zhang, Y.; Cui, X.J.; Liu, B.M.; Xie, J.H.; Zhang, S.L.; Lin, G. Understanding of the geological and geodynamic controls on the formation of the South China Sea. A numerical modelling approach. *J. Geodyn.* **2006**, *42*, 63–84. [CrossRef]
9. Xia, B.; Cui, X.; Xie, J.; Wang, R. Thinking about the dynamics mechanism study on formation and evolution of the South China Sea. *Geotecton. Metallog.* **2004**, *28*, 221–227 (In Chinese with English Abstract) [CrossRef]
10. Ma, M.; Liu, C.; Qi, J.; Zhang, D.; Zhang, S.; Wang, J.; Miao, Q. Cenozoic subsidence history of the Pearl River Mouth Basin, northern South China Sea. *Geol. J.* **2020**, *55*, 750–770. [CrossRef]
11. Liao, F.; Wu, X.; Huang, S. Geochemical Characteristics of Natural Gas in the Yinggehai-Qiongdongnan Basin, South China Sea. *Energy Explor. Exploit.* **2010**, *28*, 1–12. [CrossRef]
12. Zhang, Y.; Zhang, L.; Mi, L.; Lu, X.; Wu, S.; Tang, L.; Zhou, J.; Xiong, X.; Zhu, J. Quantitative Analysis of Cenozoic Extension in the Qiongdongnan Basin, South China Sea: Insight on Tectonic Control for Hydrocarbon Reservoir Accumulation and Formation. *Energies* **2022**, *15*, 4011. [CrossRef]
13. China Institute of International Studies. Available online: https://www.ciis.org.cn/english/COMMENTARIES/202007/t20200715_2762.html (accessed on 11 May 2015).
14. Huang, B.; Xiao, X.; Li, X. Geochemistry and origins of natural gases in the Yinggehai and Qiongdongnan basins, offshore South China Sea. *Org. Geochem.* **2003**, *34*, 1009–1025. [CrossRef]
15. Zhao, Z.; Sun, Z.; Sun, L.; Wang, Z.; Sun, Z. Cenozoic tectonic subsidence in the Qiongdongnan Basin, northern South China Sea. *Basin Res.* **2016**, *30*, 269–288. [CrossRef]
16. Li, X.; Fairweather, L.; Wu, S.; Ren, J.; Zhang, H.; Quan, X.; Jiang, T.; Zhang, C.; Su, M.; He, Y.; et al. Morphology, sedimentary features and evolution of a large palaeo submarine canyon in Qiongdongnan basin, Northern South China Sea. *J. Asian Earth Sci.* **2013**, *62*, 685–696. [CrossRef]
17. Cao, L.; Jiang, T.; Wang, Z.; Zhang, Y.; Sun, H. Provenance of Upper Miocene sediments in the Yinggehai and Qiongdongnan basins, northwestern South China Sea. Evidence from REE, heavy minerals and zircon U–Pb ages. *Mar. Geol.* **2015**, *361*, 136–146. [CrossRef]
18. Shi, X.; Burov, E.; Leroy, S.; Qiu, X.; Xia, B. Intrusion and its implication for subsidence. A case from the Baiyun Sag, on the northern margin of the South China Sea. *Tectonophysics* **2005**, *407*, 117–134. [CrossRef]
19. Zhu, W.; Mi, L. *Atlas of Oil and Gas Basins, China Sea*; Petroleum Industry Press: Beijing, China, 2010; pp. 478–480. (In Chinese)
20. All China Commission of Stratigraphy. *The Instruction of China Regional Stratigraphy (Geological Time Scale) Chart*; Geological Publishing House: Beijing, China, 2002; pp. 235–237. (In Chinese)
21. Wang, X.; Chen, X. *Stratigraphy Division and Correlation of Each Geological Period in China*; Geological Publishing House: Beijing, China, 2005; pp. 278–283. (In Chinese)
22. Chen, Y.; Wang, Z.; Huang, Z. *The Pandect of Stratigraphical of China*; Geological Publishing House: Beijing, China, 2009; p. 557. (In Chinese)
23. Liu, X.; Xie, J.; Zhang, H.; He, W. Chronostratigraphy of planktonic foraminifera in the Yinggehai-Qiongdongnan Basin. *Acta Micropalaeontoligical Sin.* **2009**, *26*, 181–192. (In Chinese with English Abstract)
24. Gao, H.; Du, D.; Zhong, G. Quantitative simulation of subsidence history and analysis of Pearl River Mouth Basin in South China Sea. *Res. Geol. South China Sea* **2006**, *13*, 11–20. (In Chinese with English Abstract)
25. Athy, L.F. Density, Porosity, and Compaction of Sedimentary Rocks. *AAPG Bull.* **1930**, *14*, 1–24. [CrossRef]
26. Airy, G.B. On the Computation of the Effect of the Attraction of Mountain-Masses, as Disturbing the Apparent Astronomical Latitude of Stations in Geodetic Surveys. *Philos. Trans. R. Soc. Lond.* **1855**, *145*, 101–104. [CrossRef]
27. Steckler, M.S.; Watts, A.B. Subsidence of the Atlantic-type continental margin off New York. *Earth Planet. Sci. Lett.* **1978**, *41*, 1–13. [CrossRef]
28. Mckenzie, D.P. Some remarks on the development of sedimentary basins. *Earth Planet. Sci. Lett.* **1978**, *40*, 25–32. [CrossRef]
29. Sclater, J.G.; Christie, P.A.F. Continental stretching. An explanation of the Post-Mid-Cretaceous subsidence of the central North Sea Basin. *J. Geophys. Res. Solid Earth* **1980**, *85*, 3711–3739. [CrossRef]
30. Allen, P.A.; Allen, J.R. *Basin Analysis. Principles and Applications to Petroleum Play Assessment*, 3rd ed.; John Wiley & Sons Ltd.: Chichester, UK, 2013; p. 523.
31. Madon, M.B.; Watts, A.B. Gravity anomalies, subsidence history and the tectonic evolution of the Malay and Penyu Basins (offshore Peninsular Malaysia). *Basin Res.* **1998**, *10*, 375–392. [CrossRef]
32. Watts, A.B.; Steckler, M.S. Subsidence and eustasy at the continental margin of eastern North America. In *Deep Drilling Results in the Atlantic Ocean. Continental Margings and Paleoenviroment*; Maurice Ewing Symposium Series; Talwani, M., Hay, W., Ryan, W.B.F., Eds.; American Geophysical Union: Washington, DC, USA, 1979; Volume 3, pp. 218–234.
33. Haq, B.U.; Hardenbol, J.; Vail, P.R. Chronology of fluctuating sea levels since the Triassic. *Science* **1987**, *235*, 1156–1167. [CrossRef] [PubMed]
34. Kominz, M.A.; Browning, J.V.; Miller, K.G.; Sugarman, P.J.; Mizintseva, S.; Scotese, C.R. Late Cretaceous to Miocene sea-level estimates from the New Jersey and Delaware coastal plain coreholes. An error analysis. *Basin Res.* **2008**, *20*, 211–226. [CrossRef]
35. Yin, A.; Harrison, T.M. Geologic Evolution of the Himalayan-Tibetan Orogen. *Annu. Rev. Earth Planet. Sci.* **2000**, *28*, 211–280. [CrossRef]

36. Northrup, C.J.; Royden, L.H.; Burchfiel, B.C. Motion of the Pacific plate relative to Eurasia and its potential relation to Cenozoic extension along the eastern margin of Eurasia. *Geology* **1995**, *23*, 719–722. [CrossRef]
37. Yao, B. The Tectonic evolution and sedimentary basins of South China Sea in Cenozoic. *Geol. Res. South China Sea* **1998**, *10*, 1–17. (In Chinese with English Abstract)
38. Briais, A.; Patriat, P.; Tapponnier, P. Updated interpretation of magnetic anomalies and seafloor spreading stages in the south China Sea. Implications for the Tertiary tectonics of Southeast Asia. *J. Geophys. Res. Solid Earth* **1993**, *98*, 6299–6328. [CrossRef]
39. Clift, P.D.; Sun, Z. The sedimentary and tectonic evolution of the Yinggehai-Song Hong basin and the southern Hainan margin, South China Sea. Implications for Tibetan uplift and monsoon intensification. *J. Geophys. Res. Solid Earth* **2006**, *111*, 1–28. [CrossRef]
40. Dupré, S.; Bertotti, G.; Cloetingh, S. Tectonic history along the South Gabon Basin. Anomalous early post-rift subsidence. *Mar. Pet. Geol.* **2007**, *24*, 151–172. [CrossRef]
41. Yin, X.; Ren, J.; Lei, C.; Wang, S.; Zhang, J. Post rift rapid subsidence characters in Qiongdongnan Basin, South China Sea. *J. Earth Sci.* **2011**, *22*, 273–279. [CrossRef]
42. Zhong, D.; Ding, L. The discussion of the uplift process of the Tibetan Plateau and its mechanism. *Sci. China (Ser. D)* **1996**, *26*, 289–295. (In Chinese)
43. Sun, Z.; Zhou, D.; Zhong, Z.; Zeng, Z.; Wu, S. Experimental evidence for the dynamics of the formation of the Yinggehai basin, NW South China Sea. *Tectonophysics* **2003**, *372*, 41–58. [CrossRef]
44. Lee, T.Y.; Lawver, L.A. Cenozoic plate reconstruction of Southeast Asia. *Tectonophysics* **1995**, *251*, 85–138. [CrossRef]
45. Coleman, M.; Hodges, K. Evidence for Tibetan plateau uplift before 14 Myr ago from a new minimum age for east-west extension. *Nature* **1995**, *374*, 49–52. [CrossRef]
46. Krishnappan, B.G. Reviewof a Semi-Empirical Modelling Approach for Cohesive Sediment Transport in River Systems. *Water* **2022**, *14*, 256. [CrossRef]
47. Lu, B.; Wang, P.; Zhang, G.; Wang, W. Characteristic of regional fractures in South China Sea and its basement tectonic framework. *Prog. Geophys.* **2015**, *30*, 1544–1553. (In Chinese with English Abstract) [CrossRef]

processes

MDPI

Article

Experimental Study of Acid Etching and Conductivity of High-Temperature-Resistant Cross-Linked Acid

Hai Lin [1], Tengfei Hou [2], Fuguo Wang [3], Long Yue [4], Shiduo Liu [1], Guide Yuan [3], Guoqing Wang [3], Yong Liu [1], Qing Wang [5,*] and Fujian Zhou [5,*]

1 Drilling and Production Technology Research Institute, PetroChina Qinghai Oilfield Company, Beijing 816400, China
2 CNPC Engineering Technology R&D Company Limited, Beijing 102206, China
3 The Exploration Enterprise Department of Qinghai Oilfield Company, Dunhuang 816401, China
4 Engineering and Technology Department of Qinghai Oilfield Company, Dunhuang 816401, China
5 State Key Laboratory of Oil and Gas Resources and Prospecting, China University of Petroleum, Beijing 102249, China
* Correspondence: 2021311217@student.cup.edu.cn (Q.W.); zhoufj@cup.edu.cn (F.Z.)

Abstract: Acid fracturing is one of the effective techniques for developing low-permeability carbonate reservoirs economically. With the increasing reservoir depth, the reservoir temperature and closure pressure increase, posing new challenges to the acid system. In this paper, a high-temperature-resistant cross-linked acid system is selected, which maintains a viscosity above 80 mPa·s in the temperature range of 120 °C to 140 °C and can effectively reduce acid leak-off. The acid system can not only open the reservoir and ensure the extension of the fracture, but also reduce the reaction rate between the acid and the reservoir and increase the etching distance. The rock slab acid etching and conductivity tests show that the optimum injection rate is 50 mL/min, the rock etching morphology is channel type, and the conductivity remains above 110 D·cm. However, as the acid concentration decreases, the rock slab conductivity decreases considerably, especially at 10% acid concentration, where the closure pressure rises to 15 MPa, and there is almost no conductivity. In particular, after the acid system is broken, the reacted acid can form a filter cake on the core surface, hindering further intrusion of the residue into the core and reducing reservoir damage. The study shows that high-temperature-resistant cross-linked acid systems can effectively improve the stimulation of deeply fractured carbonate reservoirs at high temperatures.

Keywords: acid fracturing; cross-linked acid; conductivity; high temperature

Citation: Lin, H.; Hou, T.; Wang, F.; Yue, L.; Liu, S.; Yuan, G.; Wang, G.; Liu, Y.; Wang, Q.; Zhou, F. Experimental Study of Acid Etching and Conductivity of High-Temperature-Resistant Cross-Linked Acid. *Processes* **2023**, *11*, 722. https://doi.org/10.3390/pr11030722

Academic Editor: Alina Pyka-Pająk

Received: 21 December 2022
Revised: 17 February 2023
Accepted: 21 February 2023
Published: 28 February 2023

1. Introduction

With the exploitation of global oil and gas resources, the exploitation depth of carbonate reservoirs is increasing. Porosity and permeability are the key parameters of the reservoir, and also determine the level of oil production. Deep carbonate reservoirs are tight (low permeability and low porosity) [1,2]. Low-permeability carbonate reservoirs have no natural productivity after drilling and need reservoir stimulation to create high-speed channels for oil and gas flow to achieve commercial development [3,4]. Acid fracturing is an effective method to improve the production of carbonate reservoirs. Acid fracturing is to crack the formation when it is higher than the formation fracture pressure, and then the injected acid reacts with the rock on the fracture surface [5,6]. Due to the heterogeneity of reservoir rock mineral distribution and acid distribution, the fracture surface is unevenly etched by acid. When the acid fracturing is completed, the rough fracture surface cannot be closed entirely under the closure pressure, forming a high conductivity channel for oil and gas flow to the wellbore [7,8].

The effect of acid fracturing on increasing oil and gas production is related to the effective length and conductivity of acid-etching fracture [9,10]. The effective length of the acid-etched fracture is mainly related to acid–rock reaction rate and acid filtration [11,12]. At reservoir temperature, the slower the reaction rate between acid and rock, the more fractures can communicate with more reservoirs. The conductivity is related to the etching form of the fracture surface and closure pressure [13,14]. In general, since the rock is composed of different minerals, the surface of the fracture surface is uneven after the rock reacts with the acid solution [15–19]. However, some carbonate reservoirs have high mineral content (>95%) and cannot spontaneously form uneven etching morphology, which requires the assistance of acid fracturing process. Therefore, the properties of acid are significant for the effect of acid fracturing. In particular, the reservoir temperature is high (>120 °C), and the permeability is low for deeply fractured carbonate reservoirs. Conventional hydrochloric acid, gelled acid [20], and organic acid [21,22] are not suitable for acidizing on deep tight carbonate reservoirs due to the problems of a fast acid–rock reaction rate, low viscosity, and extensive filtration. Compared with other acids mentioned above, cross-linked acid has the advantages of high viscosity, slow reaction rate, low filtration, and low friction. It is widely used in acid fracturing of high-temperature tight carbonate reservoirs.

Cross-linked acid refers to a high-viscosity acid system formed by the chemical cross-linking of thickener macromolecules and cross-linking agent small molecules under acidic conditions [23]. Avtar [24] developed a high-viscosity acid for the first time using polymer and metal ion cross-linking agents. Although the viscosity of the acid solution is high under low pH conditions (15% HCl), the temperature resistance is poor, and the viscosity decreases sharply at high temperatures, which cannot meet the acid fracturing of high-temperature reservoirs. At the same time, the iron-based cross-linking agent remains in the formation, causing damage to the reservoir [25]. In response to this problem, on the one hand, the researchers added a certain amount of organic acid to the acid solution to reduce the hydrochloric acid content, thereby reducing the acid–rock reaction rate. Buijse [26] used a mixture of hydrochloric acid and organic acid as a cross-linked acid–base solution, confirming that adding organic acids can reduce the acid–rock reaction rate and increase the effective distance of the acid in the fracture. On the other hand, more excellent thickeners and cross-linking agents were synthesized. Patil [27] prepared a new aluminum cross-linked acid system, which can withstand temperatures up to 135 °C in 20% HCl. The cross-linking agent has good compatibility with the corrosion inhibitor. When the pH is 1.5~3, the polymer can be effectively cross-linked, and the corrosion inhibition effect is good at 93 °C and 107 °C. Wang [28] synthesized a high-temperature-resistant cross-linked acid, at 140 °C, 170 s^{-1}, after shearing 1 h apparent viscosity stability at 100 mP·s. Fang [29] used organic zirconium cross-linking agent to improve the temperature resistance of the acid to 160 °C.

For stimulation of acid fracturing in low permeability carbonate reservoirs, cross-linked acid is the most suitable acid fluid system. The research on cross-linked acid primarily focuses on temperature resistance, but acid fracturing is a comprehensive problem of acid etching and fracture closure. It is necessary to comprehensively study the temperature resistance, acid-etching ability, and the damage of reacted acid to the reservoir [30–32]. On the basis of clarifying the change of acid viscosity with temperature, exploring the relationship between acid-etching pattern and conductivity, and adjusting acid injection parameters to form high conductivity, is of great significance to guide acid fracturing stimulation.

To solve this problem, this paper selects a new type of ground cross-linked acid system. The polymer thickener used is amphoteric polyacrylamide, and the cross-linking agent is organic zirconium. In this paper, the research on the acid fracturing stimulation of a low permeability carbonate reservoir is mainly divided into three aspects. The first is the temperature resistance and shear resistance test of the acid system to ensure that the acid's structure is not damaged at the reservoir temperature, which can reduce the acid leakage. The second is the acid etching and conductivity test of the ground cross-linked acid system and rock slab. By changing the acid injection parameters and adjusting the acid-etching pattern, the method of obtaining high conductivity is explored. In addition, the damage evaluation of reacted acid to rock is also carried out, revealing the mechanism of the formation of filter cake on the core surface by the reacted acid of cross-linked acid to reduce reservoir damage.

2. Materials and Methods

2.1. Materials

The rock slab used in this paper is made of tight carbonate rock with high calcite content. XRD mineral composition results of the rock slab are shown in Table 1 and the slabs are sized to meet API conductivity cell requirements (Figure 1). Cores were drilled in the same batch of rocks for porosity and permeability tests. The results show that core's porosity is about 10%, and the permeability distribution range is 0.074~0.305 mD, which has strong heterogeneity (Table 2). The new cross-linked acid comprises 20 wt % HCl, 1 wt % thickener, 1 wt % cross-linking agent, and 3 wt % corrosion inhibitor. The thickener is amphoteric polyacrylamide. The cross-linking agent is organic zirconium cross-linking agent. It is worth noting that the cross-linking agent is composed of A and B, according to the proportion of 1:7 before use, mixed evenly and stirred, ready to use, to avoid cross-linking agent failure.

Table 1. XRD mineral analysis results.

No.	Mineral Contents (%)		
	Quartz	Calcite	Dolomite
1-1#	0.9	96.2	2.9
1-2#	0.7	95.7	3.6
1-3#	1.2	96.3	2.5

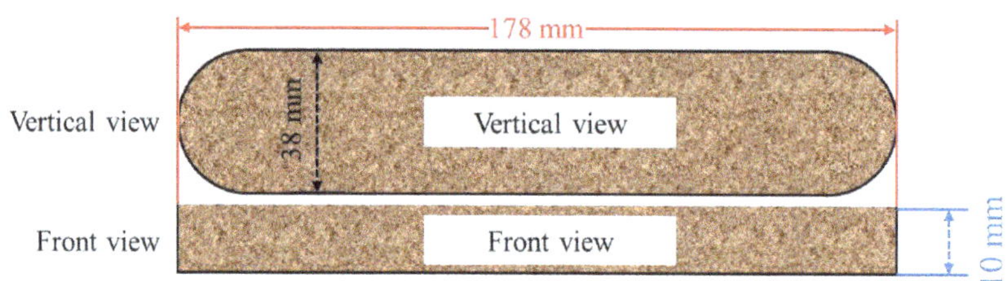

Figure 1. Rock slab size diagram.

Table 2. Porosity and permeability analysis results.

No.	Length (cm)	Diameter (cm)	Porosity (%)	Permeability (mD)
2-1#	5.979	2.506	10.85	0.149
2-2#	5.689	2.501	9.82	0.074
2-3#	5.714	2.502	11.17	0.305

2.2. Experimental Method and Equipment

2.2.1. Acid Rheological Test

The acid rheological test used the Haake Mars III rheometer coaxial cylinder test system (Figure 2). The control mode is rate control. The shear time is 120 min, the shear rate is 170 s^{-1}, and the test temperature is 60, 80, 100, 120, 140, and 160 °C. The apparent viscosity of the acid is tested with the shear time. Further, using the step-type constant temperature module in the rheometer, the control mode is set to rate control, the shear rate scanning range is set to 0.1~500 s^{-1}, the test temperature is 140 °C, and the sampling points are 200. Firstly, the cross-linked acid was preheated to the test temperature, and the variation of the apparent viscosity of the acid with the shear rate was tested. The purpose was to obtain the cross-linked acid flow index and the consistency coefficient.

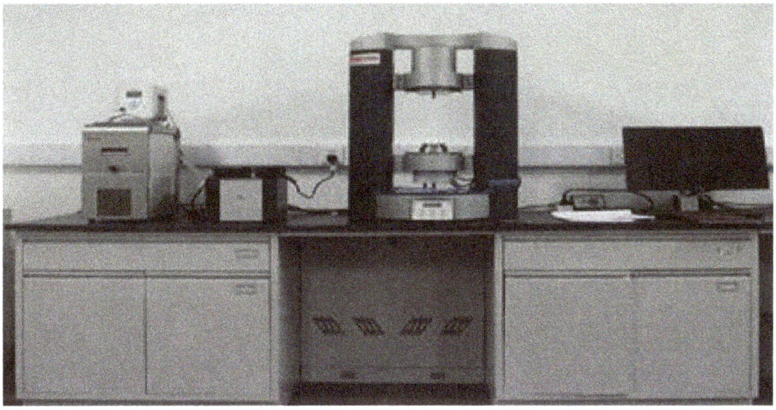

Figure 2. Haake Mars III rheometer.

2.2.2. Acid Etching and Conductivity Test

The experiment of acid-etching fracture conductivity is divided into two parts. The first part is to simulate the acid-etching process on the fracture surface by acid-etching experiment and analyze the etching morphology by X-ray computed tomography (CT scanner). The specific steps are shown in Zhang's research [33], and the experimental scheme is shown in Table 3. The second part is to simulate the seepage pattern of acid-etched fractures in underground rock mass after acid fracturing (Figure 3). The experimental results can be used to evaluate the conductivity of local acid-etched fractures. The principle of acid-etching fracture conductivity test is Darcy's law.

$$\mathrm{K} = \frac{5.555\mu Q}{\Delta P W_f} \tag{1}$$

Table 3. Acid-etching experimental scheme.

Slab No.	Acid	Injection Rate (mL/min)	Total Volume (mL)	Width (mm)
1#	20% cross-linked acid	40	1000	2
2#	20% cross-linked acid	50	1000	2
3#	20% cross-linked acid	60	1000	2
4#	15% cross-linked acid	50	1000	2
5#	10% cross-linked acid	50	1000	2

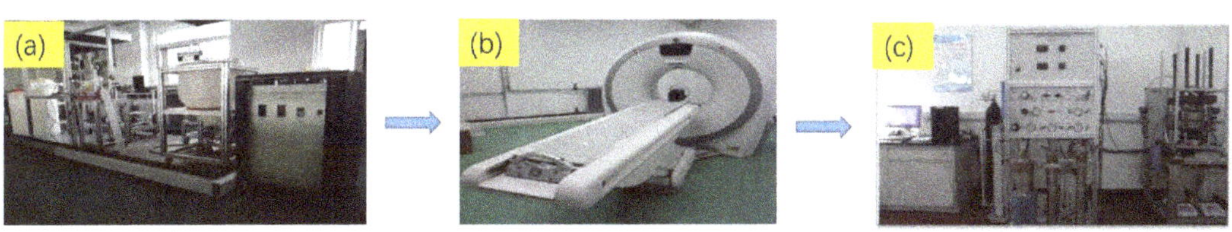

Figure 3. Equipment for acid etching and conductivity test: (**a**) Rock slab acid etching device, (**b**) CT scanner, (**c**) Fracture conductivity test.

The conductivity can be further expressed as:

$$KW_f = \frac{5.555\mu Q}{\Delta P} \quad (2)$$

where K is fracture permeability, mD; μ is viscosity, mPa·s; Q is flow rate, cm^3/min; ΔP is the pressure difference, kPa; W_f is fracture width, cm.

2.2.3. Permeability Damage Test

The core damage test is performed using a core-flooding device. Fresh cross-linked acid is prepared first, and then calcium carbonate powder is continuously added until the pH of the acid solution is 6. Before the damage, the initial permeability of the core is tested first. The reacted acid is injected to damage the core. Finally, the permeability of the core after the damage is measured again. The damage degree of reacted acid to the core is calculated through the permeability change.

$$K = \frac{\mu QL}{A\Delta P} \quad (3)$$

where K is the core permeability, mD; Q is the flow through the core, cm^3/s; a is the cross-sectional area of the fluid through the core, cm^2; ΔP is the pressure difference, 0.1 MPa; μ is fluid viscosity, mPa·s; L is the core length, cm.

After obtaining the core permeability before and after damage, the damage rate of rock is calculated by the following formula:

$$P_e = \frac{K_b - K_a}{K_a} \times 100\% \quad (4)$$

where K_b is the core permeability before reacted acid damage, mD; K_a is core permeability after reacted acid damage, mD; P_e is defined as core damage rate.

3. Result and Discussion

3.1. Rheological Test Results

Figure 4 shows the relationship between the viscosity and shear time at 60 °C and 170 s^{-1}. The average viscosity of the last stable 30 min is the viscosity of the cross-linked acid at this temperature. The viscosity change of cross-linked acid at different temperatures is shown in Figure 5. When the temperature increases from 60 °C to 120 °C, the viscosity of the acid solution decreases rapidly, and the reduction rate reaches 50%. When the temperature increases from 120 °C to 140 °C, the viscosity changes little, and the influence of the temperature range on the cross-linked acid is small. When the formation temperature reaches 140 °C, the viscosity of the acid can still be maintained at about 80 mPa·s. When the temperature rises to 160 °C, the viscosity of the acid decreases rapidly. It is inferred that the polymer breaks at this temperature, decreasing the acid viscosity. From the viscosity curve, we can find that the cross-linked acid still maintains high viscosity at high temperatures, reducing the rate of H^+ transfer to the rock, thereby reducing the acid–rock reaction rate [34]. In general, the 20% cross-linked acid system has appropriate temperature and shear resistance during the test, which can meet the technical requirements of acid fracturing in high temperature deep reservoirs of the Changqing gas field [35].

The power exponential form is used to fit the curve, indicating that the reasonable correlation is good and the fitting result is reliable. It can be seen from Figure 6 that the rheological characteristics of the cross-linked acid agree with the rheological model of the power-law fluid. The consistency coefficient is 41,255 mPa·s^n, and the power law index is 0.937 at 140 °C, which shows typical non-Newtonian fluid characteristics. The viscosity decreases with the increase in shear rate, and the fluidity increases.

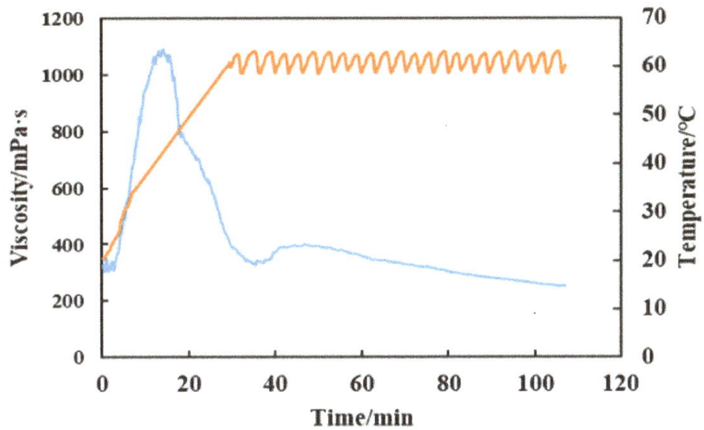

Figure 4. Viscosity–time curve of 20% cross-linked acid at 170 s^{-1} and 60 °C.

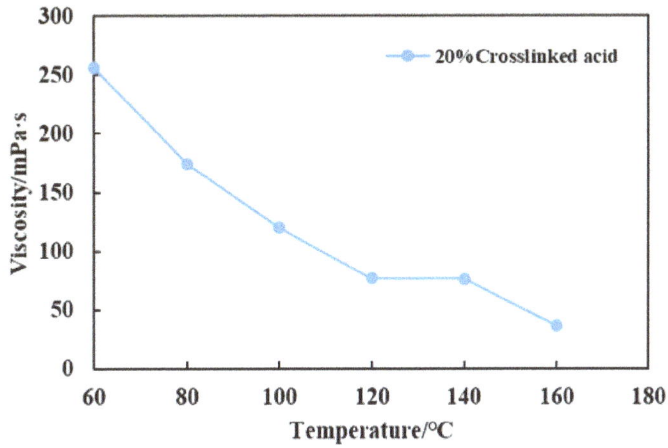

Figure 5. Viscosity test results of 20% cross-linked acid at different temperatures.

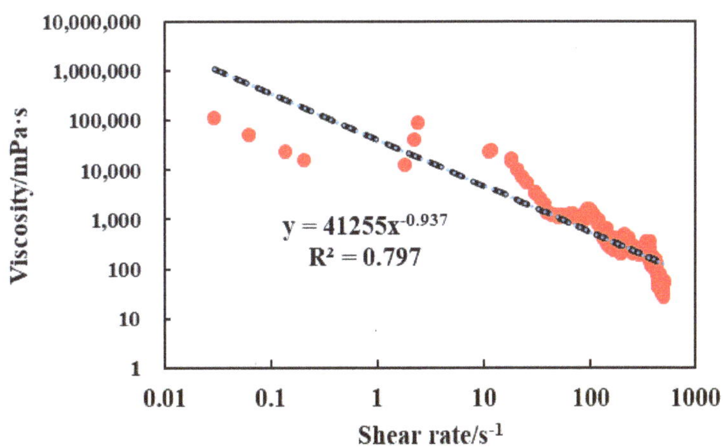

Figure 6. Relationship between viscosity of cross-linked acid and shear rate.

3.2. Acid Etching and Conductivity Test Results

3.2.1. Mass Change after Acid Etching

The first three groups of experimental injection rates were 40, 50, and 60 mL/min, and the total amount of acid injection was 1000 mL. Due to the different contact times of acid rock, with the increase in injection rate, the reaction time of acid rock becomes shorter, and the mass difference before and after the rock slab reaction decreases gradually (Figure 7). The fourth and fifth groups have 15% and 10% acid concentrations, respectively, and the injection rate is 50 mL/min. With the decrease in acid concentration, the quality of the rock slab participating in the reaction decreases rapidly. Due to the high viscosity of the acid, the rock plate is tight, the rock slab surface is etched, but the acid basically has no leak-off and no wormhole formation. Therefore, the acid–rock contact area of the five groups of experiments is the same, and the rock quality involved in the reaction is within a reasonable range.

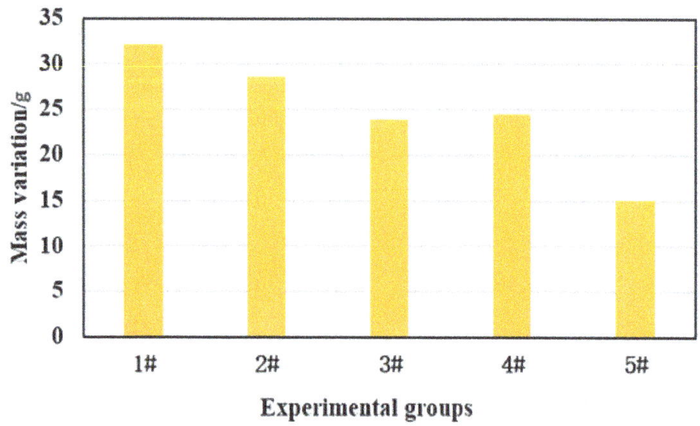

Figure 7. Mass of acid-etched rock slab.

3.2.2. Slab Etching Morphology under Different Injection Rates

Before acid etching, the rock slab is grayish white, and some areas are black. The rock surface is smooth. The red arrow represents the direction of acid injection. After acid etching, the surface color becomes darker, with some cross-linked acid residue (Figure 8). The reason may be that after the acid etching, when the water replaces the acid solution, because the acid's viscosity is much larger than that of the water, the water forms a viscosity fingering in the acid solution, causing the waterfront to quickly break through to the outlet end of the rock slab. The acid solution cannot be replaced entirely. Regarding the problem of acid residue on the surface of the rock slab and causing damage to the reservoir, the core displacement experiment analysis is carried out separately. After cleaning the rock plate and observing it again, it is found that the surface of the rock slab is uneven, and the dominant acid fluid flows through the channel on the flank. As the rock slab is tight, no wormhole is formed on the slab's surface. In subsequent experiments, the slabs before acid etching are smooth, ensuring the unity of the experiment. Due to the deep color of the slab after acid etching, the etching morphology cannot be distinguished by the naked eye, so the surface morphology is reconstructed by CT scanning. According to the difference of atomic number/density of different components in the core, the objects with different densities are divided and three-dimensional imaging is carried out to show the acid-etching morphology after the reaction of acid solution and rock slab. In subsequent experiments, the results of CT scans were displayed.

Figure 8. Photos of the first group of rock slabs before and after reaction.

After the first group of acid etching (Figure 9a), the natural pores at the outlet of the rock slab show an expanding trend. However, there are many non-uniform dissolution pits on other surfaces, and there is no apparent wormhole formation. According to the study of Pournik [36], along the acid flow direction is the formation of the dominant flow channel. After the second group of etching (Figure 9b), due to the increase in injection rate, the surface etching morphology heterogeneity is higher than that of the first group. On 2#rock slab, along the acid flow direction forms a flow channel. The convex part can support the fracture after the fracture is closed and maintain high conductivity. After acid etching, the third group (Figure 9c) also has no wormhole formation, more in the direction perpendicular to the acid flow to form a tiny flow channel, which is closely related to the mineral distribution.

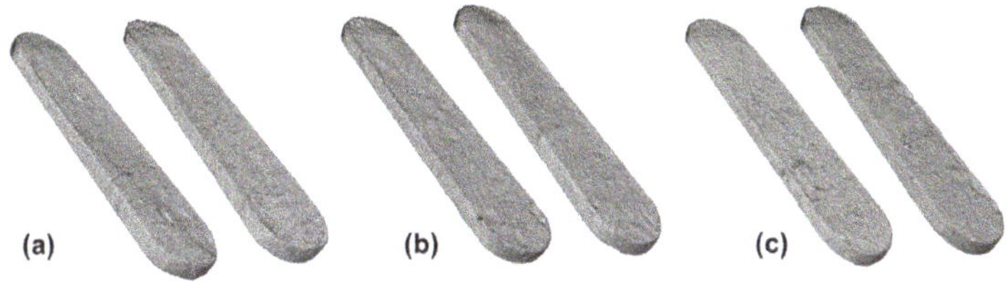

Figure 9. CT reconstruction of acid-etching morphology under different injection rates. (**a**–**c**) represents the injection rate of 40, 50 and 60 mL/min.

3.2.3. Slab Etching Morphology under Different Acid Concentrations

In the fourth group (Figure 10a), the acid concentration decreases and heterogeneous etching also appears on the slab's surface, but no prominent acid flow channel appears. Compared with high concentration, the number of pits on the surface of rock plate after etching in low concentration acid solution is more, but the area is relatively small. The fifth group (Figure 10b) uses 10% cross-linked acid. The surface etching morphology is relatively flat. From the etching morphology and the mass difference before and after the reaction, the etching effect of low-concentration acid solution is worse than that of high-concentration acid solution. This shows that the front fracture edge is etched by low-concentration acid in the actual acid fracturing process, and the effect is far less than that near the wellbore.

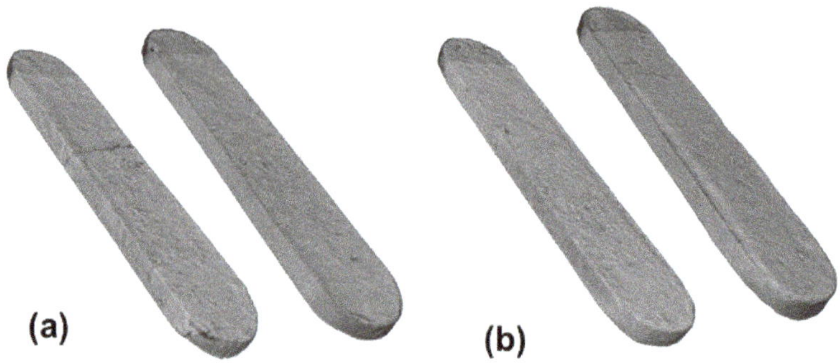

Figure 10. CT reconstruction of acid-etching morphology under different acid concentrations. (**a**,**b**) represents the acid concentration of 15% and 10%.

3.2.4. Conductivity Test Results

At different injection rates, the conductivity curve shows different trends. It can be seen from Figure 11 that the initial conductivity is very high at an injection rate of 40 mL/min. However, when the closure pressure increases to 10 MPa, the conductivity decreases rapidly to about 70 D·cm. When the closure pressure increases to 25 MPa, the conductivity decreases to 15 D·cm. When the injection rate is 50 mL/min, under different closure pressures, the conductivity remains basically unchanged at about 110 D·cm, and the effect is the best. Under 60 mL/min injection rate, the initial fracture conductivity is high. When the closure pressure is increased to 25 MPa, the conductivity decreases rapidly, and the final conductivity is zero.

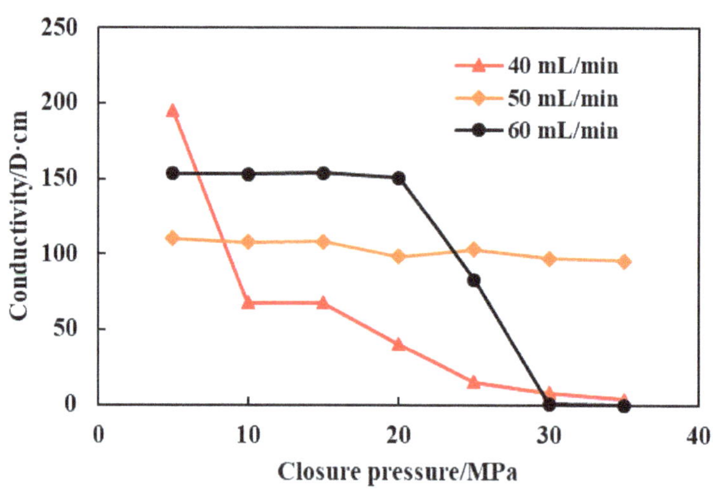

Figure 11. Variation of conductivity with closure pressure at different injection rates.

Compared with the acid-etching morphology formed at different injection rates of cross-linked acid, the acid-etching morphology at a smaller injection rate (40 mL/min) tends to be the weak channel type, with some small support points. The acid-etching morphology at the intermediate injection rate (50 mL/min) belongs to the channel type, and the support area is connected into a piece. The acid-etching morphology at higher injection rate (60 mL/min) is rough, and the support point area is dispersed. The results of conductivity show that the conductivity generated by the channel acid-etching morphology decreases slowly with the increase in closure pressure. Choosing the appropriate injection rate to form channel-type acid-etching morphology is of great significance for the development of deep and low permeability carbonate reservoirs.

It can be seen from Figure 12 that with the decrease in acid concentration, the conductivity of rock slab is greatly reduced under the same closure pressure. The conductivity produced by 20% cross-linked acid remains basically unchanged with the increase in closure pressure, and the conductivity produced by 15% cross-linked acid decreases with the increase in closure pressure. Especially when the acid concentration is 5%, the closure pressure is increased to 15 MPa, and there is almost no conductivity. From the perspective of acid-etching morphology, the quality of acid–rock reaction is obviously reduced after the acid concentration is reduced. It shows that the acid concentration is critical to the acid action distance in the actual acid fracturing stimulation. The decrease in conductivity caused by acid concentration also shows that the problem of different conductivity at different fracture distances should be fully considered in numerical simulation.

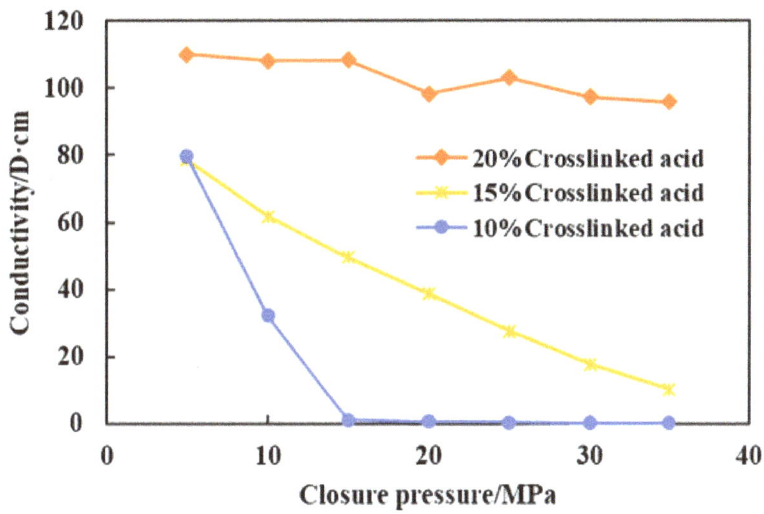

Figure 12. Variation of conductivity with closure pressure at different acid concentrations.

3.3. *Permeability Damage Test*

The reacted acid is injected into the core, and the change value of core permeability and core surface morphology judges the damage degree of reacted acid to the reservoir.

After the core is damaged by reacted acid, the photos and displacement pressure difference curves are shown in Figures 13 and 14. From the core photos after displacement, it can be found that there is a large amount of reacted acid at the entrance of the core. At the same time, there is less reacted acid at the outlet of the core. By observing the displacement pressure difference curve, it is found that the maximum displacement pressure difference can reach 12 MPa when reversing flooding kerosene after damage and decreases rapidly. This is because the reacted acid and kerosene flow in the opposite direction. In the previous stage, the reacted acid blocked the core entrance. When kerosene flows into the core for the second time, the reacted acid increases the flow resistance of kerosene, resulting in a sharp increase in the flow pressure difference. When the pressure reaches a certain value, the resistance of the reacted acid is broken through, and the subsequent kerosene flows out along the low-resistance flow channel. After the reacted acid damage, the kerosene displacement pressure difference increases slightly. The permeability damage is 6.72%, indicating that the core permeability decreases, and the reacted acid causes damage to the pore (Figure 15). Through microscopic observation (Figure 16.), there is gray–black acid slag accumulation at the inlet and glial residues on the entire core surface, which do not easily fall off after washing with water.

Figure 13. Core inlet (left) and outlet (right) after damage.

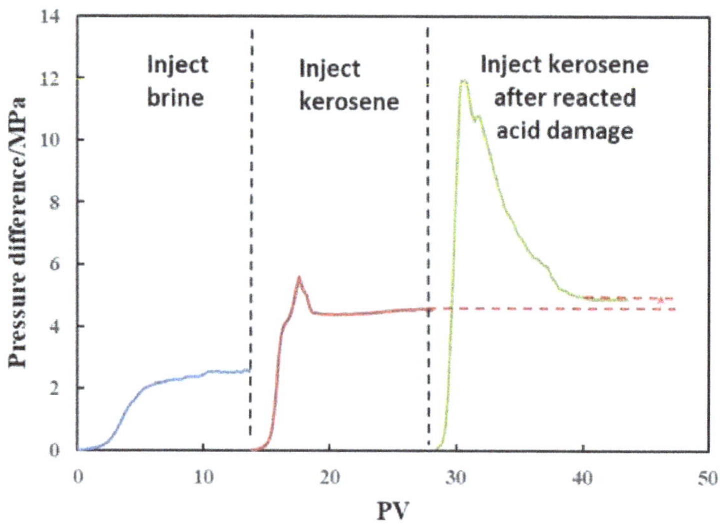

Figure 14. Pressure difference curve before and after damage.

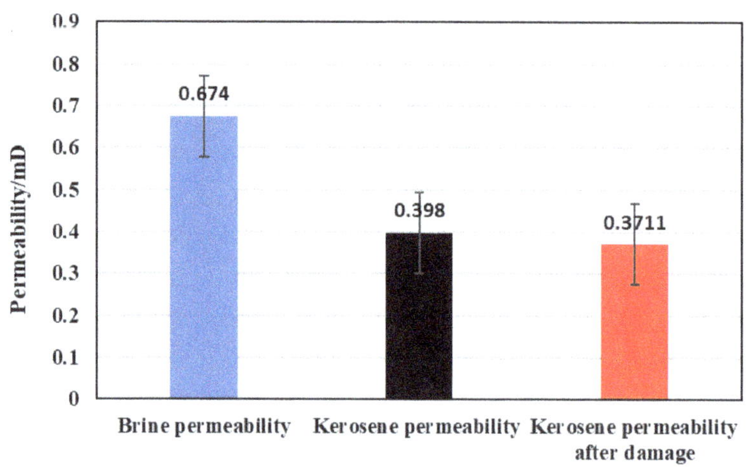

Figure 15. Core permeability in different treatment stages.

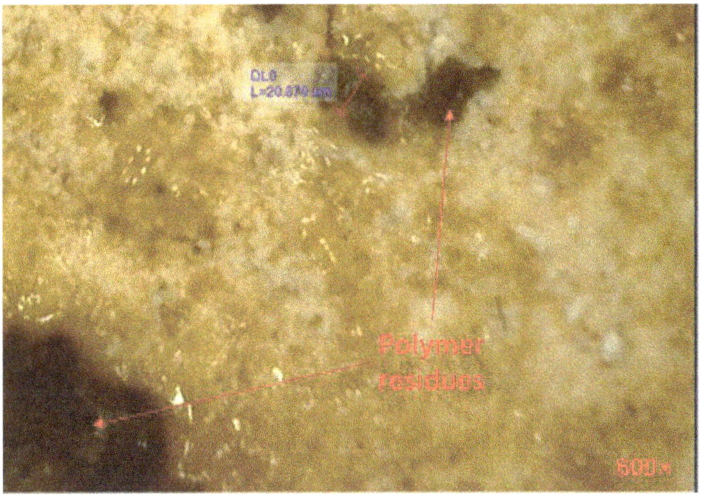

Figure 16. Core entrance observed by microscope.

4. Conclusions

This paper evaluated the performance of a ground cross-linked acid system from the aspects of rheological properties, rock slab acid etching, conductivity, and reacted acid damage with high carbonate mineral rock. The main conclusions and suggestions are as follows:

(1) The viscosity of the cross-linked acid can be maintained at about 80 mPa·s at 120~140 °C, and the viscosity at 160 °C is about 40 mPa·s, indicating that the system has good temperature resistance and shear resistance during the test time;
(2) The etching morphology and conductivity test show that when the injection rate is 50 mL/min, the conductivity is 110 D·cm. With the increase in closure pressure, the decreased conductivity rate is low. The conductivity is closely related to the acid = etching morphology. The experimental results show that the channel-type acid-etching morphology has a large conductivity and is not easy to reduce;
(3) With the decrease in acid concentration, the etching effect worsens, and the conductivity decreases rapidly. When the acid concentration is 10%, the closure pressure increases to 15 MPa, with almost no conductivity. The relationship between conductivity and closure pressure is related to the strength of fracture surface. It is the trend of subsequent research to establish the relationship of acid-etching conductivity considering strength;
(4) The reacted acid pollution mainly occurs at the inlet, forming a dense filter cake, hindering the entry of subsequent residues, and reducing reservoir damage.

The wormholes produced in the acid-etching process and natural fractures increase the acid fluid loss and reduce the acid-etching fracture's length. This paper studies the damage of reacted acid to the matrix core without considering its damage to natural fractures. These problems are essential for improving the stimulation efficiency of carbonate acid fracturing and need further research.

Author Contributions: Conceptualization, H.L.; methodology, T.H.; experiment, F.W.; validation, L.Y.; data analysis, S.L.; investigation, G.Y.; writing—original draft preparation, Q.W.; writing—review and editing, G.W. and Y.L.; funding acquisition, F.Z. All authors have read and agreed to the published version of the manuscript.

Funding: This research was financially supported by National Natural Science Foundation of China (grant No. 52174045).

Institutional Review Board Statement: Not applicable.

Informed Consent Statement: Not applicable.

Data Availability Statement: Not applicable.

Conflicts of Interest: The authors declare no conflict of interest.

References

1. Klym, H.; Karbovnyk, I.; Piskunov, S.; Popov, A.I. Positron Annihilation Lifetime Spectroscopy Insight on Free Volume Conversion of Nanostructured MgAl2O4 Ceramics. *Nanomaterials* **2021**, *11*, 3373. [CrossRef] [PubMed]
2. Luchechko, A.; Shpotyuk, Y.; Kravets, O.; Zaremba, O.; Szmuc, K.; Cebulski, J.; Ingram, A.; Golovchak, R. Microstructure and luminescent properties of Eu3+-activated MgGa2O4: Mn2+ ceramic phosphors. *J. Adv. Ceram.* **2020**, *9*, 432–443. [CrossRef]
3. Schwalbert, M.P.; Aljawad, M.S.; Hill, A.D.; Zhu, D. Decision Criterion for Acid-Stimulation Method in Carbonate Reservoirs: Matrix Acidizing or Acid Fracturing? *SPE J.* **2020**, *25*, 2296–2318. [CrossRef]
4. Zhu, D.; Wang, Y.; Cui, M.; Zhou, F.; Wang, Y.; Liang, C.; Zou, H.; Yao, F. Acid System and Stimulation Efficiency of Multistage AcidFracturing in Porous Carbonate Reservoirs. *Processes* **2022**, *10*, 1883. [CrossRef]
5. Zhao, X.; Cun, X.; Li, N.; Yu, J.; Ren, X.; Shen, F.; Zhu, J.; Zhang, Y.; Ren, Z. Simulation of Volumetric Acid Fracturing Fracture in Low Permeability Carbonate Rock. *Pet. Sci. Technol.* **2022**, *40*, 2336–2360. [CrossRef]
6. Dong, R.; Alpak, F.O.; Wheeler, M.F. Accurate Two-Phase Flow Simulation in Faulted Reservoirs by Combining Two-Point Flux Approximation and Mimetic Finite Difference Methods. *SPE J.* **2022**, *28*, 111–129. [CrossRef]
7. Aljawad, M.S.; Aljulaih, H.; Mahmoud, M.; Desouky, M. Integration of Field, Laboratory, and Modeling Aspects of Acid Fracturing: A Comprehensive Review. *J. Pet. Sci. Eng.* **2019**, *181*, 106158. [CrossRef]
8. Antelo, L.F.; Pournik, M.; Zhu, D.; Hill, A.D. Surface Etching Pattern and Its Effect on Fracture Conductivity in Acid Fracturing. In Proceedings of the SPE Hydraulic Fracturing Technology Conference, The Woodlands, TX, USA, 19–21 January 2009.
9. Gou, B. Acid-Etching Fracture Morphology and Conductivity for Alternate Stages of Self-Generating Acid and Gelled Acid during Acid-Fracturing. *J. Pet. Sci. Eng.* **2021**, *200*, 108358. [CrossRef]
10. Jafarpour, H.; Aghaei, H.; Litvin, V.; Ashena, R. Experimental Optimization of a Recently Developed Matrix Acid Stimulation Technology in Heterogeneous Carbonate Reservoirs. *J. Pet. Sci. Eng.* **2021**, *196*, 108100. [CrossRef]
11. Shabani, A.; Jamshidi, S.; Jahangiri, H.R. Application of a Mathematical Method in Calculation of the Skin Variation during a Real Field Acidizing Operation. *Chem. Eng. Sci.* **2018**, *192*, 829–839. [CrossRef]
12. Karimi, M.; Shirazi, M.M.; Ayatollahi, S. Investigating the Effects of Rock and Fluid Properties in Iranian Carbonate Matrix Acidizing during Pre-Flush Stage. *J. Pet. Sci. Eng.* **2018**, *166*, 121–130. [CrossRef]
13. Deng, J.; Mou, J.; Hill, A.D.; Zhu, D. A New Correlation of Acid-Fracture Conductivity Subject to Closure Stress. *SPE Prod. Oper.* **2012**, *27*, 158–169. [CrossRef]
14. Asadollahpour, E. The Etching and Hydraulic Conductivity of Acidized Rough Fractures. *J. Pet. Sci. Eng.* **2018**, *166*, 704–717. [CrossRef]
15. Trejos, D.Y.; Valverde, J.C.; Venturino, E. Dynamics of infectious diseases: A review of the main biological aspects and their mathematical translation. *Appl. Math. Nonlinear Sci.* **2022**, *7*, 1–26. [CrossRef]
16. Sabir, Z.; Umar, M.; Raja, M.A.Z.; Fathurrochman, I.; Hasan, H. Design of Morlet wavelet neural network to solve the non-linear influenza disease system. *Appl. Math. Nonlinear Sci.* **2022**. ahead of print. [CrossRef]
17. Yan, L.; Sabir, Z.; Ilhan, E.; Asif Zahoor Raja, M.; Gao, W.; Mehmet Baskonus, H. Design of a Computational Heuristic to Solve the Nonlinear Liénard Differential Model. *Comput. Model. Eng. Sci.* **2023**, *136*, 201–221. [CrossRef]
18. Dubey, R.S.; Goswami, P.; Baskonus, H.M.; Gomati, A.T. On the existence and uniqueness analysis of fractional blood glucose-insulin minimal model. *Int. J. Model. Simul. Sci. Comput.* **2022**, 2350008. [CrossRef]
19. Veeresha, P.; Ilhan, E.; Prakasha, D.G.; Baskonus, H.M.; Gao, W. Regarding on the fractional mathematical model of Tumour invasion and metastasis. *Comput. Model. Eng. Sci.* **2021**, *127*, 1013–1036. [CrossRef]
20. Sui, Y.; Cao, G.; Guo, T.; Li, Z.; Bai, Y.; Li, D.; Zhang, Z. Development of Gelled Acid System in High-Temperature Carbonate Reservoirs. *J. Pet. Sci. Eng.* **2022**, *216*, 110836. [CrossRef]
21. Alhamad, L.; Alrashed, A.; Al Munif, E.; Miskimins, J. Organic Acids for Stimulation Purposes: A Review. *SPE Prod. Oper.* **2020**, *35*, 952–978. [CrossRef]
22. Al-Douri, A.F. A New Organic Acid to Stimulate Deep Wells in Carbonate Reservoirs. Master's Thesis, Texas A&M University, College Station, TX, USA, 2014.
23. Wang, Q.; Zhou, F.; Zou, Y.; Yao, E. Study on the Performance of High-Temperature Resistant Cross-Linked Acid for Acid Fracturing of Tight Carbonate Reservoirs. In *All Days*; ARMA: Santa Fe, NM, USA, 2022; p. ARMA-2022-0318. [CrossRef]
24. Avtar, S.P.; David, L.H. *A New Stimulation Technique: High Strength Crosslinked Acid*; Society of Petroleum Engineers; SPE Production & Operations: Richardson, TX, USA, 1980; pp. 49–68.
25. Li, N.; Dai, J.; Liu, P.; Luo, Z.; Zhao, L. Experimental Study on Influencing Factors of Acid-Fracturing Effect for Carbonate Reservoirs. *Petroleum* **2015**, *1*, 146–153. [CrossRef]

26. Buijse, M.; Maier, R.; Casero, A.; Fornasari, S. *Successful High-Pressure/High-Temperature Acidizing with In-Situ Crosslinked Acid Diversion*; Society of Petroleum Engineers; SPE Production & Operations: Richardson, TX, USA, 2000; pp. 803–810.
27. Patil, P.; Sarda, A.; George, S.; Choudhary, Y.K.; Kalgaonkar, R. Non-Iron-Based Composition for In-Situ Crosslinked Gelled Acid System. In *All Days*; SPE: Doha, Qatar, 2012; p. SPE-156190-MS. [CrossRef]
28. Wang, Z.; Fu, M.; Song, Q.; Wang, M.; Zhao, X.; Wang, Y. Preparation and Characteristic of Cross-linked Acid for High-temperature Acid Fracturing System of Deep Carbonate Reservoirs. *Oilfield Chem.* **2016**, *33*, 601–606.
29. Fang, H.; Mou, J.; Wang, Y.; Li, C.; Luo, P. Development and evaluation of new high-temperature crosslinked acid. *Fault-Block Oil Gas Field* **2018**, *25*, 815–818.
30. Mehrjoo, H.; Norouzi-Apourvari, S.; Jalalifar, H.; Shajari, M. Experimental Study and Modeling of Final Fracture Conductivity during Acid Fracturing. *J. Pet. Sci. Eng.* **2021**, *208*, 109192. [CrossRef]
31. Chacon, O.G.; Pournik, M. Matrix Acidizing in Carbonate Formations. *Processes* **2022**, *10*, 174. [CrossRef]
32. Wang, Y.; Zhou, F.; Zhang, Y.; Wang, Y.; Su, H.; Dong, R.; Wang, Q.; Bai, H. Numerical studies and analysis on reaction characteristics of limestone and dolomite in carbonate matrix acidizing. *Geoenergy Sci. Eng.* **2022**, *222*, 211452. [CrossRef]
33. Zhang, L.; Zhou, F.; Mou, J.; Xu, G.; Zhang, S.; Li, Z. A New Method to Improve Long-Term Fracture Conductivity in Acid Fracturing under High Closure Stress. *J. Pet. Sci. Eng.* **2018**, *171*, 760–770. [CrossRef]
34. Jiang, Q.; Yang, X.; Gong, F. Development and evaluation of high temperature resistant cross-linked acid fracturing fluid system. *Chem. Ind. Eng.* **2022**, *39*, 51–57. [CrossRef]
35. Hu, J.; Chen, T.; Cheng, F.; Cheng, X.; Zhao, C. Preparation and Application of Hydrophobic Associating Crosslinking Acid Thickener. *Oilfield Chem.* **2020**, *37*, 403–408. [CrossRef]
36. Pournik, M. Laboratory-Scale Fracture Conductivity Created by Acid Etching. Ph.D. Thesis, Texas A&M University, College Station, TX, USA, 2008.

processes

Article

Case Study on the Effect of Acidizing on the Rock Properties of the Mahu Conglomerate Reservoir

Lifeng Wang [1,2], Wenting Jia [3], Yajun Xu [1,2], Jianye Mou [3,*], Ze Liao [1,2] and Shicheng Zhang [3]

1 DownHole Service Company, CNPC XiBu Drilling Engineering Company Limited, Karamay 834000, China
2 Shale Oil Reservoir Reformation Sub-Laboratory, CNPC Key Laboratory for Reformation of Oil and Gas Reservoirs, Karamay 834000, China
3 College of Petroleum Engineering, China University of Petroleum (Beijing), Beijing 102249, China
* Correspondence: moujianye@cup.edu.cn

Abstract: The development of the Mahu tight reservoir has adopted horizontal wells with staged fracturing. In the fracturing, there is a problem of a high fracturing pressure. Acid treatment is often used to lower the fracturing pressure on site. At present, the impact of this acid treatment on the physical parameters of the rocks of the reservoir in the Mahu region has not been systematically studied. Aiming to solve this problem, this paper conducted an experimental study on how acid dissolution affects the physical properties of the Mahu conglomerate, including its porosity, permeability, triaxial rock mechanical parameters, tensile strength, and mineral composition. First, the experimental scheme was designed. Next, a series of experiments were conducted. Finally, the experiment results were analyzed comparatively before and after acidizing. The acid composition, concentration, and contact time were the main factors for the analysis, based on which the acid system and related parameters were recommended. This study showed that the Mahu conglomerate exhibited brittle plasticity characteristics under stress. The carbonate content in this region was low, while the feldspar content was high, so it was necessary to use mud acid to effectively dissolve feldspar, clay, and other silicates. After acidizing, the porosity was 200% of the original value. The permeability increased by up to 14 times. The tensile strength decreased significantly by up to 84%. The value of Young's modulus of the rock decreased by up to 63.6%. The value of Poisson's ratio was reduced by up to 40.7%. A combination of 6% HF + 15% HCl is recommended, with an effective acid treatment time of over 60 min for the Mahu conglomerate. Acidizing could significantly change the mechanical properties and permeability of the rock of the Mahu conglomerate reservoir, thus effectively reducing the formation fracturing pressure. This research provides technical support for Mahu acid dipping in horizontal well fracturing.

Keywords: experiment; Mahu; acidizing; physical properties; acid system

Citation: Wang, L.; Jia, W.; Xu, Y.; Mou, J.; Liao, Z.; Zhang, S. Case Study on the Effect of Acidizing on the Rock Properties of the Mahu Conglomerate Reservoir. *Processes* **2023**, *11*, 626. https://doi.org/10.3390/pr11020626

Academic Editor: Yong Yuan

Received: 6 December 2022
Revised: 12 January 2023
Accepted: 16 January 2023
Published: 18 February 2023

1. Introduction

Sandstone and conglomerate reservoirs are widely distributed in the Songliao Basin, the Tarim Basin, the Bohai Bay Basin, and the Junggar Basin in China. The Junggar Basin, with its sizable proven oil reserves and huge development potential [1,2], is located in Xinjiang province in the northwest of China. It is the primary development area of conglomerate reservoirs in Xinjiang and China [3–5]. The Mahu conglomerate oil reservoir in the northwest edge of the Junggar Basin, with an area of nearly 5000 km^2, is an unconventional reservoir with poor physical properties and low permeability [6]. This kind of tight formation generally uses horizontal wells with a staged fracturing stimulation. Horizontal well fracturing in the Mahu area has the problem of that it is hard to break the formations due to the high fracturing pressure required. The main methods of lowering the fracturing pressure include acid pretreatment, hydro jetting, high-energy gas fracturing, and perforation [7–12]. At present, acid pretreatment technology is commonly used to reduce the fracturing pressure in the Mahu area along with perforation.

Li [13] derived an analytical formula of the formation fracture pressure P_f for horizontal well perforation completion based on the formula of formation fracture pressure for vertical well perforation completion.

$$P_f = \frac{9\sigma_h - 3\sigma_V - \sigma_H + 2v(\sigma_V - \sigma_h) - \varphi\frac{1-2v}{1-v}p_0 + \sigma_f}{4 + \varphi_c - \varphi\frac{1-2v}{1-v}} \quad (1)$$

where P_0 is the original formation pore pressure, MPa; φ is the rock porosity, %; φ_c is the rock contact porosity, %; σ_f is the uniaxial tensile strength of rock, MPa; and v is Poisson's ratio of rock, which is dimensionless. It can be seen from Equation (1) that increasing the pore pressure and decreasing the tensile strength can lower the breaking pressure. Acidizing can significantly increase the permeability so as to increase the pore pressure before initiating the fracture. Acidizing also dissolves some minerals so as to decrease the tensile strength. This is why acidizing can decrease the fracturing pressure.

Some researchers have studied the influence of acid pretreatment on the mechanical properties of rock, but this research has mostly been conducted on shale instead of conglomerate. Wang et al. [14] used a triaxial rock mechanics testing system to conduct a comparative study on the compressive strength of reservoir rocks before and after sandstone acidizing, and they found that a high concentration of hydrofluoric acid can greatly reduce the compressive strength of rocks. Wang et al. [15] proposed using a rock mechanics testing system to determine the mechanical properties of rocks before and after acidizing and systematically studied the effects of the acid composition and concentration on the mechanical properties of sandstone rocks. Guo et al. [16] conducted a comparative study on the rock elastic modulus, Poisson's ratio, and rock compressive strength before and after acidizing in ultra-deep reservoirs with a high fracturing pressure by using reservoir rock samples, and the results showed that the acid rock reaction reduced the rock strength, thus reducing the formation fracturing pressure. Deng et al. [17] studied the mechanical properties of sandstone before and after acidizing for high-pressure, high-temperature, and low-permeability reservoirs, and they elaborated the mechanism of acid pretreatment in changing the mechanical properties of rocks and effectively lowering the fracturing pressure from a microscopic perspective. Their results showed that mineral dissolution changed the shape of the intergranular pores and crystal solution pores and improved the connectivity of the pores, thus reducing the rock strength.

Acid pretreatment has been applied to horizontal well fracturing in the Mahu fracturing reservoir for some time, but some treatments take a long time, and some are not effective. Some problems of the acid pretreatment in the Mahu reservoir remain unclear, such as the effective acid composition and concentration, the acid–rock contact time, and the variation in the physical properties of the Mahu conglomerate. Therefore, in this paper, we conducted a systematic experimental study on the interaction between the acid and minerals and the rock properties, and we recommended an acid system and its related parameters and provided guidance for the in-field acidizing pretreatment in the Mahu area.

2. Experimental Apparatus and Approach

2.1. Experimental Apparatus

The acid flooding test was accomplished using apparatus composed of a flooding pump, containers, a thermostat, and Hastelloy acid-resistant holders. The maximum experimental temperature could reach 200 °C, meeting the requirements of simulating the formation temperature. Figure 1 shows the schematic drawing of the acid flooding apparatus.

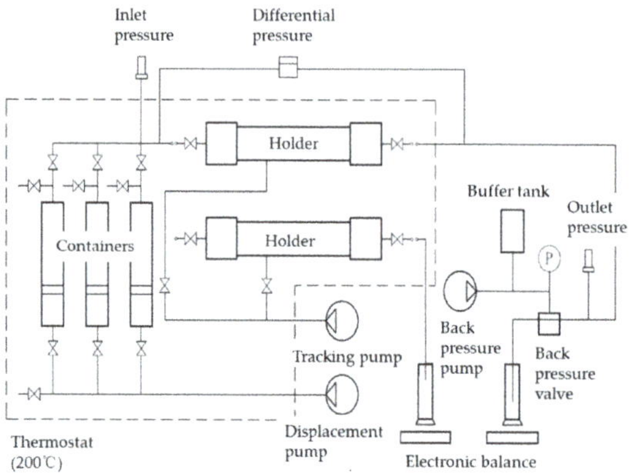

Figure 1. Experimental platform schematic drawing.

2.2. Experimental Samples

This paper took the downhole cores of Well Ma 132, 136, and 139 as the research object. The rock samples are conglomerate with a high gravel content and strong heterogeneity [18]. For triaxial experiment, the cores taken were 25.4 mm in diameter and 50 mm in length. For the tensile strength test, the cores taken were 25.4 mm in diameter and 13 mm in length. The perimeter of the rock sample was ensured to be smooth, keeping the parallelism of the two end faces within 0.02 mm and the perpendicularity of the section to the axis within 0.05 mm. Ensure the cylinder surface is smooth, and avoid uneven or gravel peeling on the cylinder surface.

2.3. Experimental Approach and Procedures

The temperature of acid flooding test was set as 90 °C based on the formation temperature in the Mahu region; the test pressure was 0.101325 MPa.

The specific experimental steps are as follows: ① carry out experimental tests on the original rock samples, including an XRD mineral composition test, porosity and permeability test, rock mechanics triaxial test, and tensile strength test. Obtain the fundamental physical properties of rock before acidizing. ② Conduct an acid flooding experiment with a fixed rate of 0.2 mL/min. Set up different acidizing parameters, including acid composition, acid concentration, and acid–rock contact time to conduct the acid flooding test. ③ Clean the acidized rock samples, place them for drying, and conduct mineral composition and mechanical parameters tests again for acidized rock samples. A complete study workflow is exhibited in Figure 2 below.

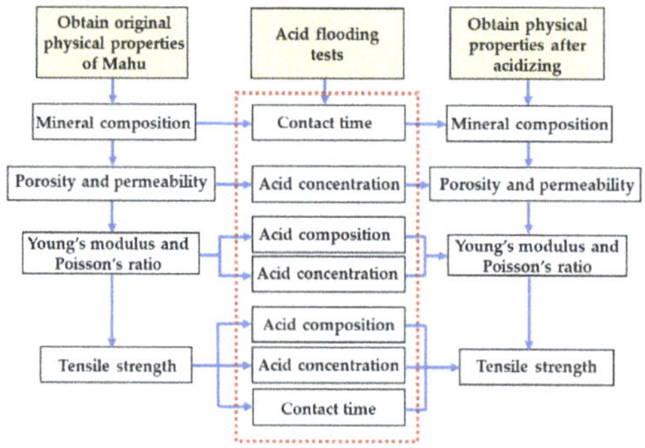

Figure 2. Workflow process chart of acidizing factors study.

The acidizing parameters were designed as follows: ① the mineral composition experiment was conducted in 13 groups of rock samples, including nine unacidified rock samples used to obtain the original mineral content. The acid composition was 6% HF + 15% HCL, and the acid treatment time was set as 30 min, 60 min, 90 min, and 120 min, respectively. ② The porosity and permeability comparative experiments were conducted in 18 groups of rock samples, including 9 groups of unacidified rock samples used to obtain the original porosity and permeability. Others were treated with mud acid, with concentrations of 1.5% HF + 15% HCL, 3% HF + 15% HCL, and 6% HF + 15% HCL. ③ In the triaxial test comparison experiment, 15 groups of rock samples were tested, including 3 groups of unacidified rock samples. Hydrochloric acid and mud acid were selected for treatment. Acid concentrations were set as 15% HCL, 20% HCL, 1.5% HF + 15% HCL, and 6% HF + 15% HCL. In total, 22 tensile strength tests were performed, including 11 groups of unacidified rock samples. First, the effects of hydrochloric acid and mud acid were compared. Next, concentrations of 10% HCL, 15% HCL, 20% HCL, 1.5% HF + 15% HCL, 3% HF + 15% HCL, and 6% HF + 15% HCL were considered. Finally, different contact times of 10 min, 20 min, 30 min, 60 min, and 120 min were considered. Table 1 represents the overall experiment scheme.

Table 1. Experiment scheme.

Experiment	Acid Composition	Acid Concentration	Contact Time	Experiment Groups	Control Groups
Mineral content test	HF + HCL	6% HF + 15% HCL	30 min	1	9
	HF + HCL	6% HF + 15% HCL	60 min	1	
	HF + HCL	6% HF + 15% HCL	90 min	1	
	HF + HCL	6% HF + 15% HCL	120 min	1	
Porosity and permeability test	HF + HCL	1.5% HF + 15% HCL	-	3	9
	HF + HCL	3% HF + 15% HCL	-	3	
	HF + HCL	6% HF + 15% HCL	-	3	
Triaxial rock mechanics test	HCL	15% HCL	-	3	3
	HCL	20% HCL	-	3	
	HF + HCL	1.5% HF + 15% HCL	-	3	
	HF + HCL	6% HF + 15% HCL	-	3	
Tensile strength test	HCL	10% HCL	60 min	1	11
	HCL	15% HCL	60 min	1	
	HCL	20% HCL	60 min	1	
	HF + HCL	1.5% HF + 15% HCL	60 min	1	
	HF + HCL	3% HF + 15% HCL	10 min	1	
	HF + HCL	3% HF + 15% HCL	20 min	1	
	HF + HCL	3% HF + 15% HCL	30 min	1	
	HF + HCL	3% HF + 15% HCL	60 min	2	
	HF + HCL	3% HF + 15% HCL	120 min	1	
	HF + HCL	6% HF + 15% HCL	60 min	1	

3. Mineral Content Variation by Acidizing

The acid composition selection and acid rock reaction rate depend on the mineral composition. Therefore, it is necessary to test and obtain the mineral composition of the target reservoir. Meanwhile, the carbonate rock content determines how to recommend the acid composition. The MiniFlex II benchtop X-ray diffractometer was used for mineral composition testing. Figure 3 illustrates how the apparatus tests.

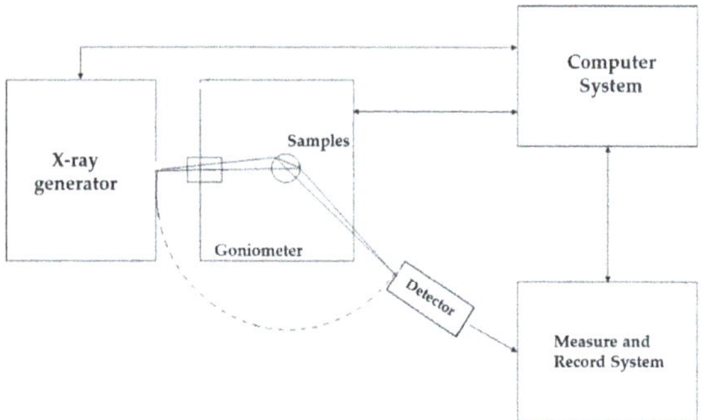

Figure 3. X-ray diffraction structure diagram.

The mineral composition of Mahu was obtained by the test; the results are revealed in Table 2, which contains the original mineral component of Mahu sag, and mineral composition after 60 min acidizing. The original value statistics are shown in Figure 4. As can be seen from the results, the mineral composition of Mahu sag is mainly dominated by quartz with an average content of 46.54%, carbonate rock with an average content of 6.08%, clay with an average content of 8.02%, potassium feldspar with an average content of 13.84%, and plagioclase with an average content of 20.82%. The clay and carbonate rock content are low, and the feldspar content is high at more than 30%. The rock samples were treated with 6% HF + 15% HCL for 1 h; calcite was dissolved completely. Hydrochloric acid cannot dissolve mud or quartz, while hydrofluoric acid reacts not only with carbonate, but also with silicate. Clay and feldspar were dissolved; they were reduced by almost 34%.

Table 2. Mineral composition variation after 60 min acidizing.

Well	Clay (%)		Quartz (%)		Potash Feldspar (%)		Plagioclase (%)		Calcite (%)		Dolomite (%)		Augite (%)	
	Without Acidizing	60 min Acidizing	Without Acidizing	60 min Acidizing	Without Acidizing	60 min Acidizing	Without Acidizing	60 min Acidizing	Without Acidizing	60 min Acidizing	Without Acidizing	60 min Acidizing	Without Acidizing	60 min Acidizing
MA132	8.87	4.44	43	65.76	11.53	5.12	24.13	18.03	4.47	0.00	2.73	0.95	5.27	5.7
MA136	6.23	3.32	47.17	62.45	15.33	10.1	20.65	16.05	-	-	4.95	1.72	5.67	6.36
MA139	8.97	4.55	49.43	67.2	14.67	10.55	17.67	11.68	3.63	0.00	2.47	1.05	3.17	4.97
Average	8.02	4.11	46.54	65.13	13.84	8.59	20.82	15.25	2.7	0.00	3.38	1.24	4.70	5.68

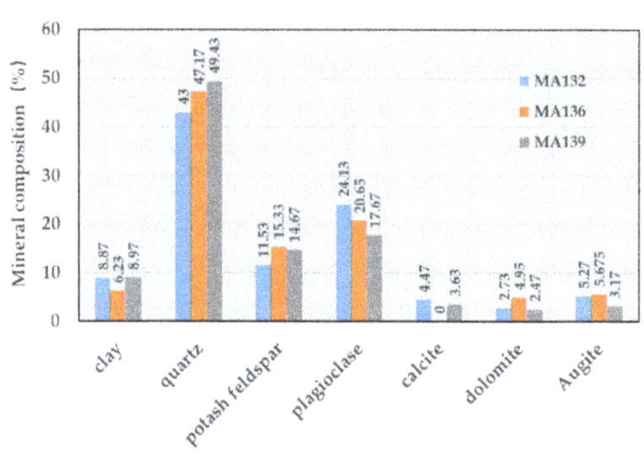

Figure 4. Mineral component distribution in Mahu reservoir.

Figure 5 is a graph which presents the image of rocks before and after acidizing. The rocks were dissolved partially, which can be seen from the edge of the rocks clearly. The micro-cracks and corrosion holes were irregularly distributed over the whole surface of the rocks.

Figure 5. Mineral dissolution by the acid.

Figure 6 shows a mineralogy change section through a rock sample. The acid flowed through and left distinct marks. There are several manifest cracks which can be observed inside the rock after acidizing.

Figure 6. Acid marks and obvious fractures after acidizing.

The rock samples were treated with 6% HF + 15% HCL. We tested and obtained the mineral composition of the rock samples after acidizing at different treatment times to analyze the variation in the carbonate minerals and clay and feldspar minerals.

Figures 7 and 8 reflect variations in the clay, feldspar minerals, and carbonate minerals when the acid treatment time was increased. Under the condition of the same acid concentration, with the increase in the acid treatment time, the content of calcite, dolomite, and other carbonate minerals in Mahu conglomerate decline continuously. After 30 min acidizing, the average content of carbonate rock reduces by 37%. In contrast, the content of quartz sees an opposite trend. The content of clay and feldspar also drop significantly; they drop by 19.7% for 30 min, 34.4% for 1 h, 54.3% for 90 min, and 61.7% for 120 min. Furthermore, the clay minerals grow down more than feldspar. Mud acid can effectively dissolve feldspar and clay. Therefore, the mixture of hydrofluoric acid and hydrochloric acid is chosen to have a better acidizing effect. Mud acid was mainly used in this study, and some experiments considered hydrochloric acid for comparative experiments.

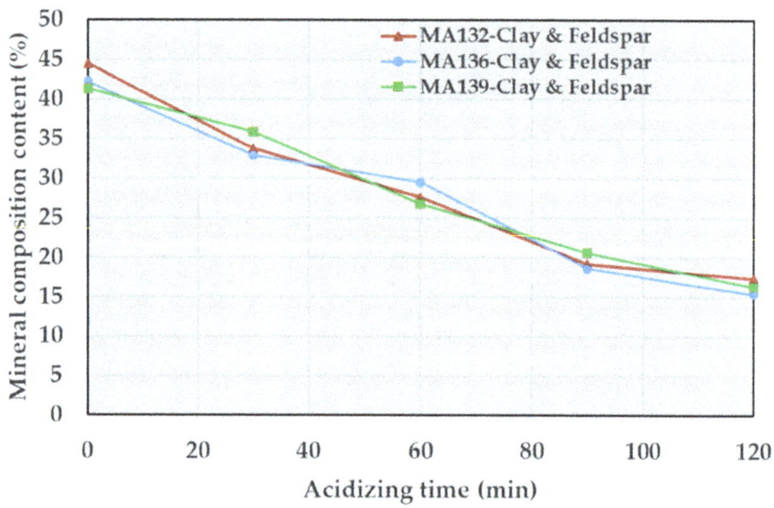

Figure 7. Changes in clay and feldspar minerals of Mahu conglomerate under different acid treatment times.

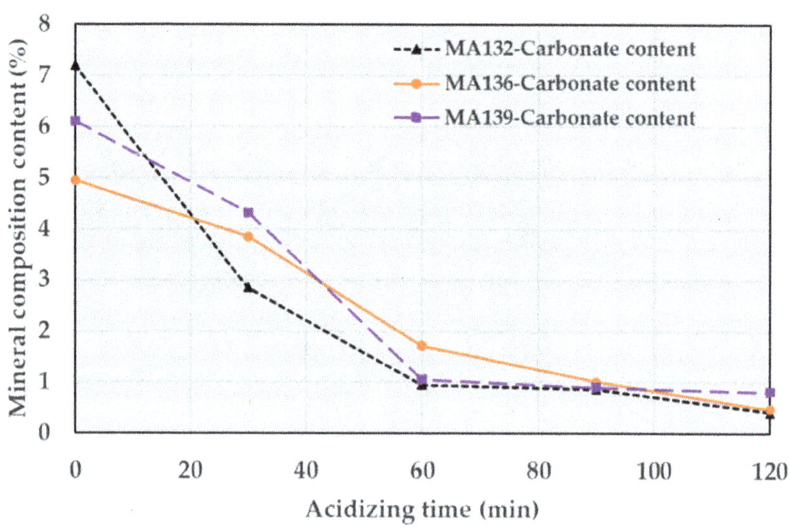

Figure 8. Changes in carbonate minerals of Mahu conglomerate under different acid treatment times.

4. Porosity and Permeability Increase after Acidizing

The volume in which a rock can store fluid and the ability of fluid transport in it are usually characterized by porosity and permeability. The ratio of the pore volume to the total rock volume is defined as porosity [19,20]. Based on the connecting status of pores in rocks, the porosity can be divided into effective and total porosity. Effective porosity is used to measure the volume of interconnected pores in rocks, which was measured in this paper. The quantity, indicating the ease of fluid flow in the medium, is called permeability. The permeability test was conducted by a 38 mm core holder and micro gas flow meter, which is the gas testing method. It is obtained by reading the micro gas flow meter, and the permeability can be calculated by combining the formula. The weight difference calculates the effective porosity before and after the core is saturated. The equations for calculating the permeability and porosity are given below.

$$K = \frac{2P_2Q_0\mu L}{A(P_1^2 - P_2^2)} \times 1000 \tag{2}$$

where K is the permeability, 10^{-3} um^2; A is the sectional area of the core, cm^2; L is the rock length, cm; P_1 is the absolute pressure at the inlet of the rock sample, 0.1 MPa; P_2 is the absolute pressure at the outlet of the rock sample, 0.1 MPa; μ is the gas viscosity, mPa.s; and Q_0 is the flow rate at the atmospheric pressure condition, cm^3/s.

$$\phi = \frac{V_p}{V_f} = \frac{V_f - V_g}{V_f} \tag{3}$$

where ϕ is the porosity, %; V_f is the gross rock volume, cm^3; V_p is the rock pore volume, cm^3; and V_g is the aggregate particles volume, cm^3.

The average porosity and permeability of unacidified rock samples were obtained by the tests. The average porosity of the Mahu region is 8.74%, and the average permeability is 4.18×10^{-4} μm^2. At the same time, the control groups were treated with different concentrations of mud acid. After acidizing, the rock samples were tested again to obtain the porosity and permeability.

As can be seen from Figure 9, the curves reflect the change in the trends of rock porosity and permeability after acid treatment. Different acidizing parameters on rock porosity and permeability arouse different results. Figure 9a shows the effect of acid concentrations on the porosity. The porosity is 200% of the original value. Figure 9b represents the effect of acid concentrations on the permeability, which increases by 5 times to 14 times. In general, the porosity and permeability see an upward trend with the increase in the acid concentration, but there is no obvious linear rule.

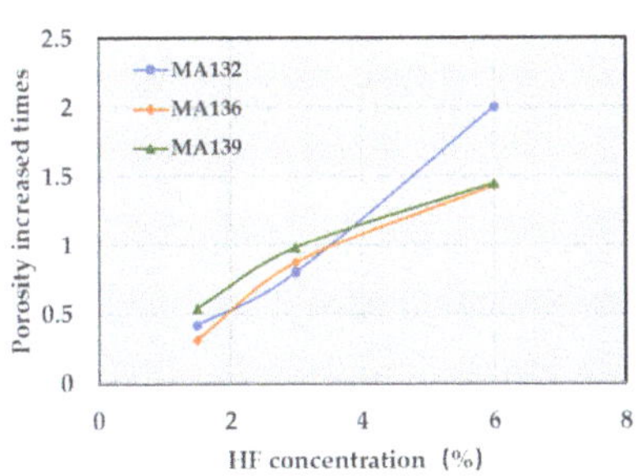

(a) Porosity

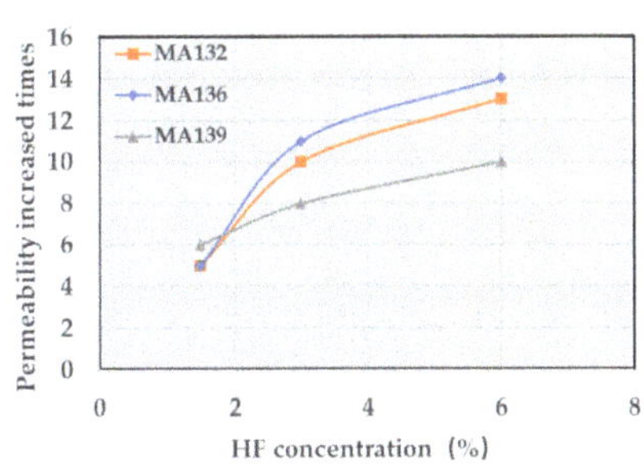

(b) Permeability

Figure 9. Growth multiples under different acid concentrations. (**a**) Growth multiples of porosity. (**b**) Growth multiples of permeability.

5. Triaxial Rock Mechanics Parameter Variation by Acidizing

The triaxial rock mechanics test is carried out with the GCTS RTR-1500 test system. The axial stress is 2000 KN, the maximum allowable confining pressure is 120 MPa, and the maximum pore pressure is 70 MPa. The sample produces longitudinal displacement and transverse displacement under the action of external force. The deformation of the sample can be calculated by the displacement value and the geometric size of the sample. Young's modulus and Poisson's ratio can be calculated by measuring the longitudinal and transverse deformation of the rock sample with a regular shape under axial and transverse pressure. The calculation formula is as follows:

$$E = \sigma_{(50)} / \varepsilon_{h(50)} \tag{4}$$

where E is Young's modulus, MPa; $\sigma_{(50)}$ is 50% of the maximum principal stress difference, MPa; and $\varepsilon_{h(50)}$ is the axial compression strain at $\sigma_{(50)}$, dimensionless.

$$v = \left| \frac{\varepsilon_{d(50)}}{\varepsilon_{h(50)}} \right| \tag{5}$$

where ν is Poisson's ratio, dimensionless; $\varepsilon_{h(50)}$ is the axial compression strain at $\sigma_{(50)}$, dimensionless; and $\varepsilon_{d(50)}$ is the radial compression strain at $\sigma_{(50)}$, dimensionless.

Under the conditions of reservoir temperature and stress, the confining pressure was set at 25 MPa; the measured average Young's modulus of unacidified cores was 26 GPa. The average Poisson's ratio was 0.29. The triaxial test was conducted again on the rock samples after acidizing with different experiment parameters. The results are shown in Table 3.

Table 3. Young's modulus and Poisson's ratio at different acid concentrations.

	Core Number	Unacidified	15% HCL	20% HCL	15% HCL + 1.5% HF	15% HCL + 6% HF
Young's modulus (GPa)	136-(6)	27.09	20.74	15.06	17.66	9.85
	136-(7)	24.33	20.11	14.12	16.24	10.57
	136-(8)	26.57	21.02	16.17	18.39	12.42
Poisson's ratio	136-(6)	0.28	0.24	0.19	0.21	0.17
	136-(7)	0.32	0.29	0.22	0.25	0.21
	136-(8)	0.27	0.23	0.18	0.21	0.16

The above table provides information about the value changes in Young's modulus and Poisson's ratio. Young's modulus has a similar tendency to Poisson's ratio after the acid treatment. The effects of the acid concentration on Young's modulus and Poisson's ratio are revealed in Figure 10.

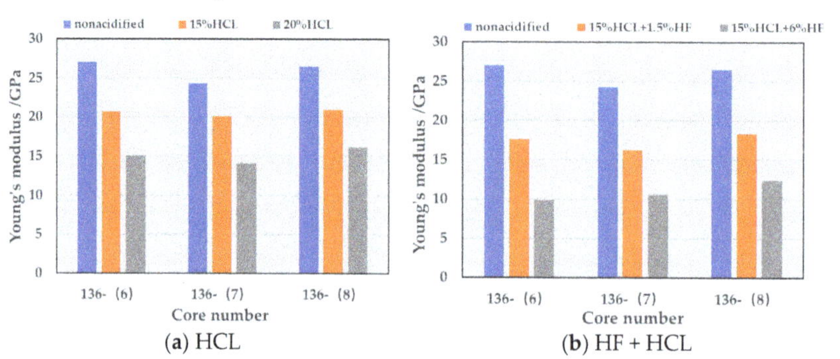

Figure 10. Influence of acid concentration on Young's modulus. (**a**) Hydrochloric acid treatment. (**b**) Mud acid treatment.

It is apparent from the diagram that Young's modulus witnesses a downward trend after being treated with both hydrochloric acid and mud acid, which declines more significantly after mud acid treatment on the condition of the same acid concentration and contact time.

Figure 11 illustrates how the acid concentration affects Poisson's ratio. There is a slight decrease after acidizing. The decrease in Poisson's ratio after hydrochloric acid treatment is less than that after mud acid treatment.

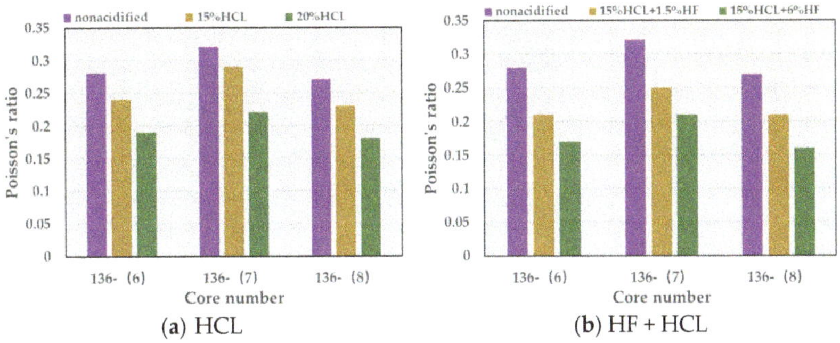

Figure 11. Influence of acid concentration on Poisson's ratio. (**a**) Hydrochloric acid treatment. (**b**) Mud acid treatment.

Figure 12 depicts the impact on Young's modulus and Poisson's ratio. Acid composition has prominent effects on Young's modulus, which shows a sharp drop, whereas Poisson's ratio only has a slight decrease.

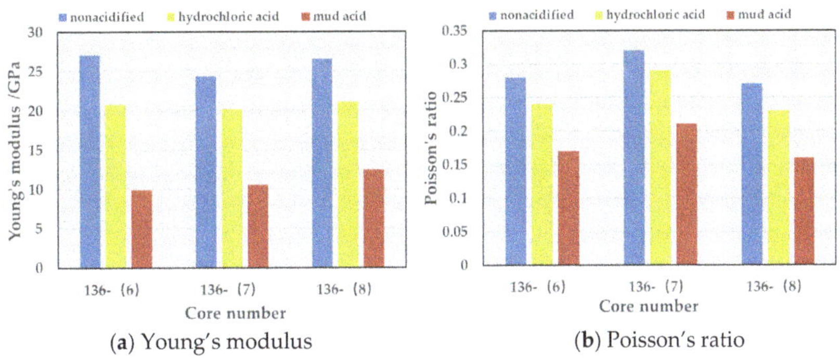

Figure 12. Effect of acid composition on Young's modulus and Poisson's ratio. (**a**) Effect of acid composition on Young's modulus. (**b**) Effect of acid type on Poisson's ratio.

The fracture modes of rocks obtained from the triaxial test are displayed in Figure 13. It shows the fracture morphology of unacidified rocks and acidized rocks under 25 MPa confining pressure. The failure modes of Mahu conglomerate are mainly tensile failure. It can be clearly seen from the graph that the fractures of acidized rocks become more obvious and the number of fractures increases after acidizing. More micro-cracks occurred around the main cracks after acidizing, and the cracks apparently increased compared with the unacidified rocks. The reasons accounting for this phenomenon can be explained that acid dissolves minerals and destroys cement, making the rock easier to crack. It was also found that the fractures propagation of unacidified rocks is mainly around the gravel, and some are wear gravel as well as acidized rocks. Moreover, acidized rocks have more wear gravel than unacidified rocks. The reason lies in that acidizing changed the rock strength.

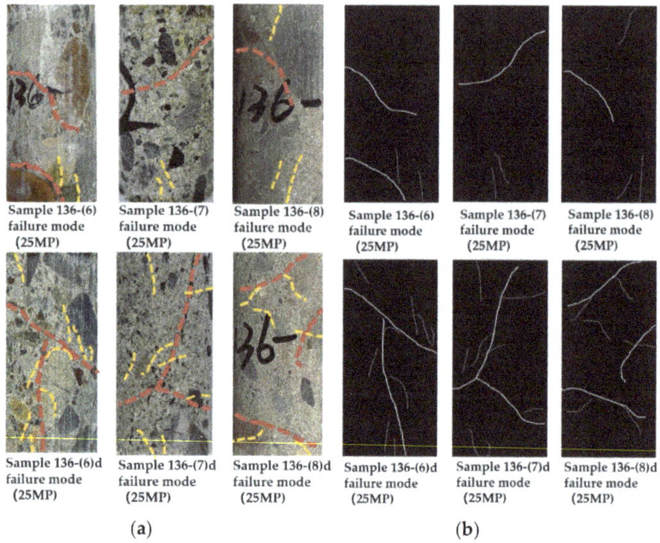

Figure 13. Failure modes of Mahu conglomerate under 25 MPa confining pressure. (**a**) Rock samples, (**b**) failure modes.

The stress–strain curve can reflect the brittle plastic characteristics of the rock. Figure 14 shows the stress–strain curves of the rock samples obtained from the triaxial experiment, and Figure 15 shows three typical stress–strain curves. Brittle rock has a good linear relationship between the stress and strain, which suddenly breaks after stress loading to the peak value. For brittle plastic rock, the initial stress–strain correlation is linear and nonlinear in the later stage. The fracture occurs after the stress reaches the peak value. Plastic rock exhibits nonlinear characteristics under low stress. The stress increases slowly with strain, without an obvious breaking point. The stress–strain curves of the Mahu conglomerate reservoir show that the stress and strain have a good linear correlation in the early stage, showing brittle characteristics. After the stress reaches the peak, there is no sudden fracture and there are nonlinear characteristics before breaking, revealing the plastic characteristics. To conclude, the Mahu conglomerate belongs to brittle plastic rock.

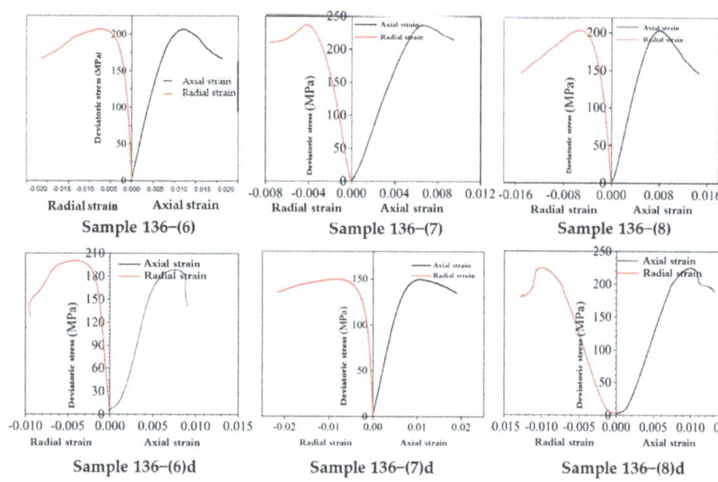

Figure 14. Stress–strain curve to determine the lithological characteristics of Mahu reservoir.

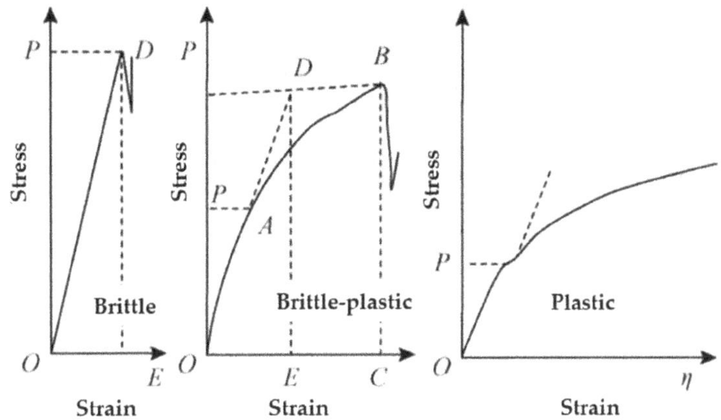

Figure 15. Typical stress–strain curve. (Reprinted with permission from Ref. [21]. 2022, Liu, G.)

6. Tensile Strength Decrease after Acidizing

The Brazilian splitting test is the most common method to obtain the rock tensile strength [22], which is determined by measuring the failure load in the diameter direction and calculating the size of the cylinder sample. The experimental diagram is shown in Figure 16.

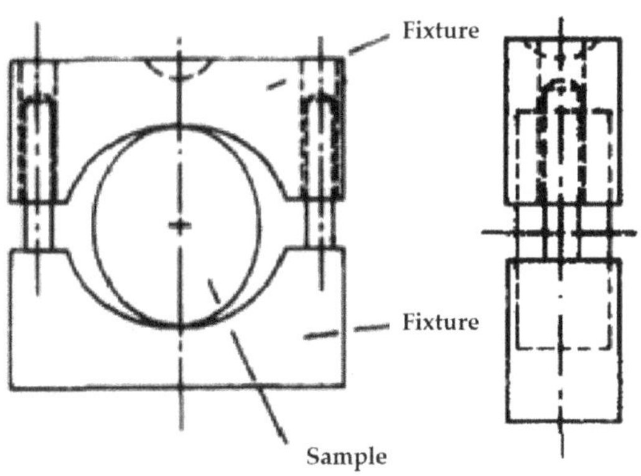

Figure 16. Schematic diagram of splitting experiment.

The tensile strength of the rock sample can be determined by Equation (6).

$$S_t = \frac{2P}{\pi DL} \tag{6}$$

where S_t is the tensile strength of rock, MPa. P is the failure load, N. D is the diameter of the specimen, cm. L is the thickness of the specimen, cm.

First, 11 rock samples as control groups were tested to obtain the original value. Second, the left rock samples were treated with different acids. The rock tensile strength test was conducted again. The influence law of acid composition, acid concentration, and acid treatment time on the tensile strength were obtained. The results are shown in Table 4. The left side is the original tensile strength value, and the right side is the tensile strength after acidizing.

Table 4. Rock tensile strength test results.

Core Number	Tensile Strength (MPa) (before)	Core Number	Tensile Strength (MPa) (after)	Acidizing Time (min)	Reduction Rate (%)	Acidizing Fluid System
136-(1)a	2.57	136-(1)b	0.41	60	84	6% HF + 15% HCL
136-(2)a	4.19	136-(2)b	1.17	60	72	3% HF + 15% HCL
136-(3)a	6.19	136-(3)b	2.02	60	67	1.5% HF + 15% HCL
136-(4)a	3.16	136-(4)b	1.07	60	66	15% HCL
136-(6)a	5.27	136-(6)b	2.69	60	49	12% HCL
136-(5)a	8.32	136-(5)b	4.74	60	43	10% HCL
136-(7)a	4.72	136-(7)b	4.53	10	4	3% HF + 15% HCL
136-(8)a	4.33	136-(8)b	3.59	20	17	3% HF + 15% HCL
136-(9)a	6.83	136-(9)b	4.64	30	32	3% HF + 15% HCL
136-(10)a	5.39	136-(10)b	3.13	60	71	3% HF + 15% HCL
136-(11)a	6.42	136-(11)b	3.08	120	83	3% HF + 15% HCL

As is demonstrated in Figure 17, the reduction rate of tensile strength soared to 84%. The red trendline represents mud acid treatment, and the blue trendline expresses hydrochloric acid treatment. The effect of mud acid acidizing is better and more obvious. The explanation is that the carbonate minerals of the Mahu area is not high, while the feldspar minerals are high. The mud acid can effectively dissolve feldspar, clay, and other minerals, so the damage to the mechanical properties of the rock is more evident.

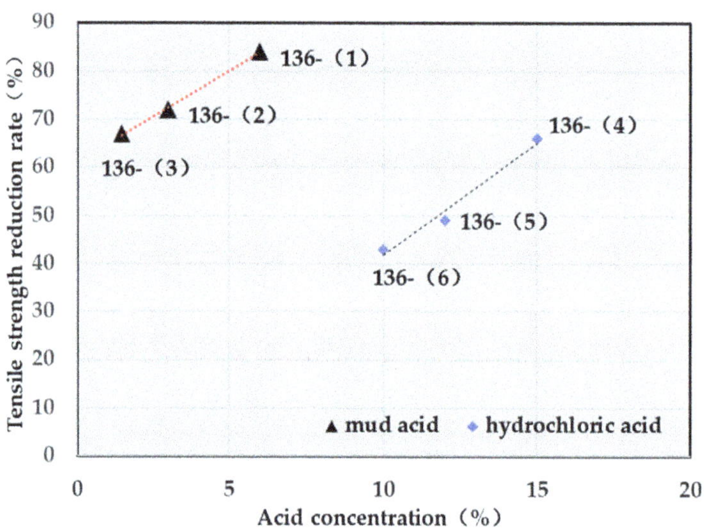

Figure 17. Effect of acid concentration and acid composition on the tensile strength.

Figure 18 represents how the acid treatment time impacts the tensile strength. The longer the acid treatment time is, the more sufficient the dissolution is, and the greater the decline proportion is. Within 60 min of acidizing, the tensile strength decreases sharply, and above 60 min of acidizing, the rate of decrease slows down, indicating that the dissolution is gradually sufficient. Therefore, it is necessary to ensure a certain effective acid–rock contact time to allow the acid to fully dissolve the rock minerals.

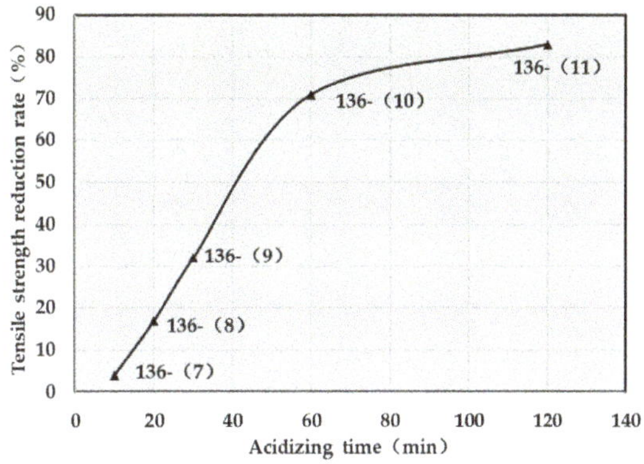

Figure 18. Effect of acid treatment time on the tensile strength.

Figure 19 provides a comparison photograph of a rock sample before and after testing. Figure 19a is the top view of the rock, which contains an unacidified rock sample image on the left and the same rock after acid treatment on the right. It is obvious that the fractures mainly propagate around the gravel. Figure 19b represents a cross section of the same rock sample after acidizing. The whole rock has been treated thoroughly, which can be seen from the color changed inside. The acidizing was sufficient, and micro-cracks and corrosion holes are also clearly observed inside.

(**a**) Top view (**b**) Section plane

Figure 19. Micro-cracks and dissolved pores after acidizing. (**a**) Top view of rock sample before and after acidizing. (**b**) Rock section plane after acidizing.

7. Conclusions

In this paper, through conducting a series of experiments before and after acid treatment, we analyzed how acid dissolution affects rock properties and the effect on fracturing pressure in the Mahu conglomerate reservoir, which provided a theoretical basis for the acid pretreatment of hydraulic fracturing in the Mahu conglomerate reservoir. The following conclusions were obtained:

1. The Mahu conglomerate has a mineral composition of 46.54% quarts, 6.08% carbonate, 34.66% feldspar, 8.02% clay, and 4.70% others. The carbonate content is relatively low, and the clay and feldspar content are relatively high. Contacted by 6% HF + 15% HCl for 1 h, carbonate was dissolved almost, and more than 40% of clay and feldspar were dissolved; other minerals were consumed only a little. Mineral dissolution by the acid was obvious in acid dipping, resulting in a significant property change in the rock, which beneficially promotes breaking the formation in hydraulic fracturing in the Mahu conglomerate reservoir.
2. Acid treatment can increase the porosity and permeability significantly, as the porosity increased by up to 2 times and the permeability increased up to 14 times.
3. After acid treatment, the tensile strength decreased by up to 84%, Young's modulus decreased by up to 63.6%, and Poisson's ratio decreased up to 40.7%

4. Mud acid has a stronger dissolution to the Mahu conglomerate than HCl. A 6% HF + 15% HCl acid composition is recommended for 60 min plus the acid–rock contact time.
5. The stress–strain curves of the Mahu conglomerate showed that the rock is a brittle-plastic type. The mechanism that acid treatment lowers the fracture pressure is that it reduces the tensile strength and increases the permeability. The increased permeability let the fluid pressure propagate into the formation to increase the pore pressure. The raised pore pressure can reduce the fracturing pressure.

In this study, the experiments were conducted with small-size standard core samples. Whether the law can be generalized to the field scale should be validated by field applications. Further study is recommended, such as acid pretreatment modeling and how the acidizing lowers the formation breaking pressure in hydraulic fracturing.

Author Contributions: Conceptualization, L.W. and J.M.; methodology, W.J.; software, W.J. and J.M.; validation, L.W., Y.X. and S.Z.; formal analysis, W.J. and J.M.; investigation, W.J.; resources, L.W.; data curation, Y.X. and Z.L.; writing—original draft preparation, W.J.; writing—review and editing, J.M.; visualization, W.J.; supervision, S.Z.; project administration, L.W., J.M., Y.X. and Z.L. All authors have read and agreed to the published version of the manuscript.

Funding: This research was funded by Efficient Fracturing Technology for Horizontal Wells in Plastic Conglomerate Reservoir Research Project (JX2021C9-34).

Data Availability Statement: The raw data required to reproduce these findings cannot be shared at this time as the data also form part of an ongoing study.

Acknowledgments: This work was supported by Efficient Fracturing Technology for Horizontal Wells in Plastic Conglomerate Reservoir Research Project.

Conflicts of Interest: The authors declare no conflict of interest.

References

1. Wang, S.; Deng, H.; Yu, H.; Zhu, J.; Tian, G.; Lin, Y.; Tang, W.; Guo, C. Rock damage and fracturing pump pressure reduction mechanism of conglomerate reservoirs in Baikouquan Formation of Mahu Sag after acidizing treatment. *Sci. Technol. Eng.* **2021**, *21*, 8841–8850.
2. Yang, Z.; He, R.; Chen, J.; Li, X.; Xie, B.; Meng, J.; Li, W. A Hybrid Model for Selecting Horizontal Candidate Wells for Re-Fracturing of Tight Oil Reservoirs—A Case Study in the Baikouquan Formation, North Mahu Oil Field, Western China. In Proceedings of the SPE Russian Petroleum Technology Conference, Moscow, Russia, 22–24 October 2019.
3. Lu, C.; Ma, L.; Guo, J.; Xiao, S.; Zheng, Y.; Yin, C. A new hydraulic fracturing method for large-thickness conglomerate reservoirs. *Nat. Gas Explor. Dev.* **2020**, *43*, 64–70.
4. Yang, L.; Gao, J.; Chang, S.; Wan, Q.; Liu, C. Study on Fracture Conductivity of Mahu Sandy Conglomerate Tight Oil Reservoir. In Proceedings of the SPE/IATMI Asia Pacific Oil & Gas Conference and Exhibition, Virtual, 12–14 October 2021.
5. Xu, J.; Li, J.; Wu, Y.; Ding, K.; Jiang, H. Exploration and practice of volume fracturing technology in horizontal well of Mahu tight conglomerate reservoirs. *China Pet. Explor.* **2019**, *24*, 241–249.
6. Yun, J.; Qin, G.; Xu, F.; Li, X.; Zhong, N.; Wu, W. Development and utilization prospects of unconventional natural gas in China from a low-carbon perspective. *Acta Pet. Sin.* **2012**, *33*, 526–532.
7. Tan, P.; Jin, Y.; Han, L.; Shan, Q.; Zhang, Y.; Chen, G.; Zhou, Y. Influencing mechanism of acidification pretreatment on hydraulic fracture for deep fractured shale reservoirs. *Chin. J. Geotech. Eng.* **2018**, *40*, 384–390.
8. Wu, Y.; Li, M.; Zhang, J.; Xu, C.; Tian, X. Research and application for fracturing technology of volcanic reservoir in Chagan Depression. *Drill. Prod. Technol.* **2016**, *39*, 60–63.
9. Zou, C.; Yang, Z.; Zhu, R.; Zhang, G.; Hou, L.; Wu, S.; Tao, S.; Yuan, X.; Dong, D.; Wang, Y.; et al. Progress in China's unconventional oil & gas exploration and development and theoretical technologies. *Acta Geol. Sinca* **2015**, *89*, 979–1007.
10. Morsy, S.S.; Hetherington, C.J.; Sheng, J.J. Effect of low-concentration HCl on the mineralogical, mechanical, and physical properties of shale rocks. In Proceedings of the SPE Eastern Regional Meeting, Pittsburgh, PA, USA, 20–22 August 2013.
11. Morsy, S.S.; Sheng, J.J.; Hetherington, C.J.; Soliman, M.Y.; Ezewu, R.O. Impact of matrix acidizing on shale formation. In Proceedings of the Nigeria Annual International Conference and Exhibition, Lagos, Nigeria, 5–7 August 2013.
12. Morsy, S.S.; Sheng, J.J.; Soliman, M.Y. Improving hydraulic fracturing of shale formations by acidizing. In Proceedings of the SPE Eastern Regional Meeting, Pittsburgh, PA, USA, 20–22 August 2013.
13. Li, P. Analytical formula of formation breakdown pressure for horizontal well. *J. Shanghai Univ. Eng. Sci.* **2011**, *25*, 41–45.

14. Wang, B.; Wang, X.; Feng, P. Determination of rock mechanical parameters of reservoir before and after sandstone matrix acidizing. *J. Xinjiang Pet. Inst.* **1999**, 11–14.
15. Wang, X.; Zou, H.; Cheng, X.; Jiang, X.; Zhang, G. Study on Rock Mechanics Properties Before and After Acidizing of Sandstone Matrix. In Proceedings of the Chinese Society for Rock Mechanics & Engineering, Wuhan, China, October 2000.
16. Guo, J.; Xin, J.; Zhao, J.; Li, G. The calculation analysis of decreasing formation fracturing pressure by acidizing pretreatment. *J. Southwest Pet. Univ. (Sci. Technol. Ed.)* **2008**, *30*, 83.
17. Deng, Y.; Xue, R.; Guo, J. The mechanism of high-pressure high-temperature and low permeability acid pretreatment to reduce fracturing pressure. *J. Southwest Pet. Univ. (Sci. Technol. Ed.)* **2011**, *33*, 125–129.
18. Yang, Y.; Li, X.; Wang, Z.; Ju, Y. Influence of Perforation Parameters on Propagation Law of Hydraulic Fracture in Heterogeneous Sandstones. *China Civ. Eng. J.* **2022**, *55*, 1–9.
19. Zhang, L.; Pan, B.; Shan, G. Progress in Experimental Research on Porosity and Permeability of Core Samples. *Prog. Geophys.* **2018**, *33*, 777–782.
20. Zhang, L.; Pan, B.; Shan, G. Experimental Study on Rock Porosity and Permeability under Overburden Pressure. In Proceedings of the Annual Meeting of Chinese Geoscience Union, Beijing, China, 20 October 2014.
21. Liu, G.; Xiao, F.K.; Guo, Z.B.; Chi, X.H.; Jiang, Y.N.; Yu, H.; Hou, Z.Y.; Zhao, R.X. Plastic characteristics of rock under point load. *Sci. Technol. Eng.* **2018**, *18*, 217–222.
22. Deng, H.; Li, J.; Zhu, M.; Wang, R.; Yuan, X.; Luo, Q. Research on Effect of Disc Thickness-To-Diameter Ratio on Splitting Tensile Strength of Rock. *Chin. J. Rock Mech. Eng.* **2012**, *31*, 792–798.

processes

MDPI

Article

Preparation of Polymer Solution for Profile Control and Displacement Using Wastewater with High Ca^{2+}/Mg^{2+} and Fe^{2+} Concentrations

Xuanran Li [1], Anzhu Xu [1], Mengqi Ma [2,*], Shanglin Liu [1], Jun Ni [1] and Lun Zhao [1]

1 Research Institute of Petroleum Exploration and Development (RIPED), Beijing 100083, China
2 College of Petroleum Engineering, China University of Petroleum, Beijing 102249, China
* Correspondence: mengqi_maggie@163.com; Tel.: +86-136-9125-2916

Abstract: In the present study, we used Kalamkas, which is a typical Kazakhstani oilfield, which produces wastewater with high Ca^{2+}/Mg^{2+} and Fe^{2+} concentrations, as a case study. We investigated a method for preparing Fe^{2+} polymer solutions without oxygen isolation under the conditions of salinity >110 × 10^3 mg/L, Ca^{2+}/Mg^{2+} concentration >7000 mg/L, and Fe^{2+} concentration >30 mg/L. Fe^{2+}-resistant groups were grafted onto the molecular chains of a hydrophobically associating polymer prepared using existing synthesis technology to overcome the decrease in apparent viscosity of the polymer solution due to the oxidation of Fe^{2+} during solution preparation. The experiments showed that PAM-IR with iron-resistant groups can be completely dissolved in the wastewater within 180 min, and can tolerate an NaCl concentration of up to 0.23 × 10^6 mg/L, a Ca^{2+} concentration of up to 10 × 10^3 mg/L, an Mg^{2+} concentration of up to 9 × 10^3 mg/L, and a Fe^{2+} concentration of up to 90 mg/L, with favorable thickening performance and resistances to NaCl, Ca^{2+}, Mg^{2+}, and Fe^{2+}. PAM-IR has good injection performance and can establish a high resistance factor (F_R) and residual resistance factor (F_{RR}) to increase the sweep efficiency. Therefore, it is potentially useful for enhancing oil recovery.

Keywords: formation water with high salinity; Fe^{2+}-resistant polymer; apparent viscosity; resistance factor; microscopic state

Citation: Li, X.; Xu, A.; Ma, M.; Liu, S.; Ni, J.; Zhao, L. Preparation of Polymer Solution for Profile Control and Displacement Using Wastewater with High Ca^{2+}/Mg^{2+} and Fe^{2+} Concentrations. *Processes* **2023**, *11*, 325. https://doi.org/10.3390/pr11020325

Academic Editors: Linhua Pan, Yushi Zou, Jie Wang, Minghui Li, Wei Feng, Lufeng Zhang and Qingbang Meng

Received: 31 October 2022
Revised: 30 December 2022
Accepted: 13 January 2023
Published: 19 January 2023

1. Introduction

It is common to extract chemical reagents from oilfield wastewater produced during polymer flooding or in-depth profile control flooding. The strategy is environmentally friendly and economical. However, this method requires a high-performance chemical displacement agent, especially if the concentrations of Ca^{2+}/Mg^{2+} and Fe^{2+} are high, which will seriously affect the performance of the chemical displacement agent [1–7].

The effect of polymer flooding varies considerably depending on the reservoir [1,8,9], and the water used for preparing the polymer solution. Wastewater contains inorganic salts, divalent and high-valence ions, sulfides, solid impurities, crude oil, and bacteria. When it is used directly to prepare a polymer solution, the viscosity of the generated flooding system is significantly reduced [2,8,9]. Previous studies [3–5,10–14] have indicated that cations affect the stability of a polymer solution in the order: $Fe^{2+} > Fe^{3+} > Mg^{2+} (Ca^{2+}) > Na^{+} (K^{+})$. When the polymer is diluted with wastewater containing Fe^{2+} at a concentration of 1700 mg/L, the viscosity of the polymer solution measured at 70 °C is less than 5 mPa·s, and the viscosity retention rate is less than 10%, both of which seriously impede the displacement effect. A water quality survey has shown that the Fe^{2+} concentration of the wastewater is greater than 0.5 mg/L. According to the laboratory results reported in the literature [3–6,10–15], the Fe^{2+} concentration of the injected water is the main factor that affects the viscosity of the polymer solution during in-depth profile control [9]. Therefore, the polymer solution is often prepared directly using fresh water (e.g., at the Xinjiang oilfield in China), or prepared using fresh water that is then diluted with wastewater (e.g., at the Daqing oilfield in China) [11]. Many researchers have proposed adding chemicals (e.g., alcohols, thiourea, formaldehyde, or sodium borohydride) to the polymer to eliminate the influence of Fe^{2+} through reactions, wherein the molecular chain of the polymer is unaffected by the oxidation reactions of Fe^{2+}. This method can reduce the solid content of the polymer, but it significantly increases its cost. Other researchers have also mentioned the use of physical oxygen isolation methods (e.g., nitrogen production and oxygen isolation, and the improved design of fluid transfer systems) [16–18]. However, these methods also have poor economic benefits, and impose high demands on the injection equipment. Therefore, they are not conducive to long-term use. Moreover, the results from laboratory tests and from actual field application differ. The methods described above may require large amounts of clean water or involve the treatment of wastewater. The consumption of clean water resources and the cost of wastewater treatment both increase the input of oilfields and the risk of environmental pollution while reducing the overall economic benefits of polymer flooding. Therefore, only if the comprehensive effect of reducing wastewater treatment and increasing oil production can maximize the benefits to the oilfield, will they promote the application of the technology.

The Kalamkas oilfield In the North Uschut Basin of Kazakhstan is a stratified unintegrated reservoir with a gas cap and edge water. It is one of the high-aquifer sandstone reservoirs that are typical of Kazakhstan, and has an average reservoir thickness of 14 m, an average porosity of 0.27, and an average permeability of 441 millidarcy (mD). It has been developed for 42 years. Currently, most wells have a water cut of more than 95%, and the vertical production degree of the oilfield is only approximately 50%. Ineffective water injection is a serious problem, and production efficiency is low, which adversely affects recovery. The problem of ineffective water circulation during the development process needs to urgently be solved. In light of the experience gained through the development of medium–high-permeability sandstone oilfields throughout the world, the Kalamkas oilfield is believed to be suitable for in-depth profile control. However, the oilfield is developed by re-injecting the wastewater, which has a salinity of >110 × 10^3 mg/L, a Ca^{2+}/Mg^{2+} concentration of >7000 mg/L, an oil content of >70 mg/L, and, most importantly, a Fe^{2+} concentration of 30–70 mg/L. In addition, there is no water treatment system. These conditions are technically and economically far from the requirements for polymer solution preparation and injection; therefore, it is difficult to screen out economically feasible polymers, which makes the preparation of a flooding agent challenging [19]. Therefore, it is necessary to develop proper Fe^{2+}-resistant polymers to enhance oil recovery, lower costs, and improve economic performance at the Kalamkas oilfield.

The present paper describes the development of a new Fe^{2+}-resistant polymer for the preparation of a polymer solution using oilfield wastewater with a high Fe^{2+} concentration and high salinity. According to ShiraziM1 and other authors [20–26], a hydrophobically associating polymer solution is stable under conditions of high salinity and high temperature and, therefore, is useful as a flooding agent. During wastewater treatment, complex compounds are generally used to stabilize Fe^{2+} [27]. Dong and Lin et al. adopted oxygen isolation to mitigate the reduction in polymer viscosity due to Fe^{2+} oxidation [28]. Since 2014, a nitrogen blanketing system has been used to isolate air at the Kalamkas oilfield to eliminate the degradation of the polymer solution caused by Fe^{2+} oxidization [17]. Cao et al. and Xiong et al. have indicated that certain hydrophobically associating polymers can form three-dimensional network morphologies that improve thickening performance. They further divulged that such hydrophobically associating polymers are water-soluble and are formed by introducing a small amount (generally < 2%) of an associating functional monomer into a polyacrylamide molecular chain. Polymers synthesized from various functional monomers have distinct molecular structures and functions. The hydrophobically associating polymers described have potential for use in polymer flooding, as viscosity modifiers for fracturing fluids, or for heat-resistant, salt-tolerant filtrate reducers for water-based drilling fluids [29–33]. In the present study, we attempted to graft Fe^{2+}-resistant groups onto the molecular chain of a hydrophobically associating polymer to form a kind of 3D network that can trap Fe^{2+} while maintaining the viscosity of the polymer solution. This polymer ensures that the high viscosity and long-term stability of a solution prepared using oilfield sewage are retained.

2. Materials and Methods

2.1. Materials and Instruments

The chemicals used to prepare the experimental samples comprised: NaCl, $CaCl_2$, $MgCl_2 \cdot 6H_2O_2$, and $(NH_4)_2Fe(SO_4)_2 \cdot 6H_2O$, which were all of analytical purity (i.e., >99%); acrylamide (AM), acrylic acid (AA), NaOH, ammonium persulfate ($(NH_4)_2S_2O_8$), sodium bisulfite ($NaHSO_3$), octadecyl dimethyl allyl ammonium chloride (DMAAC-18), and *N,N*-methylene bisacrylamide (MBA), which were all industrial products (with purities > 98%); hydrolyzed polyacrylamide (HPAM), with a molecular weight of 23×10^6, a solid content of 90.0%, and a hydrolysis degree of 22.5%, which is commonly used for flooding in China's Daqing and Xinjiang oilfields, and was provided by Xinjiang Keli New Technology Development Co., Ltd. Karamay, China, and a hydrophobically associating water-soluble polymer (HAWP) comprising a Fe^{2+}-resistant polyacrylamide with a molecular weight of 12×10^6, a solid content of 90.0%, and a hydrolysis degree of 20%.

Simulated brine, with a salinity of 116,700 mg/L, was prepared by adding 90.17 g of NaCl, 15.00 g of $CaCl_2$, and 24.68 g of $MgCl_2 \cdot 6H_2O$ to 1 L of distilled water.

The wastewater sample from the Kalamkas oilfield was turbid, yellow, and oily, with a salinity of 124,991.1 mg/L, a density of 1.084 g/cm^3, a pH of 6.67, a K^++Na^+ concentration of 40,427.1 mg/L, a Ca^{2+} concentration of 5611.2 mg/L, an Mg^{2+} concentration of 1702.4 mg/L, a Fe^{3+} concentration of 3.81 mg/L, a Fe^{2+} concentration of 33.46 mg/L, a Cl^- concentration of 77,103.75 mg/L, a HCO^{3-} concentration of 164.64 mg/L, a CO_2 concentration of 145.76 mg/L, and a total hardness of 420 mg/L. No ${SO_4}^{2-}$, ${CO_3}^{2-}$, or dissolved O_2 were found.

The core comprised of an epoxy resin cement (with approximate dimensions of 3.8 cm × 30 cm), with a N_2-logging permeability of 390–410 mD, taken from the Kalamkas oilfield, where the average reservoir permeability is 400 mD.

The instruments comprised: a FD240-Binder convection oven; a ME2002 electronic balance (Mettler Toledo Instruments); an MCR302 advanced rheometer (Anton Paar); a scanning electron microscope (SEM); a VERTEX 70 Fourier-transform infrared spectroscopy (FTIR) spectrometer; a constant-temperature and constant-pressure displacement unit (Changzhou Yiyong, China); and a Fe-HX-10 quick iron-measuring tube (Beijing Huaxing, China). The other commonly available test devices comprised a heating mantle, Erlenmeyer flasks, and a cylinder containing 99.6% pure nitrogen.

2.2. Experimental Methods and Contents

In this study, the experiments on the preparation of polymer solution, the influence of metal ions on the viscosity of the polymer, the characteristic viscosity number, the molecular weight, and others were performed with the methods given in the Recommended Practices for Evaluation of Polymers Used in Enhanced Oil Recovery (SYT6576-2003), which is equivalent to API RP 63:1990. The experiments on resistance factor (F_R) and residual resistance factor (F_{RR}) were completed in accordance with the Technical Criterions of Polymer for Oil Displacement (SY/T5862-2020). Other experiments were conducted according to the literature [2,4,11–13,26,32]. The experiments are described as follows:

(1) Preparation of polymer solution

Simulated brine was used for preparing the polymer solution at normal temperature. We took a certain volume of simulated brine and stirred it with a stand blender at (400 ± 20) r/min. Then, we weighed the polymer with a mass required for preparing the solution with the designed polymer concentration and added it into the simulated brine being stirred. We further stirred (usually for 2–3 h) and observed the solution until the particulate matters were completely dissolved, that is, until there was no "fisheye" in the solution and the insoluble particulate matters disappeared.

(2) Effect of Fe^{2+} on the viscosity of the polymer solution

We prepared a 1000 mg/L Fe^{2+} solution from $(NH_4)_2Fe(SO_4)_2 \cdot 6H_2O$ and added the simulated brine (116,700 mg/L) with Fe^{2+} concentrations of 1, 2, 5, 10, 20, 30, or 50 mg/L to form three 0.3% polymer solutions. We determined the solubility, and measured the apparent viscosity of each of the three polymer solutions at 3 h, then placed them in an incubator set to 40 °C. At 40 °C, and a shear rate of 10 s^{-1}, we measured the apparent viscosity changes at various time-points to determine the effect of Fe^{2+} on the various polymers and to assess the tolerance of these polymers with regard to Fe^{2+}.

(3) Measurement of Fe^{2+} concentration in the polymer solution

We measured the Fe^{2+} concentration before and after solution preparation using the quick iron-measuring tube to determine the changes in Fe^{2+} concentration, and to investigate the stability of Fe^{2+} in the polymer and assess the tolerance of the new polymer with regard to Fe^{2+}.

(4) Determination of the basic physical property parameters of the polymers

(a) Characteristic viscosity and polymer molecular weight: The characteristic viscosity number $[\eta]$ of the polymer and weak gel was obtained by measurement of the Uhler vissimeter, and the average molecular weight of the polymer and weak gel was calculated using the formula

$$[\eta] = K\overline{M}_{\eta}{}^{a},$$

where K and a are empirical constants, $K = 6.31 \times 10^{-3}$, and $a = 0.8$ in the present paper.

(b) Degree of hydrolysis: The degree of hydrolysis of the partially hydrolyzed polyacrylamide was determined by the method described in GB/T 12005.6-1989.

(5) Polymer characterization

The polymer microstructure was characterized by FTIR (VERTEX 70), and the infrared spectra of the synthesized samples was obtained. The microstructure of the polymer solution was characterized by SEM; hydrophobic association polymer (1500 mg/L) and Fe^{2+}-resistant polymer solutions were prepared with distilled water, and the microstates of the solutions were examined under a microscope.

(6) Performance evaluation of the PAM-IR solution We prepared a 0.5% PAM-IR solution using simulated brine solutions with various concentrations of NaCl, Ca^{2+}, Mg^{2+}, and Fe^{2+}. We determined the state of the polymer in the solution at 3 h and measured its apparent viscosity to assess the resistance of the polymer solution to salinity and Fe^{2+}. The apparent viscosity was measured using the MCR302 rheometer at 40 °C and a shear rate of 10 s^{-1}. The experiments are performed on:

(a) Salt sensitivity: various concentrations of NaCl were added to distilled water to form the simulated brines, which were then used to prepare polymer solutions.

(b) Ca^{2+} influence: various concentrations of Ca^{2+} were added to a 100×10^3 mg/L NaCl aqueous solution to form simulated brines, which were then used to prepare polymer solutions.

(c) Mg^{2+} influence: various concentrations of Mg^{2+} were added to a 100×10^3 mg/L NaCl aqueous solution to form simulated brines, which were then used to prepare polymer solutions.

(d) Fe^{2+} influence: various concentrations of Fe^{2+} were added to a 100×10^3 mg/L NaCl aqueous solution to form simulated brines, which were then used to prepare polymer solutions.

(7) Verification experiment with wastewater

The water sample was prepared using the injected water collected from the Kalamkas oilfield to determine the apparent viscosity and apparent viscosity-polymer concentration relationship of the developed PAM-IR solution, with the aim of providing a basis for subsequent field application. We quickly prepared a 0.5% polymer mother solution using the injected water sample, stirred it for 3 h, then determined its solubility and measured its apparent viscosity after diluting it to various concentrations. We also took a sample of injected water and stirred it for 3 h for dynamic aeration, then exposed it to air for 24 h for static aeration. Finally, we used this sample to prepare a 0.5% polymer solution. We measured its viscosity with the DVIII rotational viscometer (18 rotors) at 40 °C and a shear rate of 10 s^{-1}.

(8) Resistance factor and residual resistance factor test

We used the constant-temperature and constant-pressure displacement unit. First, we injected the polymer solution into the core at a rate of 1 mL/min under stable pressure after injecting it with the simulated brine. When the stable injection pressure p_p was reached, the resistance factor (F_R) of the polymer solution was calculated using the following equation:

$$F_R(P) = \left(\frac{p_p}{p_w}\right)_q \tag{1}$$

Then, we injected the simulated brine into the core and injected the polymer solution at the same rate until a stable injection pressure P'_W was reached. We calculated the residual resistance factor (F_{RR}) of the polymer solution using the following equation:

$$F_{RR}(P) = \left(\frac{p'_W}{p_w}\right)_q \tag{2}$$

If the injection pressure continued to rise, it indicated that the polymer had blocked the core.

3. Results and Discussion

3.1. PAM-IR Synthesis

The synthesis mechanism comprised attaching lipophilic and hydrophilic groups to a skeleton consisting of flexible hydrocarbon chains (See Figure 1). This allowed them to generate a supramolecular effect in the solution owing to the presence of a corresponding quantity of cationic functional groups and non-ionic structural units. The strong interactions between the chain links and between adjacent molecules eliminated the influence of the metal ions, crude oil, and solid impurities in the wastewater on the viscosity of the solution.

a+b+c+d

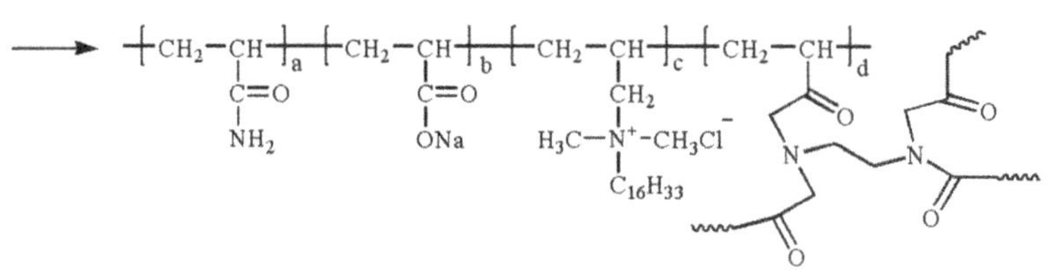

Figure 1. Formula of PAM-IR.

First, we synthesized Fe^{2+}-resistant functional monomers. Then, we dissolved a certain quantity of Fe^{2+}-resistant functional monomer in distilled water, added a certain amount of AM and AA, adjusted the pH with aqueous NaOH, and added a certain amount of hydrophobic (functional) monomer. Next, we stirred the solution evenly with a magnetic stirrer, and injected nitrogen for 30 min to remove the oxygen. Finally, we added a certain amount of $(NH_4)_2S_2O_8$/$NaHSO_3$ and placed the reactants in a constant-temperature water bath for 6 h to produce PAM-IR.

3.2. Results of Basic Physical Properties Determination

The synthesized samples were analyzed according to the experimental method described in Section 2.2 (3), and the results are shown in Table 1.

Table 1. Determination of the basic physical properties of the PAM-RI polymer.

Polymer	Characteristic Viscosity (dL/g)	Viscose Average Molecular Weight (10^6)	Degree of Hydrolysis (mol%)
PAM-RI	31.5	29.2	23.9

3.3. Characterization of the New Polymer

(1) FTIR analysis results

Figure 2 shows the infrared spectrum of the new polymer. The peak at 3500–3100 cm^{-1} is attributable to free NH_2 and associated NH_2; the peak at 2934 cm^{-1} is attributable to the antisymmetric stretching vibration of methylene; and the peak at 2333 cm^{-1} is attributable to the symmetric stretching vibration of methylene. The peak at 1681 cm^{-1} is attributable to the carbonyl group corresponding to amide (C=O stretching vibration); the peak at 1560 cm^{-1} is attributable to the amide (N-H bending vibration); the peak at 1406 cm^{-1} is attributable to methylene deformation; and the peak at 770–550 cm^{-1} is attributable to the oscillation stretching vibration of N-H. All the profiles featured the characteristic absorption peaks of partially hydrolyzed polyacrylamide, proving the new polymer had been successfully prepared.

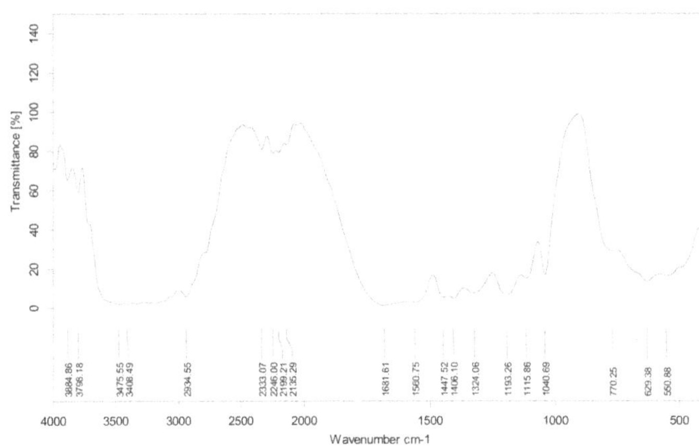

Figure 2. Infrared spectrum of the new polymer.

(2) SEM characterization results

As shown in Figure 3, we used SEM to investigate the 1500 mg/L HAWP and PAM-IR solutions prepared with distilled water for signs of the aggregation state and micromorphology. HAWP exhibited a comb-shaped structure with functional groups, and a relatively regular three-dimensional network morphology, which is consistent with the findings described in the literature [7,32–34]. Similar to HAWP, PAM-IR presented a relatively regular three-dimensional network morphology, because the Fe^{2+}-resistant groups were grafted onto the branch chains of the hydrophobically associating polymers synthesized to avoid the viscosity reduction in the polymer solution while capturing the iron ions, so that the polymer solution prepared with oilfield wastewater has a high viscosity and a satisfactory long-term stability. Therefore, the morphology of PAM-IR, as shown in the SEM image, can allow the solution to be Fe^{2+}-resistant, and inherit the salt-resistant performance of HAWP.

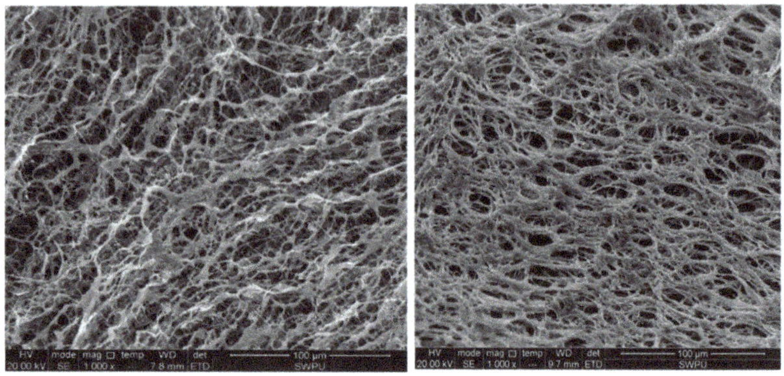

(**a**) HAWP (**b**) PAM-IR

Figure 3. Microstructures of the polymer solutions. (**a**) Microstructure of HAWP; (**b**) Microstructure of PAM-IR.

3.4. Effect of Fe^{2+} on the Apparent Viscosity of the Polymer Solution

Three 0.3% polymer solutions were prepared using simulated Fe^{2+}-containing brine solutions, and their solubilities at 3 h and apparent viscosities after dissolution were determined. The results are shown in Table 2 and Figure 4.

Table 2. Measured apparent viscosities of three polymers dissolved in simulated Fe^{2+}-containing brine solutions.

Fe^{2+} (mg/L)	Viscosity (mPa·s)								
	3 h			1 d			2 d		
	HPAM	HAWP	PAM-IR	HPAM	HAWP	PAM-IR	HPAM	HAWP	PAM-IR
0	28.0	60.7	47.5	27.7	60.3	47.5	28.0	60.5	47.2
1	26.5	55.2	47.2	26.5	58.2	47.2	25.3	55	47
2	24.2	45.3	47.3	24.2	46.5	47.3	20.5	45.2	46.5
5	13.8	40.5	47.2	13.5	20.5	46.2	12.5	11.2	43.2
10	3.5	20.3	46.5	11.3	3.2	45.5	3.2	4.5	42.1
20	2.5	15.2	46.3	10.3	2.1	45.3	2.1	3.5	41.5
30	2.5	11.2	45.3	5.2	2.3	44.1	2.2	3.5	41
50	2.3	8.5	42.5	3.2	2.2	42	1.5	2.5	40.8

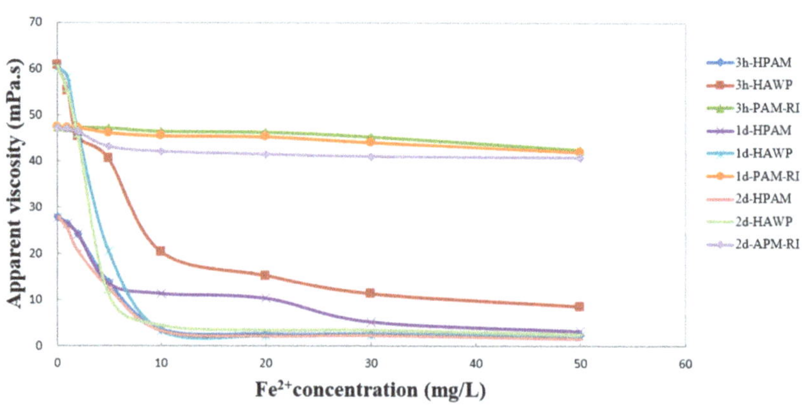

Figure 4. Fe^{2+} concentration versus apparent viscosity of the polymer solutions.

According to the results of the experiments involving dissolution at 3 h and the viscosity measurements, when the Fe^{2+} concentration of the simulated brine was 5 mg/L, the apparent viscosity of HPAM decreased to below 50%, the apparent viscosity of HAWP decreased to below 70%, and the apparent viscosity of PAM-IR remained almost unchanged, compared with that when the Fe^{2+} concentration of the simulated brine was 0. When the Fe^{2+} concentration of the simulated brine was 10 g/L, it was not possible to completely dissolve the HPAM or HAWP. When the Fe^{2+} concentration of the simulated brine was 50 mg/L, the apparent viscosity of HPAM was less than 10%, the apparent viscosity of HAWP was less than 20%, and the viscosity retention of PAM-IR was higher than 90%. The results are shown in Figure 5.

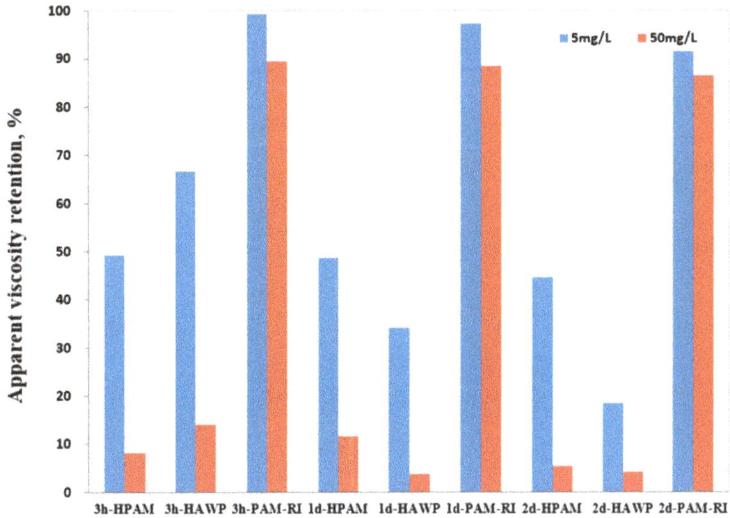

Figure 5. Apparent viscosity retention of polymer solutions with Fe^{2+} concentrations of 5 mg/L and 50 mg/L.

According to the results obtained after aging the polymer solutions at 40 °C after dissolution, the viscosity retention of PAM-IR remained above 90% over time, demonstrating favorable resistance to iron ions. In contrast, the apparent viscosity of HPAM decreased to less than 10%, which was close to the viscosity of the water sample. The apparent viscosity of HAWP decreased to less than 10%, also suggesting degradation.

3.5. *Fe^{2+} Concentration of the Polymer Solution*

The Fe^{2+} concentrations of the 0.3% polymer solutions were measured using the quick iron-measuring tube, and the color and Fe^{2+} concentration of the simulated water sample changed. The results are shown in Figures 6 and 7.

3.6. *Evaluation of the Fe^{2+}-Resistant Polymers*

We determined the state of the polymer at 3 h and measured the apparent viscosity of the 0.5% PAM-IR solution prepared using simulated brine solutions containing various concentrations of NaCl, Ca^{2+}, Mg^{2+}, and Fe^{2+}. As shown in Figures 8–12, PAM-IR can tolerate an NaCl concentration of up to 230×10^3 mg/L, a Ca^{2+} concentration of up to 10×10^3 mg/L, an Mg^{2+} concentration of up to 9×10^3 mg/L, and a Fe^{2+} concentration of up to 90 mg/L, respectively, demonstrating favorable thickening performance and resistance to NaCl, Ca^{2+}, Mg^{2+}, and Fe^{2+}.

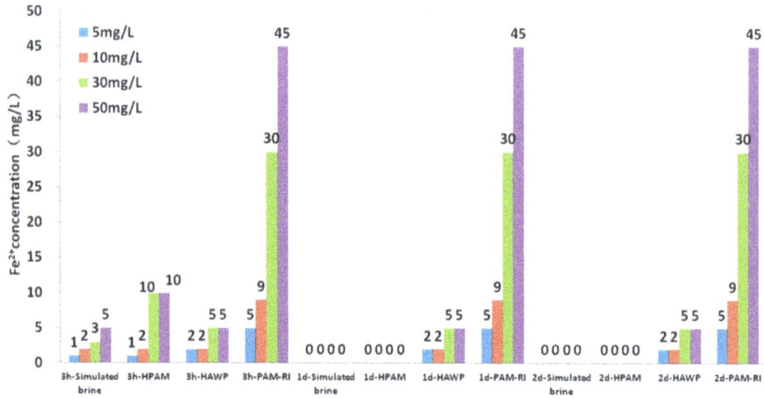

Figure 6. Measured Fe^{2+} concentrations of the polymer solutions.

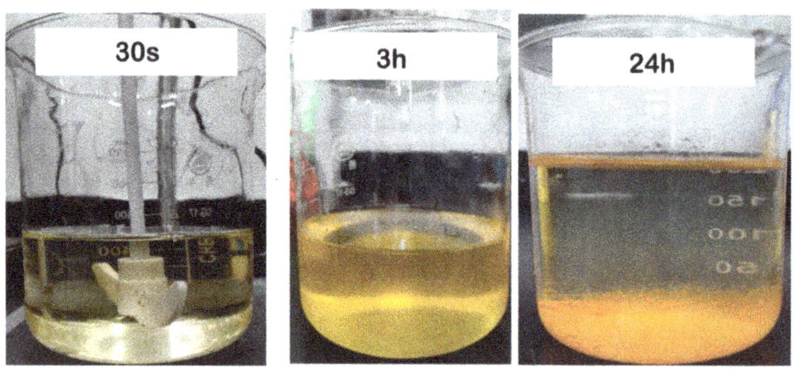

Figure 7. Changes in the simulated brine containing 50 mg/L Fe^{2+}.

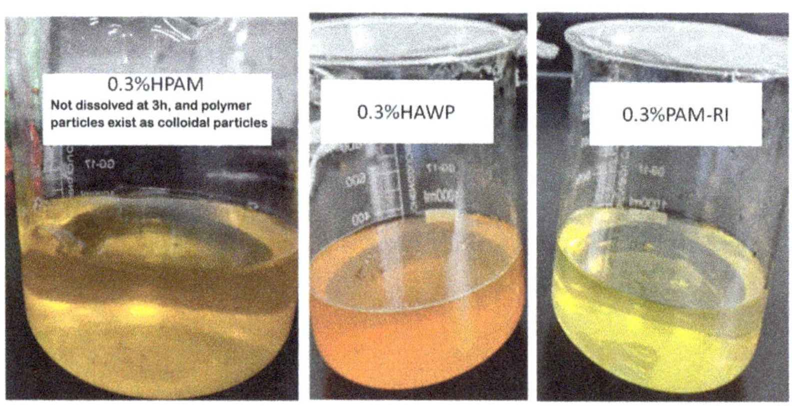

Figure 8. Changes in the three polymer solutions prepared with the simulated brine containing 50 mg/L Fe^{2+} at 3 h.

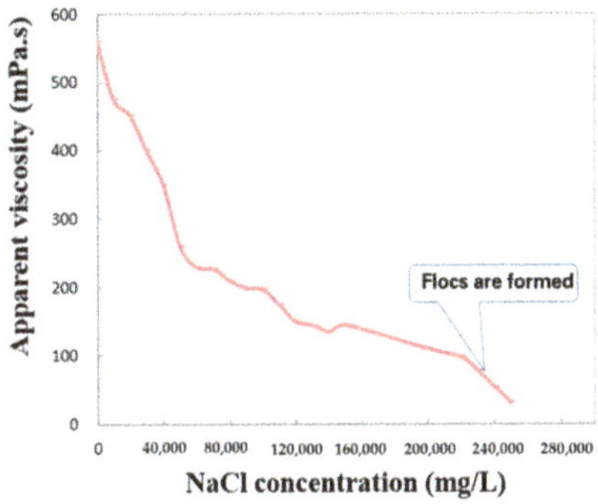

Figure 9. NaCl concentration versus apparent viscosity.

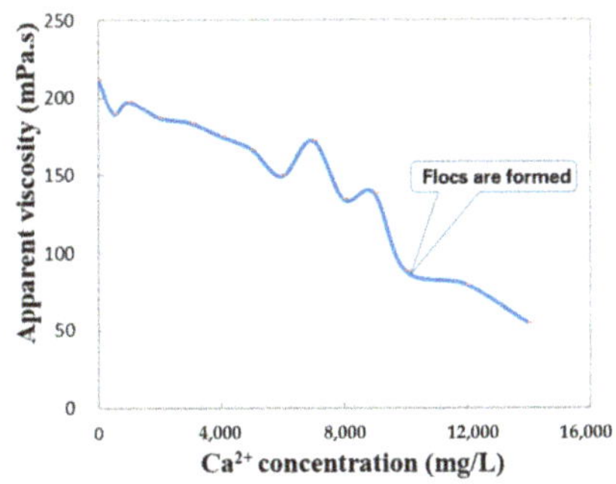

Figure 10. Ca^{2+} concentration versus apparent viscosity.

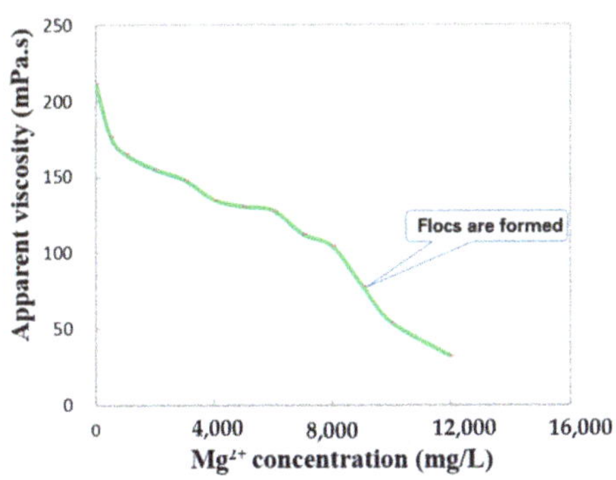

Figure 11. Mg^{2+} concentration versus apparent viscosity.

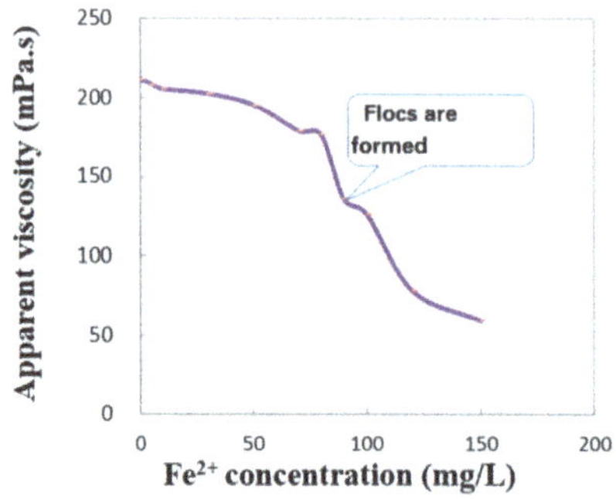

Figure 12. Fe^{2+} concentration versus apparent viscosity.

3.7. *Verification with the Field Water Sample*

The 0.5% polymer solutions were prepared using water collected from the Kalamkas oilfield and aerated water (the Fe^{2+} concentration was determined to be 0 after filtration). We determined the solubilities of the solutions and diluted them to polymer concentrations of 0.1–0.5%. As illustrated in Figures 13 and 14, the color of each polymer solution had not changed after 3 h, the polymer particles were completely dissolved, the apparent viscosity-polymer concentration curve was linear, and the viscosity reduction rate in the polymer solution prepared with raw water was 20% smaller than that of the polymer prepared with aerated water. Aeration completely removes the effect of iron ions on the viscosity of the polymer solution; however, high-salinity water loses calcium when it is aerated to remove iron ions. Thus, the concentrations of metal ions (e.g., Ca and Mg) in the aerated water were lower than those in the raw water, which produced different experimental results.

Figure 13. Injected water sample and polymer solution at 3 h.

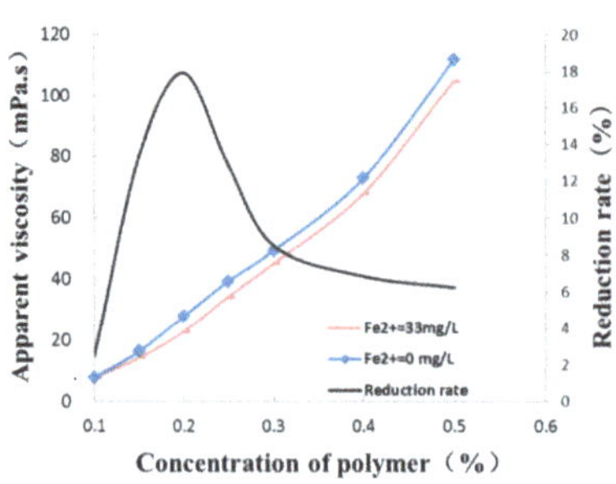

Figure 14. Apparent viscosity–polymer concentration curve.

3.8. Resistance Factor and Residual Resistance Factor Tests

The resistance factor and residual resistance factor are critical parameters for evaluating the concentration of a polymer solution in polymer flooding and are key indicators of the capacity of a polymer solution to improve the mobility ratio and reduce reservoir permeability [15,29,30]. In the present study, we determined the resistance factor (F_R) and residual resistance factor (F_{RR}) using a cement core with a permeability of approximately 400 mD after injecting polymer solutions of various concentrations at a rate of 1 mL/min at 40 °C. The test results are shown in Table 3 and Figure 15. As shown, the polymer solution was injected smoothly. During the injection process, the equilibrium pressure was reached, and no blockage formed. As the concentration of the injected polymer increased, F_R and F_{RR} increased. In actual applications, the polymer concentration can be selected depending on the oil viscosity and permeability of the reservoir.

Table 3. Test results showing the F_R and F_{RR} values of the polymer solutions.

Core	Gas Logging Permeability (mD)	Water Logging Permeability (mD)	Injected Polymer Concentration (%)	F_R	F_{RR}
1#	405.1	210.9	0.1	0.53	0.34
2#	410.5	226.5	0.2	3.39	1.57
3#	398.2	200.8	0.3	8.85	4.77
4#	402.8	210.5	0.4	19.29	11.64
5#	395.8	215.3	0.5	24.00	19.64

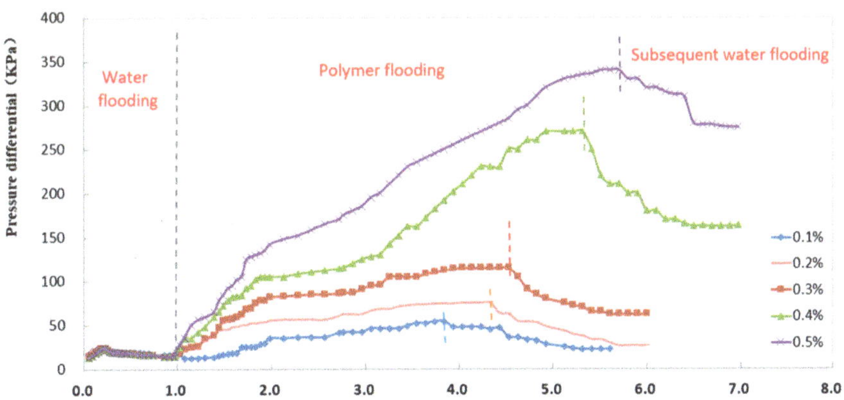

Figure 15. PV versus pressure differential.

4. Conclusions

(1) The new polymer, PAW-IR, exhibits a relatively regular three-dimensional network morphology, and has a good solubility and a favorable thickening performance with regard to oilfield wastewater. It can be completely dissolved in oilfield wastewater within 180 min. It can tolerate an NaCl concentration of up to 23 × 10^4 mg/L, a Ca^{2+} concentration of up to 1 × 10^4 mg/L, an Mg^{2+} concentration of up to 0.9 × 10^4 mg/L, and a Fe^{2+} concentration of up to 90 mg/L, demonstrating favorable thickening performance and resistance to salt and Fe^{2+}.

(2) PAM-IR is prepared by grafting the Fe^{2+}-resistant groups onto the branch chains of the traditional hydrophobically associating polymers synthesized to avoid the viscosity reduction in the polymer solution while capturing the iron ions. Compared with traditional physical methods (e.g., nitrogen production and oxygen isolation, and better design of fluid transfer) and chemical methods (e.g., the precipitation of Fe^{2+} by the addition of alcohol, formaldehyde, and sodium borohydride), the

preparation of PAM-IR does not require additional investment and complex process design, which will improve the overall efficiency of polymer flooding in reservoirs with high concentrations of Fe^{2+} and Ca^{2+}/Mg^{2+} in the formation water.

(3) PAW-IR has a good injection performance and can establish high resistance factor and residual resistance factor, indicating a satisfactory plugging performance. It is promising for enhancing oil recovery by flooding with high-salinity and high-Fe^{2+} polymers.

Author Contributions: Conceptualization, X.L. and S.L.; methodology, X.L.; software, S.L.; validation, J.N., X.L. and M.M.; formal analysis, X.L. and J.N.; investigation, X.L.; resources, X.L.; data curation, M.M. and A.X.; writing—original draft preparation, X.L. and A.X.; writing—review and editing, M.M.; visualization, J.N.; supervision, L.Z.; project administration, J.N.; funding acquisition, L.Z. All authors have read and agreed to the published version of the manuscript.

Funding: This research was funded by the Major Science and Technology Projects of CNPC (grant number 2021DJ3201).

Institutional Review Board Statement: Not applicable.

Informed Consent Statement: Not applicable.

Data Availability Statement: Not applicable.

Conflicts of Interest: The authors declare no conflict of interest.

References

1. Alfazazi, U.; AlAmeri, W.; Hashmet, M.R. Screening of New HPAM Base Polymers for Applications in High Temperature and High SalinityCarbonate Reservoirs. In Proceedings of the Abu Dhabi International Petroleum Exhibition and Conference, Abu Dhabi, United Arab Emirates, 12–15 November 2018. SPE-192805-MS. [CrossRef]
2. Fan, J.; Wei, L.; Luo, W.L. Influencing Factors of the Viscosity Decrease on Polymer Sewage Solution. *Oilfield Chem.* **2011**, *28*, 251–253.
3. Chen, Y.; Li, D.; Song, H.; Song, Y. Research progress of the effect of metal cations on the viscosity of polyacrylamide solution. *J. Chem. Ind. Eng.* **2013**, *34*, 36–41.
4. Li, M.R.; Liu, Z.; Song, X.W.; Ma, B.D.; Zhang, W. Effect of metal ions on the viscosity of polyacrylamide solution and the mechanism of viscosity degradation. *J. Fuel Chem. Technol.* **2012**, *40*, 43–47.
5. YU, Y.; Sun, C.; Huang, J.; Yan, Y.; Li, X.; Xu, W.; Xing, L.; Wu, X.; Wang, J. Reconstruction Project of Iron Removal Process from Oilfield Produced Wastewater for Polymer Injection. *China Water Wastewater* **2012**, *10*, 66–69.
6. Wang, D.; Seright, R.S. Examination of literature on colloidal dispersion gels for oil recovery. *Pet. Sci.* **2021**, *18*, 1097–1114. [CrossRef]
7. Zaitoun, A.; Dupuis, G. Conformance control using SMG microgels: Laboratory evaluation and first field results. In Proceedings of the SPE Europec Featured at 79th EAGE Conference and Exhibition, Paris, France, 12–15 June 2017. [CrossRef]
8. Zhou, X.; Zhang, D. Analysis on Influence Factors of Formulating Polymer Solution with Sewage and Thickening Effect. *Contemp. Chem. Ind.* **2016**, *45*, 272–275.
9. Kang, W.L.; Meng, L.W.; Niu, J.G. Mechanism of the Effect of Salinity on HPAM Solution Viscosity. *Polym. Mater. Sci. Eng.* **2006**, *22*, 175–181.
10. Bin, C.; Xiaoyan, W.; Shanshan, W.; Shijia, C.; Qingquan, Z. Effect of Fe^{2+} on the apparent viscosity of polymer solution and controlling methods. *Chem. Eng. Oil Gas* **2014**, *43*, 168–173.
11. Wang, Z.; Wang, F.; Zong, H. Influence of iron ions on viscosity of polymer solution and its removal. *Ind. Water Wastewater* **2017**, *48*.
12. Ding, Y.; Zhang, J.; Ma, B.; Zhou, M.; Huang, M.; Li, M.; Hao, Q. Tackification Measures and Mechanism of Polymer Solution Prepared by Sewage. *Oilfield Chem.* **2013**, *32*, 123–127.
13. Yang, H.J.; Luo, P.Y. Factors for and Control of Polymer Degradation in Recycled Produced Water Solutions. *Oilfield Chem.* **2005**, *22*, 158–162.
14. Wang, Z.; Wang, Z. Discussion on Effective Reinjection Method of Produced Wastewater in Polymer Flooding. *Environ. Prot. Oil Gas Fields* **2018**, *22*, 40–42.
15. Yao, C.; Xu, X.; Wang, D.; Lei, G.; Xue, S.; Hou, J.; Steenhuis, T.S. Research and application of micron-size polyacrylamide elastic microspheres as a smart sweep improvement and profile modification agent. In Proceedings of the SPE Improved Oil Recovery Conference, Tulsa, OK, USA, 11–13 April 2016. [CrossRef]
16. Xing, E.; Tao, X.; Jin, L. Determination and Influence Factors of Resistance Coefficient and Residual Resistance Coefficient. *Liaoning Chem. Ind.* **2016**, *45*, 284–287.

17. Sagyndikov, M.; Salimgrayev, I.; Ogay, E.; KSeright, R.S.; Kudaibergnov, S.E. Assessing polyacrylamide solution chemical stability during a polymer flood in the Kalamkas field, Western Kazakhstan. *Bull. Univ. Kargnda-Chem.* **2022**, *105*, 99–112. [CrossRef]
18. Temizel, C.; Putra, D.; Zhang, M.; Moreno, R. Smart nanoparticles for conformance improvement in waterflooding. In Proceedings of the SPE Annual Caspian Technical Conference and Exhibition, Baku, Azerbaijan, 1–3 November 2017. [CrossRef]
19. Xiong, Z.; Deng, T.; Xu, H.; Jia, F.; Yu, X.; Ma, C. Performance evaluation on a hydrophobically associating polymer. *J. Guangdong Univ. Petrochem. Technol.* **2022**, *32*, 24–30.
20. Shirazi, M.; Kord, S.; Jamialahmadi, M.; Tamsilian, Y. Polymer Enhanced Oil Recovery Process: An Updated, Narrowed Review. *Acad. J. Polym. Sci.* **2018**, *1*, 555572. [CrossRef]
21. Mohsenatabar Firozjaii, A.; Zargar, G.; Kazemzadeh, E. An investigation into polymer flooding in high temperature and high salinity oil reservoir using acrylamide based cationic co-polymer: Experimental and numerical simulation. *Orig. Pap.-Prod. Eng.* **2019**, *9*, 1485–1494. [CrossRef]
22. Rellegadla, S.; Prajapat, G.; Agrawal, A. Polymers for enhanced oil recovery: Fundamentals and selection criteria. *Appl. Microbiol. Biotechnol.* **2017**, *101*, 4387–4402. [CrossRef]
23. Skauge, T.; Ormehaug, P.A.; Alsumaiti, A.; Masalmeh, S.; Skauge, A. Polymer Stability at Harsh Temperature and Salinity Conditions. In Proceedings of the SPE Conference at Oman Petroleum & Energy Show, Muscat, Oman, 21–23 March 2022.
24. Reichenbach-Klinke, R.; Langlotz, B.; Wenzke, B.; Spindler, C.; Brodt, G. Hydrophobic Associative Copolymer with Favorable Properties for the Application in Polymer Flooding. In Proceedings of the SPE International Symposium on Oilfield Chemistry, Woodlands, TX, USA, 11–13 April 2011.
25. Shi, L.; Chen, L.; Ye, Z.; Zhou, W.; Zhang, J.; Yang, J.; Jin, J. Effect of polymer solution structure on displacement efficiency. *Pet. Sci.* **2012**, *9*, 230–235. [CrossRef]
26. Zhi, L.I.U. Thickening Mechanism of Hydrophobically Associating Polyacrylamide and Polyacrylamide. *Acta Pet. Sin. (Pet. Process. Sect.)* **2012**, *28*, 1037–1042.
27. Wang, S.; Zhu, G.; Yu, Z.; Li, C.; Wang, D.; Cao, X. Ozonation Degradation of Fiber Water Catalyzed by Fe^{2+} Pyrophosphate Dodium Complex. *Spec. Petrochem.* **2020**, *37*, 15–21.
28. Dong, Z.X.; Lin, M.Q.; Xin, J.; Li, M.Y. Influence of dissolved oxygen content on oxidative stability of linked polymer solution. *Pet. Sci.* **2009**, *6*, 421–425. [CrossRef]
29. Zhang, H.; Wang, D.; Wang, L. Flow law of polymer solution in porous media and mechanism of improving oil displacement efficiency. *Daqing Pet. Geol. Dev.* **2002**, *21*, 57–61.
30. Xulong, C.A.O.; Yanfeng, J.I.; Yangwen, Z.H.U. Research advance and technology outlook of polymer flooding. *Reserv. Eval. Dev.* **2020**, *10*, 8–16.
31. Sagyndikov, M.; Mukhambetov, B.; Orynbasar, Y.; Nurbulatov, A.; Aidarbayev, S. Evaluation of Polymer Flooding Efficiency at brownfield development stage of giant Kalamkas oilfield, Western Kazakhstan. In Proceedings of the SPE Annual Caspian Technical, Conference and Exhibition, Astana, Kazakhstan, 31 October–2 November 2018. SPE-192555-MS.
32. Seright, R.S.; Wavrik, K.E.; Zhang, G.; AlSofi, A.M. Stability and Behavior in Carbonate Cores for New Enhanced-Oil-Recovery Polymers at Elevated Temperatures in Hard Saline Brines. *SPE Reserv. Eval. Eng.* **2003**, *24*, SPE-200324-PA. [CrossRef]
33. RSeright, R.S.; Skjevrak, I. Effect of Dissolved Iron and Oxygen on Stability of Hydrolyzed Polyacrylamide Polymers. *SPE J.* **2015**, *20*, 433–441. [CrossRef]
34. Jouenne, S. Polymer flooding in high temperature, high salinity conditions: Selection of polymer type and polymer chemistry, thermal stability. *J. Pet. Sci. Eng.* **2020**, *195*, 107545. [CrossRef]

processes

MDPI

Article

Change Characteristics of Heavy Oil Composition and Rock Properties after Steam Flooding in Heavy Oil Reservoirs

Ting Huang [1,*], Kai Peng [2], Wenzhi Song [3], Changpeng Hu [4] and Xiao Guo [5]

[1] Hubei Cooperative Innovation Center of Unconventional Oil and Gas, Yangtze University, Wuhan 430100, China
[2] PetroChina Jidong Oilfield Company, Tangshan 063200, China
[3] Xinjiang Oilfield Heavy Oil Development Company, Karamay 834000, China
[4] China Petroleum Technology and Development Corporation, Beijing 100028, China
[5] State Key Laboratory of Oil and Gas Reservoir Geology and Exploitation, Southwest Petroleum University, Chengdu 610500, China
* Correspondence: huangting331@126.com

Abstract: The thermal recovery method of steam flooding is one of the most common development methods for heavy oil reservoirs. However, after multiple rounds of steam injection development, the composition of crude oil and reservoir rock properties have changed greatly, which is unfavorable for the subsequent enhanced oil recovery. It is necessary to study the distribution of the remaining oil after the thermal recovery of heavy oil reservoirs, and clarify the change characteristics of the components of the crude oil under different steam injection conditions. At the same time, the change of porosity and the permeability of the rocks after steam flooding, and its influence on oil recovery, are investigated. In this paper, the composition changes of heavy oil before and after steam flooding are studied through experiments and numerical simulation methods. A numerical model is established to study the retention characteristics of heavy components in heavy oil reservoirs by the CMG software. The effects of different steam injection conditions, and heavy oil with different components on the residual retention of heavy components, are compared and studied. The changes of rock physical properties in heavy oil reservoirs after steam flooding is clarified. The results show that after steam flooding, the heavy components (resin and asphaltenes) of the recovered oil decrease, and the heavy components in the formation increase in varying degrees. With the increase of heavy components in the crude oil, the remaining oil in the formation increases after steam flooding, and the retention of heavy components increases; after steam flooding, the stronger the rock cementation strength, the higher the degree of reserve recovery, and it is difficult to form breakthrough channels; the greater the steam injection intensity, the earlier to see steam breakthrough in the production well, and the lower the degree of reserve recovery. The research reveals the changes of heavy oil components and rock properties after steam flooding, providing support for the subsequent enhanced oil recovery.

Keywords: heavy oil; steam flooding; asphaltenes; residual oil; rock property

Citation: Huang, T.; Peng, K.; Song, W.; Hu, C.; Guo, X. Change Characteristics of Heavy Oil Composition and Rock Properties after Steam Flooding in Heavy Oil Reservoirs. *Processes* **2023**, *11*, 315. https://doi.org/10.3390/pr11020315

Academic Editors: Linhua Pan, Yushi Zou, Jie Wang, Minghui Li, Wei Feng and Lufeng Zhang

Received: 28 September 2022
Revised: 22 November 2022
Accepted: 29 November 2022
Published: 18 January 2023

1. Introduction

The thermal oil recovery method is widely used in many heavy oil fields in the world. It is generally believed that the role of steam in the process of thermal oil recovery is to reduce the viscosity of heated heavy oil [1–4]. In the process of steam injection, heavy oil will not only undergo physical changes under the combined action of water and heat, but also may be accompanied by chemical reactions, such as hydrothermal pyrolysis. However, after long-term thermal injection development, the properties of unconsolidated sandstone heavy oil reservoirs, such as porosity and permeability, have changed greatly. Due to the increase of porosity and permeability, it is easy to form a high-permeability zone and an ultra-high-permeability zone, which is called the breakthrough channel. When the steam flows through the breakthrough channel in the formation, which makes the thermal efficiency of steam low, the heat spreads to a small range. The steam forms an inefficient or ineffective circulation along the breakthrough channel, which seriously affects the oil displacement efficiency of the steam and makes the distribution of the remaining oil in the reservoir vary significantly.

In response to the physicochemical changes during the thermal recovery of heavy oil, Hyne et al. [5] called all chemical changes that occur when oil sands are heated in the presence of water, as an aquathermolysis reaction. The aquathermolysis reaction of heavy oil [6] was first reported in Canada, which was mainly used to describe the lightening of asphaltenes. Cao et al. [7] experimentally studied the composition changes of Liaohe super-heavy oil before and after the aquathermolysis reaction (280 °C, 24 h). After the aquathermolysis reaction, the content of aromatics (the sum of light aromatic, medium aromatic and heavy aromatic), saturates and asphaltenes in super-heavy oil increased, while the content of resins decreased, and the resins had the tendency to transform into aromatics and asphaltenes. Wang [8] found that the composition changes of crude oil during thermal recovery mainly showed an increase of heavy hydrocarbons and a decrease of light hydrocarbons, and paraffin and asphaltene precipitated in large quantities, causing serious blockage of oil wells. Khormali et al. [9,10] studied the influencing factors of asphaltene precipitation and its effect on the permeability of carbonate core samples. The above existing studies found that the changes of crude oil properties in the formation after steam flooding in heavy oil reservoirs are still very complicated. Not only the aquathermolysis reaction should be considered, but also the influence of light components flowing underground and being recovered by steam extraction on the composition and viscosity of heavy oil.

Many researches have been conducted on rock property (mainly permeability) changes during the thermal recovery of heavy oil. As early as the 1970s, a method was proposed for predicting breakthrough flow in a five-spot pattern by using the flow characteristics of tracer. Abbaszadeh and Brigham [11] used well-to-well tracer data to interpret the heterogeneity of reservoirs, and the prediction and description of the high permeability zone between wells in oil fields have progressed toward quantitative direction. Ge [12] thought that the breakthrough of the injected fluid is caused by the high permeability zone, and the degree of breakthrough was determined by the pressure field of the fluid and the flow resistance of the fluid.

Later researchers have carried out a large number of studies on large pore sealing and blocking agent. However, it is necessary to further explore the formation mechanism, identification, description and performance characteristics of the breakthrough channel in the oil reservoir, so as to have a deeper understanding of the breakthrough channel. Chen [13] argued that due to the change of cementation capacity of the unconsolidated sandstone, the dragging effect of the viscous crude oil and the long-term scouring of the injected fluid, the problem of sand production in formation occurred and a large pore-like high permeability zone was formed. Zhao [14] discussed the tracer method to describe and identify the large pores, and qualitatively judge whether there is a high permeability zone in the formation and quantitatively calculate the relevant reservoir parameters. Zeng et al. [15,16] considered that different sedimentary characteristics, reservoir physical properties, the development process and other factors will affect the formation of breakthrough channel. A fluid-solid coupling mathematical model for the formation of large pores was established; the large pores in formation were identified and described by using conventional dynamic data and gray correlation theory; and the formation of large pores and the sand production were combined to study the changes of formation characteristics.

Researchers have carried out a lot of work to study the changes of reservoir physical properties in the process of the water flooding development of oil fields, which provides a powerful basis for the formation mechanism of breakthrough channel during heavy oil thermal recovery. Guo et al. [17] studied the swelling, dispersion and migration of clay by microscopic seepage experiments, which paved the way for the study of the formation mechanism of breakthrough channel. Few studies have been conducted on the breakthrough channel and steam breakthrough formed by the thermal recovery of heavy oil reservoirs, but it has certain similarity and reference with the formation of the breakthrough channel after water injection development.

The change of fluid and rock physical properties of the heavy oil reservoir after steam flooding, affects the ultimate recovery of heavy oil to a certain extent. Therefore, this paper studies the change characteristics of fluid and rock physical properties after the steam flooding of the heavy oil reservoir. The residual retention properties of heavy oil with different compositions in porous media are studied by physical simulation experiments. After steam flooding, the physical properties of the heavy oil reservoir will have certain changes. According to the mechanism of physical property change, the residual retention model of the heavy component and the mechanism model of changes in physical property of rock are established by the numerical simulation method. Specifically, they include: (1) prepare the heavy oil with the same viscosity and different components, and study the composition change of the heavy oil before and after steam flooding; (2) establish the residual retention model of heavy components in heavy oil thermal recovery by the CMG software, and study the residual retention characteristics of heavy components after steam flooding; (3) the effects of different steam injection parameters on the residual retention of heavy components are compared and studied; and (4) the change characteristics of rock physical properties in the heavy oil reservoir after steam flooding are investigated. The mechanism model is established by the numerical simulation method, so as to study the change of rock physical properties after steam flooding and to clarify the influence of cementation strength, crude oil viscosity and other factors on the change of rock physical properties.

2. Physical Simulation Experiment of Heavy Oil Composition Change

2.1. Experimental Materials and Methods

2.1.1. Experimental Materials

Crude oil with the same viscosity and different heavy components was prepared in the laboratory, and the composition change of crude oil before and after steam flooding was analyzed by taking component as a single factor. Three sets of sand-filled pipe (Table 1) were used for the physical simulation experiments. Two sand-filled pipes were filled with 160 mesh glass micro-sand to compare the composition changes of crude oil with different components before and after steam flooding under the same permeability; the other sand pack was mainly filled with 40 mesh glass micro-sand to make the permeability different, and the replacement efficiency of the crude oil (the same oil sample with the same composition) before and after the steam flooding was compared under different permeability conditions. The oil used for experiments is No. 1 and No. 2 compound crude oil (Table 2).

Table 1. Basic physical property of sand-filled pipes.

Number of Sand-Filled Pipe	Length of Sand-Filled Pipe (mm)	Diameter (mm)	Permeability ($\times 10^{-3}$ μm^2)
1	300	38	958.305
2	300	38	1023.168
3	300	38	5977.387

Table 2. Content of heavy components and viscosity of compound oil.

Compound Oil Samples	Light Oil Mass Fraction (%)	Residual Oil Mass Fraction (%)	Compound Oil Viscosity (30 °C) (mPa.s)	Mass Fraction of Resins and Asphaltenes in Compound Oil (%)
1	10	90	4075	35.2890
2	23	77	4000	31.4107

2.1.2. Experimental Methods

The following process was adopted in the experiment: (1) The gas tightness of the device was tested with high-pressure nitrogen gas, and the pressure was kept at 5 MPa for 30 min. (2) Saturate the sand-fill pipe with water at a flow rate of 1 mL/min. After saturation, calculate the volume of saturated water in the sand-fill pipe, that is, the pore volume of the sand pack; the pressure difference between the two ends of the sand-fill pipe was measured, and the water-phase permeability of the two sand-fill pipes was calculated according to Darcy's formula. (3) Saturate the crude oil at a flow rate of 0.2 mL/min, and set the whole set of experimental equipment to the corresponding temperature. (4) Steam flooding is carried out after the sand-fill pipe is saturated with oil. The steam temperature is controlled to 300 °C and the flow rate is 2 mL/min. (5) The experiment is stopped when the sand-fill pipe is flooded to a stage when the water cut reaches 98%.

2.2. Experimental Results and Analysis

2.2.1. The Displacement Efficiency under Different Conditions

No. 1 sand-filled pipe and No. 2 sand-filled pipe are saturated with No. 1 compound oil and No. 2 compound oil, respectively, to compare the displacement pressure (Figure 1) and displacement efficiency (Figure 2) of crude oil with different components during steam flooding under the same permeability conditions.

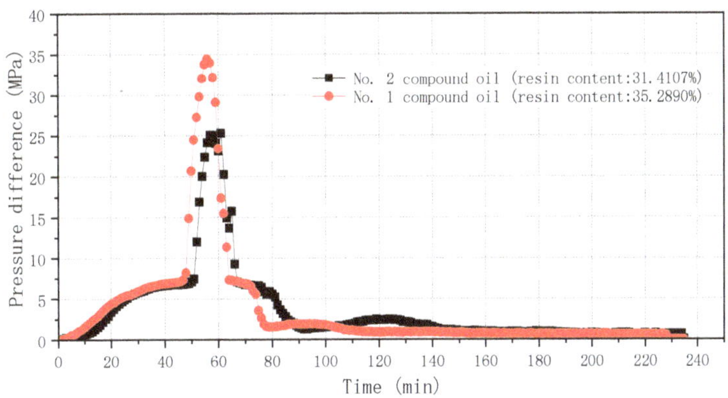

Figure 1. Displacement pressure of two different compound oil.

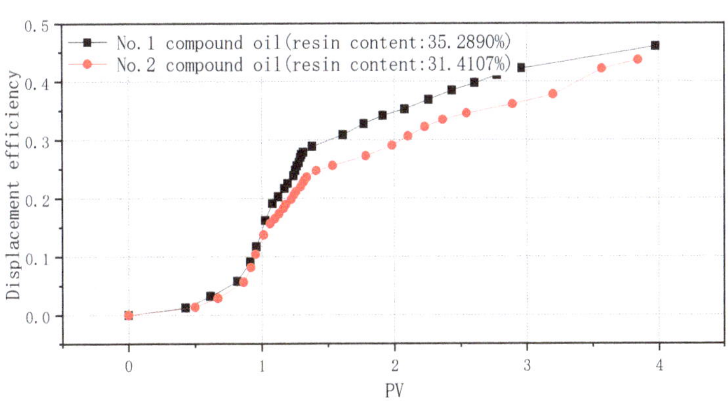

Figure 2. Displacement efficiency of two different compound oil.

As can be seen from Figure 1, under the same permeability condition (1 μm^2), the pressure difference required to displace oil samples with high resin content is larger than that required to displace oil samples with low resin content. This is because the high resin content makes it easier for the heavy components in the heavy oil to form a network spatial structure, which requires more pressure to deform the heavy oil and start to flow during displacement.

As can be seen from Figure 3, under the condition that different permeability sand-filled pipes (1 μm^2 and 6 μm^2) are saturated with the same oil sample, the water-free oil recovery period of the high-permeability pipe (HPP) is shorter than that of the low-permeability pipe (LPP). However, once the low-permeability pipe produces water, the water content rises especially fast, and the displacement efficiency of the high-permeability pipe is higher than that of the low-permeability pipe, but the pressure required to displace the crude oil in the high-permeability pipe is also smaller than that in the low-permeability pipe, as shown in Figure 4.

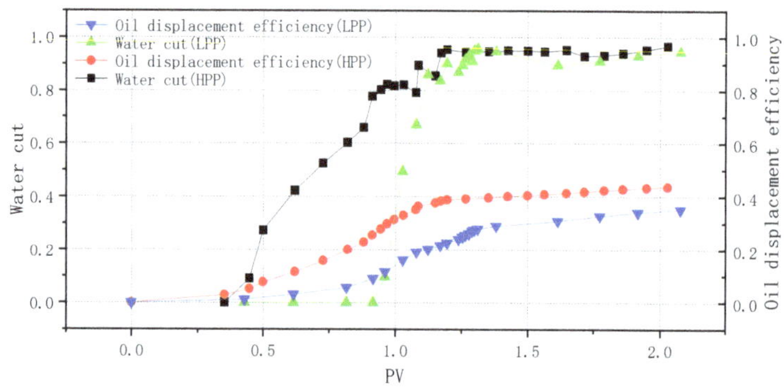

Figure 3. Comparison curves of oil repelling efficiency and water content of high and low permeability pipes (same oil sample).

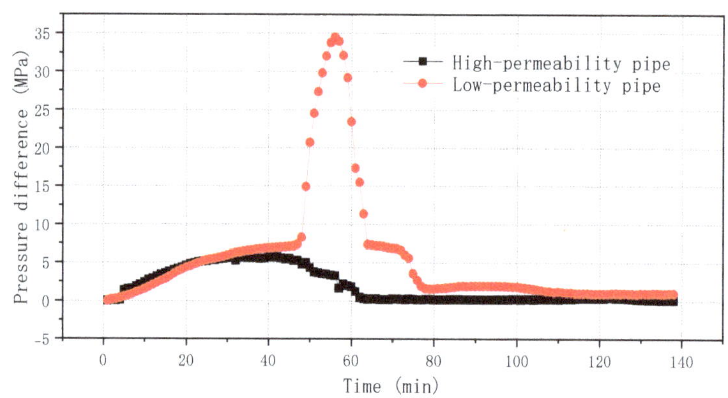

Figure 4. Displacement pressure of two different permeability sand pack.

2.2.2. Change of Heavy Oil Composition

The composition changes of different compound oil before and after steam flooding are shown in Table 3. When the water cut of the two sand-filled pipes reaches 98% at the outlet of the sand-filled pipe by steam flooding, the resin content of the No. 1 oil sample displaced from the sand-filled pipe decreased from 34.92% to 34.01%, and the resin content of the No. 2 oil sample decreased from 31.10% to 27.99%. This indicates that the heavy component of oil produced after steam flooding is reduced, so the heavy fraction remaining in the sand-filled pipe are increased to varying degrees. This is because during the steam flooding, the high-temperature steam distills the light components of crude oil and makes it easier to be recovered. However, the heavy components have a high distillation temperature, and the resins and asphaltenes in it are easier to be adsorbed on the rock wall, which are more difficult to be recovered than the light components and thus remain in the formation [18].

Table 3. SARA composition of two different compound oil before and after steam flooding.

Test	No. 1 oil Sample (before Flooding)	No.1 oil Sample (after Flooding)	No.2 oil Sample (before Flooding)	No.2 oil Sample (after Flooding)
Asphaltenes, %	0.37	0.14	0.32	0.58
Resins, %	34.92	34.01	31.10	27.99
Aromatics, %	27.34	23.40	25.23	21.00
Saturates, %	32.99	30.34	39.37	34.71
Light component + loss, %	4.38	12.11	3.98	15.72

3. Numerical Simulation of Heavy Oil Properties Change

The change of fluid properties after steam flooding is mainly reflected in the change of components before and after steam flooding. According to the results of physical simulation experiments, the light components in crude oil are easier to be recovered after steam flooding because they are heated and volatilized, so the heavy components of the remaining oil in the formation will increase [19–22].

3.1. Numerical Simulation Model

The average geological parameters of the steam flooding pilot test well group in the Qi40 block of the Liaohe oilfield are selected to establish a homogeneous geologic model. The buried depth in the middle of the reservoir is 900 m, the well spacing is 140 m, the porosity is 27% and the permeability is 1500×10^{-3} μm^2. In the model, the crude oil was divided into light and heavy components, and the viscosity-temperature relationships of the two components are shown in Table 4. In the geological model, the mass fraction ratio of the light component and the heavy component is 1:1.

Table 4. Viscosity-temperature relationship.

Temperature (°C)	Light Components (mPa·s)	Heavy Components (mPa·s)
20	23.39	20624.00
40	15.31	4178.59
60	10.84	1552.49
80	7.45	576.81
100	5.28	257.65
200	0.95	10.47
300	0.17	0.91
400	0.040	0.076
450	0.020	0.022

The STARS module of the CMG software was applied to establish a model with a grid number of $20 \times 13 \times 5$. The grid step is 7 m in X and Y directions and 4 m in Z direction, and the well spacing is 140 m. In the steam flooding stage, there is a steam injection well and a production well. The steam injection temperature is 300 °C, the dryness of steam is 0.6, the steam injection intensity of 1.6 $m^3/d/m/ha$, and the production-injection ratio is 1.2.

3.2. Changes in Fluid Properties after Steam Flooding

As can be seen from Figure 5, there is less remaining oil in the formation between the injection and production wells. This is because the rocks between the injection and production wells have experienced long-term steam flooding and the sand-carrying effect of heavy oil, resulting in a high permeability zone. Therefore, the steam will displace to the production well along the high permeability zone with low resistance, which makes the oil washing efficiency in the high permeability zone high, resulting in less oil remaining in the high permeability zone and more oil remaining in the formation outside the high permeability zone.

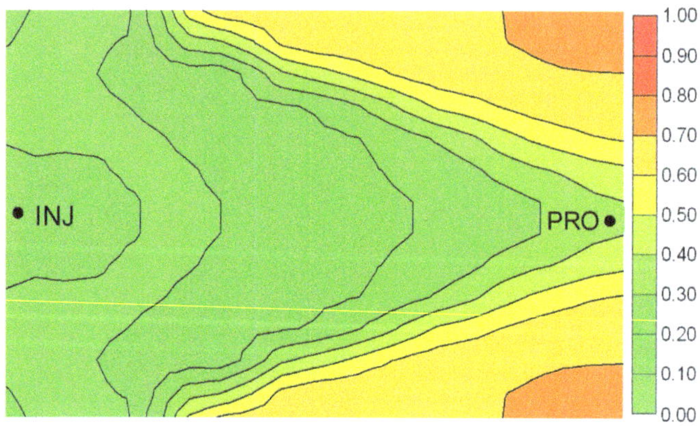

Figure 5. Oil saturation distribution of reservoir after steam flooding.

It can be found from Figure 6 that in the high permeability zone with low oil saturation, the mass fraction of heavy components in the remaining oil is 100% and the mass fraction of light components is 0%. This indicates that the light components in the heavy oil is easier to be volatilized and recovered than the heavy components after long-term steam flooding, while the heavy components remain in the formation.

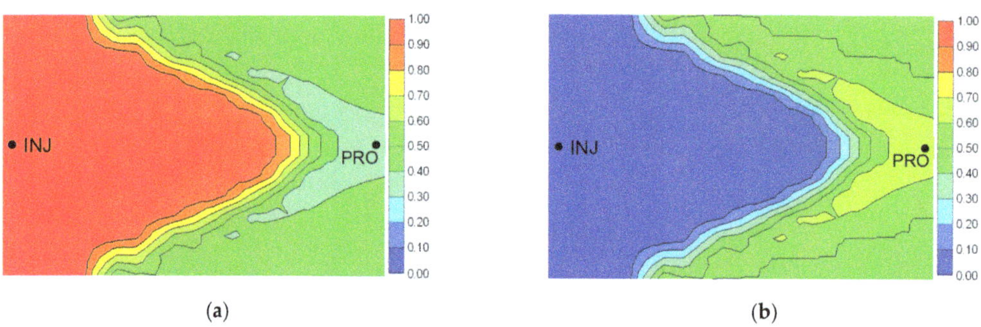

Figure 6. Mass fraction distribution of components of oil after steam flooding. (**a**) Mass fraction distribution of heavy components. (**b**) Mass fraction distribution of light components.

3.3. *Results and Analysis*

3.3.1. Different Ratios of Light and Heavy Components

In the geological model, the light and heavy components were proportioned with different mass fractions to make the heavy components accounted for 0%, 25%, 50%, 75% and 100% of the crude oil mass fraction. Then, the effects of different ratios on the quantity of remaining oil and the retention of different components after steam flooding were analyzed.

As shown in Figure 7, the total cumulative oil production decreases, which indicates that the heavier the crude oil, the lower the degree of recovery. We can see from the curves in Figures 8 and 9, that the cumulative production of light components gradually decreases as the mass fraction of heavy components increases, the production of heavy components increases and the rising trend gradually slows down. The decrease of recovered heavy components indicates that the more heavy components are retained in the formation.

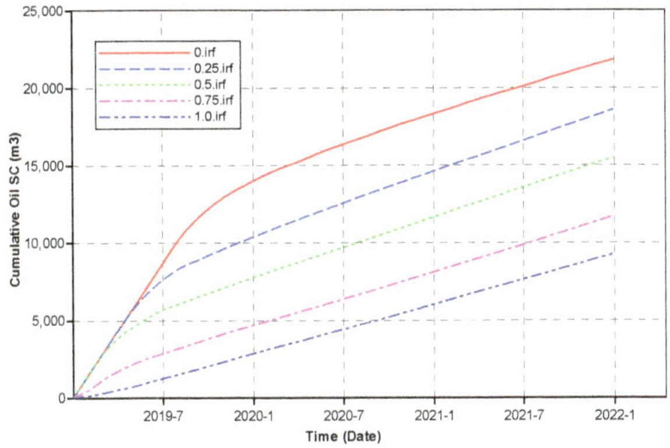

Figure 7. Cumulative oil production after steam flooding at different proportion of components.

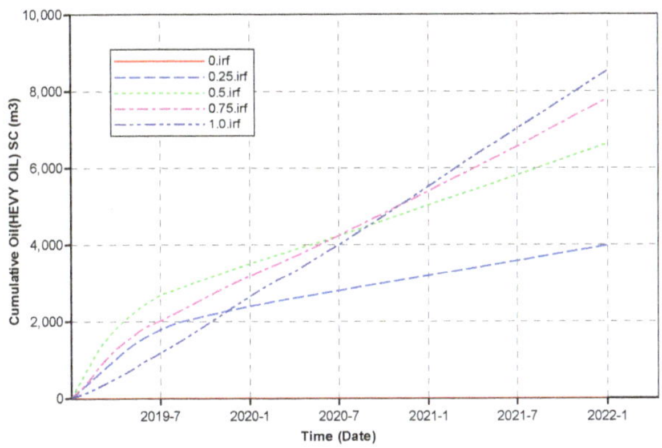

Figure 8. Cumulative production of heavy component after steam flooding at different proportion of components.

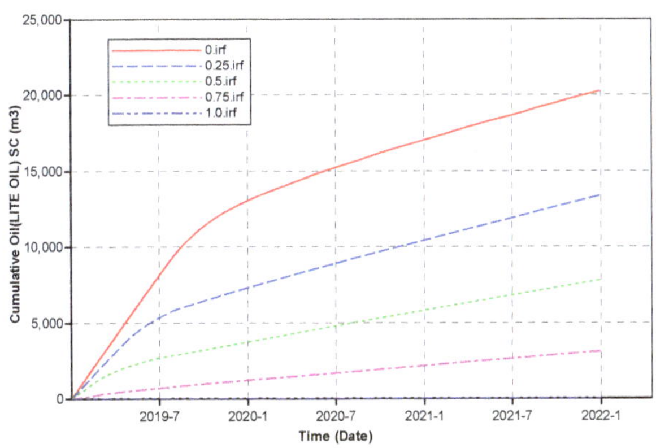

Figure 9. Cumulative production of light component after steam flooding at different proportion of components.

Figures 10 and 11 show the distribution of oil saturation and the mass fraction of heavy component after steam flooding under different ratios of light component and heavy component, respectively. It can be seen from Figure 10 that with the increase of heavy component, the amount of remaining oil in the formation after steam flooding increases. This is because when the mass fraction of the light component in the crude oil is larger, the oil-water viscosity ratio is smaller, thus making the steam flooding have a large planar sweep area in the formation and therefore less oil remaining in the plane.

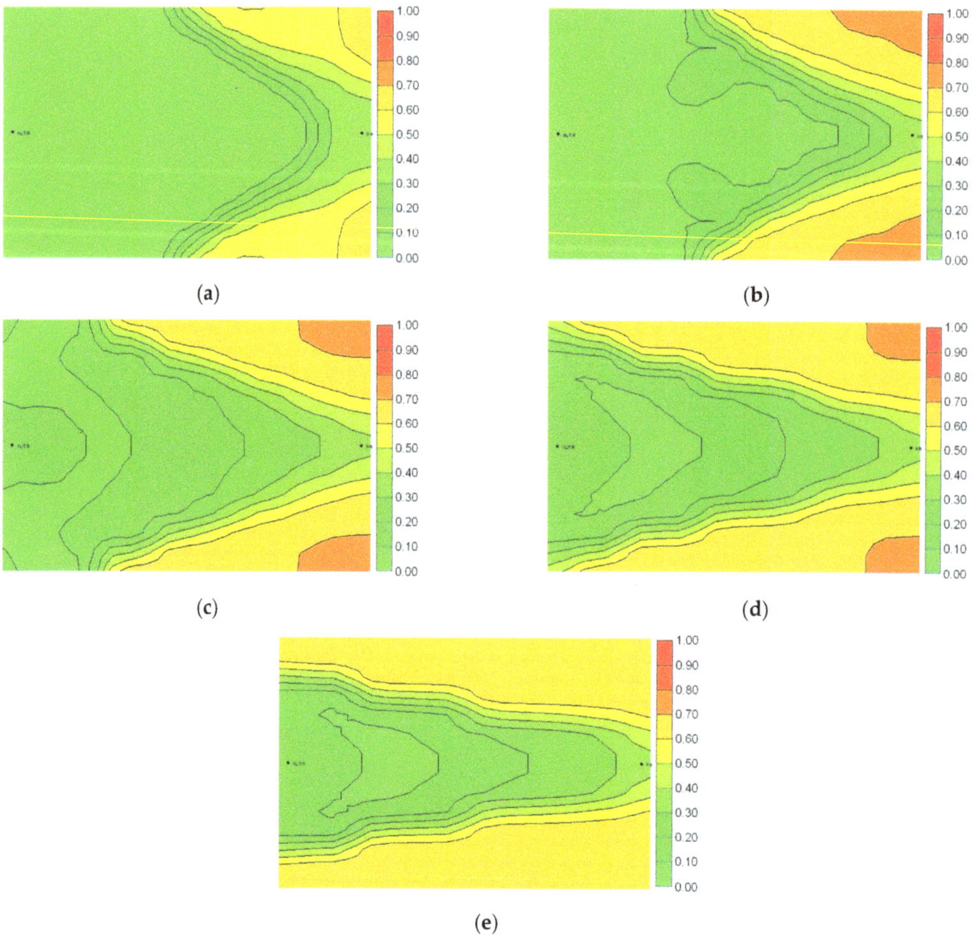

Figure 10. Oil saturation distribution after steam flooding. (**a**) mass fraction of heavy components: 0. (**b**) mass fraction of heavy components: 0.25. (**c**) mass fraction of heavy components: 0.5. (**d**) mass fraction of heavy components: 0.75. (**e**) mass fraction of heavy components: 1.0.

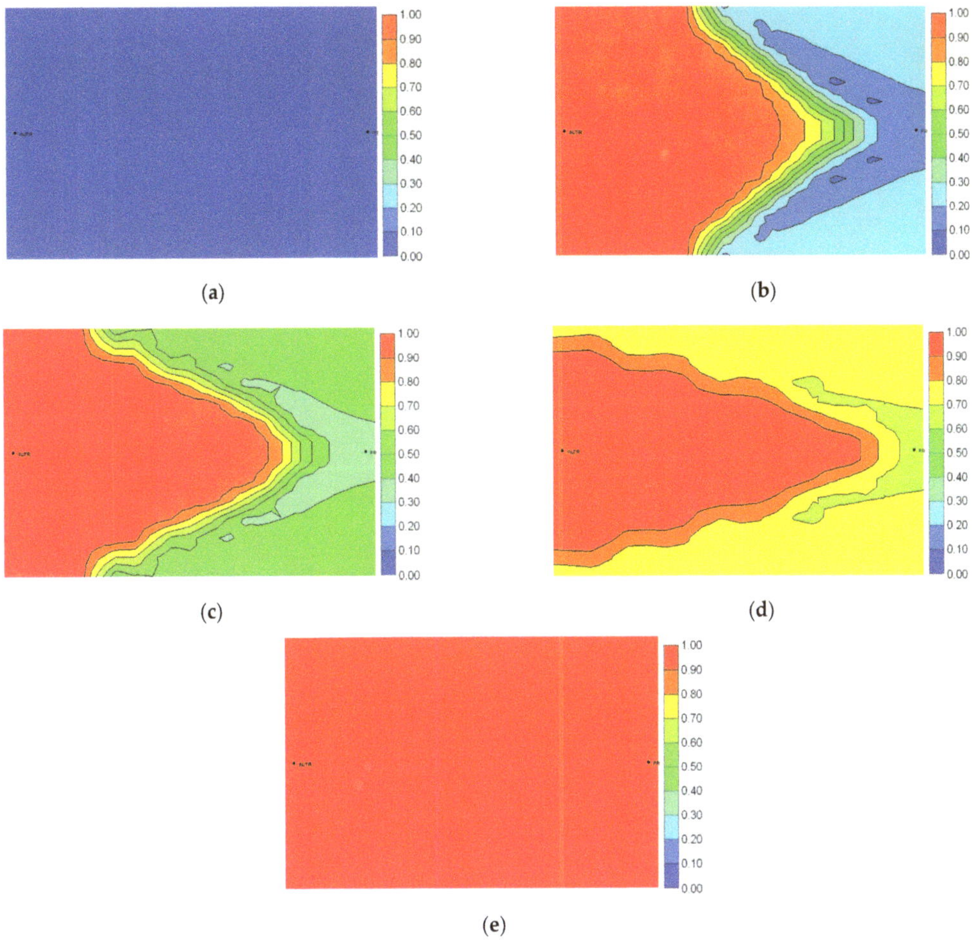

Figure 11. Mass fraction distribution of heavy component after steam flooding. (**a**) mass fraction of heavy components: 0. (**b**) mass fraction of heavy components: 0.25. (**c**) mass fraction of heavy components: 0.5. (**d**) mass fraction of heavy components: 0.75. (**e**) mass fraction of heavy components: 1.0.

It can be seen from Figure 11 that as the initial mass fraction of heavy component in crude oil increases, the greater the mass fraction of heavy component in the formation after steam flooding, that is, the more the heavy component is retained. Moreover, in formations with less oil saturation, the remaining oil contains more heavy components; in formations with high oil saturation, the more heavy components remain after steam flooding, as the initial mass fraction of heavy components in crude oil increases.

3.3.2. Steam Injection Intensity

Steam was injected at steam injection intensities of 1.2, 1.4, 1.6 and 1.8 ($m^3/(d \cdot m \cdot ha)$ under the condition that the mass fraction of heavy components was 50% and the production-injection ratio is unchanged.

It can be seen from Figure 12 that the greater the steam injection intensity, the shorter the time to see steam breakthrough in the production well. Figure 13 shows the cumulative oil production of the wells at the time of steam breakthrough under the four steam injection intensities. We can find that the cumulative oil production of the wells increases with the increase of the steam injection intensity, but decreases after the steam injection intensity reaches 1.6 $m^3/(d \cdot m \cdot ha)$. Therefore, the cumulative oil production is the highest at the time of steam breakthrough with a steam injection intensity of 1.6 $m^3/(d \cdot m \cdot ha)$.

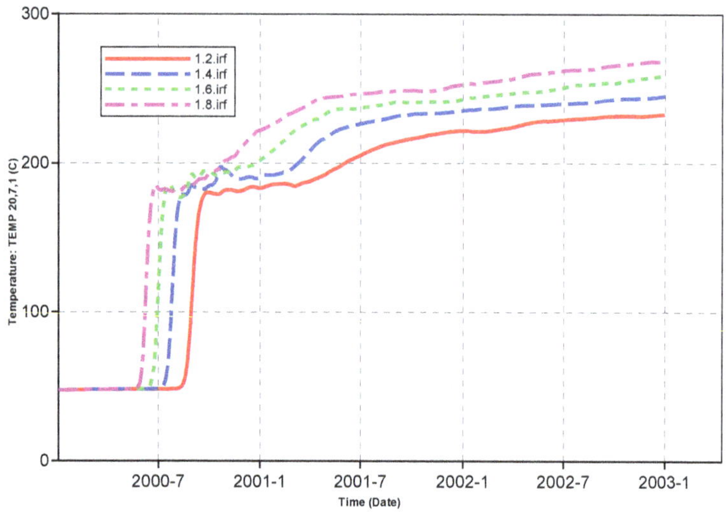

Figure 12. The change of temperature of the well head of production well.

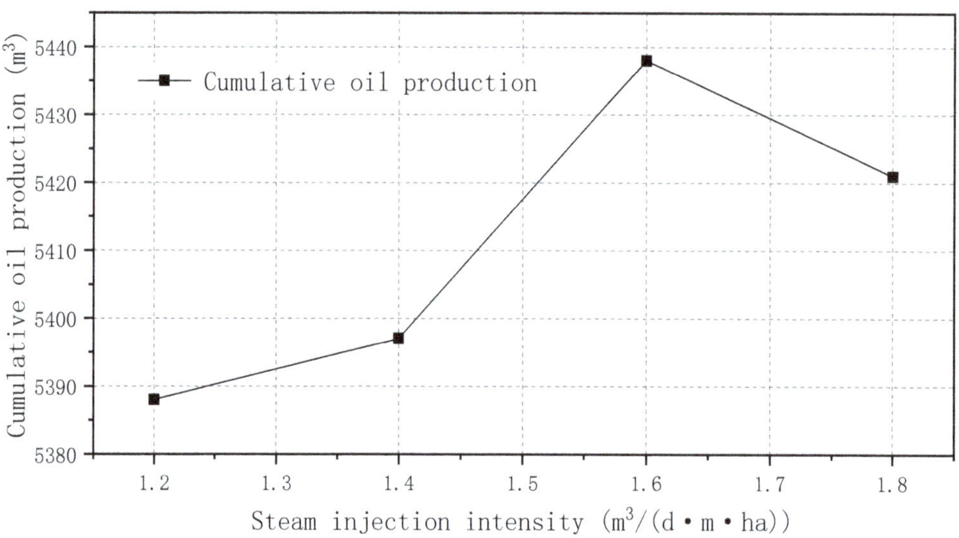

Figure 13. Cumulative oil production at different steam injection intensity.

It can be seen from Figure 14 that the mass fraction distribution of heavy components retained in the formation did not change much with increasing steam injection intensity with steam breakthrough in the production well, that is, different steam injection intensities have little effect on the retention of heavy components.

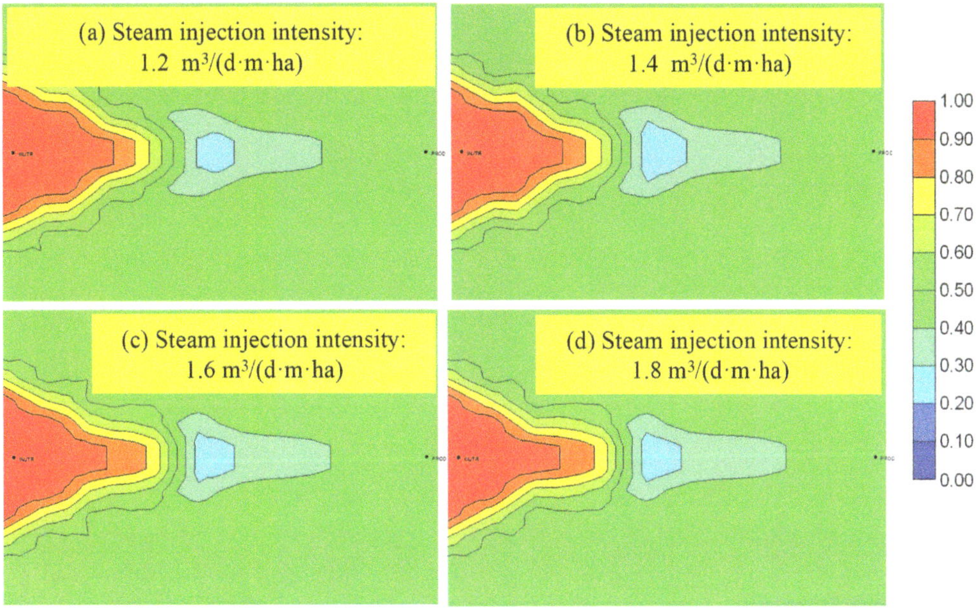

Figure 14. Mass fraction distribution of heavy component.

4. Numerical Simulation of Reservoir rock Physical Properties Change

After long-term thermal injection development, it is easy to form breakthrough channel in the formation, due to the increase in the pore radius and permeability of the formation [23–25]. When the steam flows along the breakthrough channel in the formation, the thermal efficiency of the steam is low and the thermal sweep range is small. The inefficient or ineffective circulation of steam along the breakthrough channel seriously affects the oil displacement efficiency of the steam flooding. Therefore, it is necessary to study the physical property changes of reservoir rocks after steam flooding, mainly focusing on the changes of rock porosity and permeability, and the factors leading to the changes of rock physical properties after steam flooding [26–29].

4.1. Numerical Simulation Model

The average geological parameters of the steam flooding pilot test well group in the Qi40 block of the Liaohe oilfield are selected to establish a homogeneous geologic mode. The buried depth in the middle of the reservoir is 900 m, the well spacing is 140 m, the porosity is 27% and the permeability is 1500×10^{-3} μm^2. In the model, the following crude oil viscosity-temperature curve was used to analyze the influence of crude oil viscosity on the thermal recovery effect.

The STARS module of the CMG software was applied to establish a model with a grid number of $20 \times 13 \times 5$. The grid step is 7 m in X and Y directions and 4 m in Z direction, and the well spacing is 140 m. The geological model was established to simulate steam huff and puff and then steam flooding. The distribution characteristics and influencing factors of rock physical property changes after steam flooding were studied. In the steam flooding stage, there is a steam injection well and a production well, the steam temperature is 300 °C, the steam dryness is 0.6, the steam injection intensity is 1.6 m^3/ha/m/d, and the production-injection ratio is 1.2; the viscosity-temperature relationship of the heavy oil is shown in Table 5.

Table 5. Viscosity-temperature relationship [30].

Temperature (°C)	Viscosity (mPa·s)
20	9136.2
40	2980.9
65	953.4
80	403.4
100	170.7
200	7.3
300	2.1
400	1.6

4.2. Changes in Reservoir Physical Properties after Steam Flooding

The porosity and permeability changes of a steam injection well and a production well in homogeneous heavy oil reservoir were studied to summarize and analyze the changes of reservoir physical properties during the steam flooding stage. Figure 15 shows the distribution of porosity and permeability between the two wells after steam flooding. It can be seen that a high permeability zone was formed between the two wells due to factors such as sand production.

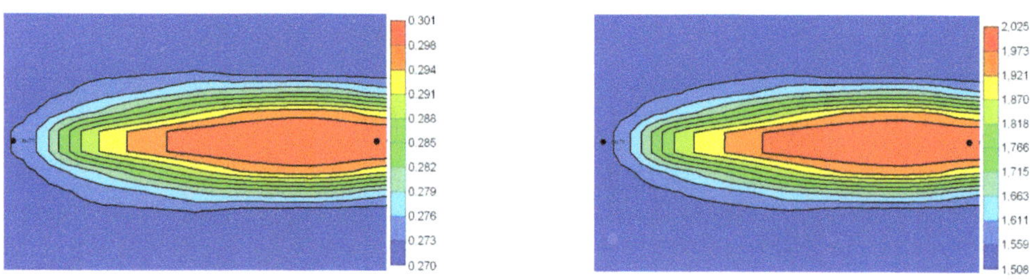

(**a**) (**b**)

Figure 15. Physical property distribution between two wells after steam flooding. (**a**) Porosity distribution; (**b**) permeability distribution.

Meanwhile, we established a numerical model to study the changes of porosity and permeability in the injection-production system of five-spot well pattern, as shown in Figure 16. The center well is a steam injection well and the corner wells are oil production wells. After steam flooding, the porosity and permeability between the injection and production wells become larger, which makes it easy to form cross-flow between the wells.

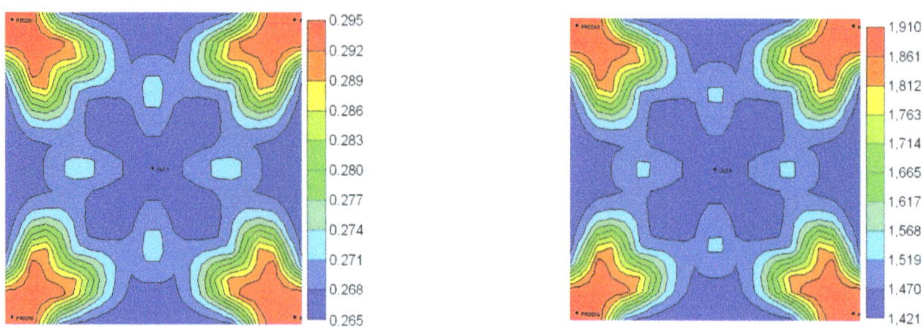

(**a**) (**b**)

Figure 16. Physical property distribution in 5-spot well pattern after steam flooding. (**a**) Porosity distribution; (**b**) permeability distribution.

4.3. Results and Analysis

In the absence of inter-well high permeability zones or natural fractures, it is difficult to form steam breakthrough during the steam huff and puff stage. However, in the steam flooding stage, due to the continuous steam injection, the thermal front is continuously advancing from the injection well to the production well, which improves the plane sweep volume of the steam flooding. At the same time, due to the heterogeneity of formation, steam breakthrough occurs when the thermal front reaches the production well along the formation with less resistance.

4.3.1. Effect of Cementation Degree

The degree of cementation is determined by the change rate of porosity and the relationship between porosity and permeability. The larger the value of the cementation degree, the faster the change rate of porosity and the greater the change of permeability, and the weaker the degree of cementation.

The porosity changes of formations with different degrees of cementation after steam flooding are shown in Figure 17. It can be seen that the stronger the degree of cementation, the less sand production, and the porosity changes in a small range; however, formations with a weak degree of cementation has serious sand production, and the range of porosity changes is large. The porosity of formation affected by steam between the two wells becomes larger in different degrees, and the porosity of formation around the production well changes to a greater extent than that around the steam injection well.

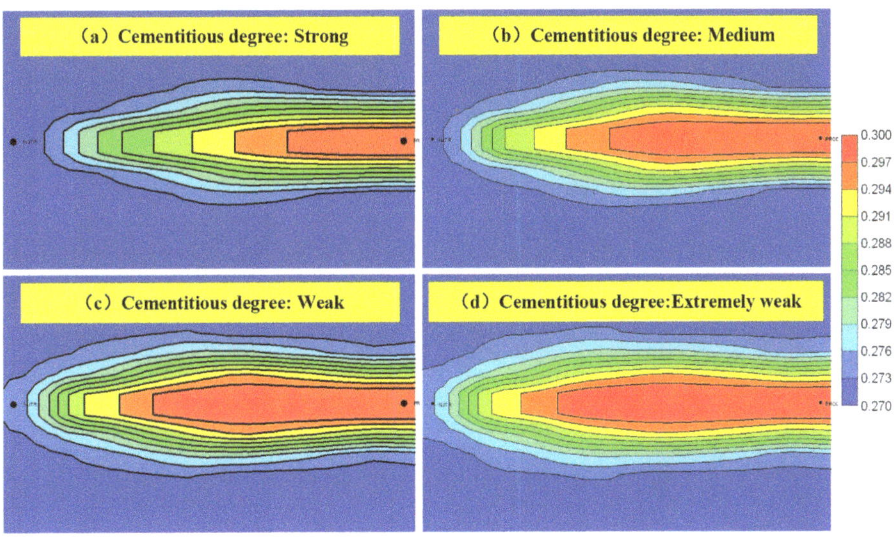

Figure 17. Porosity distribution after steam flooding at different cementation degrees.

Heavy oil formations often show the characteristics of high permeability, high porosity and weak cementation. The clay minerals, siliceous minerals, aluminum and boron minerals and other cementation minerals in heavy oil formations will change under high temperature or alkali liquor. The structure, distribution and occurrence of clay minerals will be damaged. Silica, aluminum and boron minerals will be dissolved. The aquathermolysis reaction will cause the non-expansive clay to transform into expansive clay, which will provide a basis for the increase of porosity to a certain extent [31–33]. Therefore, both high-temperature steam and alkalinity will affect porosity, while loosely cemented formation is more susceptible to temperature and alkalinity in terms of mineral type and contact area.

It can be seen from Table 6, Figures 18 and 19 that the stronger the degree of cementation, the less likely it is to form breakthrough channels, the longer the steam breakthrough time, and the higher the recovery degree. The weaker the degree of cementation, the easier the breakthrough flow will occur, the shorter the steam breakthrough time, and the lower the recovery degree.

Table 6. Steam breakthrough time and recovery degree for different cementation degree.

Cementation Degree	Steam Breakthrough Time (d)	Recovery Degree (%)
0.01 (Strong)	434	15.07
0.03 (Medium)	363	13.96
0.05 (Weak)	337	13.62

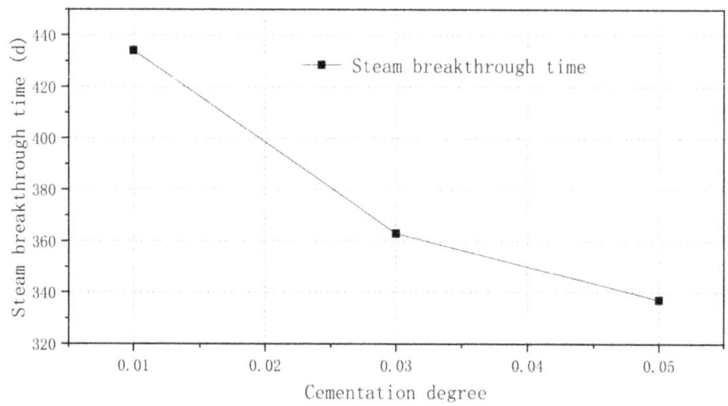

Figure 18. Steam breakthrough time for different cementation degree.

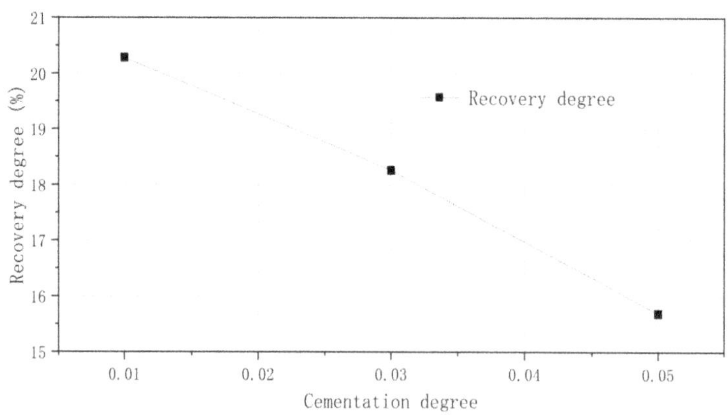

Figure 19. Recovery degree for different cementation degree.

4.3.2. Effect of Steam Injection Strength

In the model, the production-injection ratio is 1.2, and the steam injection intensity is the daily injection volume per unit area and per unit formation thickness, the unit of steam injection intensity is t/(d·m·ha). The effect of steam injection intensity on the change of rock properties (porosity and permeability) after steam flooding can be obtained by changing the steam injection intensity of the numerical model.

As shown in Figure 20, as the steam injection intensity increases, the production pressure difference between injection and production wells gradually increases, and the flow rate increases. The greater the change of formation porosity around the production well, the faster the steam reaches the production well, and the steam breakthrough occurs in production well in a shorter time.

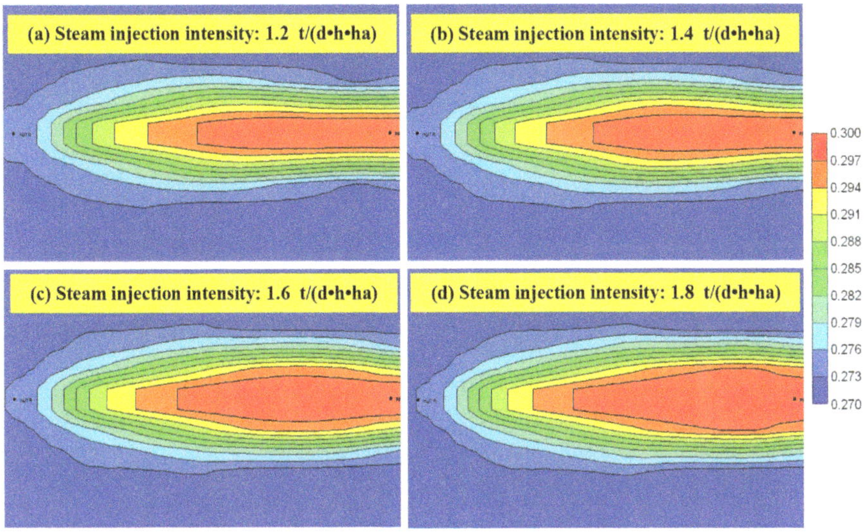

Figure 20. Porosity distribution after steam flooding under different steam injection intensity.

As can be seen from Table 7, Figures 21 and 22, the greater the steam injection intensity, the earlier the breakthrough channel is formed, the easier it is to form breakthrough flow, the shorter the steam breakthrough time, and the lower the recovery degree.

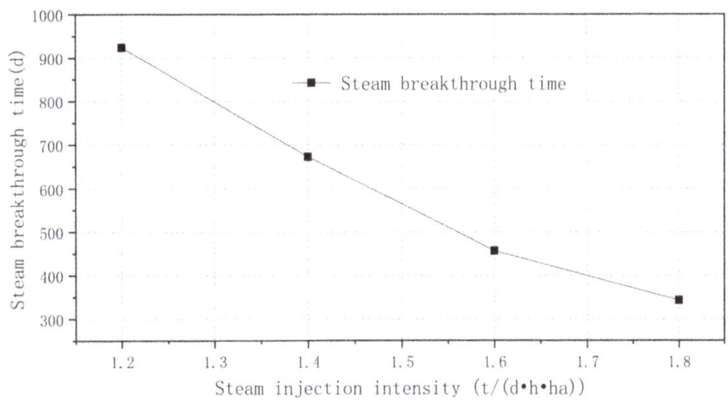

Figure 21. Steam breakthrough time for different steam injection intensity.

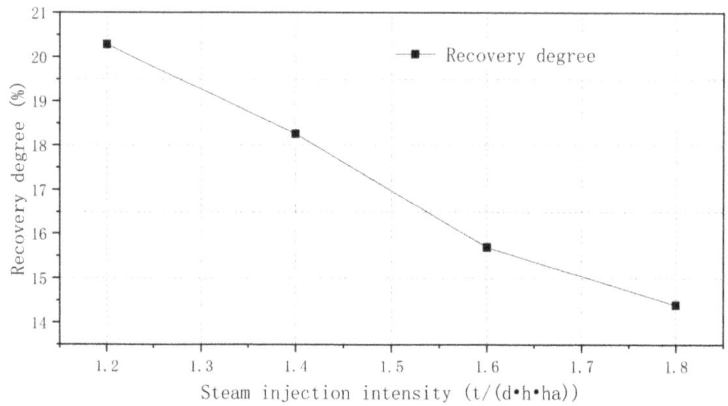

Figure 22. Recovery degree for different steam injection intensity.

Table 7. Steam breakthrough time and recovery degree for different steam injection intensity.

Steam Injection Intensity (t/(d·h·ha))	Steam Breakthrough Time (d)	Recovery Degree (%)
1.2	924	20.28
1.4	673	18.26
1.6	456	15.69
1.8	343	14.39

5. Conclusions

This paper studied the residual retention properties of heavy oil with different compositions in porous media by physical simulation experiments, and established the residual retention model of heavy component and the mechanism model of changes in the physical property of rock by the numerical simulation method. The following conclusions can be drawn:

(1) Heavy oil samples with the same viscosity and different components were prepared for steam flooding experiments. Results showed that the pressure difference required to displace oil samples with high resin content is larger than that required to displace oil samples with low resin content. By comparing the SARA composition of compound oil before and after steam flooding, we can find that the resin content of the recovered oil after steam flooding decreased.

(2) The CMG software was used to establish the residual retention mechanism model of heavy components, and the effects of different ratios of heavy components and light components, steam injection parameters and other factors on the residual retention of heavy components, were studied. The simulation shows that the residual oil after steam flooding is mainly composed of heavy components. In the formation with less residual oil saturation after steam flooding, the heavy components are retained more. With the increase of remaining oil saturation, the mass fraction of heavy components in crude oil decreases. Under the condition of gas injection intensity of 1.6 $m^3/(d \cdot m \cdot ha)$, oil production is the largest.

(3) The mechanism model of rock properties (porosity and permeability) change after steam flooding is established by using the CMG software. After long-term thermal injection development, the porosity and permeability of the reservoir become larger, forming a breakthrough channel. The weaker the cementation degree is, the easier it is to form a breakthrough channel, shorten the steam breakthrough time and reduce the recovery degree. The increase of steam injection intensity will cause the breakthrough channel to appear earlier, and the steam breakthrough time and recovery factor will also decrease.

Author Contributions: Methodology, X.G.; software, T.H.; investigation, K.P.; writing, W.S.; editing, C.H. All authors have read and agreed to the published version of the manuscript.

Funding: This research was funded by Open Fund (PLN2020-02) of State Key Laboratory of Oil and Gas Reservoir Geology and Exploitation (Southwest Petroleum University). The authors would like to acknowledge the financial support of National Natural Science Foundation of China (Grant Nos. 51704032).

Data Availability Statement: All data included in this study are available upon request by contacting with the corresponding author.

Acknowledgments: The authors would like to thank Hubei Cooperative Innovation Center of Unconventional Oil and Gas for supporting our work.

Conflicts of Interest: The authors declare no conflict of interest.

References

1. Fan, H.F.; Liu, Y.J.; Zhao, X.F. Study on composition changes of heavy oils under steam treatment. *J. Fuel Chem. Technol.* **2001**, *29*, 269–272.
2. Hongfu, F.; Yongjian, L.; Liying, Z.; Xiaofei, Z. The study on composition changes of heavy oils during steam stimulation processes. *Fuel* **2002**, *81*, 1733–1738. [CrossRef]
3. Zhao, X.F. Development of Catalyst and Its Support for Super Heavy Oil Hydrothermal Cracking Reaction. Ph.D. Thesis, Daqing Petroleum Institute, Daqing, China, 2006.
4. Fan, H.F.; Liu, Y.J.; Zhong, L.G. The composition and viscosity changes of heavy oils after aquathermal cracking at the presence of reservoir minerals. *Oilfield Chem.* **2001**, *18*, 299–301.
5. Liu, W.Z.; Tang, Y.W. *Selected Papers of the International Conference on Heavy Crude Oil Exploitation*; Petroleum Industry Press: Beijing, China, 1988; Volume II.
6. Hepler, L.G.; Aostr, H. *Heavy Oil Technical Manual*; Petroleum Industry Press: Beijing, China, 1992.
7. Cao, D.D.; Yi, Y.F.; Ding, F.C. The Influence of Hydrothermal Reaction to Heavy Oil Component and the Character in Liaohe. *Oil-Gas Field Surf. Eng.* **2009**, *28*, 13–14.
8. Wang, H.T. Variations of Physical Properties and Control Measures of Formation Damages during Steam Injection in Heavy Oil Reservoirs. Ph.D. Thesis, China University of Geosciences, Beijing, China, 2009.
9. Khormali, A.; Sharifov, A.R.; Torba, D.I. Experimental and modeling analysis of asphaltene precipitation in the near wellbore region of oil wells. *Pet. Sci. Technol.* **2018**, *36*, 1030–1036. [CrossRef]
10. Khormali, A.; Sharifov, A.R.; Torba, D.I. The control of asphaltene precipitation in oil wells. *Pet. Sci. Technol.* **2018**, *36*, 443–449. [CrossRef]
11. Abbaszadeh-Dehghani, M.; Brigham, W.E. Analysis of well-to-well tracer flow to determine reservoir layering. *J. Pet. Technol.* **1984**, *36*, 1753–1762. [CrossRef]
12. Ge, J.L. *Seepage Flow in Porous Media*; Petroleum Industry Press: Beijing, China, 1982.
13. Chen, Y.S. *Reservoir Seepage Field*; Petroleum Industry Press: Beijing, China, 1998.
14. Zhao, G. The application of interwell tracer method in oilfield production. *Pet. Explor. Dev.* **1999**, *26*, 92–95.
15. Zeng, L.F. *Study on Formation Mechanism and Seepage Law of Large Pore in Unconsolidated Sandstone Reservoir*; Petroleum Industry Press: Beijing, China, 2002.
16. Zeng, L.F.; Zhao, G.; Zhang, Z. The macroscopic throats forming mechanism of unsolidated sand-reservior and their identyfying method. *J. Basic Sci. Eng.* **2002**, *3*, 266–268.
17. Guo, S.P.; Huang, Y.Z.; Zhou, J.; Hu, Y.R.; Zhou, Y.R.; Yu, D.S. *Microscopic Mechanism of Physical-Chemical Percolation*; Science Press: Beijing, China, 1990.
18. Gray, M.R. *Upgrading Oilsands Bitumen and Heavy Oil*; University of Alberta Press: Edmonton, AB, Canada, 2015.
19. Alshaikh, M.A.I. Impact of Ionic and Polar Interaction between Heavy Crude Oil and Surfactants during Surfactant-Steam Flooding Processes. Ph.D. Thesis, Texas A&M University, College Station, TX, USA, 2020.
20. Crivelaro, K.C.O.; Damacena, Y.T.; Andrade, T.H.F.; Lima, A.G.B.; Farias Neto, S.R. Numerical simulation of heavy oil flows in pipes using the core-annular flow technique. WIT Trans. *Eng. Sci.* **2009**, *63*, 193–203.
21. Ado, M.R.; Greaves, M.; Rigby, S.P. Numerical simulation of the impact of geological heterogeneity on performance and safety of THAI heavy oil production process. *J. Pet. Sci. Eng.* **2019**, *173*, 1130–1148. [CrossRef]
22. Jing, J.; Zhang, S.; Qin, M.; Luo, J.; Shan, Y.; Cheng, Y.; Tan, J. Numerical simulation study of offshore heavy oil desanding by hydrocyclones. *Sep. Purif. Technol.* **2021**, *258*, 118051. [CrossRef]
23. Rabiu Ado, M. Numerical Simulation of Heavy Oil and Bitumen Recovery and Upgrading Techniques. Ph.D. Thesis, University of Nottingham, Nottingham, UK, 2017.
24. Yuan, Z.; Liu, P.; Zhang, S.; Li, X.; Shi, L.; Jin, R. Experimental study and numerical simulation of nitrogen-assisted SAGD in developing heavy oil reservoirs. *J. Pet. Sci. Eng.* **2018**, *162*, 325–332. [CrossRef]

25. Arhuoma, M.; Yang, D.; Dong, M.; Li, H.; Idem, R. Numerical simulation of displacement mechanisms for enhancing heavy oil recovery during alkaline flooding. *Energy Fuels* **2009**, *23*, 5995–6002. [CrossRef]
26. Shafiei, A.; Dusseault, M.B. Geomechanics of thermal viscous oil production in sandstones. *J. Pet. Sci. Eng.* **2013**, *103*, 121–139. [CrossRef]
27. Dong, X.; Liu, H.; Chen, Z.; Wu, K.; Lu, N.; Zhang, Q. Enhanced oil recovery techniques for heavy oil and oilsands reservoirs after steam injection. *Appl. Energy* **2019**, *239*, 1190–1211. [CrossRef]
28. Gurgel, A.; Moura, M.C.P.D.A.; Dantas, T.N.D.C.; Barros Neto, E.L.; Dantas Neto, A. A Review on Chemical Flooding Methods Applied in Enhanced Oil Recovery. *Braz. J. Pet. Gas* **2008**, *2*, 29–41.
29. Theune, U. Seismic Monitoring of Heavy Oil Reservoirs: Rock Physics and Finite Element Modelling. Ph.D. Thesis, University of Alberta, Edmonton, AB, Canada, 2004.
30. Feng, X. The Formation Mechanism of Channeling after Heavy Oil Thermal Recovery. Ph.D. Thesis, China University of Petroleum, Beijing, China, 2010.
31. Su, X.; Lian, Z.; Li, Z.; Xiong, H.; Lv, Z.; Min, L. Temperature-sensitive damage theory of thermal recovery of super-heavy oil reservoir. *Fault-Block Oil Gas Field* **2018**, *15*, 227–231.
32. Dong, X.; Liu, H.; Chen, Z. *Hybrid Enhanced Oil Recovery Processes for Heavy Oil Reservoirs*; Elsevier: Cambridge, MA, USA, 2021.
33. Zhow, T.; Li, C.; Guo, S.S. Experiment of Aquathermolysis reaction between heavy oil and over-heated steam. *J. Southwest Pet. Univ. Sci. Technol. Ed.* **2009**, *31*, 89–92.

processes

MDPI

Article

Numerical Simulation Research on the Effect of Artificial Barrier Properties on Fracture Height

Jian Zou [1,*], Ying Zhang [2,*], Liping Zhang [1], Jiyun Jing [1], Yangyang Fu [1], Yunjin Wang [2], Guchang Zhang [2] and Fujian Zhou [2]

[1] Bohai Petroleum Research Institute, Tianjin Branch, CNOOC (China) Co., Ltd., Tianjin 300452, China
[2] State Key Laboratory of Petroleum Resources and Prospecting, China University of Petroleum (Beijing), Beijing 102249, China
* Correspondence: zoujian@cnooc.com.cn (J.Z.); wubaili_zhang@126.com (Y.Z.)

Abstract: Hydraulic fracturing is an important measurement for the stimulation of oil and gas wells and is widely used in the development of low-permeability and ultra-low-permeability reservoirs. However, fractures can pass through barriers with poor properties during fracturing, resulting in fractures that do not reach the pre-designed length. In a worse situation, it is possible to communicate with the water layer and cause sudden water flooding, resulting in the failure of the fracturing construction. In order to improve the efficiency of fracturing construction, an effective way to control the height of fractures is by laying diverting agents to form artificial barriers. In this study, we established a three-dimensional numerical calculation model of fracture propagation, considering artificial barriers in the finite element analysis framework; the fracture propagation is governed by a cohesive zone model. The influence of artificial barriers with different Young's modulus and different permeability on the fracture height was simulated and calculated. Different fracture geometries under different pumping injection rates were also considered. The simulation results show that the smaller the Young's modulus of the artificial barrier, the smaller the extension in the direction of the fracture height: when its Young's modulus is 28 GPa, the half fracture height is about 25 m, while when Young's modulus increases to 36 GPa, the half fracture height increases by about 10m. When the fracture does not penetrate the artificial barrier area, the larger the Young's modulus, the smaller the fracture width and the larger the fracture height. With the change in the permeability of the artificial barrier, the change in the fracture width direction of the fracturing fracture is only about 0.5 m, but the inhibition on the fracture height direction is more obvious; in the case of maximum permeability and minimum permeability, the fracture height change is 10 m. The influence of pumping injection rates on the width and height of the fracture is obvious: with the increase in the pumping rates, both the height and width of the fractures increase. However, when the pumping rate increases from 0.12 m^3/s to 0.14 m^3/s, the change in the direction of fracture height is no longer significant, and the increase is only 0.6 m. This study investigates the role of artificial barrier properties and pumping rates in controlling fracture height extension, clarifies the feasibility of artificial barriers to control fracture height technology, and provides guidance for the selection of diverting agents and the determination of the pumping rate in the process of fracturing construction.

Keywords: artificial barrier; cohesive zone model; finite element; hydraulic fracturing

Citation: Zou, J.; Zhang, Y.; Zhang, L.; Jing, J.; Fu, Y.; Wang, Y.; Zhang, G.; Zhou, F. Numerical Simulation Research on the Effect of Artificial Barrier Properties on Fracture Height. *Processes* **2023**, *11*, 310. https://doi.org/10.3390/pr11020310

Academic Editor: Linhua Pan

Received: 27 December 2022
Revised: 15 January 2023
Accepted: 16 January 2023
Published: 17 January 2023

1. Introduction

Currently, the hydrocarbon industry encompasses a significant part of the world economy and is the primary energy source [1,2]. Fracturing stimulation technology is the key technology for the effective development of oil and gas reservoirs. By pumping high-pressure fluid, the reservoir rock is broken to produce fractures and injected with proppant to prevent fracture closure, providing a channel with high conductivity for reservoir fluid [3,4]. However, in the fracturing stimulation construction of thin poor layers, the fracture height often exceeds the production layer and continues to extend [5,6]. In this case, the excessive extension of fracture height will result in low efficiency of fracturing fluid and inefficient laying of proppant, so that the fracturing fractures cannot reach the designed fracture length, thereby reducing oil and gas production [7,8]. However, when the oil layer is close to the water layer, the excessive expansion of the fracture height will communicate with the water layer, causing water channeling, causing the water content of the oil well to rise sharply, and directly leading to the failure of the fracturing construction [9,10]. Therefore, effective measures must be taken to control the height of fracturing fractures to achieve the purpose of oil and gas stimulation during the fracturing and stimulation construction of thin poor reservoirs.

At present, there has been a lot of research on the control of fracture height [11,12], mainly optimizing the construction parameters, preferred fracturing fluid, and laying diverting agents to form the artificial barrier. Among them, artificial barrier technology is the most widely used. Its basic principle is to use the floating or sinking diverting agent to form a certain thickness of low-permeability or impermeable artificial barrier at the top or bottom of the fracture, increase the impedance of the fracture tip, eliminate the stress concentration at the fracture tip, and achieve the purpose of delaying the vertical expansion of the fracture. Nguyen proposed for the first time that during the fracturing of the gas–water barrier, in order to prevent the loss of control in the fracture height direction caused by the press through the upper and lower barriers, it was proposed to use artificial barriers to control the growth of fractures in the height direction [13]. Morales added radioactive tracers to the diverting agent and tested the gamma curves before and after fracturing to verify the feasibility of controlling fracture height with artificial barriers [14]. Arp discussed the problem of fracture height control in several production layers in eastern Texas [15]. Some researchers have studied the placement of artificial barriers and recommended that viscous fracturing fluid be used first to pre-treat and keep the fracture tip open, and then 5–10 cP low-viscosity fracturing fluid be used to carry a lighter proppant to seal the top and bottom of the fracture to achieve a fracture control effect [16]. Many scholars have also carried out experiments and numerical simulation studies on the law of fracture propagation. Khodaverdian demonstrated through experiments that fracture propagation is mainly controlled by fluid invasion and shear failure propagation in the fracture tip region [17]. Now, most of the numerical models used to simulate fracture are based on the theory of linear elasticity, such as the tensile fracture model, PKN model, KGD model, P3D model, and so on [18–20]. In order to study more complex fracture morphology propagation problems, many numerical simulation calculation models have been established [21,22]. Simoni and Secchi used a re-meshing algorithm and a staggered solving algorithm to model cohesive fracture propagation under fluid pressure. They considered fluid exchanges between the fracture and the porous medium [23]. Lacampion investigated the extended finite element method to model hydraulic fracture in an impermeable medium [24]. However, this work neither considers the longitudinal flow of fluids nor the propagation of fractures. Chen adopted an interface element governed by a cohesive law to model fracture propagation in an impermeable medium, and their numerical simulation results were in good agreement with the toughness-dominated numerical solution [25]. Sun used the finite element method to investigate the parameters of the hydraulic fracturing process in porous media, and the results showed that as the fracture extension length increased and the fracture fluid pressure decreased [26]. Wan has studied the effects of stress difference, tensile strength, and Young's modulus on the fracture propagation law in view of the difficulty in controlling the width and height of hydraulic fracturing fractures in thin poor layers [27]. Zhao carried out engineering application research on the basis of numerical simulation and studied the potential influencing factors of the difficulty of hydraulic fracture penetration and expansion in continental shale reservoirs [28]. There are also some studies that conduct sensitivity analyses on main parameters such as fluid viscosity, distribution of natural fractures, differential stress of reservoir and interlayer, and pump injection rate to explore their influence on fracture extension or diversion [29–31]. In summary, previous studies on artificial barriers are mainly based on laboratory or oilfield experiments and the results from these studies further validate the feasibility of this technology and optimize the design of actual fracturing construction parameters. However, such studies often lack quantitative results on the characterization of the actual fracture morphology, which is often obtained through numerical simulation studies. Additionally most scholars' numerical simulation studies about fracture propagation do not consider the effect of artificial barriers on fracture morphology. Therefore, it is a challenging task to accurately establish a three-dimensional numerical calculation model to simulate the effect of artificial barriers in control fracture height.

In this study, in order to explore the influence of the properties of artificial barriers on the fracture height, a numerical calculation model was established to simulate fracture propagation after the placement of diverting agents in the framework of finite element analysis, which is based on the bilinear traction-separation criterion and uses stiffness degradation to describe the damage evolution of the unit. This study focuses on carrying out a numerical simulation to obtain experimental results. Compared with experimental studies, numerical simulation studies have lower costs, faster results, and an advantage in terms of efficient resources [32]. Many studies that cannot be performed in experiments can be well studied in numerical simulation software, and numerical simulation studies are more suited to fault tolerance. Experiments on artificial barrier control fracture height technology must be carried out after a large number of numerical simulation experiments to draw some conclusions, and then combined with laboratory experiments in the oil field. Otherwise, it may face problems such as high investment and low return, or more seriously affect the normal production operation of the oilfield. Different from previous studies, this study sets up an artificial barrier area between the reservoir and the caprock in the finite element software for the first time, and discusses in detail the differences in the morphological changes in fractures before and after penetrating the artificial barrier, which is not only a simple study of the influence of heterogeneity between layers on fracture height propagation. Therefore, a zero-thickness element is used to simulate fracture propagation. In this work, the fluid-solid coupling in the hydraulic fracturing process is especially considered. The simulated calculation unit with pore pressure seepage is used to simulate and calculate the degree of influence of artificial barriers with different Young's modulus and different permeability on the expansion of fracture height. In addition, under different pumping rates, the change in fracture morphology with the pumping rate is also simulated.

2. Mathematical Model

2.1. Fluid Flow Equation in Fractures

Fluid flow in the damaged area of the cohesive unit is divided into tangential flow along the cohesive unit, which can be characterized by Newton or power-law fluid models, and the normal flow perpendicular to the upper and lower surfaces of the cohesive unit can reflect the resistance caused by agglomeration and scaling, such as Figure 1 shows. The basis of fluid flow in this study is CFD (computational fluid dynamics), which is a hybrid finite volume method and finite element solution method to calculate incompressible laminar and turbulent flow problems with high solution accuracy [33].

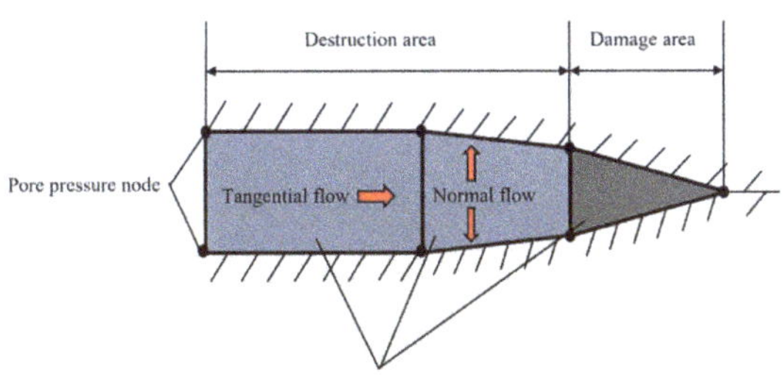

Figure 1. The fluid flow schematic within a damaged unit.

Assuming that the fluid inside the cohesive unit is an incompressible Newtonian fluid, the calculation formula for the tangential flow is [34]:

$$q = -\frac{w^3}{12\mu}\nabla p \tag{1}$$

where q is the tangential flow rate, m^3/s; ∇p is the pressure gradient along the length direction of the cohesive unit, Pa/m; w is the fracture width, m; μ is the viscosity of the fracturing fluid, Pa s.

The normal leak-off loss on the upper and lower surfaces of the Cohesive unit can be described as [34]:

$$\begin{cases} q_t = c_t(p_f - p_t) \\ q_b = c_b(p_f - p_b) \end{cases} \tag{2}$$

where p_t and p_b are the pore pressure at the upper and lower surfaces of the fracture, Pa; p_f is the fluid pressure in the fracture, Pa; c_t, c_b are the fluid loss coefficients at the upper and lower surfaces, $m^3/(Pa \cdot s)$; q_t, q_b are the normal volume flow of the upper and lower surfaces, respectively, m^3/s.

The fluid mass conservation equation of the Cohesive unit is [35]:

$$\frac{\partial w}{\partial t} + \nabla \cdot q + (q_t + q_b) = Q(t)\delta(x, y) \tag{3}$$

where Q(t) is the injection velocity of fracturing fluid, m^3/s.

2.2. *Criteria for Fracture Initiation and Propagation*

At present, the commonly used finite element methods for numerical simulation of hydraulic fracturing include the finite element method based on the cohesive model and the extended finite element method [19]. The former is adopted in this paper, and the problem of infinite stress at the fracture tip of linear elastic fracture mechanics can be effectively avoided by inserting cohesive units between cells for describing nonlinear fracture problems. Cohesive elements provide a solution without refining the mesh, which can effectively reduce the number of meshes and improve the efficiency of operations [36].

2.2.1. Cohesive Unit Damage Model

The relationship between the tensile stress and the distance between the upper and lower surfaces is defined in the damage model of the Cohesive unit, and the definition of the damage evolution consists of two parts. A bilinear T-S criterion was proposed by V. Tomar [37]. As shown in Figure 2, the first part involves determining the effective displacement δ_m^f at full damage relative to the effective displacement δ_m^0 at the onset of damage, or the energy dissipation G_C due to failure. The second component defines the evolutionary nature of the damage variable D between initial and complete failure, which can be specified directly in the form of a table of effective displacement, effective displacement versus damage, by means of a linear or exponential softening law; the expression for the damage variable D when the linear displacement expansion criterion is used is:

$$D = \frac{\delta_m^f\left(\delta_m^{max} - \delta_m^0\right)}{\delta_m^{max}\left(\delta_m^f - \delta_m^0\right)} \tag{4}$$

where δ_m^{max} is the maximum displacement of the unit; δ_m^f is the displacement of the unit opening; δ_m^0 is the displacement of the unit starting damage.

The Cohesive unit uses stiffness degradation to describe the damage evolution of the unit.

$$T = (1 - D)\overline{T} \tag{5}$$

where T_n, T_{s1}, T_{s2} are the stresses obtained in the three directions of the cohesive unit according to the linear elastic deformation in the undamaged phase, respectively, and D is the damage variable in Equation (4).

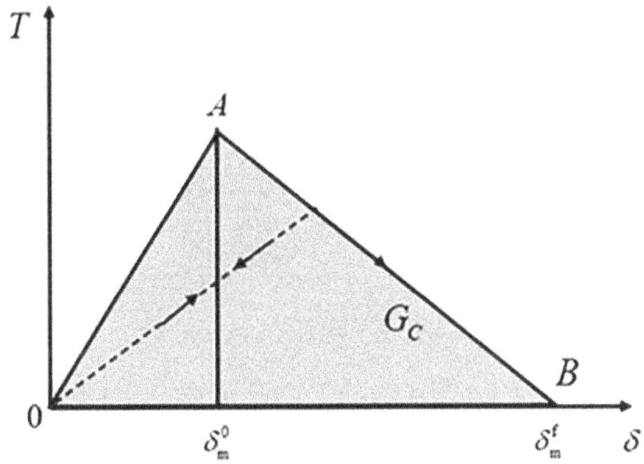

Figure 2. Bilinear traction-separation curve.

The T-S criterion is based on the tensile stress of the cohesive unit as the damage criterion; the unit is very stiff before the damage, the stress is proportional to the displacement and can be recovered after unloading, but when the tensile stress exceeds the tensile strength of the material, the stress it can withstand decays linearly with the increase in displacement and is not recoverable, which is more suitable for the fracture expansion process in hydraulic fracturing.

2.2.2. Cracking and Expansion of Cohesive Units

The following criteria are often used in the finite element method to determine fracture initiation: maximum positive stress criterion, maximum positive strain criterion, quadratic stress criterion and quadratic strain criterion [38,39]. In this study, after many attempts and simulations, it is considered that the maximum positive stress criterion is used, which is more stable as the basis for judging fracture initiation, and the calculation process is easy to converge and more compatible. That is, as soon as the stress in either direction of the unit reaches its critical stress, the unit starts to crack. The expression of the maximum positive stress criterion is as follows:

$$\max\left\{\frac{\sigma_n}{\sigma_n^{max}}, \frac{\tau_s}{\tau_s^{max}}, \frac{\tau_t}{\tau_t^{max}}\right\} = 1 \tag{6}$$

where σ_n^{max} refers to the maximum tensile stress that the unit can withstand in the vertical direction, i.e., the tensile strength of the reservoir, and τ_s^{max} and τ_t^{max} refer to the maximum shear stress that the unit can withstand in both directions, i.e., the shear strength of the reservoir.

For the expansion of composite fractures after fracture initiation, this paper adopts the B-K criterion, which is the criterion of critical energy release rate of crack expansion proposed by Benzeggagh and Kenane [40].

$$G_n^C + \left(G_s^C - G_n^C\right)\left\{\frac{G_S}{G_T}\right\}^{\eta} = G^C \tag{7}$$

In Equation (7), $G_S = G_s + G_t, G_T = G_n + G_s$, G_n^C is the normal fracture critical strain energy release rate; G_s^C, G_t^C are the two tangential fracture critical energy release rates, respectively, and the B-K criterion considers $G_s^C = G_t^C$; η is a constant related to the properties of the material itself; G^C is the composite fracture critical energy release rate.

Figure 3 shows a schematic diagram of the crack expansion of the cohesive unit. When the calculated energy release rate at the fracture tip node is greater than the B-K critical energy release rate, the current fracture tip node of the cohesive unit will unbind the bound part, and then the fracture will open and continue to expand along the next cohesive unit.

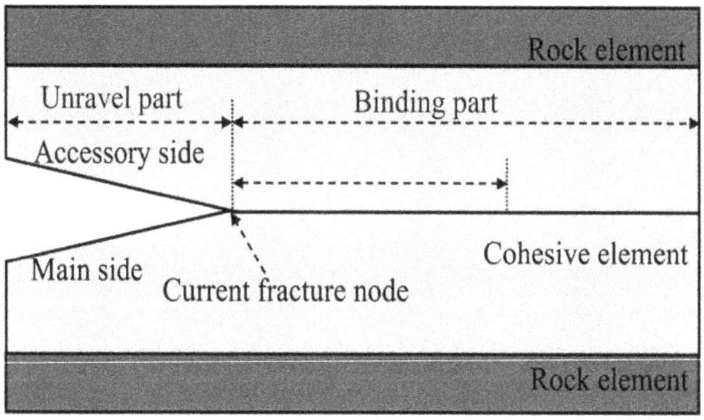

Figure 3. Fracture propagation in the cohesive unit.

2.3. Fluid-Solid Coupling Control Equation

In the process of hydraulic fracturing, the fluid percolation pressure acting on the fracture wall increases continuously with the increase in pump pressure, which leads to the increase in fluid leak-off loss into the formation, resulting in the change in stress state in the rock pore space, and the change in stress in the rock will cause the change in parameters such as reservoir porosity and fluid percolation velocity, which in turn affects the change in pore pressure in the percolation field on the fracture wall. This relationship between fluid seepage and rock deformation is called fluid-solid coupling [41].

The rock skeleton stress balance equation for a unit body with volume V and surface area S is:

$$\int_V \sigma\delta\varepsilon dV = \int_V t\cdot\delta v S + \int_V f\cdot\delta v dV \tag{8}$$

where σ and δ_ε are the stress matrix and the imaginary strain rate matrix, respectively; t, f and δ_v are the surface force vector, the volume force vector and the imaginary velocity vector, respectively.

Fluid seepage is required to satisfy the continuity equation:

$$\frac{1}{J}\frac{\partial}{\partial t}(J\rho_w n_w) + \frac{\partial}{\partial X}(\rho_w n_w v_w) = 0 \tag{9}$$

where J is the variation ratio of porous media body; ρ_w is the fluid density; n_w is the pore ratio; v_w is the fluid percolation velocity; X is the spatial vector.

Fluid flow in the cohesive unit satisfies Darcy's law:

$$v_w = -\frac{1}{n_w g \rho_w}\mathbf{k}\left(\frac{\partial p_w}{\partial x} - \rho_w g\right) \tag{10}$$

where **k** is the permeability matrix of porous media; g is the acceleration of gravity.

2.4. Fracture Wall Stress Distribution Model

This paper focuses on the upper propagation of the fracture. Generally, the model of numerical simulation calculation will be different from the actual situation, and some simplifications will be made [42]. To simplify the model, it is assumed that the upper and lower interlayer stresses in the reservoir are the same and symmetrically distributed along the centerline of the fracture. The model has the following assumptions: ①: the reservoir is an ideal thick oil formation, the rock is an ideal linear elastic material, and the ground stress is linearly distributed; ②: the fracture flanks are symmetrically distributed with the wellbore as the axis; ③: the fracturing fluid is an incompressible power-law fluid when it flows inside the fracture, and the two-dimensional flow of the fluid is considered; ④: the matrix permeability is low, and the leak-off of the fracturing fluid to the fracture wall is ignored; ⑤: the inertia effect is ignored.

According to the above assumptions, the stress distribution at the fracture wall is shown in Figure 4. Assuming that the reservoir thickness is 2H and the height of fracture penetration into the barrier $h_b = H - h$, let the minimum horizontal principal stress at the center of the fracture be σ_b, the minimum horizontal principal stress gradient be g_s, the upper and lower reservoir interlayer stresses be the same, the reservoir interlayer stress difference be $\sigma_b - \sigma_r$, the pressure drop of fracturing fluid gravity in the direction of the fracture height be ρgy, and the pressure drop generated by the artificial interlayer in the direction of the fracture height be $k_a h_a$.

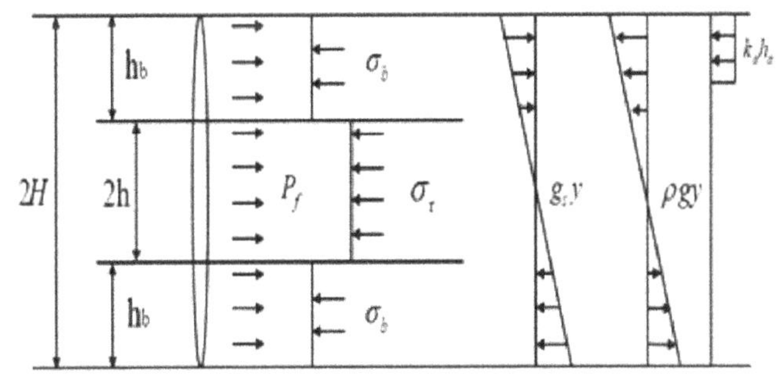

Figure 4. Fracture wall stress distribution.

2.5. Mode Description

The numerical calculation model established in this paper is shown in Figure 5 for a three-dimensional model with dimensions of 40 m × 100 m × 100 m. The following assumptions are made for the model material: the reservoir is an ideal thick oil layer and the rock is an ideal linear elastic homogeneous material. The model is dissected into four areas, the thickness of both upper and lower layers as barriers is 20 m, the thickness of the middle area as the target reservoir is 60 m, and the height of the artificial barrier area formed by the diverting agents placement is 5 m. The (a) and (b) figures in Figure 6 show the profiles of the xoy plane and yoz plane, respectively. The perforation tunnel is chosen to be the center of the xoy plane, y is the direction of vertical principal stress, z is the direction of maximum horizontal principal stress, and x is the direction of minimum horizontal principal stress. The built-in Cohesive cell is embedded in the yoz plane as the fracture initiation and propagation path, perpendicular to the direction of the minimum horizontal principal stress, and the shot hole position is parallel to the direction of the maximum horizontal principal stress. The fracture propagation path and the basic morphology at different moments (t_1, t_2, t_3) are depicted in Figure 6b. The type of solid element is C3D8P, which is a three-dimensional calculation unit with 8 nodes. It has the advantage of more accurate results for the displacement and the accuracy of the analysis will not be affected too much when there is distortion and deformation of the mesh. The mesh size of the numerical calculation model is chosen to be 2 m, and hexahedral elements are used completely in the mesh. For the three-dimensional geometric model, the mesh is first generated on the face and then stretched along the sweep path to obtain the three-dimensional mesh, and a total of 50,000 meshes are divided. The initial saturation of the target layer and the interlayer is set to 1; both are saturated seepage, and the constant pore pressure (20 MPa) is maintained at the left, right and upper and lower boundaries. The initial pore pressure is 20 MPa, the initial porosity of the reservoir is 0.3, and the initial porosity of the interlayer is 0.2. The fracturing fluid pumping discharge is set to 0.06 m^3/s for 600 s, the fracturing fluid viscosity is 100 mPa·s, and the other initial values of the main calculated parameters are shown in Table 1.

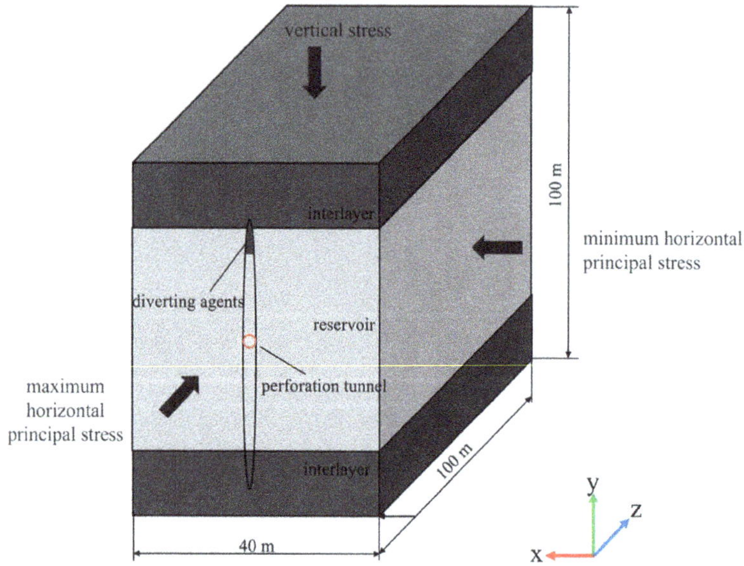

Figure 5. Three-dimensional numerical calculation model.

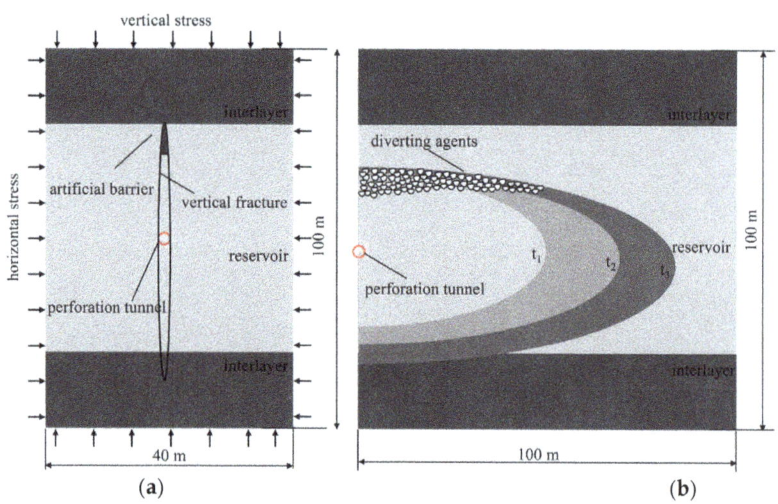

Figure 6. Numerical calculation model (**a**): xoy profile chart, (**b**): yoz profile chart.

Table 1. The rock mechanics parameters of the stratum.

Parameters	Units	Value of Reservoir Rock	Value of Interlayer
Young's modulus	GPa	30	28
Poisson's ratio	/	0.25	0.3
Leak-off coefficient	m/(Pa·s)	1×10^{-7}	1×10^{-8}
Sy	MPa	32	32
Sx	MPa	25	25
Sz	MPa	30	30

Figure 7 shows the workflow of this study. The first step is to establish the fluid-solid coupling control equation on the basis of computational fluid dynamics. The second step is to establish a three-dimensional numerical calculation model considering artificial barriers in the finite element simulation software. On the basis of this model, combined with the previous laboratory experimental results, the calculation parameters are determined and the numerical simulation of fracture propagation is carried out. After the simulation, 3D fracture morphology data, model stress, and pore pressure clouds are extracted for parameter sensitivity analysis and discussion.

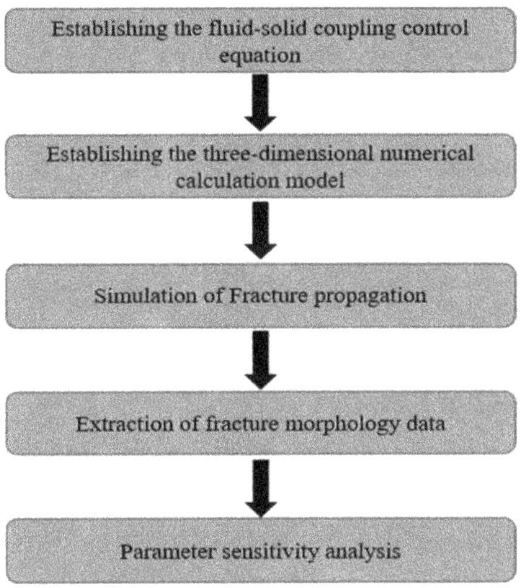

Figure 7. Workflow of this study.

3. Result and Discussion

3.1. Young's Modulus

The Young's modulus of the reservoir rock is 30 GPa, the Young's modulus of the interlayer is 28 GPa, and the Young's modulus of the artificial barrier formed by diverting agents is 36, 34, 32 and 28 GPa, respectively. The values of other calculated parameters are shown in Table 2. The effect of different Young's modulus of the artificial barrier on the fracture morphology is simulated and studied. Figure 8 shows the graphs of the fracture width and half-slit height under different Young's modulus. It can be seen from Figure 8 that the larger the Young's modulus of the artificial barrier, the more easily the fracture extends in the direction of the fracture height. Additionally, the change in the direction of the fracture width is not obvious and the range of the fracture width is distributed from 4.7 to 5.2 mm. However, there is a significant increase in the fracture width compared to the case without artificial barrier. It is analyzed that this is because the Young's modulus will make the width of the fracture narrower, and the fracture height will increase correspondingly for a certain volume of injected fluid and leak-off. On the other hand, the higher the Young's modulus of the artificial barrier, the more likely it is to develop high stresses at the fracture tip, and the more likely it is for the fracture to expand forward under the same tensile strength conditions.

It is noteworthy that the fracture breaks through the area of each artificial barrier to extend into the area of the interlayer under the condition of Young's modulus of the artificial barrier of 34 GPa or 36 GPa. Figure 8 shows that when the fracture penetrates the intersection and extends in the artificial barrier with a smaller Young's modulus (28 GPa), the fracture width will increase slightly due to the sudden increase in the amount of leak−off when the fracture penetrates the dividing line between the artificial barrier and the interlayer, which leads to a slight increase in the fracture width and also an increase in the fracture height. Figure 9 shows the variation curve of pore pressure at the injection point with the injection time during the fracturing process, and the pore pressure variation when the fracture expands to the artificial barrier area is labeled. From Figure 9, it can be seen that the pore pressure at the injection point decreases slightly with the increase in Young's modulus of the artificial barrier; the higher the pore pressure at the injection point, the greater the impediment to fracture extension. This indicates that the small Young's modulus of the artificial barrier does have a hindering inhibiting effect on the vertical propagation of the fracture.

Table 2. The rock mechanics parameters of the stratum.

Parameters	Units	Value of Reservoir Rock	Value of Interlayer	Value of Artificial Barrier
Young's modulus	GPa	30	28	28~36
Poisson's ratio	/	0.25	0.3	0.3
Leak-off coefficient	m/(Pa·s)	1×10^{-7}	1×10^{-8}	1×10^{-10}
Sy	MPa	32	32	32
Sx	MPa	25	25	28
Sz	MPa	30	30	36

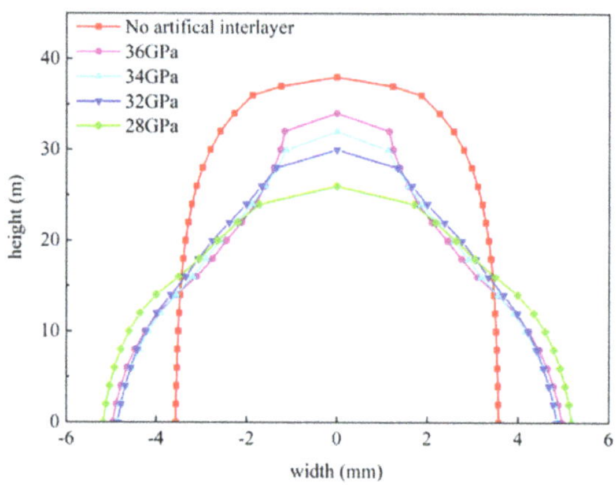

Figure 8. Fracture morphology at different Young's modulus of artificial barrier.

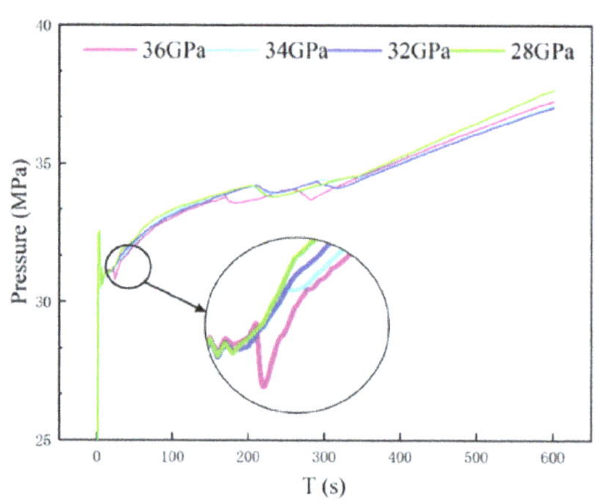

Figure 9. Pore pressure at injection point with different Young's modulus.

Figure 10 shows the stress distribution clouds of the fractured reservoir under different Young's modulus conditions of the artificial barrier. It can be seen from the figure that the stress concentration area of the fracture changes with the increase in Young's modulus, which is due to the different Young's modulus of the artificial barrier, resulting in different fracture morphology. The larger Young's modulus is the more unfavorable to control the fracture height.

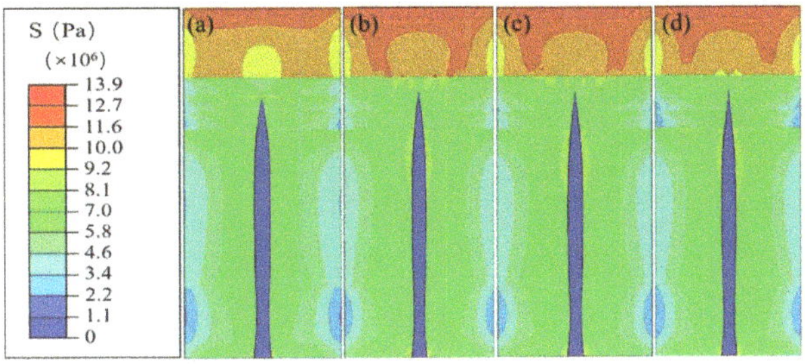

Figure 10. Stress distribution cloud after fracturing under different Young's modulus conditions: (**a**) 28 GPa, (**b**) 32 GPa, (**c**) 34 GPa, (**d**) 36 GPa.

Figure 11 shows the pore pressure distribution clouds of the reservoir after fracturing under different Young's modulus conditions of the artificial barrier, from which it can be seen that: the reservoir pore pressure increases with the increase of Young's modulus; the pore pressure shows a concentration phenomenon after fracturing; and a higher stress concentration zone is formed around the fracture.

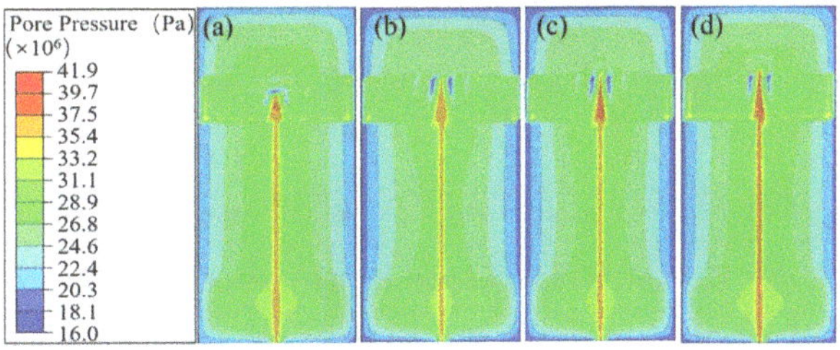

Figure 11. Pore pressure distribution cloud after fracturing under different Young's modulus conditions: (**a**) 28 GPa, (**b**) 32 GPa, (**c**) 34 GPa, (**d**) 36 GPa.

3.2. Leak-Off Coefficient

The size of the formation permeability is an important factor affecting fracture expansion, with low permeability reservoirs prone to fractures of a smaller horizontal extent and larger width, and the opposite for high permeability reservoirs [43]. The values of other calculated parameters are shown in Table 3. Since the fracture is extended on the preplaced cohesive cell, the fluid normal leak-off coefficient for the cohesive cell pore material defines the pressure-flux relationship between the intermediate nodes of the bonded cell and their adjacent surface nodes, and thus this coefficient can be interpreted as the cohesive bonded cell surface permeability of a finite material layer. Changing the leak-off coefficient of artificial barrier (1×10^{-7}, 1×10^{-10}, 1×10^{-13}, 1×10^{-16}) and studying the effect of the change in permeability of the artificial barrier on the fracture height, Figure 12 shows the graphs of the fracture width and half-slit height under different leak-off coefficients. With the decrease in the permeability of the artificial barrier, the morphology of the fracture changes greatly, and the change in fracture width direction is not significant. However, the control of permeability on the direction of fracture height of the fractured fracture is more obvious, in which the change in fracture height under the condition of maximum and minimum leak-off coefficient reaches about 10 m. When the leak-off coefficient of the artificial barrier is 1×10^{-16}, the half fracture height is 16 m, when the leak-off coefficient increases to 1×10^{-7}, the half fracture height increases by 10 m, but the width of the fracture decreases by about 0.2 m. This shows that the artificial barrier with low leak-off coefficient has significant control on the fracture height. Therefore, the placement of the artificial barrier with good sealing properties and a low leak-off coefficient is an important technical means to effectively control the fracture height.

Table 3. The rock mechanics parameters of the stratum.

Parameters	Units	Value of Reservoir Rock	Value of Interlayer	Value of Artificial Barrier
Young's modulus	GPa	30	28	26
Poisson's ratio	/	0.25	0.3	0.3
Leak-off coefficient	m/(Pa·s)	1×10^{-7}	1×10^{-8}	$1 \times 10^{-16\sim-7}$
Sy	MPa	32	32	32
Sx	MPa	25	25	28
Sz	MPa	30	30	36

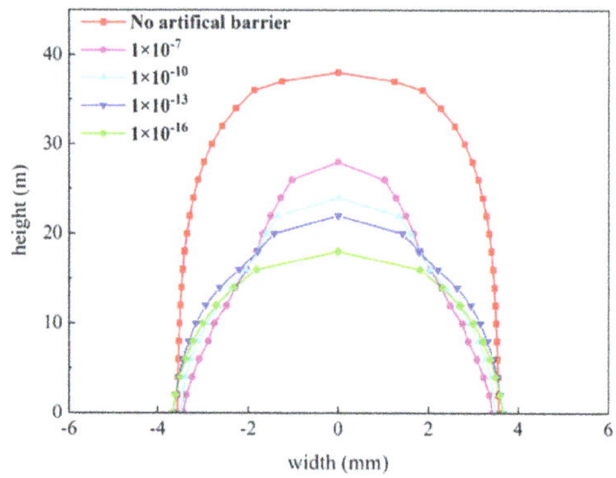

Figure 12. Fracture morphology at different leak-off coefficient of artificial barrier.

Figure 13 shows the pore pressure distribution clouds of the fracture after fracturing under different leak-off coefficients, from which it can be seen that the formation pore pressure increases slightly with the decrease in the leak-off coefficient of the artificial barrier. When the leak-off coefficient is 1×10^{-7} and 1×10^{-10}, the height of the fracture is smaller, but the fracture extension height above the injection point is greater than that of the leak-off coefficient of 1×10^{-13} and 1×10^{-16}. The analysis suggests that because the total amount of fracturing fluid pumped remains unchanged, more fracturing fluid is transported to the lower end of the fracture due to the decrease in the leak-off coefficient of the artificial barrier, and the pore pressure at the lower end of the fracture is greater (c, d), thus achieving the purpose of controlling the upward expansion of the fracture.

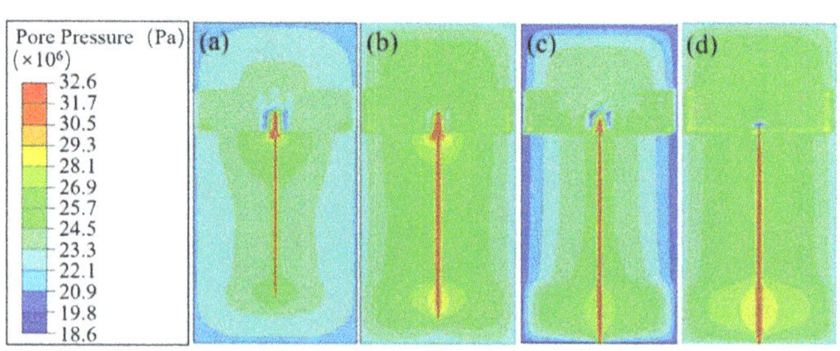

Figure 13. Pore pressure distribution cloud after fracturing under different leak-off coefficient conditions: (**a**) 1×10^{-7}, (**b**) 1×10^{-10}, (**c**) 1×10^{-13}, (**d**) 1×10^{-16}.

3.3. Pumping Rate

In this section, the effect of different pumping rates on the fracture height extension in the presence of artificial barrier conditions is investigated. Since the model is plane strain, the pumping rates in this model are discounted accordingly to the field discharge, and five gradients are set: 0.04, 0.08, 0.1, 0.12, and 0.14 m^3/s. Other parameters of the stratum are shown in Table 4, and the fracture morphology calculated by a numerical simulation is shown in Figure 14. With the increase in pumping rate, the fracture width and fracture height increase, and when the displacement increases from 0.12 m^3/s to 0.14 m^3/s, the extension in the fracture height direction is not obvious, and the fracture width continues to increase. The reason is that at this time, the factors affecting the fracture height are not only the size of pumping volume, but also the Young's modulus and permeability of the artificial barrier.

Table 4. The rock mechanics parameters of the stratum.

Parameters	Units	Value of Reservoir Rock	Value of Interlayer	Value of Artificial Barrier
Young's modulus	GPa	30	28	26
Poisson's ratio	/	0.25	0.3	0.3
Leak-off coefficient	m/(Pa·s)	1×10^{-7}	1×10^{-8}	1×10^{-10}
Sy	MPa	32	32	32
Sx	MPa	25	25	28
Sz	MPa	30	30	36

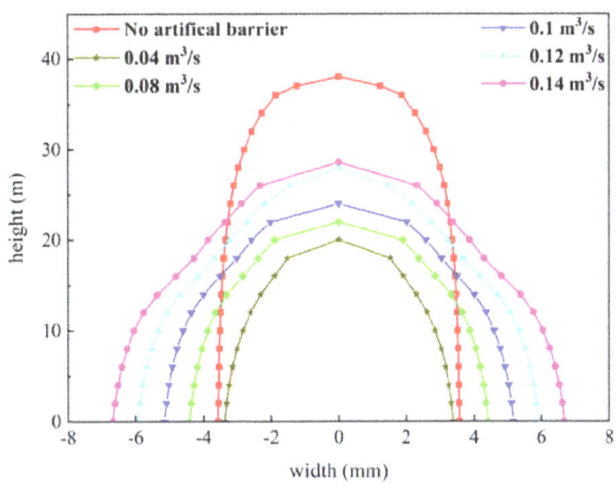

Figure 14. Fracture morphology under different pumping rate.

Figure 15 shows the cloud plot of pore pressure distribution after fracturing under a different pumping rate. From the figure, it can be seen that the forming pore pressure increases with the increase in pumping rate, indicating that the displacement increases and the fracturing fluid leak–off loss also increases, while the pore pressure results in the pressure concentration phenomenon after fracturing, forming a high stress concentration zone around the fracture. When the pumping displacement is 0.04 m^3/s, the lower end of the fracture is not enough to extend to the lower reservoir, and a high pore pressure concentration is formed at the lower end of the fracture.

Through the sensitivity analysis of the effects of the above three different parameters on the fracture morphology, the effects of Young's modulus and the permeability of the artificial barrier on the fracture height were clarified, and the effects of pumping rates on the fracture propagation under the artificial barrier conditions were investigated. However, there are still some parts of this study that need to be improved. The three-dimensional numerical calculation model used in this study is homogeneous material and does not consider the effect of natural fractures and stratigraphic bedding on fracture morphology [44]. It should also be noted that the numerical simulation results of this study cannot be verified in the laboratory experiment. We will continue to refine this three-dimensional numerical model and add more different parameters to study the artificial barrier control fracture height technique, combined with microseismic monitoring techniques to verify the simulation results of this study.

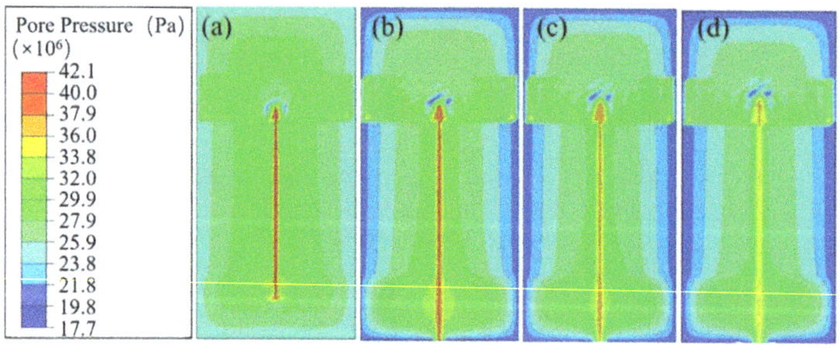

Figure 15. Pore pressure distribution cloud after fracturing under different pumping rate: (**a**) 0.04 m^3/s, (**b**) 0.08 m^3/s, (**c**) 0.1 m^3/s, (**d**) 0.12 m^3/s.

4. Conclusions

In this study, a three-dimensional numerical calculation model of fracture propagation based on the bilinear traction-separation criterion is established, which considers the relationship between the mutual constraints and effects of fluid seepage and rock deformation. The effects of Young's modulus, leak-off coefficient of the artificial barrier and pumping rate on fracture morphology are studied, and the following conclusions are drawn. The Young's modulus of the artificial barrier has a significant effect on fracture morphology. Overall, the smaller the Young's modulus of the artificial barrier, the smaller the fracture height formed by the fracturing construction, while the fracture width size does not vary much. When the Young's modulus of the artificial barrier is 36 GPa, the half fracture height decreases by 4 m compared with the case without an artificial barrier. Similarly, the leak-off coefficient of the artificial barrier has great influence on fracture height, but has little influence on fracture width. When the leak-off coefficient decreases from 1×10^{-7} to 1×10^{-16}, the half fracture height decreases by 10 m. The pore pressure cloud diagram shows that the artificial barrier with small leak-off coefficient can effectively organize the pressure in the fracture to transfer to the fracture end, so as to achieve the purpose of controlling the fracture height. The effect of pumping rates on fracture width and height are both more significant, which is different from the first two factors. With the increase in pumping rates, the height and width of the fracture increases. When the pumping rates increase from 0.04 m^3/s to 0.12 m^3/s, the half fracture height increases by 8 m, and when the pumping rate is 0.14 m^3/s, the fracture height does not change much compared with 0.12 m^3/s. In a subsequent study, we will continue to refine this three-dimensional numerical model to simulate the effects of other different properties of the artificial barrier on the fracture morphology and investigate the mechanism of the artificial barrier to control the fracture height. It will also be combined with field tests, such as using microseismic monitoring technology to verify the feasibility of artificial barrier control fracture height technology, and investigate the impact of artificial barriers formed by different types of diverting agents on oil and gas production.

Author Contributions: Writing—original manuscript, J.Z. and Y.Z.; methodology, Y.Z. and J.Z.; software, Y.Z. and L.Z.; writing—original draft preparation, J.J. and Y.F.; supervision, Y.W. and F.Z.; investigation, J.J.; data curation, Y.W. and G.Z.; formal analysis, Y.F. All authors have read and agreed to the published version of the manuscript.

Funding: This research was funded by the National Natural Science Foundation of China (No. 52174045).

Conflicts of Interest: The authors declare no conflict of interest.

References

1. Ahmadi, Y.; Mohammadi, M.; Sedighi, M. Chapter 1—Introduction to Chemical Enhanced Oil Recovery. In *Chemical Methods*; Hemmati-Sarapardeh, A., Schaffie, M., Ranjbar, M., Dong, M., Li, Z., Eds.; Enhanced Oil Recovery Series; Gulf Professional Publishing: Houston, TX, USA, 2022; pp. 1–32. ISBN 978-0-12-821931-7.
2. Mansouri, M.; Parhiz, M.; Bayati, B.; Ahmadi, Y. Preparation of Nickel Oxide Supported Zeolite Catalyst (NiO/Na-ZSm-5) for Asphaltene Adsorption: A Kinetic and Thermodynamic Study. *Iran. J. Oil Gas Sci. Technol.* **2021**, *10*, 63–89. [CrossRef]
3. Ahmadi, Y. Relationship between Asphaltene Adsorption on the Surface of Nanoparticles and Asphaltene Precipitation Inhibition During Real Crude Oil Natural Depletion Tests. *Iran. J. Oil Gas Sci. Technol.* **2021**, *10*, 69–82. [CrossRef]
4. Lin, H.; Tian, Y.; Sun, Z.; Zhou, F. Fracture Interference and Refracturing of Horizontal Wells. *Processes* **2022**, *10*, 899. [CrossRef]
5. Zhu, D.; Wang, Y.; Cui, M.; Zhou, F.; Zhang, Y.; Liang, C.; Zou, H.; Yao, F. Effects of Spent Viscoelastic-Surfactant Acid Flow on Wormholes Propagation and Diverting Performance in Heterogeneous Carbonate Reservoir. *Energy Rep.* **2022**, *8*, 8321–8332. [CrossRef]
6. Dong, R.; Wheeler, M.F.; Su, H.; Ma, K. Modeling Acid Fracturing Treatments in Heterogeneous Carbonate Reservoirs. In *Proceedings of the SPE International Conference on Oilfield Chemistry, The Woodlands, TX, USA, 29 December 2021*; OnePetro: Richardson, TX, USA, 2021.
7. Zeng, H.; Jin, Y.; Wang, D.; Yu, B.; Zhang, W. Numerical Simulation on Hydraulic Fracture Height Growth across Layered Elastic–Plastic Shale Oil Reservoirs. *Processes* **2022**, *10*, 1453. [CrossRef]
8. Merzoug, A.; Ellafi, A.; Rasouli, V.; Jabbari, H. Anisortopic Modeling of Hydraulic Fractures Height Growth in the Anadarko Basin. *Appl. Mech.* **2023**, *4*, 44–69. [CrossRef]
9. Cao, Y.; He, Q.; Liu, C. Numerical Investigation of Fracture Morphology Characteristics in Heterogeneous Reservoirs. *Processes* **2022**, *10*, 2604. [CrossRef]
10. Zhu, D.; Wang, Y.; Cui, M.; Zhou, F.; Wang, Y.; Liang, C.; Zou, H.; Yao, F. Acid System and Stimulation Efficiency of Multistage Acid Fracturing in Porous Carbonate Reservoirs. *Processes* **2022**, *10*, 1883. [CrossRef]
11. Dong, R.; Wheeler, M.F.; Su, H.; Ma, K. Modeling Multistage Acid Fracturing Treatments in Carbonate Reservoirs. In *Proceedings of the SPE Hydraulic Fracturing Technology Conference and Exhibition, Virtual, TX, USA, May 2021*; OnePetro: Richardson, TX, USA, 2021.
12. Dong, R.; Wheeler, M.F.; Su, H.; Ma, K. Modeling Acid Fracturing Treatments with Multi-Stage Alternating Injection of Pad and Acid Fluids. In *Proceedings of the SPE Reservoir Simulation Conference, On-Demand, TX, USA, October 2021*; OnePetro: Richardson, TX, USA, 2021.
13. Nguyen, H.X.; Larson, D.B. Fracture Height Containment by Creating an Artificial Barrier With a New Additive. In *Proceedings of the SPE Annual Technical Conference and Exhibition, San Francisco, CA, USA, October 1983*; OnePetro: Richardson, TX, USA, 1983.
14. Morales, R.; Fragachan, F.E.; Prado, E.; Santillan, J. Production Optimization by an Artificial Control Of Fracture Height Growth. In *Proceedings of the SPE European Formation Damage Conference, The Hague, The Netherlands, June 1997*; OnePetro: Richardson, TX, USA, 1997.
15. Arp, M.E.; Hilton, R.E. Case History Study of Fracture Height Containment in East Texas. In *Proceedings of the SPE East Texas Regional Meeting, Tyler, TX, USA, April 1986*; OnePetro: Richardson, TX, USA, 1986.
16. Mukherjee, H.; Paoli, B.F.; McDonald, T.; Cartaya, H.; Anderson, J.A. Successful Control of Fracture Height Growth by Placement of Artificial Barrier. *SPE Prod. Facil.* **1995**, *10*, 89–95. [CrossRef]
17. Khodaverdian, M.; McElfresh, P. Hydraulic Fracturing Stimulation in Poorly Consolidated Sand: Mechanisms and Consequences. In *Proceedings of the SPE Annual Technical Conference and Exhibition, Dallas, TX, USA, October 2000*; OnePetro: Richardson, TX, USA, 2000.
18. Peng, S.; Zhang, X.; Tang, X.; Xu, J.; Jiao, F.; He, M. Experimental Investigation of Non-Linear Seepage Characteristics in Rock Discontinuities and Morphology of the Shear Section in the Shear Process. *Processes* **2022**, *10*, 2625. [CrossRef]
19. Carrier, B.; Granet, S. Numerical Modeling of Hydraulic Fracture Problem in Permeable Medium Using Cohesive Zone Model. *Eng. Fract. Mech.* **2012**, *79*, 312–328. [CrossRef]
20. Yan, J.I.N.; Xu-dong, Z.; Mian, C. Initiation Pressure Models for Hydraulic Fracturing of Vertical Wells in Naturally Fractured Formation. *Acta Pet. Sin.* **2005**, *26*, 113. [CrossRef]
21. Adachi, J.; Siebrits, E.; Peirce, A.; Desroches, J. Computer Simulation of Hydraulic Fractures. *Int. J. Rock Mech. Min. Sci.* **2007**, *44*, 739–757. [CrossRef]
22. Xu, B.; Liu, Y.; Wang, Y.; Yang, G.; Yu, Q.; Wang, F. A New Method and Application of Full 3D Numerical Simulation for Hydraulic Fracturing Horizontal Fracture. *Energies* **2019**, *12*, 48. [CrossRef]
23. Simoni, L.; Secchi, S. Cohesive Fracture Mechanics for a Multi-phase Porous Medium. *Eng. Comput.* **2003**, *20*, 675–698. [CrossRef]
24. Lecampion, B. An Extended Finite Element Method for Hydraulic Fracture Problems. *Commun. Numer. Methods Eng.* **2009**, *25*, 121–133. [CrossRef]

25. Chen, Z.; Bunger, A.P.; Zhang, X.; Jeffrey, R.G. Cohesive Zone Finite Element-Based Modeling of Hydraulic Fractures. *Acta Mech. Solida Sin.* **2009**, *22*, 443–452. [CrossRef]
26. Sun, S.; Zhou, M.; Lu, W.; Davarpanah, A. Application of Symmetry Law in Numerical Modeling of Hydraulic Fracturing by Finite Element Method. *Symmetry* **2020**, *12*, 1122. [CrossRef]
27. Wan, B.; Liu, Y.; Zhang, B.; Luo, S.; Wei, L.; Li, L.; He, J. Investigation of the Vertical Propagation Pattern of the 3D Hydraulic Fracture under the Influence of Interlayer Heterogeneity. *Processes* **2022**, *10*, 2449. [CrossRef]
28. Zhao, Y.; Wang, L.; Ma, K.; Zhang, F. Numerical Simulation of Hydraulic Fracturing and Penetration Law in Continental Shale Reservoirs. *Processes* **2022**, *10*, 2364. [CrossRef]
29. Behnia, M.; Goshtasbi, K.; Zhang, G.; Mirzeinaly Yazdi, S.H. Numerical Modeling of Hydraulic Fracture Propagation and Reorientation. *Eur. J. Environ. Civ. Eng.* **2015**, *19*, 152–167. [CrossRef]
30. Dahi Taleghani, A.; Gonzalez-Chavez, M.; Yu, H.; Asala, H. Numerical Simulation of Hydraulic Fracture Propagation in Naturally Fractured Formations Using the Cohesive Zone Model. *J. Pet. Sci. Eng.* **2018**, *165*, 42–57. [CrossRef]
31. Li, J.; Dong, S.; Hua, W.; Li, X.; Pan, X. Numerical Investigation of Hydraulic Fracture Propagation Based on Cohesive Zone Model in Naturally Fractured Formations. *Processes* **2019**, *7*, 28. [CrossRef]
32. Putra, R.U.; Basri, H.; Prakoso, A.T.; Chandra, H.; Ammarullah, M.I.; Akbar, I.; Syahrom, A.; Kamarul, T. Level of Activity Changes Increases the Fatigue Life of the Porous Magnesium Scaffold, as Observed in Dynamic Immersion Tests, over Time. *Sustainability* **2023**, *15*, 823. [CrossRef]
33. Tauviqirrahman, M.; Jamari, J.; Susilowati, S.; Pujiastuti, C.; Setiyana, B.; Pasaribu, A.H.; Ammarullah, M.I. Performance Comparison of Newtonian and Non-Newtonian Fluid on a Heterogeneous Slip/No-Slip Journal Bearing System Based on CFD-FSI Method. *Fluids* **2022**, *7*, 225. [CrossRef]
34. Economides, M.J.; Nolte, K.G. *Reservoir Stimulation*; Prentice Hall: Englewood Cliffs, NJ, USA, 1989; Volume 2.
35. Peirce, A.; Detournay, E. An Implicit Level Set Method for Modeling Hydraulically Driven Fractures. *Comput. Methods Appl. Mech. Eng.* **2008**, *197*, 2858–2885. [CrossRef]
36. Neves, L.F.R.; Campilho, R.D.S.G.; Sánchez-Arce, I.J.; Madani, K.; Prakash, C. Numerical Modelling and Validation of Mixed-Mode Fracture Tests to Adhesive Joints Using J-Integral Concepts. *Processes* **2022**, *10*, 2730. [CrossRef]
37. Tomar, V.; Zhai, J.; Zhou, M. Bounds for Element Size in a Variable Stiffness Cohesive Finite Element Model. *Int. J. Numer. Methods Eng.* **2004**, *61*, 1894–1920. [CrossRef]
38. Campilho, R.D.S.G.; de Moura, M.F.S.F.; Domingues, J.J.M.S. Modelling Single and Double-Lap Repairs on Composite Materials. *Compos. Sci. Technol.* **2005**, *65*, 1948–1958. [CrossRef]
39. Camanho, P.P.; Davila, C.G. *Mixed-Mode Decohesion Finite Elements for the Simulation of Delamination in Composite Materials*; No. NAS 1.15: 211737; NASA Center for AeroSpace Information: Hanover, MD, USA, 2002. Available online: https://ntrs.nasa.gov/citations/20020053651 (accessed on 10 December 2022).
40. Benzeggagh, M.L.; Kenane, M. Measurement of Mixed-Mode Delamination Fracture Toughness of Unidirectional Glass/Epoxy Composites with Mixed-Mode Bending Apparatus. *Compos. Sci. Technol.* **1996**, *56*, 439–449. [CrossRef]
41. Yang, K.; Gao, D. Numerical Simulation of Hydraulic Fracturing Process with Consideration of Fluid–Solid Interaction in Shale Rock. *J. Nat. Gas Sci. Eng.* **2022**, *102*, 104580. [CrossRef]
42. Dong, R.; Wheeler, M.F.; Ma, K.; Su, H. A 3D Acid Transport Model for Acid Fracturing Treatments With Viscous Fingering. In *Proceedings of the SPE Annual Technical Conference and Exhibition, Virtual, TX, USA, October 2020*; OnePetro: Richardson, TX, USA, 2020.
43. Ren, L.; Lin, R.; Zhao, J.; Yang, K.; Hu, Y.; Wang, X. Simultaneous Hydraulic Fracturing of Ultra-Low Permeability Sandstone Reservoirs in China: Mechanism and Its Field Test. *J. Cent. South Univ.* **2015**, *22*, 1427–1436. [CrossRef]
44. Wu, J.; Li, X.; Wang, Y. Insight into the Effect of Natural Fracture Density in a Shale Reservoir on Hydraulic Fracture Propagation: Physical Model Testing. *Energies* **2023**, *16*, 612. [CrossRef]

processes

MDPI

Article

Study on Salt Dissolution Law of High Salinity Reservoir and Its Influence on Fracturing

Liyan Pan [1], Lei Wang [2,3,*], Weijie Zheng [1], Feipeng Han [1], Ariya Zibibula [1], Zhenlong Zhu [2,3] and Shengxiang Li [2,3]

1 Engineering and Technology Research Institute, Xinjiang Oilfield Company, CNPC, Karamay 834000, China
2 School of Petroleum Engineering, Yangtze University, Wuhan 430100, China
3 Hubei Key Laboratory of Oil and Gas Drilling and Production Engineering, Wuhan 430100, China
* Correspondence: wang-lei@yangtzeu.edu.cn; Tel.: +86-189-7116-8785

Abstract: For the high-salt reservoir of the Fengcheng Formation in the Mahu area, the production decreases rapidly due to the conductivity decrease after fracturing. The analysis shows that this has a great relationship with the special salt dissolution characteristics of the High salinity reservoir. In order to study the problem of salt dissolution pattern, the effect of different temperatures, the salt concentration of fracturing fluid, the viscosity of fracturing fluid, and injection rate on the rate of salt dissolution was evaluated by using the dynamic experimental evaluation method of salt dissolution. Through the grey correlation analysis of salt rock dissolution rate, it can be found that the degree of influence is from large to small which the influence of temperature is greater than fracturing fluid velocity, followed by fracturing fluid viscosity and, finally, fracturing fluid salt concentration. The results of compressive strength tests on salt-bearing rocks after dissolution show that the compressive strength is greatly reduced after salt dissolution by more than 60%. At the same time, the test results of proppant-free conductivity showed that the conductivity increased first and then decreased sharply after salt dissolution. This shows that in the early stage of salt dissolution, the flow channel will increase through dissolution. The rock strength decreases greatly with the increase of salt dissolution. As a result, collapse leads to a sharp reduction in the facture conductivity. Therefore, it is necessary to choose saturated brine fracturing fluid. In the proppant conductivity experiments, by optimizing the use of saturated brine fracturing fluid with 30/50 mesh or 20/40 mesh ceramic proppant with a sand concentration of 5 Kg/m^2, a high facture conductivity can be achieved under high closure pressure conditions. Based on the above study, directions and countermeasures for improving high saline reservoirs are proposed, which point the way to improve the fracturing conductivity.

Keywords: high salinity reservoir; influencing factors of salt dissolution; rock strength evaluation; fracture conductivity; fracturing process improvement direction

Citation: Pan, L.; Wang, L.; Zheng, W.; Han, F.; Zibibula, A.; Zhu, Z.; Li, S. Study on Salt Dissolution Law of High Salinity Reservoir and Its Influence on Fracturing. *Processes* **2023**, *11*, 304. https://doi.org/10.3390/pr11020304

Academic Editor: Youguo Yan

Received: 8 October 2022
Revised: 11 January 2023
Accepted: 13 January 2023
Published: 17 January 2023

1. Introduction

Junggar basin is an important petroliferous basin in China. Mahu sag is a secondary tectonic unit in the basin. The sedimentary environment of the Fengcheng Formation in Mahu Sag is an alkaline lake environment [1]. Compared with the freshwater environment, it has the characteristics of strong cementation and early hydrocarbon generation. Therefore, the diagenetic evolution of fine-grained sedimentary rocks in salt lakes is more complicated. The Fengcheng Formation in Mahu Sag is a set of typical ancient alkaline lake sediments with various rock types. These include terrigenous sedimentary sandstone (conglomerate), siltstone, mudstone, and saline rocks formed by alkali lake deposition. The lithological characteristics of different depths represented by the central area of the alkaline lake are controlled by salinity vertically. P1f1 layer: volcanic rocks, dolomitic argillaceous siltstone, alkali minerals, low salinity; p1f2 layer: the edge area is dominated by cloud mudstone and cloudy siltstone, and the salt rock strata in the center of the alkali lake appear in large quantities with high salinity; p1f3 layer: dolomitic rocks, limestone, etc. alkali minerals gradually decreased to disappear. The main salt layer is the P1f2 layer [2]. The development of salt minerals in this area indicates that the climate was dry and hot during the deposition of dolomitic rocks in the Fengcheng Formation of Mahu Sag. As a result, warm-phase alkaline minerals and wide-temperature-phase salt minerals were precipitated [3].

In the early stage, the Fengcheng Formation of Mahu Sag was fractured for 11 wells. Fracturing mainly uses a reverse mixed fracturing process. The implementation wells generally have high oil saturation, and the initial effect of fracturing is good so that industrial oil flow can be obtained. The initial production capacity is 15~20 t/d, but the stable production period is generally within 1 month. The later production decreased rapidly due to the decrease in fracture conductivity. Through analysis, it is believed that this has a great relationship with the special lithology of the High salinity reservoir. Studies have shown that one of the important characteristics of high salinity reservoirs is the high content of soluble salts (mainly sodium bicarbonate, sodium silicate, calcium carbonate, etc.). In the process of stimulation measures, since the fracturing fluid is a water-based system, salt will be dissolved in large quantities. This excessive salt dissolution will cause a decrease in strength and eventually lead to rock collapse. At the same time, the conductivity decreases and directly leads to a sharp decrease in production. Therefore, it is of great significance for the high salinity reservoir to study the influence factors of salt dissolution and take measures to maintain effective conductivity [4].

Water dissolution is a basic characteristic of rock salt. From a physicochemical point of view, the contact between salt rock and water leads to two opposite effects at the solid-liquid interface and in an aqueous solution: the dissolution of rock salt and the crystallization of liquid solution [5–7]. From the point of view of chemical kinetics, the dissolution process of salt rock can be regarded as the heterogeneous reaction at the interface between salt and water (i.e., solid-liquid phase interface). The reaction includes three basic processes: water entering the surface of rock salt, the interaction between water and rock salt, and the diffusion of dissolved rock salt into water [8,9]. In recent years, scholars in China and abroad have carried out a series of related research on the dissolution characteristics of salt rocks. Durie [10] studied the surface characteristics of dissolution surface during salt dissolution. He found that the inhomogeneity of rock composition caused the uneven dissolution of salt corrosion surfaces. The dissolution area increased, thereby increasing the dissolution rate. Saberian [11] analyzed the influence of flow rate, temperature, and other factors on the dissolution rate of salt rock through a large number of experiments. Hans Ulrich Rohr [12] studied the effect of salt composition on the dissolution rate through a large number of basic experiments and gave the dissolution rate of different kinds of salt. The results showed that in the multi-component brine system dominated by NaCl, the dissolution rate of most salts decreased with the increase of NaCl concentration. Manvan Alkattan [13] found through experiments that Co, Cr, Cd, and Pb metals could reduce the dissolution rate of rock salt, while Fe and Zn metals had little effect on the dissolution rate. Xiao Changfu [14] et al. obtained the change rule of the dissolution rate of salt rock along with the main influencing factors and the corresponding calculation formula through the experimental study on dissolution characteristics of salt rock. Liang Weiguo [15], Wang Chunrong [16], Tang Kang [17], Li Xiaokang [18], and Zhang Zhandong [19] et al. studied the influence of different temperatures, pressure, concentration, and flow velocity on the dissolution rate of salt rock through the salt rock dissolution test.

Although the above researchers have completed a series of research related to the dissolution characteristics of salt rock. However, most of them considered the dissolution characteristics of salt rock in the process of salt cavitating. Researchers have not studied the law of salt dissolution in the process of fracturing. Taking the high salinity reservoir in Mahu Sag, Junggar Basin as an example, this paper studies the salinity dissolution of the highly saline reservoir and its influence on rock strength and conductivity through a salt dissolution experiment, compressive experiment, and conductivity experiment.

In order to make the experimental data more representative, the P_1f_2 layer with the highest salt content in the typical well XY1 in this block was selected to carry out a whole-rock mineral analysis. As can be seen from the whole-rock mineral analysis results (Table 1), it is generally characterized by "low clay minerals, low quartz, low feldspar, and high salt". The salt-containing types are mainly sodium bicarbonate (easily soluble in water) and sodium silicate boron-type (insoluble in water), and the overall salt content of the core is between 62.9 and 100%, of which the soluble salt content is 0–100%. In order to clarify the law of salt dissolution and its influence on the change of rock strength and flow conductivity, experiments on influencing factors of salt dissolution, laboratory of mechanical strength changes before and after salt dissolution, and experiments on the influence of salt dissolution on the change of flow conductivity were carried out. In the experiments, we selected the core with a buried depth of 5,381.5 m and soluble salt content of 78% to carry out salt dissolution experiments. In the conductivity experiment, an 80% salt plate was made.

Table 1. Mineral analysis results.

Depth of Sample (m)	Mineral Content (%)								
	Quartz	Potassium Feldspar	Plagioclase	Sodium Silicate Boronite	Carbonite	Sodium Hydrocarbon Stone	Pyrite	Amphibole	Clay
5372.0	7.6	6.1	15.1	55	7.9	/	3.8	2	2.5
5375.5	/	/	/	22	/	78	/	/	/
5378.0	2.8	2.9	7.1	81.1	/	/	/	2.8	3.3
5380.0	/	/	/	86.6	/	13.4	/	/	/
5381.0	/	/	/	73.8	/	26.2	/	/	/
5381.5	/	/	/	/	/	100	/	/	/

2. Mechanism and Influencing Factors of Fracturing Fluid Salt Dissolution

2.1. Influencing Factors of Salt Dissolution Law in the Fracturing Process

In this paper, a series of dissolution experiments of salt-containing rock samples were carried out by laboratory experiments. The most important factors affecting the dissolution rate of rock salt were selected for the experiment: formation temperature, the salinity of fracturing fluid water, the viscosity of fracturing fluid, and the pumping rate. The influence of these factors on the dissolution rate of rock salt will be observed in the experiment, and the research results will be helpful in guiding the optimization of fracturing technology.

This reservoir has a wide range of salt content values. In order to make the experimental results more representative, the core used in the salt dissolution experiment in this paper is taken from the main P1f2 layer of well XY1. The content of sodium bicarbonate in this section is 78% at 5375.5.5 m. The salt dissolution characteristics of this kind of reservoir were studied by using a core with high salt content.

Before the experiment, the core was processed into a standard cylindrical specimen with a diameter of 25 mm and a length of 20 mm. The cutting process ensures that no water is touched. The processing of rock samples is carried out in strict accordance with the specifications. The processed rock sample is shown in Figure 1. During the experiment, both the upper surface and the side of the sample were sealed to keep only the lower surface of the sample in contact with the liquid.

The experiment was carried out in SD-II reactive mechanics experimental facility, Figure 2. During the experiment, the liquid was first added to the set temperature and pressure, and then the core was put into the solution for reaction. After the reaction for 30 min, the core was taken out for drying and weighing. The mass change of the rock sample before and after the reaction was measured, and the dissolution rate was represented by the mass change of the rock sample per unit area within a unit of time. See Equation (1).

$$\text{dissolution rate}: \ v = \frac{\Delta m}{At} \tag{1}$$

Δm—the quality difference of rock samples before and after the experiment, g;
A—the dissolved area, cm^2;
t—the experiment time, min;
v—the dissolving rate, $g/(cm^2 \cdot min)$.

The experimental design refers to the actual reservoir environment and fracturing parameters. The formation temperature is 110 °C, the formation pressure is 40 Mpa, and the injecting rate is 4–8 m^3/min. Four factors were considered during the experiment. The three experimental temperatures were 70, 90, and 110 °C, respectively. The three viscosity liquids were clear water with a viscosity of 1 mPas, low viscosity fracturing fluid with a viscosity of 5 mPa·s, and high viscosity fracturing fluid with a viscosity of 55 mPa·s. The three kinds of salinity fracturing fluid are 0% clear water with salt content, 5% salt content, and 16.19% salt water with salt content. The injection rates of the three kinds are 4 m^3/min, 6 m^3/min, and 8 m^3/min, respectively. The above experiments were carried out in 12 groups, It is shown in Table 2, the experimental results were analyzed, and the primary and secondary factors of salt dissolution were identified.

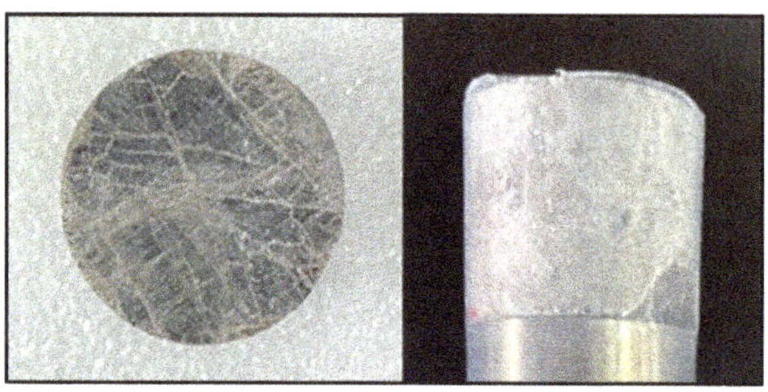

Figure 1. Rock sample for the experiment.

Figure 2. SD-II-type reaction kinetics experimental setup.

Table 2. Summary of the influences of various factors on dissolution rate.

Factor of Experiment	Experimental Serial Number	Temperature of Solution/°C	Concentration of Solution/%	Viscosity of Solution/mPa·s	Flow Rate of Solution/(m^3/min)	Corresponding Speed of Rotation (r/min)
Reservoir temperature	1	70	5	15	6	153
	2	90	5	15	6	153
	3	110	5	15	6	153
Salt concentration in fracturing fluid	4	110	0	15	6	153
	5	110	5	15	6	153
	6	110	16.19 (saturated)	15	6	153
Fracturing fluid viscosity	7	110	5	1	6	153
	8	110	5	15	6	153
	9	110	5	55	6	153
Fracturing fluid flow rate	10	110	5	15	4	102
	11	110	5	15	6	153
	12	110	5	15	8	204

2.1.1. Influence of Temperature on Dissolution Rate of Salt Rock

The present experimental plan was to vary the solution temperature. The change in dissolved mass of salt rock samples at different temperatures and the effect on dissolution rate were studied. The solutions were low-viscosity fracturing fluids. The solution temperatures were 70, 90, and 110 °C. Firstly, the treated cores were put into the solution. Then the pressure and speed were adjusted to reach the set value. After dissolving for 30 min, the rock samples were weighed and recorded, and the data were processed, as shown in Figure 3. From the experimental results, it can be seen that the dissolution rate of the rock samples all increased with the increase in temperature. This is because as the temperature increases, the molecular activity in the solution becomes stronger. This leads to an increase in the chance of collision between the solution molecules and the salt rock molecules, which leads to an increase in the dissolution rate. Finally, the data were fitted, and the results of the fit showed an exponential relationship between the dissolution rate and the solution temperature.

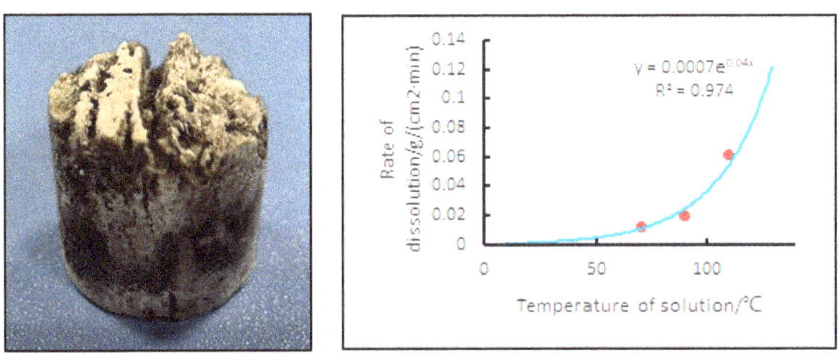

Figure 3. Dissolution of the core at different solution temperatures. (**Left**) salt-dissolved morphology of fracturing fluid after salt dissolution at 90 °C; (**Right**) Influence curve of dissolution rate at different temperatures.

2.1.2. Influence of Fracturing Fluid with Different Salinity on Dissolution Rate of Salt Rock

This group of experiments studied the effect of different salinity fracturing fluids on the dissolution rate of rock samples. The solution was 1 L of 0% salt water, 5% sodium bicarbonate, and 16.19% sodium bicarbonate low-viscosity fracturing fluid. The treated cores were placed in the three solutions. Then adjust the temperature, pressure, and speed to the set value. After 30 min of dissolution, the rock samples were taken out, weighed, recorded, and processed, as shown in Figure 4. Figure 4 shows that the dissolution rate gradually decreases with the increase in concentration, and the salt rock specimen is no longer dissolved in a saturated solution. From the point of view of chemical kinetics, the difference between the concentration of the solution and the concentration of the saturated solution is one of the chemical potentials for the dissolution reaction of salt rock. The greater the difference between the two, the greater the dissolution rate. When the solution concentration is 0%, the difference is the largest, and the rock salt dissolution rate reaches the maximum. When the solution concentration is equal to the solution saturation concentration, the dissolution rate of rock salt is zero. The data were fitted, and the fitting results showed that the dissolution rate was linear with the solution concentration.

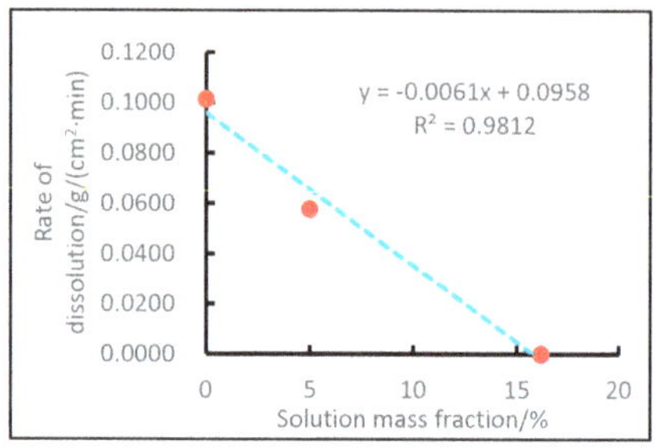

Figure 4. Dissolution of cores in solutions of different concentrations. (**Left**) salt–dissolved morphology of fracturing fluids containing 5% sodium bicarbonate; (**Right**) Influence curve of solution with a different salt concentration on dissolution rate.

2.1.3. Effect of Fracturing Fluid Viscosity on Dissolution Rate

In this experiment, by changing the viscosity of the solution, the dissolution quality of salt rock samples under different viscosity and its influence on the dissolution rate were studied. Clean water, low viscosity fracturing fluid, and high viscosity fracturing fluid were selected for 1 L each (the three liquids were 55 mpa·s high viscosity fracturing fluid, 15 mpa·s low viscosity fracturing fluid, and 1 mpa·s clear water, respectively). The treated core was put into the solution, and the temperature, pressure, and rotational speed were adjusted to reach the set values. Figure 5. It can be seen from Figure 5 that the dissolution rate of rock samples decreases with the increase of viscosity. This is because the higher the viscosity of the solution, the greater the resistance to the movement of molecules in the solution, the more difficult the diffusion of solute molecules in the solution, and the lower the diffusion rate of solute to the surface of the rock samples and the lower the dissolution rate of rock samples.

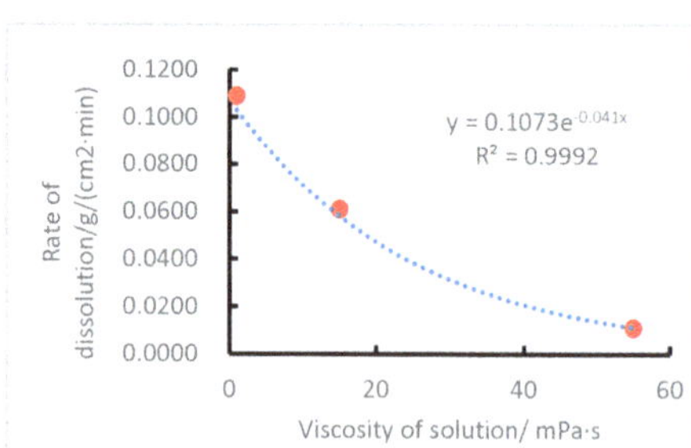

Figure 5. Dissolution of rock samples in solutions of different viscosities within 30 min. (**Left**) salt-dissolved morphology of low-viscosity fracturing fluid; (**Right**) Influence curve of solution with a different salt concentration on dissolution rate.

2.1.4. Effect of Fracturing Fluid Velocity on Dissolution Rate

In this experiment, the dissolution of salt rock under the simulated dynamic condition under a certain flow state was studied, and the change of dissolution mass of salt rock under different flow rates was studied. In this experiment, based on the principle of the same linear velocity, the salt dissolution characteristics of 121, 153, and 204 r/min were adopted when the flow rate was 4, 6, and 8 m^3/min, respectively. Put the treated core into the solution, adjust the temperature and pressure to reach the set value, take out the rock sample weighing record and process the data after 30 min of dissolution. Figure 6 shows that within a certain flow velocity range, the dissolution rate of salt rock increases with the increase in flow velocity. The motion of the solution accelerates the convection and diffusion of the salt solution and the salt particles in the solution, which can accelerate the dissolution rate of minerals to a certain extent. The dissolution rate of minerals will be different under different flow rates. The data were fitted, and the fitting results showed that there was an exponential relationship between the dissolution rate and the velocity of the solution.

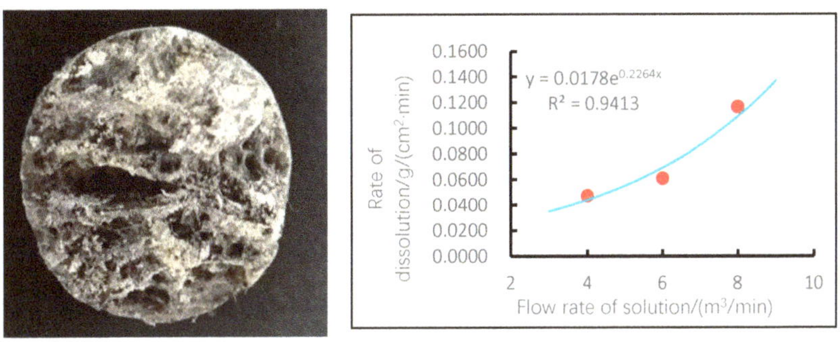

Figure 6. Dissolution of rock samples in solutions of different flow rates in 30 min. (**Left**) salt-dissolved morphology of 6 m^3/min flow rate; (**Right**) Influence curve of solution with different flow rate on dissolution rate.

2.2. Analysis of Primary and Secondary Factors Affecting Salt Dissolution

Grey correlation analysis is a method to measure the degree of correlation between factors according to the degree of similarity or dissimilarity of development trends between factors. Therefore, grey correlation analysis is used to analyze the correlation degree of each influencing factor so as to find out the main factors affecting the dissolution rate of salt rock. The main idea is as follows: Firstly, the dissolution rate is determined as the reference sequence, and the sub-factors (the above-influencing factors) for the correlation degree compared with the reference sequence are taken as the comparison sequence, as shown in Table 3. Before the analysis, it is necessary to average the data so as to find out the grey correlation value between the solving parent sequence and the feature sequence and then calculate the correlation degree between them. The greater the value, the closer the relationship between the sub-factor and the parent factor.

Through the grey correlation analysis of salt rock dissolution rate, it can be found that each influencing factor restricts the other, but the degree of influence is very different. The degree of influence is arranged from large to small: solution temperature > solution flow rate > solution viscosity > solution concentration, as shown in Table 4. Therefore, in the actual fracturing process, how to reduce the solution temperature, using low displacement injection, and using brine high viscosity fracturing fluid are the main directions to reduce the impact of salt dissolution.

Table 3. Summary of the influences of various factors on dissolution rate.

Factor of Experiment	Experimental Serial Number	The Temperature/°C	Solution Salt Concentration/%	Viscosity of Solution/mPa·s	Flow Rate of Solution/(m^3/min)	Rate of Dissolution/g/(cm^2·min)
Temperature	1	70	5	15	6	0.0123
	2	90	5	15	6	0.0199
	3	110	5	15	6	0.0609
fracturing fluid with different salinity	4	110	0	15	6	0.1089
	5	110	5	15	6	0.0654
	6	110	16.19	15	6	0.0078
fracturing fluid viscosity	7	110	5	1	6	0.1089
	8	110	5	15	6	0.0609
	9	110	5	55	6	0.011
fracturing fluid velocity	10	110	5	15	4	0.0471
	11	110	5	15	6	0.0609
	12	100	5	15	8	0.1165

Table 4. Results of correlation degree.

Item of Evaluation	Degree of Correlation	Ranking
Temperature/°C	0.8472	1
Salinity of fracturing fluid/%	0.7415	4
Viscosity of fracturing fluid/mPa·s	0.7546	3
Flow rate of fracturing fluid/(m^3/min)	0.8345	2

3. Effect of Salt Dissolution on the Strength of Rocks

3.1. Simulation of Dynamic Salt Solution Process

The salt dissolution process was simulated by core displacement. First, the salt-containing rock core with a buried depth of 5375.5.5 m and sodium bicarbonate content of 78% was still selected. The core size was φ25 × 50 mm. Water was used to simulate the dynamic dissolution process. The injection rate was 0.1 mL/min, and the temperature was 110·°C. The pressure changes were then observed during the injection process. When the pressure dropped sharply at 175 min, the dissolution passed through the core, indicating the end of the dynamic dissolution process (Figure 7). After the core was removed, it was found that there were dissolved pores through the front and back (Figure 8).

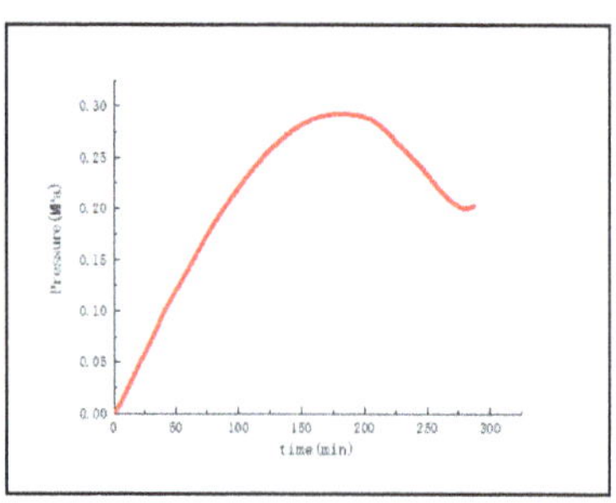

Figure 7. Dynamic dissolution curve of the core.

Figure 8. Morphology of the core after dissolution.

3.2. Change of Compressive Strength before and after Salt Solution

Uniaxial compression tests were carried out on cores before and after dissolution in the same layer. The experimental results show that the uniaxial compressive strength decreased by 65% from 53.6 MPa to 18.6 MPa (Figure 9). It shows that salt dissolution will greatly reduce the strength of the core and have a great impact on the maintenance of fracture conductivity.

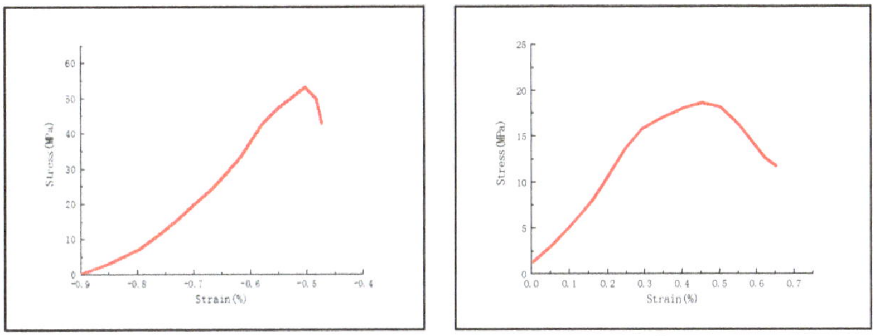

(a) (b)

Figure 9. Uniaxial compressive strength test before and after salt solution. (**a**) Uniaxial compressive strength before salt solution; (**b**) Uniaxial compressive strength after salt solution.

4. Effect of Salt Dissolution on the Change of Fracture Conductivity

The fracture conductivity largely reflects the effectiveness of hydraulic fracturing. As the fracture conductivity increases, the hydraulic fracturing effect is better, and the production of fractured wells is better. In view of the prominent characteristics of salt dissolution in high salinity reservoirs, it is necessary to study the trend of change of fracture conductivity under salt dissolution [20,21].

4.1. Test Pieces and Equipment

Due to the small number of cores in the Fengcheng Formation of Mahu, it is difficult to obtain natural cores to make rock slabs. Therefore, this experiment refers to the mineral analysis results of well XY1 at 5375.5.5 m and makes artificial rock with a sodium bicarbonate content of 78%. The rock slab with the required size of API conductivity was cut and processed by emery wire cutting, as shown in Figure 10.

The HXDL-2C proppant conductivity system device was used in the fracture conductivity test. The experimental instrument mainly includes a closure pressure system, API flow chamber, displacement monitoring and measurement system, back pressure control, measurement system, and data acquisition and processing system, as shown in Figure 11. The instrument can be used to evaluate the performance of different proppant fracture conductivity by simulating downhole closure pressure, temperature, and other reservoir-related conditions under standard experimental conditions. The main parameters are the maximum temperature of 200 °C, the maximum closure pressure of 120 MPa, and the fluid flow rate in the range of 0–20 mL/min.

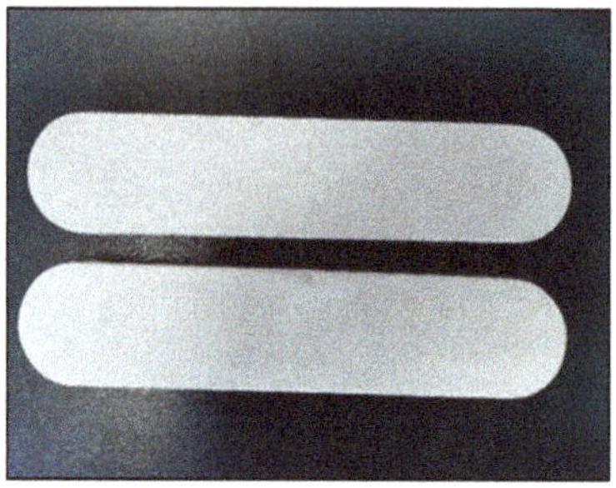

Figure 10. Artificial high salinity slab.

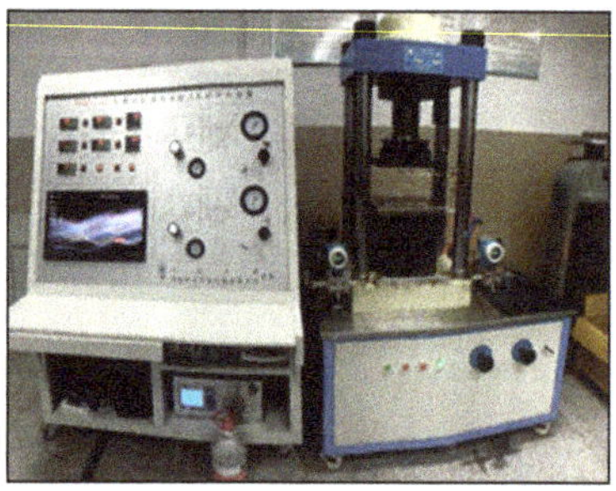

Figure 11. HXDL-2 system for conductivity.

4.2. The Experimental Method

In this conductivity experiment, the influence of salt dissolution on the change of fracture conductivity under no proppant condition was first considered. And the suitable fluid type was selected through the experiment. Secondly, experiments were carried out under different proppant particle sizes and proppant placement concentrations. 8 groups were carried out in the experiment, as shown in Table 5.

Table 5. Experimental design.

Group	Proppant	Ceramsite Particle Size (mesh)	Ceramsite Concentration (Kg/m^2)	Temperature (°C)	Solution Medium
1	/	/	/	110	Saturated sodium bicarbonate fracturing fluid
2	/	/	/	110	Clear water fracturing fluid
3	Yes	20/40	5	110	Saturated sodium bicarbonate fracturing fluid
4		30/50			
5		40/70			
6	Yes	30/50	2.5	110	Saturated sodium bicarbonate fracturing fluid
7			5		
8			10		

The corresponding fracture conductivity can be calculated by measuring the flow rate and pressure difference during the experiment and combining them with the fluid viscosity.

$$k \cdot W_f = \frac{5.555 \mu Q}{\Delta P} \quad (2)$$

$k \cdot W_f$—propped fracture conductivity, D·cm;
Q—the flow rate through the fracture, cm^3/s;
μ—fluid viscosity, mpa·s;
ΔP—pressure difference between the two ends of the propped fracture, kPa.

4.3. Analysis of Experimental Results

Firstly, the experiments with saturated sodium bicarbonate fracturing fluid without proppant were conducted. At the beginning of the experiment, the fracture width of the rock slab was small due to the lack of proppant. The flow path was the gap generated by the rough surface of the upper and lower rock slabs, and the measured initial fracture conductivity was 16.12 cm·D. As shown in Figure 12, when the solution passed through the rock slab, the conductivity decreased rapidly and dropped to 1.54 cm·D at 30 min. Moreover, there was almost no change in the subsequent measurement time. By weighing the rock slab after this group of diversion experiments, it was found that the weight loss of the rock slab was very small. This result indicates that salt dissolution can be neglected. As can be seen from the picture of the rock slab after the experiment, the surface of the rock slab is smooth and flat, and there is no sign of dissolution. The saturated sodium bicarbonate fracturing fluid is closer to the change law of conductivity of conventional rock in the process of conductivity test. Under a certain closing pressure, the conductivity decreases rapidly in the initial stage and gradually reaches a stable value in the later stage.

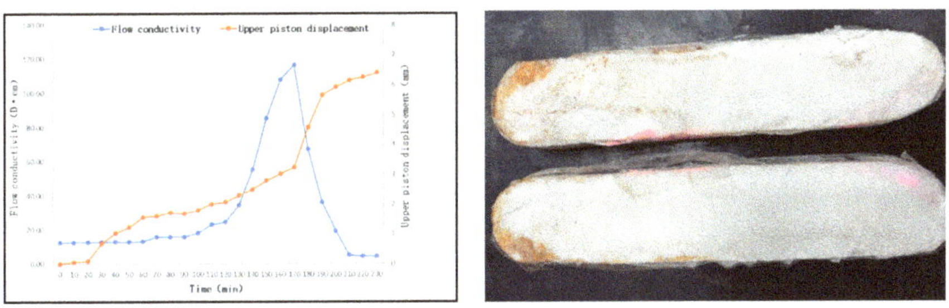

Figure 12. Variation curve of clearwater fracturing fluid conductivity and dissolution pattern of the rock slab.

Through the conductivity experiment of clean water fracturing fluid without proppant, it was found that the initial conductivity measured at the beginning of the experiment was 12.48 cm D. After that, in the process of the solution passing through the rock slab, the conductivity began to increase gradually due to the salt dissolution of the solution to the rock slab. The flow channel is also gradually increasing, and the conductivity is rising faster and faster, up to 116.39 cm D. When the experiment was carried out for 170 min, the flow channel generated by the salt solution collapsed under the action of long-term closure pressure. The fracture closes quickly and gradually blocks the rock slab gap. At the same time, the conductivity decreased sharply and finally decreased to 2.52 cm D. Subsequent observation of the rock slab after the conductivity experiment (Figure 13), the surface collapse corrosion was significant. After weighing, it was found that the weight loss of the rock slab was 124.07 g, and the salt solution reduction effect was significant. Summarizing the above two sets of experimental results, combined with the previous salt solution experiment and compression experiment, the final conductivity of the water fracturing fluid is slightly higher. However, considering reservoir stability and conductivity, saturated brine fracturing fluid is more suitable for High salinity reservoirs.

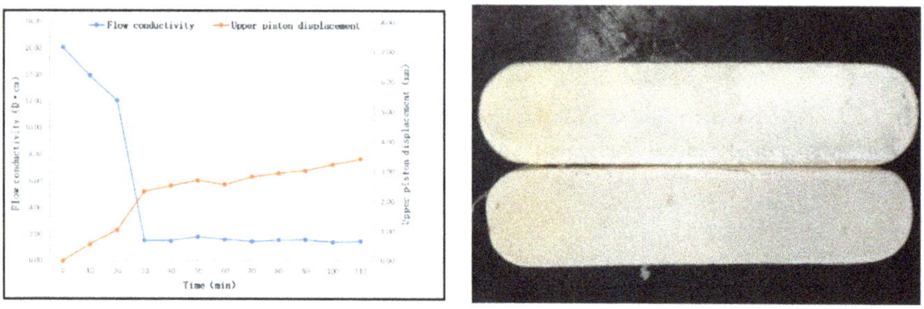

Figure 13. Variation curve of saline fracturing fluid conductivity and dissolution pattern of the rock slab.

The conductivity experiments of different particle sizes of proppant were carried out using saturated saline fracturing fluid and Ceramic proppant with a sand concentration of 2.5–10 kg/m^2 and a closing pressure of 10–70 MPa. As shown in Figure 14, it can be seen that the larger the particle size of the proppant, the higher the fracture conductivity. The larger the particle size of the proppant, the smaller the embedding effect; so that the larger the effective fracture width can be provided, the higher the fracture conductivity.

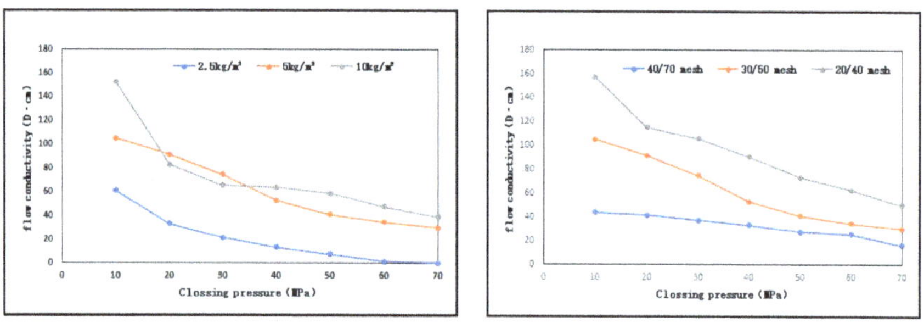

Figure 14. Proppant fracture conductivity curve under different particle sizes and sand concentration of saturated saline fracturing fluid.

Using saturated saline fracturing fluid and 30/50 Ceramic proppant with a closing pressure design of 10–70 MPa, three groups of different sand concentrations of 10 kg/m^2, 5 kg/m^2, and 2.5 kg/m^2 were carried out, respectively. With the increase of proppant concentration, the conductivity changes significantly. Final fracture conductivity size 10 kg/m^2 greater than 5 kg/m^2 greater than 2.5 kg/m^2. At high closing pressure of 2.5 kg/m^2 proppant, it is embedded in the rock slab part more, which can hardly provide fracture width that fracture conductivity is very low. There is little difference between 10 kg/m^2 and 5 kg/m^2 fracture conductivity.

5. Conclusions

(1) The high-salinity reservoir of Mahu sag is a special reservoir with a salt content of 62.9–100%, among which sodium bicarbonate is easily soluble in water. It has a significant impact on the development of this type of reservoir.

(2) In the fracturing process, due to the salt solubility, the strength of the rock and the maintenance of the fracture width will have a great impact, so how to control the salt solution is very important to this type of reservoir.

(3) For fracturing, the use of saturated brine to configure high-viscosity fracturing fluid, higher sand concentration, and increasing proppant particle size the above methods are very necessary for the fracturing of high salt reservoirs.

Author Contributions: Conceptualization, L.P. and L.W.; methodology, L.P. and L.W.; software, L.P.; validation, L.P. and W.Z.; formal analysis, L.P. and L.W.; investigation, L.P.; resources, L.W.; data curation, A.Z. and F.H.; writing—original draft preparation, L.W. and L.P.; writing—review and editing, Z.Z. and S.L.; visualization, L.W.; supervision, L.W.; project administration, L.W.; funding acquisition, L.W. and L.P. All authors have read and agreed to the published version of the manuscript.

Funding: This study is supported by the major project "Policy Research on Effective Development of Salt Shale Oil" funded by the Ministry of Science and Technology of the People's Republic of China, and the approval number is 2017ZX05049-005-010.

Data Availability Statement: Not applicable.

Acknowledgments: Thanks to other authors for his support of the experimental methods, experimental materials and amendments to the content of the article.

Conflicts of Interest: The authors declare no conflict of interest.

References

1. Lei, D.; Tang, Y.; Chang, Q. *Genetic Types and Main Controlling Factors of Permian Reservoirs in the Northwestern Margin of Junggar Basin*; Petroleum Industry Press: Beijing, China, 2013.
2. Zhengxiang, L.; Zheyuan, L.; Yuefeng, L. Diagenesis of alkaline lake dolomitic reservoirs in the Permian Fengcheng Formation in Mahu Sag. *Lithol. Reserv.* **2022**, *34*, 26–37.
3. Zhang, X.; Zheng, M. Research progress of salt minerals in the Qinghai-Tibet Plateau. *Sci. Technol. Rev.* **2017**, *35*, 72–76.
4. Gao, T. Microscopic Pore Structure Characteristics and Seepage Mechanism of Inter-Salt Shale Oil Reservoirs. Master's Thesis, University of Chinese Academy of Sciences, Beijing, China, 2019.
5. Liang, W. Research on Multi-Field Coupling Theory and Application of Hydraulic Fracturing and Water-Soluble Mining of Salt Deposits. Ph.D. Thesis, Taiyuan University of Technology, Taiyuan, China, 2004.
6. Liang, W. Research on laboratory test of water-soluble mining of salt mines. *J. Liaoning Univ. Eng. Technol.* **2003**, *22*, 54–57.
7. Zhao, Z. Research on the Theory of Fluid Transport in the Water-Soluble Cavity of the Salt Rock Gas Storage and the Change Law of the Cavity Shape. Ph.D. Thesis, Chinese Academy of Sciences, Beijing, China, 2003.
8. Wallner, M. *Rock Mechanics in Petroleum Engineering*; OnePetro: Richardson, TX, USA, 1994; pp. 761–768.
9. Langer, M. *Rock Mechanics in Petroleum Engineering*; OnePetro: Richardson, TX, USA, 1994; pp. 697–703.
10. Durie, R.W.; Jessen, F.W. The influence of surface features in the salt dissolution process. *Soc. Pet. Eng. J.* **1964**, *4*, 275–281. [CrossRef]
11. Saberian, A. Cavity development in a three layer beddedsalt model. *Soc. Pet. Eng. J.* **1977**, *12*, 58–98.
12. Rohr, H.U. Rates of dissolution of salt minerals leaching caverns in salt—Fundamentals and practical application. *Fifth Int. Symp. Salt—North. Ohio Geol. Soc.* **1979**, *19*, 78–84.
13. Alkattan, M.; Oelkers, E.H.; Dandurand, J.-L.; Schott, J. Experimental studies of halite dissolution kinetics: II. The effect of the presence of aqueous trace anions and K,Fe(CN)6.L. *Chem. Geol.* **1997**, *143*, 17–26. [CrossRef]
14. Xiao, C.; Yang, Y.; Wu, G.; Li, C.; Qiu, X. Study on dissolution characteristics and mass transfer process of halite. *J. Chongqing Univ. Nat. Sci. Ed.* **1993**, *2*, 51–57. [CrossRef]
15. Liang, W.; Li, Z.; Zhao, Y. Study on laboratory test of water solution Mining of Salt Mine. *J. Liaoning Tech. Univ.* **2003**, *1*, 54–57.
16. Wang, C. Experimental Study on Influencing Factors of Dissolution Rate of Rock Salt. Master's Thesis, Chongqing University, Chongqing, China, 2009.
17. Tang, K. Research on Expansion and Stability of Water-Soluble Cavity in Horizontal Well. Master's Thesis, Chongqing University, Chongqing, China, 2017.
18. Li, X. Study on Flow Field Migration and Cavity Expansion Law of Water-Soluble Forming Cavity in Salt Rock Double Well. Master's Thesis, Chongqing University, Chongqing, China, 2018.
19. Zhang, Z. Research on Effectiveness Evaluation of Artificial Fractures in Intersalt Shale Oil Reservoir. Master's Thesis, China University of Petroleum, Beijing, China, 2018.
20. Chen, M. Experimental study on long-term conductivity of fractures supported by inter-salt non-sandstone formations. *J. Xi'an Shiyou Univ. Nat. Sci. Ed.* **2008**, *23*, 51–54.
21. Wu, L. Study on fracturing fluid suitable for inter-salt non-sandstone reservoir stimulation. *J. Jianghan Pet. Work. Univ.* **2008**, *21*, 44–51.

processes

Article

A New Model of Temperature Field Accounting for Acid–Rock Reaction in Acid Fracturing in Shunbei Oilfield

Jianye Mou [2,*], Jiayuan He [1], Haiqian Zheng [2], Rusheng Zhang [1], Lufeng Zhang [1,*] and Budong Gao [2]

1 Key Laboratory of Marine Oil & Gas Reservoirs Production, Sinopec, Petroleum Exploration and Production Research Institute, Beijing 100083, China
2 State Key Laboratory of Petroleum Resources and Engineering, China University of Petroleum, Beijing 102249, China
* Correspondence: moujianye@cup.edu.cn (J.M.); zlfcupb@163.com (L.Z.)

Abstract: The Shunbei oil formation is a deep, high-temperature carbonate reservoir. Acid fracturing is an effective technology to stimulate this formation. For acid fracturing, the temperature field is fundamental information for the acid system selection, acid–rock reaction, live acid penetration distance prediction, acid fracturing design, etc. Therefore, in this paper, we conduct a numerical study on the temperature field in acid fracturing to account for the acid–rock reaction in the Shunbei formation. Firstly, a new mathematical model of the fracture temperature field during acid fracturing is established based on the laws of mass and energy conservation and acid–rock reaction kinetics. The fracture model is based on a PKN model, which accounts for a few factors, such as the acid–rock reaction heat, acid–rock reaction rate dependence on the temperature, and the fracture width change with acid erosion. Then, the numerical mode is developed. Next, an extensive numerical study and a parameter analysis are conducted based on the model with the field data from the Shunbei formation. The study shows that the acid–rock reaction in acid fracturing has obvious effects on the temperature field, resulting in a 10~20 °C increase in the Shunbei formation. The acid–rock reaction dependence on temperature is a factor to be accounted for. The rock dissolution increases first and then decreases from the inlet to the tip of the fracture, unlike the monotonous decrease without temperature dependence. The temperature gradient is high near the inlet and then decreases gradually. Beyond half of the fracture, the temperature is close to the formation temperature. The temperature drops fast in the initial injection stage and tends to stabilize at about 50 min.

Keywords: temperature field; acid fracturing; acid–rock reaction; numerical simulation; Shunbei oilfield

Citation: Mou, J.; He, J.; Zheng, H.; Zhang, R.; Zhang, L.; Gao, B. A New Model of Temperature Field Accounting for Acid–Rock Reaction in Acid Fracturing in Shunbei Oilfield. *Processes* **2023**, *11*, 294. https://doi.org/10.3390/pr11010294

Academic Editor: Yidong Cai

Received: 30 December 2022
Revised: 9 January 2023
Accepted: 12 January 2023
Published: 16 January 2023

1. Introduction

Acid fracturing is primarily used in the stimulation of carbonate formations. The key to evaluating the effectiveness of acid fracturing is the effective acid fracture length and acid fracture conductivity. The effective acid fracture length depends on the live acid penetration distance. Among the parameters affecting the live acid penetration distance, the temperature is an important factor affecting the acid–rock reaction, rock dissolution, and acid viscosity and determining the effective acid action distance [1,2]. The Shunbei carbonate reservoir is characterized by a deep burial, high reservoir temperature, and high fracture pressure, compared with acid fracturing in other oilfields, and the high reservoir temperature and faster acid–rock reaction rate in the Shunbei oilfield result in a short effective acid penetration distance. Therefore, studying the temperature distribution within the fracture is essential to guide the design of the acid fracturing in deep formations under high-temperature and high-pressure conditions, which has far-reaching significance in selecting the optimal acid system and the treatment parameters [3–6].

The first model for calculating the fracture temperature field was proposed by Wheeler [7], who used an analytical solution to establish the relationship between the dimensionless temperature and the factors such as the fracture width, injection rate, injection time, leak-off rate, and fluid and formation properties for a fixed fracture width and fluid loss. After summarizing the Wheeler model, Dysart and Whitsitt [8] first proposed a one-dimensional mathematical model (the D-W model) to calculate the fluid temperature distribution in a hydraulic fracture at a given distance from the wellbore with a fixed fracture width and a fixed fracture height. Nevertheless, they neglected the difference in the heat transfer between the fracturing fluid and the formation and the energy change caused by the heat convection of the fracturing fluid loss. Kamphuis et al. [9] researched the temperature field in a fracture and found that the heat transfer in the direction of the fracturing fluid flow is controlled by a convective heat transfer. In contrast, thermal diffusion controls the heat transfer in the direction perpendicular to the fracture walls. This led to the development of a mathematical model for the hydraulic fracture temperature based on the average fracture width (the K-D-R model), which is still widely used. Based on this, Ji [10] established a mathematical model for calculating the fracture temperature field of acid fracturing and considered the acid–rock reaction heat as a constant value by adding the acid–rock reaction heat to the energy equation, but he did not consider the fracture dimension change or the dependence of the acid–rock reaction on the temperature. Hu [11] developed a temperature model considering the acid–rock reaction during the acid fracturing of naturally fractured carbonate reservoirs, where wormholes form and thus affect the temperature distribution [12–15].

In this paper, based on the K-D-R model, a new temperature field model in acid fracturing is developed for the Shunbei carbonate formation. Compared to conventional models, this model considers the acid–rock reaction heat generated during acid fracturing and the dependence of the reaction rate on the temperature. Based on the model, extensive numerical simulations are conducted for a parameter analysis with the Shunbei formation properties. The finding in this paper can improve the acid system selection and parameter design of the acid fracturing treatment [16–19].

2. Mathematical Model

2.1. Physical Model

Considering the propagation of hydraulic fractures in the acid fracturing process, a physical model is established (Figure 1), which divides the fracture and peri-fracture temperature field during acid fracturing into three parts, namely the fluid temperature field in the fracture, temperature field in the leak-off zone, and temperature field in the formation. To facilitate the model solving, the following assumptions are adopted as other acid fracturing models [9,11] did: (1) The hydraulic fracture propagation conforms to the PKN model, i.e., the fracture height is constant and equal to the reservoir thickness. (2) The reservoir and injection fluid thermodynamic parameters are constant and do not vary with temperature, flow state, and other conditions. (3) The fluid flow within the fracture is stable and incompressible. (4) The reaction system's volume does not change during the acid–rock reaction, i.e., the volumes of the reactants and products are equal. (5) Energy changes due to fluid friction and volume changes are ignored [16]. (6) The formation is considered as 100% of calcite or dolomite.

The energy-coupling process between the fluid in the fracture and the bedrock of the formation is characterized by the following energy changes:

(1) Low-temperature acid fluid is continuously injected through perforations and leaks into the rock matrix. The energy of the formation around the fracture and the leak-off zone is constantly taken away, resulting in heat dissipation by leak-off.

(2) Heat conduction from the high-temperature formation to the low-temperature fluid in the fracture goes through the following processes: from the formation to the fracture leak-off zone and finally to the fluid in the fracture.

(3) Acid–rock reactions occur in the fracture wall when acid is injected, and a portion of the energy is released [20,21].

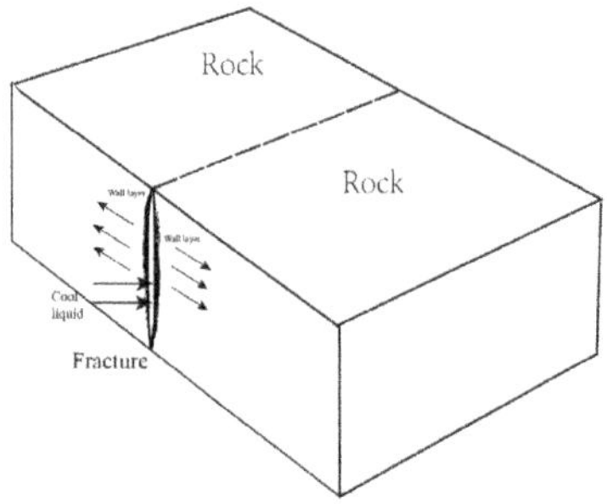

Figure 1. Physical model of the fracture temperature field for acid fracturing.

2.2. Governing Equations

(1) The continuity equation

Based on the established physical model, the acid flow within the fracture satisfies the law of conservation of mass. An element within the fracture is assumed according to the flow process, as shown in Figure 2.

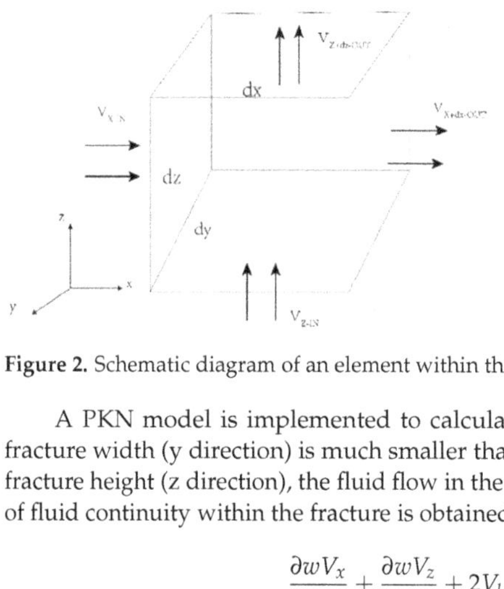

Figure 2. Schematic diagram of an element within the fracture.

A PKN model is implemented to calculate the fracture dimensions. Because the fracture width (y direction) is much smaller than the fracture length (x direction) and the fracture height (z direction), the fluid flow in the y direction can be neglected. The equation of fluid continuity within the fracture is obtained based on the law of conservation of mass:

$$\frac{\partial wV_x}{\partial x} + \frac{\partial wV_z}{\partial z} + 2V_{leak-off} = -\frac{\partial w}{\partial t} \tag{1}$$

where w is the fracture width (m), $V_{z,x}$ is the velocity along the direction of the fracture length and height (m/s), $V_{leak-off}$ is the acid leak-off rate (m/s), and t is the treatment time (s).

(2) Acid balance equation

The flow of the injected acid through the fracture is divided into two main parts: diffusion and leak-off. The diffusion part of the acid is the main source of acid to dissolve the rock wall, which contributes to the acid fracture wall dissolution and the fracture width widening. The liquid leak-off part of the acid flows into the formation, forming a leak-off zone, which is usually considered not to participate in the acid–rock reaction at the fracture wall, but the heat of the acid–rock reaction released from the leak-off zone is one of the energy sources in the fracture [22,23].

The reaction of carbonate rock with HCl is controlled by the mass transfer rate. The acid concentration at the fracture wall is much lower than that at the center of the fracture as the acid flows within the fracture. Therefore, the concentration gradient in the fracture width direction can be simplified to the extent that the acid concentration at the fracture wall can be approximated as zero. Therefore, the concentration gradient is expressed in terms of the effective mass transfer coefficient of hydrogen ions and the average acid concentration. That is,

$$-D\frac{\partial C}{\partial y}\Big|_{y=\frac{w}{2}}=k_g(C-C_0)=k_gC \tag{2}$$

C and C_0 denote the average acid concentration from the center of the fracture to the fracture wall and the acid concentration at the fracture wall, respectively. k_g is the mass transfer coefficient (m/s).

According to the law of conservation of mass, the equation for the distribution of acid concentration in the fracture is as follows:

$$-\frac{\partial(Cv_xw)}{\partial x}-\frac{\partial(Cv_zw)}{\partial z}-2(V_{leak-off}+k_g)C=\frac{\partial(Cw)}{\partial t} \tag{3}$$

(3) Fracture width variation

The acid reacting with the carbonate is believed to be mainly transported to the fracture wall by diffusion. In contrast, only a tiny amount of the leak-off acid is involved in the acid–rock reaction at the fracture wall. Most leak-off acid flows into the formation through wormholes, which is believed to make a small contribution to the fracture width increasement.

The mass conservation equation for the change in acid-etched fracture width per unit of time due to the dissolution of the fracture wall by the acid–rock reaction is as follows:

$$(2k_gC+2\eta V_{leak-off}C)\frac{\beta}{\rho_r(1-\phi)}=\frac{\partial w}{\partial t} \tag{4}$$

where k_g is the H^+ mass transfer coefficient (m/s), β is the dissolving power of the acid, fractional, ρ_r is the density of the rock (kg/m^3), and ϕ is the porosity of the formation. η is the proportion of the acid that dissolves the rock on the surface of the fracture before the acid leaks into the formation, often taken as 30%.

(4) Energy conservation equation in the fracture

When the acid flows through the fracture, it is assumed that the temperature inside the element has reached thermal equilibrium and the temperature at the center of the fracture is T. The acid inside the fracture flows in the direction of the fracture length, on the one hand, and leaks off along the direction perpendicular to the fracture wall, on the other.

Considering the incompressible fluid, ignoring the effect of kinetic energy on the internal energy of the fluid, and assuming that heat conduction is thermally homogeneous, the energy conservation equation for the fluid in the fracture is as follows:

$$\rho_fc_f\frac{\partial wT_f}{\partial t}=-\rho_fc_f\left(\frac{\partial wT_fv_x}{\partial x}+\frac{\partial wT_fv_z}{\partial z}\right)+k_f\left(\frac{\partial^2wT_f}{\partial x^2}+\frac{\partial^2wT_f}{\partial z^2}\right)+Q_r-Q_{leak-off} \tag{5}$$

where c_f is the specific heat capacity of the acid (J/(kg·°C)), ρ_f is the density of the acid (kg/m^3), k_f is the acid thermal conductivity (W/(m·°C)), T_f is the acid temperature (°C), w is the fracture width (m), and T_d is the temperature of the leak-off zone (°C). v is the velocity of the acid in the fracture (m/s).

$$Q_r=2h_d\left(T_d-T_f\right) \tag{6}$$

$$Q_{leak-off}=2c_f\rho_fV_{leak-off}T_f \tag{7}$$

where $V_{leak-off}$ is the acid leak-off rate (m/s) and h_d is the convective heat transfer coefficient. The convective heat transfer coefficient satisfies the following equation under the condition that the heat transfer at the fracture surface conforms to Fourier's law.

$$h_d \approx \frac{k_f}{\delta_T} \tag{8}$$

The thickness of the temperature boundary layer decreases as the velocity increases and is more difficult to describe accurately, so the average thickness of the boundary layer is usually assumed. Meanwhile, Plante [10] suggests that if the acid velocity along the object's surface is sufficiently high, the temperature boundary layers are very thin, and the boundary layer approximation can be used. The convective heat transfer coefficient for a fully developed temperature field is calculated as follows.

$$h_d = \frac{k_f Nu}{w} \tag{9}$$

where Nu is the Nussle number, and it usually takes 4~5.

(5) Energy conservation equation in leak-off zone

Compared with the fracture length, the leak-off zone is small, and the temperature distribution of the leak-off zone in the direction perpendicular to the fracture wall can be ignored. It is assumed that the temperature gradient of the leak-off zone is considered in the direction of the fracture length and fracture height. As a result, the energy conservation of an element in the leak-off zone can be described as follows: the sum of the energy flowing into the element and the energy introduced by the acid–rock reaction minus the energy flowing out of the element is equal to the total energy change in the element in the leak-off zone. The energy equation for the leak-off zone is written as follows:

$$\begin{aligned} &\xi\left[c_f\rho_f\phi + c_r\rho_r(1-\phi)\right]\frac{\partial T_d}{\partial t} \\ &= c_f\rho_f V_{leak-off}\left(T_f - T_d\right) + k_{ef}(T_{res} - T_d)\frac{1}{\sqrt{\pi D\tau}}\frac{e^{-\frac{F^2}{D}}}{1+erf\left(\frac{F}{\sqrt{D}}\right)} \\ &+ \frac{k_f Nu}{w}\left(T_f - T_d\right) + (2k_g\overline{C} + V_{leak-off}\phi k_g\overline{C}\Delta t) \times \Delta H_{HCl} \end{aligned} \tag{10}$$

$$F = \frac{c_f\rho_f C_t}{c_f\rho_f\phi + c_r\rho_r(1-\phi)} \tag{11}$$

$$D = \frac{K_{ef}}{c_f\rho_f\phi + c_r\rho_r(1-\phi)} \tag{12}$$

$$k_{ef} = k_r(1-\varphi) + k_f\varphi \tag{13}$$

where T_d is the temperature of the leak-off zone (°C); c_f is the specific heat capacity of the acid (J/(kg·°C)); c_r is the specific heat capacity of formation rock (J/(kg·°C)); ρ_r is the density of formation rock (kg/m^3); ρ_f is the density of the acid (kg/m^3); ϕ is the porosity, fractional; ξ is the leak-off zone thickness (m); $V_{leak-off}$ is the acid leak-off rate (m/s); C_t is the integrated leak-off factor ($m/\sqrt{\text{min}}$); Δt is the time step; and T_f is the acid temperature (°C).

The energy equation for the leak-off zone consists of four components: convective heat exchange, heat transfer between the fluid in the fracture and the leak-off zone, heat transfer between the in situ rock and the leak-off zone, and heat generation by the acid–rock reaction [17]. The equation can be solved by coupling the energy changes in these four components to achieve an energy balance.

The following equation gives the enthalpy of the reaction between hydrochloric acid and limestone during acid fracturing, considering the temperature and pressure conditions [24].

$$\begin{aligned}\Delta H_m^{\ominus}(T,p) &= -13.692 + \tfrac{1}{1000}(-6.443 \times 10^{-3}T^2 + 16.075T - \tfrac{17.406\times10^5}{T}) \\ &+ \textstyle\int_{1atm}^{p} V_{co_2}[V_{co_2} - T(\tfrac{\partial V_{co_2}}{\partial T})_p]_T dp\end{aligned} \tag{14}$$

The enthalpy of the reaction of dolomite with hydrochloric acid is calculated as follows:

$$\begin{aligned}\Delta H_m^{\ominus}(T,p) &= -43.272 + \tfrac{1}{1000}(-21.508 \times 10^{-3}T^2 + 49.552T - \tfrac{1.46\times10^5}{T}) \\ &+ \textstyle\int_{1atm}^{p} V_{co_2}[V_{co_2} - T(\tfrac{\partial V_{co_2}}{\partial T})_p]_T dp\end{aligned} \tag{15}$$

where $\Delta H_m^{\ominus}(T,p)$ is the standard molar reaction enthalpy of limestone (kJ/mol) and V_{co_2} is the molar volume of CO_2, determined by the relevant equation of state; P and T are pressure (atm) and temperature (°C) at which the acid–rock reaction takes place.

(6) Initial conditions and boundary conditions

Initial conditions for the model:

$$\begin{aligned} T_d(x,z,t)|_{t=0} &= T_{res} \\ T_f(x,z,t)\Big|_{t=0} &= T_{res} \end{aligned} \tag{16}$$

Boundary conditions for the model:

$$\begin{aligned} T(x,z,t)|_{x=0} &= T_{inject} \\ T(x,z,t)|_{x=L} &= T_{res} \\ C(x,z,t)|_{x=0} &= C_{inject} \\ \frac{\partial p}{\partial x}\Big|_{x=0} &= -\frac{12\mu q}{Sw^2} \\ p_{x=L} &= p_e \end{aligned} \tag{17}$$

where T_{inject} is the temperature of injected acid (°C), C_{inject} is the acid concentration (wt.%) and q is the injection rate, S is the cross-sectional area of inlet (m^2), μ is the acid viscosity (Pa·s), and w is the fracture width of inlet (m).

3. Numerical Implementation Process

The numerical model of the temperature field is discretized according to the finite difference method, and a block-centered grid is used to mesh the fracture along the fracture length direction. The overall numerical implementation process of the temperature field model is summarized as follows.

(1) Input the formation parameters, treatment parameters, physical properties of the injected fluid, etc.

(2) The hydraulic fracture dimensions (fracture length and width) at the current time step are derived from the input parameters based on the PKN model, and then the fracture meshes into girds.

(3) Calculate the continuity equation based on the law of conservation of mass to obtain the pressure and velocity fields within the fracture.

(4) Based on the pressure and velocity fields, the acid concentration distribution in the fracture at the current time step is calculated, and the acid-etched fracture width is calculated based on the acid–rock reaction. The fracture dimension parameters are updated.

(5) Considering the heat generated by the acid–rock reaction and assuming that the distribution of the fluid temperature field in the fracture is T_{f0}, based on the updated fracture parameters and the initial conditions of the formation, calculate the temperature field in the leak-off zone under the current temperature–pressure conditions.

(6) Calculate the fluid temperature field in the fracture T_f at the current temperature and pressure condition based on the temperature field in the leak-off zone derived in step (5). Compare the magnitude of T_f with the assumed T_{f0}. If the calculated result T_f does not meet the error requirement $\left|T_{f0} - T_f\right| \leq \varepsilon$, then set $T_{f0} = T_f$ and repeat steps (5) to (6) until the computed result meets the error requirement. The temperature field of the fluid and the leak-off zone in the current time step are obtained.

(7) Repeat steps (2) to (6) until the entire acid fracturing process is completed.

4. Results and Analysis

4.1. Data for Simulation

To analyze the factors affecting the fracture temperature field during acid fracturing, a simulation study was conducted using actual data from the Shunbei carbonate reservoir. Taking Shunbei well #T as an example, the well was designed to stimulate with acid fracturing in an open-hole section ranging from 6550.00 to 6781.63 m to create acid-etched fractures. The reservoir lithology is mainly micrite, in which the content of calcite is more than 95%, quartz about 4%, and clay about 1%. In addition, the pressure gradient in the area is 1.10, which is a normal pressure reservoir. The regional stratigraphic temperature gradient is 2.26 °C/100 m, with an expected reservoir temperature of 145.2 °C at a mid-depth of 6424.0 m.

The treatment parameters, rock mechanical properties, and thermodynamic properties are given in Table 1 below, and the injection parameters are given in Table 2 below.

Table 1. Treatment parameters and stratigraphic parameters for Shunbei well #T.

Parameters (Units)	Value	Parameters (Units)	Value
Reservoir depth (m)	6500	Geothermal gradients (°C/m)	0.0226
Reservoir temperature (°C)	145.2	Porosity (%)	0.2
Fracture height (m)	50	Temperature of injected fluid (°C)	25
Injection rate (m^3/min)	5~7	Young's modulus of the rock (MPa)	36,700
Poisson ratio	0.26	Thermal conductivity of the rock (W/(m·K))	5.2
Density of the rock (kg/m^3)	2700	Specific heat capacity of the reservoir crude oil (J/(kg·K))	1981
Specific heat capacity of the reservoir rock (J/(kg·K))	999	Thermal conductivity of crude oil (W/(m·K))	0.339
Density of crude oil (kg/m^3)	840	-	-

Table 2. Injection parameters for Shunbei well #T.

Parameters (Units)	Value	Parameters (Units)	Value
Density (kg/m^3)	1090–1110	Specific heat capacity (J/(kg·K))	4180
Coefficient of thermal conductivity (W/(m·K))	0.65	Acid concentration (%)	15
Acid–rock reaction rate constant (mol/cm^3)	0.9849×10^{-6}	The heat generated by the acid–rock reaction (kJ/mol)	$\Delta H_m^{\ominus}(T, p)$
Order of the reaction (dimensionless)	0.88	Mass transfer coefficient (m/s)	1.4693×10^{-5}

4.2. Effect of Acid–Rock Reaction Heat on Temperature Distribution

To investigate the effect of the reaction heat on the temperature distribution in the fracture, two simulation cases are performed, one considering the reaction heat and the other not. Viscous acid with an injection rate of 6 m^3/min and a temperature of 25 °C is injected for 90 min. Figures 3 and 4 show the simulation results. The temperature inside the fracture gradually increases along the fracture and eventually approaches the formation temperature. Comparing the distribution of the fluid temperature and the leak-off zone temperature with and without considering the reaction heat, it can be found that the acid–rock reaction heat has a significant effect on the distribution of the temperature in the fracture. It results in a maximum temperature difference of 15 °C. The temperature difference is slight at the two ends of the fracture but significant in the middle. The potential reason is that the temperature at the inlet of the fracture is low due to the continuous injection of the cool acid, which leads to a slow acid–rock reaction. At the fracture tip, the temperature is close to the formation temperature, and the concentration of the acid is low, sometimes becoming residual acid, which also has little effect on the reaction heat. The acid concentration in the middle of the fracture is high, and the fracture temperature is relatively high, so the acid–rock reaction rate is fast, and the effect of the reaction heat on the temperature distribution is significant (Figure 5). Therefore, the acid–rock reaction heat cannot be ignored in the fracture temperature calculation.

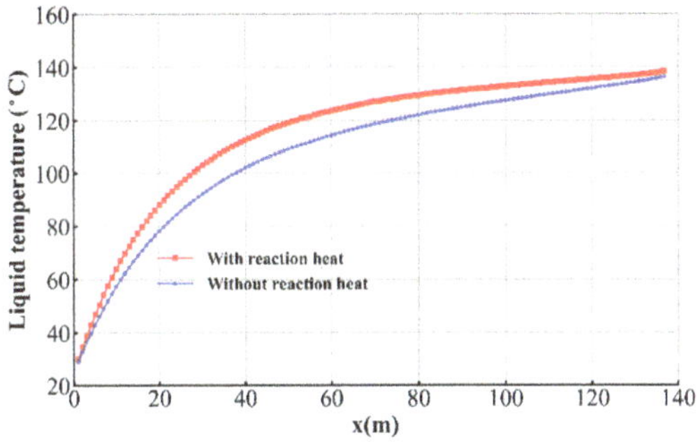

Figure 3. Temperature distribution of the fluid in the fracture (blue line: without reaction heat; red line: considering reaction heat).

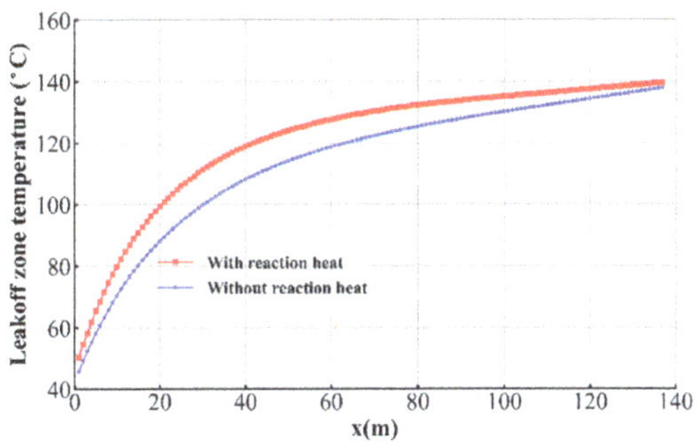

Figure 4. Temperature distribution of the leak-off zone (blue line: without reaction heat; red line: considering reaction heat).

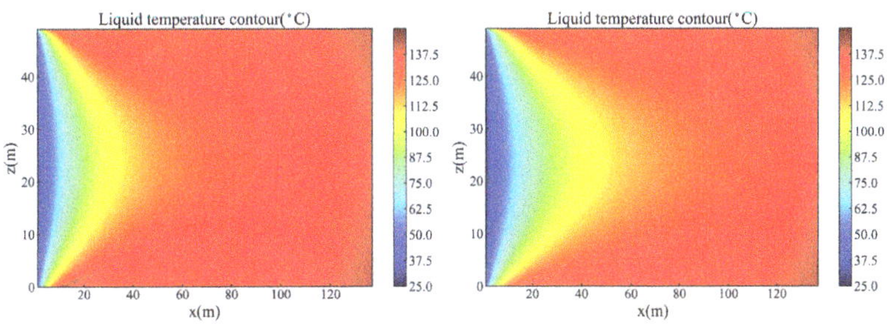

Figure 5. Thermal map of fluid temperature distribution in the fracture leak-off zone with and without considering acid–rock reaction heat. (**a**) considering reaction heat, (**b**) without considering reaction heat.

4.3. Effect of Acid Properties on Temperature Field

The reaction rates between different acids and carbonate rocks vary, among which the reaction rate between limestone and the acid is mainly determined by the mass transfer rate. The effect of the temperature variation on the mass transfer coefficients of different acid species is discussed in this section. Table 3 and Figure 6 show the variation in the mass transfer coefficients of three commonly used acids in the oilfield at different temperatures. In our paper, plain acid refers to HCl with a mass concentration of 15%, gelled acid is obtained from a plain acid mixing with additives such as gelling agents, and crosslinked acid refers to the acid obtained by adding an organic polymer crosslinking agent to plain acid (HCl of 15%wt). Because the viscosity of different acid species differs, the fracture dimensions formed vary. To eliminate the influence of the fracture dimensions, the fracture dimension created by three acid species was set the same in the model. The mass transfer coefficient of plain acid is significantly affected by the temperature, while the gelled acid and crosslinked acid were less affected. Figure 7 shows the distribution of the acid-etched fracture width of the injecting plain acid, gelled acid, and crosslinked acid, considering the variation in the mass transfer coefficient with the temperature. Figure 8 demonstrates a two-dimensional fracture width distribution of the injecting plain acid and crosslinked acid. Compared with the acid-etched fracture width under the constant mass transfer coefficient shown in Figure 9, it is concluded that when considering the variation in the mass transfer coefficient with the temperature, the acid–rock reaction rate decreases at the inlet of the fracture due to the cooling effect of the continuous injection, and the maximum acid-etched fracture width offsets the fracture inlet. Meanwhile, when considering the cooling effect, the maximum etched fracture width is less, and the live acid penetration distance is longer than those without considering the cooling effect.

Table 3. Variation in mass transfer coefficients with temperature for three commonly used acids in the field.

Temperature (°C)	Acid Concentration (mol/L)	Mass Transfer Coefficient (10^{-5} m/s)		
		Plain Acid	Gelled Acid	Plain Acid
20	4.5 (15%wt)	1.348	0.451	0.058
60	4.5 (15%wt)	3.254	1.086	0.321
100	4.5 (15%wt)	5.89	1.685	0.532
140	4.5 (15%wt)	7.956	2.351	0.776
160	4.5 (15%wt)	9.034	2.649	0.918

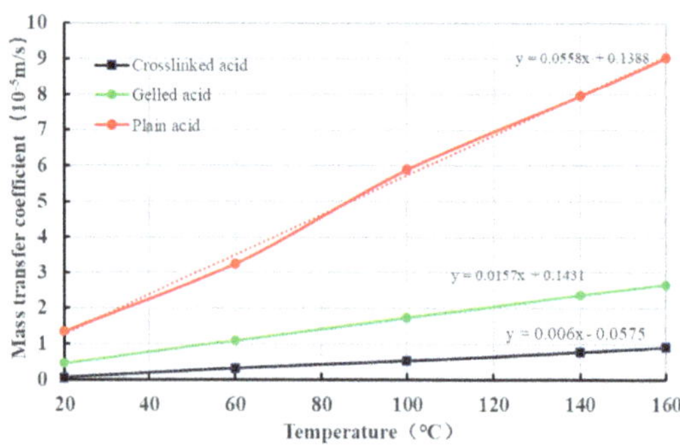

Figure 6. Modeling the temperature dependence of mass transfer coefficients for three acid species.

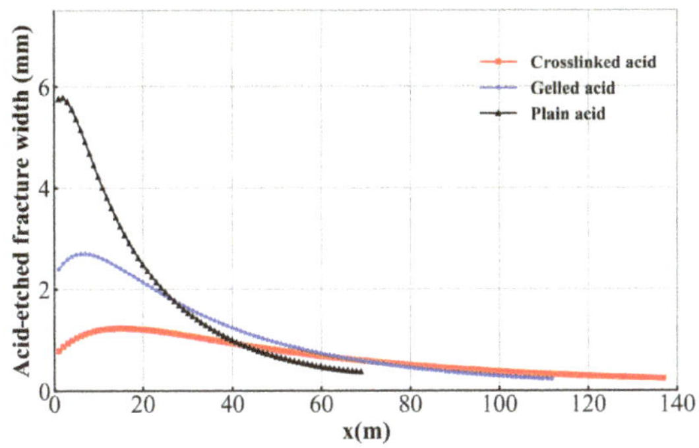

Figure 7. Acid-etched fracture width distribution of different acid solutions when considering the variation in mass transfer coefficient with temperature.

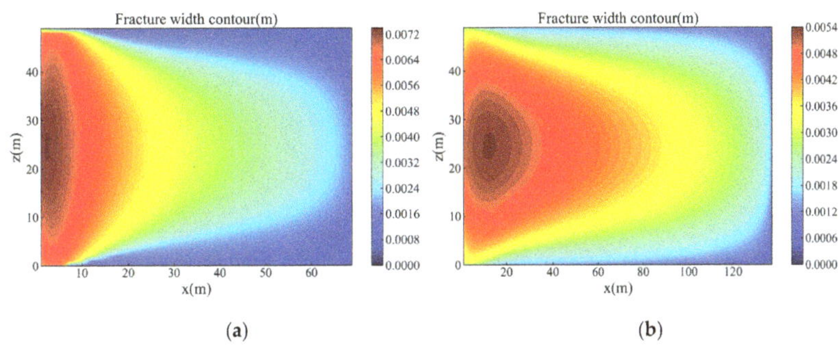

Figure 8. Two-dimensional acid-etched fracture width distribution by injection of plain acid and crosslinked acid (considering the variation in mass transfer coefficient with temperature). (**a**) plain acid, (**b**) crosslinked acid.

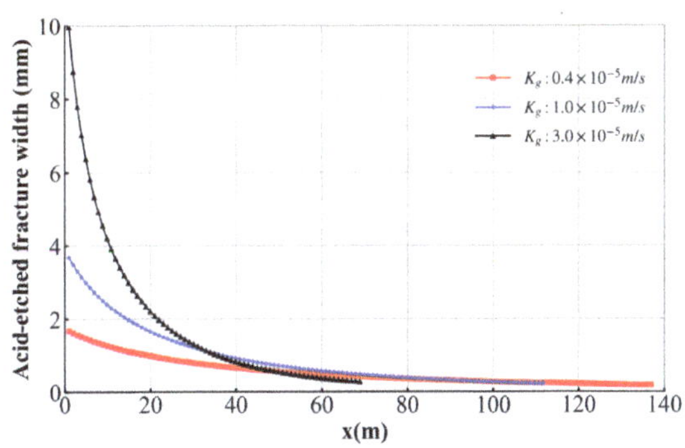

Figure 9. Acid-etched fracture width distribution under constant mass transfer coefficient (without considering the variation in mass transfer coefficient with temperature).

4.4. Temperature Field in Time Sequence

The cooling effect comes into play mainly in the first third of the fracture, where the temperature decreases sharply with a large reduction, whereas the temperature of the other part of the fracture is close to the initial formation temperature. Comparing the temperature changes after different injection times (Figures 10 and 11), it can be seen that as the injection time increases, the fracture temperature at a given location decreases, but the cooling rate gradually becomes slower, and the cooling area is concentrated in the middle front of the fracture. This is because as the cool fluid is injected, the fracture temperature continues to decrease. With the temperature difference decreasing, it becomes more difficult to lower the fracture temperature, and the temperature in the fracture gradually reaches a dynamic equilibrium state. Figure 12 shows the range, degree, and distribution pattern of the overall cooling effect after a period of injection.

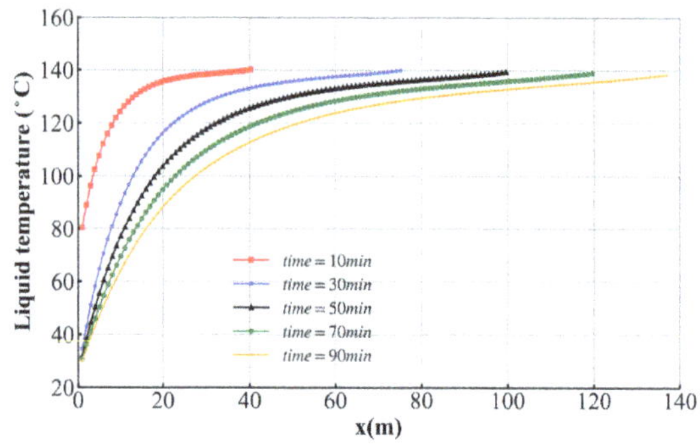

Figure 10. Fluid temperature distribution in the fracture at different injection times.

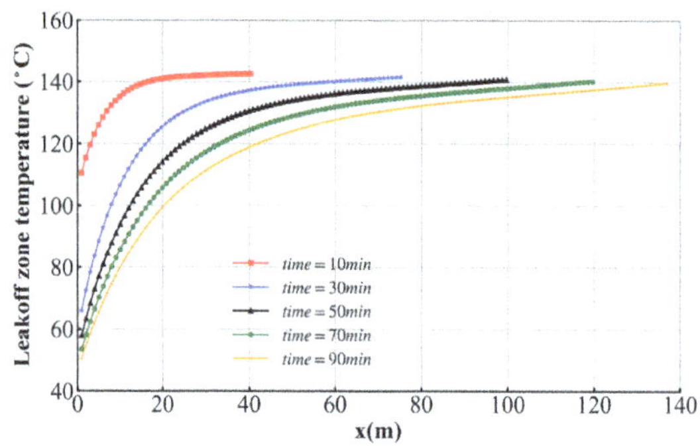

Figure 11. Temperature distribution in the leak-off zone at different times.

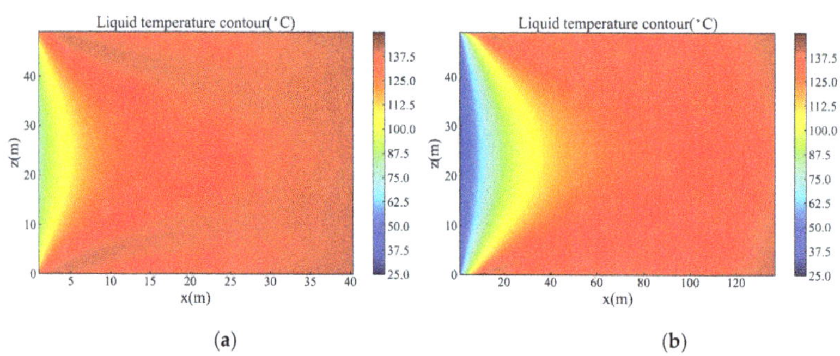

Figure 12. Thermal maps of the fluid in the fracture at 10 and 90 min. (**a**) 10 min, (**b**) 90 min.

5. Conclusions

This paper develops a new temperature field model of acid fracturing in the Shunbei carbonate formation. The model accounts for the acid–rock reaction heat and the dependence of the acid–rock reaction rate on the temperature. Based on the extensive numerical simulations, the following conclusions can be reached:

(1) The acid–rock reaction in acid fracturing obviously affects the temperature field, resulting in about a 10~20 °C increase in the Shunbei formation.

(2) The acid–rock reaction rate dependence on the temperature is a vital factor that should be accounted for. The rock dissolution increases first and then decreases from the inlet to the tip of the fracture, unlike the monotonous decrease without the temperature dependence.

(3) The temperature gradient is high near the inlet and then decreases gradually. Beyond half of the fracture, the temperature is close to the formation temperature.

(4) The temperature decreases fast in the initial injection stage and tends to stabilize at about 50min.

Author Contributions: Conceptualization, J.M.; Formal analysis, H.Z.; Funding acquisition, J.H.; Methodology, J.M. and H.Z.; Project administration, R.Z.; Resources, J.H. and L.Z.; Supervision, J.M. and L.Z.; Writing—original draft, H.Z.; Writing—review and editing, B.G. All authors have read and agreed to the published version of the manuscript.

Funding: This research was funded by Key Laboratory of Marine Oil & Gas Reservoirs Production, Sinopec, grant number: 33550000-22-ZC0613-0311 and by Science and Technology Department, Sinopec, grant number: P21063-3.

Data Availability Statement: Not applicable.

Acknowledgments: This work is supported by the Open Fund Project for Numerical Simulation of Multiple-Cluster Fracture Propagation in Horizontal Wells during Acid Fracturing of a Carbonate Formation (No. 33550000-22-ZC0613-0311) provided by Key Laboratory of Marine Oil & Gas Reservoirs Production, Sinopec and by the Project Research on Key Engineering Technologies of Unused Reserves Development in Tahe Oilfield (P21063-3) provided by Science and Technology Department, Sinopec.

Conflicts of Interest: The authors declare no conflict of interest.

Nomenclature

C_t	the integrated leak-off factor ($m/\sqrt{\min}$)
C_{inject}	the acid concentration (wt.%)
C	the average acid concentration from the center of the fracture to the fracture wall (wt.%)
C_0	the acid concentration at the fracture wall (wt.%)
$\Delta H_m^{\ominus}(T,p)$	the standard molar reaction enthalpy of limestone (kJ/mol)
S	the cross-sectional area of inlet (m^2)
T_f	the acid temperature (°C)
T_d	the temperature of the leak-off zone (°C)
T_{inject}	the temperature of injected acid (°C)
V_{co_2}	the molar volume of CO_2
$V_{z,x}$	the velocity along the direction of the fracture length and height (m/s)
$V_{leak-off}$	the acid leak-off rate (m/s)
Nu	the Nussle number, and it usually takes 4~5
c_f	the specific heat capacity of the acid (J/(kg·°C))
c_r	the specific heat capacity of formation rock (J/(kg·°C))
h_d	the convective heat transfer coefficient
k_g	the mass transfer coefficient (m/s)
k_f	the acid thermal conductivity (W/(m·°C))
q	the injection rate (m^3/min)
t	the treatment time (s)
Δt	the time step
w	the fracture width (m)
β	the dissolving power of the acid, fractional
ρ_f	the density of the acid (kg/m^3)
ρ_r	the density of formation rock (kg/m^3)

ϕ	the porosity of the formation
η	the proportion of the acid that dissolves the rock on the surface of the fracture before the acid leaks into the formation, often taken as 30%
ξ	the leak-off zone thickness (m)
μ	the acid viscosity (Pa·s)

References

1. Aljawad, M.S.; Zhu, D.; Hill, A.D. Temperature and geometry effects on the fracture surfaces dissolution patterns in acid fracturing. In Proceedings of the SPE Europec featured at 80th EAGE Conference and Exhibition, Copenhagen, Denmark, 11–14 June 2018. SPE-190819-MS.
2. Jawad, M.S.H.A. Development of a Fully Integrated Acid Fracture Model. Ph.D. Thesis, The University of Texas at Austin, College Station, TX, USA, 2018.
3. Guo, J.C.; Ren, J.C.; Wang, S.B. Numerical simulation and application of multi-field coupling of acid fracturing in fractured tight carbonate reservoirs. *Acta Pet. Sin.* **2020**, *41*, 1219.
4. Li, T.; Wang, S.B. Research on Acid Rock Reaction Model in High Temperature Carbonate Fracture. In Proceedings of the IPPTC Organizing Committee, Shanghai, China, 26 August 2020; p. 10.
5. Luo, P.D.; Li, H.Y.; Zhai, L.J.; Li, C.Y.; Lv, X.R.; Mou, J.Y. Supercritical CO_2 fracturing wellbore and fracture temperature field in Tahe Oilfield. *Fault-Block Oil Gas Field* **2019**, *26*, 225–230.
6. Khormali, A.; Sharifov, A.R.; Torba, D.I. The control of asphaltene precipitation in oil wells. *Pet. Sci. Technol.* **2018**, *36*, 443–449. [CrossRef]
7. Wheeler, J.A. Analytical calculations for heat transfer from fractures. In Proceedings of the SPE Improved Oil Recovery Symposium, Tulsa, OK, USA, 13–15 April 1969. SPE-2494-MS.
8. Dysart, G.R.; Whitsitt, N.F. Fluid temperature in fractures. In Proceedings of the fall meeting of the society of petroleum engineers of AIME, New Orleans, LA, USA, 1–4 October 1967. SPE-1902-MS.
9. Kamphuis, H.; Davies, D.R.; Roodhart, L.P. A new simulator for the calculation of the in-situ temperature profile during well stimulation fracturing treatments. *J. Can. Pet. Technol.* **1993**, *32*, 38–47. [CrossRef]
10. Ji, W. Research of Acid Fracturing Temperature Field Model with Considering Acid-Rock Reaction Heat. Ph.D. Thesis, Chengdu University of Technology, Chengdu, China, 2010.
11. Hu, J.Y. Modeling the Acidification Temperature Field of Horizontal Wells in Fractured Carbonate Reservoirs. Ph.D. Thesis, Southwest Petroleum University, Chengdu, China, 2016.
12. Biot, M.A.; Masse, L.; Medlin, W.L. Temperature analysis in hydraulic fracturing. *J. Pet. Technol.* **1987**, *39*, 1389–1397. [CrossRef]
13. Liu, W.; Zeng, Q.; Yao, J. Numerical simulation of elasto-plastic hydraulic fracture propagation in deep reservoir coupled with temperature field. *J. Petrol. Sci. Eng.* **2018**, *171*, 115–126. [CrossRef]
14. Ben, N.K.; Stephenson, P. Models of heat transfer in hydraulic fracturing. In Proceedings of the SPE/DOE Low Permeability Gas Reservoirs Symposium, Denver, CO, USA, 19 May 1985. SPE-13865-MS.
15. Wang, T.; Yang, Y.; Peng, Y.; Zhao, J.; Qi, T.; Zeng, J. Simulation and evaluation for acid fracturing of carbonate reservoirs based on embedded discrete fracture model. *Nat. Gas Ind. B* **2021**, *8*, 637–649. [CrossRef]
16. Guo, J.; Liu, H.; Zhu, Y.; Liu, Y. Effects of acid–rock reaction heat on fluid temperature profile in fracture during acid fracturing in carbonate reservoirs. *J. Petrol. Sci. Eng.* **2014**, *122*, 31–37. [CrossRef]
17. Liu, H.F.; Baletabieke, B.; Wang, G. Influences of real-time acid-rock reaction heat on etched fracture dimensions during acid fracturing of carbonate reservoirs and field applications. *Heliyon* **2022**, *8*, e11659. [CrossRef] [PubMed]
18. Dong, C. Acidizing of Naturally Fractured Carbonate Formations. Ph.D. Thesis, The University of Texas at Austin, Austin, TX, USA, 2001.
19. Zhai, W.B. Mechanism of Fractures Propagation of Multi-Stage Fracturing in Heterogeneous Shale Reservoir. Ph.D. Thesis, China University of Petroleum (Beijing), Beijing, China, 2020.
20. Xian, C. Study on the Temperature of Self-Generated Solid Phase Chemical Fracturing. Ph.D. Thesis, Southwest Petroleum University, Chengdu, China, 2018.
21. Wang, L.; Shen, X.Y.; Wang, R. Effects of acid-rock reaction heat on fracture temperature field and effective distance of live acid. *J. Chongqing Univ. Sci. Technol.* **2014**, *16*, 21–24.
22. John, D. Measuring diffusion coefficients in acid fracturing fluids and their application to gelled and emulsified acids. In Proceedings of the SPE Annual Technical Conference and Exhibition, New Orleans, LA, USA, 25–28 September 1994; SPE-28552-MS.
23. Xu, H.R.; Cheng, J.R.; Zhao, Z.H. Numerical study and application of acid-fracturing in the carbonate geothermal reservoirs from North China. *Acta Geol. Sin.* **2020**, *94*, 2157–2165.
24. Liu, H.F.; He, C.M.; Zhu, Y.Q. Method to calculate acid-rock reaction heat during acid fracturing of carbonate reservoir. *Nat. Gas Explor. Dev.* **2011**, *34*, 84–87+91+102–103.

processes

MDPI

Article

Case Study: Successful Application of a Novel Gas Lift Valve in Low Pressure Wells in Fuling Shale Gas Field

Qiaoping Liu [1], Jingfei Tang [2,*], Wenqi Ke [2], Haibo Wang [2] and Uzezi Davis Orivri [3]

1 Sinopec Chongqing Fuling Shale Gas Exploration and Development Co., Ltd., Chongqing 408014, China
2 Petroleum Exploration & Production Research Institute, Sinopec, Beijing 102206, China
3 Schlumberger Oilfield Services Ltd., Lagos 100283, Nigeria
* Correspondence: tangjf.syky@sinopec.com; Tel.: +86-10-5660-8078

Abstract: The Fuling shale gas field is facing a rapid gas production decline due to heavy liquid loading issues. Given the condition that most wells are located at remote areas in the mountains, the traditional gas lift methods that require either fixed compressor or skid-mounted gas lift trucks do not seem feasible and occur high operation costs. A new type of gas lift valve that can be opened or closed at a low valve dome pressure indicates the high sensitivity to low production pressure. Thus, the piping line pressure can be utilized to activate the valve due to its new advantages. In addition, the specially designed structure of the gas lift valve can be activated via pressure increases in the tubing to create a channel between the tubing and annulus. The valve that previously functioned as a dummy valve was then switched to a gas lift valve. Field application results show that all wells were successfully restarted by only utilizing the low piping pressure, and loaded liquid was lifted with gas production at an incremental rate that reached up to 27.4 × 10 kscm/d per well. Fewer slickline operations were conducted to replace the dummy valve. The result of the application shows that the new type of gas lift system has a wide range of application prospect for low pressure wells, especially for shale gas wells.

Keywords: shale gas; low pressure; gas lift; piping line

Citation: Liu, Q.; Tang, J.; Ke, W.; Wang, H.; Orivri, U.D. Case Study: Successful Application of a Novel Gas Lift Valve in Low Pressure Wells in Fuling Shale Gas Field. *Processes* **2023**, *11*, 19. https://doi.org/10.3390/pr11010019

Academic Editor: Nicolas Dietrich

Received: 26 October 2022
Revised: 20 November 2022
Accepted: 22 November 2022
Published: 22 December 2022

1. Introduction

The Fuling shale gas field, which is located in southwest China [1], announced the first gas production since 2013, and it is now the largest gas field in China (Figure 1) in which most wells are horizontal wells and completed with a hydraulic multi-fracturing method [2,3]. The annual gas production increased dramatically from 5 bcm to 10 bcm with an increased number of new tied-in wells and effective stimulation methods such as hydraulic fracturing. The gas reservoirs are widely distributed with a large variety of production characteristics [4–6]. For instance, wells in the Jiangdong block are produced with average water production of 25 scm/d and peak production of 200 scm/d. With average well depths of 3571 m, wells tend to load liquid due to hydraulic fracturing and well undulation. The statistics show that 85% of gas wells' production has started to decline, and the annual gap between real production and actual demand is around 0.35 bcm, (Figure 2) which imposes a big challenge to the field development.

Gas lift systems have proven to be an extremely flexible and economical means of lifting fluid from deep wells, particularly from those with high-flowing bottom-hole pressures [7]. Whereas the required high gas lift pressure on the surface of the traditional gas lift method may not only require a high gas compressor capacity but also the high CAPEX, studies also show that there might be flow assurance issues—e.g., salt precipitation—in wells produced under high pressure and high temperature [8]. This may cause a negative impact on the production layer because it may force the loaded liquid into the reservoir of wells with depths more than 3657 m. Since the average well depth in the Fuling gas field is more than 3657 m, most wells are loaded with a heavy water column that imposes high requirements of compressor capacity to lift water. The wellsite compressor with a capacity of 25 MPa can hardly meet the requirement and eventually a big portion of wells were shut in. A novel gas lift valve has been successfully applied in the Fuling shale gas field with more than 40 gas wells that produce a new solution to handle the liquid loading problem by running holes in multiple special gas lift valves with tubing to unload the liquid in the tubing by staged gas lifting methods. This new type of gas lift valve was designed under a low valve dome pressure, thus it is sensitive to low pressure which can be opened and closed by injecting the piping line pressure to activate the valve and create a channel between the tubing and annulus. With the staged gas lift operation process, the gas lift system restored the gas wells' production rapidly after hydraulic fracturing by fully utilizing the piping line pressure (Figure 3). This method in general reduces the operation pressure and operation time in field applications and promote the application efficiency in shale gas reservoirs.

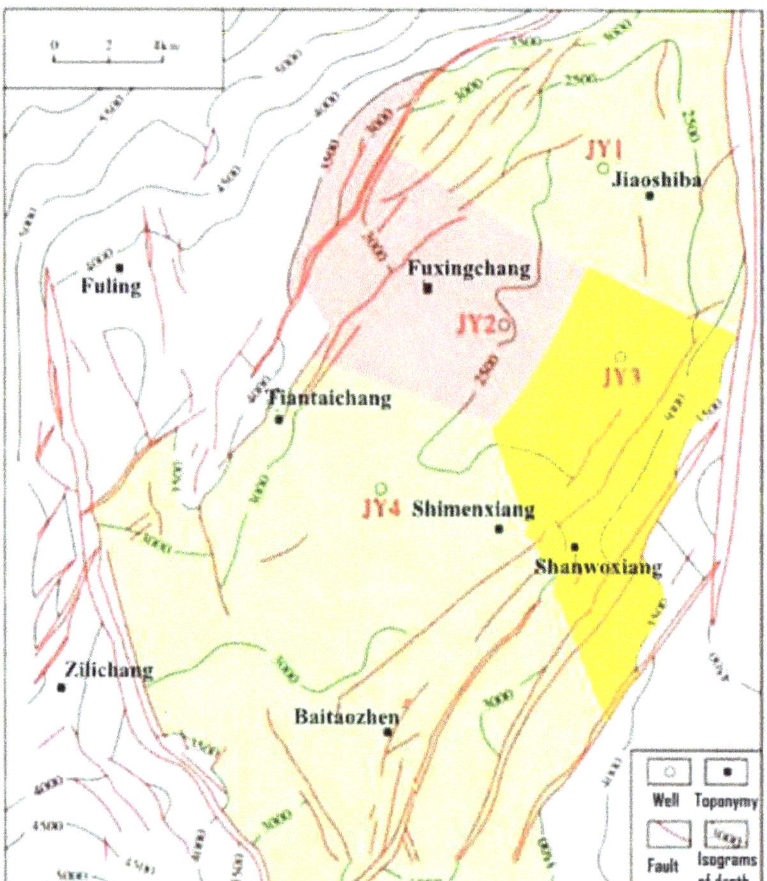

Figure 1. The main district of the Fuling shale gas field.

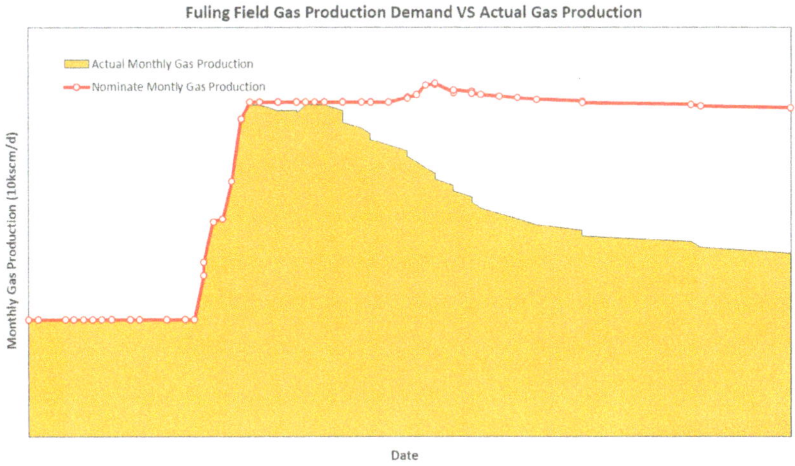

Figure 2. Fuling field gas production demand vs. actual gas production.

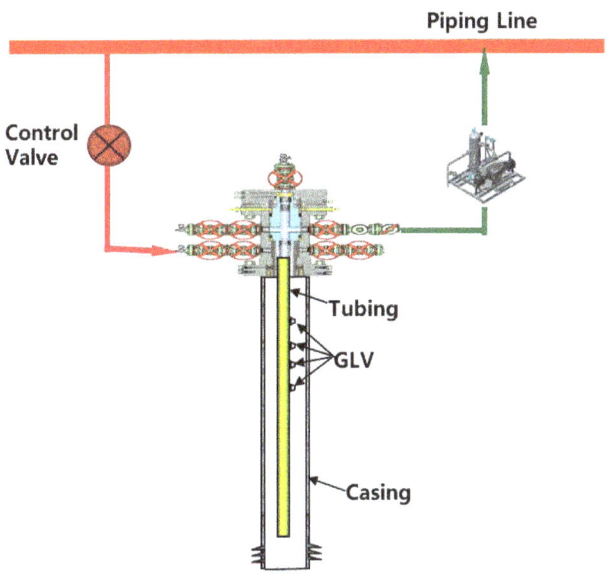

Figure 3. Low pressure wells' gas lift operation loop.

2. Gas Lift Valves

This fixed gas lift valve is a newly designed valve that contains 3 main parts (Figure 4): upper part 1, middle part 2 and low part 3; the parts are connected with thread. The first through hole 4 is located in the middle part, and the second through hole 5 is located in the hydraulic moving piston 6. On top of the valve is the fishing neck 7. The advantages of the valve compared with the traditional valve is list as follows.

1. The pressure control components are made of three layers of monel alloy and the test rack opening pressure is very sensitive to low pressure, thus it meets the accuracy of low-pressure well activation.
2. The new type of gas lift valve can connect/disconnect the first and second through-hole of the internal structure of the air lift valve by installing a hydraulic piston that can move axially inside the valve body. This technology connects the space of the

annulus and tubing by injecting liquid inside the tubing to activate the gas lift valve, thus it greatly reduces the operation frequency of replacing the dummy valve and as a result reduces the cost and risks and saves time.

3. The most import function of this new type of valve can greatly reduce the required gas lift pressure to the level of piping line pressure which is approximately around 5 MPa. By utilizing piping line pressure, the gas wells' production can be restored instantaneously after a hydraulic fracturing operation or being heavily water-loaded.

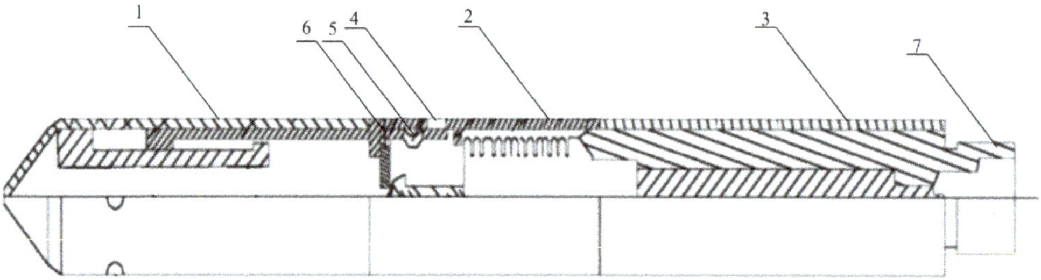

Figure 4. New type of fixed gas lift valve structure.

3. Geological Overview

The Fuling shale gas field is located in the Sichuan Basin of the south region of China (Figure 5) [8], which is the first major commercial discovery of shale gas outside North America in the world [3]. Most gas occurs in the lower Paleozoic Wufeng–Longmaxi marine shale play which consists of black shale and has a steady thickness of about 328 ft buried between 2150 mTVD and 3150 mTVD. This gas field was once screened out as shale gas potential due to its high complexity. Nevertheless, the structure of this field, the Jiaoshiba Structure, is a diamond faulted anticline with axial trending in a northeast direction, which is controlled by two main groups of faults trending northeast and near north, respectively.

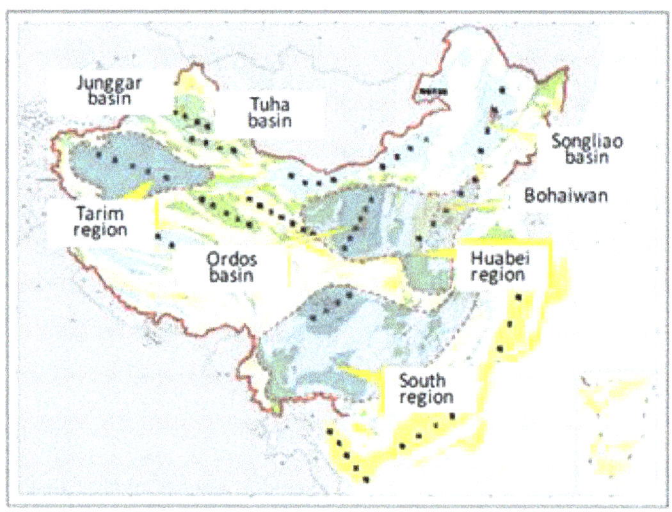

Figure 5. Distribution map of favorable exploration areas for shale gas in China.

Since the discovery of the Fuling field after a successful drilling of well Jiaoye 1, the production formation of Wufeng–Longmaxi was proven to be a prospective formation of an industrial gas flow at 0.2 bm/d. By the end of July 2014, the proven gas reserves climbed to 106 bcm [3]. The targeting Wufeng–Longmaxi Fm. is short for the Upper Ordovician Wufeng Formation and the first section of the Lower Silurian Longmaxi Formation (Figure 6). The formation is the main prospective formation of the lower Paleozoic marine shale of the Sichuan Basin. Based on the seismic date, Wufeng–Longmaxi Fm. is proven to consist of sandstone, whereas the Longmaxi formation mainly consists of limestone.

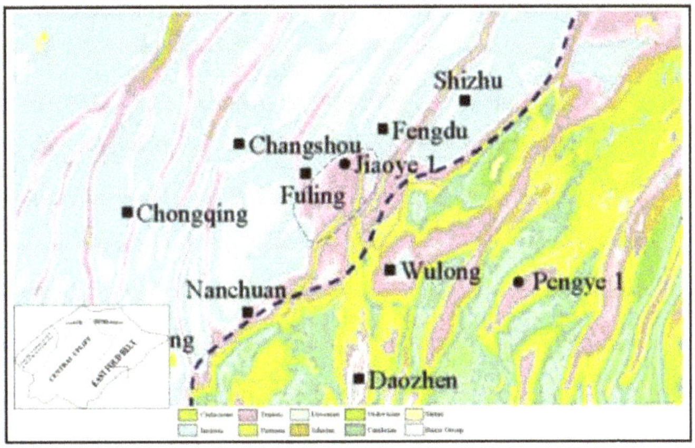

Figure 6. The location of the Fuling shale gas field.

The Wufeng–Longmaxi Fm. can be divided into three type of layers (Figure 7) which are: (1) section I, carbonaceous siliceous shale; (2) section II, gray black silty mudstones; (3) section III, gray argillaceous mudstones and gray black carbonaceous shale.

Sichuan Basin					
ERA	PERIOD	EPOCH	FORMATION		AGE(Ma)
PALEOZOIC	Permian	Lower	Maokou	TP_1m	230-270
			Qixia	TP_1q	
			Liangshan	TP_1l	
	Carboniferous	Middle	Huanglong	TC_2h	270-320
	Silurian	Middle	Hanjiadian	TS_2h	320-570
		Lower	Xiaoheba	TS_1x	
	The Wufeng-longamxi Fm.		Longmaxi -3	TS_1l^3	
			Longmaxi -2	TS_1l^2	
			Longmaxi -1	TS_1l^1	
	Ordovician	Upper	Wufeng	TO_3w	
			Jianchaogou	TO_3j	
		Middle	Baota	TO_2b	
		Lower	Meitan	TO_1m	
			Honghuayuan	TO_1h	
			Tongzhi	TO_1t	

Figure 7. Stratigraphy of the Paleozoic of the Sichuan Basin, highlighting the Wufeng-Longmaxi Formation.

4. Candidates Selection

The selection of an appropriate gas well de-liquification method depends on several factor [9,10], such as well and reservoir characteristics, field locations, operational conditions (power availability, temp, facility constraints, water production etc.), long term reservoir performance, change in production over the life of the well and economics. The table listed below (Table 1) shows the specifications needed for a well to conduct a gas lift operation [11].

Table 1. Candidate selection criteria for different gas well de-liquification methods.

Comparison	Selection Criteria	SRP	PCP	ESP	Jet Pump	GL
System Condition	System Complexity	Simple	Simple	Downhole Complex	Surface Complex	Surface Complex
	Initial Investment	Low	Low	High	High	Highest
	Operation Cost	Low	Low	High	Low	Low
Fluid Range,	Fluid Rate Range	1–100	10–200	80–700	10–500	30–3180
m^3/d	Maximum Range	300	250	1400	1590	–7945
Lifting Depth, m	Lifting Depth	<3000	<1500	<2000	<2000	<4000
	Maximum Depth	4421	1700	3084	3500	4500
Downhole Condition	Small Tubing Size	Not Applicable	Not Applicable	Not Applicable	Not Applicable	Not Applicable
	Multiple Layer Production	Not Applicable	Not Applicable	Applicable	Applicable	Applicable
	Slanted Well	Normal Wear	Normal Wear	Applicable	Applicable	Applicable
	Degree of Hollowing Out	High	Very High	High	Very High	High
Surface Condition	Offshore	Not Applicable	Very Applicable	Applicable	Applicable	Very Applicable
	Remote Area	Normal	Normal	Applicable	Applicable	Applicable
Operation	High GOR	Very Applicable	Normal	Not Applicable	Applicable	Very Applicable
	Heavy Oil	Applicable	Applicable	Not Applicable	Very Applicable	Not Applicable
	Sand Production	Very Applicable	Applicable	Not Applicable	Applicable	Very Applicable
	Corrosion	Applicable	Applicable	Applicable	Applicable	Applicable
	Scale	Applicable	Not Applicable	Not Applicable	Applicable	Applicable
	Adjust Working System	Convenient	Convenient	Not Convenient	Convenient	Convenient
	Power Supply	Electricity, Oil, NG	Electricity, Oil, NG	Electricity	Electricity, Oil, NG	Electricity, Oil, NG
	Power Medium Requirements	None	None	None	Hydrodynamic Fluid	Anti-hydrate
Maintenance Management	Pump Inspection	Tubing WO	Tubing WO	Tubing WO	Slickline Ops	Slickline Ops
	Avg. WO Period, a	2	1	1.5	0.5	3
	Auto-control	Applicable	Applicable	Applicable	Applicable	Applicable

Based on the selection criteria above, a workflow has been created to screen most candidate wells in the Fuling gas field. The tubing sizes are 48.26 mm and 60.33 mm with an average well depth of 4738 m and surface piping pressure of 4.87 MPa. All wells are horizontal wells that are completed with hydraulic fracturing. Six wells are new tied-in wells that are not activated due to a heavy fracturing liquid column and the rest of the wells are heavily water-loaded (Table 2). To fit the slickline operation, the gas lift valves should be set within the inclination of 38° and the gas lift valves were assembled on tubing during tubing run in hole operations.

Table 2. Well candidates for low pressure gas lift operation.

Series No.	Well Name	Tubing Size, mm	Casing Size, mm	Well Depth (MD), m	Piping Line Pressure, MPa	Tubing Setting Depth, m	New Tied in Well
1	FL-1HF	48.26	139.70	4576	5.5	3067	N
2	FL-2HF	48.26	139.70	4157	6.5	2620	Y
3	FL-3HF	48.26	139.70	5176	6.0	3564	Y
4	FL-4HF	48.26	139.70	4840	6.1	3277	N
5	FL-5HF	48.26	139.70	4840	6.3	3212	N
6	FL-6HF	48.26	139.70	5335	5.9	3527	N
7	FL-7HF	48.26	139.70	4471	5.7	2977	N
8	FL-8HF	48.26	139.70	4640	5.9	2506	Y
9	FL-9HF	48.26	139.70	4545	6.7	2976	N
10	FL-10HF	60.33	139.70	4324	6.2	2568	N
11	FL-11HF	48.26	139.70	4341	6.1	2717	Y
12	FL-12HF	48.26	139.70	4966	6.5	3452	Y
13	FL-13HF	48.26	139.70	5030	6.1	3248	Y

5. Project Validation

5.1. Unloading

As described in Section 4, the new gas lift valves were installed on tubing to restart heavily water-loaded wells. The gas lift valves that have been used are specially designed for low pressure wells due to its high sensitivity to low pressure. The setting depth was designed by PIPESIM by considering well completion, reservoir pressure and compressor capacity. The assumption is to determine the gas injection point depth under constant wellhead pressure. The curve below shows the basic principles of how to calculate the injecting point (Figure 8).

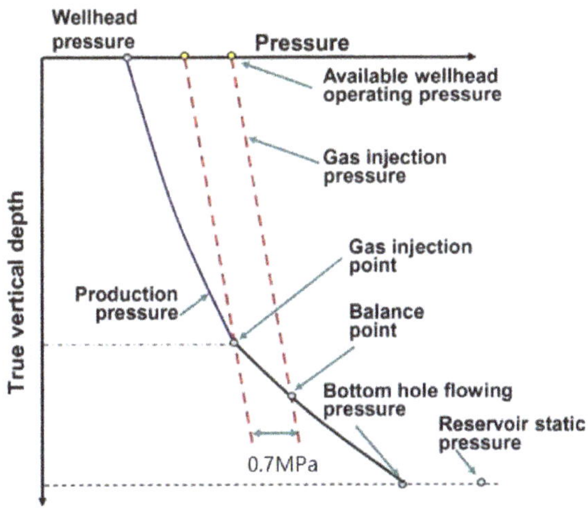

Figure 8. Depth of the operating gas injection point.

With variable sensitive data of GLR, the lifting liquid rate can be obtained at a different pressure gradient curve (Figure 9).

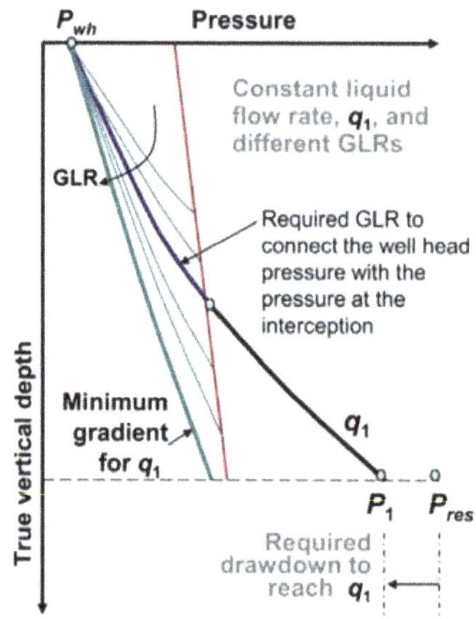

Figure 9. Procedure to find the liquid production at different pressure gradients.

Note that the liquid production here is the liquid that should be lifted to restart the loaded gas well.

5.2. Field Trail Result and Data Validation

The gas lift valves' specifications are listed as follows. Total length of the valve is 300 mm with OD of 18.8 mm (Table 3), which was assembled with a fixed mandrel that aligned with the tubing size. For example, with FL-2HF, the well was vented to the atmosphere after hydraulic fracturing, though no production was observed, and the tubing head pressure dropped to zero right after the opening of the well. With the proper design of the gas lift, six stages of gas lift valves were installed on tubing (Table 4) to lift the loaded water. The loaded liquid was gradually lifted to the surface, and gas production reached 33.8×10 kscm/d at the peak production rate. Although the well was shut in after the operation, the gas rate was restored to a higher number, mostly due to the pressure build up effect (Figure 10).

Table 3. Parameters of gas lift valve.

Type	SKY-GLV-18.8
OD, mm	18.8
Length, mm	300
Effective area of bellows, sq. mm	110
Connection Thread	1/4 NPT
Pressure Rating, MPa	50
Valve hole size, mm	3.17, 4.76

Table 4. Setting depth of gas lift valve—FL-2HF.

Kick off Pressure: 15 MPa Operating Pressure: 5.5 MPa Injecting Gas Rate: 0.8 10 kscm/d								
Stages	**Setting Depth (MD), m**	**Setting Depth (TVD), m**	**Inclination Angle, °**	**Valve Hole Size, mm**	**Pro, MPa**	**Pv-Open, MPa**	**Pv-Close, MPa**	**Mandrel Type**
1	470.18	470.13	0.6	3.2	5.032	5.098	4.954	SKY-FGLM-90
2	907.39	907.29	0.8	3.2	4.602	4.975	4.851	SKY-FGLM-90
3	1294	1293.9	0.7	3.2	4.382	4.852	4.747	SKY-FGLM-90
4	1629.6	1629.4	0.4	3.2	4.193	4.73	4.644	SKY-FGLM-90
5	1913.3	1913.1	6	3.2	4.037	4.61	4.541	SKY-FGLM-90
6	2157	2146.7	32.2	3.2	3.913	4.492	4.437	SKY-FGLM-90

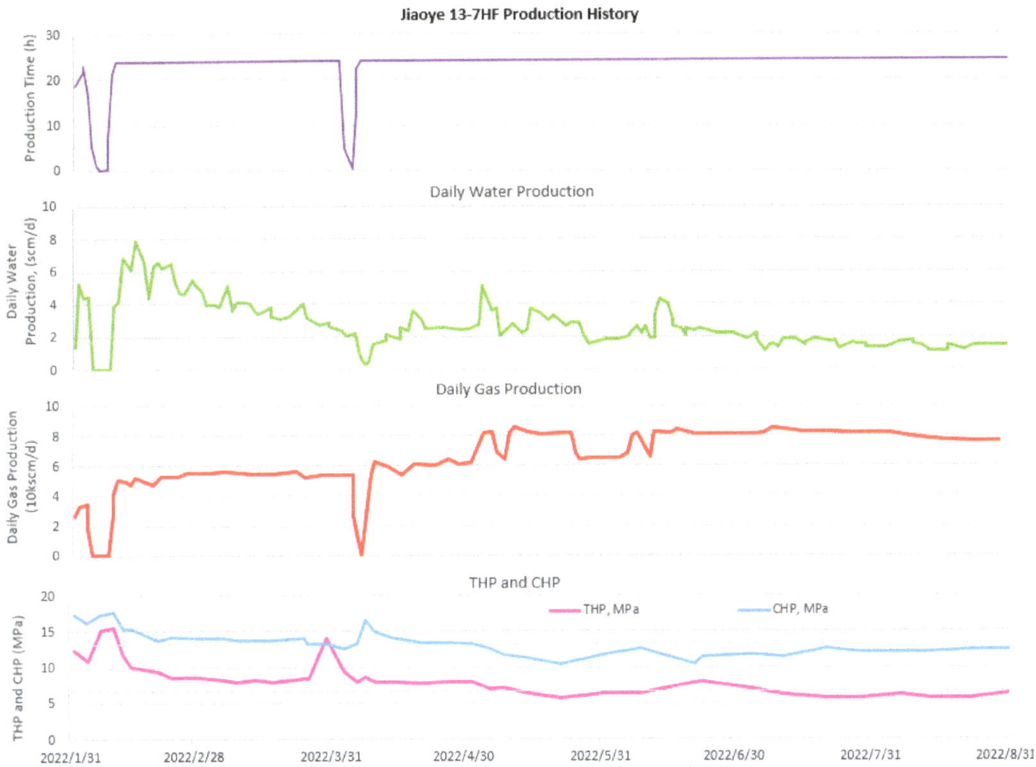

Figure 10. FL-2HF's production history.

A total of 13 wells were assembled with gas lift valves and 8 of them were operated successfully; the total incremental of gas rate is 219 × 10 kscm/d with a total lifted liquid volume of 104.7 scm (Table 5). The rest of the wells were waiting to operate once the equipment was ready. The field operation showed the great potential of the new gas lift valves in the Fuling gas field.

Table 5. Gas lift result with the new gas lift valves.

Well Name	Lifted Liquid Volume, scm	Gas Rate before GL Operation, 10 kscm /d	Gas Rate after GL Operation, 10 kscm/d	Incremental of Gas Rate, 10 kscm /d	Remarks
FL-1HF	52	0.3	1.83	1.53	Wells loaded with liquid
FL-2HF	4.6	0.1	3.48	3.38	New tied-in well shut in due to loaded liquid
FL-3HF	2	0.3	1.4	1.1	Wells loaded with liquid
FL-4HF	/	/	/	/	Waiting to be operated
FL-5HF	/	/	/	/	Waiting to be operated
FL-6HF	/	/	/	/	Waiting to be operated
FL-7HF	6	1.7	2.1	0.4	Wells loaded with liquid
FL-8HF	5.5	0.1	5.8	5.7	New tied-in well shut in due to loaded liquid
FL-9HF	23	0.05	2.4	2.35	Wells loaded with liquid
FL-10HF	6	0.1	4.3	4.2	Wells loaded with liquid
FL-11HF	5.6	0.1	3.2	3.1	New tied-in well shut in due to loaded liquid
FL-12HF	/	/	/	/	Waiting to be operated
FL-13HF	/	/	/	/	Waiting to be operated
Total	104.7			21.76	

6. Conclusions

A new type of gas lift valve that was specially designed for shale gas reservoir low-pressure wells has been introduced in this paper. On the basis of field operation, we derived conclusions as follows.

The new type of gas lift valve is very sensitive to low piping pressure due to its low valve dome pressure, and as a result it was used as an effective tool to re-activate the heavy water-loaded wells by staged gas lift operations. Field operation results show that the piping line pressure that is required for the gas lift operation can be as low as 5.5 MPa, which shows a prospective application for the shale gas field in the future. The low pressure gas wells gas lift operation process was created by the candidates selection criteria, and eight wells were selected to conduct the operation using the noval gas lift valves. Eight wells in the Fuling gas field was successfully restarted by this new type of well in tandem with the staged gas lift operation to unload water in wells. The total lifted water for the eight wells of one operation is 13 scm, and the total of gas production incremental is 218 $\times$ 10 kscm/d. The field operation cost can be greatly reduced due to its special function in which the dummy valve can be swtiched to the gas lift valve via pressure changes in the tubing. As a result, the operation risks can also be reduced to a low level.

Author Contributions: Conceptualization, Q.L. and J.T.; methodology, J.T.; validation, W.K. and Q.L.; formal analysis, J.T.; investigation, H.W.; resources, H.W.; data curation, Q.L.; writing—original draft preparation, J.T.; writing—review and editing, J.T.; visualization, J.T.; supervision, W.K.; project administration, U.D.O.; funding acquisition, W.K. All authors have read and agreed to the published version of the manuscript.

Funding: This research was funded by the project from Sinopec Chongqing Fuling Shale Gas Exploration and Development Co., Ltd. Grant number 33550000-21-FW0425-0001.

Institutional Review Board Statement: Not applicable.

Informed Consent Statement: Not applicable.

Data Availability Statement: Not applicable.

Conflicts of Interest: The authors declare no conflict of interest.

References

1. Tingxue, J.; Dehua, Z.; Changgui, J.; Haitao, W.; Xiaobing, B.; Shuangming, L.; Xiaobo, X.; Weiran, W.; Suyuan, S. The Study and Application of Multi-Stage Fracturing Technology of Horizontal Wells to Maximize ESRV in the Exploration & Development of Fuling Shale gas Play, Chongqing, China. In Proceedings of the SPE Asia Pacific Hydraulic Fracturing Conference, Beijing, China, 24–26 August 2016. Available online: https://onepetro.org/speaphf/proceedings/16APHF/2-16APHF/Beijing,%20China/185205 (accessed on 2 November 2022).
2. Yaowen, L.; Yuanzhao, L.; Chi, Z.; Yue, M.; Jialin, X.; Rong, H.; Zichao, W.; Jiao, Z.; Wu, C. First Successful Application of Casing in Casing CiC Refracturing Treatment in Shale Gas Well in China: Case Study. In Proceedings of the Abu Dhabi International Petroleum Exhibition & Conference, Abu Dhabi, United Arab Emirates, 15–18 November 2021. Available online: https://onepetro.org/SPEADIP/proceedings/21ADIP/1-21ADIP/D011S006R002/473700 (accessed on 2 November 2022).
3. Guo, T.; Li, J.; Lao, M.; Li, W. Integrated Geophysical Technologies for Unconventional Reservoirs and Case Study within Fuling Shale Gas Field, Sichuan Basin, China. In Proceedings of the Unconventional Resources Technology Conference, San Antonio, TX, USA, 20–22 July 2015; p. 10.
4. Zhang, L. Application and Prospect of Gas Lift Technology with Gas Lift Valves in Fuling Shale Gas Field. *J. Jianghan Pet. Univ. Staff. Work.* **2022**, *35*, 20–21, 60.
5. Di, D.; Pang, W.; Mao, J.; Ai, S.; Ying, H. Production Logging Application in Fuling Shale Gas Play in China. In Proceedings of the SPE Asia Pacific Oil & Gas Conference and Exhibition, Perth, Australia, 25–27 October 2016. Available online: https://onepetro.org/SPEAPOG/proceedings/16APOG/All-16APOG/Perth,%20Australia/185405 (accessed on 5 November 2022).
6. Wei, P.; Zuqing, H.; Cuiping, X.; Juan, D.; Sun, Z. SRV Analysis of Shale Gas Wells in China. In Proceedings of the SPE Asia Pacific Unconventional Resources Conference and Exhibition, Brisbane, Australia, 9–11 November 2015. Available online: https://onepetro.org/SPEURCE/proceedings/15URCE/All-15URCE/Brisbane,%20Australia/183964 (accessed on 5 November 2022).
7. Demoss, E.E.; Ellis, R.C.; Kingsley, G.S. New Gas-Lift Concept-Continuous-Flow Production Rates from Deep, Low-Pressure Wells. *J. Pet. Technol.* **1974**, *26*, 13–18. [CrossRef]
8. Azizollah, K.; Petrakov, D.G.; Benson, L.A.B.; Rastegar, R.A. Prevention of Calcium Carbonate Precipitation during Water Injection into High-Pressure High-Temperature Wells. In Proceedings of the SPE European Formation Damage Conference and Exhibition, Budapest, Hungary, 3–5 June 2015. Available online: https://onepetro.org/SPEEFDC/proceedings/15EFDC/All-15EFDC/Budapest,%20Hungary/183005 (accessed on 5 November 2022).
9. Wang, Y.; Tian, Z.; Yang, L.; Yan, X.; Yi, X.; Lu, H.; Yaoyao, D. What We Have Learned on Shale Gas Fracturing During the Past Five Years in China. In Proceedings of the SPE Asia Pacific Oil & Gas Conference and Exhibition, Perth, Australia, 25–27 October 2016. Available online: https://onepetro.org/SPEAPOG/proceedings/16APOG/All-16APOG/Perth,%20Australia/185283 (accessed on 11 November 2022).
10. Priscilla, E.; Ademola, A.; Oluwafemi, O.; Nchekwube, L.; Emmanuel, M.; Fred, O.; Fatoke, O. Evaluating Alternate Artificial Lift Methods in the Niger Delta. In Proceedings of the SPE Nigeria Annual International Conference and Exhibition, Lagos, Nigeria, 2–4 August 2021. Available online: https://onepetro.org/SPENAIC/proceedings/21NAIC/3-21NAIC/D031S019R005/465696 (accessed on 11 November 2022).
11. Wei, P.; Qiong, W.; Ying, H.; Juan, D.; Tongyi, Z.; Christine, A. Production Analysis of One Shale Gas Reservoir in China. In Proceedings of the SPE Annual Technical Conference and Exhibition, Houston, TX, USA, 28–30 September 2015. Available online: https://onepetro.org/SPEATCE/proceedings/15ATCE/2-15ATCE/Houston,%20Texas,%20USA/180473 (accessed on 11 November 2022).

processes

MDPI

Article

Study on the Flow Pattern and Transition Criterion of Gas-Liquid Two-Phase Flow in the Annular of Shale Gas Fractured Horizontal Wells

Yu Lei [1], Zhenghua Wu [1], Wei Wang [1,2,*], Jian Wu [1] and Bin Ma [1]

1 Gas-Lift Technology Center of Tuha—CNPC, Shanshan, Turpan 838200, China
2 College of Petroleum Engineering, Yangtze University, Wuhan 430100, China
* Correspondence: llliuzq@163.com

Abstract: Improving the accuracy of pressure prediction in the wellbore annular is of great importance for the design in oil and gas production. However, due to the existence of double-layer liquid membrane and the lack of relevant experiments, the existing correlations fail to the field application. In this study, a new model of flow pattern transition in inclined annulus pipe is proposed by using a mechanistic approach to classify the flow patterns. Firstly, a gas-liquid two-phase flow experiment in annulus pipe was carried out in a pipe with an outer diameter of 73.02 mm and an inner diameter of 121.36 mm, and then the influence of inclined angle on the transition boundary of flow pattern is discussed. Finally, a hydrodynamic transition criterion for the flow pattern model of inclined annulus pipe is established and verified in detail. The experimental results show that bubble flow, slug flow, churn flow and annular flow were observed under different inclination angles, and the results indicate that the slug flow will be shifted to the larger gas-liquid superficial flow rate region with the smaller inclination angle, and the annular flow will appear in the higher gas superficial flow rate region. Compared to the performance of the existing correlations (Kelessidis and Zhang) and the present model using the experimental data, the accuracy of the new model reached 83%, significantly higher than the other two models, and the new correlation was better in predicting the transition from slug flow to churn flow and churn flow to annular flow.

Keywords: annulus pipe; gas-liquid two-phase flow; flow pattern transition; slug flow

Citation: Lei, Y.; Wu, Z.; Wang, W.; Wu, J.; Ma, B. Study on the Flow Pattern and Transition Criterion of Gas-Liquid Two-Phase Flow in the Annular of Shale Gas Fractured Horizontal Wells. *Processes* **2022**, *10*, 2630. https://doi.org/10.3390/pr10122630

Academic Editors: Linhua Pan, Yushi Zou, Jie Wang, Minghui Li, Wei Feng and Lufeng Zhang

Received: 24 October 2022
Accepted: 1 December 2022
Published: 7 December 2022

1. Introduction

Hydraulic fracturing can effectively improve the recovery rate of shale gas and has become the main development method of shale gas fields. However, the flowback of shale gas fracturing fluid is complex and easy to accumulate fluid, which seriously restricts gas well productivity [1,2]. Prediction of pressure drop in annulus pipe is the basis for determining shale gas wellbore fluid accumulation and drainage process preferences. In addition, the conventional oil and gas extraction process, such as underbalanced drilling and spontaneous injection wells at high production rates, will involve the flow of two-phase flow in annulus pipe [3].

The division of flow pattern boundaries is a prerequisite for the calculation of annular air-water two-phase flow parameters. However, the flow patterns, especially slug flow and annular flow, in the annulus pipe are different from those in the circular pipe. There are two liquid films in the annular two-phase flow, one is the casing film in contact with the casing wall (or well wall), and the other is the tubing film in contact with the tubing wall. In addition, according to the experimental observations of Caetano [4], Taylor bubbles are no longer symmetrical and there is a region of high turbulence behind the Taylor bubble, due to the presence of a channel connecting the tubular membrane to the casing membrane in the dorsal direction. The annular flow in the annulus pipe occurs at very high gas flow rates, the gas phase contains liquid droplets, and a very thin liquid film around the core.

Due to the structure of the annular, two liquid films also exist, an inner film in contact with the tubing wall and an outer film in contact with the casing wall, the outer film being thicker than the inner film. At high superficial liquid-phase flow rates, Taitel [5] suggested that turbulent disturbance forces cause the gas phase in a bubbling or elastomeric flow to shatter into smaller bubbles, which are dispersed inside the continuous liquid phase. When the turbulence intensity can be sufficient to disperse Taylor bubbles into smaller bubbles than the critical bubble size, it will prevent them from re-aggregating, and the dispersed bubble flow will occur.

Few studies are devoted to its flow pattern and flow pattern distribution diagram. The accuracy of transition prediction for slug flow and annular flow is poor. There are even fewer studies on the flow pattern of inclined pipe annular two-phase flow. In 1982, Sadatomi studied two-phase flow in a non-circular pipe with air and water [6]. He studied the flow pattern and pressure drop of gas-liquid two-phase flow in a vertical concentric annulus pipe with a diameter ratio N of 0.5. However, he classified the types of flow patterns roughly and did not give a specific model for the transition between flow patterns. In 1989, Kelessidis measured the local gas content of in annulus pipe by the electric probe method and studied on the transition of flow patterns, in which the bubbly flow was predicted more accurately and the slug flow, and annular flow were predicted with larger errors [7]. In 1992, Caetano gave a more detailed description of the types of flow patterns and the characteristics different from those of circular pipe flow patterns [4]. Taitel's circular pipe flow pattern transition model was modified and applied to vertical annulus pipe. The flow pattern distribution of vertical concentric and vertical eccentric annulus pipe was given. In 1992, Hasan applied the drift flow model to investigate the gas content rate of the annular flow pattern. A correction was presented to the calculation equation for the rise velocity of Taylor bubble, and a more accurate equation for the calculation of the gas content rate of inclined pipe annular was obtained [8]. In 2003, Zhang established a vertical annular hollow pipe section slug flow and annular flow transition model, and studied the vertical annular air-liquid two-phase flow with inner pipe rotation [9]. In 2014, Yin established a hydrodynamic model applicable to vertical and inclined tubular annular air-liquid two-phase flow, and modified the annular flow pattern transition criterion, which has better accuracy compared to the previous models [10].

At present, the studies for the annular air-liquid two-phase flow in wellbore are based on vertical and horizontal pipes [11–21]. Mostly, the concept of equivalent diameter is used to analyze and calculate with the help of the circular pipe method. There are fewer studies on the two-phase flow in the annular of inclined pipes. Due to the influence of pipe inclination on the flow pattern of annular two-phase flow, and the existence of a double liquid film, the flow state in the annulus pipe is more complex than the circular pipe flow. Therefore, the direct application of the vertical pipe flow pattern transition criterion may lead to large errors.

Due to the complex characteristics of annular air liquid two-phase flow and the lack of relevant experiments, in this study, a gas-liquid two-phase flow experiment in annulus pipe was carried out, and then the influence of inclined angle on the transition boundary of flow pattern is discussed. The experimental results indicate that the slug flow will be shifted to the larger gas-liquid superficial flow rate region with the smaller inclination angle, and the annular flow will appear in the higher gas superficial flow rate region. A hydrodynamic transition criterion for the flow pattern model of inclined annulus pipe is established and verified in detail, the results show that the new correlation has better performance in predicting the transition from slug flow to churn flow, and churn flow to annular flow.

2. Experiment and Analysis

2.1. Experimental Setup and Process

The experiments were conducted in the multiphase flow experimental platform of CNPC (China National Petroleum Corporation), as shown in Figure 1. The platform can carry out the study of multiphase pipe flow dynamics in single-phase, gas-liquid two-phase, and oil-gas-water three-phase under various conditions such as different inclination angles and different gas-liquid velocities. The platform is composed of nine parts, including simulated wellbore, oil-water stabilization system, gas stabilization system, and cooling water system. It is equipped with independent metering modules such as gas and liquid flow standard devices and piston type manometer standard devices, respectively. The accuracy of liquid flow meter is 0.3%, gas flow meter is 1%, and the accuracy of differential pressure sensor can be up to 0.25%, and high-speed camera frequency is 500 fps with a resolution of 1920 × 1080, which provides guarantee for the accuracy of measurement of experimental results.

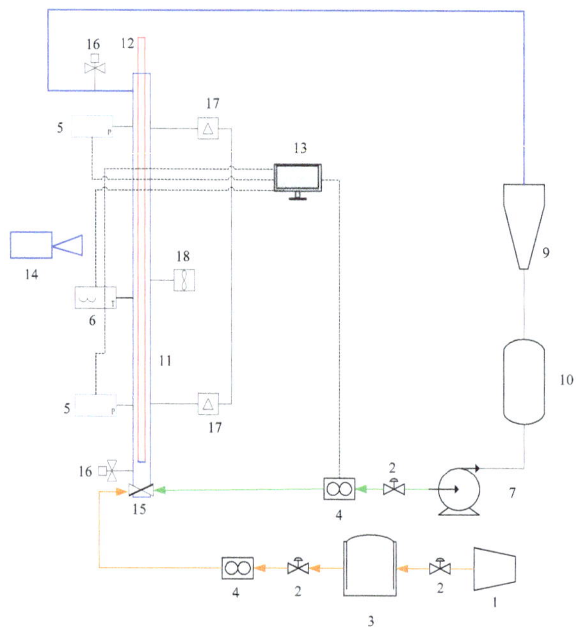

Figure 1. Flow chart of the experimental device for circulating air-liquid two-phase flow. 1—Air compressor; 2—Valves; 3—Air storage tank; 4—Flow meter; 5—Pressure sensor; 6—Temperature sensor; 7—Pump; 9—Gas-liquid separator; 10—Oil-water mixing tank; 11—Casing; 12—Oil pipe; 13—Data acquisition system; 14—High-speed camera; 15—Mixer; 16—Quick-closing valve; 17—Differential pressure sensor; 18—Capacitance probe.

The experiment was conducted in an annulus pipe with an outer diameter of 73.02 mm and an inner diameter of 121.36 mm. The experimental length of the pipe is 7.55 m, and the distance of the quick-closing valve is 11.32 m. The liquid-phase is water, and the gas-phase is air. In the room temperature of 11~14 °C environment, the specific experimental operations are as follows. The range of experimental parameters as shown in Table 1.

Table 1. Range of experimental parameters.

Angle (°)	30, 45, 60, 90
Liquid superficial velocity (m/s)	0~0.63
Gas superficial velocity (m/s)	0~56.83

2.2. *Analysis of Experimental Results*

In the experiment, bubble flow, slug flow, churn flow, and annular flow can be observed, and the flow patterns were captured by high-speed camera. The experimental phenomena and results of different inclination angles are shown in Figures 2–5.

(**a**) Bubble flow (**b**) Slug flow (**c**) Churn flow (**d**) Annular flow

Figure 2. Flow pattern observed in the experiment at an inclination angle of 30°.

(**a**) Bubble flow (**b**) Slug flow (**c**) Churn flow (**d**) Annular flow

Figure 3. Flow pattern observed in the experiment at an inclination angle of 45°.

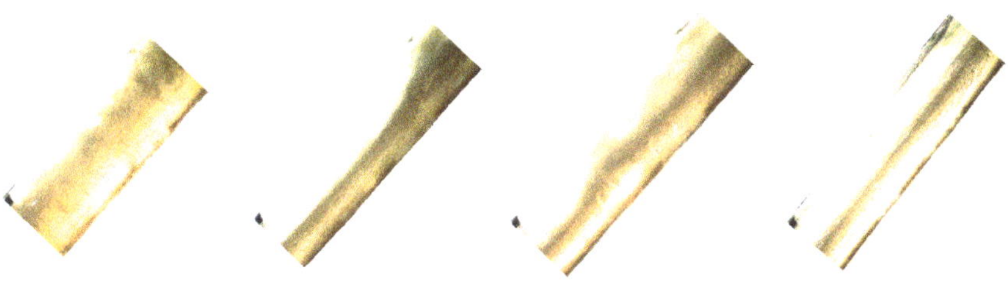

(**a**) Bubble flow (**b**) Slug flow (**c**) Churn flow (**d**) Annular flow

Figure 4. Flow pattern observed in the experiment at an inclination angle of 60°.

(**a**) Bubble flow (**b**) Slug flow (**c**) Churn flow (**d**) Annular flow

Figure 5. Flow pattern observed in the experiment at an inclination angle of 90°.

In the experiment, the slug flow consists of the air embolism zone and the liquid embolism zone. The Taylor bubble and the liquid film around it can be observed in the segment plug flow gas bolus region. When the gas flow velocity (range from 0.4 m/s to 7.8 m/s) is small, the liquid film in the gas embolus area will show obvious liquid fall back due to the effect of gravity. The returned liquid flows along the underside of the annulus pipe to form a flow channel. When the gas flow velocity continues to increase (greater than 7.8 m/s), the liquid backflow phenomenon weakens, and the segment plug frequency accelerates and gradually transforms to churn flow. After the superficial gas flow velocity is greater than 15 m/s, it can be observed that the flow pattern gradually transforms to annular flow. The gas is in the middle of the annular space, and the liquid adhere to the annulus pipe wall to form a double-layer liquid film.

When the physical parameters of medium, pipe diameter, gas and liquid superficial flow velocities are determined, the flow patterns of Kelessidis [7], Caetano [4], Hasan [8], and Zhang [9] can be plotted, as shown in Figure 6. From the figure, it can be found that all the above four flow pattern prediction models can accurately predict the transition boundary between the bubbly flow and the slug flow. However, there are large errors in the transition from slug flow to churn flow, and churn flow to annular flow. Among them, the transition boundaries predicted by Hasan [8] and Caetano [4] models differ greatly from the experimental data. The experimental data show that the superficial gas velocity increases with the increase in the superficial liquid velocity during the transition from the churn flow to the annular flow. The superficial gas velocity is not constant as predicted by the Hasan [8] and Caetano [4] model. Compared to the Kelessidis [7] model, the boundary trend of the transition from slug flow to churn flow is more accurately predicted by the Zhang [9] model, but the superficial gas velocity is shifted in a smaller direction. The Kelessidis [7] model is more consistent with the present experimental data in the boundary prediction of the transition from churn flow to slug flow.

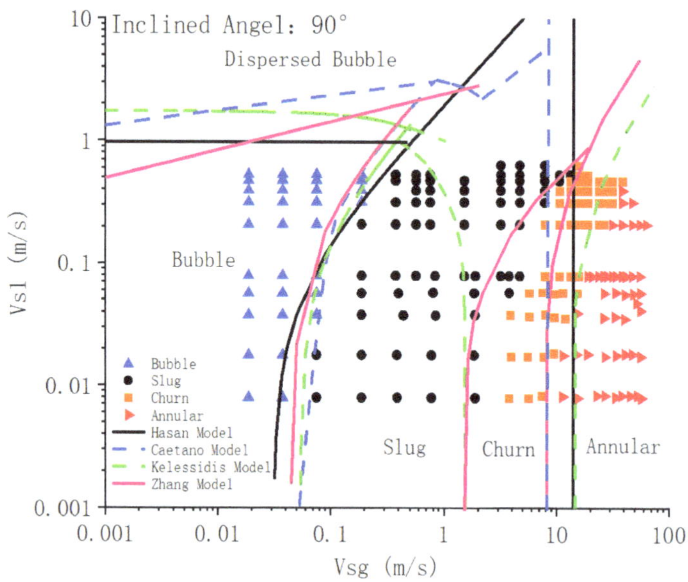

Figure 6. Comparison between flow pattern diagrams of different models and experiments with 90° inclination angle.

The flow pattern observed in the experiment at different angles were plotted, as shown in Figures 7–9. From the experiment, it can be observed that when the tilt angle is changed, the transition boundary from slug flow to churn flow and annular flow will be shifted significantly. When the tilt angle is changed from 90° to 30°, the transition boundary of the slug flow to the churn flow and the annular flow also shifts to the direction of greater superficial gas flow velocity. Since the Kelessidis [7] and Zhang [9] model only gives the flow transition boundary of the vertical annulus pipe, it cannot accurately predict the flow transition of the inclined pipe. Therefore, it can be seen that when the annulus pipe is inclined, the liquid film at the bottom becomes thicker due to the influence of gravity. The shapes of bubbles in complex flow patterns such as slug flow are changed. There is also a special double liquid film in the annulus pipe, the flow state in the annulus pipe is more complicated than the circular pipe flow. Therefore, the direct application of the vertical pipe flow pattern transition formula may lead to large errors.

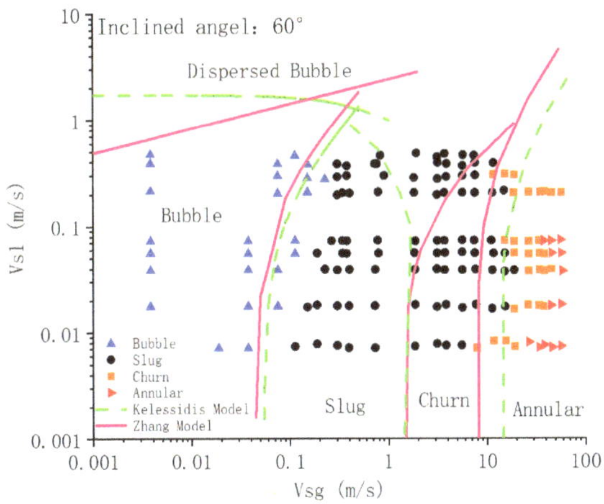

Figure 7. Comparison between flow pattern diagrams of different models and experiments with 60° inclination angle.

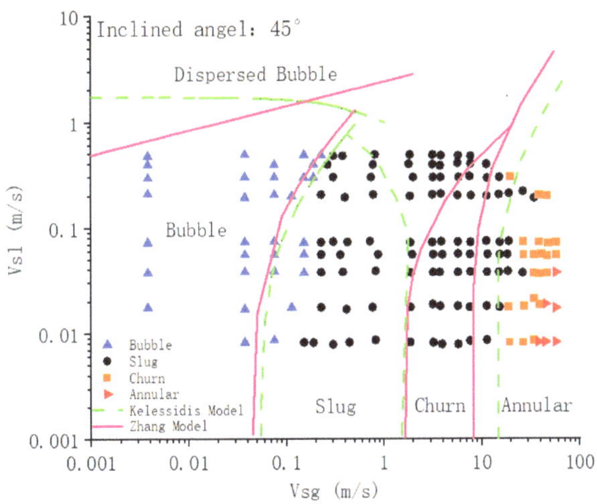

Figure 8. Comparison between flow pattern diagrams of different models and experiments with 45° inclination angle.

In this paper, the flow pattern discriminant diagrams of Kelessidis [7] and Zhang [9] models will be modified by combining experimental phenomena. The flow pattern is classified by using the mechanism approach and a new flow transition model for the slug flow in the inclined annulus pipe section is proposed.

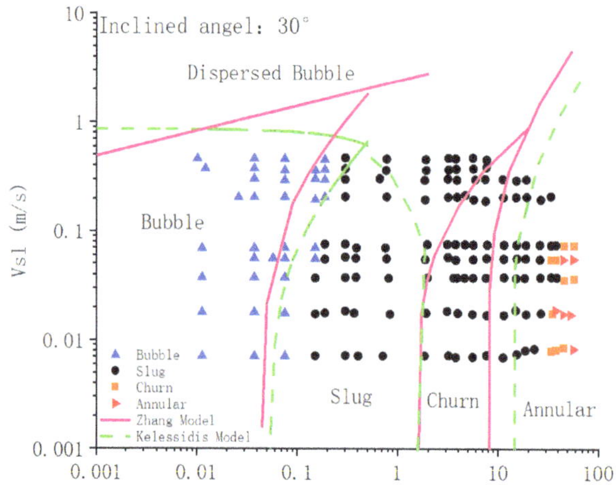

Figure 9. Comparison between flow pattern diagrams of different models and experiments with 30° inclination angle.

3. Model Building and Discussion

3.1. Establishment of the Flow Model for the Slug Flow in The Annulus Pipe Section

Most of the existing annular flow pattern transition guidelines are established in vertical pipe conditions. However, due to the influence of gravity, the gas-liquid distribution of the inclined pipe slug flow is very different compared to that of the vertical pipe. From the experimental observation, it is clear that the liquid phase of the slug flow in the inclined annulus pipe becomes asymmetrically distributed. In the middle of the double liquid film of the annulus pipe, a preferential channel is formed. This paper is based on the dynamic model of the slug flow in the annulus pipe. The effects of gas-liquid slip and negative frictional pressure drop are considered. The mass conservation and momentum conservation analysis is carried out for the annular slug unit body. Figure 10 shows the schematic diagram of the velocity and force analysis of the slug flow in the annulus pipe.

The casing film is much thicker than the oil pipe film in the gas embolism area, the oil pipe film H_{LFd} is much smaller than the casing film H_{LFc}. The oil pipe membrane is much smaller than the casing membrane. Therefore, the oil pipe film is ignored in the slug flow model of this paper.

According to the experimental observation, the slug flow unit in steady state can be composed of a Taylor bubble and a section of liquid bolus. In the middle of the Taylor bubble, a preferential channel is formed. The fluid in the air bolus area will flow backward through this channel. It is assumed that the fluid in the control body of the cell is incompressible. The mass of the liquid phase flowing in from the bottom boundary of the liquid film is equal to the mass of the liquid phase flowing out from the top boundary of the liquid film.

$$(v_{\mathrm{T}} - v_{\mathrm{LLS}})H_{\mathrm{LS}} = (v_{\mathrm{T}} + v_{\mathrm{LFc}})H_{\mathrm{LFc}} + (v_{\mathrm{T}} - v_{\mathrm{c}})H_{\mathrm{Lc}} \quad (1)$$

where: v_{T} is the liquid slug transport velocity, m/s. v_{LLS} is the liquid phase flow velocity in the liquid slug area, m/s. H_{LS} is the liquid holding rate in the liquid slug area, %. v_{LFc} is the casing membrane flow velocity, m/s. H_{LFc} is the casing membrane holding rate, %. v_{c} is the Taylor bubble flow velocity, and m/s. H_{Lc} is the Taylor bubble holding rate, %.

Similarly, the mass of the gas phase entering the liquid membrane is equal to the mass of the gas phase leaving the liquid membrane. The mass conservation equation can be established.

$$(1 - H_{\mathrm{LS}})(v_{\mathrm{T}} - v_{\mathrm{GLS}}) = (v_{\mathrm{T}} - v_{\mathrm{c}})(1 - H_{\mathrm{LFc}} - H_{\mathrm{Lc}}) \quad (2)$$

where, v_{GLS} is the gas phase flow velocity in the liquid bolus region, m/s.

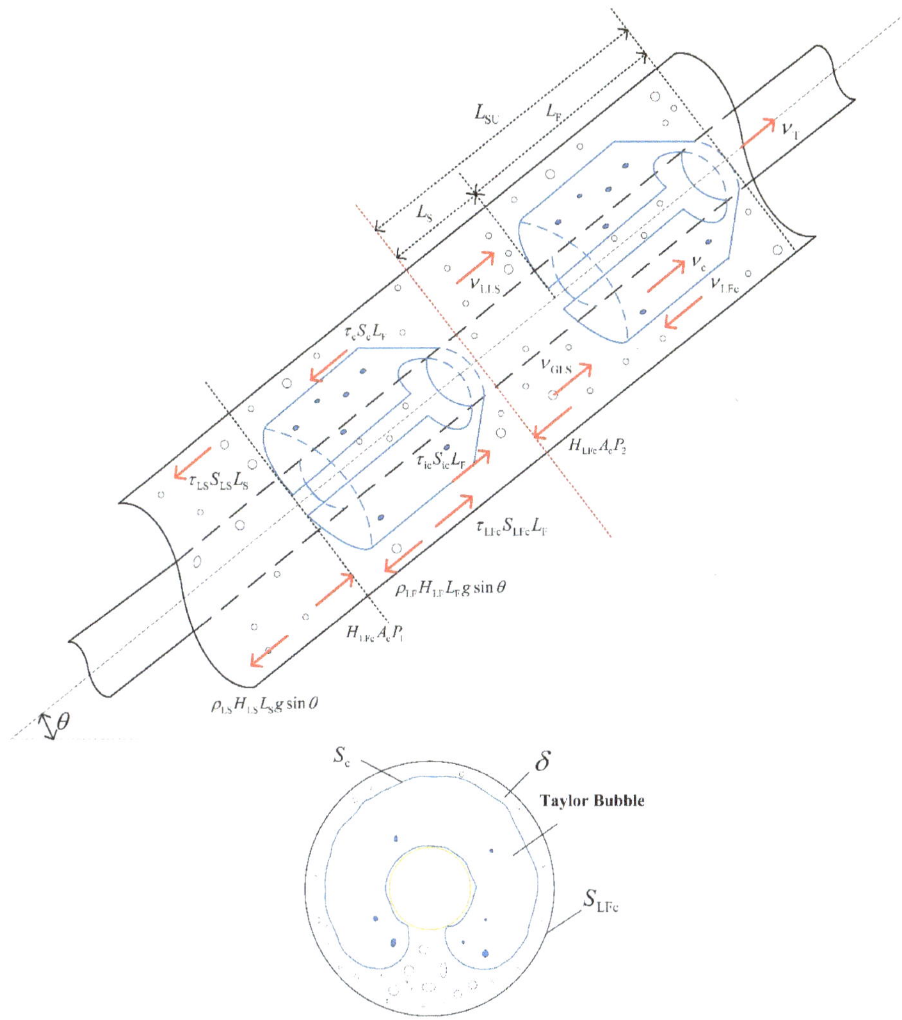

Figure 10. Schematic diagram of slug flow velocity and force analysis in the annular section.

Consider the liquid and gas in the liquid bolus region there is a slippage situation, the gas flow rate v_{GLS} should be greater than the liquid flow rate v_{LLS}. Combined with the slip velocity defined by Caetano [4] in the following equation.

$$v_{\mathrm{s}} = v_{\mathrm{GLS}} - v_{\mathrm{LLS}} = 1.53\left[\frac{(\rho_{\mathrm{l}} - \rho_{\mathrm{g}})g\sigma}{\rho_{\mathrm{l}}^{2}}\right]^{0.25}(H_{\mathrm{LS}})^{0.5} \tag{3}$$

$$v_{\mathrm{LLS}} = (v_{\mathrm{sl}} + v_{\mathrm{sg}}) - 1.53\left[\frac{(\rho_{\mathrm{l}} - \rho_{\mathrm{g}})g\sigma}{\rho_{\mathrm{l}}^{2}}\right]^{0.25}(H_{\mathrm{LS}})^{0.5}(1 - H_{\mathrm{LS}}) \tag{4}$$

where, v_{s} is the liquid slug flow rate, m/s. v_{sl} is the liquid superficial flow rate, m/s. v_{sg} is the gas superficial flow rate, m/s. ρ_{l} is the liquid phase bulk density, and kg/m^3. ρ_{g} is the gas phase bulk density, kg/m^3.

Therefore, combining the above equation yields.

$$v_{\mathrm{LFc}} H_{\mathrm{LFc}} = v_{\mathrm{s}} H_{\mathrm{LS}} + (1 - H_{\mathrm{LFc}}) v_{\mathrm{c}} - v_{\mathrm{GLS}} \tag{5}$$

According to Zhang [9], the equation for the gas content in the liquid bolus region is known.

$$\alpha_{\mathrm{s}} = \frac{v_{\mathrm{sg}}}{C_2 + C_3 v_{\mathrm{m}}} \tag{6}$$

where, for the values of C_2 and C_3, Schmidt [22] gave values of 0.331 and 1.25. Zhang [9] suggested that 0.425 and 2.65 are more applicable for the annular, and the liquid holding rate H_{LS} in the liquid bolus region can be calculated.

$$H_{\mathrm{LS}} = 1 - \alpha_{\mathrm{s}} \tag{7}$$

Since the liquid content rate and the tubular membrane holding rate in Taylor bubble are much smaller than the casing membrane. Therefore, the casing membrane liquid-holding rate H_{LFc} can be approximately equal to the liquid-holding rate H_{LF} in the gas embolism zone.

$$H_{\mathrm{LF}} = \frac{4D_{\mathrm{C}}}{D_{\mathrm{C}} + D_{\mathrm{T}}} \left(\frac{\delta}{D_{\mathrm{e}}} \right) - \frac{4D_{\mathrm{e}}}{D_{\mathrm{C}} + D_{\mathrm{T}}} \left(\frac{\delta}{D_{\mathrm{e}}} \right)^2 \tag{8}$$

$$\frac{L_{\mathrm{LS}}}{L_{\mathrm{SU}}} = \frac{v_{\mathrm{sl}} + v_{\mathrm{LFc}} H_{\mathrm{LF}}}{v_{\mathrm{LLS}} H_{\mathrm{LS}} + v_{\mathrm{LFc}} H_{\mathrm{LF}}} \tag{9}$$

where, D_{C} is the casing diameter, m. D_{T} is the tubing diameter, m. D_{e} is the annular hollow pipe hydraulic diameter, m. δ is the casing liquid film thickness, m. L_{LS} is the liquid plug zone length, and m. L_{SU} is the segment plug unit length, m.

From the cross-sectional diagram of the slug flow in Figure 10, it can be seen that Taylor bubble do not occupy the total cross-sectional area of the pipe. Instead, a preferential channel is formed. The liquid film around the Taylor bubble wets the tubing and casing wall through the priority channel. The liquid film thickness is calculated using the average casing liquid film converted thickness with the following equation.

$$\delta = 9.06 \times 10^{-9} \left[\frac{3 v_{\mathrm{l}} v_{\mathrm{TB}}}{2g} (D_{\mathrm{e}}/2 - \delta) \right]^{-0.733} \tag{10}$$

where, v_{l} is the liquid phase kinematic viscosity, Pa · s. v_{TB} is the Taylor bubble kinematic velocity, m/s.

According to the empirical equation of Zhang [9], the equation of the thickness of the falling liquid film around the Taylor bubble is shown in the following equation.

$$\delta \left[\frac{(\rho_{\mathrm{l}} - \rho_{\mathrm{g}}) g}{\rho_{\mathrm{l}} {v_{\mathrm{l}}}^2} \right]^{1/3} = k \mathrm{Re}_f^{\mathrm{m}} \tag{11}$$

where, k and m are related to the flow pattern, and the values taken in this paper are k = 0.0682 and m = 2/3.

$$\mathrm{Re}_f = \frac{4 v_{\mathrm{LFc}} \delta}{v_{\mathrm{l}}} \tag{12}$$

From Equations (10)–(12), the liquid film flow velocity can be found. Combined with the mass conservation equation, the Taylor bubble flow velocity v_{c} can be calculated.

The force analysis of the casing film and Taylor bubble in the air embolism area in the slug flow unit of the annular section is carried out separately, which is known according to the law of conservation of momentum.

$$H_{\mathrm{LF}} A_{\mathrm{c}} \frac{\Delta P}{L_{\mathrm{LF}}} = H_{\mathrm{LF}} A_{\mathrm{c}} \rho_{\mathrm{l}} g \sin\theta - \tau_{\mathrm{LFc}} S_{\mathrm{LFc}} - \tau_{\mathrm{ic}} S_{\mathrm{ic}} \tag{13}$$

$$(1-H_{\mathrm{LF}})A_{\mathrm{c}}\frac{\Delta P}{L_{\mathrm{LF}}}=(1-H_{\mathrm{LF}})A_{\mathrm{c}}\rho_{\mathrm{g}}g\sin\theta+\tau_{\mathrm{c}}S_{\mathrm{c}}+\tau_{\mathrm{ic}}S_{\mathrm{ic}} \tag{14}$$

Similarly, the conservation of momentum within the liquid bolus.

$$H_{\mathrm{LS}}A_{\mathrm{c}}\frac{\Delta P}{L_{\mathrm{LS}}}=H_{\mathrm{LS}}A_{\mathrm{c}}\rho_{\mathrm{l}}g\sin\theta+\tau_{\mathrm{LS}}S_{\mathrm{LS}} \tag{15}$$

where, τ_{LFc} is the shear force at the contact interface between the casing wall and the casing membrane, Pa. τ_{c} is the shear force at the contact interface between the casing liquid membrane and the Taylor bubble, Pa. τ_{LS} is the shear force at the contact surface between the liquid slug and the casing membrane tubular membrane, Pa. S_{LFc} is the wet perimeter of the casing wall, m. S_{c} is the wet perimeter of the Taylor bubble, m. S_{ic} is the wet perimeter of the contact interface between the casing membrane and the Taylor bubble, and m. S_{id} is the wet perimeter of the contact interface between the tubular membrane and the Taylor bubble, m. According to Yin [10], the wall shear force in the flow conservation equation of the annular section slug is calculated as follows.

$$\tau_{\mathrm{LFc}}=f_{\mathrm{LFc}}\rho_{\mathrm{LF}}v_{\mathrm{LFc}}^{2}/2 \tag{16}$$

$$\tau_{\mathrm{c}}=f_{\mathrm{c}}\rho_{\mathrm{c}}v_{\mathrm{c}}^{2}/2 \tag{17}$$

$$\frac{\tau_{\mathrm{LS}}S_{\mathrm{LS}}}{A_{\mathrm{c}}}=\frac{2f_{\mathrm{s}}}{D_{\mathrm{e}}}\rho_{\mathrm{s}}(v_{\mathrm{sg}}+v_{\mathrm{sl}})^{2} \tag{18}$$

Friction factors:

$$f_{\mathrm{LFc}}=C_{1}\mathrm{Re}_{\mathrm{LFc}}^{a} \tag{19}$$

$$f_{\mathrm{c}}=C_{2}\mathrm{Re}_{\mathrm{c}}^{b} \tag{20}$$

$$f_{\mathrm{s}}=C_{3}\mathrm{Re}_{\mathrm{s}}^{c} \tag{21}$$

where, f_{LFc} is the friction factor of casing film. f_{c} is the friction factor of tubing film. f_{s} is the friction factor of Taylor bubble.

The hydrodynamic equivalent diameter method is used. The Reynolds number at the interface between the casing film, the tubing film, and the Taylor bubble can be expressed as the following equation.

$$\mathrm{Re}_{\mathrm{LFc}}=\frac{v_{\mathrm{LFc}}\rho_{\mathrm{LF}}d_{\mathrm{LFc}}}{\mu_{\mathrm{LF}}},\ d_{\mathrm{LFc}}=\frac{4H_{\mathrm{LF}}A_{\mathrm{c}}}{(S_{\mathrm{LFc}}+S_{\mathrm{c}})} \tag{22}$$

$$\mathrm{Re}_{\mathrm{c}}=\frac{v_{\mathrm{c}}\rho_{\mathrm{c}}d_{\mathrm{c}}}{\mu_{\mathrm{c}}},\ d_{\mathrm{c}}=\frac{4(1-H_{\mathrm{LF}})A_{\mathrm{c}}}{S_{\mathrm{c}}} \tag{23}$$

$$\mathrm{Re}_{\mathrm{s}}=\frac{v_{\mathrm{LS}}\rho_{\mathrm{LS}}d_{\mathrm{LS}}}{\mu_{\mathrm{LS}}},\ d_{\mathrm{LS}}=\pi(D_{\mathrm{C}}+D_{\mathrm{T}}) \tag{24}$$

where, $\mathrm{Re}_{\mathrm{LFc}}$ is the Reynolds number of casing film. Re_{c} is the Reynolds number of tubing film. Re_{s} is the Reynolds number of Taylor bubble.

Wet Week:

$$S_{\mathrm{LFc}}=\pi D_{\mathrm{C}} \tag{25}$$

$$S_{\mathrm{c}}=S_{\mathrm{Ic}}+S_{\mathrm{Id}}=(D_{\mathrm{C}}-2\delta+D_{\mathrm{T}})\pi+D_{\mathrm{e}}-2\delta \tag{26}$$

3.2. Inclined Annulus Pipe Flow Pattern Transition Criterion

Based on the formation mechanism of different flow patterns, this paper establishes the transition criterion of flow pattern in the annulus pipe on the basis of previous research results.

3.2.1. Transition from Bubble Flow to Slug Flow

The decisive role in the transition from bubble flow to slug flow in the inclined annulus pipe is the cross-sectional gas content, the proportion of gas volume in the whole cross-section. According to Kelessidis [7], $\alpha_G = 0.25$, while Caetano's experimental data and Zhang's experimental data show that the cross-sectional gas content in the transition from bubble flow to slug flow is lower than that in the circular pipe, which is about $\alpha_G = 0.2$. In this paper, we take $\alpha_G = 0.2$, and the transition criterion of bubble flow to slug flow is as follows.

$$v_{SG} = 0.25 v_{SL} + 0.306 \left[\frac{(\rho_L - \rho_G) g \sigma}{\rho_L^2} \right]^{1/4} \sin\theta \tag{27}$$

3.2.2. Transition from Slug Flow to Churn Flow

The transition from slug flow to churn flow is complicated, and the definition of churning flow when it occurs is different. Therefore, the transition boundaries predicted by the above models differ greatly from each other. In this paper, the liquid film at the bottom of the pipe is thicker in the inclined annular hollow pipe. When the superficial flow velocity of gas phase reaches the critical value of transition from slug flow to churn flow, the gas in the Taylor bubble will enter the liquid plug zone in the previous segment. The boundary between the Taylor bubble and the liquid plug region starts to disappear, causing the appearance of the churn flow. In this paper, the true gas content is considered to be about 0.69 at the critical transition from the slug flow to the churn flow, and the gas content in the Taylor bubble region can be taken as the total gas content of the flow unit. The transition criterion of slug flow to churn flow is given by the following equation.

$$v_{SG} = [v_{SL}(1/\alpha - 1.2) - v_{TB} \sin\theta]/1.2 \tag{28}$$

where, v_{TB} is the rate of rise of Taylor bubble, and in this paper we use Hasan's relation.

$$v_{TB} = (0.345 + 0.1N) \sqrt{g \frac{\rho_L - \rho_G}{\rho_L} d_2} \tag{29}$$

3.2.3. Churn Flow to Annular Flow Transition

The transition from churn flow to annular flow occurs when the gas flow rate in the annular is very high. The liquid phase moves up the pipe wall as a liquid film and may also exist as small droplets in the gas core. When the liquid film in the churn flow becomes infinitely long, the transition from the stirred flow to the annular flow occurs, making the momentum exchange between the liquid plug and the liquid film zero. In this paper, the criterion for the transition from churn flow to annular flow is defined as $\alpha_G = 0.75$.

$$H_{LFc} + H_{LFd} = C_1 \frac{[H_{LS}(v_T - v_S) + v_{SL}](v_{SG} + v_{SL} F_E) - v_T v_{SL} F_E}{v_T v_{SG}} \tag{30}$$

3.3. *Model Validation*

In this paper, the present flow pattern transition criterion model was used to predict the flow pattern at different liquid phase superficial velocities and gas phase superficial velocities, and compared to the experimental results. Figures 11–14 show the flow pattern distribution of the annulus pipe under each inclination angle condition. Tables 2–5 shows the comparison of the model in this paper, Kelessidis [7] model, Zhang [9] model, and the experimental results. Table 6 shows the accuracy statistics of different models. It can be seen that the model established in this paper can predict the flow pattern of the inclined annulus pipe more accurately. Further, the prediction of slug flow to churn flow, and churn flow to annular flow is more accurate than that of the previous models.

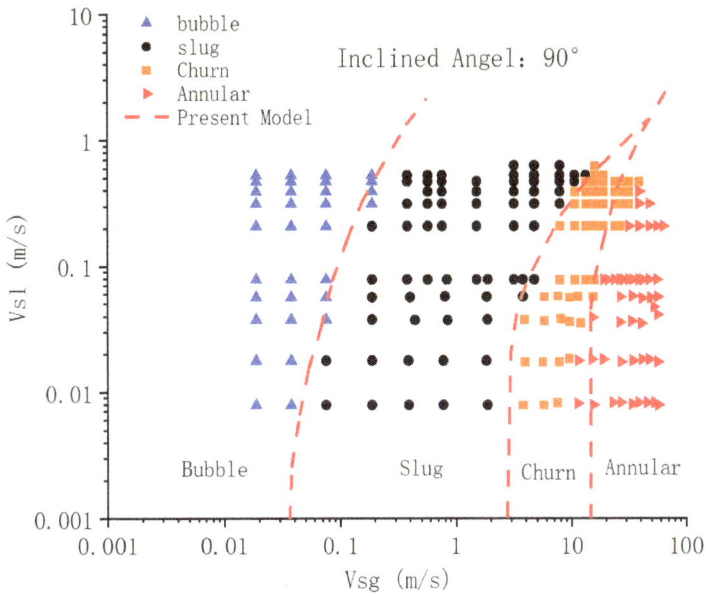

Figure 11. Flow pattern distribution of gas-liquid two-phase flow in the annulus pipe at an inclination angle of 90°.

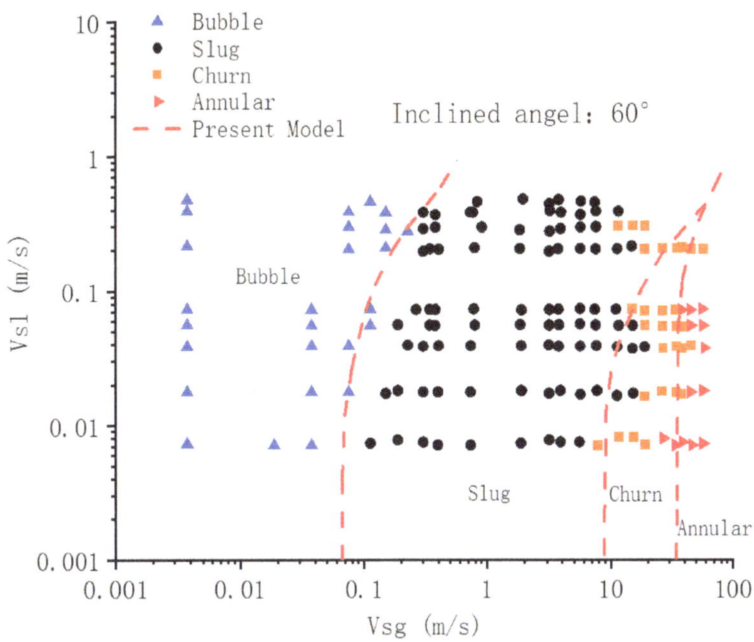

Figure 12. Flow pattern distribution of gas-liquid two-phase flow in the annulus pipe at an inclination angle of 60°.

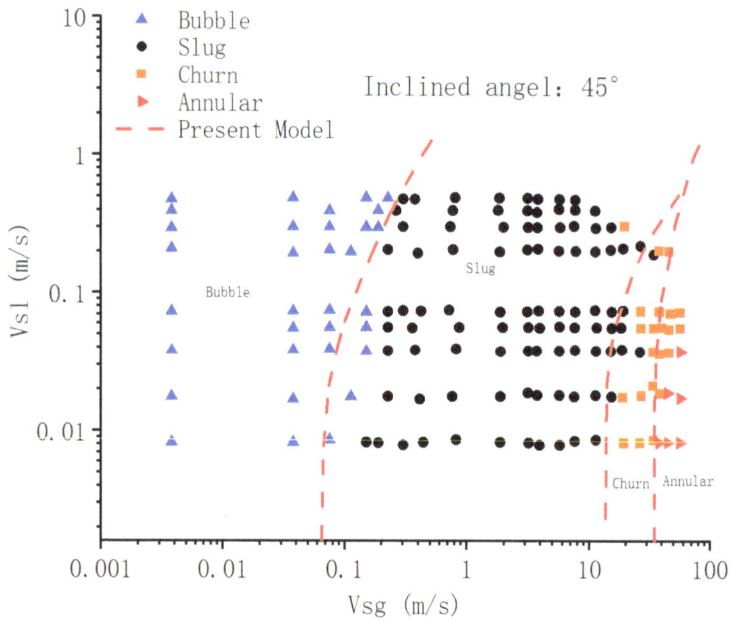

Figure 13. Flow pattern distribution of gas-liquid two-phase flow in the annulus pipe at an inclination angle of 45°.

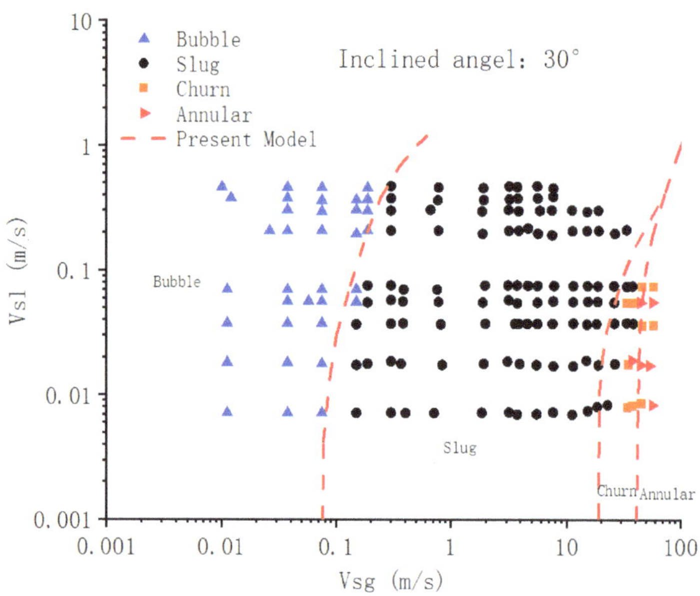

Figure 14. Flow pattern distribution of gas-liquid two-phase flow in the annulus pipe at an inclination angle of 30°.

Table 2. Comparison of model predictions with experimental results at an inclination angle of 90°.

No.	v_{SL} ($m{\cdot}s^{-1}$)	v_{SG} ($m{\cdot}s^{-1}$)	Flow Pattern of Experimental	The Present Model	Kelessidis Model	Zhang Model
1	0.0177	0.0376	BB	BB	BB	BB
2	0.0564	0.0752	BB	BB	BB	SL
3	0.0376	0.0752	BB	BB	SL	SL
4	0.392	0.188	BB	SL	BB	SL
5	0.470	0.188	BB	BB	BB	SL
6	0.008	1.876	SL	SL	CH	CH
7	0.037	1.870	SL	SL	CH	CH
8	0.207	3.162	SL	SL	CH	SL
9	0.313	1.506	SL	SL	CH	SL
10	0.628	7.830	SL	SL	CH	SL
11	0.008	7.393	CH	CH	CH	CH
12	0.037	3.859	CH	CH	CH	CH
13	0.035	11.833	CH	CH	CH	AN
14	0.056	9.538	CH	CH	CH	AN
15	0.313	18.822	CH	CH	CH	AN
16	0.018	14.912	AN	AN	AN	AN
17	0.017	11.350	AN	CH	CH	AN
18	0.078	18.822	AN	AN	CH	AN

Note: BB—Bubble flow; SL—Slug flow; CH—Churn flow; AN—Annular flow.

Table 3. Comparison of model predictions with experimental results at an inclination angle of 60°.

No.	v_{SL} ($m{\cdot}s^{-1}$)	v_{SG} ($m{\cdot}s^{-1}$)	Flow Pattern of Experimental	The Present Model	Kelessidis Model	Zhang Model
1	0.007	0.037	BB	BB	BB	BB
2	0.017	0.075	BB	BB	SL	SL
3	0.039	0.075	BB	BB	SL	SL
4	0.055	1.112	BB	SL	SL	SL
5	0.208	0.150	BB	BB	SL	SL
6	0.007	0.731	SL	SL	SL	SL
7	0.017	3.117	SL	SL	CH	CH
8	0.055	0.790	SL	SL	SL	SL
9	0.196	3.122	SL	SL	CH	SL
10	0.465	5.713	SL	SL	CH	SL
11	0.306	11.338	CH	SL	CH	CH
12	0.073	11.928	CH	CH	CH	CH
13	0.007	18.996	CH	CH	CH	AN
14	0.018	25.981	CH	CH	AN	AN
15	0.072	26.307	CH	CH	CH	AN
16	0.008	26.667	AN	CH	AN	AN
17	0.037	58.376	AN	AN	AN	AN
18	0.073	58.341	AN	AN	AN	AN

Note: BB—Bubble flow; SL—Slug flow; CH—Churn flow; AN—Annular flow.

Table 4. Comparison of model predictions with experimental results at an inclination angle of 45°.

No.	v_{SL} $(m{\cdot}s^{-1})$	v_{SG} $(m{\cdot}s^{-1})$	Flow Pattern of Experimental	The Present Model	Kelessidis Model	Zhang Model
1	0.008	0.037	BB	BB	BB	BB
2	0.039	0.075	BB	BB	BB	SL
3	0.037	0.151	BB	SL	SL	SL
4	0.199	0.113	BB	BB	BB	SL
5	0.486	0.226	BB	BB	BB	SL
6	0.017	0.411	SL	SL	SL	SL
7	0.206	0.775	SL	SL	SL	SL
8	0.299	1.976	SL	SL	CH	SL
9	0.304	7.446	SL	SL	CH	SL
10	0.220	26.369	SL	SL	CH	AN
11	0.008	19.093	CH	CH	AN	AN
12	0.307	19.281	CH	SL	CH	AN
13	0.203	27.369	CH	CH	AN	AN
14	0.056	33.714	CH	CH	AN	AN
15	0.037	45.406	CH	AN	AN	AN
16	0.008	37.705	AN	AN	AN	AN
17	0.018	44.482	AN	AN	AN	AN
18	0.037	56.553	AN	AN	AN	AN

Note: BB—Bubble flow; SL—Slug flow; CH—Churn flow; AN—Annular flow.

Table 5. Comparison of model predictions with experimental results at an inclination angle of 30°.

No.	v_{SL} $(m{\cdot}s^{-1})$	v_{SG} $(m{\cdot}s^{-1})$	Flow Pattern of Experimental	The Present Model	Kelessidis Model	Zhang Model
1	0.007	0.075	BB	BB	SL	SL
2	0.037	0.075	BB	BB	BB	BB
3	0.195	0.150	BB	BB	BB	BB
4	0.206	0.188	BB	BB	BB	SL
5	0.297	0.188	BB	BB	BB	SL
6	0.017	0.369	SL	SL	SL	SL
7	0.054	1.873	SL	SL	CH	SL
8	0.037	7.564	SL	SL	CH	CH
9	0.302	7.437	SL	SL	CH	SL
10	0.074	33.755	SL	CH	CH	AN
11	0.008	33.829	CH	CH	AN	AN
12	0.008	45.101	CH	AN	AN	AN
13	0.055	34.480	CH	CH	AN	AN
14	0.074	45.438	CH	CH	AN	AN
15	0.074	57.370	CH	AN	AN	AN
16	0.018	37.448	AN	CH	AN	AN
17	0.008	56.361	AN	AN	AN	AN
18	0.055	56.689	AN	AN	AN	AN

Note: BB—Bubble flow; SL—Slug flow; CH—Churn flow; AN—Annular flow.

Table 6. Results of model prediction.

Model	Accuracy				
	Angle of 90°	Angle of 60°	Angle of 45°	Angle of 30°	Total
Kelessidis model	56%	56%	56%	44%	53.00%
Zhang model	50%	56%	44%	44%	48.50%
The present model	89%	83%	83%	78%	83.25%

4. Conclusions

In this work, we reported an experimental investigation of gas and water flow in an inclined annulus pipe, and presented a new model of flow pattern transition in inclined annulus pipe by using a mechanistic approach to classify the flow patterns. This work can provide experimental proof for field application, which are lacking in existing studies. The specific conclusions are shown as follows.

1. The experimental results indicate that the slug flow will be shifted to the larger gas-liquid superficial flow rate region with the smaller inclination angle, and the annular flow will appear in the higher gas superficial flow rate region.
2. Using the experimental data, the performance of 4 existing correlations did not agree well with the transition from slug flow to churn, and churn to annular flow in the inclined annulus pipe. Based on the formation mechanism of different flow patterns, this establishes the transition criterion of flow pattern in the annulus pipe on the basis of previous research results. The results indicate that predictions of the correlation agree very well with the experimental data and performs better than the existing correlations we considered (Kelessidis [7] and Zhang [9]), and the new correlation was better in predicting the transition from slug flow to churn flow and churn flow to annular flow.

In this study, only one oil/casing size combination experiment was carried out, however, the applicability of the new model under more combinations in the field needs to be verified. In addition, there is a wide range of pressure changes from the bottom to the wellhead, it is necessary to carry out relevant research on the impact of pressure on the transition criterion to ensure the accuracy of flow pattern prediction under different pressure conditions.

Author Contributions: Methodology, Y.L.; validation, Z.W.; investigation, W.W.; resources, W.W.; data curation, Y.L.; writing—original draft preparation, J.W.; writing—review and editing, B.M. All authors have read and agreed to the published version of the manuscript.

Funding: This research received no external funding.

Institutional Review Board Statement: Not applicable.

Informed Consent Statement: Not applicable.

Data Availability Statement: Data available on request from the corresponding author if necessary.

Conflicts of Interest: The authors declare no conflict of interest.

Nomenclature

BB	Bubble flow
SL	Slug flow
CH	Churn flow
AN	Annular flow

References

1. Li, Y.W.; Peng, G.B.; Chen, M.; Zhang, J.; Wei, J.G.; Yang, C.G. Gas-liquid-solid three phase flow model of CO_2 foam fracturing in wellbore. *J. Pet.* **2022**, *43*, 386–398.
2. Yao, L.; Li, H.X.; Jiao, W.; Zhou, J.; Ke, C.Y.; Zhang, Q.Z. Progress and prospect of reuse technology of fracturing flowback liquid. *Oilfield Chem.* **2022**, *39*, 548–553.
3. Falavand-Jozaei, A.; Hajidavalloo, E.; Shekari, Y.; Ghobadpouri, S. Modeling and simulation of non-isothermal three-phase flow accurate prediction in underbalanced drilling. *Pet. Explor. Dev.* **2022**, *49*, 358–365. [CrossRef]
4. Caetano, E.F.; Shoham, O.; Bril, J.P. Upward Vertical Two-Phase Flow through an Annular—Part I: Single-Phase Friction Factor, Taylor Bubble Rise Velocity, and Flow Pattern Prediction. *J. Energy Resour. Technol.* **1992**, *114*, 1–13. [CrossRef]
5. Taitel, Y.; Barnea, D.; Dukler, A.E. Modelling flow pattern transitions for steady upward gas-liquid flow in vertical pipes. *Aiche J.* **2010**, *26*, 345–354. [CrossRef]

6. Sadatomi, M.; Sato, Y. An Examination of One Dimensional Two-Phase Flow Models Using the Date for Horizontal Isosceles-Triangular Channels. *Tech. Rep. Kumamoto Univ.* **1983**, *32*, 21–28.
7. Kelessidis, V.C.; Dukler, A.E. Modeling flow pattern transitions for upward gas-liquid flow in vertical concentric and eccentric annuli. *Int. J. Multiph. Flow* **1989**, *15*, 173–191. [CrossRef]
8. Hasan, A.R.; Kabir, C.S. Two-phase flow in vertical and inclined annuli. *J. Multiph. Flow* **1992**, *18*, 279–293. [CrossRef]
9. Zhang, J.; Chen, T.K.; Luo, Y.S.; Wang, H.J. Transition from slug flow to churn flow for upward gas liquid in vertical concentric annular. *Chem. Eng.* **2002**, *53*, 360–363.
10. Yin, B.T.; Li, X.F.; Sun, B.J.; Zhang, H.Q. Hydraulic model of state multiphase flow in wellbore annuli. *Pet. Explor. Dev.* **2014**, *41*, 399–407. [CrossRef]
11. Sukamta; Sudarja; Sundari, S. Comparison of Void Fraction on Annular and Churn Two-Phase Flow Patterns in Mini Pipe with a Slope of 15° against Horizontal Position. *IOP Conf. Ser. Mater. Sci. Eng.* **2021**, *1063*, 012011. [CrossRef]
12. Yu, T.T.; Zhang, H.Q.; Li, M.X.; Sarica, C. A mechanistic model for gas/liquid flow in upward vertical annuli. *SPE Prod. Oper.* **2010**, *25*, 285–295. [CrossRef]
13. Hernández, L.; Enrique Julia, J.; Ozar, B.; Hibiki, T.; Ishii, M. Flow regime identification in boiling two-phase flow in a vertical annulus. *J. Fluids Eng.* **2011**, *133*, 091304. [CrossRef]
14. Tazoe, H.; Nishioka, D.; Kawahara, A. Effects of grid spacer with mixing vane on liquid film thickness for gas-liquid annular flow in a vertical pipe. *Jpn. Soc. Mech. Eng.* **2018**, *71*, A34. [CrossRef]
15. Uche, O. Evaluation of Liquid Film Thickness in Gas-Liquid Annular Flow in Horizontal Pipes Using Three Methods. *Int. J. Energy Environ. Sci.* **2020**, *5*, 57. [CrossRef]
16. Li, H.; Sun, B.; Guo, Y.; Gao, Y.; Zhao, X. Recognition and measurement gas-liquid two-phase flow in a vertical concentric annulus at high pressures. *Heat Mass Transf.* **2018**, *54*, 353–362. [CrossRef]
17. Sethu, A.; Jayanti, S. Studies of Upward and Downward Gas-Liquid Annular Flow. In Proceedings of the 9th ICMF 2016, Florence, Italy, 22–27 May 2016.
18. Morenko, I.V. Mathematical Modeling of Gas–Liquid Flow Patterns in an Annular Space with a Rotating Inner Cylinder. *Theor. Found. Chem. Eng.* **2022**, *56*, 99–106. [CrossRef]
19. Freitas, L.; Alves, M.; Francisco, R.R. Comparison of numerical simulations with experimental data for vertical annular and churn gas-liquid flows. In Proceedings of the 18th Brazilian Congress of Thermal Sciences and Engineering, Online, 16–20 November 2020.
20. Aliyu, A.M.; Baba, Y.D.; Lao, L.; Yeung, H.; Kim, K.C. Interfacial friction in upward annular gas–liquid two-phase flow in pipes. *Exp. Therm. Fluid Sci.* **2017**, *84*, 90–109. [CrossRef]
21. Basha, M.; Shaahid, S.M.; Alhems, L.M. Effect of Viscosity on Pressure Drop of Oil-Water Two Phase Flow in 6″ Horizontal and Inclined Stainless Steel Annulus Pipe. *J. Adv. Res. Fluid Mech. Therm. Sci.* **2020**, *69*, 156–167. [CrossRef]
22. Schmidt, J.; Giesbrecht, H. Pressure Drop in Emergency Vent Lines with Two-Phase Flow of Gas and Viscous Liquid. *Chem. Eng. Technol. Ind. Chem. Plant Equip. Process Eng. Biotechnol.* **2018**, *41*, 853–859. [CrossRef]

processes MDPI

Article

Study on the Influence of Pressure Reduction and Chemical Injection on Hydrate Decomposition

Lei Wang [1,2], Zhikang Song [1,2,*], Xin Huang [3], Wenjun Xu [1,2] and Zhengbang Chen [1,2]

1 School of Petroleum Engineering, Yangtze University, Jingzhou 434000, China
2 Key Laboratory of Drilling and Production Engineering for Oil and Gas, Jingzhou 434000, China
3 Sinopec Petroleum Exploration and Development Research Institute, Beijing 122206, China
* Correspondence: szkyangtzeu@163.com; Tel.: +86-18056012832

Abstract: This study simulated seabed high pressure and low temperature conditions to synthesize natural gas hydrates, multi-stage depressurization mode mining hydrates as the blank group, and then carried out experimental research on the decomposition and mining efficiency of hydrates by depressurization and injection of different alcohols, inorganic salts, and different chemical agent concentrations. According to the experimental results, the chemical agent with the best decomposition efficiency is preferred; the results show that: the depressurization and injection of a certain mass concentration of chemical agents to exploit natural gas hydrate is more effective than pure depressurization to increase the instantaneous gas production rate. This is because depressurization combined with chemical injection can destroy the hydrate phase balance while effectively reducing the energy required for hydrate decomposition, thereby greatly improving the hydrate decomposition efficiency. Among them, depressurization and injection of 30% ethylene glycol has the best performance in alcohols; the decomposition efficiency is increased by 52.0%, and the mining efficiency is increased by 68.2% within 2 h. Depressurization and injection of 15% calcium chloride has the best performance in inorganic salts; the decomposition efficiency is increased by 46.3%, and the mining efficiency is increased by 61.1% within 2 h. In the actual mining process, the appropriate concentration of chemical agents should be used to avoid polluting the environment.

Keywords: natural gas hydrate; depressurization and chemical injection combined method; hydrate decomposition efficiency; hydrate recovery efficiency

Citation: Wang, L.; Song, Z.; Huang, X.; Xu, W.; Chen, Z. Study on the Influence of Pressure Reduction and Chemical Injection on Hydrate Decomposition. *Processes* **2022**, *10*, 2543. https://doi.org/10.3390/pr10122543

Academic Editors: Ofelia de Queiroz Fernandes Araujo and Vladimir S. Arutyunov

Received: 6 October 2022
Accepted: 22 November 2022
Published: 29 November 2022

1. Introduction

As a new type of strategic energy in the future, the exploitation of natural gas hydrate is of great significance. The mining principle is to change the stable phase equilibrium condition of natural gas hydrate, and decompose natural gas hydrate to obtain natural gas. The mining methods proposed by domestic and foreign hydrate mining experts mainly include the pressure reduction method, the chemical injection method, the heat injection method and the carbon dioxide replacement method [1,2]. Judging from the current indoor test research and test mining, the pressure reduction method is more preferred.

The depressurization method reduces the pressure of the hydrate reservoir to below the equilibrium pressure of the hydrate phase, and destroys the stability of the natural gas hydrate [3]. Almenningen et al. [4] considered the depressurization method to be the most cost-effective method through field examples and the status quo. However, the experimental results of Li et al. [5], Wang et al. [6], and Zhou et al. [7] show that the depressurization amplitude has a great influence on depressurization mining efficiency. Zhao et al. [8] found that excessive pressure reduction may lead to reservoir instability. Song et al. [9] simulated vertical well pressure reduction mining through ABAQUS and found that reducing production pressure can effectively increase hydrate production, but formation subsidence is also more likely to occur. Konno et al. [10] found that the excessive production of a pressure difference could lead to problems such as ice plugging or secondary hydrate formation in the wellbore. The single depressurization method still faces the problems of the rapid decay of the mining rate, low heat transfer efficiency, and low energy utilization rate, but the use of stepped depressurization (i.e., multi-stage depressurization) can effectively alleviate such problems. Therefore, using only the depressurization method is not suitable for the long-term development of natural gas hydrate.

The chemical agent injection method is to inject chemical agents into the formation to move the phase equilibrium curve of natural gas hydrate to a higher pressure and lower temperature, so that the natural gas hydrate is decomposed [11,12]. The advantage of the chemical injection method is that the heat required for the decomposition of natural gas hydrate can be reduced, and the decomposition rate of natural gas hydrate can be effectively increased in a short time. Sung et al. [13], Fan et al. [14] studied the effects of the pressure reduction method and chemical injection method on the heat of hydrate decomposition, and found that the chemical injection can significantly reduce the heat required for natural gas hydrate decomposition and increase the decomposition efficiency. Sun et al. [15] found through experiments that, compared with the conventional depressurization method, ethylene glycol injection can significantly improve the recovery efficiency of natural gas hydrate. Sun et al. [16] used a small three-dimensional device to deepen the study of chemical injection to extract methane hydrate, and explored from the aspects of injection concentration and injection rate. To sum up, chemical injection can effectively make up for the shortage of depressurization mining.

In recent years, the combined depressurization and extraction technologies of natural gas hydrate mainly include the depressurization and heat injection combined method, the depressurization and replacement combined method, the depressurization and ground decomposition combined method, etc. [17]. Among them, Sun et al. [18] established a mathematical model, and the combined method of depressurization and heat injection can greatly reduce the water cut and improve the recovery factor. Li S X et al. [19] and Bai Y H et al. [20] found through physical and numerical simulations, that although the combined method of heat injection and depressurization can effectively improve energy utilization, long-term depressurization may lead to reservoir instability. Gupta et al. [21] found that the depressurization and CO_2 replacement method is efficient and economical, and can effectively solve the permeability problem, but this method must strictly limit the large amount of CO_2 overflow, so it has extremely high technical requirements and is not suitable for conventional mining. The combined method of depressurization and ground decomposition can easily cause formation instability due to the instability of shallow hydrate reservoirs on the seabed and the simultaneous action of pressure drop and shallow mining, so the mining efficiency of this method is low. However, depressurization and chemical injection can avoid the secondary formation of hydrate under the action of chemical agents [22], and greatly improve the recovery efficiency in a short period of time, obtain considerable economic benefits, and maintain reservoir stability. There are also other technologies that increase the production rate, such as asphaltene control [23].

At present, there is a serious lack of basic data on the effect of that multi-stage depressurization and chemical injection combined method on hydrate recovery efficiency. It is necessary to verify the feasibility of the multi-stage depressurization and chemical injection method through experimental simulation, and at the same time, it is necessary to evaluate the performance of efficient decomposition promotion. It can provide reasonable guidance for the efficient exploitation of natural gas hydrate.

In this paper, self-made equipment is used to simulate the temperature and pressure conditions of submarine hydrate reservoirs to generate hydrates with high saturation, and a multi-stage depressurization and chemical injection combined exploration of natural gas hydrate is carried out. The influence of the recovery efficiency, instantaneous gas production dynamics, and the cumulative gas production changes during the hydrate decomposition process were studied.

2. Materials and Methods

2.1. Experimental Materials and Equipment

Experimental materials: deionized water, made in the laboratory; methane gas (purity 99.99%), Wuhan Huaxin Gas Company; quartz sand (0.82 mm, porosity 0.45), Shanghai Sinopharm Reagent Chemical Factory; methanol (≥99%), ethylene glycol (≥99.5%), ethanol (≥95%), calcium chloride (≥96%), sodium chloride (≥99.5%), potassium chloride (≥99.5%), Longxi Science.

Experimental equipment: A XFH-II type natural gas hydrate synthesis and decomposition simulation device was used in the experiment. The schematic diagram of the composition of the gas hydrate synthesis and decomposition device is shown in Figure 1. The device consists of an injection system, a pressurization system, a constant temperature control system, a reactor system, a gas metering system and a data acquisition system. Among them, the high-pressure stainless steel reaction kettle has a volume of 500 mL, a pressure resistance of 20 MPa, and a working temperature of −20 to 50 °C; the temperature sensor range is greater than 400 °C, and the accuracy is 0.1 °C; the gas flow meter range is 1000 mL/min, and the accuracy is 0.1 mL/min.

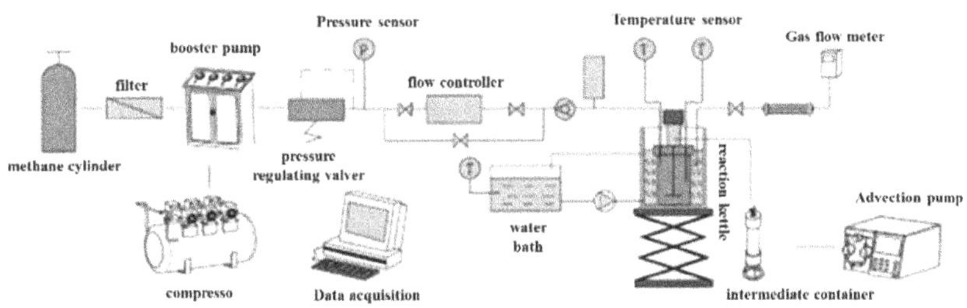

Figure 1. Schematic diagram of the composition of the gas hydrate synthesis and decomposition device.

2.2. Experimental Steps

2.2.1. Natural Gas Hydrate Synthesis Steps

More than 90% of natural gas hydrates exist in seabed sediments with a temperature of 0–10 °C and a pressure of more than 10 MPa [24]. The main composition of hydrate deposits is silt in the South China Sea, with a porosity of 33% to 55% [25] and an average gas hydrate saturation of 13.7% to 45.2% [26]. Therefore, this experiment is based on the above research to ensure that the basic requirements of the natural gas hydrate reservoir environment are met. First, ensure that methane hydrate with high saturation is generated, and then ensure that the pressure in the kettle is about 10 MPa after the methane hydrate is completely generated. After many experiments and tests, when the experimental temperature is 4 °C and the initial pressure is about 15 MPa, and the quartz sand with particle size of 0.82 mm and porosity of 0.45 is selected to simulate the reservoir, basic methane hydrate reservoir conditions can be achieved.

Natural gas hydrate synthesis steps: ① Rinse the reactor with distilled water 3 times to remove other chemical reagents left in the reactor from the previous experiment; ② After connecting the pipeline, check the airtightness of the device, and inject 4 MPa of gas into the reaction kettle. If the pressure remains stable within 24 h, the airtightness is good; ③ Fill the reaction kettle with sand to 350 mL; ④ Add distilled water to predetermined mass (160 g); ⑤ Open the exhaust valve, and quickly and repeatedly replace the air in the reaction kettle with high-purity methane gas sample twice, so as to ensure that the gas in the kettle has the same composition as the prepared gas as much as possible; ⑥ Use a constant temperature water bath to adjust the temperature in the kettle to 4 °C, and in the process of waiting for the temperature to reach a constant value, prepare a decomposer solution for later use; ⑦ Methane gas was injected into the reaction kettle, and the injection of methane gas was stopped after the pressure in the kettle reached 15 MPa. When the pressure in the reactor is stable (observation for 2 h), the hydrate reaction is complete.

2.2.2. Natural Gas Hydrate Decomposition Steps

① After the hydrate is completely formed, multi-stage depressurization is carried out, the pressure in the reactor is reduced to the specified decomposition and production pressure for many times (the pressure shall be maintained after each stage of depressurization, and the mining shall be continued for 1 h), and the chemical agent solution prepared in advance is passed through the advection pump. The chemical agent solution was injected into the reactor at the injection rate of 8 mL·min−1, so that the hydrate began to decompose; ② When the gas production rate dropped to 0, the hydrate decomposition reaction was considered to be over; ③ The data acquisition system was used to record the temperature and pressure changes in the reaction kettle during the whole experiment process, save the data after the experiment was completed, and clean the reaction kettle.

3. Results

3.1. Multi-Stage Depressurization and Alcohol Injection to Promote Hydrate Decomposition

At 4 °C, the equilibrium pressure of the methane hydrate phase is 3.9 MPA [27]. In the actual production process of hydrate reservoirs, an excessive wellhead pressure difference may impact the wellbore and cause it to be damaged or collapse. Therefore, the three-stage depressurization mode of -4 MPa-3 MPa-2 MPa was used in the experiments. The use of the third-order depressurization mode can not only reduce the risk of reservoir collapse, but also shorten the effective production time of natural gas hydrate and increase the decomposition efficiency of natural gas hydrate. The mining conditions of multi-stage depressurization +20%/30% ethylene glycol, multi-stage depressurization +20%/30% methanol, and multi-stage depressurization +20%/30% ethanol were investigated respectively, and they were decomposed with pure depressurization. Methods were compared. The experiment was repeated 3 times, and a set of data closest to the average was taken. The experimental results are shown in Figures 2 and 3 and Table 1.

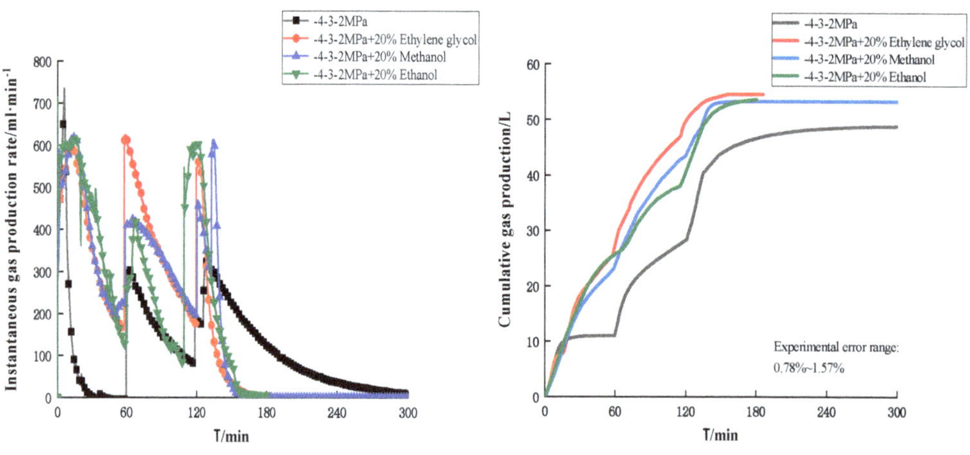

Figure 2. Decomposition curve of multi-stage depressurization and injection of 20% alcohol hydrate: (**a**) Instantaneous gas production rate; (**b**) Cumulative gas production.

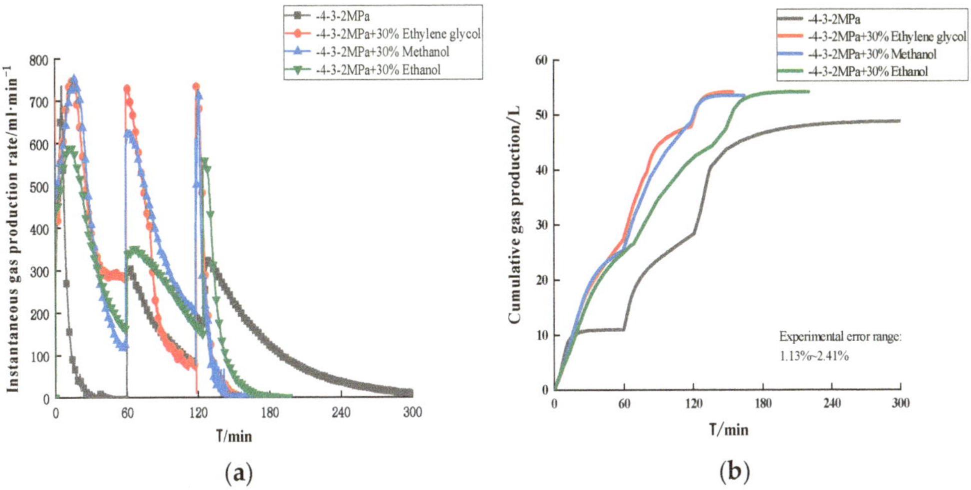

Figure 3. Decomposition curve of multi-stage depressurization and injection of 30% alcohol hydrate: (**a**) Instantaneous gas production rate; (**b**) Cumulative gas production.

Table 1. Relevant experimental parameters of gas hydrate depressurization and alcohol injection production.

Index	Pure Multistage Buck	Multi-Stage Depressurization + 20% Ethylene Glycol	Multi-Stage Depressurization + 30% Ethylene Glycol	Multi-Stage Depressurization + 20% Methanol	Multi-Stage Depressurization + 30% Methanol	Multi-Stage Depressurization + 20% Ethanol	Multi-Stage Depressurization + 30% Ethanol
Experimental pressure/MPa	14.85	14.82	14.80	14.81	14.83	14.81	14.83
Experimental temperature/°C	4	4	4	4	4	4	4
Experiment end pressure/MPa	10.27	10.12	10.30	10.14	10.24	10.34	10.24
Hydrate Saturation/%	48.0	48.5	46.5	48.5	48.0	46.1	48.2
Injection rate/(mL·min^{-1})	8	8	8	8	8	8	8
Buck mode/MPa	-4-3-2	-4-3-2	-4-3-2	-4-3-2	-4-3-2	-4-3-2	-4-3-2
Cumulative gas production/L	48.83	54.51	54.14	53.25	53.47	53.76	53.45
Decomposition complete time/min	300	163	144	163	152	194	179
Average gas production rate/(mL·min^{-1})	161.7	334.4	376.0	326.7	351.8	277.1	298.6

It can be seen from Figures 2 and 3 and Table 1 that in the pure depressurization mode, when the pressure is reduced to 4.00 MPa for constant pressure production, the instantaneous gas production rate increases rapidly, and then decreases rapidly. This is because the decomposition occurs when the pressure is reduced to 4.00 MPa. The driving force is insufficient, the decomposition gas production is small, the instantaneous gas production rate cannot be supported, and the gas production is basically free gas; the migration in the hydrate layer causes the decomposition of methane hydrate, and the instantaneous gas production rate is greatly improved, which is obviously better than pure depressurization production. Among them, the decomposition performance of multi-stage depressurization + 20%/30% ethylene glycol mode is the best, and the time required for multi-stage depressurization + 20% ethylene glycol and multi-stage depressurization + 30% ethylene glycol to completely decompose natural gas hydrate is 163 min and 144 min, respectively. Compared with pure multi-stage depressurization, the decomposition efficiency is increased by 45.7% and 52.0%, respectively. When multi-stage depressurization + 30% ethylene glycol is produced at a constant pressure of 4.00 MPa, the early instantaneous gas production rate exceeds 750 mL/min, the gas production rate is still 292 mL/min in the later stage, which is obviously better than other alcohols; when the multi-stage depressurization + 30% ethylene glycol is produced at a constant pressure of 3.00 MPa, the instantaneous gas production rate in the final stage drops to 78 mL/min, this is because the hydrate has basically completely decomposed during the constant pressure production of 4.00 and 3.00 MPa in the multi-stage depressurization + 30% ethylene glycol mode, and the decomposed gas is insufficient, so the instantaneous gas production rate drops rapidly.

Since the initial pressure is about 14.85 MPa, and multi-stage depressurization is reduced to 2.00 MPa, the final cumulative gas production should be basically the same within the allowable error range. In order to understand the production efficiency of the combined method in a certain period of time, the gas production at the end of production at a constant pressure of 4.00 and 3.00 MPa (i.e., 2 h of production) was selected for comparison. It can be seen from Figures 2 and 3 that the production efficiency of multi-stage depressurization and alcohol injection in the first 2 h of production time is much greater than that of the pure depressurization mode. The mining efficiency of +30% ethylene glycol is the best; compared with the pure multi-stage depressurization mode, the multi-stage depressurization + 20% ethylene glycol mode increases the production efficiency by 65.3%, and the multi-stage depressurization + 30% ethylene glycol mode produces Efficiency increased by 68.2%.

From the experimental results, multi-stage depressurization and alcohol injection can effectively improve the extraction and decomposition efficiency of hydrate reservoirs compared with simple depressurization. In addition, the chemical concentration of the combined method is much lower than that of Wang et al. [28], which is more efficient and economical. This is because the multi-stage depressurization destroys the stability of the hydrate, and, secondly, the organic reagents such as ethylene glycol, methanol, ethanol and other alcohol molecules, and the hydrophilic hydroxyl group destroys the crystal structure of the hydrate and promotes the decomposition of the hydrate [29]. At the same time, with the increase of alcohol concentration, the hydrate decomposition efficiency also increased gradually. This is because the high-concentration alcohol solution has a greater impact on the phase equilibrium conditions of methane hydrate, and adding high-concentration alcohol solution can reduce the decomposition heat of methane hydrate [15], which improves the decomposition efficiency of hydrate.

The experiment found that ethylene glycol had the best effect on promoting the decomposition of hydrate. This is because, compared with methanol and ethanol, the decomposition heat in ethylene glycol solution is lower, and the cluster structure formed by methanol molecules and ethanol molecules reduces the number of active molecules, which is not conducive to the decomposition of hydrates [30].

3.2. Multi-Stage Depressurization and Salt Injection to Promote Hydrate Decomposition

High-quality salts may crystallize and cause blockage of the injection pipeline. Therefore, multi-stage depressurization + 10%/15% calcium chloride, multi-stage depressurization + 10%/15% sodium chloride, multi-stage depressurization + 10%/15% potassium chloride are the best. The mining conditions of decompression + 10%/15% potassium chloride were compared with pure decompression decomposition. The experiment was repeated three times, and the group closest to the average was taken. The experimental results are shown in Figures 4 and 5 and Table 2.

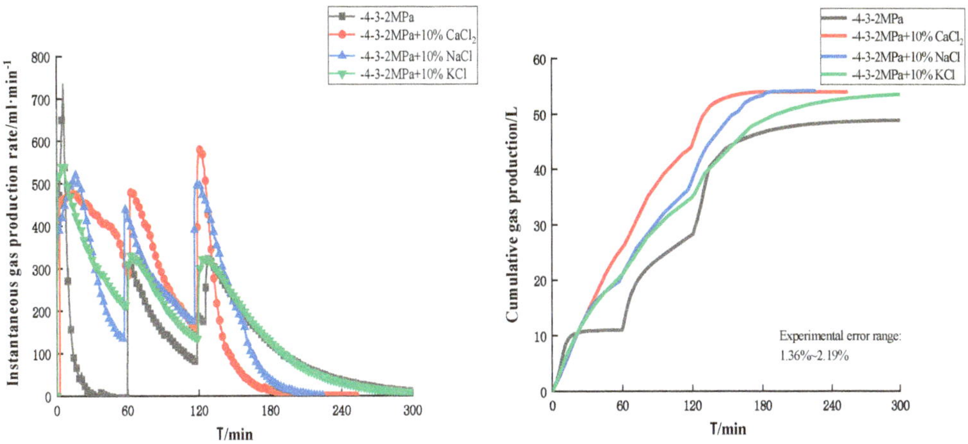

(a) (b)

Figure 4. Decomposition curve of multi-stage depressurization and injection of 10% salt hydrate: (**a**) Instantaneous gas production rate; (**b**) Cumulative gas production.

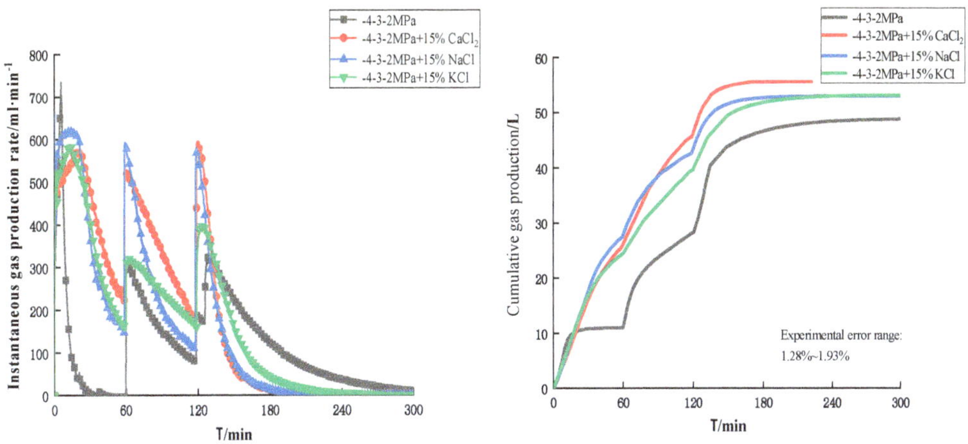

(a) (b)

Figure 5. Decomposition curve of multi-stage depressurization and injection of 15% salt hydrate: (**a**) Instantaneous gas production rate; (**b**) Cumulative gas production.

Table 2. Relevant experimental parameters of gas hydrate depressurization and salt injection production.

Index	Pure Multistage Buck	Multi-Stage Depressurization + 10%$CaCl_2$	Multi-Stage Depressurization + 15% $CaCl_2$	Multi-Stage Depressurization + 10% NaCl	Multi-Stage Depressurization + 15% NaCl	Multi-Stage Depressurization + 10% KCl	Multi-Stage Depressurization + 15% KCl
Experimental pressure/MPa	14.85	14.92	15.07	14.91	14.86	14.85	14.86
Experimental temperature/°C	4	4	4	4	4	4	4
Experiment end pressure/MPa	10.27	10.46	10.40	10.24	10.48	9.33	10.07
Hydrate Saturation/%	48.0	46.5	48.2	49.1	45.7	57.2	50.5
Injection rate/($mL \cdot min^{-1}$)	8	8	8	8	8	8	8
Buck mode/MPa	-4-3-2	-4-3-2	-4-3-2	-4-3-2	-4-3-2	-4-3-2	-4-3-2
Cumulative gas production/L	48.83	53.84	55.20	54.17	52.92	53.90	53.11
Decomposition complete time/min	300	174	161	211	189	289	240
Average gas production rate/($mL \cdot min^{-1}$)	161.7	309.4	342.9	256.7	280.0	186.5	221.3

As can be seen from Figures 4 and 5 and Table 2, after the combined method of multi-stage depressurization and salt injection is adopted, the instantaneous gas production rate is also significantly improved, which is better than that of pure multi-stage depressurization; multi-stage depressurization and injection of 10% Calcium chloride, multi-stage depressurization and injection of 15% calcium chloride all show excellent decomposition performance, the instantaneous gas production rate at the constant pressure stage of 4 and 3 MPa is significantly higher than that of other salt injections as the hydrate is rapidly decomposed. Therefore, the natural gas hydrate decomposes completely at a relatively fast rate during the constant pressure decomposition of 2 MPa; the time required for the multi-stage depressurization + 10% calcium chloride injection and the multi-stage decompression + 15% calcium chloride injection to completely decompose the natural gas hydrate are, respectively, 174 min and 161 min, and the decomposition efficiency was increased by 42.0% and 46.3%, respectively compared with pure multistage depressurization.

It was found from the experimental process that the mining efficiency of multi-stage depressurization + 10% calcium chloride injection and multi-stage depressurization + 15% calcium chloride injection is high, and much larger than the pure multi-energy depressurization mode; compared with the pure multi-energy depressurization mode, the mining efficiency of multi-stage depressurization + 10% calcium chloride injection is improved by 54.6%, and the mining efficiency of multi-stage depressurization + 15%, while calcium chloride injection is improved 61.1%.

From the experimental results, multi-stage depressurization and salt injection can effectively improve the extraction and decomposition efficiency of hydrate reservoirs compared with simple depressurization, and it increases with the increase of mass concentration, which is consistent with the experimental results of Ding et al. [31]. This is due to the strong coulomb force of inorganic salt reagents, and it is easy to break the hydrate crystal structure through the combination of hydrogen bonds and water molecules, thereby promoting the decomposition of hydrates. This ability increases with the increase of particle concentration; inorganic salts can change the self-diffusion effect of water molecules in the system. As the mass concentration of inorganic salt increases, the activity of water molecules in the solution decreases and the self-diffusion coefficient decreases, while the self-diffusion coefficient of water molecules in hydrates increases [32], thereby promoting the decomposition of hydrates.

It was also found from the experiment that calcium chloride has the best effect on promoting the decomposition of natural gas hydrate. This is because the addition of salt solution reduces the energy required for hydrate decomposition; in calcium chloride solution, the energy required for natural gas hydrate decomposition is very low, and the ability to destroy the natural gas hydrate lattice is stronger [33] and more favorable for hydrate decomposition.

4. Conclusions

The experimental study found that multi-stage depressurization combined with the chemical agent injection method can increase mining efficiency by more than 68% in the first two hours after chemical agent injection, compared with the single depressurization method, which has the advantage of significantly increasing the mining efficiency. However, considering the field exploitation of natural gas hydrates, this method also has some limitations. Multi-stage depressurization combined with the injection of thermodynamic chemicals with a higher mass concentration (such as methanol, ethylene glycol, etc.) may pollute the environment and cost more.

From the perspective of economic and environmental benefits, the combined method needs to reasonably optimize the injection amount of thermodynamic chemicals or use multiple chemicals together to reduce costs and increase gas production. Secondly, in the future, it is necessary to constantly develop green, efficient, and economic kinetic chemicals to replace traditional thermodynamic chemicals. I believe that the reasonable application of multi-stage depressurization combined with chemical injection method is expected to be an efficient method for the on-site exploitation of natural gas hydrate.

Author Contributions: Conceptualization, Z.S. and L.W.; methodology, Z.S. and L.W.; software, Z.S.; validation, Z.S. and Z.C.; formal analysis, Z.S. and L.W.; investigation, Z.S. and W.X.; resources, L.W.; data curation, Z.S. and Z.C.; writing—original draft preparation, Z.S.; writing—review and editing, Z.S.; visualization, L.W.; supervision, L.W.; project administration, L.W.; funding acquisition, L.W. and X.H. All authors have read and agreed to the published version of the manuscript.

Funding: This research was funded by the project group of the Ministry of Science and Technology of Sinopec "Research on the Key Technologies of Effective Exploitation of Natural Gas Hydrate", grant number P20040-4.

Institutional Review Board Statement: The study does not involve humans or animals.

Informed Consent Statement: The study does not involve humans or animals.

Data Availability Statement: Not applicable.

Acknowledgments: Thanks to other authors for his support of the experimental methods, experimental materials and amendments to the content of the article.

Conflicts of Interest: The authors declare no conflict of interest.

References

1. Huang, X.; Wang, H.B.; Zhang, L.; He, J.Y.; Cen, X.Q. Research progress on production stimulation technology of natural gas hydrate reservoir exploitation. *Sci. Technol. Eng.* **2022**, *22*, 3405–3415.
2. Zhang, X.B.; Lu, X.B.; Liu, L.L. Research progress on natural gas hydrate extraction methods. *Adv. Geophys.* **2014**, *29*, 858–869.
3. Collett, T.S.; Ginsburg, G.D. Gas hydrates in the Messoyakha gas field of the west Siberian basin—A re-examination of the geologic evidence. *Int. J. Offshore Polar Eng.* **1998**, *8*, 22–29.
4. Almenningen, S.; Flatlandsmo, J.; Fernø, M.A.; Ersland, G. Production of sedimentary methane hydrates by depressurization. SPE Bergen One Day Seminar. *OnePetro* **2016**.
5. Li, G.; Li, B.; Li, X.S.; Zhang, Y.; Wang, Y. Experimental and numerical studies on gas production from methane hydrate in porous media by depressurization in pilot-scale hydrate simulator. *Energy Fuels* **2012**, *26*, 6300–6310. [CrossRef]

6. Wang, Y.; Feng, J.C.; Li, X.S.; Zhang, Y.; Li, G. Analytic modeling and large-scale experimental study of mass and heat transfer during hydrate dissociation in sediment with different dissociation methods. *Energy* **2015**, *90*, 1931–1948. [CrossRef]
7. Zhou, Y.; Castaldi, M.J.; Yegulalp, T.M. Experimental investigation of methane gas production from methane hydrate. *Ind. Eng. Chem. Res.* **2009**, *48*, 3142–3149. [CrossRef]
8. Zhao, J.; Shi, D.; Zhao, Y. Mathematical model and simulation of gas hydrate reservoir decomposition by depressurization. *Oil Gas Sci. Technol.* **2012**, *67*, 379–385. [CrossRef]
9. Song, B.; He, Y.; Fan, Q.; Wu, Z.; Yang, B.; Zhou, Y. Numerical Study on Gas Production and Stratum Response to Marine Hydrate Dissociation Directed by Depressurization. In Proceedings of the 31st International Ocean and Polar Engineering Conference, Rhodes, Greece, 20 June 2021.
10. Konno, Y.; Jin, Y.S.; Shinjou, K.; Nagao, J. Experimental evaluation of the gas recovery factor of methane hydrate in sandy sediment. *RSC Adv.* **2014**, *4*, 51666–51675. [CrossRef]
11. Quan, H.P.; Li, Q.; Chen, T.D.; Ma, Q.; Zhai, X.E. Synthesis of a kinetic gas hydrate inhibitor. *Sci. Technol. Eng.* **2013**, *13*, 3986–3989.
12. Kawamura, T.; Sakamoto, Y.; Ohtake, M.; Yamamoto, Y.; Haneda, H.; Yoon, J.H.; Komai, T. Dissociation behavior of hydrate core sample using thermodynamic inhibitor. *Int. J. Offshore Polar Eng.* **2007**, *16*, 5–9.
13. Sung, W.; Kang, H. Experimental investigation of production behaviors of methane hydrate saturated in porous rock. *Energy Sources* **2003**, *25*, 845–856. [CrossRef]
14. Fan, S.S.; Tian, G.L.; Liang, D.Q.; Liang, D.; Li, D. Natural gas hydrate dissociation by presence of ethylene glycol. *Energy Fuels* **2006**, *20*, 324–326. [CrossRef]
15. Sun, Y.F. *Experimental Study on Gas Hydrate Exploitation Combined with Depressurization and Other Technologies*; China University of Petroleum: Beijing, China, 2019.
16. Sun, Y.F.; Zhong, J.R.; Li, W.Z.; Ma, Y.M.; Li, R.; Zhu, T.; Ren, L.L.; Chen, G.J.; Sun, C.Y. Methane Recovery from Hydrate-Bearing Sediments by the Combination of Ethylene Glycol Injection and Depressurization. *Energy Fuels* **2018**, *32*, 7585–7594. [CrossRef]
17. Zhang, J.; Guan, F.J.; Zhao, H. Feasibility analysis of natural gas hydrate exploitation by binary composite technology. *Contemp. Chem. Ind.* **2018**, *47*, 309–312.
18. Sun, T.; Chen, Y.; Zhao, Y. Combining DWS Depressurization Method and Heat Injection to Improve the Hydrate Recovery. In Proceedings of the Offshore Technology Conference, Richardson, TX, USA, 30 April 2018.
19. Li, S.X.; Chen, Y.M.; Zhang, W.W.; Xia, X.R. Experimental study on heat injection and depressurization production of natural gas hydrate in porous media. *Exp. Mech.* **2011**, *26*, 202–208.
20. Bai, Y.H.; Li, Q.P. Simulation of gas hydrate reservoirs combined with warm water injection and pressure reduction method. *Sci. China Tech. Sci.* **2011**, *41*, 262–268.
21. Gupta, A.; Aggarwal, A. Gas hydrates extraction by swapping-depressurisation method. In Proceedings of the Offshore Technology Conference-Asia, Kuala Lumpur, Malaysia, 25 March 2014.
22. Wang, B.; Fan, Z.; Wang, P.F.; Liu, Y.; Zhao, J.; Song, Y. Analysis of depressurization mode on gas recovery from methane hydrate deposits and the concomitant ice generation. *Appl. Energy* **2018**, *227*, 624–633. [CrossRef]
23. Khormali, A.; Sharifov, A.R.; Torba, D.I. The control of asphaltene precipitation in oil wells. *Pet. Sci. Technol.* **2018**, *36*, 443–449. [CrossRef]
24. Liu, D.J.; Qi, L.P. Research progress on formation mechanism and reserve prediction of seabed sandstone hydrate. *J. Petrochem. Univ.* **2021**, *34*, 76–84.
25. Wang, D.D. *Geological Characteristics and Physical Properties of Low-Permeability Weakly Consolidated Hydrate Reservoirs in the South China Sea*; China University of Geosciences: Wuhan, China, 2021. [CrossRef]
26. Dai, S.J. Study on the remodeling of gas hydrate sediment samples in the South China Sea. *China Pet. Chem. Stand. Qual.* **2022**, *42*, 129–131.
27. Gayet, P.; Dicharry, C.; Marion, G.; Graciaa, A.; Lachaise, J.; Nesterov, A. Experimental determination of methane hydrate dissociation curve up to 55MPa by using a small amount of surfactant as hydrate promoter. *Chem. Eng. Sci.* **2005**, *60*, 5751–5758. [CrossRef]
28. Wang, Y.F.; Sun, C.Y.; Yu, X.C.; Wang Qing Li, Q.; Chen, G. Decomposition rule of hydrate injection inhibitor in pilot plant. *Chem. Prog.* **2022**, *41*, 1–13. [CrossRef]
29. Zhang, G.Q. Mechanism analysis of the effect of electrolytes and alcohols on the phase balance of natural gas hydrate. *Guangzhou Chem. Ind.* **2020**, *48*, 17–20.
30. Sun, Z.N. *Molecular Simulation of Hydrate Decomposition Mechanism under the Action of Inhibitors*; China University of Petroleum: Qingdao, China, 2015.
31. Ding, T.; Wang, R.; Xu, J.; Wang, X.; Yu, Z.; Cheng, Y.; Wang, Z. Experimental study and modeling of methane hydrate dissociation by depressurization and chemical injection. In Proceedings of the 29th International Ocean and Polar Engineering Conference, Honolulu, HI, USA, 16 June 2019.
32. Peng, L. *Study on the Decomposition Law of Methane Hydrate in the Presence of Inorganic Salts*; China University of Petroleum: Qingdao, China, 2017.
33. Zhang, Z.X.; Xu, L.; Zhang, T.J. Molecular simulation of the mechanism of hydrate decomposition of inorganic salts. *Inn. Mong. Petrochem. Ind.* **2020**, *46*, 16–22.

processes MDPI

Article

Investigation of the Vertical Propagation Pattern of the 3D Hydraulic Fracture under the Influence of Interlayer Heterogeneity

Bingqian Wan [1], Yancheng Liu [2], Bo Zhang [1], Shuai Luo [1], Leipeng Wei [3], Litao Li [2] and Jiang He [4,*]

1 PetroChina Xinjiang Oilfield Development Company, Karamay 834000, China
2 China United Coalbed Methane Corporation Ltd., Beijing 100016, China
3 CNPC Western Drilling Engineering Co., Ltd., Karamay 834000, China
4 School of Geoscience and Technology, Southwest Petroleum University, Chengdu 610500, China
* Correspondence: swpu_hejiang@126.com

Abstract: The low permeability and thinly interbedded reservoirs have poor physical properties and strong interbedded heterogeneity, and it is difficult to control the hydraulic fracture (HF) height and width during hydraulic fracturing, which affects the effect of HF penetration and sand addition. In this work, a three-dimensional fluid–solid fully coupled HF propagation model is established to simulate the influence of interlayer heterogeneity on vertical HF height and HF width, and the relationship between HF length and HF width under different treatment parameters is further studied. The results show that, in thin interbedded strata, the high interlayer stress contrast, high tensile strength, and low Young's modulus will inhibit the vertical propagation of HFs. The interlayer heterogeneity results in the vertical wavy distribution of HF width. Under the high interlayer stress contrast, Young's modulus, and tensile strength, the HF width profile becomes narrow and the variation amplitude decreases. The HF length decreases and the HF width increases as the injection rate and fracturing fluid viscosity increase. This study is of great significance for clarifying the vertical propagation pattern in thinly interbedded reservoirs, optimizing the treatment parameters, and improving the effect of cross fracturing and proppant distribution.

Keywords: hydraulic fracturing; HF vertical propagation; finite element method; interlayer heterogeneity; thin interlayers

Citation: Wan, B.; Liu, Y.; Zhang, B.; Luo, S.; Wei, L.; Li, L.; He, J. Investigation of the Vertical Propagation Pattern of the 3D Hydraulic Fracture under the Influence of Interlayer Heterogeneity. *Processes* **2022**, *10*, 2449. https://doi.org/10.3390/pr10112449

Academic Editors: Linhua Pan, Yushi Zou, Jie Wang, Minghui Li, Wei Feng and Lufeng Zhang

Received: 17 October 2022
Accepted: 16 November 2022
Published: 18 November 2022

1. Introduction

Jimsar shale oil reservoir in Xinjiang has high oil reserves and poor reservoir physical properties (porosity of 10%, permeability of 0.05 mD). Large-scale hydraulic fracturing is a necessary technology for the profitable development of this reservoir. However, the target reservoir presents the typical characteristics of thin vertical layers, with strong interbedded heterogeneity (Qi et al., 2022) [1]. It is difficult to predict and control the height and width of hydraulic fractures (HFs) during the process of hydraulic fracturing, which affects the processes of the through-layer fracturing and the sand-adding fracturing (Dan et al., 2015; Wang et al., 2021; Xie et al., 2022) [2–4]. Therefore, clarifying the vertical propagation pattern of HFs in thinly interbedded reservoirs is of great significance to improve the pertinence of fracturing construction and the fracturing effect.

Scholars at home and abroad have studied the pattern of HF propagation in layered strata based on physical model experiments and numerical simulations [5–15]. Mukhtar et al. (2022) [5] present a coupled multiphysics 3D generalized finite element method to simulate hydraulic fracture propagation. This method is validated against a hydraulic fracture experiment on PMMA. There are two types of factors affecting vertical HF propagation: formation factors and treatment parameters. The former include interlayer stress contrast, petrophysical property, and tensile strength, while the latter include pump injection rate and fracturing fluid viscosity (Tan et al., 2021; Sun et al., 2020; Tan et al., 2019) [6–8]. Miskinmins et al. (2003) [9] studied the influence of interlayer properties of sand and mud on vertical HF propagation and analyzed the influence of interlayer interface shear slip, rock mechanical properties difference, and pore elasticity on HF height propagation. Li et al. (2014) [10] studied the HF propagation pattern in layered strata based on the finite element method, and the results showed that the high stress, high strength, and low modulus significantly inhibited the vertical HF propagation, thus the HF height can be controlled. Sun et al. (2020) [7] established a multi-layer HF propagation model based on the cohesive element theory, and the variation in the angle of the relative vertical extension of the bedding plane and the tensile strength of the bedding plane on the vertical extension of the HF is analyzed. The simulation results show that reservoirs with low vertical stress differences and nearly horizontal bedding planes with low dip angles are found to be favorable for opening the bedding planes, and reservoirs with high vertical stress differences and bedding plane dip angles are favorable for the longitudinal expansion of HFs. Wang et al. (2021) [11] conducted triaxial fracturing simulation experiments on full-diameter shale cores with different horizons to study the vertical propagation patterns of HFs in different reservoirs and established a pseudo-three-dimensional HF propagation model of multi-layer shale oil reservoirs. The results show that the HF height of shale is smaller than that of sandstone owing to bedding limitations. When sandstone and shale are fractured at the same time, the HF extension height reaches the maximum after the two layers are connected. When the sandstone–shale interface is strong, increasing the pumping rate can increase the HF height. Fu et al. (2021) [12] studied the effect of permeability difference, in situ stress condition, and lithological interface on hydraulic propagation by building a 3D HF propagation numerical model. The results show that the ability of HF to cross the lithologic boundary decreases with the increase in the high permeability difference coefficient. In addition, when the stress difference between layers is small, the fracture penetration boundary has a higher probability to expand to the additional layer. Liu et al. (2022) [13] considered the Hancheng area, Ordos Basin, China, as an example to understand macrolithotype differences, established finite element numerical models of the cohesive zone, and evaluated the HF initiation and vertical propagation behavior of laminated coal reservoirs. Mukhtar et al. (2022) [14] present a comprehensive study on multiple hydraulic fracture propagation and their interactions under different treatment conditions. A systematical parametric study was carried out and the considered factors included fracture spacing, injection fluid viscosity, number of fracture clusters, and the stress conditions.

In conclusion, the present study has clarified the main controlling factors of HF height propagation in stratified strata and their influencing patterns. However, owing to the limitation of the sample scale, the boundary effect in stress loading cannot be eliminated in the model experiment, which leads to a certain gap between the HF height and the reality. In addition, the HF width in the model experiment is generally tens of microns, which cannot capture the vertical distribution of HF width. The existing numerical simulation work mainly studies the vertical propagation pattern of HFs in three-layer strata. However, the Jimsar shale oil reservoir in Xinjiang presents the characteristics of thin interbedding, while the reservoir and the interbedding appear alternately. The variation pattern of vertical HF height and width needs to be further studied. Based on the finite element method and cohesive zone method, a fluid–solid fully coupled three-dimensional HF propagation model is established for thin interlayer formation and the influence pattern of strong interlayer heterogeneity on vertical HF propagation is studied. The research in this paper has certain theoretical guiding significance for the formulation of the fracturing scheme of Jimsar shale oil.

2. Physical Process and Mathematical Models

Three-dimensional HF propagation involves multiple complex physical processes: tangential/normal flow of fluid in the HF, matrix seepage, rock stress/strain, fluid loss at the

HF surface, and initiation and propagation at the HF tip. In addition, these processes affect each other and need to be solved in a coupled manner (Gonzalez-Chavez et al., 2015) [15].

2.1. Rock Deformation and Fluid Flow Equations

The matrix pore pressure is changed by fluid filtration in the HF, which affects the effective stress of the rock and then affects the deformation of the rock. Considering the coupling effect of fluid filtration, pore pressure change, and rock deformation, the governing equation of rock deformation and fluid is provided. Assuming the characteristics of homogeneity, isotropy, and linear elasticity of rock, the governing equation of rock deformation is as follows (Wang et al., 2018) [16]:

$$\begin{cases} \nabla \sigma + \mathbf{f} = 0 \\ \varepsilon = (\nabla \mathbf{u} + (\nabla \mathbf{u})^T)/2 \\ \sigma = \mathbf{D}\varepsilon \end{cases} \tag{1}$$

where σ is the stress tensor, Pa; $\mathbf{D}$ is the stiffness matrix, Pa; $\mathbf{f}$ is the force vector per unit volume, N/m^3; ε is the strain tensor, dimensionless; and $\mathbf{u}$ is the displacement vector, m.

Assuming that the fluid is incompressible, the governing equation of tangential flow in the HF is based on Poiseuille's cubic law:

$$\mathbf{q_f} = -\frac{w^3}{12\mu}\nabla p_f \tag{2}$$

where $\mathbf{q_f}$ is the tangential flow rate of fluid in HF, m^3/s; w is HF width, m; μ is fluid viscosity, Pa·s; and p_f is fluid pressure in the HF, Pa.

The continuity equation of the fluid flow in the HF is as follows:

$$\nabla \mathbf{q_f} - \frac{\partial w}{\partial t} + q_b + q_t = 0 \tag{3}$$

where q_b and q_t are the normal fluid filtration velocity (m/s) of the upper and lower surfaces of the HF, respectively.

The normal filtration equation of the HF surface is as follows:

$$\begin{cases} q_t = c_t(p_i - p_t) \\ q_b = c_b(p_i - p_b) \end{cases} \tag{4}$$

where c_t and c_b are the leak-off coefficient into the top and bottom HF surfaces, respectively; p_i, p_t, and p_b are the pore pressures within the HF, top HF surface, and bottom HF surface, respectively; and q_t and q_b are the normal flow rates into the top and bottom HF surfaces, respectively.

2.2. Hydraulic Fracture (HF) Tip Initiation and Propagation Equation

Based on the theory of linear elastic fracture mechanics, there exist stress singularities at the HF tip, and the calculation is huge. Mukhtar et al. (2020) [17] combined the generalized finite element method with mesh adaptivity for the robust and computationally efficient simulation of HF propagation. To eliminate singularity calculation of the HF tip, a cohesive zone model was used to characterize HF initiation and propagation, and there exists a process zone at the HF tip. In this region, there is a bilinear relationship between the interface force and the interface distance (T–S criterion). Before the interface distance reaches the initial damage distance, the interface force and interface distance meet the linear elastic relationship. When the interface distance reaches the initial damage distance, the interface stiffness degrades gradually. When the interface distance reaches the complete damage distance, the interface gravity is 0 and a new HF element is generated. In addition, the cohesive zone model can accurately simulate the tangential flow in Equation (2) and normal fluid loss in Equation (4), which gradually becomes an effective method to simulate HF propagation.

The secondary stress criterion is used to control the initiation of HF elements; when the sum of the squares of the ratio of the actual stress in three directions to the critical stress in the corresponding direction is equal to 1, the initial damage of HF elements occurs (Wang et al., 2021) [18]:

$$\left\{\frac{\langle t_n\rangle}{t_n^0}\right\}^2+\left\{\frac{\langle t_{s1}\rangle}{t_{s1}^0}\right\}^2+\left\{\frac{\langle t_{s2}\rangle}{t_{s2}^0}\right\}^2=1 \quad (5)$$

where t_n, t_{s1}, and t_{s2} are the real nominal stress in the normal, first, and second shear nominal directions, respectively; t_n^0, t_{s1}^0, and t_{s2}^0 are the peak nominal stress purely in the normal, first, and second shear directions, respectively. $\langle\,\rangle$ is the Macaulay bracket.

After initial damage, the interface stress of the HF element can be calculated by the following equation:

$$\sigma_n=\begin{cases}(1-D)\overline{\sigma}_n, \overline{\sigma}_n\geq 0\\ \overline{\sigma}_n\end{cases} \quad (6)$$

$$\begin{aligned}\sigma_s&=(1-D)\overline{\sigma}_s\\ \sigma_t&=(1-D)\overline{\sigma}_t\end{aligned} \quad (7)$$

where σ_n, σ_s, and $\overline{\sigma}_t$ are the stresses calculated in the normal direction and two tangential directions according to the elastic criterion of the undamaged front. D is a dimensionless damage factor, whose value lies between 0 and 1. The material is not damaged when $D = 0$ and completely damaged when $D = 1$.

The expression of damage factor D is as follows:

$$D=\frac{\delta_m^f(\delta_m^{\max}-\delta_m^0)}{\delta_m^{\max}(\delta_m^f-\delta_m^0)} \quad (8)$$

where δ_m^0, δ_m^f, and $\delta_m^{\max}$ are the effective displacement at the stage of HF initiation, complete HF formation, and the maximum value during the loading history, respectively.

The energy criterion is used to characterize the condition of complete damage of the HF element. G^c represents HF energy, which is equal to the area enclosed by the triangle, kN/m. The expression of the energy criterion is as follows:

$$G^c=G_n^c+(G_s^c-G_n^c)\left\{\frac{G_S}{G_T}\right\}^\eta \quad (9)$$

where G_n^c and G_s^c are Mode I and Mode II critical HF energies, respectively, kN/m. $G_S = G_s + G_t$, $G_T = G_n + G_S$. G_n and G_s are the HF tensile (Mode I) and shear (Mode II) energy components, respectively. η is the material parameter.

3. Model Establishment and Input Parameters

In the Jimsar shale oil reservoir, multistage fracturing is commonly used in horizontal wells. This paper focuses on a perforation cluster within one stage and focuses on vertical HF propagation in layered formations. Based on the finite element method and cohesive zone model, a three-dimensional (3D) fluid–solid coupling model is established in Abaqus (a commercial software) to study the vertical propagation pattern of 3D HFs under the influence of interlayer heterogeneity. As shown in Figure 1a, the length × width × height of the model is 150 m × 30 m × 60 m, and there are five simulated pay zones (red zone) and six simulated interlayers (blue zone). The thickness of the single pay zone and interlayer in the middle reservoir is 5 m and the interlayer thickness at both ends is 7.5 m. The in situ stresses are perpendicular to each other and the injection point is located in the middle of the model. Considering the symmetry of the model, the upper half of the 1/2 model is used for the calculation to shorten the calculation time (Figure 1b). The 3D eight-node displacement and pore pressure element (C3D8P) was used for the matrix rock, and the 3D 12-node displacement and pore pressure cohesive element (COH3D8P) was used for the HF propagation path. In this way, the nodes of the matrix element and the side nodes of the cohesive elements have both displacement and pore pressure degrees of freedom, while the middle nodes of the cohesive element only have the pore pressure degrees of freedom. The mesh around the HF element is refined to improve the calculation accuracy. Considering multi-cluster fracturing within one stage, the injection rate of a single HF was 4 m^3/min and the injection time was 30 min. The displacement degrees of freedom of the model boundary were fixed and the constant pore pressure boundary was set. Table 1 shows the basic input parameters of the model. Figure 2 presents the fracture profile at different times during HF propagation.

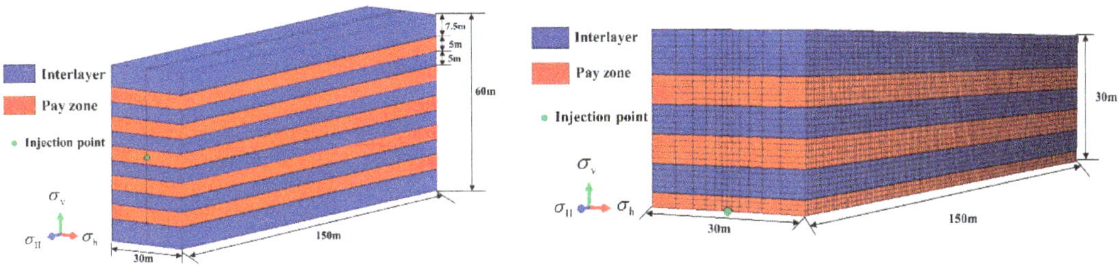

(a) (b)

Figure 1. Model establishment: (**a**) geometry model; (**b**) finite element model.

Table 1. The base input parameters for the model.

Item	Parameter	Pay Zone	Interlayer
Rock parameter	Young's modulus (GPa)	20	25
	Possion ratio	0.25	0.2
	Filtration coefficient (m/s)	3×10^{-4}	1.2×10^{-5}
Cohesive element property	Tensile strength/(MPa)	3	6
	Energy release rate (N/m)	12,000	16,000
	Leak off coefficient (m^3/(Pa·s))	10^{-13}	10^{-14}
In situ stress	The effective minimum horizontal principal stress (MPa)	20	23
	The effective maximum horizontal principal stress (MPa)	28	31
	The effective vertical principal stress (MPa)	38	38
Fluid property	Fluid viscosity (mPa·s)	100	100
	Injection rate (m^3/min)	4	4
Initial condition	Pore pressure (MPa)	37	35
	Porosity ratio	0.12	0.1

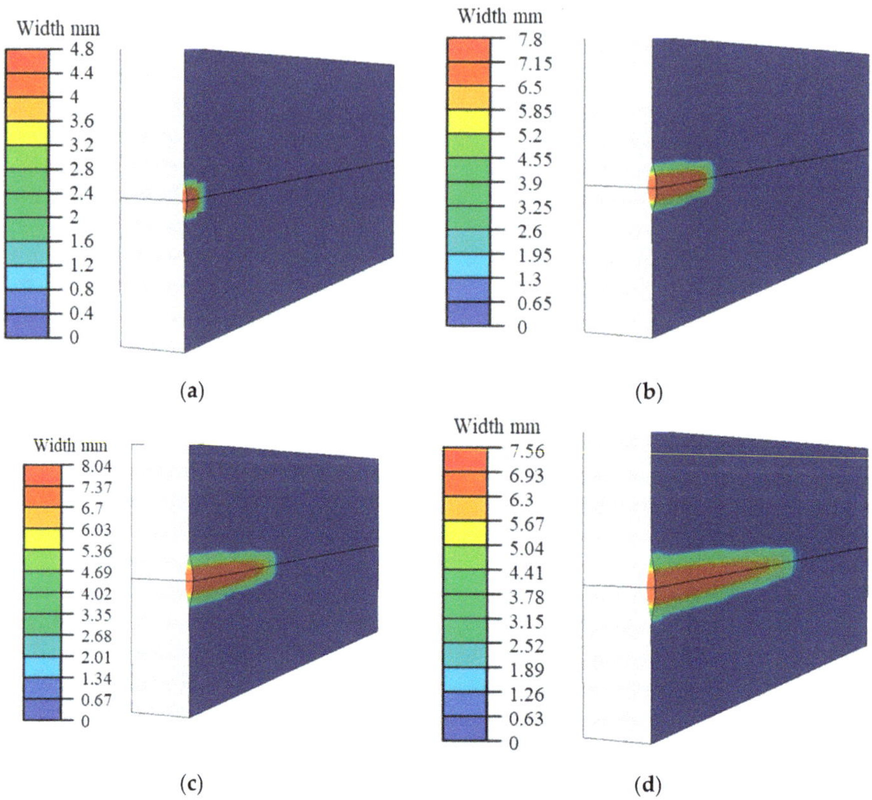

Figure 2. Fracture profile at different times during HF propagation (PFOPEN denotes the HF width, m). (**a**) 1 min; (**b**) 8 min; (**c**) 16 min; (**d**) 24 min.

4. Results and Analysis

The research results show that the properties of the interlayer interface affect the vertical HF propagation pattern; when the strength of the interlayer interface is weak, interface slip or HF propagates along the interface during fracturing, and T-shaped or fishbone HFs are generated longitudinally (Du et al., 2022) [19]. Laboratory rock mechanical parameters tests show that the mechanical strength of the interface is high, thus the fracturing fluid cannot penetrate into the interface between layers in the Jimsar shale oil reservoir. To focus on the effect of interlayer heterogeneity on HF vertical propagation, this paper assumes that the cementation between the pay zone and the interlayer is intact. The variation rule of 3D HF height and width under the condition of a thin interlayer is quantitatively studied with different in situ stress, elastic modulus, and tensile strength values of the interlayers.

4.1. Interlayer Stress Contrast

Based on the parameters in Table 1, the minimum horizontal principal stress of the pay zone was set as 20 MPa and kept unchanged, the minimum horizontal principal stress of the interlayer (23 MPa, 27 MPa, 31 MPa) was set successively, and other parameters were kept consistent. The variation law of HF height and width was studied under the interlayer stress contrasts of 3 MPa, 7 MPa, and 11 MPa, respectively. As shown in Figure 3 (injection time of 30 min), the vertical distribution profile of HF width is drawn with the HF center as the coordinate origin. The ordinate represents the half-HF height and the abscissa represents the HF width. It can be seen that, as the stress contrast between layers increases, the HF height decreases. When the stress contrast between layers is 3 MPa, 7 MPa, and 11 MPa, the corresponding HF heights are 25.80 m, 14.18 m, and 10.80 m, respectively, which decrease by 58.14%. The reason is that, the higher the stress in the interlayer, the more difficult the vertical HF propagation, and the higher the net pressure in the HF. For the same volume of injected fluid, the HF tends to propagate along the HF length in the pay zone owing to the strong containment of the interlayers. In addition, compared with the interlayer stress contrast of 3 MPa and 7 MPa, the difference in HF height is 45.04%. Compared with the stress contrast of 7 MPa and 11 MPa, the difference in HF height is 23.84%. Therefore, with the increase in stress contrast between layers, the decrease in HF height becomes slower. The reason is that, the higher the stress contrast between layers, the higher the energy consumed when the HF propagates vertically through the interlayer and the easier it is to stop the HF in the interlayer, and the smaller the HF height, the smaller the gap.

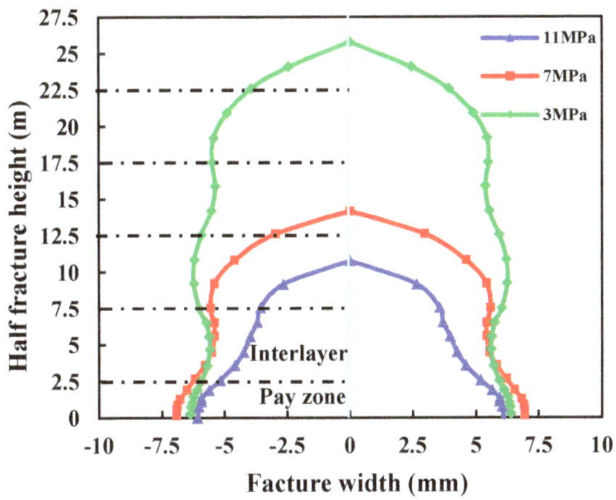

Figure 3. HF morphology under different interlayer stress contrast.

As far as HF width is concerned, owing to the alternating occurrence of the pay zone and the interlayer (the dashed line represents the interface between the pay zone and the interlayer), the horizontal minimum principal stress contrast between the zones constantly changes the HF width and the longitudinal distribution of HF width is wavy (Figure 3). The smaller the horizontal minimum principal stress, the smaller the resistance and the larger the central HF width. In addition, the HF width in the pay zone where the HF center is located is less affected by the interlayer stress. The width of the HF center corresponding to the interlayer stress contrast of 3 MPa, 7 MPa, and 11 MPa is 12.7 mm, 13.8 mm, and 12.1 mm, respectively.

4.2. Young's Modulus

Young's modulus of the reservoir rock reflects the deformation resistance of the rock and then affects the HF height and width. Young's modulus of the pay zone was kept unchanged at 20 GPa, and Young's modulus of the interlayer was set to be 25 GPa, 30 GPa, and 35 GPa, respectively. Other parameters were kept consistent. The variation law of the HF height and width of the HF was studied under Young's modulus differences of 5 MPa, 10 GPa, and 15 GPa. As shown in Figure 4 (injection time of 30 min), the HF heights corresponding to Young's modulus of 25 GPa, 30 GPa, and 35 GPa are 15.92 m, 24.11 m, and 27.80 m, respectively, with an increase of 42.73%. The higher the Young's modulus of the interlayer, the stronger the vertical propagation ability and the higher the HF height. The reason is that, in the interlayer with a high Young's modulus, the rock deformation is small, the HF width is small, the net pressure in the HF is high, and the vertical propagation ability of the HF is strong. In addition, compared with 30 GPa and 25 GPa, the gap in height was 33.97%. The gap between Young's modulus of 30 GPa and 35 GPa was 13.27%. Therefore, with the increase in Young's modulus of the interlayer, the increase in the HF height becomes slower.

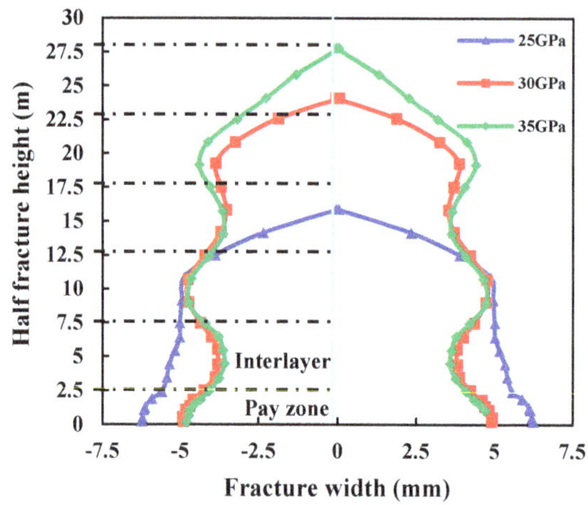

Figure 4. HF morphology under different Young's modulus of the interlayer.

In terms of HF width, the Young's modulus of the pay zone is small and the Young's modulus of the interlayer is large. Therefore, the width gap of the pay zone is large, the width gap of the interlayer is small, and the longitudinal distribution of the width gap is wavy. In addition, with the decrease in the Young's modulus of the interlayer, the ability of the rock to resist deformation is weakened, and the HF width at the HF center increases significantly. When the Young's modulus of the interlayer is 25 GPa, 30 GPa, and 35 GPa, the corresponding central HF width is 12.5 mm, 9.8 mm, and 9.58 mm, respectively. Therefore, with the increase in the Young's modulus of the interlayer, the variation range of the HF width profile decreases.

4.3. Rock Tensile Strength

Rock tensile strength reflects the ability of the rock to resist tensile failure. The higher the tensile strength, the more difficult the HF initiation and propagation, which significantly affects the vertical HF propagation pattern. The tensile strength of the pay zone was kept unchanged at 3 MPa and the tensile strength of the interlayer was set as 6 MPa, 12 MPa, and 18 Mpa, respectively. Other parameters were kept consistent. The variation law of vertical HF height and width of HFs was studied under the difference of the tensile strength between the layers of 3 MPa, 9 MPa, and 15 MPa. As shown in Figure 5 (injection time of 30 min), the HF heights corresponding to the tensile strength of interlayer of 6 MPa, 12 MPa, and 18 MPa are 24.12 m, 15.73 m, and 12.41 m, respectively. The HF height of the first case decreases by 48.55% compared with the last case. The reason is that, the higher the tensile strength of the interlayer, the more difficult it is for the HF to cross the interlayer, so it tends to propagate along the HF length. In addition, compared with the interlayer tensile strength of 6 MPa and 12 MPa, the difference in HF height is 34.78%. Compared with the interlayer tensile strength of 12 MPa and 18 MPa, the difference in HF height is 21.11%, indicating that the change range of HF height decreases with the increase in interlayer rock tensile strength.

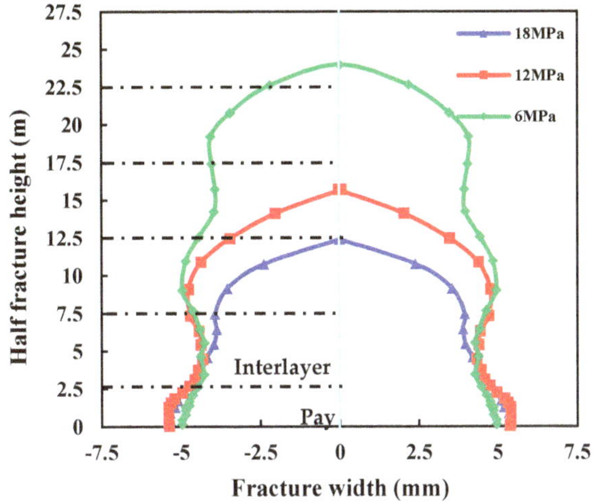

Figure 5. HF morphology under different tensile strengths of the interlayer.

As far as HF width is concerned, the difference in tensile strength between the pay zone and interlayer results in wavy distribution of HF width in the vertical direction. The higher the tensile strength of the interlayer, the more difficult it is to penetrate the interlayer, the smaller the HF width in the interlayer, and the smaller the width gap at the HF center. The reason is that the main reservoir factors affecting HF width are in situ stress and Young's modulus, and the tensile strength of the interlayer will not affect the width of the HF in the pay zone. The tensile strength determines the HF initiation pressure; after the generation of a new HF element, the HF net pressure remains a constant value, thus the HF width has a neglectable difference.

4.4. Treatment Parameters

This paper focuses on the influence of interlayer heterogeneity on HF vertical propagation. In addition to the two-dimensional extension of HF height and width, there is also the extension of HF length. Because of the influence of reservoir factors and engineering factors, HF height, width, and length affect each other. In the fracturing process, the larger the HF width, the lower the difficulty of sand addition. The longer the HF length, the larger the contact area between the wellbore and the reservoir and the higher the single well productivity. The viscosity and injection rate of fracturing fluid are the key parameters that can be controlled artificially in the field fracturing process. This section focuses on the influence of fluid injection rate and fluid viscosity on the length and width of HF.

Based on the numerical model in Figure 1 and the basic parameters in Table 1, the injection rate is set as 3, 4, 5, 6, and 7 m^3/min, respectively, and the injection time is adjusted to ensure that the total volume of fluid is consistent. Figure 6 shows the relationship between HF length and width in the intermediate pay zone under different injection rates. It can be seen that, with the increase in the injection rate, the HF length decreases and the HF width increases. The reason is that a higher injection rate produces a higher net pressure in the HF and the HF width is proportional to the net pressure in the HF, thus increasing the width of the HF, and taking into account the volume conservation (the total liquid pumping volume is the same), thus reducing the length of the HF.

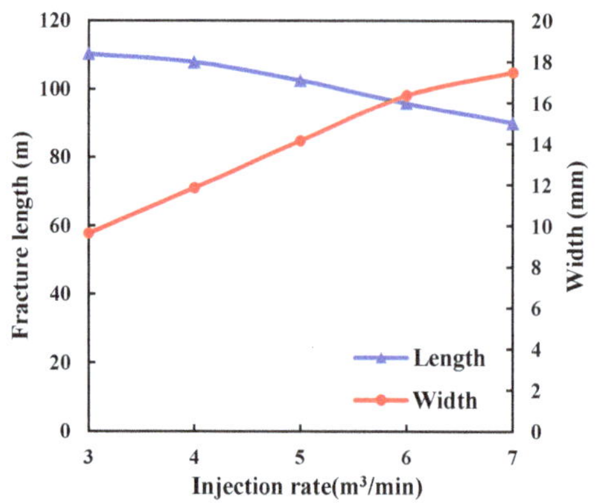

Figure 6. HF length and aperture curves under various flow rates after injecting for 30 min.

Based on the numerical model in Figure 7 and the basic parameters in Table 1, the value range of fracturing fluid viscosity is 1–300 mPa·s, and the relationship curve between the HF length and HF width under different fracturing fluid viscosity is studied. As shown in Figure 7, with the increase in fracturing fluid viscosity, the HF length decreases and the HF width increases slowly. The reason is that, with the increase in fracturing fluid viscosity, the fluid flow resistance in the HF increases, the fluid loss rate decreases, and the net pressure in the HF increases, thus the HF width increases with the HF net pressure. According to the volume conservation, the HF length decreases significantly.

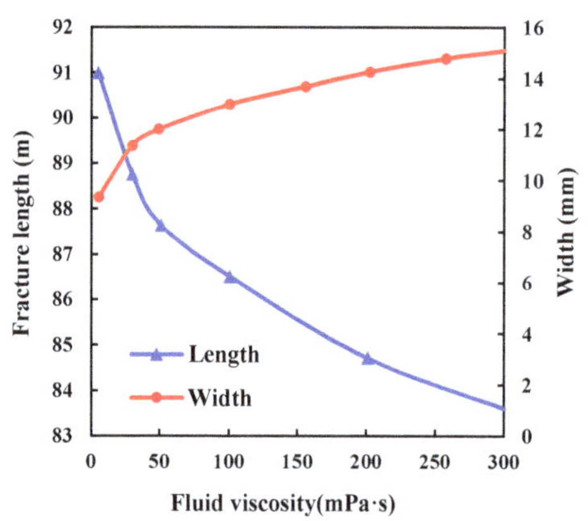

Figure 7. HF length and aperture curves under various fracturing fluid viscosity after injecting for 30 min.

Therefore, for the Jimsar thin interbedded reservoir, 3D HF morphology characteristics can be effectively controlled by controlling the fluid injection rate and fracturing fluid viscosity. Under certain reservoir conditions, pumping excessive viscosity liquid limits the HF length; pumping too low viscosity liquid tends to produce narrow HFs, which is not conducive to sand addition. Therefore, it is necessary to consider all factors to achieve efficient fracturing in thinly interbedded formations.

5. Conclusions

Based on the cohesive zone model and finite element method, this work established a 3D fluid–solid coupling HF propagation model and investigated HF vertical propagation patterns under the influence of interlayer heterogeneity. Moreover, this work presented the methods to control HF vertical propagation by adjusting the injection rate and fracturing fluid viscosity. The main conclusions are as follows:

1. The larger the stress contrast between the pay zone and interlayer, the smaller the HF height. The stress contrast between layers increases from 3 MPa to 11 MPa, and the HF height decreases by 58.14%. The larger the stress contrast, the narrower the HF width profile and the smaller the variation range of the HF width profile.
2. The higher the Young's modulus of the interlayer, the higher the HF height of the interlayer. The Young's modulus of interlayer increases from 25 GPa to 35 GPa, and the HF height increases by 42.73%. The larger the Young's modulus of the interlayer, the narrower the HF width profile and the smaller the variation range of the HF width profile.
3. The higher the interlayer tensile strength, the smaller the HF height. The interlayer tensile strength increases from 6 MPa to 18 MPa, and the HF height decreases by 48.55%. The larger the interlayer tensile strength, the narrower the HF width profile and the smaller the variation range of the HF width profile. The HF width in the middle pay zone is nearly the same.
4. In a thin interbedded reservoir, the HF width profile is wavy in the longitudinal direction. The influence of the HF width profile on proppant longitudinal placement should be considered in the process of sand-adding fracturing to improve the stimulation effect.
5. The larger the fluid injection rate and fracturing fluid viscosity, the smaller the HF length and the larger the HF width. When designing the fracturing construction, the sand-adding effect and reservoir contact area should be taken into account to determine the best fluid injection rate and fracturing fluid viscosity.

Author Contributions: Methodology, Y.L.; software, B.Z.; formal analysis, L.W.; investigation, S.L.; data curation, L.L.; writing—review and editing, B.W.; supervision, J.H.; project administration, J.H.; funding acquisition, J.H. All authors have read and agreed to the published version of the manuscript.

Funding: This research was funded by the Science and Technology Cooperation Project of the CNPC-SWPU Innovation Alliance (2020CX020000).

Data Availability Statement: Data available on request from the corresponding author if necessary.

Conflicts of Interest: The authors declare no conflict of interest.

References

1. Qi, H.; Su, J.; Hu, X.; Ma, A.; Dong, Y.; Li, A. Study on Well Logging Technology for the Comprehensive Evaluation of the "Seven Properties" of Shale Oil Reservoirs—An Example of Shale Oil in the Lucaogou Formation in the Jimsar Sag, Junggar Basin. *Front. Earth Sci.* **2022**, *9*, 827380. [CrossRef]
2. Dan, X.U.; Ruilin, H.U.; Wei, G.A.O.; Jiaguo, X. Effects of laminated structure on hydraulic fracture propagation in shale. *Pet. Explor. Dev.* **2015**, *42*, 573–579. [CrossRef]
3. Wang, L.; Sheng, Z.; Zhao, Z.; Song, D.; Wang, L.; Wang, G. Large-Section and Multi-Cluster Fracturing Technology for Horizontal Wells in the Jimsar Shale Oil Reservoir. *Pet. Drill. Tech.* **2021**, *49*, 106–111. [CrossRef]
4. Jianyong, X.; Xinjiang, C.; Wenbo, L.; Jingsheng, Z.; Xiaohu, W.; Yanjie, C.; Jinfeng, Z. Exploration and practice of benefit development of shale oil in Jimsar Sag, Junggar Basin. *China Pet. Explor.* **2022**, *27*, 99–110. [CrossRef]
5. Mukhtar, F.M.; Duarte, C.A. Coupled multiphysics 3-D generalized finite element method simulations of hydraulic fracture propagation experiments. *Eng. Fract. Mech.* **2022**, 108874. [CrossRef]
6. Tan, P.; Jin, Y.; Pang, H. Hydraulic fracture vertical propagation behavior in transversely isotropic layered shale formation with transition zone using XFEM-based CZM method. *Eng. Fract. Mech.* **2021**, *248*, 107707. [CrossRef]
7. Sun, C.; Zheng, H.; Liu, W.D.; Lu, W. Numerical simulation analysis of vertical propagation of hydraulic fracture in bedding plane. *Eng. Fract. Mech.* **2020**, *232*, 107056. [CrossRef]
8. Tan, P.; Jin, Y.; Yuan, L.; Xiong, Z.Y.; Hou, B.; Chen, M.; Wan, L.M. Understanding hydraulic fracture propagation behavior in tight sandstone–coal interbedded formations: An experimental investigation. *Pet. Sci.* **2019**, *16*, 148–160. [CrossRef]
9. Miskimins, J.L.; Barree, R.D. Modeling of hydraulic fracture height containment in laminated sand and shale sequences. In *SPE Production and Operations Symposium*; OnePetro: Oklahoma City, OK, USA, 2003.
10. Yang, L.; Jingen, D.; Baohua, Y.; Liu, W.; Chen, J. Effects of Reservoir Rock/Barrier and Interfacial Properties on Hydraulic Fracture Height Containment. *Pet. Drill. Tech.* **2014**, *42*, 80–86.

11. Wang, Y.; Hou, B.; Wang, D.; Jia, Z. Features of fracture height propagation in cross-layer fracturing of shale oil reservoirs. *Pet. Explor. Dev.* **2021**, *48*, 402–410. [CrossRef]
12. Fu, S.; Hou, B.; Xia, Y.; Chen, M.; Wang, S.; Tan, P. Hydraulic fracture height growth law for deep coal measures shale reservoir. *Fault Block Oil Gas Field* **2021**, *28*, 555–560.
13. Liu, Y.; Tang, D.; Xu, H.; Zhao, T.; Hou, W. Effect of interlayer mechanical properties on initiation and propagation of hydraulic fracturing in laminated coal reservoirs. *J. Pet. Sci. Eng.* **2022**, *208*, 109381. [CrossRef]
14. Mukhtar, F.M.; Shauer, N.; Duarte, C.A. Propagation mechanisms and parametric influence in multiple interacting hydraulic fractures: A 3-DG/XFEM hydro-mechanical modeling. *Int. J. Numer. Anal. Methods Geomech.* **2022**, *46*, 2033–2059. [CrossRef]
15. Gonzalez-Chavez, M.; Dahi Taleghani, A.; Olson, J.E. A Cohesive Model for Modeling Hydraulic Fractures in Naturally Fractured Formations. In Proceedings of the Spe Hydraulic Fracturing Technology Conference, The Woodlands, TX, USA, 3–5 February 2015.
16. Wang, B.; Zhou, F.; Wang, D.; Liang, T.; Yuan, L.; Hu, J. Numerical simulation on near-wellbore temporary plugging and diverting during refracturing using XFEM-Based CZM. *J. Nat. Gas Sci. Eng.* **2018**, *55*, 368–381. [CrossRef]
17. Mukhtar, F.M.; Alves, P.D.; Duarte, C.A. Validation of a 3-D adaptive stable generalized/eXtended finite element method for mixed-mode brittle fracture propagation. *Int. J. Fract.* **2020**, *225*, 129–152. [CrossRef]
18. Wang, B.; Liu, X.; Hu, J. Numerical simulation method of in-fracture temporary plugging and diversion fracturing. *Bull. Pet. Sci.* **2021**, *6*, 10.
19. Du, J.; Chen, X.; Liu, P.; Zhao, L.; Chen, Z.; Yang, J.; Miao, W. Numerical Modeling of Fracture Height Propagation in Multilayer Formations Considering the Plastic Zone and Induced Stress. *ACS Omega* **2022**, *7*, 17868–17880. [CrossRef] [PubMed]

processes

MDPI

Article

Research on Gas Channeling Identification Method for Gas Injection Development in High-Pressure Heterogeneous Reservoir

Juan Luo [1,2] and Lei Wang [1,2,*]

1 College of Petroleum Engineering, Yangtze University, Wuhan 430100, China
2 Key Laboratory of Drilling and Production Engineering for Oil and Gas, Wuhan 430100, China
* Correspondence: wang-lei@yangtzeu.edu.cn; Tel.: +86-18971168785

Abstract: In a typical ultra-deep high-temperature and high-pressure heterogeneous reservoir in Xinjiang, gas channeling quickly occurs during gas injection because of the heterogeneity of the reservoir, the low viscosity of gas injection, and the high gas-oil fluidity ratio. The identification and prediction methods of gas channeling in gas injection development were studied. First, gas channeling discrimination parameters were determined by the numerical simulation method. According to the ratio of gas to oil produced and the composition of oil and gas produced, the flow stages of formation fluid were divided into five regions: gas phase zone, two-phase zone, miscible zone, dissolved gas and oil zone, and original oil zone. The basis for gas channeling identification (namely, the field characterization parameters for gas channeling discrimination) was discovered through analysis and the knowledge of the operability of field monitoring data as the following two parameters: (1) the C1 content rising again on the previous platform when the trailing edge of the two-phase zone is produced and (2) the continuous rise of the gas-oil ratio in production. Then, considering the original high-pressure characteristics of the reservoir, the field characterization parameters of gas channeling under different formation pressures in the exploitation process (namely, C1 content and gas-oil ratio) were simulated and determined. Thus, a gas channeling discrimination method was established for gas injection development in ultra-deep high-temperature and high-pressure heterogeneous reservoirs. According to this gas injection approach, a gas channeling discrimination method was developed, and the field gas channeling judgment was carried out for a gas injection effective D1 well. The results of gas tracer detection were compared to verify the accuracy of this method, leading to strong support for this method in slowing down the gas channeling.

Keywords: high-pressure reservoir; heterogeneous; numerical simulation; gas channeling discrimination; identification parameters

Citation: Luo, J.; Wang, L. Research on Gas Channeling Identification Method for Gas Injection Development in High-Pressure Heterogeneous Reservoir. *Processes* **2022**, *10*, 2366. https://doi.org/10.3390/pr10112366

Academic Editors: Linhua Pan, Yushi Zou, Jie Wang, Minghui Li, Wei Feng and Lufeng Zhang

Received: 6 October 2022
Accepted: 2 November 2022
Published: 11 November 2022

1. Introduction

Gas injection has become an important method to improve oil recovery in ultra-deep heterogeneous reservoirs. However, in the process of gas injection development, due to the heterogeneity of the formation, with the increase of the injected pore volume multiple (PV), the leading edge of the injected gas will form a dominant flow channel along the highly permeable layer in the formation and break through to the bottom of the production well (gas crossing phenomenon), which reduces the sweep coefficient of the injected gas and leads to the poor development effect of gas injection.

Gas channeling is generally judged by a large rise in gas-oil ratio [1]. Studies on gas injection and gas channeling in gas cap reservoirs and condensate gas reservoirs have been conducted in China, mainly in the analysis of influencing factors of gas channeling and prevention of gas channeling [1–11]. However, there are few studies on the definition of gas channeling in gas injection drive. At present, only Si Yong studied the identification of gas injection and gas channeling in a fractured buried hill reservoir in 2020 [12]. He used numerical simulation to analyze the characteristic curves of production gas-oil ratio in different gas channeling directions, and concluded that there were different gas-oil ratio derivative curve characteristics in three different gas channeling types: top, bottom, and transverse. However, for ultra-deep and high-pressure heterogeneous complex reservoirs, the gas-oil ratio has several upward changes in different stages of the gas injection production process, so the gas channeling cannot be judged solely from the gas-oil ratio or its derivative change, and the variation characteristics of gas-oil ratio and other gas channeling characterization parameters in different fluid phase zones and pressure drop process should be considered.

The current judgment methods of gas channeling mainly include the empirical parameter method [2,3,7,13], plate method [14], numerical simulation method [15], and dynamic monitoring method [6,16]. Among them, the first two methods are mainly applied in the judgment of gas channeling in condensate gas reservoirs [13,17], and dynamic monitoring discriminant method is commonly used in gas channeling discrimination [8–10], such as the gas flooding front technology monitored by microseismic monitoring or the dynamic tracer monitoring method, while the numerical simulation method is more mature in the judgment of gas channeling in hydrocarbon reservoir injection. The numerical simulation method can establish the 3D geological model corresponding to the actual reservoir situation and set up a corresponding component model scheme for injected gas. Therefore, the numerical simulation method has strong adaptability and practicability in judging gas channeling in reservoir gas injection.

Taking an oil reservoir in Xinjiang as an example, the author used the numerical simulation method to analyze and find the characterization parameters of gas channeling, due to the heterogeneity in the gas injection process, and finally established a gas channeling discrimination method for ultra-deep high-temperature and high-pressure heterogeneous reservoirs.

2. Materials and Methods

2.1. Reservoir Profile

In the study of an ultra-deep high-temperature and high-pressure heterogeneous reservoir in Xinjiang, there are intercalations between and within each sand group. The average porosity of the reservoir was 15.1% and the average permeability was 68.1 mD. The 0 and 1 sand groups belonged to low porosity and low permeability reservoirs, and the 2, 3, and 5 sand groups belonged to medium porosity and medium permeability reservoirs, and the overall performance of the reservoir was medium porosity and medium permeability reservoirs. The surface oil density was 0.8547–0.8778 g/cm^3, the viscosity was 5.23–12.47 MPa·s (20 °C), the original formation pressure was 62.38 MPa, the pressure coefficient was 1.12, and the original formation temperature was 140 °C. Since 2014, major development tests of top natural gas injection assisted gravity flooding have been carried out. By the end of December 2021, 13 wells had been opened, with a daily oil level of 379 T at the wellhead and a comprehensive water cut of 48.84%. The gas viscosity was small, the gas-oil fluidity ratio was high, and the gas channeling phenomenon occurred easily, which lead to the low sweep coefficient, especially reservoir heterogeneity which aggravated the gas channeling. There were four gas injection wells in this reservoir. At present, with the progress of gas injection, the overall gas-oil ratio has kept rising, which has led to the deterioration of gas injection development effect. Therefore, it is urgent to identify gas channeling to prevent gas channeling.

2.2. Numerical Model Establishment and PVT Fitting of Crude Oil

In this study, a single well mechanism numerical simulation model was established.

With the PVTi phase calculation software of Eclipse [18] and the combination of reservoir geological parameters, a three-dimensional three-phase component mechanism model was established to fit multiple indexes, such as crude oil density under the crude oil formation condition, gas-oil ratio in flash test, saturation pressure, and PV relationship in the isocomponent expansion test. By analyzing the variation of gas-oil ratio and molar content of each component of produced gas in different fluid phase zones, and combining with the operability of field monitoring data, the characterization parameters of gas channeling discriminating in the field can be determined.

The PR equation of state was selected for simulation, and then appropriate critical parameters of state equation were selected for regression to fit the experimental results. The critical parameters of each component after regression were obtained for the prediction calculation and simulation study of fluid. A Cartesian grid system was used in the model (Table 1), with a total of 51 × 31 × 119 grids, with a grid size of 800 × 400 × 100 and injection-production well spacing of 400 m.

Table 1. Main parameters of the numerical simulation mechanism model.

Parameter	Value
Model size (m)	800 × 400 × 100
Grid system (x × y × z)	51 × 31 × 119
Porosity (%)	15
Permeability (mD)	According to the permeability (30–80 md) of a well in this area, and considering the heterogeneity in X, Y, and Z directions
Injection-production well spacing	One injection and one pick
Injection-production well spacing (m)	400

The setting parameters of the numerical simulation mechanism model are shown in the following table.

The reservoir in Xinjiang had the advantage of rich natural gas resources and had oil displacement properties under ultra-high-temperature conditions; in addition, the physical properties of crude oil was good, and so was suitable for the injection of natural gas to enhance oil recovery via the injection of more gas components (mainly methane).

The intermediate hydrocarbon content in the well flow of a reservoir in Xinjiang was 0.1182 and the content above C7 was 0.75931. At 140 °C and 47.9 MPa, the gas-oil ratio was 15.1 m^3/m^3. The bubble point pressure of crude oil was low at 6.072 MPa. The volume coefficient of crude oil was 1.0904 m^3/m^3. The shrinkage rate of crude oil was 8.29% and its shrinkage was low; the average gas dissolution coefficient was 2.487 m^3/m^3/MPa and the average gas dissolution coefficient was low. Crude oil density was low at 0.8549 g/cm^3. The above showed that the reservoir had good physical properties of crude oil, suitable for gas injection to enhance oil recovery.

By adjusting the thermodynamic parameters of heavy components, mainly C11+, the above main indexes were fitted. After fitting, the calculated values of the software were compared with the measured values (Table 2). Overall, the fitting errors of several main indicators were below 5%.

Table 2. Matching comparison of main properties of formation crude oil fluid.

High-Pressure Physical Property	Measured Value	Software Calculated Value	Error (%)
Saturation pressure (MPa)	6.072	6.045	−0.44
Single degassing oil ratio (m^3/m^3)	15.1	15.13	0.20
Underground crude oil density (g/cm^3)	0.7989	0.799	0.01
Degassed crude oil density (g/cm^3)	0.8515	0.8233	−3.31

The software calculated that the value of the relative volume fitted by PV relationship was basically consistent with the experimental measured value. The pressure-density relationship was also very close, the error was generally below 1%, the fitting effect was good. The fitted critical properties and thermodynamic parameters of each component could be used for further phase analysis.

2.3. *Gas Channeling Discrimination Method Research*

In the multi-stage contact process of gas injection, mass transfer between injected gas and formation crude oil causes the changes of phase characteristics and physical parameters (composition, interfacial tension, point degree, density, etc.). Mass transfer between components is the main mechanism of miscible displacement.

The characteristics of fluid flow in porous media and the change of phase state of each component can be expressed by a three-dimensional, three-phase, multi-component mechanical mathematical model. At the same time, the phase equilibrium calculation of an oil and gas system can be used to accurately evaluate the phase characteristics of reservoir fluid and the variation of phase characteristics in the extraction-dissolution mass transfer process between injected gas and formation fluid [19].

Research was conducted regarding the numerical simulation method on the mechanism model. In the process of gas injection, the fluid produced by the production well is in different phase zones in its formation, and its properties vary. In immiscible flooding, the injected gas at the displacement front is partially dissolved in oil to reduce viscosity and improve flow capacity. In the near-miscible flooding, the mass transfer effect is more obvious, the viscosity of crude oil is further reduced, and the near-miscible zone is formed. During miscible flooding, the injected dry gas continuously contacts with the formation crude oil and gradually forms a miscible zone. Under the three modes of immiscible flooding, near-miscible flooding, and miscible flooding, the formation fluid can be divided into five regions: gas-phase zone, two-phase zone, miscible zone, dissolved gas-oil zone, and original oil zone (Figure 1).

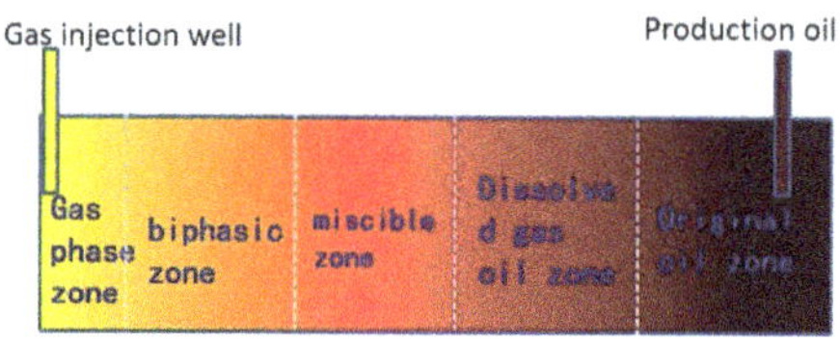

Figure 1. Division diagram of formation fluid flow stages.

For miscible flooding, the output of miscible zone is the most favorable. Natural gas is fully dissolved in crude oil, which plays a role in reducing its viscosity, reducing two-phase interfacial tension, increasing crude oil volume and supplementing formation energy. When the miscible zone fluid is displaced, it enters the two-phase zone production stage. At this time, the oil production decreases and the gas-oil ratio increases. The oil saturation in the front of two-phase zone is relatively high, and the oil production can be maintained at a certain level. The gas saturation at the trailing edge of the two-phase zone rises to a higher level. At this time, the injected gas no longer plays a displacement role, so the state corresponding to this time point is considered to be gas channeling. After the occurrence of gas channeling, the P-T phase diagram, C1 molar content, gas-oil ratio, density, and viscosity of the produced well flow will change.

3. Results

3.1. *Determination of Gas Channeling Identification Parameters*

In terms of dynamic simulation of the above model, the main production indicators with the production time relationship were as follows (Figures 2 and 3).

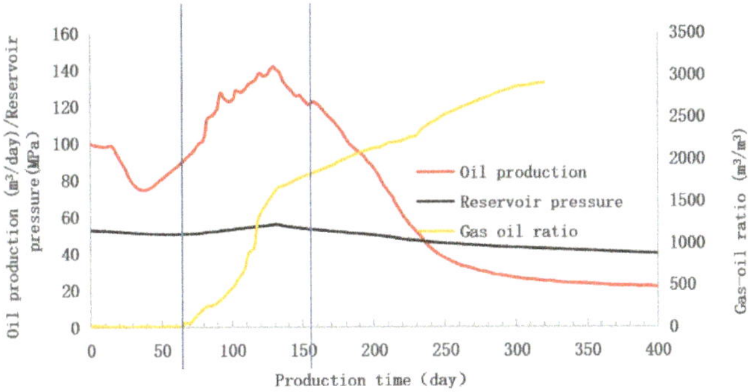

Figure 2. Prediction curve of oil production and gas-oil ratio of production well.

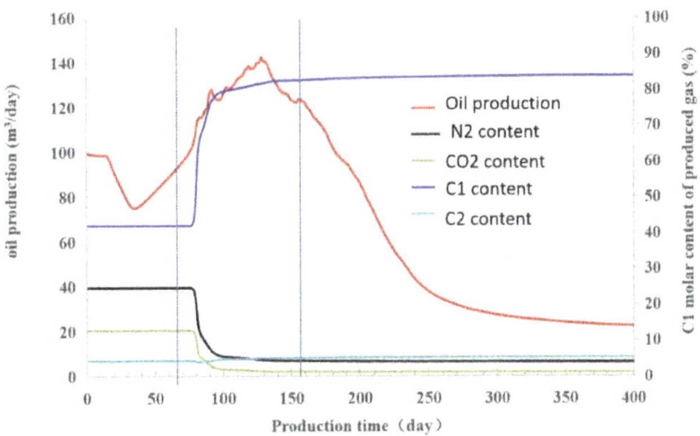

Figure 3. Analysis of gas channeling parameter selection.

Based on the analysis of changes in oil production, gas-oil ratio, and molar content of C1 component of produced gas, combined with the distribution of oil saturation in different periods, it can be divided into three main production stages:

The first stage was the original oil phase output stage, which lasted for 0~66 days. As the formation pressure drops in the early stages, oil production declined quickly. The increase of oil production in the later period was due to the replenishment of formation energy by injected gas. The main discrimination method at this stage was as follows: the produced natural gas is dissolved gas of original crude oil, so the gas production was small and the gas-oil ratio was constant at about 14 m^3/m^3.

From the oil saturation distribution (Figure 4), when the production time was 30 days, the oil saturation around the injection well changed significantly, while the oil saturation around the production well did not change significantly. However, because the injection wells continuously inject gas, the injected gas plays a role in replenishment of the formation energy to a certain extent, and the oil production of the production wells is increased by constantly flooding the crude oil around the production wells. However, at this time, the flow of the produced wells was still the original formation crude oil, so the gas-oil ratio index was constant.

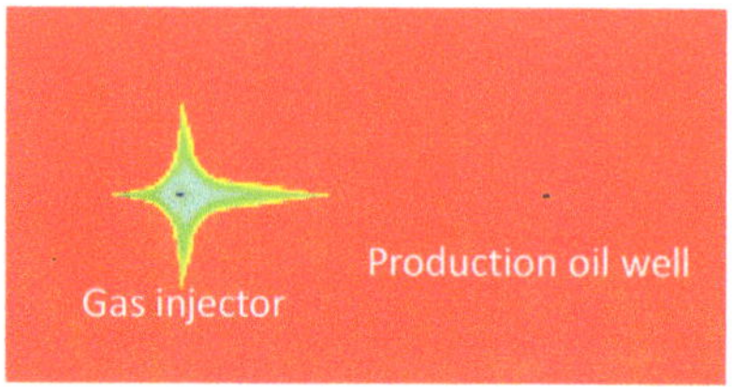

Figure 4. Distribution of formation oil saturation at 30 days of production.

The second stage was the output stage of dissolved gas-oil zone, miscible zone, and front edge of the two-phase zone, which lasted for 66–154 days. The production characteristics of this stage were comprehensively analyzed in terms of oil production, gas-oil ratio, and C1 content of produced gas.

Of the corresponding relationships of the three main indicators in each phase zone, the initial oil production, gas-oil ratio, and content of produced gas C1 increased rapidly, which was the well performance of the output stage of the dissolved gas-oil area. In the middle, the oil production in the miscible flooding stage peaked (the platform production was due to the result of a period of time in the miscible zone), the gas-oil ratio appeared in steps, and the C1 content of the produced gas increased steadily. In the later stage, the oil production decreased rapidly, the gas-oil ratio continued to rise, and C1 content of the produced gas began to rise again. The later stage should involve the two-phase drive front reaching the production well. However, in order to accurately divide the miscible zone, the early dissolved gas oil zone, and the later two-phase drive front zone, there is no accurate criterion from the morphological characteristics of each index. In this study, from the perspective of gas channeling judgment, the time when C1 content rose to the high value platform again after the production of the two-phase flooding front was marked as the end of production at this stage, and it was divided into the mixed zone in the formation.

In terms of oil saturation distribution, at 100 and 130 days of production (Figures 5 and 6), the fluid saturation around the injection well was basically the same, although the saturation around the producing well was different. Therefore, no matter when miscible, dissolved gas-oil zone or the front of the two-phase zone was produced, there was little difference in oil saturation near the production well, which made it difficult to accurately classify different stages. Due to the limited volume of injected gas and the late gas channeling, the reservoir pressure generally showed a downward trend during the gas injection stage. In the miscible phase, the injected gas not only played the role of oil displacement but also replenished the reservoir energy, resulting in a peak of oil production and a certain recovery of reservoir pressure. However, due to the limited injected gas, the pressure continued to decline after the miscible phase (Figure 2).

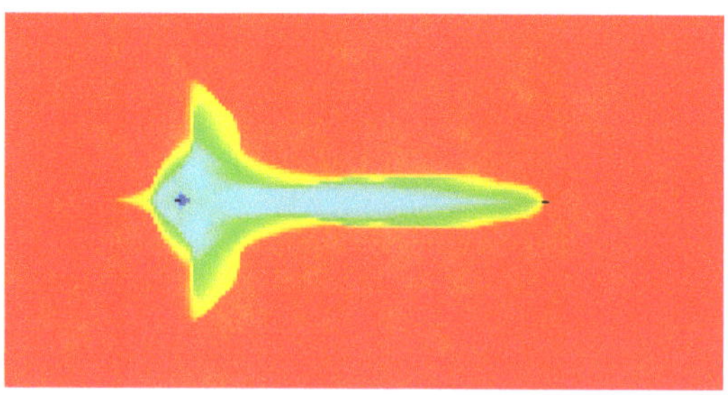

Figure 5. Distribution of formation oil saturation at 100 days of production.

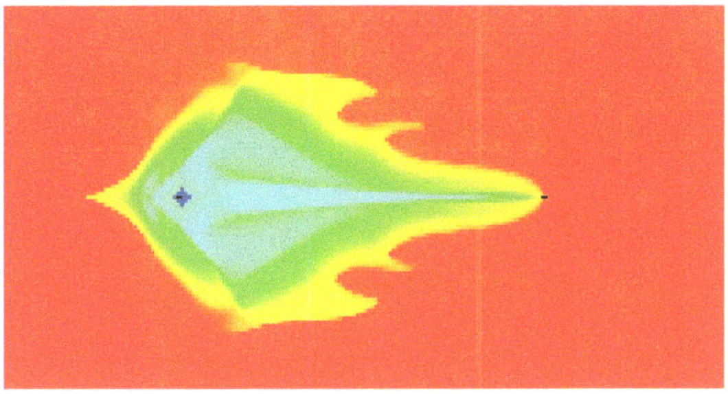

Figure 6. Distribution of formation oil saturation at 130 days of production.

The third stage was the output stage of the trailing edge of the two-phase band, 154 days later (Figure 7). After the mixing zone fluid was produced, the producing well began to enter the trailing edge of the two-phase zone. This stage had the following characteristics: oil production decreased significantly, gas-oil ratio showed an obvious upward trend, and the content of produced gas C1 showed a continuous upward trend again on the basis of relatively stable in the early stage.

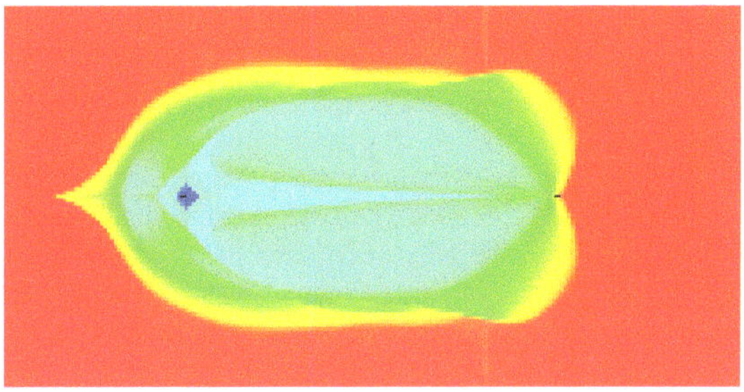

Figure 7. Distribution of formation oil saturation at 400 days of production.

From the perspective of oil saturation distribution, when the production time was 400 days, the fluid saturation around the production well changed significantly during this period, and the gas and oil phases were dominant in a large range, and the gas saturation near the wellbore increased significantly. At this time, the production well produced gas and oil two-phase which cannot form miscible or near miscible. At this stage, the proportion of injected gas output increased continuously, the gas-oil ratio increased continuously, the oil production decreased continuously, and the molar content of C1 component of produced gas increased continuously.

From the analysis of the molar content of different components (Figure 3), C1 always had the highest molar content of produced gas, which was much higher than other components such as N2 and C2. In addition, from the variation trend of C1, the value after gas channeling was obviously much higher than that in the previous stage, which had a good background value basis. Because of the low content of other components, the change in values before and after gas channeling were relatively small. In practice, it is appropriate to choose C1 content with high background value and high change value as the judgment index.

In summary, for the purpose of identifying gas channeling, C1 content will rise again on the previous platform in the third stage; that is, when the trailing edge of the two-phase zone is produced and the gas-oil ratio at this time point is taken as the discriminant index to comprehensively discriminate gas channeling in actual production wells (Table 3).

Table 3. Production index characteristics corresponding to each production phase zone of miscible flooding mode.

Serial Number	Production Phase Zone	Characteristics of Production Indicators		
		Oil Production	Gas Oil Ratio	C1 Content of Produced Gas
1	Original oil phase	Continued decline, later may be slightly increased	The original gas-oil ratio is low, about dozens	Basic constant
2	Mixed zone (dissolved gas-oil zone + miscible zone + front edge of two-phase zone)	It rises rapidly at the beginning, reaches the highest level for a period of time, and then declines rapidly	Early rapid rise, midway steady rise, and rapid rise later	It rises sharply in the early stage, slows down in the middle, and forms a relatively stable platform, and rises again in the later stage
3	Two-phase band trailing edge	Continued to decline	Continue to rise	Continue to rise slowly from a high plateau

3.2. Determination of Gas Channeling Identification Parameter Values under Different Formation Pressures

Second, considering the high pressure of the reservoir, different formation pressures are set in the injection and production mechanism model to predict the production well performance. The gas-oil ratio and C1 molar content of produced gas were analyzed and the field characterization parameters of gas channeling under different formation pressures were determined.

Considering the actual formation pressure of the reservoir, the model formation pressure was set as follows: the maximum pressure was the original pressure of 62 MPa, and then decreased to 41 MPa at every 3 MPa point. Therefore, the formation pressure was set as P = 62, 59, 56, 53, 50, 47, 44, and 41 MPa, and the model was dynamically predicted to obtain the relationship between main production indicators under different formation pressures and production time (Table 4), and the following gas breakthrough identification standards were obtained as reference values for field discrimination.

Table 4. Criteria for gas channeling.

Serial Number	Formation Pressure	Miscible Degree P_{ave}/MM% (Miscible Pressure 43.24 MPa)	Gas Channeling Judgment Eigenvalue
1	P > 50 MPa	>1.23	gas-oil ratio 1650, C1 Content 82.5%
2	47 MPa < P ≤ 50 MPa	1.09–1.23	gas-oil ratio 1450, C1 Content 82.5%
3	P < 47 MPa	<1.09	gas-oil ratio 1800, C1 Content 83.3%

When the formation pressure was high (P > 50 MPa), the recommended values were: gas-oil ratio 1650 m^3/m^3, C1 content 82.5%.

When the formation pressure was moderate (47 MPa < P ≤ 50 MPa), the recommended values were: gas-oil ratio 1450 m^3/m^3, C1 content 82.5%.

When the formation pressure was low (P < 47 MPa), there was no miscibility and near-miscibility, and the gas injection effect is poor. The recommended values were: gas-oil ratio greater than 1800 m^3/m^3, C1 content 83.3%.

This paper presented a method of gas channeling identification, which has not been studied before. It is aimed at the characteristics of high-pressure heterogeneous reservoirs, and its gas-oil ratio had increased and changed several times in different stages of the gas injection and production process. Therefore, the gas-oil ratio proposed by Si Yong [16] or its derivative change cannot be used to judge the gas flow. Instead, the identification parameters of gas channeling at the trailing edge of the two-phase zone and their values under different pressure conditions were specified. For example, when the formation pressure was greater than 50 MPa, the C1 content produced at the trailing edge of the two-phase zone rose again on the previous platform, and the gas-oil ratio was about 1650, and the C1 content value was 82.5%, which was judged to be the occurrence of gas channeling.

3.3. Application and Verification of Gas Channeling Identification Method

According to the determined results of the above gas channeling identification parameters, combined with the actual production well oil production, gas-oil ratio, and produced gas C1 content, the field gas channeling judgment was carried out for a gas injection effective D1 well and then compared with the detection results of the gas tracer method to verify the accuracy of the method.

Before June 2014, the oil production and gas-oil ratio of well D1 with a gas injection effect were basically constant (Figures 8 and 9) and the gas-oil ratio was low at only about 10 m^3/m^3, which was the production stage of the original oil zone. In June 2014, the gas-oil ratio gradually increased and C1 content initially increased rapidly. This stage was analyzed as the production stage of dissolved gas-oil area. From April 2017 to July 2019, oil production showed an obvious upward trend, and the gas-oil ratio decreased slightly during the same period, but basically remained at a plateau. Compared with the early stage, the increase trend of C1 content was obviously slowed down, and even there was a significant low value in the middle (around September 2019). Generally speaking, this stage is the production stage of miscible zone. Then it entered the leading edge production stage of the two-phase zone, and the gas-oil ratio started to rise slowly and C1 content increased to a high value. Then it entered the two-phase zone trailing edge production stage. At this time, the oil production decreased, the gas-oil ratio increased, and the C1 content of produced gas entered a high value platform.

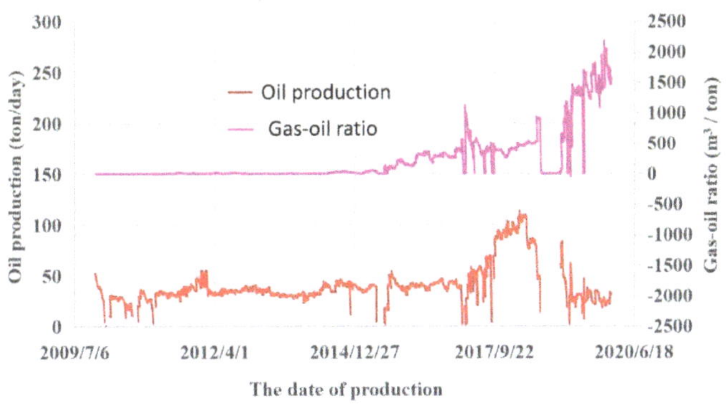

Figure 8. Production curve of oil production and gas-oil ratio of a gas injection affected well D1.

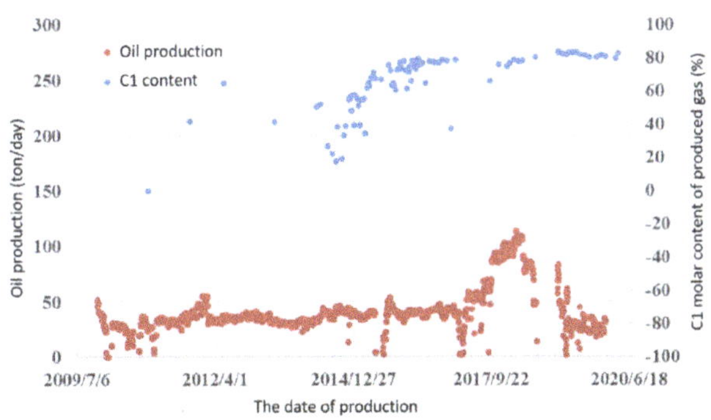

Figure 9. Production curve of oil production and gas C1 content in a gas injection well D11.

According to the gas channeling discriminant parameter and its value (the pressure is 53 MPa at this time), it can be judged that gas channeling occurred around September 2019, namely the output stage at the trailing edge of the two-phase zone. At this time, the oil production decreased, the gas-oil ratio increased, and the C1 content of produced gas entered a high value platform. The corresponding gas-oil ratio was 1642 m^3/m^3, and C1 content was more than 82.6%. The gas-oil ratio decreased in early 2020, but the oil production was significantly lower, so it was accurate to predict that the well had gas channeling.

In addition, gas tracer monitoring in late 2019 showed that the gas in the well was mainly from injection gas from the adjacent D2 well. Therefore, it was determined that the gas channeling source of the production well was the D2 well and that gas channeling had indeed occurred. The comparison results between the two methods show that the identification results based on gas channeling characterization parameters are consistent with the actual results, which prove the accuracy and reliability of gas-oil ratio and C1 molar content as gas channeling discrimination parameters. Compared with the gas tracer monitoring method, the investment cost is lower and the workload is less.

In conclusion, according to the gas-oil ratio in production and the composition of oil and gas in production, the flow stages of formation fluid can be divided into five zones: gas-phase zone, two-phase zone, miscible zone, dissolved gas-oil zone, and original oil zone. For the purpose of gas channeling identification, the time when C1 content rises again at the trailing edge of the two-phase zone is taken as the cut-off point for gas channeling identification. Combined with the gas-oil ratio at this time point as a discriminant index, the results of the gas tracer method were compared to prove its accuracy and reliability.

4. Conclusions

(1) When the produced fluid is in different phase zones in the formation, it exhibits different properties. The injected gas is dry gas with high C1 content. From the prediction curve, the gas-oil ratio and C1 content are the most sensitive to the produced fluid. Considering the convenience of field parameter selection, the gas-oil ratio and C1 content of produced gas were selected as the field characterization parameters for gas channeling discrimination.

(2) According to the variation characteristics of output parameters under different pressure conditions, the gas channeling judgment standards under different pressures were established from several parameters, such as oil production, gas-oil ratio, and C1 content of produced gas.

(3) The gas channeling discrimination method proposed in this paper was based on oil production, gas-oil ratio, and C1 content of produced gas to judge gas channeling in production wells on site, and the results of gas tracer detection were compared to prove its accuracy and reliability. The method has low cost and simple operation.

Author Contributions: J.L.: Data sorting, gas channeling discrimination method, writing—original draft preparation; L.W.: Investigation, verification, writing—editing. All authors have read and agreed to the published version of the manuscript.

Funding: This research received no external funding.

Institutional Review Board Statement: Not applicable.

Informed Consent Statement: Not applicable.

Data Availability Statement: Not applicable.

Acknowledgments: This work was supported by many technicians who have contributed to the research, and we hope to show our sincere thanks to them.

Conflicts of Interest: The authors declare no conflict of interest.

References

1. Zhao, X.; Shi, L.; Wang, W.; Bai, Y.; Tian, F. Study on gas channeling law of CO_2 flooding in heterogeneous ultra-low permeability reservoir. *J. Southwest Pet. Univ. (Sci. Technol. Ed.)* **2017**, *39*, 131–139.
2. Wu, J. *Study on Gas Channeling Law and Flow Mechanism of Fractured-Vuggy Carbonate Reservoir*; China University of Petroleum: Beijing, China, 2016.
3. Li, J.; Gao, B.; Gong, N.; Xie, Z.; Chen, Y.; Han, Y. A new evaluation method of gas channeling resistance based on fuzzy mathematics and its application. *Drill. Fluid Complet. Fluid.* **2017**, *34*, 69–74.
4. Zhang, Y.; Li, X.; Sun, Z.; Li, Z.; Liu, W.; Zhang, Y. Quantitative evaluation method and application of gas channeling in gas injection development of condensate gas reservoir. *Oil Drill. Prod. Technol.* **2017**, *39*, 667–672.
5. Peng, S. Study on gas channeling law of CO_2 flooding in ultra-low permeability reservoir of Zhenglizhuang Oilfield, Shengli. *J. Oil Gas Technol.* **2013**, *35*, 147–169.
6. Li, C. Influencing factors and rules of carbon dioxide flooding gas channeling in ultra-low permeability reservoirs. *Spec. Oil Gas Reserv.* **2018**, *25*, 82–86.
7. Gao, Y.; Zhao, M.; Wang, J.; Zong, C. Production characteristics and gas channing law of CO_2 immiscible flooding in ultra-low permeability reservoirs. *Pet. Explor. Dev.* **2014**, *41*, 79–85. [CrossRef]
8. Wei, Q.; Hou, J.; Hao, H.; Song, Z. Study on gas channeling law of CO_2 drive in ultra-low permeability reservoir. *Pet. Sci. Bull.* **2019**, *4*, 145–153.
9. Li, C.; Han, H. Identification of CO_2 gas channeling channel in the early stage of gas injection by grey fuzzy comprehensive evaluation method. *Pet. Geol. Oilfield Dev. Daqing* **2018**, *37*, 116–120.
10. Li, J. Study on gas channeling characteristics and influencing factors of reservoir developed by CO_2 injection. *J. Oil Gas Technol.* **2012**, *34*, 153–169.
11. Saifullin, E.; Zhanbossynova, S.; Zharkov, D.; Varfolomeev, M.A.; Shanbosinova, S.K.; Zharkov, D.A.; Nazarychev, S.A.; Malakhov, A.O.; Sagirov, R.N. Laboratory Studies for Design of a Foam Pilot for Reducing Gas Channeling from Gas Cap in Production Well in Messoyakhskoye Field. *SPE Reserv. Eval. Eng.* **2022**, *25*, 472–485. [CrossRef]
12. Si, Y.; Zhao, J.; Liang, F. Establishment and application of gas channeling identification method in Buried hill reservoir. *J. Shaanxi Univ. Sci. Technol.* **2020**, *38*, 94–101.
13. Liu, D.; Zhang, J.; Wang, Y.; Yan, Z. Distinguishing method and application of gas channeling in condensate gas reservoirs. *Nat. Gas Explor. Dev.* **2008**, *31*, 27–83.
14. Zhu, Y.; Li, B.; Song, W.; Tang, M.; Zhang, F.-E. Discussion on the method of distinguishing gas channeling in condensate gas reservoirs by using charts. *Pet. Geol. Recovery Effic.* **2004**, *6*, 53–85.
15. Wei, Y.; Li, Z.; Wang, X.; Cui, T.; Ding, G.; Mou, W.; Kang, J. Judgment, adjustment and effect of gas injection front breakthrough in Kekeya condensate gas reservoir. *Nat. Gas Ind.* **2006**, *6*, 97–99.
16. Pan, X. *Numerical Simulation of Nitrogen Injection Foam for Controlling Water Channeling Technology*; Northeast Petroleum University: Daqing, China, 2011.
17. Wang, B.; Zhu, Y. Gas cap channeling in a gas cap field. *Nat. Gas Ind.* **2000**, *3*, 79–84.
18. Schulumberger. *PVTI and ECLIPSE 300*; Schulumberger: Beijing, China, 2005.
19. Yuan, H.; Johns, R.T.; Egwuenu, A.M.; Dindoruk, B. Improved MMP Correlations for CO_2 Floods Using Analytical Gasflooding Theory. *SPE Reserv. Eval. Eng.* **2005**, *8*, 418–425. [CrossRef]

processes

Article

Numerical Simulation of Hydraulic Fracturing and Penetration Law in Continental Shale Reservoirs

Yanxin Zhao [1,2,*], Lei Wang [1,2,*], Kuo Ma [3] and Feng Zhang [4]

1 School of Petroleum Engineering, Yangtze University, Wuhan 430100, China
2 Key Laboratory of Drilling and Production Engineering for Oil and Gas, Wuhan 430100, China
3 Gudao Oil Production Plant of Shengli Oilfield Co., Ltd., Dongying 257231, China
4 Petroleum Engineering Technology Research Institute of Jianghan Oilfield, Sinopec, Wuhan 430000, China
* Correspondence: zyx19816099256@163.com (Y.Z.); wang-lei@yangtzeu.edu.cn (L.W.); Tel.: +86-19816099256 (Y.Z.); +86-18971168785 (L.W.)

Abstract: The vertical heterogeneity of continental shale reservoirs is strong, the difference between lithology and stress between layers is large, the weak interface between layers develops, and the hydraulic fracture penetration and expansion are difficult, resulting in poor fracturing transformation effect. In view of this, based on the finite element and cohesive element method, this paper established a fluid-solid coupling model for the hydraulic fracture propagation through the continental shale and studied the control mechanism and influence law of various geological and engineering parameters on the hydraulic fracture propagation through the continental shale reservoir using single factor and orthogonal test analysis methods. Interfacial cementation strength between high layers, high vertical stress difference, low interlaminar stress difference, low tensile strength difference, low elastic modulus difference, high pressure fracturing fluid viscosity, and high injection displacement are conducive to the penetration and expansion of hydraulic fractures. The primary and secondary order of influence degree of each factor is: interlaminar interface cementation strength > interlaminar stress difference/tensile strength difference > fracturing fluid viscosity/injection displacement > vertical stress difference > elastic modulus. In addition, engineering application research has also been carried out, and it is recommended that the injection displacement during early construction should not be less than 3 m^3/min, and the fracturing viscosity should not be less than 45 mPa·s. The field application effect is good, which verifies the engineering application value of the model.

Keywords: continental shale; hydraulic fracturing; penetration propagation; numerical simulation; engineering applications

Citation: Zhao, Y.; Wang, L.; Ma, K.; Zhang, F. Numerical Simulation of Hydraulic Fracturing and Penetration Law in Continental Shale Reservoirs. *Processes* **2022**, *10*, 2364. https://doi.org/10.3390/pr10112364

Academic Editors: Linhua Pan, Yushi Zou, Jie Wang, Minghui Li, Wei Feng and Lufeng Zhang

Received: 6 October 2022
Accepted: 1 November 2022
Published: 11 November 2022

1. Preface

The exploration and development of shale gas have a long history of nearly 200 years. Shale gas in North America has developed rapidly, realizing efficient, economic and large-scale development, becoming an important source of natural gas supply in North America, and causing significant changes in the global natural gas supply pattern [1–3]. Countries in Europe, Oceania, South America and other regions have fully recognized the value and prospects of shale gas resources and have started extensive shale gas research, exploration and development, such as basic theoretical research, resource potential evaluation, and industrial production tests [4–6]. China is rich in shale gas resources, with recoverable reserves of about 25.08 × 10^{12} m^3 and huge development potential [7]. After more than 10 years of development, the theory and key technologies for the effective large-scale development of shallow marine shale gas above 3500 m have been basically mastered, but few breakthroughs have been made in the development of continental shale gas [8]. Compared with marine shale, continental shale reservoirs have strong vertical heterogeneity, large differences in interlayer lithology and stress, developed weak interfaces between layers, and it is difficult for hydraulic fractures to propagate through layers, resulting in ineffective fracturing stimulation [9]. Therefore, it is urgent to carry out research on the propagation law of hydraulic fractures in continental shale.

At present, many scholars at home and abroad have carried out a series of studies from laboratory experiments [10–20] and numerical simulations [21–28] on the propagation law of hydraulic fractures in stratified reservoirs. Some scholars have carried out a large number of laboratory experiments on multi-lithologic combination layered rock samples such as concrete, sandstone and coal rock using a true triaxial large-scale fracturing physical simulation device and studied the influence of various geological and engineering parameters on the propagation of hydraulic fractures through layers. The experimental results show that the hydraulic fractures in the layered rock samples exhibit asymmetric and non-planar expansion characteristics in the vertical direction. After encountering the interface between layers, it presents a variety of complex expansion modes such as stopping, turning, forking and penetrating dot difference in the elastic modulus of the reservoir/barrier rock does not have a significant inhibitory effect on the propagation of hydraulic fractures through the interlayer, the horizontal minimum stress difference between layers, the vertical stress difference (the difference between the vertical stress and the horizontal minimum stress), the properties of the interlayer interface, the injection displacement and the viscosity of the fracturing fluid are the key factors that determine whether a hydraulic fracture can penetrate the layer; the smaller the minimum horizontal stress difference between layers, the greater the vertical stress difference, the higher the interface cementation strength, the higher the injection displacement and the higher the viscosity of the fracturing fluid, the more favorable the hydraulic fractures to achieve through-layer expansion [14–20]. Laboratory experiments have initially revealed the propagation law of hydraulic fractures through layers, but they can only provide some qualitative understanding and have limited guiding significance for the optimal design of hydraulic fracturing. Therefore, some scholars based on the displacement discontinuity method (DDM) [21,22], finite element (FEM) and cohesive element (cohesive element) method [23–25], extended finite element method (XFEM) [26], and numerical methods such as discrete element method (DEM) [27,28] have established a series of numerical models of hydraulic fracture penetration through layers, and carried out a large number of studies on influencing factors, further revealing the law of hydraulic fracture penetration through layers, which is a quantitative prediction method. The propagation pattern of hydraulic fracture through the layer provides an effective means.

The above studies show that various geological and engineering parameters have a significant impact on the propagation law of hydraulic fractures through layers, but the primary and secondary relationship between these factors has not been elucidated, and most of these studies are carried out on sandstone or coal rock reservoirs, which cannot systematically reveal the propagation law of hydraulic fractures in continental shale reservoirs. Therefore, in view of the development characteristics of continental shale reservoirs, this paper establishes a fluid-solid coupling model of continental shale hydraulic fractures spreading through layers based on the finite element and cohesive element method, and the model is verified by analytical solutions and experimental laboratory results. Carry out a single-factor analysis and orthogonal test research to systematically reveal the propagation law of hydraulic fractures in continental shale reservoirs.

2. Mathematical Model

In layered reservoirs, the hydraulic fracture height is usually much smaller than its length, and the fluid pressure in the fracture changes little along the fracture length except for the fracture tip. The strain model can give reasonable results [29]. Based on this, a two-dimensional plane strain model of continental shale hydraulic fractures spreading through layers is established in this paper.

2.1. Fluid-Structure Interaction Governing Equation

In the process of hydraulic fracturing, the deformation of the rock skeleton interacts with and influences the fluid flow in the pore space. The effective stress principle can be used to characterize the stress-seepage coupling relationship in the rock. Take the control volume V and its surface as S, then the rock. The coupled governing equation of solid skeleton deformation and fluid flow is [29]:

$$\int_V (\overline{\sigma} - p_w I)\cdot\delta_\varepsilon d_V = \int_S t\cdot\delta_v d_S + \int_V f\cdot\delta_v d_V \tag{1}$$

In the formula, $\overline{\sigma}$ is the effective stress matrix, Pa; p_w is the pore pressure, Pa; I is the unit matrix, dimensionless; δ_ε is the virtual strain rate matrix, s^{-1}; δ_v is the virtual velocity vector, m/s; t is the surface force vector, N/m^2; f is the body force vector, N/m^3.

The mass conservation equation of fluid seepage in the pores of the rock skeleton is:

$$\int_V \frac{1}{J}\frac{d}{dt}(J\rho_w\varphi_w)d_V + \int_S \rho_w\varphi_w n^T\cdot v_w d_S = 0 \tag{2}$$

Among them, the flow velocity, v_w, of the fluid in the rock satisfies Darcy's law [17,19,24,29]:

$$v_w = -\frac{1}{\varphi_w g\rho_w}k\left(\frac{\partial p_w}{\partial x} - \rho_w g\right) \tag{3}$$

In the formula, J is the rock volume change rate, dimensionless; ρ_w fluid density, kg/m^3; φ_w is the porosity, dimensionless; n^T is the outer normal direction vector of the surface S, dimensionless; x is the space vector, m; g is the gravitational acceleration vector, m/s^2; k is the rock skeleton permeability tensor, m/s.

2.2. Criteria for Crack Initiation and Propagation

In this paper, the cohesive element is used to simulate the initiation and propagation of hydraulic fractures and weak bedding planes, and the secondary nominal stress criterion is used to determine whether the cracks initiate or not:

$$\left\{\frac{\langle\sigma_n\rangle}{\sigma_n^o}\right\}^2 + \left\{\frac{\tau_s}{\tau_s^o}\right\}^2 = 1 \tag{4}$$

where

$$\langle\sigma_n\rangle = \begin{cases} \sigma_n & \sigma_n \geq 0 \\ 0 & \sigma_n < 0 \end{cases} \tag{5}$$

where σ_n and τ_s are the normal and tangential stresses actually borne by the cohesive element, Pa; ${\sigma_n}^o$ and ${\tau_s}^o$ are the tensile strength and shear strength of the rock, Pa.

The crack propagation process is described by the stiffness degradation criterion of the cohesive element, and its expression is as follows:

$$\begin{cases} \sigma_n = \begin{cases} (1-D)\overline{\sigma_n} & \overline{\sigma_n} \geq 0 \\ \overline{\sigma_n} & \overline{\sigma_n} < 0 \end{cases} \\ \tau_s = (1-D)\overline{\tau_s} \end{cases} \tag{6}$$

In the formula, $\overline{\sigma_n}$ and $\overline{\tau_s}$ are the stress calculated by the normal and tangential direction of the cohesive element under the current strain according to the undamaged front-line elasticity criterion, respectively; D is the damage factor, dimensionless, ranging from 0 to 1, when D = 0, the material is not damaged. When D = 1, the material is completely damaged;

that is, the hydraulic crack begins to expand (as shown in Figure 1). The calculation formula is as follows:

$$D = \frac{\delta_f(\delta_m - \delta_o)}{\delta_m\left(\delta_f - \delta_o\right)} \tag{7}$$

In the formula, δ_o and δ_f are the displacement at the initial damage and the displacement when the element is completely damaged, m; δ_m is the maximum displacement reached during the loading process, m.

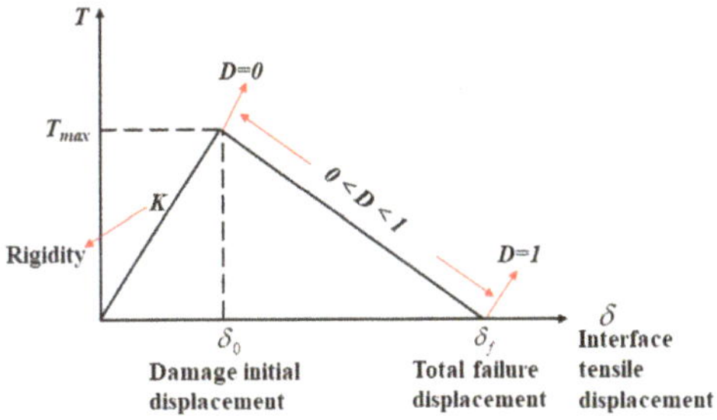

Figure 1. The traction-separation law of cohesive element.

2.3. Fluid Flow Equation in Fractures

After the cohesive cell is completely damaged, fluid will enter the damaged cell. As shown in Figure 2, there are two flow states of fluid in the damaged unit, which are divided into tangential flow and normal flow.

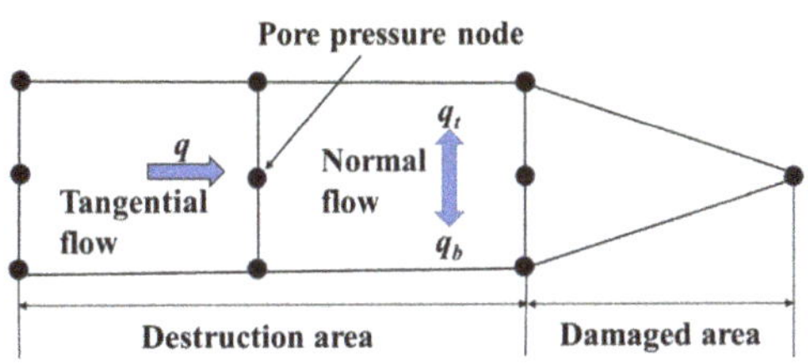

Figure 2. The fluid flow schematic within a damaged unit.

In this paper, the fracturing fluid is assumed to be an incompressible Newtonian fluid, and the formula for its tangential flow is:

$$q = -\frac{w^3}{12\mu}\nabla p \tag{8}$$

The formula for calculating normal flow is:

$$\begin{cases} q_t = c_t\left(p_f - p_t\right) \\ q_b = c_b\left(p_f - p_b\right) \end{cases} \tag{9}$$

where q is the tangential flow rate, m^3/s; μ is the fracturing fluid viscosity, Pa·s; w is the fracture width, m; ∇p is the tangential pressure gradient of the cohesive unit, Pa/m; q_t, q_b is the normal flow into the upper and lower surfaces of the unit, m^3/s; c_t, and c_b are the filtration coefficients of the upper and lower surfaces, $m^3/(Pa·s)$, p_t and p_b are the pore pressures at the upper and lower surfaces of the fracture, Pa; p_f is the fluid pressure in the fracture, Pa.

3. Model Validation

ABAQUS finite element analysis software was used to build and solve the above model, and the model solution accuracy was closely related to the mesh size. In order to determine a reasonable mesh size, the classical KGD hydraulic fracture propagation model was constructed using ABAQUS software, the mesh size was set to 1 m × 1 m, 0.2 m × 0.2 m and 0.1 m × 0.1 m, the calculation parameters are shown in Table 1, and the solution results were compared with the analytical solution. The calculation formula of the analytical solution is [30]:

$$\begin{cases} L(t) = \left(\frac{1}{4\pi}\right)^{\frac{1}{3}} \left(\frac{Eqt}{(1-v^2)K_{IC}}\right)^{\frac{2}{3}} \\ W(t) = \left(\frac{32}{\pi^2}\right)^{\frac{1}{3}} \left(\frac{K_{IC}{}^2 qt\left(1-v^2\right)^2}{E^2}\right)^{\frac{1}{3}} \end{cases} \quad (10)$$

where E is elastic modulus, MPa; v is Poisson's ratio, dimensionless; K_{IC} is rock fracture toughness, MPa·m$^{0.5}$; q is displacement per unit fracture height, m^2/s; $L(t)$ is the half-foil length of the hydraulic fracture at time t, m; $W(t)$ is the width of the hydraulic fracture at the injection point at time t, m; t is the injection time of the fracturing fluid, s. As shown in Figure 3, When the grid size is 1 m, the difference between the simulation results and the analytical solution is large, and the curve has large fluctuations. When the grid size is 0.2 and 0.1, and the half length of the hydraulic fracture exceeds 5 m, that is, 25× the side length of the grid element, the numerical simulation results agree well with the analytical solution. Based on this, the construction and meshing of the subsequent numerical model adopt the above-mentioned mesh size setting standard; that is, the edge length of the mesh element is less than 50× the size of the simulated fracture.

Table 1. Model validation calculation parameters.

Elastic Modulus/ GPa	Poisson's Ratio	Viscosity/ (mPa·s)	Fracture Toughness/(MPa·m$^{1/2}$)	Displacement/ (m^2/s)
15	0.2	1	4	0.001

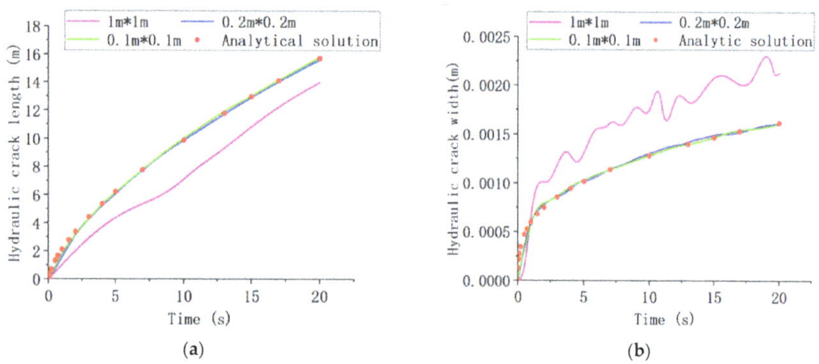

Figure 3. Comparison between analytical solution and numerical simulation results: (**a**) Comparison of hydraulic crack length; (**b**) Comparison of hydraulic crack width. The purple curve, blue curve and green curve represent the simulation results with grid sizes of 1 m × 1 m, 0.2 × 0.2 m and 0.1 × 0.1 m, respectively. The red dot represents the analytical solution result.

In order to further verify that the numerical simulation method has the ability to simulate the propagation behavior of continental shale hydraulic fractures through layers, the laboratory experiment results were used to verify it [20]. Referring to the above meshing standards, the simulation is carried out using the parameters in Table 2. As shown in Figure 4, under the conditions of different fracturing fluid viscosity, the hydraulic fracture morphology obtained by numerical simulation is basically consistent with the experimental laboratory results, which verifies the accuracy of this numerical simulation method.

Table 2. Laboratory experimental parameters [20].

Specimen Number	σh/σH/σv/ (MPa)	Displacement/ (mL/min)	Viscosity/ (mPa·s)	Elastic Modulus/ Gpa	Fracture Toughness/(Mpa·m$^{1/2}$)
RG-1	8/20/20	60	5	7.1/13.2/7.1/	0149/0.225/0.149/
RG-2	8/20/20	60	50	16/7.1	0.376/0.149

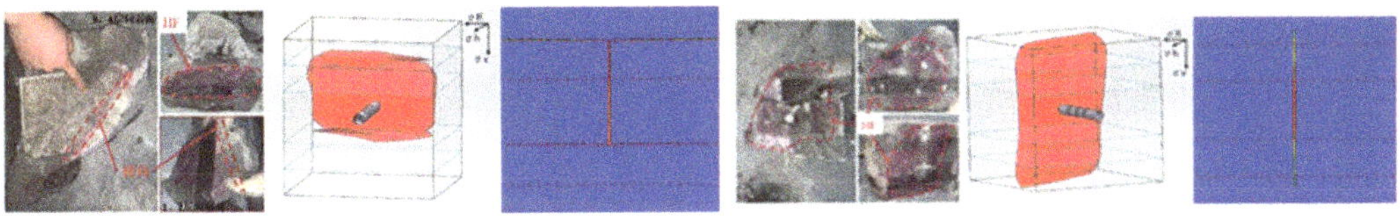

Figure 4. The results of physical simulation experiments and numerical simulation: (**a**) Anatomical map and fracture reconstruction map and numerical simulation results of RG-1 rock sample after the experiment; (**b**) Anatomical map and fracture reconstruction map and numerical simulation results of RG-2 rock sample after the experiment.

4. Analysis of Influencing Factors of Hydraulic Fractures through Layer Propagation

4.1. Model Establishment and Parameters

As shown in Figure 5a, assuming that the hydraulic fracture shape is symmetrical about the injection point, in order to improve the calculation efficiency, a 40 m × 40 m semi-model of continental shale hydraulic fracture penetration through layers is established. The upper part of the model is an interlayer, the lower part is the thickness of the reservoir is 20 m, and the grid unit size is 0.2 m × 0.2 m. The basic input parameters are shown in Table 3. Based on this model, the single-factor analysis method was used to study the influence of various geological and engineering parameters on the propagation of hydraulic fractures through layers. The simulation results are shown in Figure 5b–f.

Table 3. The rock mechanics parameters of the stratum.

Parameter Type	Specific Parameters	Reservoir/Interlayer	Interlayer Program
formation rock	Elastic Modulus/GPa	20	/
	Poisson's ratio	0.2	/
	Permeability/mD	5	/
	Minimum horizontal crustal stress/Mpa	35	/
	Maximum horizontal crustal stress/Mpa	45	/
	Vertical geostress/Mpa	39	/
	pore pressure/Mpa	27	/
	Fluid density/(N/m^3)	9800	/
Cohesive elements	Rigidity/(Gpa/m)	20,000	20,000
	Tensile strength/Mpa	4	2
	Shear strength/Mpa	40	3.6
	Filtration coefficient/(m^3·Pa^{-1}·s^{-1})	10^{-14}	
	Damage displacement/mm	0.03	0.03
Construction parameters	Displacement/(m^3/s)	3	
	Viscosity/(mPa·s)	50	

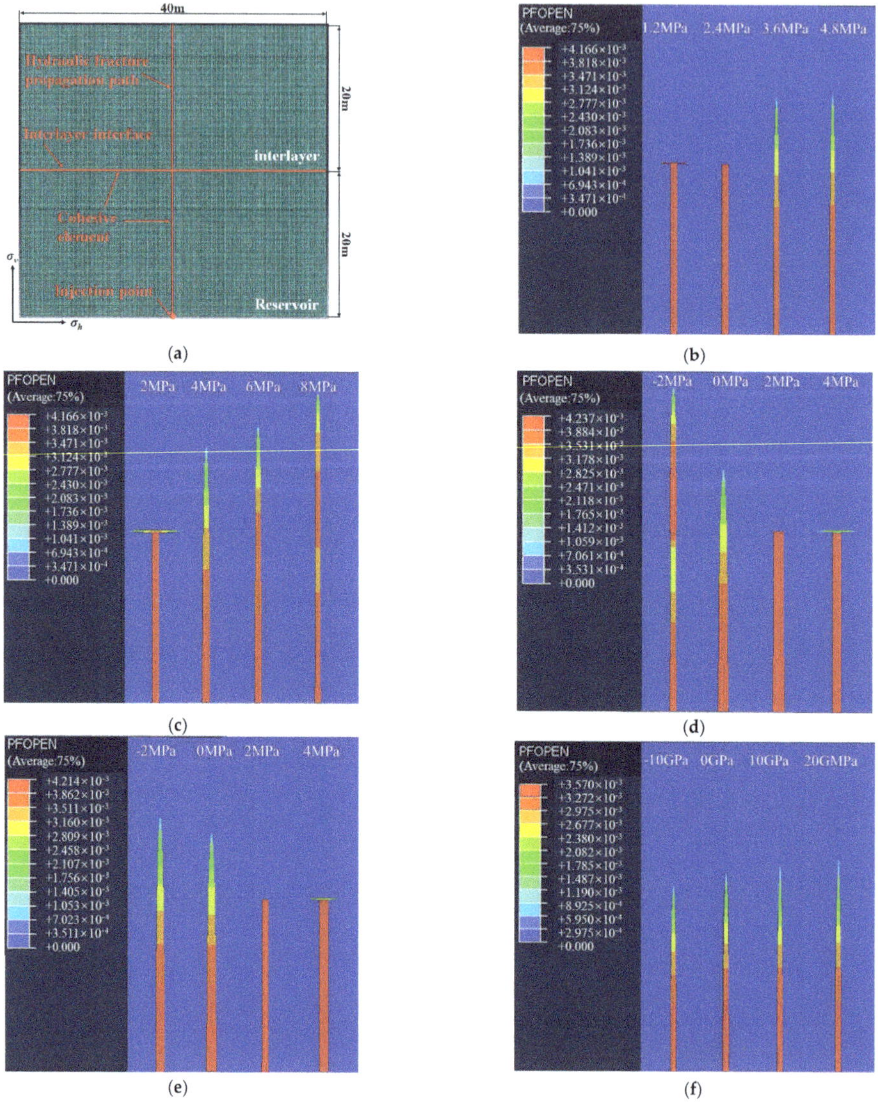

Figure 5. Numerical simulation results under different geological parameters: (**a**) Schematic diagram of model meshing; (**b**) Interlayer interface strength; (**c**) Vertical stress difference; (**d**) Interlayer stress difference; (**e**) Tensile strength difference; (**f**) Elastic modulus difference.

4.2. Influence of Formation Parameters

4.2.1. Bonding Strength of Interlayer Interface

The interfacial cementation strength of continental shale is closely related to the cement type (quartz, calcite or pyrite, etc.) and content and is usually quantitatively characterized by the interface shear strength [9,24]. The higher the cementation strength, the higher the interface shear strength. The shear strengths of the interlayer interface in the four groups of simulation examples in Figure 5b are 1.2 MPa, 2.4 Mpa, 3.6 Mpa and 4.8 Mpa, respectively. All show the simulation results under the same injection volume (the same as below). The simulation results show that the bonding strength of the interlayer interface has an important influence on the vertical propagation path of hydraulic fractures. When the bonding strength of the interlayer interface is low, the hydraulic fracture activates the interlayer interface and extends along it, and the vertical expansion is hindered; on the contrary, the hydraulic fracture penetrates the interlayer interface, enters the interlayer and continues to expand, and the interlayer interface is cemented at this time. The strengths (3.6 MP and 4.8 Mpa) have basically no effect on the vertical propagation shape of hydraulic fractures. The reason is that during the fracturing process, the interface between weakly cemented layers is more prone to shear slip under the combined action of the induced stress at the tip of the hydraulic fracture and the filtration of the fracturing fluid along the interface, resulting in the rapid release of fracturing energy and the failure of the hydraulic fracture penetrating the interlayer interface [21].

4.2.2. Vertical Stress Difference

The vertical stress difference is defined as the difference between the vertical in-situ stress and the minimum horizontal in-situ stress. The vertical in-situ stresses of the four groups of calculation examples in Figure 5c are 37 Mpa, 39 Mpa, 41 Mpa and 43 Mpa, respectively; that is, the vertical stress differences are 2 Mpa, 4 Mpa, 6 Mpa and 8 Mpa, respectively. The simulation results show that with the increase of the vertical stress difference, the ability of hydraulic fractures to penetrate through layers is significantly enhanced, and the height of hydraulic fractures increases. According to the classic Renshaw & Pollard criterion [31], when the minimum horizontal in-situ stress remains unchanged, the greater the vertical stress difference, the stronger the ability of the interlayer interface to resist shear failure, thereby avoiding shear slip caused by the interlayer interface. The vertical expansion of hydraulic fractures is hindered. In the above case, the critical vertical stress difference for hydraulic fractures to penetrate the interlayer interface is about 4 Mpa.

4.2.3. Interlayer Stress Difference

The interlayer stress difference is defined as the difference between the minimum horizontal in-situ stress of the interlayer and the reservoir. In Figure 5d, the minimum horizontal in-situ stresses of the interlayers of the four groups of calculation examples are 33 Mpa, 35 Mpa, 37 Mpa and 39 Mpa, respectively; that is, the interlayer stress differences are −2 Mpa, 0 Mpa, 2 Mpa and 4 Mpa, respectively. The simulation results show that the smaller the interlayer stress difference, the stronger the ability of hydraulic fractures to penetrate through the layers and the greater the fracture height. According to the classic Renshaw & Pollard criterion [31], the greater the interlayer stress difference, the more difficult it is for hydraulic fractures to penetrate the interlayer interface. Even if the interlayer interface penetrates, the minimum horizontal in-situ stress of the high interlayer means high extension resistance, and the height of the hydraulic fractures will also be significantly suppressed, forming "short and wide" fractures.

4.2.4. Poor Tensile Strength

The tensile strength difference is defined as the difference in the tensile strength of the barrier and the reservoir. In Figure 5e, the interlayer tensile strength of each calculation example is 2 Mpa, 4 Mpa, 6 Mpa and 8 Mpa, respectively; that is, the interlayer tensile strength difference is −2 Mpa, 0 Mpa, 2 Mpa and 4 Mpa, respectively. The simulation results show that the effect of the difference in tensile strength on the propagation of hydraulic fractures through layers is basically consistent with the difference in interlayer stress. Because in addition to the minimum horizontal in-situ stress, the tensile strength of the rock needs to be overcome in the process of hydraulic fracture extension; that is, when hydraulic fractures start from high-strength layers, it is easier to penetrate the interlayer interface into low-strength layers and propagate, which is consistent with the published numerical simulation [18] and laboratory experiments [24] results. Based on this, the wellbore traversing horizon and perforation horizon of the fracturing well can be optimized.

4.2.5. Elastic Modulus Difference

The elastic modulus difference is defined as the difference between the elastic moduli of the barrier and the reservoir. The elastic moduli of the interlayers of the four groups of calculation examples in Figure 5f are 10 Gpa, 20 Gpa, 30 Gpa, and 40 Gpa, respectively; that is, the interlayer elastic moduli differences are −10 Gpa, 0 Gpa, 10 Gpa, and 20 Gpa, respectively. The simulation results show that the influence of the elastic modulus difference on the vertical propagation path of hydraulic fractures is not significant, and this understanding has been verified by the laboratory and mine experimental results [32]; In the case of good interlayer interface bonding strength, high elastic modulus difference is a favorable factor for hydraulic fractures to achieve through-layer expansion, because when the hydraulic fracture enters the high elastic modulus interlayer, its width will be suppressed, and it is more likely to form "high and narrow" fractures. However, if the elastic modulus of the interlayer is too high, the width of the hydraulic fracture is too narrow, and the proppant migration is hindered. Even if the dynamic fracture extends to this point, it is difficult to obtain effective support after the fracture is closed, and the improvement of the fracturing effect is limited. Overall, the effect of elastic modulus on the propagation behavior of hydraulic fractures is not as significant as that of other geological factors.

4.3. Influence of Construction Parameters

4.3.1. Fracturing Fluid Viscosity

The fracturing fluid viscosities of the four groups of examples in Figure 6a are 10 mPa·s, 30 mPa·s, 50 mPa·s, and 70 mPa·s, respectively. The simulation results show that high-viscosity fracturing fluid is a favorable factor for hydraulic fractures to achieve through-layer propagation. The higher the viscosity of the fracturing fluid, the smaller the filtration loss, the greater the net pressure in the fracture, and the more sufficient the hydraulic energy, which is conducive to the continuous expansion of hydraulic fractures through the interlayer interface [33]. On the contrary, the lower the viscosity of fracturing fluid is, the lower the net pressure in the fracture is, and the weaker the hydraulic energy is. In addition, low-viscosity fracturing fluid is more likely to invade the bedding interface to produce lubrication and pressure holding, which will induce the shear slip of the interlayer interface, resulting in the rapid release of hydraulic energy, which will lead to the hydraulic fracture being captured by the interlayer interface and the vertical expansion being blocked. In addition, when the viscosity of the fracturing fluid exceeds a certain critical value (30 mPa·s), with further increases in the viscosity, the width of the fracture increases and the height decreases, forming a "short and wide" fracture. Therefore, although the high-viscosity fracturing fluid is beneficial to achieve interlayer expansion, excessively increasing the viscosity of the fracturing fluid can certainly achieve the ideal fracture height.

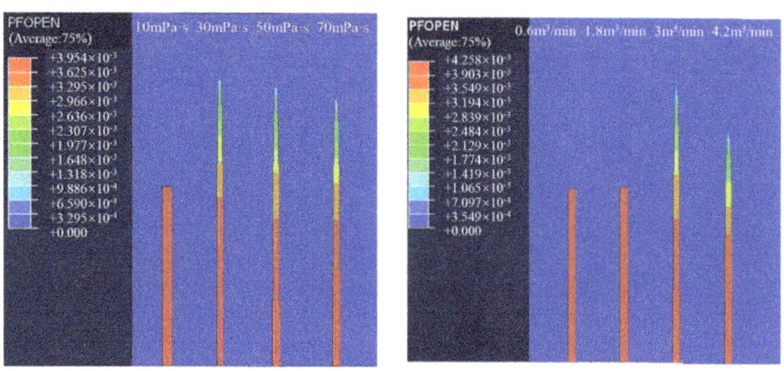

(**a**) (**b**)

Figure 6. Numerical simulation results under different construction parameters: (**a**) Viscosity; (**b**) Displacement.

4.3.2. Injection Displacement

The injection displacements of the four groups of calculation examples in Figure 6b are 0.6 m^3/min, 1.8 m^3/min, 3 m^3/min and 4.2 m^3/min, respectively. The simulation results show that the influence of the injection displacement on the propagation behavior of hydraulic fractures is similar to that of the fracturing fluid viscosity because increasing the injection displacement is also beneficial to increase the net pressure in the fractures and enhance the ability of hydraulic fractures to penetrate the interface between layers. Theoretical studies show that, in the case of ignoring fracturing fluid filtration, the net pressure in the fracture is a function of the product of the injected displacement and the fracturing fluid viscosity when the fluid injection volume is the same, that is, increasing the injection displacement or fracturing fluid viscosity by the same multiple should have the same enhancement effect on the ability of hydraulic fractures to penetrate through layers [34]. However, the simulation results in Figure 6 do not support the above conclusion. Increasing the injection displacement does not improve the penetration effect of hydraulic fractures as much as increasing the viscosity of the fracturing fluid. The main reason is that the above calculation example does not ignore the filtration behavior of fracturing fluid, which is more in line with the real situation. And in the actual fracturing construction process, due to the limitation of construction equipment, the injection displacement cannot be increased by tens or even hundreds of multiples like the viscosity of the fracturing fluid. Therefore, it is recommended to increase the viscosity of the fracturing fluid to enhance the ability of hydraulic fractures to penetrate through layers during on-site construction.

4.4. Primary and Secondary Relationship of Key Influencing Factors

To sum up, the mechanism of the hindered propagation of hydraulic fractures through layers is: (1) The shear slip at the interlayer interface changes the vertical expansion path of hydraulic fractures, limiting the growth of fracture height; (2) The width of hydraulic fractur.es is large, which weakens the growth of fracture height's ability. The former has a more significant effect and is mainly controlled by factors such as interlayer interface strength, vertical stress difference, interlayer stress difference, tensile strength difference, and fracturing fluid viscosity. In order to further reveal the primary and secondary relationship of these influencing factors, an orthogonal numerical simulation experiment was carried out based on the above numerical model [35]. Since the effect of tensile strength difference on the propagation of hydraulic fractures through interlayer is basically the same as that of interlayer stress difference, in order to reduce the number of orthogonal experiment groups, this paper introduces the extension resistance difference to characterize the composite effect of these two factors on the propagation of hydraulic fractures through the layer, which is defined as the sum of the difference in tensile strength and the difference in interlayer stress. Based on this, an orthogonal experiment with four factors and three levels was designed. The specific scheme and results are shown in Tables 4 and 5.

Table 4. Orthogonal test scheme table.

Program	Vertical Stress Difference/MPa	Shear Strength of Interlayer Interface/MPa	Extension Resistance Difference/MPa	Viscosity/(mPa·s)	Half Seam Height/m
1	2	2	0	10	20
2	2	4	4	30	
3	2	6	2	50	27.8
4	4	2	4	50	20
5	4	4	2	10	20
6	4	6	0	30	32.2
7	6	2	2	30	20
8	6	4	0	50	31.4
9	6	6	4	10	26.4

Table 5. Analysis table of orthogonal test results.

Factor Level	Average Value of Hydraulic Fracture Height under Different Influence Factors/m			
	Vertical Stress Difference	Shear Strength of Interlayer Interface	Extension Resistance Difference	Viscosity
I	$I_{IA} = 22.6$	$I_{IB} = 20$	$I_{IC} = 27.87$	$I_{ID} = 22.13$
II	$I_{IIA} = 24.07$	$I_{IIB} = 23.8$	$I_{IIC} = 22.6$	$I_{IID} = 24.07$
III	$I_{IIIA} = 25.93$	$I_{IIIB} = 28.8$	$I_{IIIC} = 22.13$	$I_{IIID} = 26.4$
Very poor crack height	$T_A = 3.33$	$T_B = 8.8$	$T_C = 5.74$	$T_D = 4.27$

It can be seen from Table 5 that $I_{IIIA} > I_{IIA} > I_{IA}$, $I_{IIIB} > I_{IIB} > I_{IB}$, $I_{IC} > I_{IIC} > I_{IIIC}$, $I_{IIID} > I_{IID} > I_{ID}$, indicating that the greater the vertical stress difference, the greater the shear strength of the interlayer interface, the smaller the extension resistance difference, and the greater the viscosity of the fracturing fluid, the better the hydraulic fracture propagation effect through the layer, which is consistent with the previous law; $T_B > T_C > T_D > T_A$, indicating that the priority order of the four key influencing factors is: interlayer interface shear strength > extension resistance difference (interlayer stress difference/tensile strength difference) > fracturing fluid viscosity > vertical stress difference.

5. Engineering Applications

Taking the FYH10 continental shale gas horizontal well as an example, the target interval of the well is divided into seven sublayers, and the horizontal wellbore mainly passes through the ④ sublayer, of which the ①–④ sublayers are high-quality shale gas reservoirs and are the key layers for fracturing stimulation, the rock's mechanical properties and in-situ stress of each sublayer are quite different (see Table 6), which makes the vertical expansion of hydraulic fractures difficult. Based on this, referring to the basic parameters in Table 6, the finite element and cohesive element method is used to establish a hydraulic fracture penetration expansion model, and carry out a case study to optimize the construction displacement and viscosity parameters, so as to obtain a better penetration fracturing effect. The simulation results are shown in Figure 7.

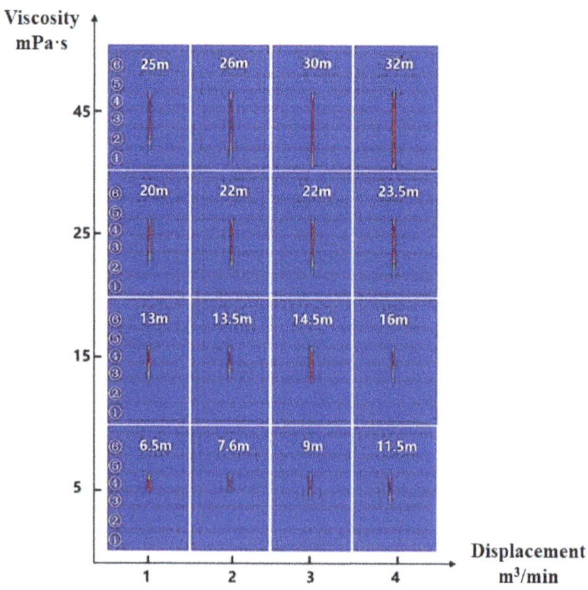

Figure 7. Example well simulation results.

Table 6. The main rock mechanics parameters of the example well.

Strata Serial Number	Formation Thickness/m	Elastic Modulus/GPa	Poisson's Ratio	Tensile Strength/Mpa	Crustal Stress/Mpa		
					Minimum Horizontal Crustal Stress	Vertical Crustal Stress	Maximum Horizontal Crustal Stress
⑦	19	15	0.25	5	63	69	71
⑥	8	15	0.25	4	62	69.2	71.5
⑤	6.5	28	0.1	8	66	69.4	72
④	6.5	18	0.2	2	60	69.8	72
③ up	3	25	0.12	6	64	70	73
③ down	3.5	23	0.13	4	62	70.2	73
②	8.5	20	0.14	4.5	63	70.4	74
①	6.5	22	0.13	5	64	70.8	74.5

It can be seen from Figure 7 that with the increase of displacement and viscosity, the ability of hydraulic fractures to penetrate layers is significantly enhanced when the displacement is not less than 3 m^3/min and the viscosity is not less than 45 mPa·s, and hydraulic fractures can achieve the effect of communicating ①–④ small layers. Based on the above optimization results, the well is divided into 32 stages for fracturing, with five to six clusters of perforations per stage, and the injection displacement is 15–18 m^3/min, 50 mPa·s high-viscosity fracturing fluid was used to realize hydraulic fracture penetration and expansion during construction, and then 5 mPa·s low-viscosity slick water was used to improve the complexity of hydraulic fractures. As shown in Figure 8, the microseismic monitoring results show that the hydraulic fractures in each fracturing section have ideal height expansion and basically achieve the effect of penetrating the small layers ①–④, and a large-scale complex fracture network is formed in the reservoir, the daily gas production was 5.58×10^4 m^3, and the daily oil production was 17.6 m^3 after the pressure test, and a good production increase effect was achieved.

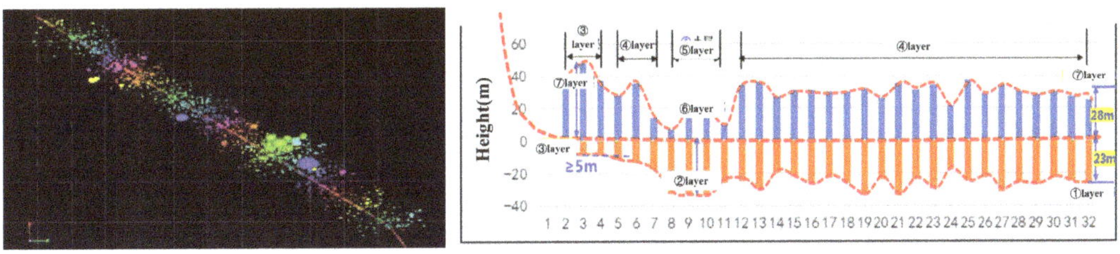

(a) (b)

Figure 8. Microseismic monitoring results of FYH10: (**a**) Vertical view; (**b**) Statistical map of the wellbore up and down sweep height.

6. Conclusions

(1) Based on the finite element and cohesive element method, a fluid-solid coupling model of continental shale hydraulic fractures spreading through layers was established, and the accuracy of the model was verified by comparing it with analytical solutions and experimental laboratory results. Based on this model, single-factor and orthogonal test analysis methods are used to reveal the control mechanism and law of various geological and engineering parameters on the propagation behavior of hydraulic fractures;

(2) The hindered mechanism of hydraulic fracture propagation through layers is: (1) The shear slip at the interlayer interface changes the vertical expansion path of hydraulic fractures, limiting the growth of fracture height; (2) The width of hydraulic fractures is large, which weakens the ability of fracture height to expand. The larger the interlayer interface strength, the larger the vertical stress difference, the smaller the interlayer stress difference, the smaller the tensile strength difference, the larger the elastic modulus difference, and the larger the fracturing fluid viscosity. The larger the injection displacement, the more favorable it is for the hydraulic fracture to achieve through-layer expansion. The primary and secondary order of the influence

degree of each factor is: shear strength of interlayer interface > interlayer stress difference/tensile strength difference > fracturing fluid viscosity > vertical stress difference > injection displacement > elastic modulus;

(3) Based on this model, engineering application research has been carried out to guide the construction parameter design of the example well. It is recommended that the injection displacement during the early construction should not be less than 3 m^3/min, and the fracturing viscosity should not be less than 45 mPa·s. The field application effect is good, realizing the purpose of cross-layer fracturing transformation, which shows that the model in this paper has high engineering application value.

Author Contributions: Conceptualization, Y.Z. and L.W.; methodology, Y.Z. and L.W.; software, Y.Z.; validation, Y.Z., L.W. and K.M.; formal analysis, Y.Z. and L.W.; investigation, Y.Z. and L.W.; re-sources, L.W.; data curation, L.W.; writing—original draft preparation, Y.Z.; writing—review and editing, Y.Z. and L.W.; visualization, L.W.; supervision, L.W.; project administration, L.W.; funding acquisition, L.W. and F.Z. All authors have read and agreed to the published version of the manuscript.

Funding: This research was funded by the Study on Control Mechanism and Optimization of Balanced Fracturing with Dense Cutting and Temporary Plugging for Continental Shale Gas Horizontal Wells, grant number PLN2021-09.

Acknowledgments: Many thanks are expressed once again to Lei Wang, Kuo Ma and Feng Zhang for their support.

Conflicts of Interest: The authors declare no conflict of interest.

References

1. Hammes, U.; Frébourg, G. Haynesville and Bossier mudrocks: A facies and sequence stratigraphic investigation, East Texas and Louisiana, USA. *Mar. Pet. Geol.* **2012**, *31*, 8–26. [CrossRef]
2. Loucks, R.G.; Ruppel, S.C. Mississippian Barnett Shale: Lithofacies and depositional setting of a deep-water shale-gas succession in the Fort Worth Basin, Texas. *AAPG Bull.* **2007**, *91*, 579–601. [CrossRef]
3. Wang, G.; Cheng, G.; Carr, T.R. The application of improved NeuroEvolution of Augmenting Topologies neural network in Marcellus Shale lithofacies prediction. *Comput. Geosci.* **2013**, *54*, 50–65. [CrossRef]
4. Dong, D.; Zou, C.; Yang, H.; Wang, Y.; Li, X.; Chen, G.; Wang, S.; Lv, Z.; Huang, Y. Shale gas resource potential and exploration and development prospect. *Geol. Bull.* **2011**, *30*, 324–336.
5. Li, S.; Qiao, D.; Feng, Z.; Liu, L.; Wang, Q.; Nie, H. The current situation of shale gas exploration and development in the world and its enlightenment to China. *Geol. Bull.* **2010**, *29*, 918–924.
6. Yassine, K. The US shale gas revolution: An opportunity for the US manufacturing sector? *Int. Econ.* **2021**, *167*, 59–77.
7. Zhao, J.; Xu, W.; Li, Y.; Hu, J.; Li, J. A new method for the evaluation of compressibility of shale gas reservoirs. *Nat. Gas Geosci.* **2015**, *26*, 1165–1172.
8. Zhao, J.; Ren, L.; Jiang, T.; Hu, D.; Wu, L.; Wu, J.; Yin, C.; Li, Y.; Hu, Y.; Lin, R.; et al. Ten Years of Shale Gas Fracturing in China: A Review and Outlook. *Nat. Gas Ind.* **2021**, *41*, 121–142.
9. Wang, R.; Hu, Z.; Liu, J.; Wang, X.; Gong, D.; Yang, T. Comparison of fracture development characteristics and main controlling factors of marine and continental shale in southern China: A case study of lower Cambrian in the Cengong region of northern Qianbei. *Oil Gas Geol.* **2018**, *39*, 631–640.
10. Jabbary, A.; Arnesa, S.R.; Samanipour, H.; Ahmadi, N. Numerical investigation of 3D rhombus designed PEMFC on the cell performance. *Int. J. Green Energy* **2021**, *18*, 425–442. [CrossRef]
11. Yaghmourali, Y.V.; Ahmadi, N.; Abbaspour-Sani, E. A thermal-calorimetric gas flow meter with improved isolating feature. *Microsyst. Technol.* **2017**, *23*, 1927–1936. [CrossRef]
12. Khormali, A.; Sharifov, A.R.; Torba, D.I. The control of asphaltene precipitation in oil wells. *Pet. Sci. Technol.* **2018**, *36*, 443–449. [CrossRef]
13. Khormali, A.; Sharifov, A.R.; Torba, D. Experimental and modeling analysis of asphaltene precipitation in the near wellbore region of oil wells. *Pet. Sci. Technol.* **2018**, *36*, 1030–1036. [CrossRef]
14. Zhang, Y.; Zhang, S.; Liu, Y.; Lu, L.; Liu, B.; Song, G.; Ma, X. Experimental study on the expansion law of hydraulic cracks in coal rock. *J. China Coal Soc.* **2012**, *37*, 73–77.
15. Meng, S.; Li, Y.; Wang, J.; Gu, G.; Wang, Z.; Xu, X. Experimental study on fracturing fracture expansion mold of coal "three gas" co-production layer group. *J. China Coal Soc.* **2016**, *41*, 221–227.
16. Gao, J.; Hou, B.; Tan, P.; Guo, X.; Chang, Z. Propagation mechanism of hydrocrack penetration between sand and coal interlayers. *J. China Coal Soc.* **2017**, *42*, 428–433.

17. Fu, H.; Wang, Z.; Xu, Y.; Liu, Y.; Xiu, N.; Yan, Y.; Guan, B. Simulation study of vertical extension of full three-dimensional hydraulic fractures. In Proceedings of the 2018 National Natural Gas Academic Annual Conference (04 Engineering Technology), Fu Zhou, China, 14 November 2018.
18. Jiang, Z.; Li, Z.; Fang, L.; Fan, Z. Propagation mechanism of segmented through-lamination fracture fractures in horizontal wells with roof plates adjacent to crushed soft coal seams. *J. China Coal Soc.* **2020**, *45*, 922–931.
19. Fu, S.; Hou, B.; Xia, Y.; Chen, M.; Tan, P.; Luo, R. Experimental study on fracture propagation of integrated fracturing in multi-rocky combined layered reservoirs. *J. China Coal Soc.* **2021**, *46*, 377–384.
20. Ma, K.; Wang, L.; Xu, W.; Zhao, Y.; Yuan, Y.; Chen, X.; Zhang, F. Physical simulation of lacustral hydraulic fracturing fracture penetration propagation law of lacustrine shale. *China Sciencepaper* **2021**, *1*, 1–7. Available online: http://kns.cnki.net/kcms/detail/10.1033.N.20210926.1041.002.html (accessed on 26 September 2021).
21. Zhang, X.; Jeffrey, R.G. Fluid-driven multiple fracture growth from a permeable bedding plane intersected by an ascending hydraulic fracture. *J. Geophys. Res. Solid Earth* **2012**, *117*, B12402. [CrossRef]
22. Xie, J.; Tang, J.; Yong, R.; Fan, Y.; Zuo, L.; Chen, X.; Li, Y. A 3D hydraulic fracture propagation model applied for shale gas reservoirs with multiple bedding planes. *Eng. Fract. Mech.* **2020**, *228*, 106872. [CrossRef]
23. Li, Y.; Deng, J.; Wei, B.; Liu, W.; Chen, J. The influence of reservoir/compartment rock and interlayer interface properties on high pressure fractures. *Pet. Drill. Tech.* **2014**, *42*, 80–86.
24. Tan, P. *Study on the Mechanical Behavior of Vertical Propagation of Hydraulic Fractures in Multi-Rocky Combined Strata Reservoirs*; China University of Petroleum: Beijing, China, 2019.
25. Sun, C.; Zheng, H.; Liu, W.D.; Lu, W. Numerical simulation analysis of vertical propagation of hydraulic fracture in bedding plane. *Eng. Fract. Mech.* **2020**, *232*, 107056. [CrossRef]
26. Abbas, S.; Gordeliy, E.; Peirce, A.; Lecampion, B.; Chuprakov, D.; Prioul, R. Limited height growth and reduced opening of hydraulic fractures due to fracture offsets: An XFEM application. In Proceedings of the SPE Hydraulic Fracturing Technology Conference, The Woodlands, TX, USA, 4–6 February 2014.
27. Fu, H.; Cai, B.; Geng, M.; Jia, A.; Weng, D.; Liang, T.; Zhang, F.; Wen, X.; Xiu, N. Three-dimensional simulation of hydraulic fracture propagation based on vertical reservoir heterogeneity. *Nat. Gas Ind.* **2022**, *42*, 56–68.
28. Zhang, F.; Wu, J.; Huang, H.; Wang, X.; Luo, H.; Yue, W.; Hou, B. Optimization of process parameters to increase the complexity of deep shale fracture propagation. *Nat. Gas Ind.* **2021**, *41*, 125–135.
29. Wang, H.; Liu, H.; Zhang, J.; Wu, H.; Wang, X. Numerical simulation study on the influence of joint height control parameters of hydraulic cracks. *J. Univ. Sci. Technol. China* **2011**, *41*, 820–825.
30. Detournay, E. Propagation regimes of fluid-driven fractures in impermeable rocks. *Int. J. Geomech.* **2004**, *4*, 35–45. [CrossRef]
31. Renshaw, C.E.; Pollard, D.D. An experimentally verified criterion for propagation across unbounded frictional interfaces in brittle, linear elastic materials. *Int. J. Rock Mech. Min. Sci.* **1995**, *32*, 237–249. [CrossRef]
32. Gu, H.; Siebrits, E. Effect of formation modulus contrast on hydraulic fracture height containment. *SPE Prod. Oper.* **2008**, *23*, 170–176. [CrossRef]
33. Yao, Y.; Wang, W.; Keer, L.M. An energy based analytical method to predict the influence of natural fractures on hydraulic fracture propagation. *Eng. Fract. Mech.* **2018**, *189*, 232–245. [CrossRef]
34. Chen, Z.; Jeffrey, R.G.; Zhang, X.; Kear, J. Finite-element simulation of a hydraulic fracture interacting with a natural fracture. *SPE J.* **2017**, *22*, 219–234. [CrossRef]
35. Liu, R.; Zhang, Y.; Wen, C.; Tang, J. Orthogonal experiment design and analytical method research. *Exp. Technol. Manag.* **2010**, *27*, 52–55.

processes

MDPI

Article

Investigation on the Injection Pattern of Intermittent Natural Gas Flooding in Ultra-Low Permeability Reservoirs

Lifei Dong [1,2,3,*], Linxiang Li [4], Wenzhuo Dong [3], Miao Wang [3] and Xiaozhi Chen [4]

1 State Key Laboratory of Shale Oil and Gas Enrichment Mechanisms and Effective Development, Beijing 102206, China
2 Research and Development Center for the Sustainable Development of Continental Sandstone Mature Oilfield by National Energy Administration, Beijing 102206, China
3 Faculty of Civil Engineering, Chongqing Three Gorges University, Chongqing 404120, China
4 Gudong Oil Production Plant of Shengli Oilfield, SINOPEC, Dongying 257237, China
* Correspondence: lfdong2012@sina.com

Abstract: Natural gas is a viable oil displacement agent in ultra-low-permeability reservoirs due to its good fluidity. It can also cause gas channeling during continuous injection, which limits its oilfield application. In order to relieve gas channeling during natural gas flooding, the injection mode should be changed. The use of intermittent natural gas injection (IGI) after the continuous natural gas injection in an ultra-low-permeability reservoir is proposed, and optimization of the injection parameters is discussed. The results show that IGI can be divided into three stages, the gas injection stage, the well shutting stage and the oil production stage. With the increase in injection time, the oil recovery enhances obviously as a result of IGI because the gas fingering can be controlled at the well shutting stage, and the gas/liquid ratio grows slowly because the gas breakthrough can be reduced at the oil production stage. The oil recovery improves with the increase in cycle time of IGI, while the increase rate reduces evidently after the cycle time reaches 360 min. The oil recovery increment is low if the cycle index exceeds 3 in the ultra-low-permeability reservoir. Thus, the optimal cycle time for each round and the appropriate cycle index of IGI are 360 min and three rounds.

Keywords: enhanced oil recovery; intermittent natural gas injection; ultra-low-permeability reservoir; cycle time; cycle index

Citation: Dong, L.; Li, L.; Dong, W.; Wang, M.; Chen, X. Investigation on the Injection Pattern of Intermittent Natural Gas Flooding in Ultra-Low Permeability Reservoirs. *Processes* **2022**, *10*, 2198. https://doi.org/10.3390/pr10112198

Academic Editor: Blaž Likozar

Received: 22 July 2022
Accepted: 30 September 2022
Published: 26 October 2022

1. Introduction

Oil development efficiency in ultra-low-permeability sandstone reservoirs is typically inadequate due to poor physical properties, such as the strong water sensitivity, intense micro heterogeneity and ultra-low permeability [1–3]. Low oil recovery ratio and high water cut are always evident during water flooding. Thus, with the acceleration of the exploration and development of ultra-low-permeability oil reservoirs, an increasing number of enhanced oil recovery technologies, such as profile control and asphaltene control, are required [4–6].

As an oil displacement agent, natural gas has many advantages compared with water. (i) The viscosity of natural gas is low. This makes the natural gas enter the small pore throat in the reservoir easily [7–11]. (ii) Natural gas can be dissolved into the oil and reduce the oil viscosity [12–14], which is beneficial for oil displacement [9]. (iii) Natural gas can be achieved at the gas reservoir, which is very convenient.

Based on the advantages mentioned above, natural gas flooding is usually considered as a major technology of enhanced oil recovery (EOR) in reservoirs, especially for the ultra-low-permeability sandstone oil reservoir. Natural gas injection has been introduced to improve oil displacement efficiency for petroleum engineers due to its property of high microscopic sweep efficiency [15–19]. However, the swept volume, controlled by the mobility ratio between the displacement agency and the oil, is unsatisfactory during natural gas injection because of its low viscosity. Gas fingering is easily formed during natural gas flooding, and the subsequent injection gas flows directly into the production well and fails to displace the oil. This limits the method's application in EOR to a great extent. In order to improve the application of gas flooding, gas fingering should be controlled. Thus, the profile control is commonly of concern.

Intermittent gas injection (IGI) and water-alternating gas injection (WAGI) are two main profile control methods currently used in gas flooding in the oilfield [20–23]. However, some studies suggest that the gas mobility control ability of IGI is better than that of water-alternating gas injection (WAGI) [24–26]. IGI has superior sweep efficiency, gas/oil ratio (GOR) stabilization and oil production.

Generally, each cycle of IGI is composed of three stages: the gas injection stage, the well shutting stage and the oil production stage. At the gas injection stage, the pressure of the reservoir increases, which is beneficial for forming an effective oil displacement pressure system. At the stage of well shutting, the natural gas has enough time to be expanded into the small pore throat and to be dissolved into the oil there. This reduces the viscosity of oil and improves its fluidity [27]. At the stage of oil production, the pressure declines rapidly because of the open output, and the reservoir has no supplementation. After these three stages, the cycle of IGI begins.

However, there is no discussion on the applicability of IGI in ultra-low-permeability reservoirs, and the technical parameters applied in the ultra-low-permeability reservoir are unknown. Experimental proof, which can provide a reference for field application, is especially inadequate.

To fill this gap, a natural gas flooding experiment was performed. First, continuous natural gas injection is simulated and was stopped when the oil could not be displaced further. Then, IGI with different cycle times and cycle indexes was performed. Finally, the optimized technical parameters of IGI were obtained by comparing on the oil recovery improvement. We hope this work can benefit the application of IGI technology.

2. Experiment

2.1. Experimental Devices and Materials

The main experimental devices are shown in Figure 1, including a high-pressure constant flow pump, a calorstat, several piston containers, several precise pressure sensors, several pressure valves, a core holder, a set of oil-water separating and measuring instrument, a computer and a gas measurer [28]. Some experimental apparatus are shown in Figure 2.

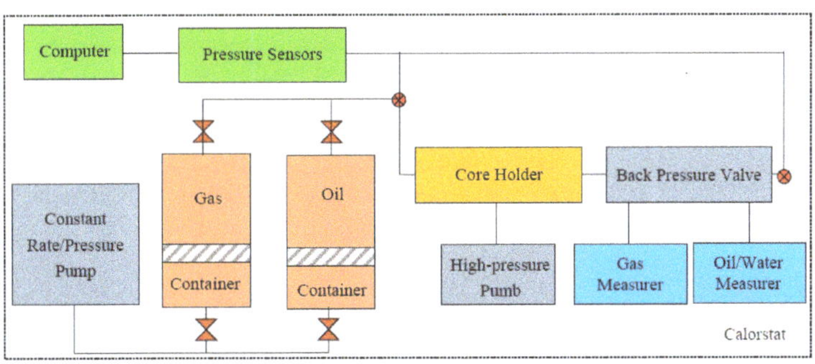

Figure 1. Experimental devices and their connection.

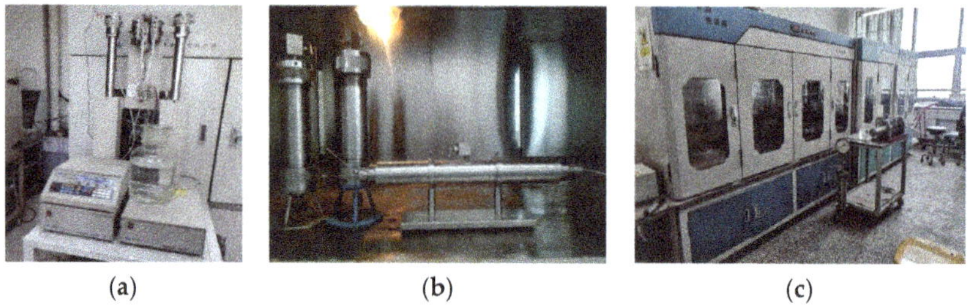

Figure 2. Experimental apparatus. (a) Pump, (b) container and core holder, (c) calorstat.

The injection rate can be controlled by the pump with the range of 0.001 to 60 mL/min. The back pressure valve is opened only when the outlet pressure exceeds the back pressure, which represents the minimum outflow pressure. The pressure sensors are set both at the beginning and the end of the core holder to test the inlet pressure and outlet pressure, respectively.

The crude oil used in the experiments was collected from Fuyu oilfield, which has considerable water channeling. It is necessary to perform gas flooding to enhance oil recovery. The viscosity of the oil is 2.04 mPa·s at the temperature of 50 °C. The oil displacement agent is natural gas, and its main components are listed in Table 1.

Table 1. Main components of natural gas.

Component	Concentration/%	Component	Concentration/%
Methane	94.08	n-Butane	0.168
Ethane	1.10	CO_2	0.532
Propane	0.326	N_2	2.81

The experimental core is a man-made core, which has the same basic physical properties as the reservoir in Fuyu oilfield. The porosity is 8.4% and the permeability is 6.88×10^{-3} μm^2.

All the oil displacement experiments were carried out under the conditions of 50 °C reservoir temperature and 10 MPa formation pressure. The protocol used to perform the experiments is GB/T 28912-2012 (Test method for two phase relative permeability in rock).

2.2. Experimental Processes

Before the oil displacement experiment, the air in the experimental cores should be vacuumized, and bound water with salinity of 4600 mg/L should be formed. The pore throat must be saturated with the experimental oil for more than 12 h. Then, natural gas flooding is performed.

The continuous natural gas flooding stops when the oil cannot be displaced further. After that, IGI is performed. In order to maximize the oil recovery ratio by IGI after continuous natural gas flood in the ultra-low-permeability reservoir, the optimized cycle time and the cycle index of IGI are discussed.

During the process of IGI, the injection pressure was kept at 12 MPa, and the back pressure was 10 MPa. In each period of IGI, the pause follows after the gas injection stage and then the oil production stage begins. At the oil production stage, the injection entrance should be closed. The cycle times of IGI, including the oil production time and pause time, were set to 90, 180, 360 and 1440 min. The cycle index ranged from 1 to 11 cycles. The injection rate of natural gas during the experiments was 0.3 mL/min, equal to 0.88 m/d. The experimental scheme is shown in Table 2.

Table 2. Experimental schemes of IGI.

Number	Permeability $/10^{-3}\ \mu m^2$	Porosity /%	Cycle Time/min			Cycle Index
			Injection Stage	Pause Stage	Total	
JX-1	6.88	8.75	60	30	90	11
JX-2		8.33	120	60	180	11
JX-3		8.09	240	120	360	11
JX-4		8.39	1080	360	1440	11

3. Experimental Results and Analysis

3.1. Analysis of EOR of IGI

The oil production dynamics of different schemes with different cycle times are shown in Figure 3.

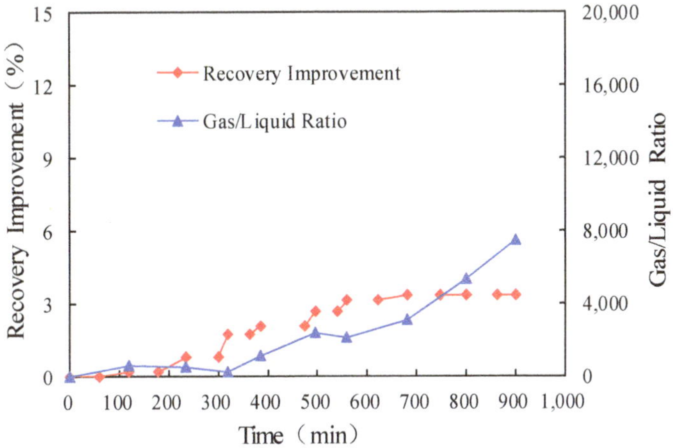

(a)

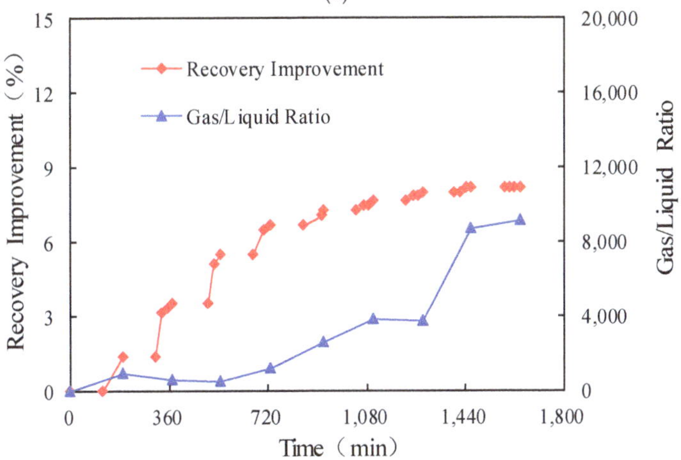

(b)

Figure 3. *Cont.*

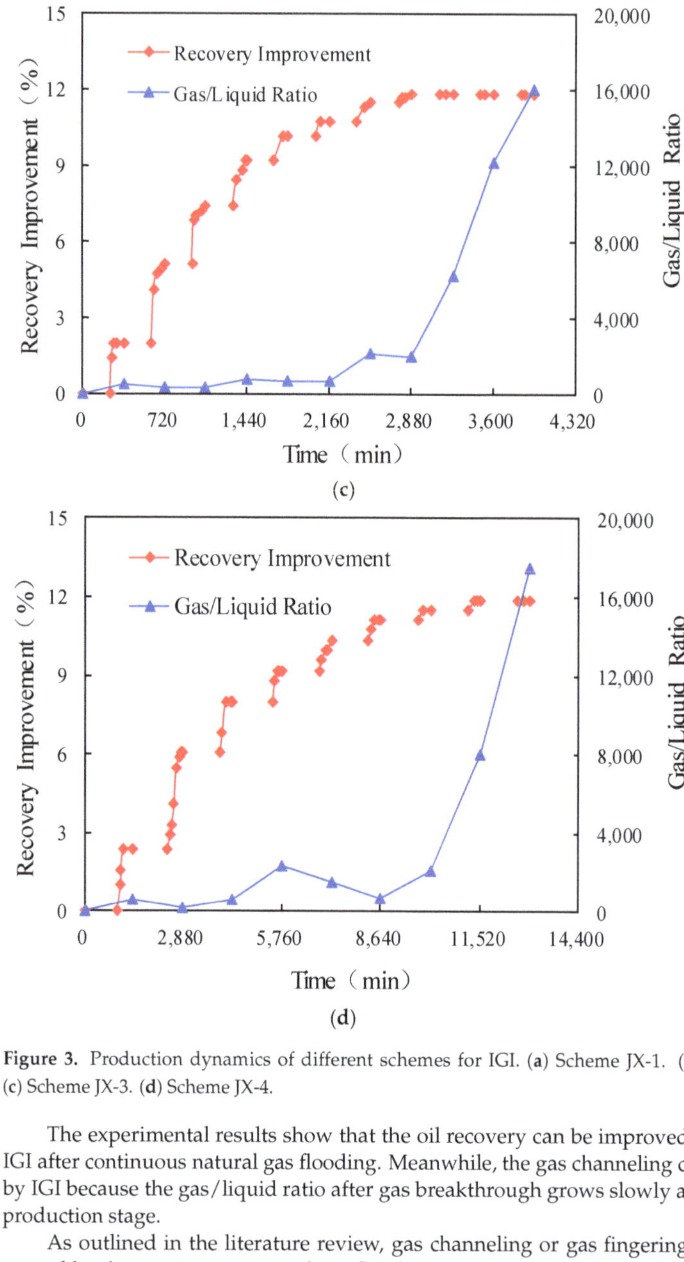

Figure 3. Production dynamics of different schemes for IGI. (**a**) Scheme JX-1. (**b**) Scheme JX-2. (**c**) Scheme JX-3. (**d**) Scheme JX-4.

The experimental results show that the oil recovery can be improved obviously by IGI after continuous natural gas flooding. Meanwhile, the gas channeling can be reduced by IGI because the gas/liquid ratio after gas breakthrough grows slowly at the initial oil production stage.

As outlined in the literature review, gas channeling or gas fingering can be easily caused by the continuous natural gas flooding. This makes the injected gas to flow into the producing well directly, which means the gas has insufficient time to make contact with or displace the oil there. Thus, the oil recovery is reduced. The pause stage of IGI can provide the redistribution time of gas and oil in the porous media, which can reduce the gas fingering obviously. After that, the gas flooding stage of IGI can displace the oil further. The longer the pause stage, the more sufficiently the gas and oil are redistributed. This is why IGI can control the gas channeling and improve the oil recovery.

3.2. *Optimization of IGI Parameters*

As seen from the analysis above, IGI has the advantages of controlling gas fingering and enhancing oil recovery. However, in practical application, the ultimate aim is to maximize benefits. Thus, a further experiment was carried out to determine the optimal technical parameters of IGI to maximize the degree of controlling gas fingering and enhancing oil recovery. An effective approach is to optimize the IGI patterns. Thus, the cycle time and the cycle index of IGI were investigated via laboratory experiments in this study.

3.2.1. The Cycle Time of IGI

As can be seen from the Scheme JX-1 to Scheme JX-4 in Figure 4, the oil recovery improves with the increase in cycle time, and the increment changes little after the cycle time reaches 360 min. This is because the natural gas needs enough time to disperse in the pore throat and to enhance the micro-swept volume. If the cycle time is insufficient, the effect of IGI mentioned above stops. Thus, the cycle time of IGI, especially for the time at the pause stage, should be long enough to accommodate the process. According to the experimental results, the suitable cycle time of IGI after the continuous natural gas flooding in the ultra-low-permeability reservoir is 360 min.

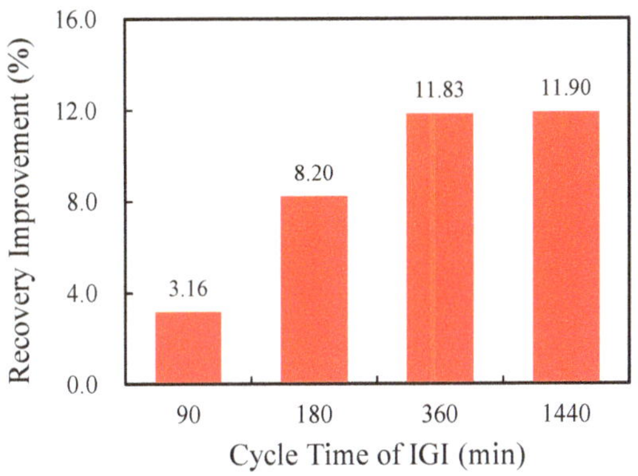

Figure 4. Oil recovery improvement for different cycle times.

3.2.2. The Cycle Index of IGI

The cycle index of IGI also affects the final oil recovery. It can be seen from Figure 5 that the oil recovery varies with the IGI cycle.

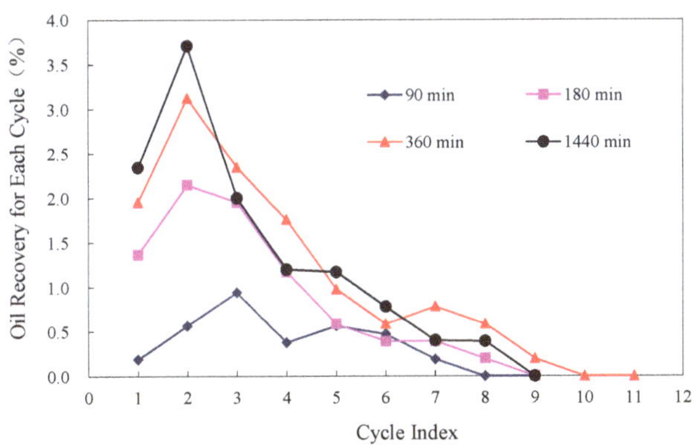

Figure 5. Oil recovery for each cycle of IGI.

As the stress sensitivity is not considered during the experiments, the permeability and porosity of cores from cycle 1 to cycle 11 are constant.

From each cycle time curve, it can be seen that the oil recovery improves obviously in the initial three cycles, but it changes little after five cycles. This is because there is an abundance of residual oil after continuous natural gas flooding, and it is displaced easily in the previous cycles of IGI. With the increase in cycle index of IGI, the remaining oil reduces, and there is little left after five cycles. Meanwhile, the higher the cycle time, the more obvious the oil recovery improves in the first three cycles, and the more apparently the curve declines after three cycles.

4. Conclusions

In this work, experiments were carried out to simulate the intermittent natural gas injection (IGI) after continuous natural gas flooding. The applicability of IGI in an ultra-low-permeability reservoir was evaluated, and the optimized technical parameters of IGI, the cycle time and the cycle index, were determined by comparing the oil recovery improvement of different experiments. This work can provide experimental proof and technical parameters for field application, which are lacking in existing studies. The specific conclusions are shown as follows.

1. Intermittent natural gas injection (IGI) has the advantage of controlling gas fingering compared with continuous natural gas flooding. It can enhance oil recovery because IGI can reduce the gas mobility and improve the swept volume effectively.
2. The oil recovery is influenced by the cycle time of IGI. It improves obviously with the increase in cycle time and changes little when the cycle time reaches 360 min in the ultra-low-permeability reservoir.
3. The oil recovery of IGI improves with the increase in the cycle index. The increment of oil recovery is obvious in the initial three cycles, while it changes little after five cycles in the ultra-low-permeability reservoir.

Author Contributions: Conceptualization, methodology and, L.D.; investigation, L.L.; project administration, L.D.; data curation, M.W.; writing—original draft preparation, W.D.; writing—review and editing, X.C. All authors have read and agreed to the published version of the manuscript.

Funding: This research was funded by (1) the project of science and technology research program of Chongqing Education Commission of China (KJQN202001203; KJZD-K202201204); (2) the Open Fund of Sichuan Oil and Gas Development Research Center (SKB21-08); (3) the Open Fund of Sustainable Development Research Center of Three Gorges Reservoir Area (2021sxxyjd03). (4) the Natural Science Foundation of Chongqing, China (2022NSCQ-MSX3909).

Data Availability Statement: Not applicable.

Acknowledgments: Funding from "Chongqing Education Commission", "Sichuan Oil and Gas Development Research Center", "Sustainable Development Research Center of Three Gorges Reservoir Area" and " Chongqing Bureau of Science and Technology" are gratefully acknowledged.

Conflicts of Interest: The authors declare no conflict of interest.

Nomenclature

IGI	intermittent natural gas injection
EOR	enhanced oil recovery
GOR	gas/oil ratio
WAGI	water-alternating gas injection

References

1. Tang, Y.; Chen, Y.; He, Y.; Yu, G.; Guo, X.; Yang, Q.; Wang, Y. An improved system for evaluating the adaptability of natural gas flooding in enhancing oil recovery considering the miscible ability. *Energy* **2021**, *236*, 121441. [CrossRef]
2. Tu, X. Comparative Analysis of CO2 Miscible Flooding and Conventional Natural Gas Flooding. *Contemp. Chem. Ind.* **2020**, *49*, 956–960.
3. Yue, X.A.; Wang, F.Y.; Wang, K.L. *Enhanced Oil Recovery Foundation*, 1st ed.; Petroleum Industry Press: Beijing, China, 2007.
4. Khormali, A.; Moghadasi, R.; Kazemzadeh, Y.; Struchkov, I. Development of a new chemical solvent package for increasing the asphaltene removal performance under static and dynamic conditions. *J. Pet. Sci. Eng.* **2021**, *206*, 109066. [CrossRef]
5. Khormali, A.; Sharifov, A.R.; Torba, D.I. Experimental and modeling analysis of asphaltene precipitation in the near wellbore region of oil wells. *Pet. Sci. Technol.* **2018**, *36*, 1030–1036. [CrossRef]
6. Nelson, P.H. Pore-throat sizes in sandstones, tight sandstones, and shales. *AAPG Bull.* **2009**, *93*, 329–340. [CrossRef]
7. Zheng, T.; Liu, X.; Yang, Z.; Luo, Y.; Zhang, Y.; Xiao, Q.; He, Y.; Zhao, X. Identification of seepage mechanisms for natural gas Huff-n-Puff and flooding processes in hydrophilic reservoirs with low and ultra-low permeabilities. *J. Energy Resour. Technol.* **2021**, *143*, 063004. [CrossRef]
8. Ding, M.; Wang, Y.; Yue, X.-A.; Chen, W.; Shi, S. The effects of initial gas content of the oil on recovery by natural gas flood. *Pet. Sci. Technol.* **2015**, *33*, 1454–1462. [CrossRef]
9. Ding, M.; Wang, Y.; Liu, D.; Wang, W.; Chen, W. Mutual interactions of CO_2/oil and natural gas/oil systems and their effects on the EOR process. *Pet. Sci. Technol.* **2015**, *33*, 1890–1900. [CrossRef]
10. Yuliang, S.; Chunxin, W.; Li, G.; Xiaodong, W. Determination of waterflooding limits for driving an ultra-Low permeability reservoir. *Pet. Drill. Tech.* **2012**, *40*, 82–86.
11. Wei, H.; Yue, X.; Zhao, Y.; Li, L.; Jia, D. The injection pattern of immiscible nitrogen displacement after water flooding in an ultra-low permeability reservoir. *Pet. Sci. Technol.* **2013**, *31*, 2304–2310. [CrossRef]
12. Lifei, D.; Miao, W.; Wei, W.; Hun, L. Investigation of natural gas flooding and its channelling prevention as enhanced oil recovery method. *Geosystem Eng.* **2021**, *24*, 137–144. [CrossRef]
13. Fu, H.T.; Wang, S.D. Research status of gas flooding. *Energy Chem. Ind.* **2015**, *36*, 44–48.
14. Sheng, J.J. Enhanced oil recovery in shale reservoirs by gas injection. *J. Nat. Gas Sci. Eng.* **2015**, *22*, 252–259. [CrossRef]
15. Tang, L. Analysis on influencing factors of natural gas flooding in low permeability reservoir. *World Pet. Ind.* **2021**, *28*, 60–64.
16. Haoguang, W.E.I.; Xiang, Y. An experimental investigation of the natural gas immiscible displacement in ultra-low permeability reservoirs. *Acta Pet. Sin.* **2011**, *32*, 307–310.
17. Zhang, Y.Y.; Cui, H.X.; Han, H.J.; Han, B.; Li, H.J. Study on numerical simulation of enhancing oil recovery using natural gas in the low-permeability oil reservoirs. *Pet. Geol. Recovery Effic.* **2005**, *12*, 61–63.
18. Guo, Y.; Yang, S.; Li, L.; Wang, G.; Zhao, W. Experiment on physical modeling of displacement oil with natural gas for long core. *FaultBlock Oil Gas Field* **2009**, *16*, 76–78.
19. Haines, H.K.; Monger, T.G. A laboratory study of natural gas huff-n-puff. In *CIM/SPE International Technical Meeting*; OnePetro: Richardson, TX, USA, 1990; ISBN 978-1-55563-480-3.
20. Phukan, R.; Gogoi, S.B.; Tiwari, P. Enhanced oil recovery by alkaline-surfactant- alternated-gas/CO_2 flooding. *J. Pet. Explor. Prod. Technol.* **2019**, *9*, 247–260. [CrossRef]
21. Bayat, M.; Lashkarbolooki, M.; Hezave, A.Z.; Ayatollahi, S. Investigation of gas injection flooding performance as enhanced oil recovery method. *J. Nat. Gas Sci. Eng.* **2016**, *29*, 37–45. [CrossRef]
22. Heidari, P.; Alizadeh, N.; Kharrat, R.; Hossein Ghazanfari, M.; Laki, A.S. Experimental analysis of secondary gas injection strategies. *Pet. Sci. Technol.* **2013**, *31*, 797–802. [CrossRef]
23. Wan, T.; Sheng, J.J. Evaluation of the EOR potential in hydraulically fractured shale oil reservoirs by cyclic gas injection. *Pet. Sci. Technol.* **2015**, *33*, 812–818. [CrossRef]
24. Ahmadi, Y.; Hasanbaygi, M.; Kharrat, R. A comparison of natural depletion and different scenarios of injection in the reservoir from the beginning of oil production. *Pet. Sci. Technol.* **2014**, *32*, 2559–2565. [CrossRef]
25. Kulkarni, M.M.; Rao, D.N. Experimental investigation of miscible and immiscible water-alternating-gas (WAG) process performance. *J. Pet. Sci. Eng.* **2005**, *48*, 1–20. [CrossRef]
26. Farsetti, S.; Farisè, S.; Poesio, P. Experimental investigation of high viscosity oil–air intermittent flow. *Exp. Therm. Fluid Sci.* **2014**, *57*, 285–292. [CrossRef]
27. Li, Y.Q. Enhanced oil recovery and application review by gas injection in low permeability oil reservoirs. *Tuha Oil Gas* **2004**, *9*, 112–116.
28. Nguyen, T.A.; Ali, S.M. Effect of nitrogen on the solubility and diffusivity of carbon dioxide into oil and oil recovery by the immiscible WAG process. *J. Can. Pet. Technol.* **1998**, *37*, 24–31. [CrossRef]

processes

MDPI

Article

Development and Performance Evaluation of Scale-Inhibiting Fracturing Fluid System

Miao Zheng [1], Lianqi Sheng [2], Hongda Ren [1], Abulimiti Yiming [1], Erdong Yao [2,*], Kun Zhang [2] and Longhao Zhao [2]

1 Engineering Technology Research Institute of PetroChina Xinjiang Oilfield Company, Karamay 834000, China
2 State Key Laboratory of Petroleum Resource and Prospecting, China University of Petroleum, Beijing 102249, China
* Correspondence: yaoed@cup.edu.cn

Abstract: The injection water and formation water in the Mahu oil field have high salinity and poor compatibility, which leads to scaling and blockage in the formation or fracture propping zone during production. In this paper, a scale-inhibiting fracturing fluid system is developed which can prevent the formation of scale in the reservoir and solves the problem of scaling in the fracture propping zone at the Mahu oil field. Firstly, based on scale-inhibition rate, the performances of six commercial scale inhibitors were evaluated, including their acid and alkali resistance and temperature resistance. Then, the optimal scale inhibitors were combined with the fracturing fluid to obtain a scale-inhibiting fracturing fluid system. Its compatibility with other additives and scale-inhibition performance were evaluated. Finally, the system's drag-reduction ability was tested through the loop friction tester. The results showed that, among the six scale inhibitors, the organic phosphonic acid scale inhibitor SC-1 has the best performance regardless of high-temperature, alkaline, and mixed scale conditions. In addition, SC-1 has good compatibility with the fracturing fluid. The scale-inhibiting fracturing fluid system can effectively prevent scaling inside the large pores in the propping zone, and a scale-inhibiting efficiency of 96.29% was obtained. The new fracture system maintained a drag-reduction efficiency of about 75%, indicating that the addition of the scale inhibitor did not cause a significant influence on the drag-reduction efficiency of the fracturing fluid.

Keywords: scale inhibitor; fracturing fluid; scaling in propping zone; compatibility; drag-reduction rate

Citation: Zheng, M.; Sheng, L.; Ren, H.; Yiming, A.; Yao, E.; Zhang, K.; Zhao, L. Development and Performance Evaluation of Scale-Inhibiting Fracturing Fluid System. *Processes* **2022**, *10*, 2135. https://doi.org/10.3390/pr10102135

Academic Editor: Jui-Yuan Lee

Received: 29 August 2022
Accepted: 16 October 2022
Published: 20 October 2022

1. Introduction

In recent years, conventional oil and gas resources have been gradually drying up, and their development has become increasingly difficult [1]. The exploration and development of unconventional oil and gas resources have shown a rapid rise [2]. Large-scale hydraulic fracturing is the most widely used and technically mature means to increase the production of unconventional oil and gas, which is the key to realizing the low-cost development of oil fields [3,4]. However, the rapid change in the temperature and pressure near the wellbore area after fracturing has already led to scaling and formation damage, especially using high-salinity fracturing fluid in dry areas. The scaling in the formation after fracturing can reduce reservoir permeability [5], severe skin damage [6], and ultimately reduced oil well production [7,8]. The Samotlor field in West Siberia of the former Soviet Union, the Foster oil field in Texas, the oil field in Louisiana, the Ujin and Retibai oil fields in the Mangyshlak region, the Changtan oil field in California, and the Burbank and Drumright oil fields in Oklahoma were all subject to scaling with various degrees, mostly of calcium carbonate and calcium sulfate [9]. In China, Daqing oil field, Zhongyuan oil field, Jilin oil field, Gaskule oil field in Qinghai, Tuha oil field, Tarim oil field, and Bohai offshore oil field had different degrees of scaling problems, which even caused them to be shut down and scrapped [10,11]. Therefore, oil field scaling can seriously affect oil and gas field development, and oil field scale prevention has attracted great attention all over the world [12,13].

In the fracturing process, there are various sources of scaling. High-salinity fracturing fluid, acid corrosion, and water–rock reaction (ion exchange between the liquid phase and solid phase) can increase the total salinity of the liquid phase at the reservoir and cause potential scaling hazards [14,15]. At present, there are various scale-inhibiting methods, mainly divided into physical and chemical methods. Although physical scale removal methods such as high-intensity shock waves have a high scale removal rate, their action area is small, mostly in the near the well zone, such as wellbore. In addition, they cannot be applied to the fracture and propping zone and their cost is high, which makes it difficult for large-scale application [16]. In contrast, the chemical scale-inhibiting method is highly effective and highly cost-effective. After the scale inhibitor is combined with fracturing fluid, it can penetrate deeply into the formation and act on the fractures, rock pores, and propping zones. If slow-release technology is used, it can achieve a long-lasting scale-inhibiting effect during the fracturing process. In addition, the operation cost is low [17].

However, as the pH, temperature, and pressure on the reservoir are unstable during the oil production process, the performance of the scale inhibitor under various conditions is different [18,19]. At present, scale inhibitors are mostly acidic and weakly acidic, which can also affect the drag-reduction or crosslinking performance of the alkaline fracturing fluid [20]. As early as the 1990s, researchers found that most scale inhibitors have poor compatibility with fracturing fluids. After adding the scale inhibitors, the fracturing fluids are either precipitation with fluids or not thermally stable at high temperatures. They proposed that qualified scale inhibitors should have features such as good compatibility, thermal stability, and acid and alkali resistance [21]. In addition, scale inhibitors need as little impact as possible on the performance of fracturing fluids [22]. Yue (2014) et al. conducted a study on the compatibility of nine commercially available scale inhibitors with fracturing fluids and concluded that only two of them could be added to the fracturing fluid [23,24].

At present, the temperature resistance and compatibility of scale inhibitors is poor, and they also have a narrow range of acid and alkali resistance. Therefore, the selection of scale inhibitors is crucial for scale-inhibiting fracturing fluids. At present, the scale inhibitors used in the Mahu oil field have poor performance in the formation environment. When combined with fracturing fluid, the stability, drag reduction, and other performance of fracturing fluid was also affected. Therefore, a new scale inhibitor should be optimized to achieve the purpose of scale inhibition while fracturing.

In this paper, an optimal scale inhibitor SC-1 was found and was combined with the fracturing fluid. Various performance indicators of scale inhibitors were evaluated, including the temperature and acid/alkali resistance as well as the mixed scale-inhibiting ability. After adding the scale inhibitor SC-1 with fracturing fluid, the scale inhibiting and the drag reduction of the new fracturing fluid on the fracturing propping zone of the scale inhibitor were studied. The results showed that the fracturing fluid system had high inhibition efficiency and high drag-reduction rate in a wide acid and alkali and temperature range. This is of great significance for improving the fracturing effect and increasing oil and gas production.

2. Materials and Methods

This section discloses the experimental details. The main experimental design ideas are: (1) first, temperature and pH resistance of scale inhibitors were evaluated, and the optimal scale inhibitor was selected; (2) then, compatibility experiments were carried out to form the new fracturing fluid formula; (3) finally, the performance of the new fracturing fluid was evaluated comprehensively, and its scale-inhibition mechanism was characterized by nuclear magnetism.

2.1. Experimental Materials

A scale inhibitor [13] has the advantages of powerful chelating ability as well as low dosage and low cost, which is the current main scale-inhibition method [6,25]. However, the existing scale inhibitors in the Mahu oil field have poor performance, and most of them have no temperature and alkali resistance. Therefore, the scaling phenomenon at the reservoir is becoming more and more serious. Here, six commercial scale inhibitors were optimized in this study. They are organic phosphoric acids and included: SC-1 (pH: 2.0–3.0, 8500 RMB/t, Hengkang Environmental Protection Technology Co., Ltd., Linyi, China), polyaspartic acid SC-5 (pH: 3.0–5.0, 8300 RMB/t Xiongguan Technology Development Co., Ltd., Tianjin, China), and composite scale inhibitors SC-2 (pH: 3.0–4.5, 8700 RMB/t Gansu Heima Petrochemical Engineering Co., Ltd., Lanzhou, China), SC-3 (pH 3.0–4.5 8200 RMB/t Gansu Heima Petrochemical Engineering Co., Ltd., Lanzhou, China), SC-4 (pH: 5.0–6.0, Xiongguan Technology Development Co., Ltd., Tianjin, China), and SC-6 (5.0–7.0, 8000 RMB/t Dongfang Chemical Co., Ltd., Yantai, China).

The formulation of the scale inhibitor fracturing fluid system is 0.08% drag-reduction agent DR800 + 0.3% scale inhibitor (preferred scale inhibitor) + 4% autogenic acid SEG-C + 0.2% clay stabilizer. The scale inhibitor fracturing fluid system is a combination of the fracturing fluid currently used at the Mahu site and the preferred scale inhibitor. DR800 is the current drag-reducing agent used in the Mahu oil field. A 0.08% proportion of drag-reducing agents is the best choice in terms of cost and effect. The reason for adding 4% the autogenic acid SEG-C is to maintain a low-pH environment in the reservoir over the long term. The autogenic acid is the organic ester, and it can slowly hydrolyze in the formation to produce the organic acid, which can be helpful for inhibiting scale precipitation [6].

2.1.1. Effect of Temperature on $CaSO_4$ Scale Inhibition

The performances of scale inhibitors on $CaSO_4$ were determined according to methods of the literature [26], which came from the China National Petroleum Corporation method (SY/T5673-93). The experimental process was as follows: prepare solution A (7.5 g/L NaCl, 11.1 g/L $CaCl_2$-H_2O), solution B (7.5 g/L NaCl, 10.66 g/L Na_2SO_4), and solution C (0.5% scale inhibitor). Mix 50 mL solution A, 50 mL solution B, 0.8 mL solution C to obtain sample 1. Mix 50 mL solution A, 50 mL solution B to obtain sample 2. Mix 50 mL solution A, 50 mL distilled water to obtain sample 3. Put the three samples in a 90 °C water bath for 24 h and filter them separately. Then, 2 mL is taken out and diluted to 100 mL. The appropriate amount of 0.1 wt% NaOH solution is added to the solution to adjust the pH between 12 and 13. Next, a little calcium indicator is added to make the solution system appear light red.

The titration was performed with 0.01 mol/L EDTA, and the finish of titration was based on the color change from red to blue. The scale-inhibition rate was calculated by Equation (1) [27]. The water bath temperature was adjusted to 70 °C, 50 °C, and 30 °C, and the above process was repeated.

$$E_f = \frac{V_1 - V_0}{V - V_0} \times 100 \quad (1)$$

where E_f is the scale-inhibition rate in %, V_0 is the volume of EDTA consumed by sample 2 in mL, V is the volume of EDTA consumed by sample 3 in mL, and V_1 is the volume of EDTA consumed by sample 1 in mL.

In order to ensure the accuracy of the experiment, each group of scale-inhibition rates was tested three times. The average value and the error were calculated, and the error bars are added in the figures.

2.1.2. Effect of Temperature on $CaCO_3$ Scale Inhibition

Prepare solution A (33.00 g/L NaCl, 12.15 g/L $CaCl_2$-H_2O, 3.68 g/L $MgCl_2$-$6H_2O$), B (33.00 g/L NaCl, 0.03 g/L Na_2SO_4, 7.38 g/L $NaHCO_3$), and C (0.5% scale inhibitor). The test method is the same as the test method in Section 2.1.1.

2.1.3. Effect of pH on $CaSO_4$ Scale Inhibition

Sample 1, sample 2, and sample 3 were prepared according to Section 2.1.2. The pH of the three solutions was adjusted to 5, 6, 7, 8, 9, 10, and 11 by 0.1 wt% HCl or 0.1 wt% NaOH solutions. Then, they were placed in a water bath at 50 °C for 24 h, and the subsequent filtration titration test was the same as Section 2.1.1.

2.1.4. Effect of pH on $CaCO_3$ Scale Inhibition

Sample 1, sample 2, and sample 3 were prepared according to Section 2.1.2. Their pH was adjusted to 5, 6, 7, 8, 9, 10, and 11. Then, they were placed in a water bath at 50 °C for 24 h, and the subsequent filtration titration test was the same as Section 2.1.1.

2.1.5. Mixed-Scale Preparation, Scale-Inhibition Ability Test, and Optimal Test of Scale Inhibitor Concentration

Prepare solution A (20.25 g/L NaCl, 11.63 g/L $CaCl_2 \cdot H_2O$, 84 g/L $MgCl_2 \cdot 6H_2O$), solution B (20.25 g/L NaCl, 5.33 g/L Na_2SO_4, 3.69 g/L Na_2CO_3), and solution C (0.5% scale inhibitor). The choice of scale inhibitor is the three best performers among the above 6 types of scale inhibitors These three samples are mixed according to Section 2.1.1 and placed in a 90 °C water bath for 24 h. The subsequent filtration titration test is the same as Section 2.1.2.

The optimal concentration of the preferred scale inhibitor is then screned. The experimental procedure is as follows: prepare solution A (33.00 g/L NaCl, 12.15 g/L $CaCl_2 \cdot H_2O$, 3.68 g/L $MgCl_2 \cdot 6H_2O$) and solution B (33.00 g/L NaCl, 0.03 g/L Na_2SO_4, 7.38 g/L $NaHCO_3$). Mix solution A, solution B, and scale inhibitor. The concentrations of the scale inhibitor are: 0.01%, 0.03%, 0.05%, 0.07%, 0.09%, and they are placed at 90 °C for 24 h. The subsequent experimental procedure is the same as Section 2.1.2. Prepare solution C (7.5 g/L NaCl, 11.1 g/L $CaCl_2 \cdot H_2O$) and solution D (7.5 g/L NaCl, 10.66 g/L Na_2SO_4); Mix solution C, solution D, and scale inhibitor. The concentrations of the scale inhibitor are 0.1%, 0.3%, 0.5%, 0.7%, and they are placed at 90 °C for 24 h. The subsequent experimental procedure is the same as Section 2.1.2

2.2. Fracturing Fluid and Its Compatibility with Scale Inhibitor

The scale inhibitor fracturing fluid was prepared according to the formula: 0.08% drag-reducing agent DR800 + 0.3% scale inhibitor (preferred scale inhibitor) + 4% autogenous acid SEG-C + 0.2% clay stabilizer. The dosed water was from Mahu injection water, and its ionic composition is shown in Table 1. The fracturing fluid system was placed in a 90°C water bath and it was observed whether there was flocculation or precipitation in the fracturing fluid system after heating for 15 min, 1 h, 3 h, 12 h, 24 h, and 48 h.

Table 1. Ion components of Mahu injection water.

Ionic Types	CO_3^{2-}	HCO_3^-	Cl^-	SO_4^{2-}	Ca^{2+}	Mg^{2+}	pH
Injected water (mg/L)	8	287.3	11,574	540	320	67	8

2.3. Performance Evaluation of Scale Inhibition

2.3.1. Rock Sample Preparation

The sandstone core with fracture and proppants was manufactured manually. Firstly, a natural tight sandstone was processed in advance, and it was split from the middle by Brazilian splitting methods. Then, a layer of 100/140 mesh quartz sand was evenly laid inside the splitting surface of the core. A porous metal screen was pasted on the upper and lower end surfaces, which can prevent the proppant in the fracture from slipping out of the fracture. Finally, it was wrapped with thermoplastic tube to form a complete core, and a strip of artificial propping fracture through the whole was formed. Two cores, named b-1 and b-2, were made according to the above methods. Table 2 shows the basic information of the cores used in the later evaluation experiment.

Table 2. Core data for the study of scale-inhibition performance.

Core No.	Length/cm	Diameter/cm	Weight/g	Porosity /%	Pore Volume /cm³	Quartz/%	Potash Feldspar/%	Plagioclase/%	Dolomite/%	Clay/%
b-1	5.71	2.38	58.11	8.22	2.09	17.23	1.10	13.20	67.20	1.27
b-2	5.42	2.31	57.68	8.17	1.85	18.57	2.13	11.80	65.90	1.60

2.3.2. Performance Evaluation of Scale-Inhibiting Fracturing Fluid

The scale formation process in the propping zone was characterized and described by using an online low-field NMR device in the laboratory. Under conditions of 5 MPa and room temperature, a 2 wt% KCl solution with flow rate of 0.2 mL/min was inject into core b-1. When the flow rate was stable at the exit end, an NMR T2 spectrum scanning was performed. Then, 1 wt% Na_2CO_3 and 1 wt% $CaCl_2$ + 0.08% DR800 were injected into core b-1 at a rate of 0.2 mL/min, and the injection time was set to 2 h. After the injection was completed, a large amount of $CaCO_3$ scale appeared in the core propping zone. Finally, all switches of the core holder were turned off and left for 6 h. The cores were scanned again by NMR T2 spectra. The relaxation times in the experimental data were collected and converted to pore size, and the formula of pore size conversion is shown in Equation (2) [28].

$$\frac{1}{T_{2s}} = \rho_2 \frac{s}{v} \tag{2}$$

where T_{2s} is the transverse relaxation time; $\frac{s}{v}$ is the ratio of the internal surface pore volume to the sample volume; ρ_2 is the transverse relaxation, which characterizes the relaxation strength of solid–fluid interactions.

The scale-inhibiting process in the propping zone was also characterized by NMR T2 spectrum. First, three solutions were prepared, including 2 wt% KCl, 1 wt% Na_2CO_3, and 1 wt% $CaCl_2$ + scale-inhibiting fracturing fluid system. Under conditions of 5 MPa and room temperature, a 2 wt% KCl solution with flow rate of 0.2 mL/min was injected into core b-2. When the flow rate was stable at the exit end, an NMR T2 spectrum scanning was performed. Then, three solutions were simultaneously injected into core b-2 at a rate of 0.2 mL/min, and the injection time was set to 2 h. After the injection was completed, all switches of the core holder were turned off and left for 6 h. Finally, the core was scanned for NMR T2 spectra, and the relaxation time in the experimental data was collected and converted into the pore size in the same way as Equation (2).

The integral area of relaxation signal has a linear dependence with the pore volume, and the change in semaphores can reflect the amount of scaling in the pore. Therefore, the core scaling rate can be calculated by Equation (3), and the scale-inhibition rate of the scale-inhibiting fracturing fluid can be calculated by using Equation (4).

$$\delta = (S_1 - S_1') \div S_1 \times 100 \tag{3}$$

$$\varnothing = [(S_1 - S_1') - (S_2 - S_2')] \div (S_1 - S_1') \times 100 \tag{4}$$

where δ is the scaling rate in %, $\varnothing$ is the scale-inhibiting rate in %; S_1 is the integral area of relaxation signal after core b-1 is saturated with 2 wt% KCl; S_1' is the integral area of the relaxation signal after scaling; S_2 is the integral area of the relaxation signal after core b-2 is saturated with 2 wt% KCl; S_2' is the integral area of relaxation signal after the new fracturing fluid system.

According to accuracy parameters from the equipment manufacturer, the magnet of the equipment is a permanent magnet, the magnetic field strength is 0.3T ± 0.05T, the magnetic field uniformity is ≤30 pm (φ110 sphere), and the magnetic field stability is ≤300 Hz/h, so the accuracy of this equipment test is quite high.

2.4. Drag-Reduction Test of Scale-Inhibiting Fracturing Fluid

2.4.1. Loop Friction Test Device

Figure 1 shows the loop friction test device, which consists of three parts, namely, the tank part, the pipe parts, and the data processing parts. The main body of the pipe parts consists of three stainless steel test tubes with different diameters, 6 mm, 8 mm, and 10 mm from top to bottom, and a mixer. The pressure difference, flow rate, and temperature of this system are automatically monitored by the computer.

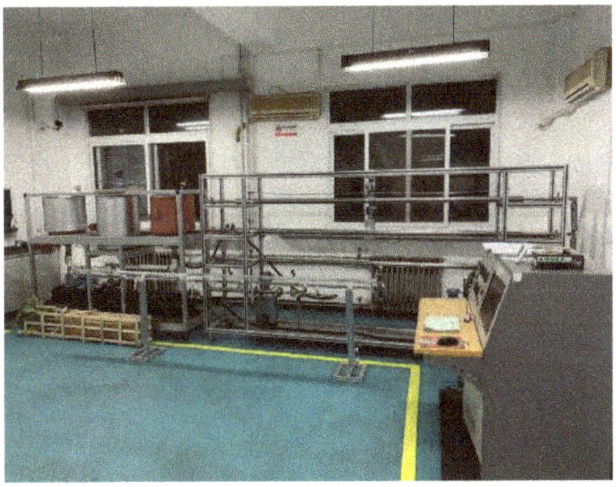

Figure 1. Picture of the loop friction test device.

2.4.2. Drag-Reduction Test

The scale-inhibiting fracturing fluid is made of slick water fracturing fluid and a scale inhibitor. Its formulation is: 0.08% scale inhibitor DR800 + 0.3% scale inhibitor SC-1 + 4% autogenic acid SEG-C + 0.2% clay stabilizer.

Testing process: (1) a 30 L measure of tap water was put into the reservoir, and the ball valve at the lower part of the reservoir was opened; (2) the injected parameters were set to 250 kg/h, 500 kg/h, 1000 kg/h, 1250 kg/h, and 1750 kg/h, and the differential pressure ΔP_0 for each was recorded, respectively; (3) in the same way, 30 L of scale-inhibiting fracturing fluid was put into the reservoir. The software was run and the differential pressure ΔP_{DR} was recorded under different injected rates. The drag-reduction rate was calculated according to Equation (5) [29]:

$$DR\% = \frac{\Delta P_0 - \Delta P_{DR}}{\Delta P_0} \times 100\% \quad (5)$$

where $DR\%$ is the drag-reduction rate; ΔP_0 is the pressure difference in Pa when the water passes through the pipe; ΔP_{DR} is the pressure difference in Pa when the fracturing fluid passes through the pipe.

The drag-reduction experiment was tested three times at each displacement and the average value was taken. Then, the error was calculated, and error bars were added to the graph.

3. Results and Discussion

3.1. Experimental Evaluation of Scale Inhibitor

3.1.1. Effect of Temperature on Scale Inhibition

$CaSO_4$ scale-inhibition rates of six commercial scale inhibitors, when the solution pH is 7 and under different temperatures, are shown in Figure 2. All scale inhibitors had a scale-inhibition rate greater than 60% at room temperature. Among them, the scale-inhibition rate of SC-1, SC-2, and SC-5 was more than 80%, which had a stronger scale-inhibition ability of $CaSO_4$. With the increase in temperature, the scale-inhibition ability of these scale inhibitors all decreased. SC-1 and SC-5 had the best temperature resistance, and the decrease in scale-inhibition rate was within 20%. The scale-inhibition rate of SC-2 decreased from 87% to about 60% at 70 °C, and the temperature resistance was average. According to an information provided by the manufacturer, SC-1 is an organophosphate scale inhibitor, and it can chelate with metal ions, which can form a quadridentate stereo ring complex. This may be the reason why it is so little-affected by the temperature [30].

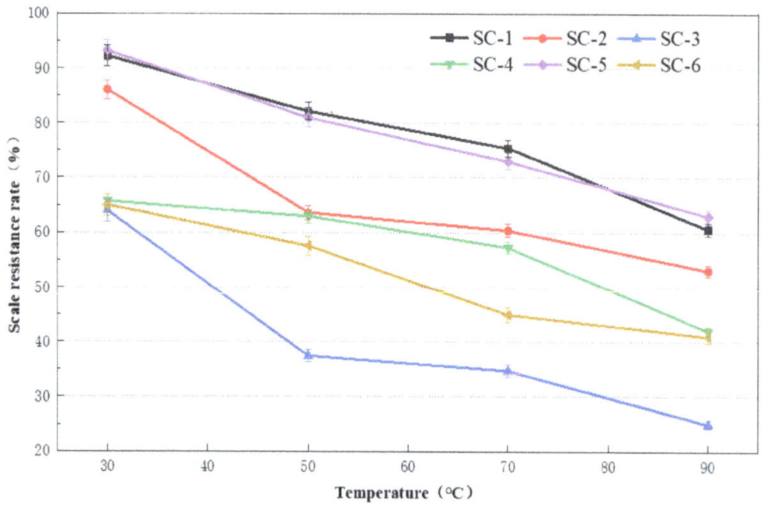

Figure 2. Effect of scale inhibitor on calcium sulfate at different temperature at pH of 7.

As shown in Figure 3, when the solution pH is 7, the scale-inhibition effect of $CaCO_3$ of six scale inhibitors under different temperature conditions is shown. The scale-inhibition rate of SC-1, SC-2, and SC-5 at room temperature was more than 70%. The scale-inhibition efficiency of SC-5 dropped sharply when the temperature was greater than 50 °C, and SC-5 was the monomer organophosphate scale inhibitor, which also makes it more susceptible to temperature effects. When the temperature rose to 90 °C, the scale-inhibition rate of SC-2 decreased to about 80%, which is lower than 82% of SC-1. Overall, SC-1 has good inhibition ability for both scales is good, and it also has better temperature stability. This is consistent with the strong chelating ability of organic phosphoric acid to calcium at different temperatures.

3.1.2. Effect of pH on Scale Inhibition

Figure 4 shows, when the temperature is 25 °C, the $CaSO_4$ scale-inhibition effect of six scale inhibitors under different pH levels. Among them, SC-1 and SC-2 have the scale-inhibition rate basically unchanged or even in an increasing trend. According to the manufacturer's information, SC-1 and SC-2 are organophosphate scale inhibitors, the chelating ability of organic phosphate for calcium ions is better under alkaline condition, they have good chemical stability, can be applied to a wide range of pH values, are not easily hydrolyzed, and have an obvious threshold value [13]. SC-5 was most seriously affected by alkalinity, and it decreased from 80% at pH 5 to 20% at pH 11. SC-5 is a polyaspartic acid scale inhibitor. With the increase in alkalinity, the scale-inhibition efficiency of polycarboxylic acid scale inhibitors was limited and weakened, which is similar to the trend reported in some literature [31].

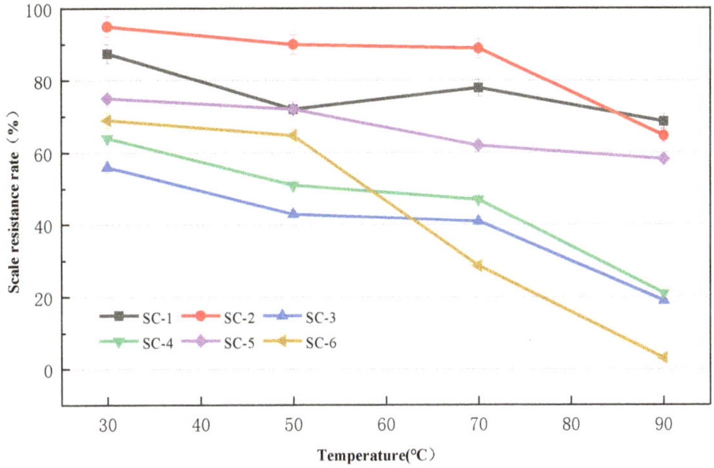

Figure 3. Effect of scale inhibitor on calcium carbonate at different temperatures at pH of 7.

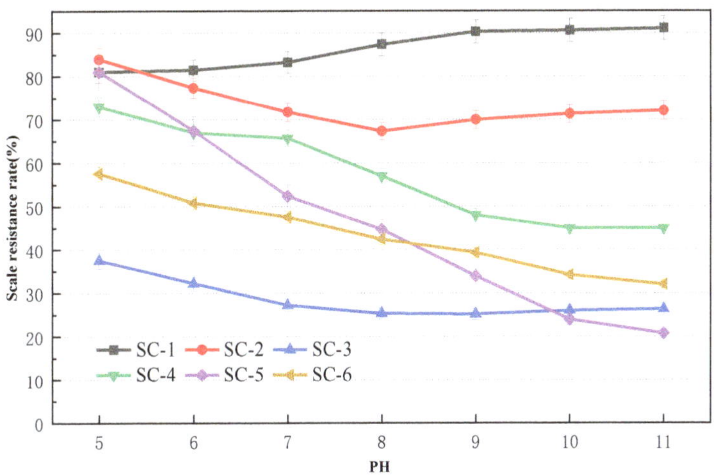

Figure 4. Effect of scale inhibitor on calcium sulfate at different pH levels and room temperature.

The $CaCO_3$ scale-inhibition effect of six scale inhibitors under different pH conditions is shown in Figure 5. In an alkaline environment, CO_3^{2-} does not hydrolyze to HCO_3^{-}, and its concentration in liquid phase is higher. Due to its low solubility product, it is easier to precipitate into scale when CO_3^{2-} and Ca^{2+} concentration is high. Therefore, the scale-inhibition effect of $CaCO_3$ is poor under alkaline environment. When the solution is weakly acidic with pH = 5, the scale-inhibition rate of SC-1 and SC-2 is larger than 75%. Although the scale-inhibition rate of SC-1 is slightly lower than that of SC-2 at lower pH, it can still maintain above 60% under strong alkaline environment. SC-1 is the most stable and strongest scale-inhibition effect compared with other five scale inhibitors. The results indicate that organic phosphoric acid has good chelating ability to calcium in both acidic and basic conditions.

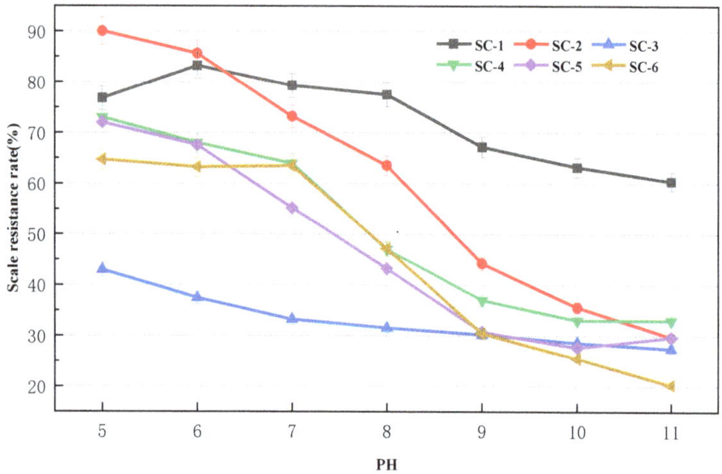

Figure 5. Effect of scale inhibitor on calcium carbonate at different pH levels and room temperature.

3.1.3. Influence of Scale-Inhibitor Concentration and Anti-Mixing Scale Evaluation Experiment

After a single scale-inhibition test, the scale inhibitors SC-1, SC-2, and SC-5 were selected, and their mixed scale-inhibition ability was further studied. The results are shown in Figure 6. Scale inhibitor SC-5 has the weakest scale-inhibition ability, and the scale-inhibition rate is only 42.8%. The scale-inhibition rate of SC-1 is 78.57% higher than that of SC-2. This is consistent with the results of a single scale-inhibition experiment. This suggests that SC-1 is preferred as the scale inhibitor for scale-inhibition fracturing fluid. Compatibility experiments are needed to verify its applicability.

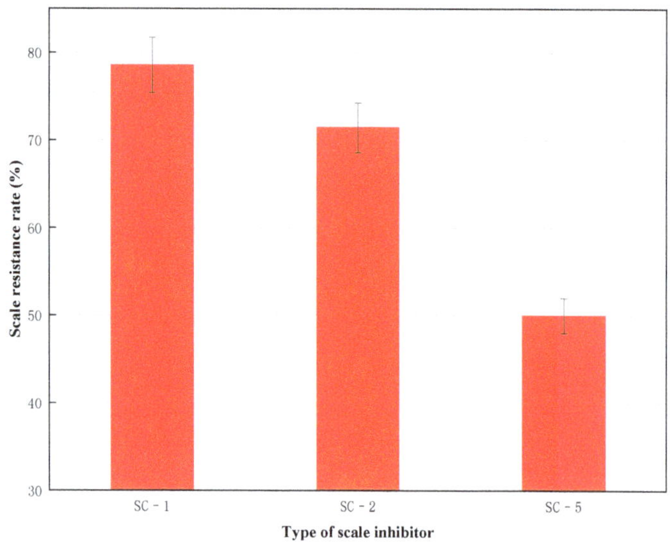

Figure 6. The mixed scale-inhibition ability of three optimal scale inhibitors at pH = 7 and room temperature.

As shown in Tables 3 and 4, the concentration of scale inhibitor SC-1 affected the scale-inhibition rate. The scale-inhibition abilities increase gradually with the scale-inhibitor concentration. However, the scale inhibitor has a threshold effect [32]. There is an economic dose achieving good scale-inhibition effect. It is generally believed that this dose can just complete chelation reaction. For $CaCO_3$, the optimal concentration of scale inhibitor is 0.07%, and the scale-inhibition rate can reach 77%. For $CaSO_4$, the optimal concentration of scale inhibitor is 0.3%, and the scale-inhibition rate is 87%. Therefore, if there is more $CaCO_3$ scale in the reservoir, the dose should be low, and if the $CaSO_4$ scale is more, the dose needs to be increased.

Table 3. Scale-inhibition rate for $CaCO_3$ of SC-1 at different concentrations.

Scale-Inhibitor Concentration (%)	**0.01%**	**0.03%**	**0.05%**	**0.07%**	**0.09%**
Scale-Inhibiting Rate (%)	70%	71%	73%	77%	78%

Table 4. Scale-inhibition rate for $CaSO_4$ of SC-1 at different concentrations.

Scale-Inhibitor Concentration (%)	**0.1%**	**0.2%**	**0.3%**	**0.5%**	**0.7%**
Scale-Inhibiting Rate (%)	65%	70%	87%	90%	91%

For $CaCO_3$, at room temperature, when the scale inhibitor is SC-1, the samples with the scale-inhibition rate under different pH conditions are counted, and the sample variance S_1^2 is calculated, and when the pH is 7, and the scale inhibitor is SC-1, the samples with the scale-inhibition rate at different temperatures are counted, and the sample variance S_2^2 is calculated. At room temperature, when pH = 7, the scale-inhibition rate samples of different scale inhibitors are counted, and the sample variance S_3^2 is calculated. S_1^2 ($CaCO_3$) = 0.0062, S_2^2 ($CaCO_3$) = 0.0043, and S_3^2 ($CaCO_3$) = 0.0219. It can be seen from the variance data that the type of scale inhibitor has the greatest impact on the scale-inhibition rate. Similarly, For $CaSO_4$, S_1^2($CaSO_3$) = 0.0016, S_2^2($CaSO_3$) = 0.0149, and S_3^2($CaSO_3$) = 0.0313, which indicates that the biggest impact on the scale-inhibitor rate is still the type of scale inhibitor.

3.2. *Compatibility Study*

As shown in Figure 7, the fracturing fluid after addition of SC-1 scale inhibitor at 90 °C has good compatibility, and no white flocculent or precipitation appeared in the sample bottle after 48 h of heating. Therefore, the scale inhibitor SC-1 can be added to the fracturing fluid system as an additive, and the autogenous acid SEG-C did not affect the compatibility of the system. However, a small number of fine bubbles appeared in the sample bottle after heating for 15 min. It is presumed that the autogenous acid generated acid which reacted with CO_3^{2-} in the fracturing fluid to produce CO_2. The autogenous acid can reduce the concentration of CO_3^{2-} in the solution, which would play an auxiliary scale-inhibition effect.

3.3. *Scale-Inhibiting Effect of Fracturing Fluid*

Figures 8 and 9 show the NMR T2 spectra of cores before and after scaling. Core b-1 applied conventional slickwater displacement, and b-2 applied scale-inhibiting effect of fracturing fluid. Scaling ions dominantly flow through the large pores, and the incompatible ions preferentially scale in the large pores of the propping zone. Although $CaCO_3$ does not completely block the pores after scaling, the scale crystals deposited and grew perpendicularly to the pore surface, resulting in a gradual decrease in the pore size of 10,000 nm large pores and an increase in the pore size of 10 nm small pores.

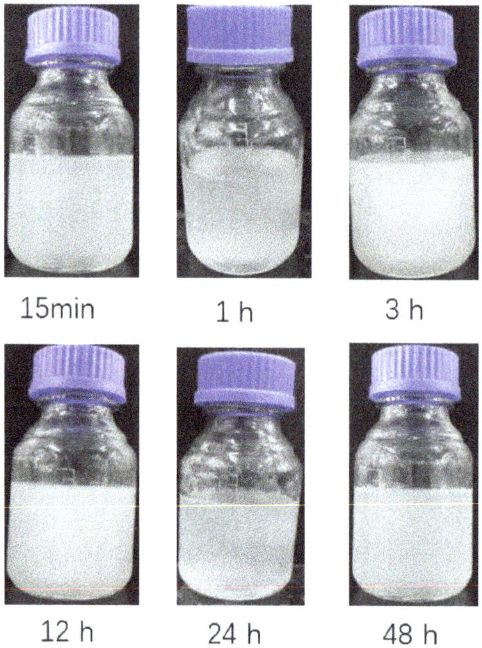

Figure 7. The compatibility of scale-inhibiting fracturing fluid system.

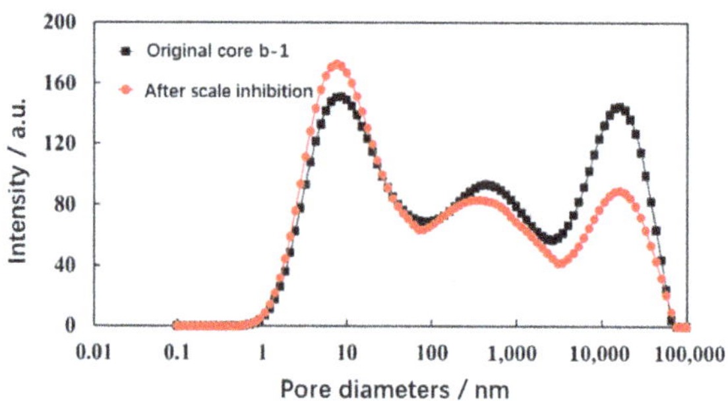

Figure 8. NMR curve of propped core b-1 before and after scaling without scale inhibitor.

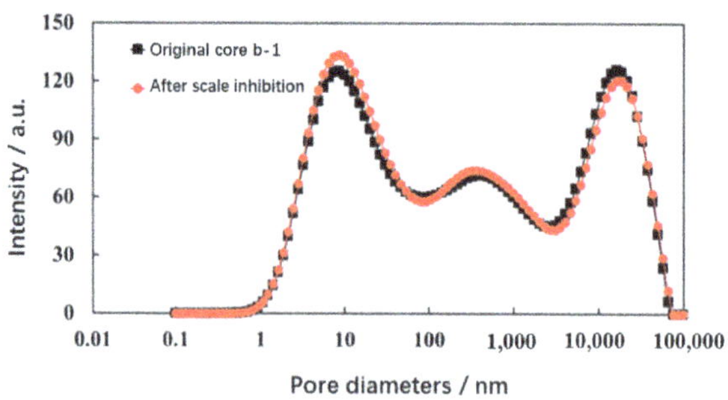

Figure 9. NMR curve of propped core b-2 before and after scaling with scale-inhibitor fracturing fluid.

After the scale-inhibiting fracturing fluid was injected into the core, the amount of Ca^{2+} and $CO_3{}^{2-}$ scaling was reduced by the chelating effect of the scale inhibitor SC-1.

In addition, a small amount of scaling in the pore space was destroyed by the lattice distortion effect [33]. Both the signal strength of 10 nm small pores and 1000 nm large pores were basically unchanged. This indicated that the scale-inhibiting fracturing fluid system effectively prevented the scaling of large pore spaces in the propping zone.

Table 5 gives the numerical results of damage rate and scale-inhibition rate. For conventional slick water, the damage rate is 31.16% after scaling. After adding the scale inhibition, the scaling damage is almost negligible. The amount of relaxation signal changes to 5864.28 from 5983.31, and the calculated scale-inhibition rate reaches 96.29%, which can effectively reduce scaling damage.

Table 5. Calculation of damage rate and scaling-inhibition rate by NMR.

	Integral Area of Core Relaxation Signal		Damage Rate/%	Scale-Inhibition Rate/%
	Original Core	Cores after Scaling		
Without scale inhibitor	10,299.63	7089.97	31.16	0
With scale inhibitor	5983.31	5864.28	1.99	96.29

3.4. Drag-Reduction Evaluation of Scale-Inhibiting Fracturing Fluid

Figure 10 shows the variation of drag reduction rate with displacement.The drag-reduction rate of conventional slickwater after adding scale inhibitor SC-1 can exceed 75%. This indicates that organic phosphate chelators do not affect the drag-reduction rate of slickwater fracturing fluid.

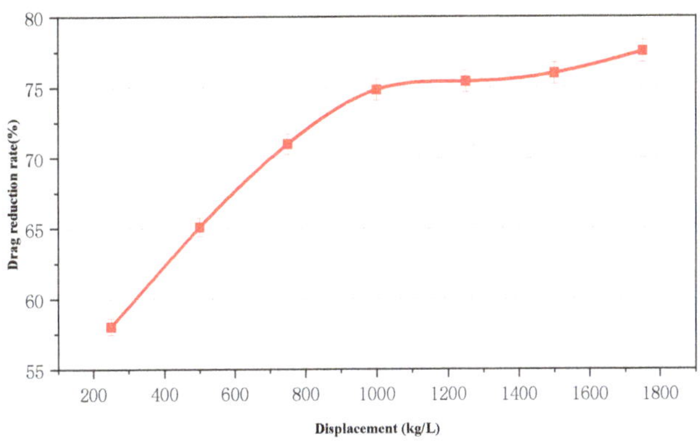

Figure 10. Drag-reduction test of scale-inhibiting fracturing fluid.

4. Conclusions

In this paper, six commercial scale inhibitors were systematically evaluated and optimized under different conditions of temperature, pH, and concentration. Then, the optimal scale inhibitor was combined with conventional slickwater fracturing fluid to produce a scale-inhibiting fracturing fluid. Finally, the compatibility, scale-inhibiting effect, and drag reduction of this new fracturing fluid were tested. The following conclusions were obtained:

(1) The scale-inhibition ability of six commercially available scale inhibitors was evaluated. Among them, organo-phosphonic acid scale inhibitor SC-1 performed well, and its scale-inhibition rate was 69% and 61% for $CaCO_3$ and $CaSO_4$ even at 90 °C. When the pH of the system is between 5 and 11, the scale-inhibition rate of SC-1 is stable around 80% for $CaSO_4$ and above 60% for $CaCO_3$.

(2) We tested the concentration of scale inhibitor for preventing CaCO3 scales, and the results showed that the most optimal concentration of the scale inhibitor was 0.07%. For inhibiting $CaSO_4$ scales, the optimal concentration of the scale inhibitor was 0.3%. Furthermore, for mixed scales, 0.3% of the scale inhibitor was also recommended. SC-1 has

optimal scale-inhibition ability for mixed scales, and the scale-inhibition rate for mixed scales can reach 78% at room temperature. Among them, the type of scale inhibitor is the most important factor affecting the scale-inhibition rate.

(3) After SC-1 was combined with the conventional fracturing fluid (0.08% drag-reducing agent DR800+0.3% SC-1+4% autogenous acid SEG-C+0.2% clay stabilizer), no white flocculent or precipitation appeared within 48 h. This indicates that the scale-inhibitor fracturing system is stable.

(4) The scale-inhibiting fracturing fluid system can effectively prevent scaling in the large pores of the core and the propping zone. Its scale-inhibition rate can reach 96.29% and the scale damage is negligible. In contrast, conventional slickwater has a 31.16% decrease in porosity.

(5) The scale-inhibiting fracturing fluid system maintains a good drag reduction, with a drag-reduction rate of 75.18% at a 1750 kg/h flow. There is no bad effect on properties of slickwater after adding the preferred scale inhibitor.

In summary, the scale-inhibiting fracturing fluid system described in the paper has good stability and scale-inhibition ability, both in different temperatures and pH environments. In addition, the new fluid maintains good compatibility and drag-reduction performance. At present, the technology can effectively solve the scaling problems of the propping zone in the fracturing process in reservoirs below 90 °C during the fracturing process, which has a positive effect on oil and gas production increase.

Author Contributions: Conceptualization, M.Z.; methodology, L.S. and E.Y.; validation, M.Z. and E.Y.; formal analysis, L.S.; investigation, H.R.; resources, H.R. and A.Y.; data curation, K.Z. and A.Y.; writing—original draft preparation, L.S.; writing—review and editing, E.Y. and L.S.; visualization, K.Z. and L.Z.; supervision, M.Z. and E.Y. All authors have read and agreed to the published version of the manuscript.

Funding: This research was financially supported by National Natural Science Foundation of China (Grant No. 52004306 and 52174045), the Strategic Cooperation Technology Projects of CNPC and CUPB (Grant No. ZLZX2020-01 and ZLZX2020-02) and the National Science and Technology Major Projects of China (Grant No. 2016ZX05030005 and 2016ZX05051003).

Data Availability Statement: The data will be available on request.

Conflicts of Interest: The authors declare no conflict of interest.

References

1. Wang, M.; Guo, Z.; Jiao, C.; Lu, S.; Li, J.; Xue, H.; Li, J.; Li, J.; Chen, G. Exploration progress and geochemical features of lacustrine shale oils in China. *J. Pet. Sci. Eng.* **2019**, *178*, 975–986. [CrossRef]
2. Qijun, H.; Haiqing, H.; Jianzhong, L.; Tao, Y. Recent progress and prospect of oil and gas exploration by PetroChina Company Limited. *China Pet. Explor.* **2018**, *23*, 1–13.
3. Wu, Q.; Xu, Y.; Wang, X.; Wang, T.; Zhang, S. Volume fracturing technology of unconventional reservoirs: Connotation, design optimization and implementation. *Pet. Explor. Dev.* **2012**, *39*, 377–384. [CrossRef]
4. Zhang, A.; Yang, Z.; Li, X.; Xia, D.; Zhang, Y.; Luo, Y.; He, Y.; Chen, T.; Zhao, X. An evaluation method of volume fracturing effects for vertical wells in low permeability reservoirs. *Pet. Explor. Dev.* **2020**, *47*, 441–448. [CrossRef]
5. Haghtalab, A.; Kamali, M.J.; Shahrabadi, A.; Golghanddashti, H. Investigation of the Precipitation of Calcium Sulfate in Porous Media: Experimental and Mathematical Modeling. *Chem. Eng. Commun.* **2015**, *202*, 1221–1230. [CrossRef]
6. Olajire, A.A. A review of oilfield scale management technology for oil and gas production. *J. Pet. Sci. Eng.* **2015**, *135*, 723–737. [CrossRef]
7. El-Said, M.; Ramzi, M.; Abdel-Moghny, T. Analysis of oilfield waters by ion chromatography to determine the composition of scale deposition. *Desalination* **2009**, *249*, 748–756. [CrossRef]
8. Kamal, M.S.; Hussein, I.; Mahmoud, M.; Sultan, A.S.; Saad, M.A.S. Oilfield scale formation and chemical removal: A review. *J. Pet. Sci. Eng.* **2018**, *171*, 127–139. [CrossRef]
9. Li, M. *Synthesis and Performance Evaluation of High-Efficiency Scale Inhibitors for Oilfields*; Ocean University of China: Qingdao, China, 2011.
10. Jing, G.; Tang, S.; Li, X.; Wang, H. The analysis of scaling mechanism for water-injection pipe columns in the Daqing Oilfield. *Arab. J. Chem.* **2017**, *10*, S1235–S1239. [CrossRef]
11. Liu, X.; Jungang, L.; Qianya, Z.; Jinlai, F.; Yingli, L.; Jingxin, S. The analysis and prediction of scale accumulation for water-injection pipelines in the Daqing Oilfield. *J. Pet. Sci. Eng.* **2009**, *66*, 161–164. [CrossRef]

12. Fan, C.; Shi, W.; Zhang, P.; Lu, H.; Zhang, N.; Work, S.; Al-Saiari, H.A.; Kan, A.T.; Tomson, M.B. Ultra-HTHP Scale Control for Deepwater Oil and Gas Production. In Proceedings of the SPE International Symposium on Oilfield Chemistry, Galveston, TX, USA, 11–13 April 2011.
13. Li, J.; Tang, M.; Ye, Z.; Chen, L.; Zhou, Y. Scale formation and control in oil and gas fields: A review. *J. Dispers. Sci. Technol.* **2017**, *38*, 661–670. [CrossRef]
14. Xiong, W.; Gill, M.; Moore, J.; Crandall, D.; Hakala, J.A.; Lopano, C. Influence of Reactive Flow Conditions on Barite Scaling in Marcellus Shale during Stimulation and Shut-In Periods of Hydraulic Fracturing. *Energy Fuels* **2020**, *34*, 13625–13635. [CrossRef]
15. Paukert Vankeuren, A.N.; Hakala, J.A.; Jarvis, K.; Moore, J.E. Mineral Reactions in Shale Gas Reservoirs: Barite Scale Formation from Reusing Produced Water As Hydraulic Fracturing Fluid. *Environ. Sci. Technol.* **2017**, *51*, 9391–9402. [CrossRef]
16. Ju, X. *Research on Multi-Layer Water Mixing Injection and Anti-Scaling Technology in Dingbian Oil Production Plant*; Xi'an Shiyou University: Xi'an, China, 2018.
17. Peng, Y.; Yue, Z.; Ozuruigbo, C.Q.; Fan, C. Carbonate Scale Control under High Level of Dissolved Iron and Calcium in the Bakken Formation. In Proceedings of the SPE International Symposium on Oilfield Chemistry, The Woodlands, TX, USA, 14 April 2015.
18. BinMerdhah, A.B.; Yassin, A.A.M.; Muherei, M.A. Laboratory and prediction of barium sulfate scaling at high-barium formation water. *J. Pet. Sci. Eng.* **2010**, *70*, 79–88. [CrossRef]
19. Cui, C.; Zhang, S. Preparation, Characterization and Performance Evaluation of a Novel Scale Inhibiting and Dispersing Copolymer Containing Natural Tannin. *J. Polym. Environ.* **2020**, *28*, 1869–1879. [CrossRef]
20. Amjad, Z.; Hooley, J. Influence of polyelectrolytes on the crystal growth of calcium sulfate dihydrate. *J. Colloid Interface Sci.* **1986**, *111*, 496–503. [CrossRef]
21. Wylde, J.J.; Mahmoudkhani, A. Development of a Scale Inhibitor for Zr-Crosslinked Seawater Systems: A Case History of Successful Testing to Failure and Field Applications. In Proceedings of the SPE International Oilfield Scale Conference and Exhibition, Aberdeen, UK, 25–26 May 2016.
22. Jing, Z.; Shao, X.; Gao, Y. Development of scale-resistant recyclable fracturing fluid systems. In Proceedings of the International Field Exploration and Development Conference 2021, Chengdu, China, 23–25 September 2021; pp. 288–294.
23. Senthilmurugan, B.; Ghosh, B.; Sanker, S. High performance maleic acid based oil well scale inhibitors—Development and comparative evaluation. *J. Ind. Eng. Chem.* **2011**, *17*, 415–420. [CrossRef]
24. Wang, X.-L.; Zhang, C.; Ouyang, P. The possibility of separating saccharides from a NaCl solution by using nanofiltration in diafiltration mode. *J. Membr. Sci.* **2002**, *204*, 271–281. [CrossRef]
25. Vazquez, O.; Herrero, P.; Mackay, E.; Jordan, M. Non-aqueous vs aqueous overflush scale inhibitor squeeze treatment in an oilfield offshore Norway. *J. Pet. Sci. Eng.* **2016**, *138*, 1–10. [CrossRef]
26. Liu, L.; Lan, X.; Li, N.; Luo, Z.; Liu, S. Study on the mechanism of acid pretreatment to reduce injury in hydraulic fracturing. *J. Southwest Pet. Univ. (Sci. Technol. Ed.)* **2016**, *38*, 150–155.
27. Li, L.; Wang, Y.; Sun, Y.; Yang, W.; Yin, X.; Chen, Y.; Liu, Y. Novel and Green Hydroxyperylene Imide Based Fluorescent Polymer for Calcium Sulfate Scale Inhibition. *J. Mol. Liq.* **2021**, *344*, 117730. [CrossRef]
28. Zhang, Z.; Lu, M.; Liu, J.; Chen, H.; Chen, Q.; Wang, B. Fluorescent-Tagged Hyper-Branched Polyester for Inhibition of CaSO4 Scale and the Scale Inhibition Mechanism. *Mater. Today Commun.* **2020**, *25*, 101359. [CrossRef]
29. Sun, Y.; Wei, L.n.; Dai, C.; Yu, Z.; Xin, Y. The carbonic acid-rock reaction in feldspar/dolomite-rich tightsand and its impact on CO2-water relative permeability during geological carbon storage. *Chem. Geol.* **2021**, *584*, 120527. [CrossRef]
30. Moayedi, H.; Aghel, B.; Vaferi, B.; Foong, L.K.; Bui, D.T. The feasibility of Levenberg–Marquardt algorithm combined with imperialist competitive computational method predicting drag reduction in crude oil pipelines. *J. Pet. Sci. Eng.* **2020**, *185*, 106634. [CrossRef]
31. Kiaei, Z.; Haghtalab, A. Experimental study of using Ca-DTPMP nanoparticles in inhibition of $CaCO_3$ scaling in a bulk water process. *Desalination* **2014**, *338*, 84–92. [CrossRef]
32. Cui, K.; Li, C.; Yao, B.; Yang, F.; Sun, G. Synthesis and evaluation of an environment-friendly terpolymer CaCO3 scale inhibitor for oilfield produced water with better salt and temperature resistance. *J. Appl. Polym. Sci.* **2019**, *137*, 48460. [CrossRef]
33. Mpelwa, M.; Tang, S.-F. State of the art of synthetic threshold scale inhibitors for mineral scaling in the petroleum industry: A review. *Pet. Sci.* **2019**, *16*, 830–849. [CrossRef]

processes

Article

Numerical Simulation Study on the Flow Properties of Materials for Plugging While Drilling in MWD

Lei Pu [1,2,3], Peng Xu [1,2,3,*], Mingbiao Xu [1,2,*], Jun Zhou [4], Qinglin Liu [1] and Jianjian Song [1,2,3]

1 School of Petroleum Engineering, Yangtze University, Wuhan 430100, China
2 Cooperative Innovation Center of Unconventional Oil and Gas, Yangtze University (Ministry of Education & Hubei Province), Wuhan 430100, China
3 Key Laboratory of Drilling and Production Engineering for Oil and Gas, Wuhan 430100, China
4 PetroChina Xinjiang Oilfield Development Company, Karamay 834000, China
* Correspondence: cdxupeng@yangtzeu.edu.cn (P.X.); xumingbiao@yangtzeu.edu.cn (M.X.)

Abstract: The method of plugging while drilling has been one of the commonly used methods to control formation loss during drilling. The damage to materials for plugging while drilling to MWD has become a complex problem. For many years, field engineers had insufficient knowledge of the passing performance of materials for plugging while drilling in measurement while drilling (MWD). In the existing research, the blocking mechanism of materials for plugging while drilling to mud screen during the flow process is still unclear. In this study, we use computational fluid dynamics coupled with discrete element method (CFD–DEM) to investigate materials' plugging mechanism while drilling. The results show that the migration process of lost circulation materials (LCMs) in the mud screen can be divided into three stages, displacement, retention, and accumulation of LCMs. The blocking mechanism of LCMs on the mud screen comes from two aspects. One is from the bridging of LCMs with larger particle size in the holes of the mud screen. Another source is the difference between the entry speed and the overflow speed of LCMs. The particle size and mass fraction of LCMs and the viscosity and displacement of the fluid affect the flow properties of LCMs from these two factors, respectively.

Keywords: mud screen; materials for plugging while drilling; numerical simulation; CFD–DEM; plugging

Citation: Pu, L.; Xu, P.; Xu, M.; Zhou, J.; Liu, Q.; Song, J. Numerical Simulation Study on the Flow Properties of Materials for Plugging While Drilling in MWD. *Processes* **2022**, *10*, 1955. https://doi.org/10.3390/pr10101955

Academic Editors: Linhua Pan, Yushi Zou, Jie Wang, Minghui Li, Wei Feng and Lufeng Zhang

Received: 21 August 2022
Accepted: 14 September 2022
Published: 28 September 2022

1. Introduction

As the oil and gas industry gradually develops into more challenging and deeper formations, this presents more significant challenges for drilling operations [1,2]. Deep formations are prone to feature narrow pressure windows and significant pressure gradient changes, and these unstable factors often lead to the loss of drilling fluids [3–5]. In order to deal with the formation leakage problem quickly and effectively, the MWD technology is widely used. Loss-while-drilling refers to the addition of circulating lost circulation materials (LCMs) to the drilling fluid. It can plug the entire well section while drilling and is a practical section to prevent complex bottom hole conditions [6]. It is especially effective in controlling the leakage of the formation with more and less clear leakage. Compared with the traditional plugging method, the measurement while drilling (MWD) method does not require lifting and unloading the MWD, significantly reducing the field operation time [7].

Measurement while drilling (MWD) is a while drilling tool included in most oil wells drilled today [8]. It is a system developed to take drilling-related measurements downhole and transmit the information to the surface while drilling. The channels they provide for drilling fluids are limited and much smaller than the inner diameter of the drill pipe [9]. Periodic movement of the valve within the tool restricts the flow of drilling fluid through the inner diameter of the tool. Debris in the drilling fluid can cause damage to the tools inside the tool, such as the pulse generator, because of the narrow flow paths [10]. Materials for plugging while drilling pose challenges to today's measurement while drilling (MWD) [11]. Improper selection of materials for plugging while drilling can bring the risk of plugging in MWD [12]. Therefore, to prevent this kind of damage, a mud screen is installed on the top of the MWD tool and connected to the drill pipe, as shown in Figure 1. This method filters out large particles in the mud that could harm the tool. A mud screen should be used whenever the circulation of the drilling fluid begins to prevent debris in the drilling fluid from flowing from the drill pipe into the tool, blocking the mud pulse generator or the drill bit jet [13].

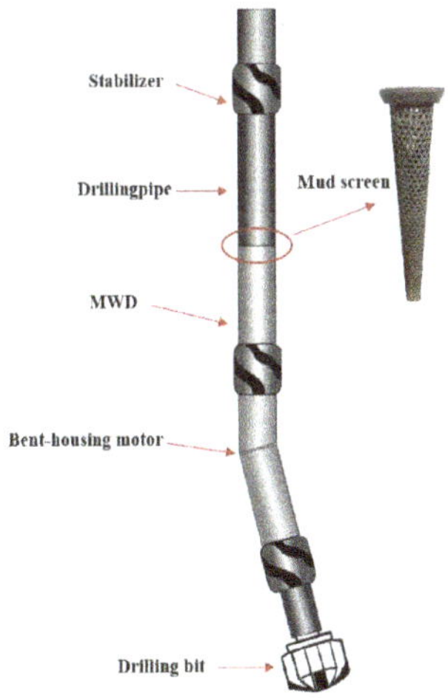

Figure 1. The mud screen placed in the MWD tool.

At present, there are few studies on the circulation of materials for plugging while drilling in MWD tools. Only a few conference papers have suggested that using LCMs for MWD poses certain risks. Hussain S et al. (2014) proposed that the concentration of LCMs is harmful to MWD [14]. Attong D J et al. (1995) suggested that MWD is highly sensitive to LCMs, especially for high concentrations of LCMs [15]. Valerio A et al. (2020) argue that the inability to know the flow-through properties of LCMs from MWD makes operators less confident in pumping large particle size or high concentrations of LCMs [10]. Klotz C et al. (2008) proposed that in a mud pulsator, adjusting the angle of the rotor entering the flow region can help LCMs to obtain larger flow channels [16]. However, such measures are still limited. At present, the operators of drilling rigs are not aware of the logic of LCMs blocking MWD, especially the mud screen. Regarding the optimal injection rate of LCM, the choice of the critical concentration of LCMs and the maximum particle size of the material is ambiguous [17]. Drilling engineers have eagerly raised this demand in the South China Sea, where leakage accidents are frequent [18].

The flow and particle properties of the drilling fluid impact the passage performance of the plugging slurry from the MWD. The flow of LCMs mixed with drilling fluid in the tool can be considered particle flow [19–21]. At present, particle flow is mainly studied by numerical simulation, most of which are coupled by the computational fluid dynamics–discrete element method (CFD–DEM). It is an Eulerian–Lagrangian method that analyzes complex problems in fluid mechanics and structural mechanics of continuum (fluids) and discontinuous media (particles) and fields. By tracking and calculating the microscopic motion between fluid, particle, and boundary in each unit, the overall flow law of LCMs in MWD is obtained. Casas G et al. (2022) discussed the flow properties of particles and fluids within the drill bit using numerical techniques of the CFD–DEM method [22]. Feng C et al. (2022) investigated the process of particles and fluids expelled from the bottom of the bit and impacted the bit body through numerical simulations and field experiments [23]. They also studied the key factors in the erosion of polycrystalline diamond compact (PDC) bits. The transport of drill cuttings in drilling operations is also a typical particle flow. Especially in horizontal wells and deviated wells, Zakerian A et al. (2018) studied cuttings migration parameters by establishing a CFD–DEM model [24]. Lin C et al. (2022) tracked the migration behavior of each particle in the fracture using numerical simulation, analyzed the sealing of the fracture by the particles in stages, and explained the evolution process of the sealing layer in the fracture [25]. The method of numerical simulation is beneficial to the understanding of the flow law between particles and fluid.

In fact, in order to ensure the safety of MWD, the selection of materials for plugging while drilling by on-site drilling engineers is often conservative. This stems from the ambiguity about the flow properties of LCMs in MWD tools. In this paper, combining the computational dynamics model with the discrete element method, the downhole mud screen tool is used as the object, and the CFD–DEM model is established simultaneously. This model simulates the entire process of the Materials for plugging while drilling from the drill string to the mud screen tool. The parameters such as the appropriate viscosity of the fluid, the injection speed of LCMs, the critical concentration of LCMs, and the maximum particle size of LCMs are further clarified. The boundary condition of materials for plugging while drilling on mud screen tool plugging is helpful for drilling engineers to expand the choice of MWD system. This facilitates dealing with a broader range of missed incidents.

2. Methods

Based on the coupling of computational fluid dynamics and discrete elements, a 3D model identical to the actual tool was established to study the flow properties of LCMs in the mud screen, in which the shape of the LCMs was set as spherical. The drilling fluid (continuous phase) is considered incompressible based on the Lagrangian method. While LCMs are considered to be the dispersed phase. The solution for the continuous phase is determined by using the Navier–Stokes (RANS) equations. The dispersed phase is solved by tracking the particles through the flow field, and each particle is numerically integrated through Newton's equations governing the translation and rotation of the particle and is individually tracked along the continuous phase by the forces acting on the particle [26].

2.1. Governing Equations of the Particle Model

2.1.1. Particle Motion

The particle is tracked based on the DEM method in Lagrangian coordinates, the velocity and acceleration of the particle are calculated based on Newton's second law, and the translation and rotation of the particle are controlled [27,28]. The governing equations are as follows:

$$m_p \frac{dv_p}{dt} = F_C + F_{f \to p} + m_p g \tag{1}$$

$$H_p \frac{dw_p}{dt} = T_C + T_{f \to p} \tag{2}$$

where m_p is the particle mass (kg), g is the gravitational acceleration (m/s^2), v_p is the particle translation velocity (m/s), and F_C is the particle–particle and particle–wall interaction force (N). $F_{f \to p}$ is the interaction force between the fluid surrounding the particle and the particle. H_p is the moment of inertia (kg·m^2), and w_p is the angular velocity of the particle (rad/s). T_C is the particle–particle and particle–wall tangential force (N) that causes particle rotation. $T_{f \to p}$ is the additional torque (kg·m^2·s^{-2}) due to the velocity gradient of the fluid.

2.1.2. Contact Force Models

In the dispersed phase, the force is mainly generated by the contact between the solid phase and the solid phase, including two particles or a particle and a wall.

Normal Force

The contact force between particle–particle and particle–wall consists of stress in the normal and tangential directions. The normal contact force model is specified by the Hertzian spring–dashpot model [29]. The Hertzian spring–dashpot model used can be written as:

$$F_n = K_H S_n^{\frac{3}{2}} + \xi_H S_n^{\frac{1}{4}} S_n \quad (3)$$

$$K_H = \frac{4}{3} E^* \sqrt{R^*} \quad (4)$$

$$\frac{1}{E^*} = \frac{1 - v_1^2}{E_1} + \frac{1 - v_2^2}{E_2} \quad (5)$$

$$\frac{1}{R^*} = \begin{cases} \frac{1}{L_1} + \frac{1}{L_2} \ for\ particle - particle\ collision \\ \frac{1}{L} \ for\ particle - boundary\ collision \end{cases} \quad (6)$$

Following the idea proposed by Tsuji Y et al. [30], the definition of the damping coefficient for the Hertz model is similar to that of the linear spring–damper model:

$$\xi_H = 2\eta_H \sqrt{m^* K_H} \quad (7)$$

Tangential Force

The model of the tangential force of the contact force uses the Mindlin–Deresiewicz model, which is called the linear friction damper model, which includes springs, dampers, and friction blocks, as shown in Figure 2. The model combines Hertz's theory to describe normal stress and Mindlin and Deresiewicz's theory [31]. The expression for the tangential force is given by:

$$F_\tau = -\mu F_n \left(1 - \zeta^{\frac{3}{2}}\right) \frac{s_\tau}{|s_\tau|} + \eta_\tau \sqrt{\frac{6\mu m^* F_n}{s_\tau, max}} \zeta^{\frac{1}{4}} s\prime_\tau \quad (8)$$

$$\zeta = 1 - \frac{\min(|s_\tau|, max)}{s_\tau, max} \quad (9)$$

where μ is the friction coefficient; F_n is the normal stress; and s_τ is the tangential displacement at the contact point. $s\prime_\tau$ is the tangential component of the relative velocity at the contact point; s_τ, max is the maximum displacement of the particle slipping at the contact point. m^* is the effective mass; η_τ is the damping ratio, which is related to the restitution coefficient of the material.

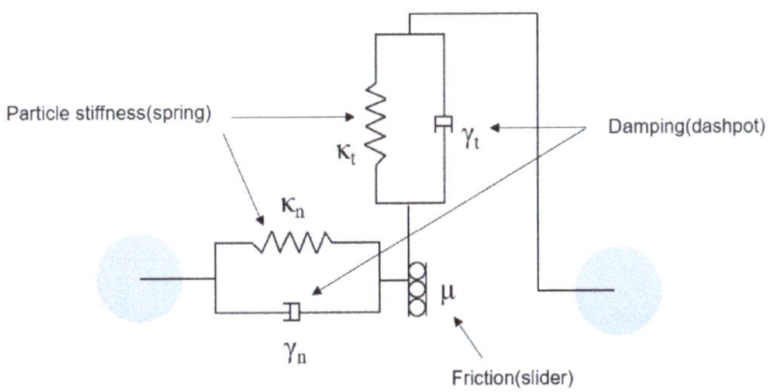

Figure 2. Spring–slider–dashpot model for Hertzian contact model [32].

2.2. *Governing Equations of the Fluid Model*

In the CFD–DEM coupled simulation, using the Multiphase Model, when using the Eulerian method, the fluid phase is regarded as a continuous phase, and the fluid phase is an incompressible fluid, which is described by the volume-averaged classical Navier–Stokes equation [33]. Its mean mass conservation equation and mean momentum conservation equation are given by Equations (10) and (11), respectively:

$$\frac{\partial}{\partial t}\left(\alpha_f\rho_f\right)+\nabla\cdot\left(\alpha_f\rho_f u\right)=0 \tag{10}$$

$$\frac{\partial}{\partial t}\left(\alpha_f\rho_f\right)+\nabla\cdot\left(\alpha_f\rho_f u\right)=-\alpha_f\nabla p+\nabla\cdot\left(\alpha_f T_f\right)+\alpha_f\rho_f g+F_{p\to f} \tag{11}$$

$$T_f=\mu f\left(\nabla_u+\nabla_u{}^T\right)+\left(\lambda_f-\frac{2}{3}\mu f\right)\nabla\cdot uI \tag{12}$$

$$F_{P\to f}=-\frac{\sum_{P=1}^{N}F_{f\to p}}{V_c} \tag{13}$$

where α_f represents the fluid volume fraction, p is the pressure, ρ_f is the fluid density, u is the fluid phase velocity, and T_f is the stress tensor of the fluid phase; the stress tensor of T_f in Equation (11) is defined by Equation (12);

In Equation (11), $F_{P\to f}$ represents the momentum source term of the interaction between particles and the fluid, which is defined according to Equation (13), where V_c is the calculation unit volume, and n is the particle in the calculation unit volume. The number, $F_{P\to f}$ describes the force the fluid exerts on the particles.

2.3. *Particle-Fluid Interaction Forces*

The interaction force $F_{f\to p}$ between the fluid and the particle is usually divided into two terms: the drag force F_D and the second term consisting of the residual (nonresistance) force F_{N-D}, in the following way [34]:

$$F_{f\to p}=F_D+F_{N-D} \tag{14}$$

The most common non-drag forces are the pressure gradient force $F_{\nabla p}$, the additional (virtual) mass force F_{VM}, and the lift force F_L, so we can also write:

$$F_{f\to p}=F_D+F_{\nabla p}+\left(F_L+F_{VM}+F_{others}\right) \tag{15}$$

In fact, depending on the flow conditions, most of these forces are negligible, and only drag and pressure gradient forces need to be considered since, in most cases, the density of particles is much greater than that of the fluid, where the drag force F_D is expressed according to Equation (16),

$$F_D=\frac{1}{2}C_D\rho_f A\left|u-v_p\right|\left(u-v_p\right) \tag{16}$$

In Formula (16), A is the projected area of the particle in the direction of the fluid; $u-v_p$ is the velocity difference between the fluid and the particle. Where the pressure gradient force $F_{\nabla p}$ is calculated according to Equation (13):

$$F_{\nabla p}=-V_p\nabla_p \tag{17}$$

where V_p is the volume of the particle and ∇_p is the local pressure.

2.4. *Coupling Algorithm*

Based on the Euler–Lagrangian framework, Figure 3 shows a diagram of the two-way coupling of CFD–DEM. It allows CFD and DEM to exchange information for the transfer of momentum and energy. First, determine the time step in the DEM solver. On the CFD solver, the flow velocity, pressure, density, and viscosity data of each grid in the computational domain are transferred to the DEM solver, and the interaction between the discrete phase and the continuous phase is calculated on the DEM solver, and the discrete phase is obtained. The speed of translation and rotation and the data are transferred to CFD for a solution again. CFD determines that the time step is an integer multiple of the DEM time step, and the initial discrete phase field in CFD is updated. The schematic diagram of CFD–DEM coupling technology is shown in Figure 3.

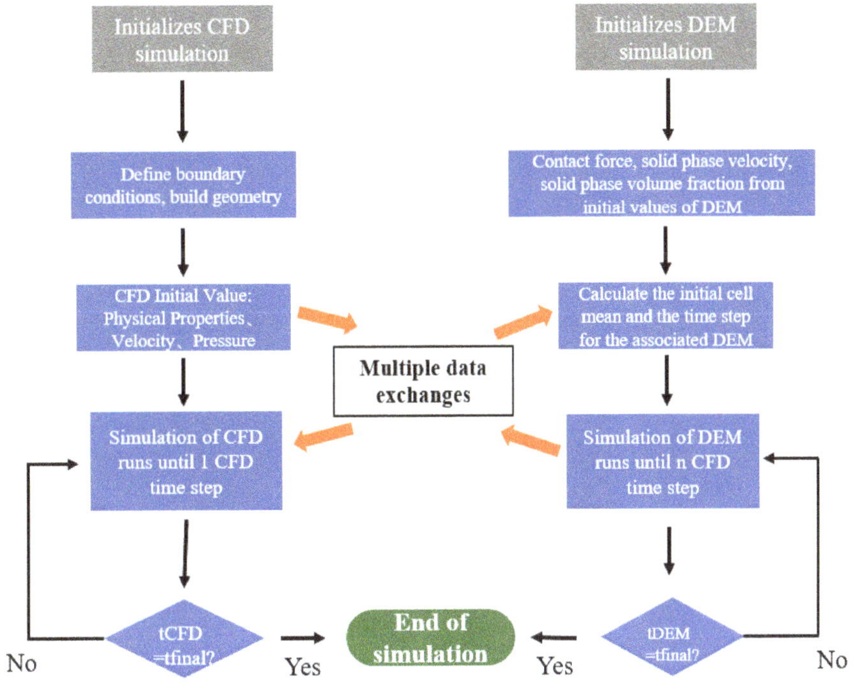

Figure 3. Schematic diagram of CFD–DEM coupling technique.

3. Numerical Model

Since the mud screen is a cone, the drilling fluid carrying LCMs flows out of the tip first, and when the particles carried by the drilling fluid are blocked at the tip, they begin to flow out from the holes on both sides. Under the action of a fluid with a specific flow rate and viscosity, the particles are not enough to block the tool, which means that the plugging slurry can safely pass through the tool. If the LCMs are trapped in the screen, the amount of LCMs coming out of the holes is less than the amount of LCMs entering the screen, and the LCMs will gradually fill the entire screen. This will cause the entire steering-while-drilling tool to be blocked, which will bring safety risks to on-site drilling construction.

3.1. Architecture of the Model

Drilling fluid enters the tool from the inlet and flows out at the screen. The particles and fluid are in the cylinder part, forming a steady flow of particles. Figure 4a depicts the process of simulating materials for plugging while drilling from the drill pipe into the mud screen tool at a steady flow rate. The geometric model established in this study consists of two parts, the stable flow area of the plugging slurry and the screen structure. The boundary conditions of the entire geometric model are shown in Figure 4b. The stable flow area of the plugging slurry is a cylindrical structure with a cross-section of 20 mm in diameter and a length of 100 mm, providing a stable flow area for the plugging slurry, simulating the upper half of the steering-while-drilling tool. Particles are randomly generated on the cylindrical structure's top face and carried into the cylindrical structure by the fluid.

The particle concentration was adjusted by the particle formation rate, and the volume fraction ranged from 7.5% to 20%. The fluid inlet also sets the top of the circular structure. The fluid velocity at the inlet is set in the range of 6.6 m/s~10 m/s to simulate the actual displacement of drilling fluid during drilling, ranging from 30 m^3/h to 50 m^3/h.

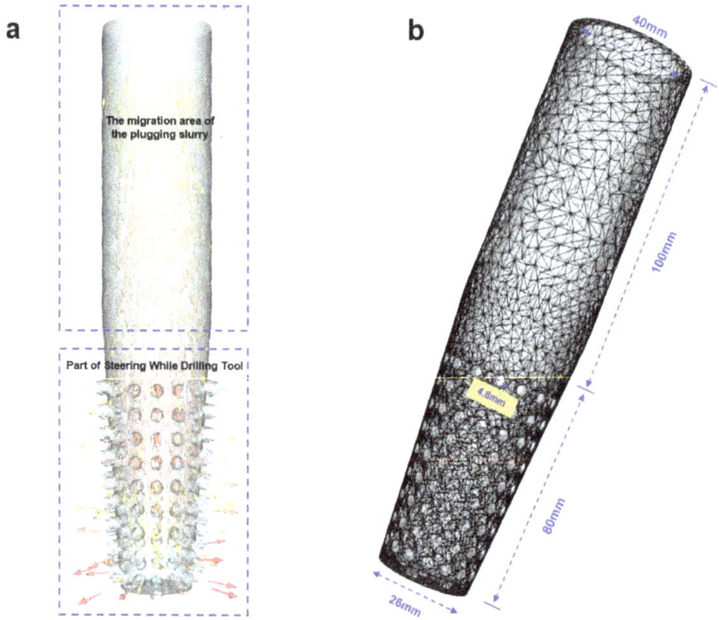

Figure 4. (**a**): Slurry migration from the mud screen tool. (**b**): Mud screen tool size and grid.

The diameter of the circle at the upper end of the mud screen tool is 20 mm, which is consistent with the cross-sectional diameter of the cylindrical structure. The diameter of the circle at the lower end is 26 mm. The overall height is 80 mm, and the cone angle is −8.58°. The conical part contains 144 holes with a diameter of 4.8 mm. The scale of the model is 1:1 with the size of the actual tool. The dimension of the entire computational domain is 0.0446 m × 0.1805 m × 0.0446 m, with a total of 54,891 boundary faces and 61,478 boundary nodes. Based on issues such as the computational stability of the solver, the accuracy of the drag force calculations, etc., the size of the mesh is kept at more than three times the diameter of the largest particle being simulated. Therefore, the maximum size of the unit is 6×10^{-3} m, and the minimum size is set to 3×10^{-3} m. The walls are set as no-slip boundary conditions.

3.2. Setting of Boundary Conditions

The inlet boundary condition of the fluid field is set to a specific velocity inlet, the outlet boundary condition is set to the pressure outlet, and the outlet pressure is standard atmospheric pressure to simulate the actual situation of drilling tools in the formation. Select the standard k-epsilon(2eqn) for the turbulence model of the fluid. For the wall function of the fluid field, select the near-wall treatment of the scalable wall function. Based on the collection and calculation of fluid parameters in the plugging operation at the drilling site, the drilling fluid in the drilling tool mostly flows in the form of laminar flow. Therefore, according to the calculated value, the turbulence intensity of the fluid is set to 1%. When leakage occurs in the formation, the leakage rate is lower than 10 m^3/h, and the method of plugging while drilling is adopted. The concentration of the MWD agent is generally controlled at 5% to 25%, and the particle size is less than 2 mm. During the drilling process, the displacement of drilling fluid is different under different working conditions. This study divides it into three types: low (55 m^3/h), medium (90 m^3/h), and high (125 m^3/h), type of displacement. Other parameter settings are the same as the actual field parameters, see Table 1 for details.

Table 1. Input parameters of DEM-CFD numerical simulation.

DEM Parameters	Values	CFD Parameters	Values
Particle Size, d (m)	$[1.1 \times 10^{-3}, 2.2 \times 10^{-3}]$	The density of the fluid, ρ_f (g/cm^3)	1.3
Particle Density, ρ_s (g/cm^3)	2.6	The viscosity of the fluid, μ_f (kg/m·s)	[0.01, 0.07]
Particle Young's Modulus, E (MPa)	1.0×10^9	Displacement of fluid, m^3/h	[55, 125]
Poisson's ratio of particles	0.3	Turbulence intensity	k-epsilon(2eqn)
Static Friction of Particle to particle, μ_s	0.8	Wall roughness constant	1%
Dynamic Friction of Particle to particle, μ_r	0.5	Wall roughness height, h (mm)	0.5
Restitution Coefficient of Particle to particle	0.3	CFD time step, Δts (s)	0.2
Static Friction of Particle to boundary, μ_s	0.9	Acceleration of gravity, g (m/s^2)	10^{-3}
Dynamic Friction of Particle to boundary, μ_r	0.5		
Restitution Coefficient of Particle to the boundary	0.2		
Tangential Stiffness Ratio	1		
DEM time step, Δts (s)	5×10^{-3}		

3.3. Validation of the Model

An experimental device for testing the flow properties of Materials for plugging while drilling in the mud screen tool was established indoors. The experimental device consists of three parts: pump, simulated wellbore, and simulated mud screen, as shown in Figure 5. The construction of the simulated mud screen is derived from the tip part of the actual mud screen. Due to the natural drilling fluid circulation environment, the tip of the mud screen is the most sensitive position for LCMs. According to feedback from field operations on the rig, whether or not it is plugged by LCMs determines whether or not the entire tool is plugged by LCMs. Therefore, considering the feasibility of indoor experiments, the most easily blocked area of the mud screen is used as the experimental model. Because the mud screen is placed inside the drill pipe close to the MWD in the design of the drilling tool, the mud screen for indoor simulation is also placed in the cylinder, which simulates the drill pipe.

Therefore, the dynamic viscosity of the drilling fluid in the room is 0.03 kg/m·s, and the density is 1.3 g/cm^3. The materials for plugging while drilling select regular spherical particles, which are as consistent as possible with the shape and properties of the LCMs used in the field. The drilling fluid is driven by a screw pump and the number of materials for plugging while drilling is set to 5%. Due to the limited performance of the indoor screw pump, the pumping displacement of the drilling fluid was set to 5 m^3/h. Under this condition, the flow properties test was carried out on materials for plugging while drilling with different particle sizes, and the particle size ranged from 0.5 mm to 1.5 mm.

Figure 6 shows the simulation and laboratory results of the flowability of materials for plugging while drilling. The simulation predictions agree well with the indoor results on a larger scale. However, when the particle size of LCMs is 0.9 mm, there is a specific difference between the two results. This is because the LCMs in the laboratory experiments do not flow in a uniform state during the cycle. The internal structure of the pump and the influence of gravity lead to the accumulation of LCMs during the circulation process, resulting in an excessively high concentration of some structures, thus making the experimental results more prone to tool blockage. Overall, the developed numerical model has high accuracy and can be used to simulate the fluid-solid flow of materials for plugging while drilling in the mud screen.

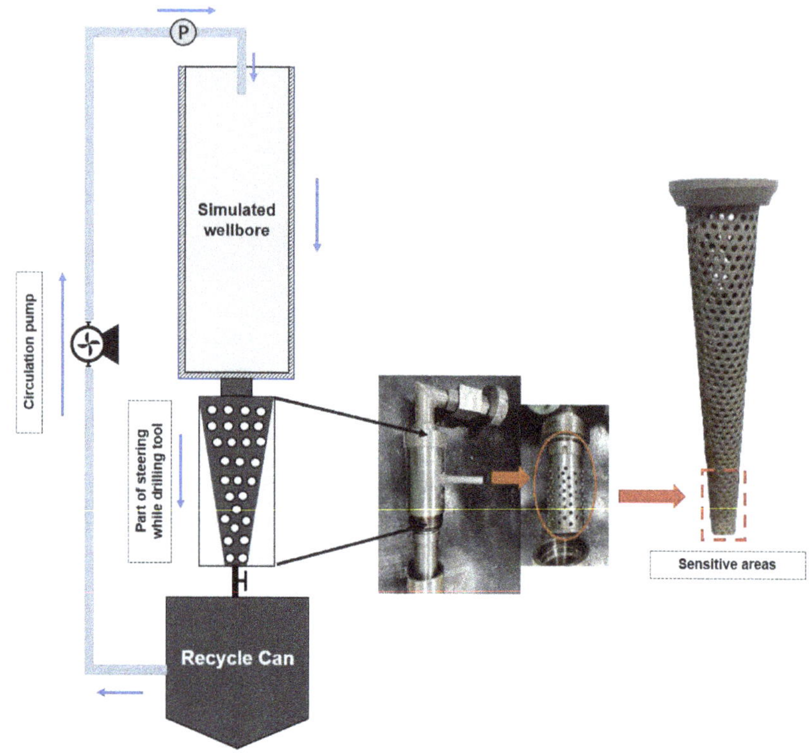

Figure 5. The experimental device for the blocked safety test of the mud screen tool.

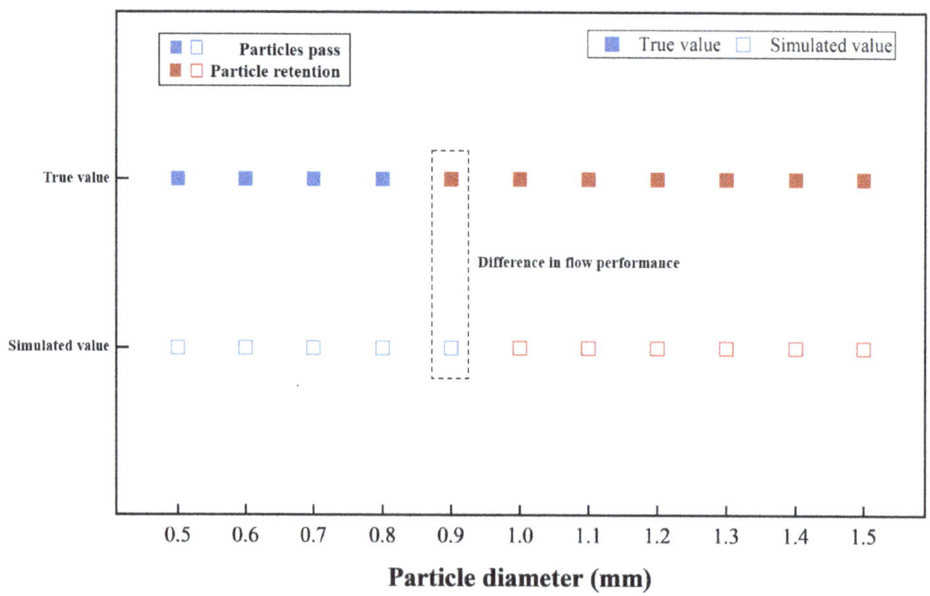

Figure 6. Flow properties of materials for plugging while drilling with different particle size in mud screen.

Figures 7 and 8 show the different results of the flow properties of the materials for plugging while drilling in the simulated mud screen in the laboratory experiment. It is evident that when the LCMs fill the entire tool, drilling fluid cannot flow through it. It is conceivable that once the LCMs clog the tool in actual drilling situations, this will be a challenging problem. The problem can only be solved once the drilling is stopped, the drill pipe is lifted, and the MWD is taken out, which is time-consuming.

Figure 7. The mud screen tool is blocked by LCMs.

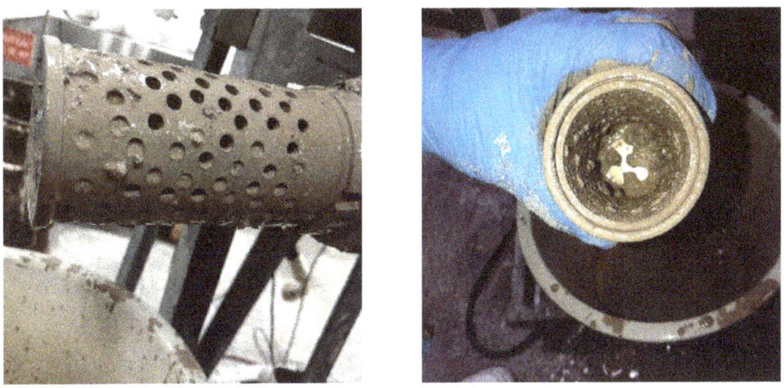

Figure 8. LCMs pass smoothly through the mud screen tool.

4. Results and Discussion

4.1. Circulation Process of LCMs in Mud Screen

It is necessary to understand how the displacement and retention of LCMs behave in the mud screen. The simulation results of the migration of LCMs in the mud screen are shown in Figure 9. The results show that the LCMs migrate smoothly from the drill pipe with a gradual increase in velocity. When the LCMs migrate into the mud screen, the speed decreases. Part of the LCMs overflowed from the holes of the mud screen, and the remaining part of the LCMs remained at the tip of the mud screen. As the number of LCMs entering the mud screen from the drill pipe gradually increased, it was much larger than the number of LCMs overflowing from the hole. At this point, the blockage develops from the tip portion of the mud screen to the entire mud screen [35]. This clogging develops rapidly. If you want to avoid clogging, you should try to control the particles and not stay at the mud screen's tip.

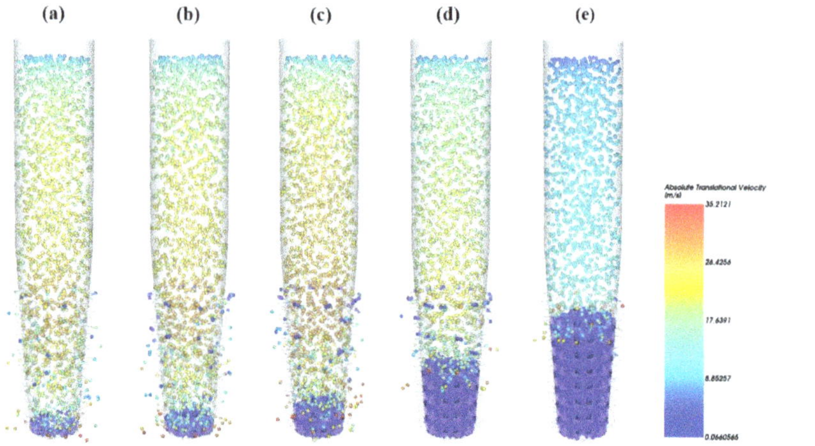

Figure 9. Displacement and retention behavior of LCMs in mud screen, (**a**–**e**): behavior changes with mud injection time.

Accordingly, the whole process can be divided into three stages, as shown in Figure 10. I, Migration of LCMS, referring to Figure 9a,b. II, LCMS retention, refer to Figure 9c,d. III, LCMS stacking, refers to Figure 9e. In the first stage, LCMS is carried by the fluid from the drill pipe to the mud screen. At this time, the entering particles are the same as the overflowing particles, and no particles block the opening of the mud screen, so the pressure of the fluid is maintained at a superficial level. In the second stage, the hole of the mud screen is gradually blocked by the remaining particles, and the passage of the fluid is blocked, resulting in an increase in the fluid pressure. In the third stage, as time increases, all the openings of the mud screen are blocked by LCMS, the main overflow channel of fluid disappears, and it can only overflow from the pores of LCMS particles. The fluid field Is restored to a new equilibrium, and the fluid pressure remains relatively stable without significant changes.

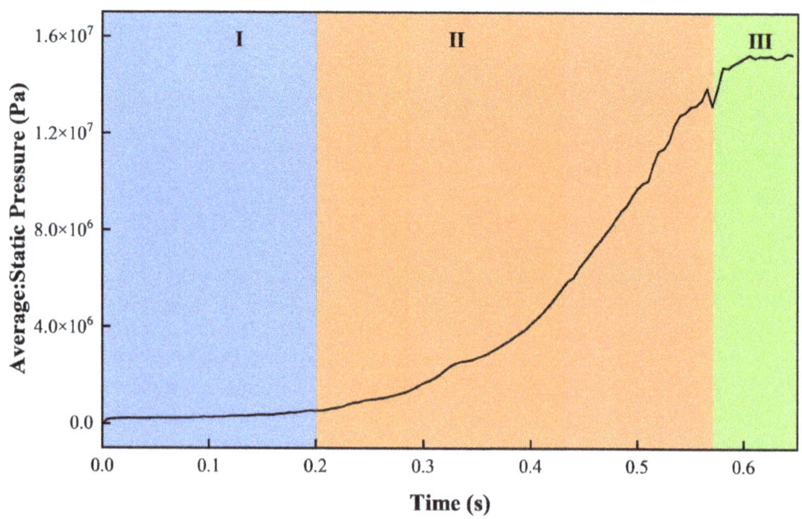

Figure 10. The pressure change of the fluid during the process of blocking the mud screen by LCMs.

4.2. Effect of Particle Size of LCMs on Flow Properties

In order to intuitively understand the effect of particle size of LCMs on the flow properties in the mud screen. The particle size range of LCMs is set at 1.1 mm~2.2 mm, and the range of mass flow is set at 1~6 kg/s. The flow-through results under different particle size ranges and mass flow conditions were tested. The differences in the occurrence time of LCMs with different particle sizes after plugging the tool were compared. As well as the particle size of LCMs, the behavior of LCMs clogging tools is classified.

The larger the particle size of the LCMs, the greater the blocking probability of the mud screen tool. According to the results given in Figure 11, when the particle size is more significant than 2.0 mm, the tool will be blocked regardless of the mass flow size pumped by the LCMs. It can therefore be concluded that for MWD, LCMs over 2.0 mm should be prohibited from being added to drilling fluids. It can cause damage to the drilling tool. At the same time, when the particle size of the particles is less than 1.2 mm, the flow properties of LCMs in the tool are good, and we think that the LCMs below 1.2 mm are in the safe range. When the particle size is between 1.2 mm and 2.0 mm, the mass concentration of LCMs needs to be considered, which is necessary to ensure the tool's safety.

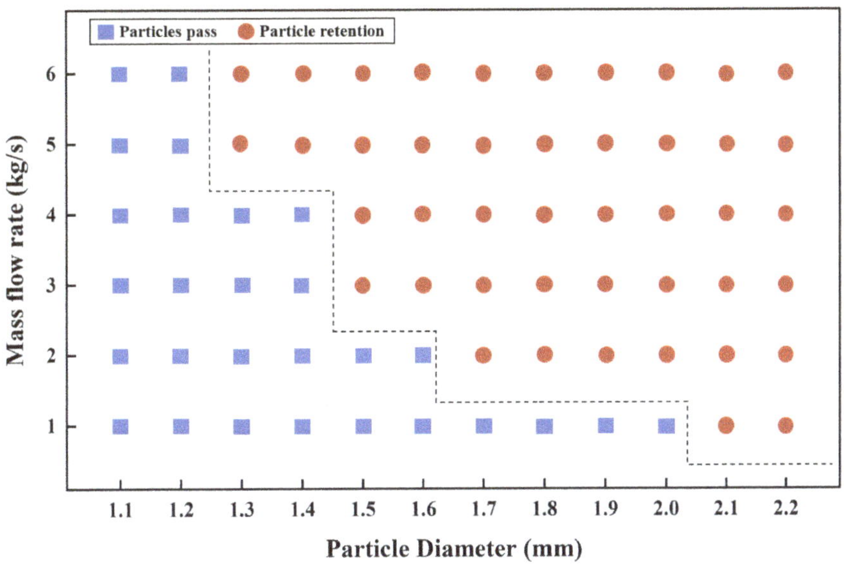

Figure 11. Effect of particle size and mass flow of LCMs on flow properties.

In Figure 12, at a mass flow of 3 kg/s, the increase in particle size accelerated the process of tool plugging by LCMs. The larger the LCMs, the faster the entire tool is filled. Figure 13 explains the difference between the two types of particle size in filling the entire tool. When the particle size of LCMs is 1.5 mm, the tool is filled with LCMs because the speed of LCMs entering the tool is greater than the speed of overflowing the tool. It can be seen that although the LCMs have blocked the entire mud screen, the small LCMs still overflow from the punched holes. Its overflow curve is also wavy. When the particle size of LCMs was 2.0 mm, the LCMs bridged in the pores, and the subsequent LCMs could not overflow from the pores. It can be seen from the overflow curve of the particles that when the number of particles overflowed reaches a peak value, it decreases rapidly. Through the study of these two different types of plugging behaviors, we learned that expanding the overflow rate of LCMs is the key to improving the flow-through performance of LCMs.

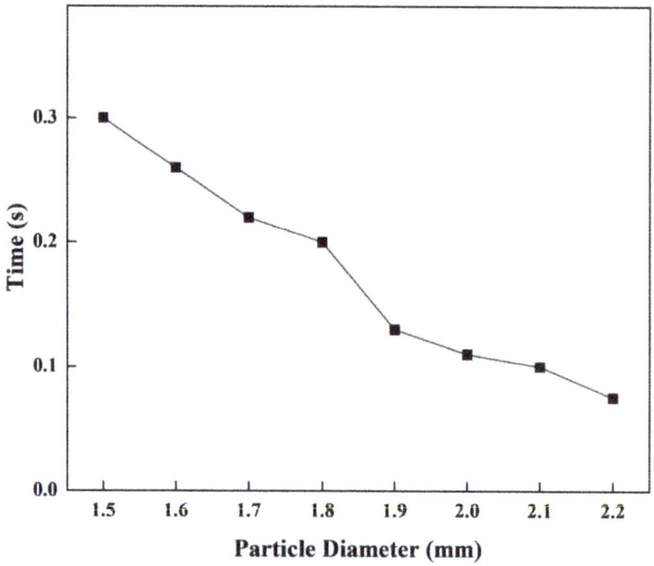

Figure 12. Effect of particle size of LCMs on the plugging time under the condition of mass flow of 3 kg/s.

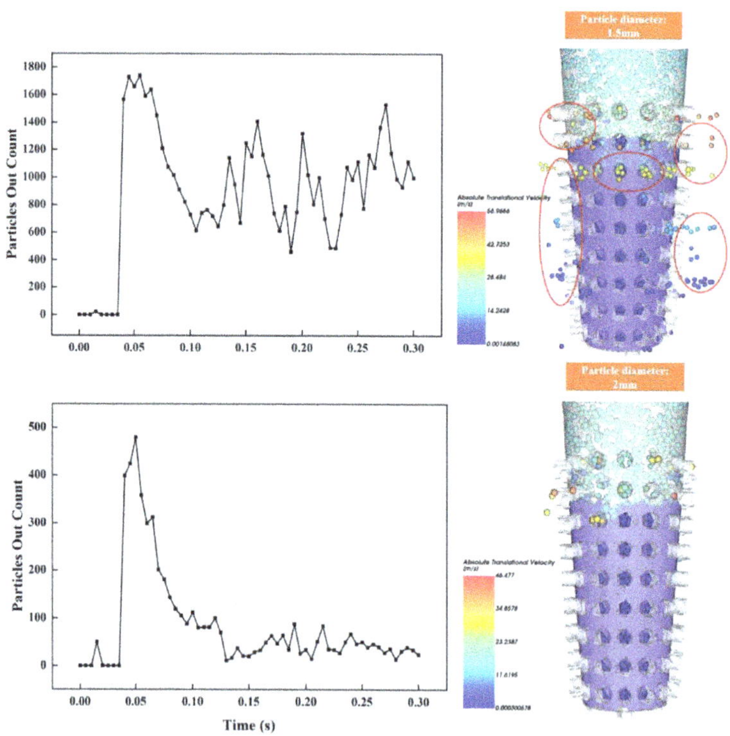

Figure 13. Behavior of LCMs with different particle size to plug the tool.

4.3. Influence of Mass Flow of LCMS on Flow Properties

As shown in Figure 14, the lower the mass flow range, the lower the probability of tool clogging by LCMs. It shows that controlling the entry speed of LCMs to keep them consistent with the speed of overflowing LCMs is beneficial to the circulation of LCMs in the mud screen. At the same time, the high mass flow makes it easier for particles to build bridges in the pores, hindering the migration of particles. Overall, LCMs are better able to ensure the safety of MWD while maintaining a lower mass flow. In the actual operation of the drilling platform, when using LCMs with larger particle size, it is necessary to ensure that the concentration of LCMs is maintained in a lower range.

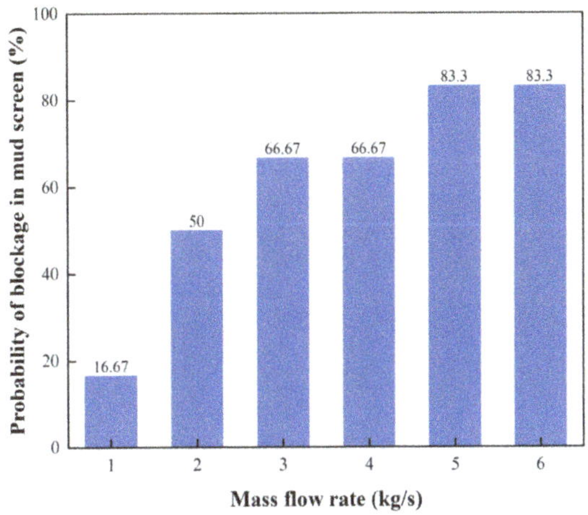

Figure 14. Influence of mass flow of LCMs on the probability of blocking in mud screen.

It can be seen from the comparison between Figures 15 and 16 that the entry velocity and the overflow velocity of LCMs with a mass flow of 2 kg/s are maintained in an equilibrium state. Therefore, LCMs maintain good liquidity. When the mass flow of LCMs rises to 3 kg/s, the entry velocity of LCMs is greater than the overflow velocity of LCMs. Meanwhile, the high concentration of LCMs had a side effect on the spillover behavior. Therefore, LCMs build up rapidly in the tool, filling the entire tool.

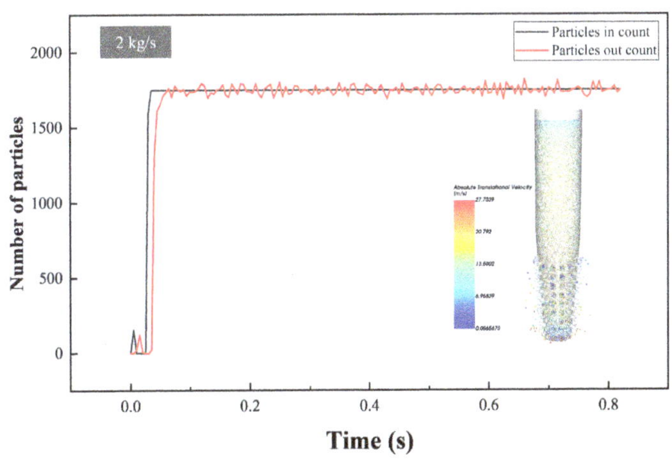

Figure 15. 1.6 mm LCMs entering at 2 kg/s mass flow.

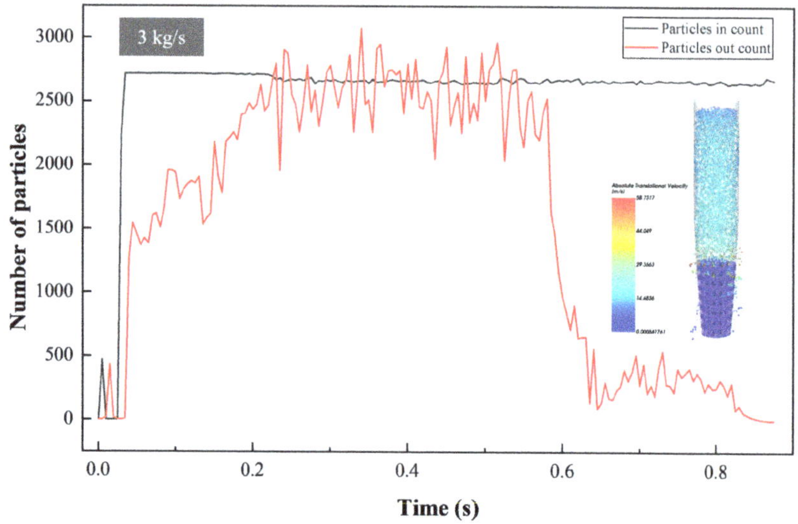

Figure 16. 1.6 mm LCMs entering at 3 kg/s mass flow.

4.4. Influence of Fluid Displacement on Flow Properties

Considering the actual working conditions in the drilling process, the displacement of the drilling fluid is different. Therefore, according to the actual parameters, the drilling fluid displacement is divided into three levels: low, medium, and high. The influence of liquid velocity was revealed by comparing the flow properties of materials for plugging while drilling under three types of displacement. Figure 17 shows the displacement velocity of LCMs in the whole tool under different displacements of fluid. It can be clearly seen that the overflow rate of LCMs increases with the increase of fluid displacement.

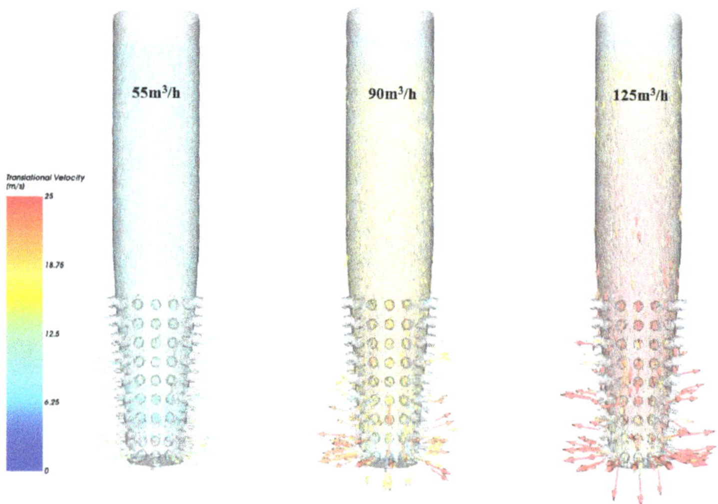

Figure 17. Velocity vector diagrams of LCMs under the action of low, medium and high displacement fluids.

Under the condition that the mass flow of LCMs is 3 kg/s, it can be found that the fluid displacement has a significant influence on the results. In Figure 18, the parameters for which the effect of fluid displacement changes the fluidity of LCMs are circled by boxes. Low displacement mainly affects the flow properties of LCMs.

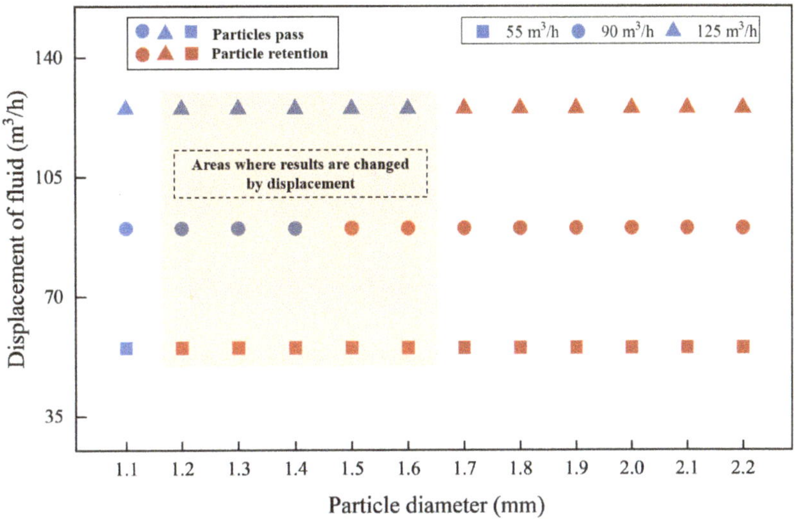

Figure 18. Flow properties of LCMs under the action of the low, medium, and high displacement fluids.

Figures 19 and 20 reflect the effect of different displacements on LCMs. The reduction of the fluid displacement will reduce the overflow speed of the particles, resulting in the deterioration of the flow properties. The drag force on the LCMs is also affected by the displacement. The larger the displacement, the greater the drag force on the LCMs. As shown in Figure 20, with a displacement of 55 m^3/h, the drag force on the LCMs suddenly increased in the subsequent period. This is because the LCMs clog the tool and reduce the fluid flow channel area, which increases the fluid velocity around the LCMs.

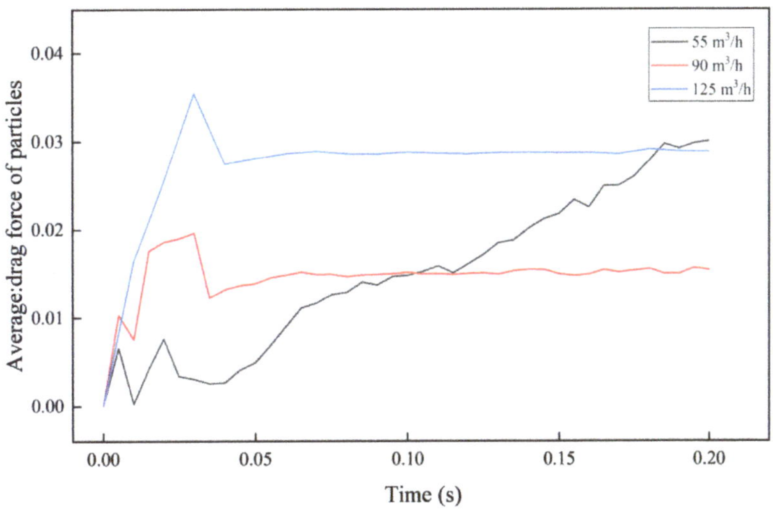

Figure 19. The magnitude of the pulling force of 1.4 mm LCMs at different displacements.

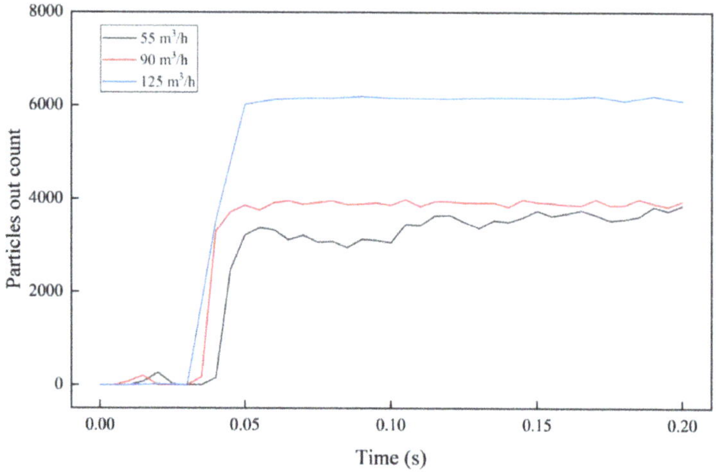

Figure 20. Spill velocity of 1.4 mm LCMs at different displacements.

4.5. Influence of Fluid Viscosity on Flow Properties

As the primary performance of drilling fluid, viscosity cannot be ignored. Drilling fluid relies on its own viscosity to suspend and carry solid phase materials. Therefore, it has a specific influence on the flow properties of LCMs.

The flow properties of LCMs with two particle size under the action of different fluid viscosities were tested. The particle size of LCMs is 1.4 mm and 1.5 mm, which are the boundary conditions for blocking the mud screen or not, respectively. Figure 21 shows that the increase in fluid viscosity effectively reduces the collision of LCMs with the tool. This weakens LCMs from being hindered by tools. This enables the LCMs to leave the mud screen in a more orderly manner from the pores, which improves the overflow rate of the LCMs. As seen in Figure 22, the increase in fluid viscosity caused the LCMs to shift from clogging the tool to a gentle spill. This shows that the fluid's viscosity can change whether the LCMs pass or not to a certain extent. In the process of plugging while drilling, the viscosity of the drilling fluid should be increased within a safe range, which is beneficial to the tool's safety.

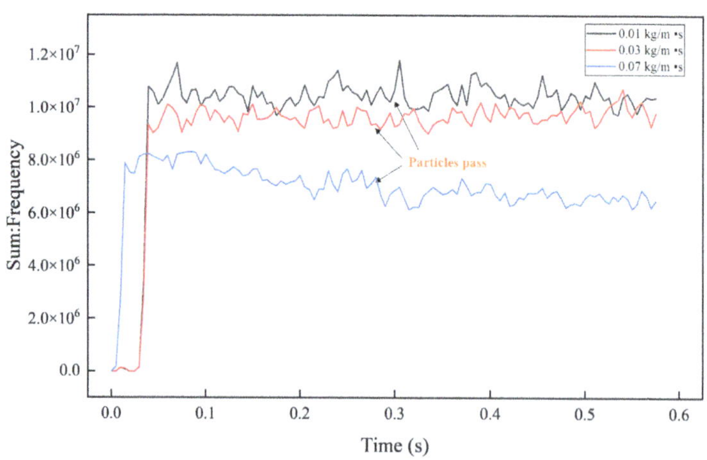

Figure 21. Flow properties of 1.4 mm LCMs under the action of drilling fluids with different viscosities.

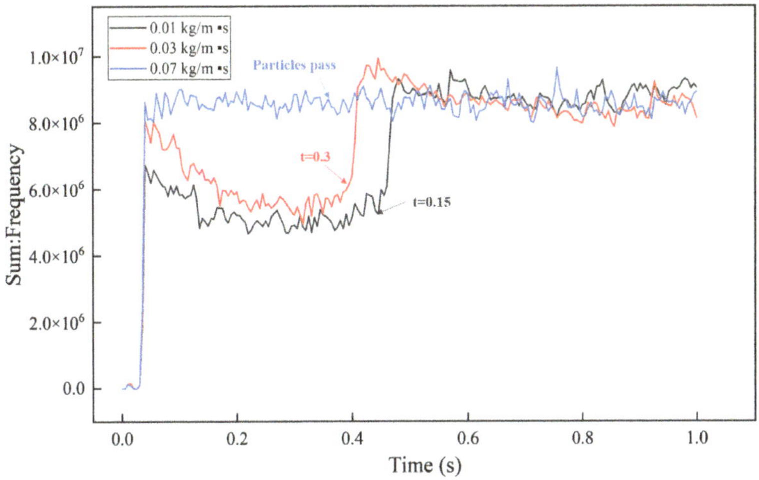

Figure 22. Flow properties of 1.5 mm LCMs under the action of drilling fluids with different viscosities.

5. Conclusions

In this study, the flow-through performance of LCMs in mud screen was investigated by using a coupled CFD–DEM method to build a scale model with the actual tool. Concluded as follow:

(1) The migration process of LCMs in the mud screen can be divided into three stages, including I, the migration of LCMS. II, Retention of LCMS. III, LCMS stacking. Among them, the retention stage of LCM dominates the clogging degree of LCMs to mud screen.
(2) The retention behavior of LCMs can be divided into two types. One is derived from the bridging of LCMs with larger particle size in the pores of the mud screen, thereby hindering the subsequent overflow of LCMs. Another source is that the difference between the entry speed and the overflow speed of LCMs causes the accumulation of LCMs in the mud screen.
(3) Mud screen requires that the particle size of LCMs needs to be controlled at 2 mm or less. When the particle size of LCMs is less than 1.2 mm, the results of the mass flow convection performance of LCMs do not change much. When the particle size of LCMs ranges from 1.2 mm to 2 mm, the mass flow of LCMs needs to be paid attention to because it dramatically affects the flow properties of LCMs.
(4) The flow properties of LCMs are favorable for fluid at high displacement. It helps LCMs achieve higher overflow speed. Low displacement results in more severe clogging of the mud screen by LCMs. The increase of fluid viscosity reduces the collision frequency of LCMs with the tool, and the overflow of LCMs from the holes is more orderly. Therefore, in the MWD construction, optimizing the viscosity and displacement of the drilling fluid can reduce the probability of the mud screen being blocked.

Author Contributions: Conceptualization, L.P. and M.X.; methodology, P.X. and M.X.; software, J.S.; validation, L.P., J.S. and M.X.; formal analysis, L.P.; investigation, P.X. and Q.L.; resources, M.X. and J.Z.; data curation, L.P.; writing—original draft preparation, L.P.; writing—review and editing, P.X.; visualization, L.P.; supervision, M.X.; project administration, P.X.; funding acquisition, M.X. All authors have read and agreed to the published version of the manuscript.

Funding: This research was supported by the Open Foundation of Cooperative Innovation Center of Unconventional Oil and Gas, Yangtze University (Ministry of Education & Hubei Province), No. UOGBX2022-02 and UOG2022-02.

Acknowledgments: We would like to thank The Bakken Laboratory of YANGTZE University for their support. At the same time, I would like to thank my teacher Mingbiao Xu for his help.

Conflicts of Interest: The authors declare no conflict of interest.

References

1. Sun, J.; Bai, Y.; Cheng, R.; Lyu, K.; Liu, F.; Feng, J.; Lei, S.; Zhang, J.; Hao, H. Research progress and prospect of plugging technologies for fractured formation with severe lost circulation. *Pet. Explor. Dev.* **2021**, *48*, 732–743. [CrossRef]
2. Pu, L.; Xu, P.; Xu, M.; Song, J.; He, M. Lost circulation materials for deep and ultra-deep wells: A review. *J. Pet. Sci. Eng.* **2022**, *214*, 110404. [CrossRef]
3. Yan, C.; Deng, J.; Yu, B.; Li, W.; Chen, Z.; Hu, L.; Li, Y. Borehole stability in high-temperature formations. *Rock Mech. Rock Eng.* **2014**, *47*, 2199–2209. [CrossRef]
4. Meng, M.; Chen, P.; Ren, R. Statistic evaluation of failure criteria in wellbore stability with temperature effects. *Fuel* **2019**, *252*, 730–752. [CrossRef]
5. Feng, Y.; Cai, W.; Deng, J.; Lin, H.; Xie, T.; Chu, M. Application of geomechanics modeling in the study of fluid loss mechanism of fractured reservoirs in Bohai Bay, China. *J. Pet. Sci. Eng.* **2022**, *212*, 110297. [CrossRef]
6. Izyurov, V.; Kharitonov, A.; Semenikhin, I.; Korsunov, E.; Gassan, A.; Tikhonov, E.; Jadan, G.; Stashko, V.; Blagonadeshniy, I.; Manikhin, A.; et al. Selecting Bridging Agents' Particle Size Distribution for Optimum Plugging While Drilling in Permeable Zones. In Proceedings of the SPE Russian Petroleum Technology Conference, Moscow, Russia, 22 October 2019; OnePetro: Richardson, TX, USA, 2019.
7. Zhai, K.; Yi, H.; Liu, Y.; Geng, Y.; Fan, S.; Zhu, D. Experimental evaluation of the shielded temporary plugging system composed of calcium carbonate and acid-soluble preformed particle gels (ASPPG) for petroleum drilling. *Energy Fuels* **2020**, *34*, 14023–14033. [CrossRef]
8. Lu, C.; Zhang, T.; Zhao, H. Improving the application depth of electromagnetic measurement while drilling (EM-MWD) systems by receiving signals from adjacent wells. *J. Appl. Geophys.* **2021**, *195*, 104468. [CrossRef]
9. Shinmoto, Y. A case study on downhole measurement to monitor drilling hazards in integrated ocean drilling program (IODP) expedition 314. In Proceedings of the OCEANS 2008-MTS/IEEE Kobe Techno-Ocean, Kobe, Japan, 28 May 2008; IEEE: Manhattan, NY, USA, 2008; pp. 1–8. [CrossRef]
10. Valerio, A.; Carvalho G, L. MWD and MPD System Combination Successfully Manages Ultra-High LCM Concentrations and Cement Slurries in Loss Scenarios. In Proceedings of the SPE International Conference and Exhibition on Formation Damage Control, Lafayette, LA, USA, 12 February 2020; OnePetro: Richardson, TX, USA, 2020. [CrossRef]
11. Reckmann, H.; Jogi, P.; Kpetehoto, F.T.; Chandrasekaran, S.; Macpherson, J.D. MWD failure rates due to drilling dynamics. In Proceedings of the IADC/SPE Drilling Conference and Exhibition, New Orleans, LA, USA, 2 February 2010; OnePetro: Richardson, TX, USA, 2010. [CrossRef]
12. Kuyken, C.; van der Plas, K.; Frederiks, K.; Williams, M.; Cockburn, C. Developing and Using Slimhole Rotary Steerable Systems in the North Sea Changing the Economics of Mature Asset Drilling. In Proceedings of the SPE Offshore Europe Oil and Gas Exhibition and Conference, Aberdeen, UK, 2 September 2003; OnePetro: Richardson, TX, USA, 2003. [CrossRef]
13. Han, G.Z.; Ai, C. In Proceedings of the Applications of Mud Pulse MWD/LWD System in Bakken Formation, North Dakota, USA. In Proceedings of the Applied Mechanics and Materials, Bangkok, Thailand, 21–22 June 2013; Trans Tech Publications Ltd.: Bäch, Switzerland, 2013; Volume 415, pp. 672–676. [CrossRef]
14. Hussain, S.; Huelvan, Y.; Adams, W. Measurement while drilling, logging while drilling, and rotary steerable systems performance, benefits, and challenges in managed pressure drilling and underbalanced drilling. In Proceedings of the SPE Bergen one day seminar, Bergen, Norway, 2 April 2014; OnePetro: Richardson, TX, USA, 2014. [CrossRef]
15. Attong, D.J.; Singh, U.B.; Teixeira, G. Successful use of a modified MWD tool in a high-concentration LCM mud system. *SPE Drill. Complet.* **1995**, *10*, 22–26. [CrossRef]
16. Klotz, C.; Bond, P.R.; Wassermann, I.; Priegnitz, S. A new mud pulse telemetry system for enhanced MWD/LWD applications. In Proceedings of the IADC/SPE Drilling Conference, Orlando, FL, USA, 4 March 2008; OnePetro: Richardson, TX, USA, 2008. [CrossRef]
17. Hache, J.M.; Till, P. New-generation retrievable MWD tool delivers superior performance in harsh drilling environments. In Proceedings of the SPE/IADC Drilling Conference, Amsterdam, The Netherlands, 27 February 2001; OnePetro: Richardson, TX, USA, 2001. [CrossRef]
18. Zhao, X.; Qiu, Z.; Sun, B.; Liu, S.; Xing, X.; Wang, M. Formation damage mechanisms associated with drilling and completion fluids for deepwater reservoirs. *J. Pet. Sci. Eng.* **2019**, *173*, 112–121. [CrossRef]
19. Zhang, T.; Zeng, X.; Guo, J.; Zeng, F.; Li, M. Numerical simulation on oil-water-particle flows in complex fractures of fractured-vuggy carbonate reservoirs. *J. Pet. Sci. Eng.* **2022**, *208*, 109413. [CrossRef]
20. Ardekani, A.M.; Rangel, R.H. Numerical investigation of particle-particle and particle–wall collisions in a viscous fluid. *J. Fluid Mech.* **2008**, *596*, 437–466. [CrossRef]
21. Wang, G.; Dong, M.; Wang, Z.; Ren, T.; Xu, S. Removing cuttings from inclined and horizontal wells: Numerical analysis of the required drilling fluid rheology and flow rate. *J. Nat. Gas Sci. Eng.* **2022**, *102*, 104544. [CrossRef]

22. Casas, G.; De-Pouplana, I.; Gandikota, R.; Oñate, E. Numerical simulation of particle impact drilling (PID) systems: A one-way coupled approach. *Comput. Part. Mech.* **2022**, *9*, 735–757. [CrossRef]
23. Feng, C.; Liu, W.; Gao, D. CFD simulation and optimization of slurry erosion of PDC bits. *Powder Technol.* **2022**, *408*, 117658. [CrossRef]
24. Zakerian, A.; Sarafraz, S.; Tabzar, A.; Hemmati, N.; Shadizadeh, S.R. Numerical modeling and simulation of drilling cutting transport in horizontal wells. *J. Pet. Explor. Prod. Technol.* **2018**, *8*, 455–474. [CrossRef]
25. Lin, C.; Taleghani, A.D.; Kang, Y.; Xu, C. A coupled CFD-DEM numerical simulation of formation and evolution of sealing zones. *J. Pet. Sci. Eng.* **2022**, *208*, 109765. [CrossRef]
26. Kloss, C.; Goniva, C.; Hager, A.; Amberger, S.; Pirker, S. Models, algorithms and validation for opensource DEM and CFD–DEM. *Prog. Comput. Fluid Dyn. Int. J.* **2012**, *12*, 140–152. [CrossRef]
27. Kanitz, M.; Grabe, J. Multiscale investigation of suffusion with coupled CFD-DEM-Influence of different drag force models. In Proceedings of the 2nd International Conference on the Material Point Method for Modelling Soil-Water-Structure Interaction, Cambridge, UK, 8–10 January 2019. Available online: https://onlinelibrary.wiley.com/doi/pdf/10.1002/nag (accessed on 20 August 2022).
28. Wang, D.; You, Z.; Wang, M.; Li, Q.; Wu, L. Numerical investigation of proppant transport at hydraulic-natural fracture intersection. *Powder Technol.* **2022**, *398*, 117123. [CrossRef]
29. Hertz, H. Über die Berührung fester elastischer Körper. *J. Für Die Reine Angew. Math.* **1882**, *92*, 22.
30. Tsuji, Y.; Tanaka, T.; Ishida, T. Lagrangian numerical simulation of plug flow of cohesionless particles in a horizontal pipe. *Powder Technol.* **1992**, *71*, 239–250. [CrossRef]
31. Mindlin, R.D.; Deresiewicz, H. Elastic Spheres in Contact Under Varying Oblique Forces. *J. Appl. Mech.* **1953**, *20*, 327–344. [CrossRef]
32. Hærvig, J. *On the Adhesive Behaviour of Micron-sized Particles in Turbulent Flow: A Numerical Study Coupling the Discrete Element Method and Large Eddy Simulations*; Aalborg University Press: Aalborg, Denmark, 2017.
33. Liu, D.; van Wachem, B.G.M.; Mudde, R.F.; Chen, X.; Ommen, J.R.V. An adhesive CFD-DEM model for simulating nanoparticle agglomerate fluidization. *AIChE J.* **2016**, *62*, 2259–2270. [CrossRef]
34. Zhou, Z.Y.; Kuang, S.B.; Chu, K.W.; Yu, A.B. Discrete particle simulation of particle–fluid flow: Model formulations and their applicability. *J. Fluid Mech.* **2010**, *661*, 482–510. [CrossRef]
35. Izyurov, V.; Kharitonov, A.; Semenikhin, I.; Korsunov, E.; Gassan, A.; Tikhonov, E.; Jadan, G.; Stashko, V.; Blagonadeshniy, I.; Manikhin, A.; et al. Lessons Learned from Formation Pressure Measurements While Drilling. In Proceedings of the SPWLA 45th Annual Logging Symposium, Noordwijk, The Netherlands, 6 June 2004; OnePetro: Richardson, TX, USA, 2004. [CrossRef]

processes

MDPI

Article

Acid System and Stimulation Efficiency of Multistage Acid Fracturing in Porous Carbonate Reservoirs

Dawei Zhu [1,*], Yunjin Wang [2,*], Mingyue Cui [1], Fujian Zhou [2], Yaocong Wang [3], Chong Liang [1], Honglan Zou [1] and Fei Yao [1]

1 Research Institute of Petroleum Exploration & Development, PetroChina, Beijing 100083, China
2 State Key Laboratory of Petroleum Resources and Prospecting, China University of Petroleum (Beijing), Beijing 102249, China
3 Oil & Gas Technology Research Institute, Petrochina Changqing Oilfield Company, Xi'an 710018, China
* Correspondence: zhudawei69@petrochina.com.cn (D.Z.); yunjingwang112@foxmail.com (Y.W.)

Abstract: With little to no natural fracture development and the high calcite content in porous carbonate reservoirs, for multistage acid fracturing, different fluids are used to form a viscous fingering in the fracture, thus enhancing the degree of nonuniform etching. However, existing studies on multistage acid fracturing mainly focused on the combination of fracturing fluid and acid, which is less specific for porous carbonate rocks. Here, the rheological properties of five fluids, including guar-based fluid, cross-linked guar, gelled acid, cross-linked acid, and diverting acid, were studied at each temperature condition, and the viscosity relationship between each fluid was clarified. Based on the rheological properties, the differences between the seven liquid combinations on the etched morphology of the fracture walls were studied and analyzed. The conductivity of the seven acid-etched fractures under different closure stress was simulated. The experimental results showed that the viscosity relationships between the fluids at different temperatures were cross-linked guar > cross-linked acid > diverting acid (spent acid) > gelled acid > guar-based liquid > diverting acid (fresh acid). Because cross-linked acid has higher viscosity than gelled acid, it can form more obvious viscous fingering with a variety of liquids, which is more suitable for acid fracturing stimulation of porous carbonate reservoirs. In addition, the combination of cross-linked and diverting acids was screened out. The multistage alternate injection of this fluid combination could form tortuous and complex etching channels, and its acid-etching fracture conductivity was significantly higher than that of other fluid combinations at different closure stress. In this study, we optimized the fluid combination of porous carbonates and clarified the effect and mechanism of nonuniform etching to provide guidance for the fluid combination selection of multistage alternate acid fracturing process for porous carbonate reservoirs.

Keywords: multistage acid fracturing; porous carbonate reservoir; morphology; conductivity; acid combination

Citation: Zhu, D.; Wang, Y.; Cui, M.; Zhou, F.; Wang, Y.; Liang, C.; Zou, H.; Yao, F. Acid System and Stimulation Efficiency of Multistage Acid Fracturing in Porous Carbonate Reservoirs. *Processes* **2022**, *10*, 1883. https://doi.org/10.3390/pr10091883

Academic Editors: Linhua Pan, Yushi Zou, Jie Wang, Minghui Li, Wei Feng and Lufeng Zhang

Received: 23 August 2022
Accepted: 11 September 2022
Published: 17 September 2022

1. Introduction

Porous carbonate reservoirs generally have low porosity and permeability, high calcite content in rock minerals, and the natural fractures are less or even not developed [1,2]. This type of reservoir is especially developed in the Middle East or central Asia [3]. Acid fracturing is one of the most effective measures for carbonate stimulation [4,5]. The acid-etching fracture conductivity after acid fracturing treatment is a key factor affecting oil and gas production, and is a key parameter in the acid fracturing design of carbonate reservoirs [6,7]. Due to the characteristics of porous carbonate with pure lithology, nonuniform etching is not easily formed by the conventional acid fracturing process [8,9]. Multistage alternate acid fracturing is widely used to enhance the degree of nonuniform etching, and was proven to be the most effective stimulation measure for porous carbonates [10].

Early multistage alternate acid fracturing technology is mainly a combination of fracturing fluid and acid fluid [11,12]. The main advantages of adding fracturing fluid to the acid fracturing process include: (1) the use of highly viscous fracturing fluid to plug the natural fractures in the reservoir and the wormhole formed in the fracture wall, thus reducing the filtration of acid through the natural fractures and fracture wall and increasing the volume of fractures during acid fracturing [13,14]; (2) the viscosity of the fracturing fluid is usually higher than that of the acid fluid, and the difference in viscosity between the alternately injected fracturing fluid and the acid fluid makes the two fluids form viscous fingering in the fracture [15]. This viscous fingering causes the distribution of the acid to be nonuniform in the fracture, thus producing nonuniform etching on the fracture wall. The formation of these nonuniform etchings prevents the fracture from completely closing under high closure stress and maintains a good oil and gas flow channel [16]. Notably, cross-linked guar is usually a chain compound formed by cross-linking of hydroxypropyl guar under alkaline conditions and the cross-link agent, thereby increasing the viscosity of the fracturing fluid [17]. Under acidic conditions, the cross-linking structure is destroyed, and the viscosity of the fluid is significantly reduced and may even be lower than that of the gelled and cross-linked acids [18]. To solve this problem, researchers have developed acid-resistant polymeric fracturing fluids that can effectively improve the fluid viscosity of fracturing fluids in highly acidic environments [19]. However, such fracturing fluids are costly, and their application in the field is limited.

Gelled and cross-linked acids are often used in multistage acid fracturing because of their higher viscosity than those of other acid systems, which can effectively reduce acid filtration and open the reservoir to form fractures [20]. In recent years, with the development of acid systems, the combination of fluid systems for multistage acid fracturing has also changed, such as the combination of fracturing fluid and gelled acid, and diverting acid and gelled acid [21,22]. The viscosity of fresh diverting acid is lower and consistent with that of water. As H^+ is consumed during the acid–rock reaction and carbonate minerals are dissolved, the pH and Ca^{2+} concentration increase in the solution. Surfactants in the diverting acid are entangled to form micelles in weakly acidic environments with metal ions, and the presence of micelles greatly enhances the viscosity of the acid solution [23]. Fresh low-viscosity acid has good flow ability in fractures, while spent high-viscosity spent acid can effectively plug the natural fractures and wormholes to reduce acid filtration [1]. Current research on porous carbonates has focused on two acid systems: gelled acid and cross-linked acid, and stimulation parameters such as injection rate and injection time. However, the stimulation effect and mechanism of the combination of diverting acid and other acid fluids on porous carbonate are still unclear.

In this study, a diverting acid whose viscosity gradually increases with reaction time was introduced. Considering the temperature variation, the rheological properties of five fluid systems at different temperatures were studied, and the viscosity relationship between the fluids was clarified. Based on the combinations of fracturing fluid and acid in conventional acid fracturing, we introduced combinations of acids. The conductivities of acid-etching fractures for different acid combinations were studied. The acid fracturing stimulation efficiencies of different acid combinations were clarified, and the fluid combinations suitable for porous carbonate reservoirs were preferably selected.

2. Experiment Preparation

In the process of acid fracturing, the acid uniformly reacts with the carbonate minerals on the fracture wall, hindering nonuniform etching on the fracture wall. With the completion of acid fracturing and the progression of oil and gas flowback, the pore pressure of the reservoir gradually decreases and the closure stress gradually increases in the reservoir. Uniformly etched fractures rapidly close under closure stress, and the conductivity is difficult to maintain, which seriously affects the acid fracturing effect. Multistage acid fracturing stimulation can reasonably use the reactions and rheological characteristics of the acid, which make the acid produce viscous fingering in the fracture, thus producing nonuniform etching on the fracture wall, finally forming a connected high-permeability oil and gas channel.

2.1. Rock Slab Preparation

Indiana limestone cores were analyzed by XRD, and the mineral composition of the cores was mainly calcite (93.7%) and dolomite (7.3%). In addition, the porosity and permeability of the Indiana limestone were 20.1% and 7.2 mD, respectively. The cores used in the experiment were processed with by wire cutting technology into a 178 mm long, 19 mm arc rock slab that was 38 mm wide and 50 mm high (Figure 1), which ensured the rock slab size fully matched the size of API acid etching conductivity chamber.

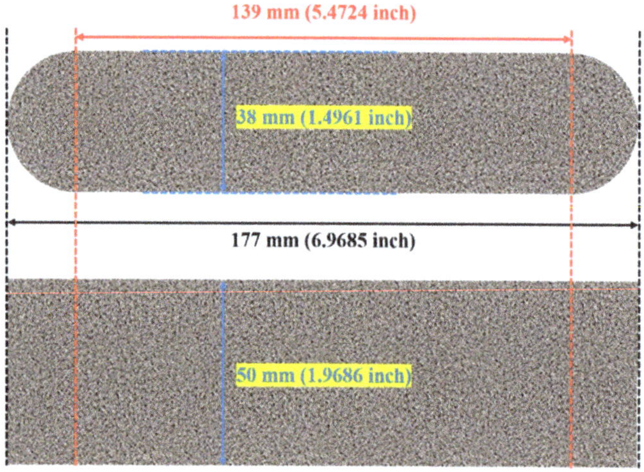

Figure 1. The size of the limestone slab.

2.2. Fluid Preparation

In this work, five kinds of liquid systems were used in the multistage acid fracturing experiment, including a guar-based solution, cross-linked guar, gelled acid, cross-linked acid, and diverting acid. The specific formulas of the liquids were as follows:

(1) Guar-based solution: 0.27 wt% HPG + 0.1 wt% fungicide, and 0.3 wt% clay stabilizer;

(2) Cross-linked guar: 0.27 wt% HPG + 0.1 wt% fungicide + 0.3 wt% clay stabilizer, and 0.3 wt% cross-link agent;

(3) Gelled acid: 20 wt% HCl + 0.6 wt% gelling agent, and 1.5 wt% corrosion inhibitor;

(4) Cross-linked acid: 20 wt% HCl + 0.6 wt% gelling agent + 0.6 wt% cross-link agent, and 1.5 wt% corrosion inhibitor;

(5) Diverting acid: 20 wt% HCl + 1.5 wt% corrosion inhibitor, and 7 wt% diverting agent.

In the reaction of diverting acid with carbonate minerals, the viscosity of the spent diverting acid solution gradually increases under environments with high pH and Ca^{2+} concentration, eventually achieving the purpose of uniform acidizing. After the chemical reaction between fresh acid and carbonate minerals at the fracture wall during acid fracturing, the highly viscous spent acid enters the porous medium through the fracture wall. The highly viscous spent acid, by entering the porous media, can enhance the percolation resistance of the subsequent acid into the fracture wall, thus reducing acid loss and improving fracturing efficiency. Therefore, in addition to the viscosity of the fresh diverting acid, the viscosity of the spent diverting acid was also tested in this study. The preparation process of the spent diverting acid was as follows: (1) 200 mL of 15 wt% HCl solution was prepared by using deionized water and 37 wt% HCl. (2) We added 1.5 wt% corrosion inhibitor to the removed liquid and 7 wt% diverting agent, which is a kind of

Gemini quaternary ammonium salt cationic surfactant. (3) Excessive Ca_2CO_3 was added to the water bath at 50 °C to ensure the HCl fully reacted, and the solution after the reaction was centrifuged at 3000 r/min to obtain the supernatant.

2.3. *Experiment Apparatus*

In this work, an acid etching fracture conductivity testing system (Figure 2) was used, which consisted of a liquid rheology testing device, an acid etching simulation device, a conductivity testing device, and a 3D laser scanning device. The rheology testing device was made of Hastelloy and was able to test the viscosity of fracturing fluid and acid at different temperatures [24]. The acid etching simulation device was used to etch the rock slab by injecting acid, thus simulating the process of etching the fracture wall during acid fracturing. The conductivity testing device was used to test the flow conductivity of the fracture at different closure stresses, thus objectively reflecting the closure of the acid-etching fracture at different closure stresses. The 3D laser scanning device was used to visualize the acid etching on the fracture walls.

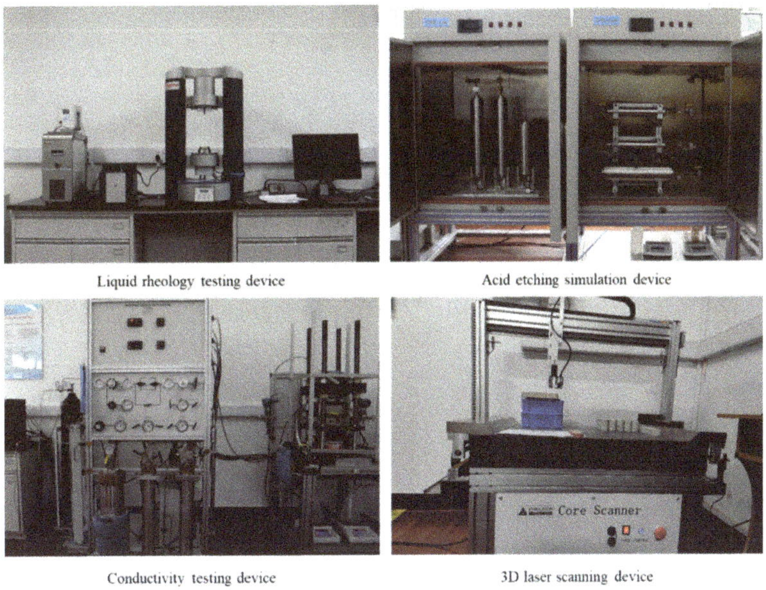

Figure 2. Acid fracturing conductivity test system.

2.4. *Experimental Procedure*

The main procedures of the acid rheology and acid etching slab conductivity testing experiments are described below:

(1) Determination of the rheological properties of the acid solution: the viscosity of the acid system used in the experiment was measured at different temperatures using the Harker rheometer to clarify the viscosity relationship of each acid system at the target temperature.

(2) Acid etching of fracture surfaces: A pair of smooth rock slabs was installed parallel to the reaction chamber of the acid-etching simulation device and began to heat up. The back pressure was set to 7 MPa during the experiment. When the ambient temperature was heated to the experimentally set value, the acid was pumped into the fracture in the reaction chamber according to the experimental scheme, and the pumped acid chemically reacted with the rock minerals on the fracture wall.

(3) Scanning the surface morphology of the fracture after acid etching: After the reaction, the acid-etched rock slab was removed from the acid etching simulation device, and the surface of the rock slab was cleaned of the acid and residue after the reaction using deionized water. The surface morphological features of the acid-etching fractures were obtained using a 3D laser scanning device.

(4) Testing the flow conductivity of acid-etching fractures: The reacted rock slabs were installed parallel in the conductivity chamber of the conductivity testing device, and the initial closure stress was applied at 5 MPa to make the two rock plates contact. Deionized water was injected at a flow rate of 5 mL/min. After the flow rate at the outlet of the device stabilized at 5 mL/min, the closure stress was increased according to the experimental plan to determine the conductivity at different closure stresses.

3. Result and Discussion

3.1. Liquid Rheology Test Result

By using a liquid rheology testing device, the viscosity of the liquids at room temperature was measured by shearing the six liquids at a constant speed for 30 min. Then, the ambient temperature was slowly raised until the temperature rose to 90 °C, which was then stabilized at 90 °C, and the liquid was continuously sheared for 60 min. After the viscosity of the liquid was stable, the average viscosity of the viscosity stable period was taken as the viscosity of the liquid at 90 °C. The shear rate was kept at 170 s^{-1} during the experiment. The rheological results of the six liquids are shown in Figure 3. As can be seen from the figure, the viscosities of the guar-based fluid, cross-linked guar, gelled acid, cross-linked acid, and spent diverting acid were 86.4, 867.3, 98.3, 394.8, and 420.6 mPa·s, respectively, at room temperature. The viscosities of all acid systems gradually decreased as the ambient temperature continued to rise. At 90 °C, the viscosities of the guar-based solution, cross-linked guar, gelled acid, cross-linked acid, and spent diverting acid were 51.6, 391.9, 28.9, 136.4, and 72.7 mPa·s, respectively. This viscosity change trend was consistent with those reported in previous studies [25,26]. The viscosity of the cross-linked guar always remained the highest at all temperature conditions, followed by that of the cross-linked acid and spent diverting acid, and the viscosities of the guar-based solution and gelled acid were the smallest and least different. The viscosity of the fresh diverting acid was 2.7 mPa·s due to the characteristics of the diverting acid, and slightly decreased as the ambient temperature kept increasing.

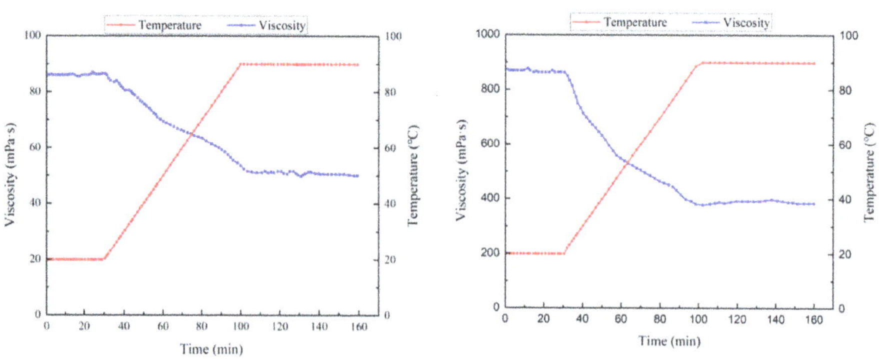

Guar based solution

Crosslinked guar

Figure 3. *Cont.*

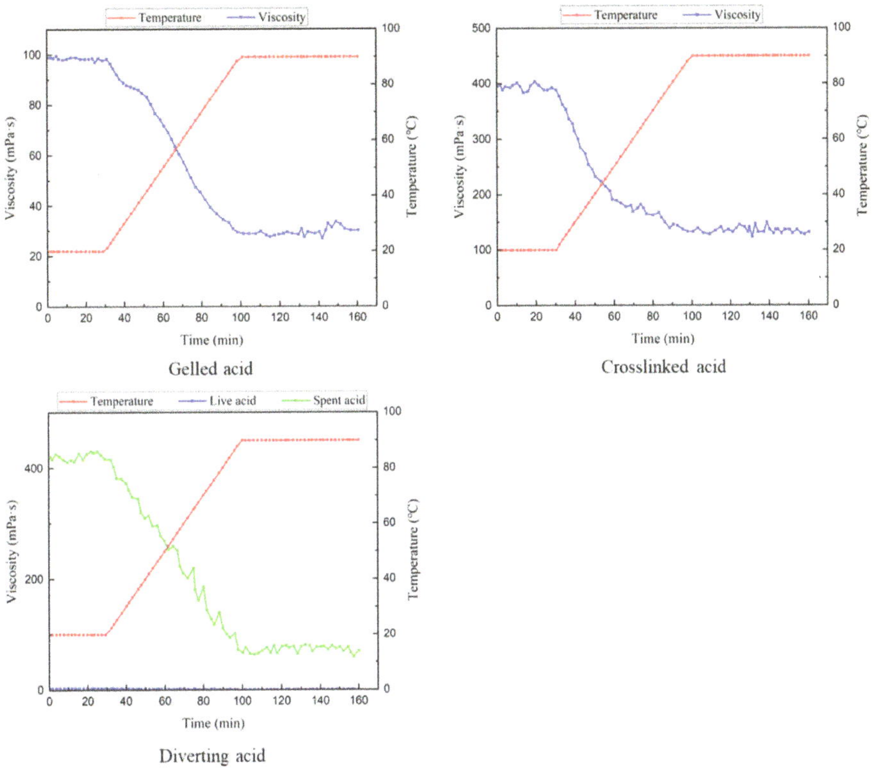

Figure 3. The viscosity of five acid systems at different temperatures.

3.2. Acid Etching Simulation Result

According to the optimization principle of multistage acid fracturing in carbonate reservoirs, seven combinations of five fluids were considered to study the effect of acid etching on the fracture wall of carbonate rocks under different fluid combinations. To ensure that the total amount of acid injected in the multistage acid fracturing process was consistent, the injection time in the multistage acid fracturing was adjusted, and the specific experimental arrangement is shown in Table 1. The ambient temperature was maintained at 90 °C during the experiments. Combined with the injection rate during acid fracturing treatment in the field and previous research results [21], the injection rate in the experiment was set to 30 mL/min.

The results of the rock slab acid etching simulation are shown in Figures 4 and 5. As can be seen from the figures, the surfaces of the rock slab after both multistage alternate injection of the guar-based solution and gelled acid and multistage alternating injection of gelled acid and diverting acid were both relatively smooth. This uniform etching completely closed the fractures under high closure stress, thus reducing the acid fracturing effect. There was nonuniform etching on the rock slab after multistage alternating injection of the cross-linked acid and guar-based solution, but the degree of nonuniform etching was lower. More obvious [27,28] etching grooves formed after multistage alternating injection of gelled acid and cross-linked guar, cross-linked acid and gelled acid, and cross-linked acid and cross-linked guar. This etched groove greatly enhanced the nonuniform etching degree of the fracture wall [29], which hindered the acid-etched fracture from completely closing under the closing pressure after the acid fracturing stimulation, thus forming a high-speed channel for the flow of oil and gas. The etched grooves formed after the multistage alternate injection of cross-linked acid and diverting acid were more obvious. These grooves greatly enhanced the degree of nonuniform etching on the fracture walls: tortuous and complex grooves were harder difficult to close under closure stress. In addition, the slab overall showed a greater degree of dissolution at the entrance, and the degree of dissolution from the entrance to the exit showed a gradually decreasing trend. This is because the acid concentration at the point of entry into the fracture was higher, which created a more intense acid–rock reaction, and the degree of etching was correspondingly higher. With the consumption of H^+ in acid by the acid–rock reaction, the concentration of acid at the exit was lower, and, thus, the degree of reaction was relatively weaker.

Table 1. Multistage acid fracturing test arrangement and conductivity test results.

Liquid Combination	Stage	Injection Time	Injection Fluid	Conductivity (D·cm)				
				15 MPa	30 MPa	45 MPa	60 MPa	75 MPa
Guar-based fluid (GB) and gelled acid (GA)	1	20 min	GB	98.5	51.3	22	7.4	1.3
	2	20 min	GA					
	3	20 min	GB					
	4	20 min	GA					
	5	20 min	GB					
	6	20 min	GA					
Cross-linked guar (CA) and gelled acid (GA)	1	20 min	CA	312.7	143.7	56.2	21.3	10.8
	2	20 min	GA					
	3	20 min	CA					
	4	20 min	GA					
	5	20 min	CA					
	6	20 min	GA					
Gelled acid and (GA) diverting acid (DA)	1	10 min	GA	154.4	81.4	24.3	11.2	6.7
	2	10 min	DA					
	3	10 min	GA					
	4	10 min	DA					
	5	10 min	GA					
	6	10 min	DA					
Guar-based fluid (GB) and cross-linked acid (CA)	1	20 min	GB	356.1	140	48.3	16.1	5.1
	2	20 min	CA					
	3	20 min	GB					
	4	20 min	CA					
	5	20 min	GB					
	6	20 min	CA					
Cross-linked guar (CG) and cross-linked acid (CA)	1	20 min	CG	217.2	97.3	49.7	20.1	5.3
	2	20 min	CA					
	3	20 min	CG					
	4	20 min	CA					
	5	20 min	CG					
	6	20 min	CA					
Cross-linked acid (CA) and gelled acid (GA)	1	10 min	CA	342.1	164.2	61.7	20.3	10.1
	2	10 min	GA					
	3	10 min	CA					
	4	10 min	GA					
	5	10 min	CA					
	6	10 min	GA					
Cross-linked acid (CA) and diverting acid (DA)	1	10 min	CA	514.2	314	148.3	73.4	41.2
	2	10 min	DA					
	3	10 min	CA					
	4	10 min	DA					
	5	10 min	CA					
	6	10 min	DA					

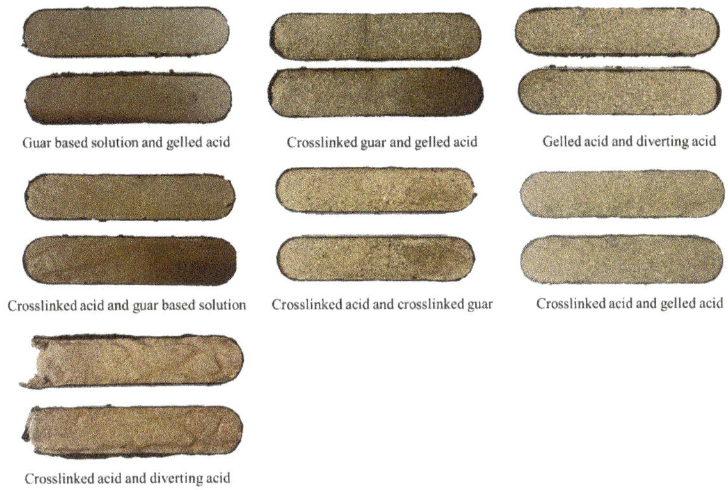

Figure 4. Surface morphology of slabs after acid etching.

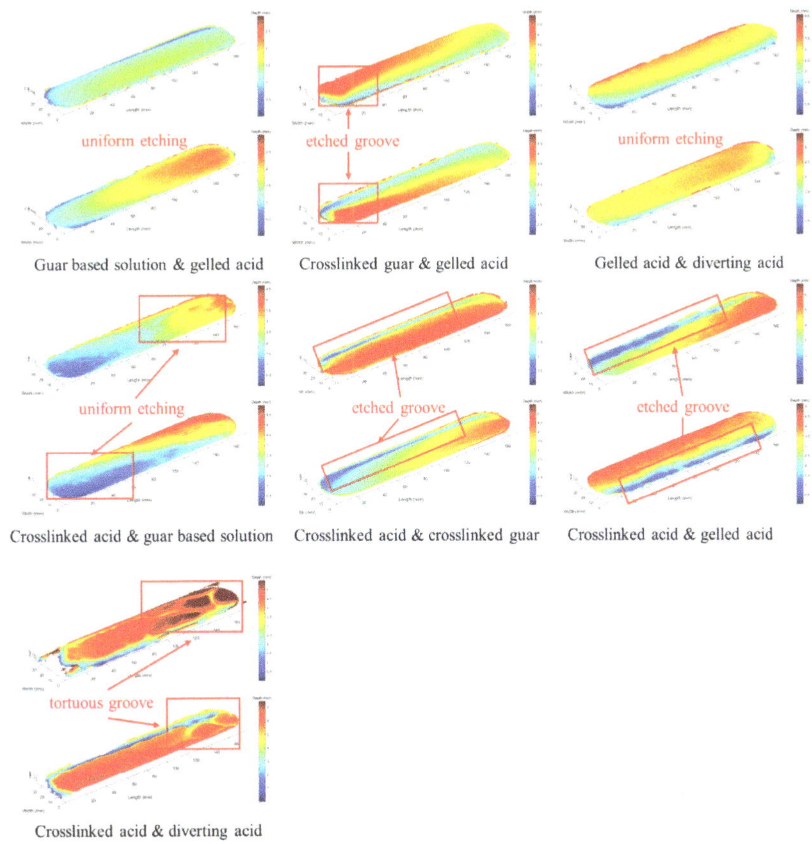

Figure 5. Surface scan of slabs after acid etching.

3.3. Conductivity Test Result

The fracture conductivity of the rock slabs, after acid etching with different acid combinations, was tested under closing pressures of 15, 30, 45, 60, and 75 MPa using the conductivity testing device, and the test results are shown in Figure 6. From the conductivity test results, we found that the fracture conductivity after cross-linked acid etching was generally higher than that after gelled acid etching. This is because the viscosity of cross-linked acid was five times higher than that of gelled acid at 90 °C, and its viscosity difference with guar-based fluid and diverting acid was even larger. Previous studies [30] showed that viscosity differences can make low-viscosity fluid form viscous fingering in high-viscosity fluid, and this viscous fingering can make the distribution of the acid nonuniform in the fracture, thus achieving nonuniform etching on the fracture wall, which greatly enhances the fracture conductivity under high closure stress. Notably, the multistage alternate injection of cross-linked acid and diverting acid had significantly higher conductivities than the other acid combinations at each closure stress. The combination of cross-linked acid and diverting acid had a conductivity of 514.2 D·cm at a closure stress of 15 MPa. The conductivity gradually decreased as the closure stress increased, and had a conductivity of 148.3 D·cm at a closure stress of 45 MPa. When the closure stress was 75 MPa, the conductivity of the acid-etched fracture was 41.2 D·cm. Combined with the acid rheology test and rock slab etching results, we determined that this was due to the large difference in viscosity between the cross-linked acid and the diverting acid, which led to more obvious viscous fingering of the acid inside the fracture. This obvious viscous fingering caused the acid etching of the fracture wall to form a tortuous and complex etching channel. Such channels cannot be completely closed even under high closure stress, which provides better flow channels for the oil and gas production process and greatly enhances the effect of acid fracturing stimulation in carbonate reservoirs.

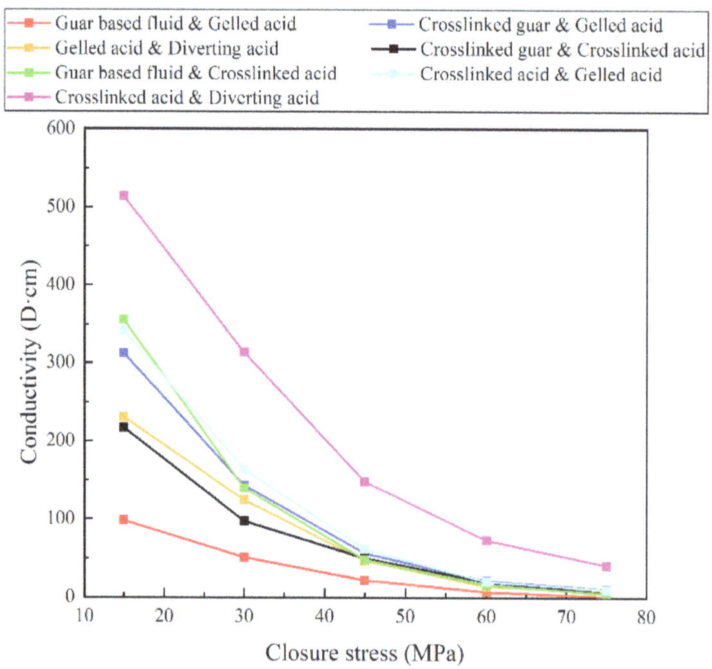

Figure 6. Conductivity under multistage alternate injection of different fluid combinations.

In summary, the fracture conductivity after etching with cross-linked acid is generally higher than that of etching with gelled acid. This is because the viscosity of cross-linked acid at 90 °C is five times higher than that of gelled acid, and its viscosity difference with guar-based fluid and diverting acid is even larger. The viscosity difference can make the acid produce viscous fingering in the fracture. This viscous fingering can make the distribution of the acid nonuniform in the fracture, thus achieving the nonuniform etching of carbonate minerals on the fracture wall. The combination of cross-linked acid and diverting acid can form tortuous and complex etching channels on the fracture wall due to their large difference in viscosity, which can improve the efficiency of acid fracturing.

4. Conclusions

In this study, the rheological properties of a guar-based solution, cross-linked guar, gelled acid, cross-linked acid, and diverting acid were investigated at different temperatures. Multistage acid fracturing experiments were carried out using different acid combinations, and the effects of different multiple-acid systems on the nonuniform etching of the fracture wall were systematically investigated. The relationship between the surface morphology and the conductivity of the acid-etched fracture was analyzed. Based on the experiment results and discussion, the following conclusions were obtained:

(1) The viscosity of all five acid fracturing working fluids gradually decreased with the increase in temperature. At different temperatures, the viscosity relationships were: cross-linked guar > cross-linked acid > diverting acid (spent acid) > gelled acid > guar-based solution > diverting acid (fresh acid).

(2) Cross-linked acid is more suitable for acid fracturing stimulation of carbonate reservoir than gelled acid. Due to the high viscosity of cross-linked acid and the large difference in viscosity with many other liquids used in acid fracturing stimulation, the combination of cross-linked acid with other liquids effectively produce viscous fingering in the fracture. This viscous fingering means the cross-linked acids generally have higher fracture conductivity than gelled acids after acid etching.

(3) The combination of cross-linked acid and diverting acid is most effective for multistage acid fracturing stimulation of carbonate reservoirs. The liquid combination of cross-linked acid and diverting acid has a large difference in viscosity and can form tortuous and complex etched channels in the fracture walls, which do not completely close even at high closure stress. When the closure stress was 75 MPa, the acid-etched fracture still had a conductivity of 41.2 D·cm.

Nonuniform acid fracturing in porous carbonatite is still a worldwide problem. In this work, we evaluated and compared the effects of various liquid combinations on multistage acid fracturing, and proposed that the combination of diverting acid and cross-linked acid can effectively improve the degree of nonuniform etching in the fracture. However, when the environmental temperature is above 90 °C, the performance of the diverting acid is seriously reduced. Therefore, the performance of diverting acid in high-temperature reservoirs must be improved. In addition, under normal conditions, the viscosity of the acid solution is controllable within a certain range, and the effect of the viscosity relationship between the diverting acid and the cross-linked acid on the acid fracturing effect still needs further study.

Author Contributions: Writing—original manuscript, D.Z.; experiment, Y.W. (Yunjin Wang) and Y.W. (Yaocong Wang); supervision, M.C. and F.Z.; investigation, C.L.; data curation, H.Z.; formal analysis, F.Y. All authors have read and agreed to the published version of the manuscript.

Funding: This research was funded by the National Natural Science Foundation of China (No. 52174045), the CNPC "Fourteenth Five Year Plan" science and technology projects (No. 2021DJ3405).

Conflicts of Interest: The authors declare no conflict of interest.

References

1. Zhu, D.; Wang, Y.; Cui, M.; Zhou, F.; Zhang, Y.; Liang, C.; Zou, H.; Yao, F. Effects of spent viscoelastic-surfactant acid flow on wormholes propagation and diverting performance in heterogeneous carbonate reservoir. *Energy Rep.* **2022**, *8*, 8321–8332. [CrossRef]
2. Li, M.-H.; Zhou, F.-J.; Liu, J.-J.; Yuan, L.-S.; Huang, G.-P.; Wang, B. Quantitative investigation of multi-fracture morphology during TPDF through true tri-axial fracturing experiments and CT scanning. *Pet. Sci.* **2022**, *19*, 1700–1717. [CrossRef]
3. Peng, Y.; Zhao, J.; Sepehrnoori, K.; Li, Y.; Yu, W.; Zeng, J. Study of the Heat Transfer in the Wellbore During Acid/Hydraulic Fracturing Using a Semianalytical Transient Model. *SPE J.* **2019**, *24*, 877–890. [CrossRef]
4. Wang, J.; Zhou, F.; Bai, H.; Wei, D.; Ma, J.; Yang, P.; Zhang, F.; Yuan, L. A new approach to study the friction-reduction characteristics of viscous/conventional slickwater in simulated pipelines and fractures. *J. Nat. Gas Sci. Eng.* **2020**, *83*, 103620. [CrossRef]
5. Ahmadpour, V.; Mirzaei, I.; Rezazadeh, S.; Ahmadi, N. Investigation of solid/liquid interface evolution in the solidification process of liquid metal in an annulus crucible at the presence of static magnetic field: Numerical study. *J. Braz. Soc. Mech. Sci. Eng.* **2019**, *41*, 404. [CrossRef]
6. Liu, Z.; Reynolds, A.C. Gradient-Enhanced Support Vector Regression for Robust Life-Cycle Production Optimization with Nonlinear-State Constraints. *SPE J.* **2020**, *26*, 1590–1613. [CrossRef]
7. Li, M.; Lv, W.; Liu, J.; Sun, Z.; Zhou, F.; Wang, B. Effect of perforation friction on 3D In-stage multiple fracture propagation: A numerical study. *Eng. Fract. Mech.* **2022**, *267*, 108415. [CrossRef]
8. Peng, Y.; Zhao, J.; Sepehrnoori, K.; Li, Z.; Xu, F. Study of delayed creep fracture initiation and propagation based on semi-analytical fractional model. *Appl. Math. Model.* **2019**, *72*, 700–715. [CrossRef]
9. Du, J.; Guo, G.; Liu, P.; Xiong, G.; Chen, P.; Liu, J.; Chen, X. Experimental Study on the Autogenic Acid Fluid System of a High-Temperature Carbonate Reservoir by Acid Fracturing. *ACS Omega* **2022**, *7*, 12066–12075. [CrossRef]
10. Dong, R.; Wheeler, M.F.; Su, H.; Ma, K. Modeling Multistage Acid Fracturing Treatments in Carbonate Reservoirs. In Proceedings of the SPE Hydraulic Fracturing Technology Conference and Exhibition, Online, 4–6 May 2021. [CrossRef]
11. Zhang, K.; Chen, M.; Zhou, C.; Dai, Y.; Liu, F.; Li, J. Study of alternating acid fracturing treatment in carbonate formation based on true tri-axial experiment. *J. Pet. Sci. Eng.* **2020**, *192*, 107268. [CrossRef]
12. Li, M.; Zhou, F.; Sun, Z.; Dong, E.; Zhuang, X.; Yuan, L.; Wang, B. Experimental study on plugging performance and diverted fracture geometry during different temporary plugging and diverting fracturing in Jimusar shale. *J. Pet. Sci. Eng.* **2022**, *215*, 110580. [CrossRef]
13. Aljawad, M.S.; Aljulaih, H.; Mahmoud, M.; Desouky, M. Integration of field, laboratory, and modeling aspects of acid fracturing: A comprehensive review. *J. Pet. Sci. Eng.* **2019**, *181*, 106158. [CrossRef]
14. Gou, B.; Guan, C.; Li, X.; Ren, J.; Zeng, J.; Wu, L.; Guo, J. Acid-etching fracture morphology and conductivity for alternate stages of self-generating acid and gelled acid during acid-fracturing. *J. Pet. Sci. Eng.* **2021**, *200*, 108358. [CrossRef]
15. Nima, A.; Sajad, R.; Mirkazem, Y.; Iraj, M. Numerical investigation of species distribution and the effect of anode transfer coefficient on the proton exchange membrane fuel cell (PEMFC) performance. *Chem. Ind.* **2012**, *46*, 71. [CrossRef]
16. Zhang, R.; Hou, B.; Zhou, B.; Liu, Y.; Xiao, Y.; Zhang, K. Effect of acid fracturing on carbonate formation in southwest China based on experimental investigations. *J. Nat. Gas Sci. Eng.* **2019**, *73*, 103057. [CrossRef]
17. Lufeng, Z.; Fujian, Z.; Shicheng, Z.; Zhun, L.; Jin, W.; Yuechun, W. Evaluation of permeability damage caused by drilling and fracturing fluids in tight low permeability sandstone reservoirs. *J. Pet. Sci. Eng.* **2019**, *175*, 1122–1135. [CrossRef]
18. You, J.; Lee, K.J. Analyzing the Dynamics of Mineral Dissolution during Acid Fracturing by Pore-Scale Modeling of Acid-Rock Interaction. *SPE J.* **2021**, *26*, 639–652. [CrossRef]
19. Mehrjoo, H.; Norouzi-Apourvari, S.; Jalalifar, H.; Shajari, M. Experimental study and modeling of final fracture conductivity during acid fracturing. *J. Pet. Sci. Eng.* **2021**, *208*, 109192. [CrossRef]
20. Shah, M.; Agarwal, J.R.; Patel, D.; Chauhan, J.; Kaneria, D.; Shah, S.N. An assessment of chemical particulate technology as diverters for refracturing treatment. *J. Nat. Gas Sci. Eng.* **2020**, *84*, 103640. [CrossRef]
21. Gou, B.; Qin, N.; Wang, C.; Ren, J.; Guo, J.; Zeng, M.; Zhou, C.; Liu, F. Acidizing Model to Couple the Closure Stress and Acid-Rock Reactive Transport in Naturally Fractured Carbonate Reservoir. In Proceedings of the 55th U.S. Rock Mechanics/Geomechanics Symposium, Online, 18–25 June 2021. Available online: https://onepetro.org/ARMAUSRMS/proceedings/ARMA21/All-ARMA21/ARMA-2021-1221/467963 (accessed on 27 November 2021).
22. Liu, Z.; Reynolds, A. Robust Multiobjective Nonlinear Constrained Optimization with Ensemble Stochastic Gradient Sequential Quadratic Programming-Filter Algorithm. *SPE J.* **2021**, *26*, 1964–1979. [CrossRef]
23. Hosseinzadeh, B.; Bazargan, M.; Rostami, B.; Ayatollahi, S. Modeling of Wormhole Propagation in Carbonate Rocks by Use of In-Situ-Gelled Acids. *SPE J.* **2017**, *22*, 2032–2048. [CrossRef]
24. Zhang, L.; He, J.; Wang, H.; Li, Z.; Zhou, F.; Mou, J. Experimental investigation on wormhole propagation during foamed-VES acidizing. *J. Pet. Sci. Eng.* **2020**, *198*, 108139. [CrossRef]
25. Nasr-El-Din, H.A.; Al-Ghamdi, A.H.; Al-Qahtani, A.A.; Samuel, M.M. Impact of Acid Additives on the Rheological Properties of a Viscoelastic Surfactant and Their Influence on Field Application. *SPE J.* **2008**, *13*, 35–47. [CrossRef]
26. Rabie, A.I.; Gomaa, A.M.; Nasr-El-Din, H.A. Reaction of In-Situ-Gelled Acids With Calcite: Reaction-Rate Study. *SPE J.* **2011**, *16*, 981–992. [CrossRef]

27. Dong, R.; Wheeler, M.F.; Ma, K.; Su, H. A 3D Acid Transport Model for Acid Fracturing Treatments With Viscous Fingering. In Proceedings of the SPE Annual Technical Conference and Exhibition, Online, 26–29 October 2020. [CrossRef]
28. Wu, B.; Zhang, M.; Deng, W.; Que, J.; Liu, W.; Zhou, F.; Wang, Q.; Li, Y.; Liang, T. Study and Mechanism Analysis on Dynamic Shrinkage of Bottom Sediments in Salt Cavern Gas Storage. *Processes* **2022**, *10*, 1511. [CrossRef]
29. Li, Y.; Zhou, F.; Wang, J.; Li, B.; Xu, H.; Yao, E.; Zhao, L. Influence of Nanoemulsion Droplet Size of Removing Water Blocking Damage in Tight Gas Reservoir. *Energies* **2022**, *15*, 5283. [CrossRef]
30. Li, Y.; Zhou, F.; Li, B.; Cheng, T.; Zhang, M.; Wang, Q.; Yao, E.; Liang, T. Optimization of Fracturing Fluid and Retarded Acid for Stimulating Tight Naturally Fractured Bedrock Reservoirs. *ACS Omega* **2022**, *7*, 25122–25131. [CrossRef]

processes

MDPI

Article

Numerical Investigation on Injected-Fluid Recovery and Production Performance following Hydraulic Fracturing in Shale Oil Wells

Kai Liao [1,2,*], Jian Zhu [2], Xun Sun [3], Shicheng Zhang [2] and Guangcong Ren [2]

1 School of Petroleum, China University of Petroleum-Beijing at Karamay, Karamay 834000, China
2 School of Petroleum Engineering, China University of Petroleum (Beijing), Beijing 102299, China
3 CNPC Engineering Technology R&D Company Ltd., Beijing 102206, China
* Correspondence: 2020592108@cupk.edu.cn

Abstract: Currently, volume fracturing of horizontal wells is the main technology for shale oil development. A large amount of fracturing fluid is injected into the formation, but the flowback efficiency is very low. Besides, the impact of fluid retention on productivity is not fully clear. There is still a debate about fast-back or slow-back after fracturing, and the formulation of a reasonable cleanup scheme is lacking a theoretical basis. To illustrate the injected-fluid recovery and production performance of shale oil wells, an integrated workflow involving a complex fracture model and oil-water production simulation was presented, enabling a confident history match of flowback data. Then, the impacts of pumping rate, slick water ratio, cluster spacing, stage spacing and flowback rate were quantitatively analyzed. The results show that the pumping rate is negatively correlated with injected-fluid recovery, but positively correlated with oil production. A high ratio of slick water would induce a quite complex fracture configuration, resulting in a rather low flowback efficiency. Meanwhile, the overall conductivity of the fracture networks would also be reduced, as well as the productivity, which indicates that there is an optimal ratio for hybrid fracturing fluid. Due to the fracture interference, the design of stage or cluster spacing is not the smaller the better, and needs to be combined with the actual reservoir conditions. In addition, the short-term flowback efficiency and oil production increase with the flowback rate. However, considering the damage of pressure sensitivity to long-term production, a slow-back mode should be adopted for shale oil wells. The study results may provide support for the design of a fracturing scheme and the optimization of the flowback schedule for shale oil reservoirs.

Keywords: complex fracture; flowback; productivity; numerical simulation; shale oil

Citation: Liao, K.; Zhu, J.; Sun, X.; Zhang, S.; Ren, G. Numerical Investigation on Injected-Fluid Recovery and Production Performance following Hydraulic Fracturing in Shale Oil Wells. *Processes* **2022**, *10*, 1749. https://doi.org/10.3390/pr10091749

Academic Editors: Linhua Pan, Yushi Zou, Jie Wang, Minghui Li, Wei Feng and Lufeng Zhang

Received: 27 July 2022
Accepted: 27 August 2022
Published: 2 September 2022

1. Introduction

In recent years, the rapid development of horizontal volume fracturing technology has made shale oil another unconventional resource with huge potential value for industrial exploitation [1]. Through large-scale staged fracturing, hydraulic fractures can communicate and expand natural fractures, or induce the creation of secondary micro-fractures. In this way, complex interconnected fracture networks are formed to increase the stimulated volume, resulting in a great expansion in productivity [2]. Unlike conventional reservoirs, shale oil reservoirs use slickwater as fracturing fluids, which reduces costs and contributes to complex fractures due to their low viscosity [3]. Field data show that tens of thousands of cubic meters of fracturing fluids are injected underground, but the flowback efficiency is extremely low, leaving a large amount of water permanently trapped [4].

Current researche shows that matrix imbibition and complex fracture closure are the main mechanisms of fracturing fluid retention in shale reservoirs: (1) in terms of matrix imbibition, Dutta observed the water phase imbibition process in tight cores by using real-time CT scanning technology and proposed that the low flowback efficiency was mainly affected by rock permeability, capillary force and heterogeneity [5]. Lin carried out NMR experiments on spontaneous imbibition of shale and found that the water retention was not only controlled by capillary force but also related to chemical osmotic pressure [6]. In addition, the factors affecting the imbibition of fracturing fluid into the deep reservoir included water-rock contact area, initial water saturation and forced pressure [7,8]. (2) In terms of fracture closure, McClure M analyzed how natural fracture closure was one of the important mechanisms for fracturing fluid to be trapped underground [9]. By using analytical methods to study the dynamic flowback characteristics of a horizontal well, Ezulike and Fu found that the hydraulic fracture networks were mainly composed of unpropped fractures, and the closure of these fractures would block the channels of fracturing fluid flowing to the wellbore [10,11].

Although the above studies have explained the reasons for low flowback efficiency in tight reservoirs from different perspectives, there is still a lack of clear understanding of the impact of water retention on shale oil production. Some scholars believed that the interaction between water and shale may be beneficial to increasing hydrocarbon production. Alamdari and Wang studied the impacts of reservoir wettability and natural fractures on imbibition and oil drainage and believed that the initial productivity could be improved through spontaneous imbibition of water in water-wet reservoirs or reservoirs with natural fractures [12,13]. Wang used a nano-oil displacement agent to achieve wettability reversal of oil-wet reservoir, and also observed oil drainage function by imbibition of water [14]. Dehghanpour found that hydration swelling of clay minerals can induce new micro-fractures and increase shale permeability [15]. However, other scholars held that the invasion and retention of fracturing fluid may cause damage to shale reservoirs: Bennion considered that matrix imbibition of unconventional reservoirs would increase water saturation near the fracture area, which could easily cause water lock and reduce hydrocarbon phase permeability [16]. Wang also drew a similar conclusion, noting that although the imbibition effect could improve initial production, it did not significantly increase ultimate recovery [17]. In addition, the reduction of rock permeability caused by scaling or other precipitations during fracturing fluid invasion could also affect production performance [18,19].

The flowback behaviors in shale wells have been mainly studied in three ways: laboratory experiment, analytical model and numerical simulation. The core imbibition experiment can simulate the fluid migration law during well shut-in and cleanup periods [5,6], but the experimental conditions are still quite different from the actual shale reservoir environment. Thus, it is difficult to clarify the flowback performance and its impact on productivity under real engineering conditions [20]. The analytical model based on rate transient analysis can predict flowback efficiency and well productivity [21], but there are limitations such as homogeneous reservoir description and simplified fracture characterization [22]. The numerical simulation method can take into account the impacts of imbibition and pressure-sensitivity and it is suitable for the mechanism study of oil and water migration in porous media [23]. Therefore, it is a practical method to study macroscopic flowback behaviors currently.

Still, the established numerical models have failed to simulate the flowback and production processes following hydraulic fracturing treatment accurately: (1) The injected fluid (then imbibed, retained or recovered) is directly contacted with complex fracture for a long time, so the properties of the fracture system are the key to determine the fluid distribution and migration. However, the fracture system formed by volume fracturing has strong heterogeneity in spatial distribution, geometric configuration, fracture conductivity and other properties, which is difficult to be successfully depicted [24]. Hydraulic fracture modelling is the deterministic approach to describing complex fractures. Compared with micro-seismic monitoring, numerical simulation has a lower operating cost and wider application scope. Besides, the simulated fracture system is based on geological data and treatment parameters, which leads to more practical and reliable fracture properties than that with the orthogonal or discrete fracture model [2]. Nevertheless, the integrated simulation combining fracturing treatment and flowback process has not been reported. The former focuses on fracture mechanics, while the latter focuses on seepage behavior. (2) It is also a problem to present the initial distribution of injected fluid after fracturing. The fluid transfer between fracture and matrix has begun from the pumping process and, at the end, the water-cut near the fracture has increased to a certain extent [25]. Moreover, the loss of fracture volume (fracture closure) during fluid recovery has also been rarely reported. We believe the pumping operation and fracture closure should be related to fluid flowback and production performance. (3) Recently, studies have mainly focused on fracture complexity and its impact on water recovery and productivity. The immediate impacts of treatment parameters such as the pumping rate and the ratio of slick water have not been reported yet.

In this paper, an integrated workflow was proposed to combine complex fracture modelling and oil-water production modelling. By complete consideration of shale geological mechanics, matrix imbibition and fracture closure, the whole migration of fracturing fluid from pumping to cleanup to production was simulated, as well as the history-match of flowback data gathered from the Jimsar shale oil reservoir in China. Then, the impacts of different engineering factors on fluid recovery and oil production were quantitatively studied. This work provides a theoretical basis for understanding the distribution of fracturing fluid, flowback efficiency and its impact on productivity, with a view to guiding a reasonable flowback schedule and promoting the efficient development of shale oil.

2. Methodology

2.1. Model Description

This section describes three parts. First, the single porosity model is used to study the fluid flow for a shale oil well. Next, complex fracture modelling is adopted to generate a fracture network. Finally, a pressure-sensitivity model is proposed to characterize the fluid distribution and fracture closure. Together they form a workflow to simulate the flowback and production processes following fracturing.

2.1.1. Governing Equations of Fluid Flow

The key assumptions for building this fluid flow model are shown here. (1) The stimulated reservoir area includes two systems (matrix and complex fracture) and outside the stimulated area is assumed to be a single media (matrix). (2) Oil and water flow in the matrix and fracture systems conform to Darcy's law, and gas exsolution from oil is negligible at reservoir conditions. (3) The matrix system considers the capillary imbibition effect, which is negligible in the fracture system. (4) Fluid and rock are both slightly compressible. The matrix system considers pressure-dependent permeability, and the fracture system considers porosity and permeability. (5) The fluid flow is an isothermal process. The governing equations of oil and water flow are as follows:

$$\frac{\partial}{\partial t}(\nabla S_o \rho_o \phi) + \nabla(\rho_o v_o) + q_o = 0 \tag{1}$$

$$\frac{\partial}{\partial t}(\nabla S_w \rho_w \phi) + \nabla(\rho_w v_w) + q_w = 0 \tag{2}$$

Auxiliary equations:

$$S_w + S_o = 1 \tag{3}$$

$$P_c = P_o - P_w \tag{4}$$

where S_o, S_w are the saturation of oil and water phase; ρ_o, ρ_w are the density of oil and water phase; Φ is the effective porosity; v_o, v_w are the volumetric velocity of oil and water phase; q_o, q_w are the sink or source term of oil and water phase per unit reservoir volume; and P_c, P_o and P_w are the capillary pressure, the pressure of oil and water phase, respectively.

2.1.2. Complex Fracture Modelling and Gridding

The unconventional fracture model (UFM) is adopted to simulate the complex fracture configuration of fractured shale from a commercial simulator with coupled geomechanics features [26]. In UFM, the generated fracture system fully considers shale physical properties, in situ stress, natural fractures and the interaction between hydraulic fractures and natural fractures [27]. The detailed properties of the basic model are derived from a typical shale oil well with 27 fracture stages in Jimsar sag, Xinjiang, China. To skip the huge calculation problem, this work is conducted with a single or partial segment of the whole horizontal interval, and the results can be scaled back to the whole stimulated well [22]. The simulated segment scale is 460 m $\times$ 500 m $\times$ 60 m. Also, the upper and lower barriers adjacent to the oil layer are set. According to logging interpretation and related reports [25,28], input parameters of shale physical and mechanical properties are summarized in Table 1, and discrete natural fracture parameters are illustrated in Table 2. In addition, the complete parameters of the typical well are shown in Table 3, to simulate the fracture propagation during the pumping process under field conditions. Based on the generated 3D fracture properties, an auto-gridding method is used to generate the unstructured grid from UFM [29].

Table 1. Basic reservoir input parameters.

Parameters	Upper Barrier	Oil Layer	Lower Barrier
Initial reservoir pressure, MPa	37	37	37
Reservoir thickness, m	20	20	20
Matrix porosity	0.03	0.11	0.03
Matrix permeability, mD	0.001	0.01	0.001
Matrix initial water saturation	0.7	0.2	0.7
Minimum horizontal stress, MPa	54	52	56
Maximum horizontal stress, MPa	60	58	62
Young's modulus, GPa	40	27	40
Poisson's ratio	0.35	0.25	0.35
Tensile Strength, MPa	3.45	3.45	3.45

Table 2. Input parameters for natural fracture generation.

Parameters	Fracture Length, m	Fracture Orientation, °	Fracture Interval, m
Mean	20	50	6.7
Standard deviation	0	10	0

Table 3. Input well completion and schedule parameters.

Parameters	Cluster Interval, m	Slickwater Viscosity, mPa·s	Gel Water Viscosity, mPa·s	Pumping Rate, m^3/min	Total Pumping Volume, m^3	Slickwater Ratio, %	Well Shut-In Time, d	Flowback Rate, m^3/d (Per Stage)
Value	10	5	200	14	1300	40	56	7.0

In this grid model, the generated fracture network is divided into the hydraulic fracture (HF) and natural fracture (NF). After that, the relative permeability and capillary pressure curves for matrix (M) and fracture systems are adopted from Chen and Jurus [25,30], respectively, as shown in Figure 1. Due to the high permeability, the capillary pressure for hydraulic fracture is ignored. And the PVT parameters of fluids are according to Liao [23]. Then, with the well schedule as shown in Table 3, the fluid flow behavior in the stimulated reservoir can be described. Here the gel breaking process of fracturing fluid is assumed to be completed immediately after pumping, so the viscosities of slick and gel water are both equal to 1 mPa·s.

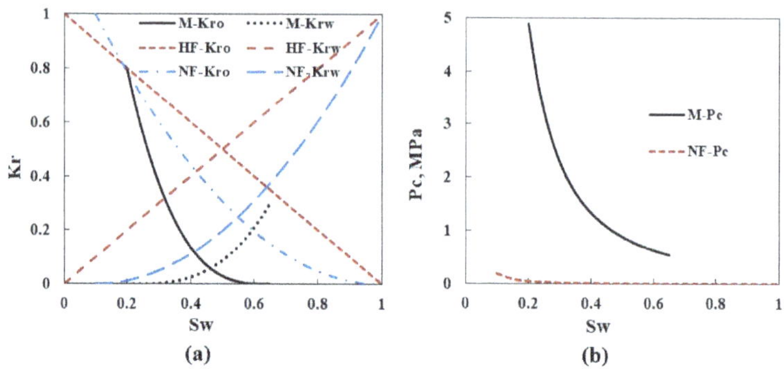

Figure 1. (**a**) Relative permeability curve. (**b**) Capillary pressure curve.

2.1.3. Pressure-Sensitivity Model Set Up

To study the fluid flow behavior, the primary task is to illustrate the distribution of reservoir pressure and saturation fields at the end of the pumping period. Therefore, oil-water production modelling needs to start with the injection of fluid. Recently, the pressure-dependent permeability of the matrix has been adopted to improve injection ability, but this method lacks rationality and overestimates the increase of shale matrix permeability [31]. Besides, it is hard to describe the key drive mechanism of fracture closure at the early period of flowback.

In this work, the decrease of matrix permeability with pressure during production is assigned according to Zhu [32]. For the pumping and flowback periods, a sensitivity model is proposed to take into account the corresponding change of porosity and permeability for fracture. This model includes two consecutive paths: (1) fracture dilation process during injection, and (2) fracture depletion process after injection. Fu indicated that fracture compressibility (C_f) could be two to three orders of magnitude higher than that of the matrix [11]. The formula Equations (5) and (6) are used to capture the significant changes in fracture porosity and permeability [23].

$$\frac{\phi_f}{\phi_{f0}} = e^{C_f P_{net}} \tag{5}$$

$$\frac{k_f}{k_{f0}} = 10^{m P_{net}} \tag{6}$$

where Φ_f is current fracture porosity; Φ_{f0} is initial fracture porosity; P_{net} is net pressure in fracture, MPa; k_f is current fracture permeability, mD; k_{f0} is initial fracture permeability, mD; and m is permeability changing factor, MPa^{-1}. The values of C_f and m can be obtained experimentally, or by data-fit if the experimental conditions are not available. In this paper, the C_f values of HF and NF obtained by the history-match of the pumping process are both 0.134, and the m values are 0.165 and 0.247, respectively. Also, for the flowback process, the matched C_f values are 0.016 and 0.033, and the m values are 0.048 and 0.099, respectively. The results of the pressure sensitivity model for complex fracture are presented in Figure 2.

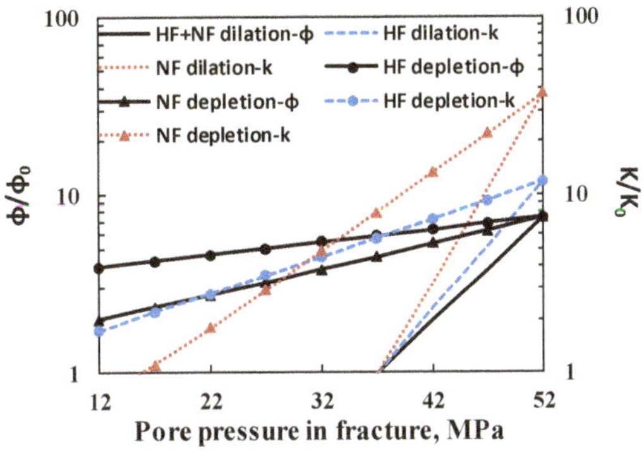

Figure 2. Pressure sensitivity curves for complex fracture from pumping to production period.

2.1.4. Integrated Workflow

As it is currently difficult to accomplish geomechanics and fluid flow mechanism within one coupled simulation at reservoir conditions, the entire workflow is divided into two steps (Figure 3). Firstly, with the input reservoir properties and treatment parameters, a complex fracture network is generated in UFM, and then finely described as an unstructured grid model. Secondly, the production modelling is conducted to forecast injected-fluid recovery and well productivity by considering capillary imbibition and pressure sensitivity in the matrix and fracture systems.

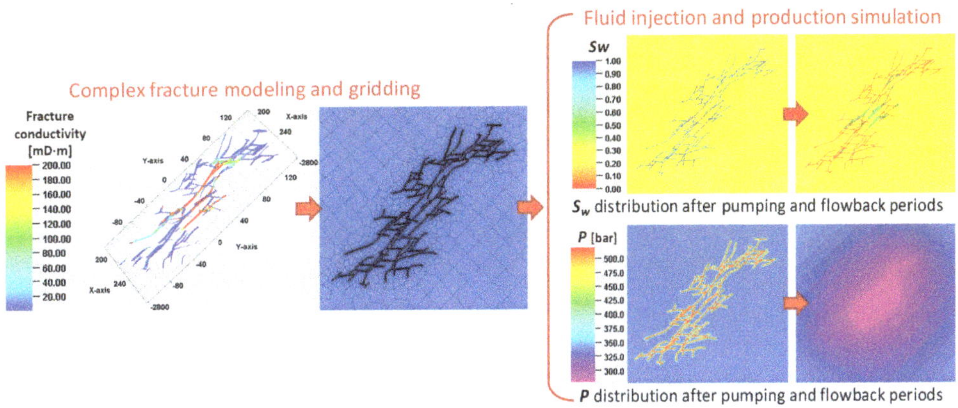

Figure 3. Numerical simulation workflow based on complex fracture modelling.

2.2. Model Validation

Due to the frequent change of choke in the early flowback stage, daily output fluctuates dramatically, which is the main challenge for quantitative validation of simulation results [22,33]. In this section, the field treatment scale (Table 3) was used to generate the complex hydraulic fracture (Figure 3). The calculated results were consistent with the micro-seismic interpretation as shown in Table 4, which verified the suitability of the proposed model to characterize complex fractures. After that, a trial-and-error method was carried out to realize the fitting between the actual and the calculated flowback performance. As the initial pressure fluctuation could easily terminate the simulation, daily liquid production data was adopted to match properly. As shown in Figure 4, the fitted results were quite acceptable, which proved the applicability and reliability of the presented workflow. Although acceptable quantitative fitting was completed, it was difficult to achieve an ideal match at the early and late stages of flowback. Since the choke size is frequently switched at the early period, higher frequency data is required to capture the production performance, which is hard for daily output. Moreover, wellhead pressure data are generally recorded rather than reliable bottom-hole pressure, which increases uncertainty for quantitative fitting.

Table 4. Comparison of hydraulic fracture geometry.

Parameters	Fracture Half-Length, m	Fracture Bandwidth, m	Fracture Height, m
Micro-seismic interpretation of each stage	60~189	50~140	33~68
Average micro-seismic values of each stage	116.4	93.6	50.7
Calculated results	125.3	89.7	44.6

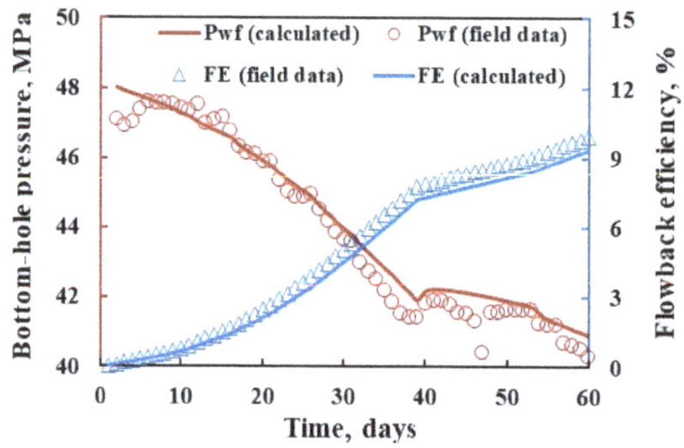

Figure 4. History-match results of bottom-hole pressure and flowback efficiency.

3. Simulation Results and Analysis

Different from conventional sandstone, due to the differences in geological features and fracturing technology, the migration of fracturing fluid in shale reservoirs is more complex. After stimulation, shale oil wells show different flowback performance from conventional ones: (1) the flowback efficiency is low and varies widely between wells; (2) there is a certain negative correlation between oil production and water recovery (flowback efficiency); (3) controlling cleanup helps increase oil production, etc. Currently, many discussions have been carried out about the impacts of shale geological factors on water imbibition and recovery, but the impacts of engineering factors on flowback behaviors are rarely studied. Therefore, this section applied the proposed models to perform a series of sensitivity analyses, which quantified the impacts of pumping rate, slick water ratio and cluster spacing. Furthermore, multi-stage fracture models were established, to investigate the impacts of stage spacing and flowback rate on water recovery and oil production.

3.1. Pumping Rate

Pumping rate is an important parameter to determine the effect of shale stimulation. Two cases of 12 and 16 m^3/min were simulated and compared with the basic model (14 m^3/min). The production lasts for 1 year, and the results of fracture configuration, pressure distribution, flowback efficiency and oil production are shown in Figures 5a and 6a.

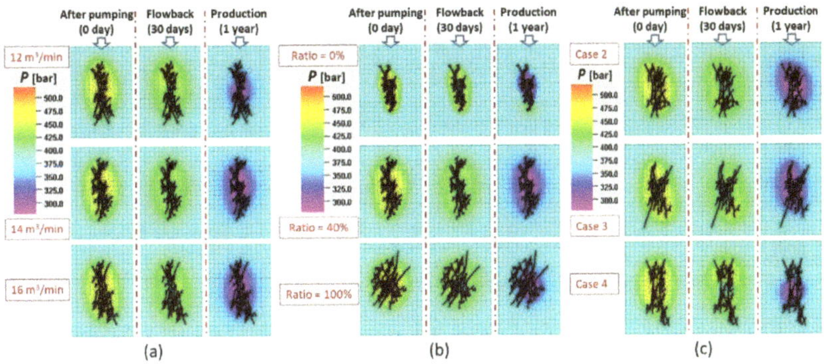

Figure 5. Variation of pressure distribution in oil layer: (**a**) pumping rate. (**b**) slick water ratio. (**c**) cluster spacing.

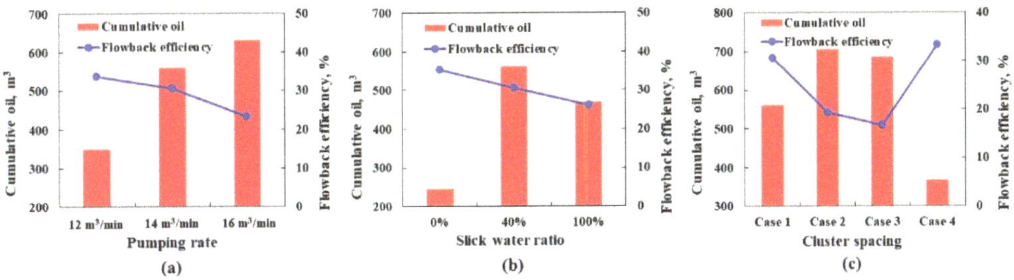

Figure 6. Cumulative oil and flowback efficiency: (**a**) pumping rate. (**b**) slick water ratio. (**c**) cluster spacing.

According to Figure 5a, with the increase in pumping rate, the fracture complexity in the stimulated zone will be improved, leading to a larger contact area between fracture and matrix. The imbibition is promoted further, resulting in more fracturing fluid remaining in the matrix pores near the fracture. So, as shown in Figure 6a, when the rate is 12, 14 and 16 m^3/min, the flowback efficiency is 33.7%, 30.5% and 23.2%, respectively, indicating that the water recovery decreases with the increase in pumping rate.

In terms of well productivity, the cumulative oil production increases with the pumping rate, but the growth slows down. As shown in Figure 5a, although the fracture complexity at 16 m^3/min is higher than that at 14 m^3/min, the diffusion area of pressure drop in the two cases is almost equal after one year of production. This is because, with the same proppant quantity, the redundant natural fractures are difficult to obtain propped and gradually inactivate.

3.2. Slick Water Ratio

Currently, hybrid fracturing fluids with the alternation of gel and slick water are widely used during the stimulation treatment of shale reservoirs. Two cases of 0% (cross-linked gel alone) and 100% (slick water alone) were simulated respectively, compared with the basic model (40%). The production lasts for 1 year, and the results of fracture configuration, pressure distribution, flowback efficiency and oil production are shown in Figures 5b and 6b.

Figure 6b reports that, when the ratio is 0%, 40% and 100%, the flowback efficiency is 35.3%, 30.5% and 26.1%, respectively, indicating that the water recovery decreases with the ratio of slick water. This is due to: (1) According to Figure 5b, the fracture complexity increases significantly with the ratio of slick water, which greatly enlarges the contact area between fracture and matrix. Thus, the amount of water imbibed and retained in the matrix increases. (2) The weak sand-carrying performance of slick water greatly reduces the fracture conductivity (F_{CD}) located in the oil layer. Figure 7 shows the F_{CD} distribution of the whole fracture system under the different ratio of slick water. Compared with the other two cases, the high permeable area near the wellbore shrinks significantly when 100% slick water is used.

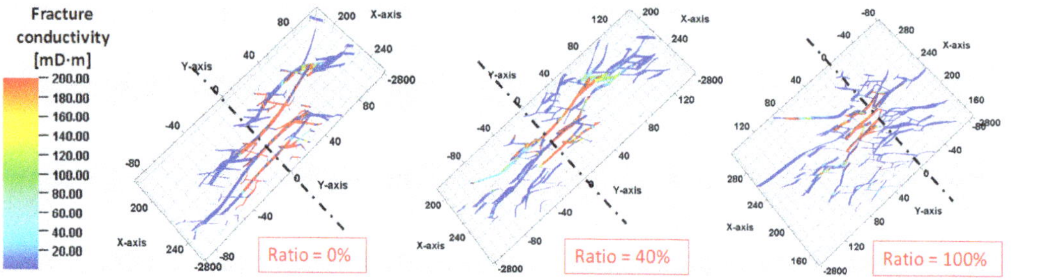

Figure 7. The F_{CD} distribution of fracture system under the different ratios of slick water (zoom scale Z = 0.2 times).

According to Figure 5b, when the slick water ratio is 40%, the diffusion area of pressure drop is the largest, followed by 100%, and 0% is the smallest. This trend is consistent with that of oil production in Figure 6b. Therefore, the slick water ratio may not be the higher the better, and just like the pumping rate, they all have an optimization interval. As long as the proppant quantity remains unchanged, the volume difference of the propped HF under different treatment parameters is often not very large. At this time, the reactivated NF connected with HF can play a role in increasing the drainage area. However, when the NFs are too complex, the connection between NFs is often closed due to the lack of effective support of proppant, resulting in the redundant NFs being unable to maintain long-term effective conductivity. Therefore, compared with blindly increasing the slick water ratio to form complex fractures, there may be an optimal ratio of hybrid fracturing fluids to improve the proppant carrying capacity and ensure that enough NFs are reactivated.

3.3. Cluster Spacing

The spacing or number of perforation clusters are important factors to determine the complexity of hydraulic fractures, which affect the fracturing fluid distribution and oil production performance. Within a single fracturing stage with a length of 45 m, three cases were simulated to compare with the basic case (case1, with 3 clusters and 10 m apart): (1) case 2, with 4 clusters and 12 m apart; (2) case 3, with 3 clusters and 20 m apart; (3) case 4, with 2 clusters and 30 m apart. The production lasts for 1 year, and the results of fracture configuration, pressure distribution, flowback efficiency and oil production are shown in Figures 5c and 6c.

As shown in Figure 5c, with the decrease in cluster spacing, the interference between fractures becomes more serious, resulting in more complex fractures. However, for multiple clusters, the expansion of the middle cluster is significantly worse than that of the two sides. Especially for case 2, the third cluster from the left failed to initiate. Therefore, from the fracture propagation results, the complexity of case 3 is not much different from that of case 2, and the stimulation near the well bore of case 4 is the worst. Likewise, the results of flowback efficiency and oil production are similar in case 2 and case 3, while the flowback efficiency in case 4 is the highest and the single well productivity is the lowest, as shown in Figure 6c.

The results indicate that the properly decrease of cluster spacing is conducive to increasing the contact area between fractures and matrix. But when the cluster spacing is less than 20 m, the interference may affect the stimulation effect. By comparing case 1 and case 3, it can also be realized that the decrease in cluster spacing is not conducive to oil production when the number of perforating clusters is the same.

3.4. Stage Spacing

Stage spacing also affects the flowback and production performance in shale oil reservoirs. Within a horizontal interval length of 225 m, three cases were simulated: (1) case 1, with 5 stages and 45 m apart; (2) case 2, with 4 stages and 56 m apart; and (3) case 3, with 3 stages and 75 m apart. Figure 8 shows the results of complex fracture propagation under different stage spacing. It can be seen that: (1) In all cases, the sum of single stage stimulation width is wider than the corresponding whole well stimulation width, indicating that the fracturing area of each stage has a certain degree of overlap. (2) With the decrease of spacing, the degree of overlap becomes higher. Moreover, compared with case 2, the overall fracture complexity of case 1 is relatively close, and only the local area near the well bore is more fully stimulated.

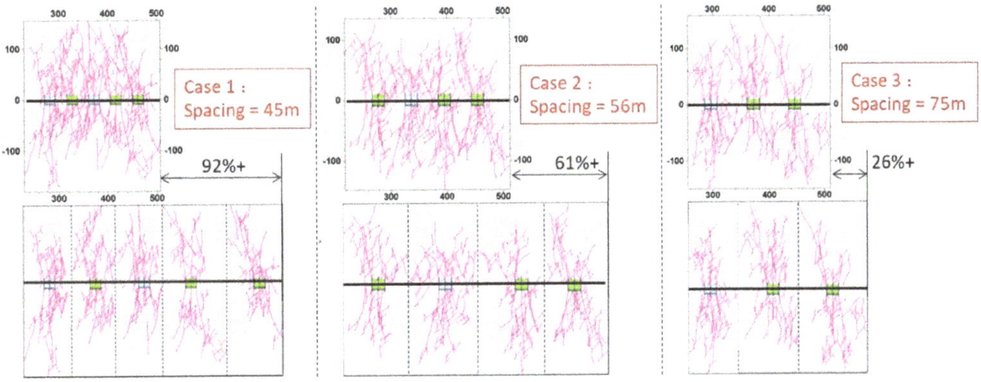

Figure 8. Complex fracture configuration under different stage spacing (global scale vs. individual scale).

Figure 9a shows that: (1) the load-recovery efficiencies of the three cases are 25.8%, 28.6% and 32.2%, respectively, which increase with the stage spacing. (2) Conversely, the oil production increases with the spacing narrowing, but the growth slows down. Furthermore, the average cumulative oil production of each individual stage is 447.7 m^3, 509.8 m^3 and 519 m^3 for the three cases, indicating that the production contribution of a single stage is positively correlated with its spacing. This is because, although the fracture complexity increases obviously with the reduction of stage spacing, the stimulation and drainage area of the individual stage is narrowed down accordingly.

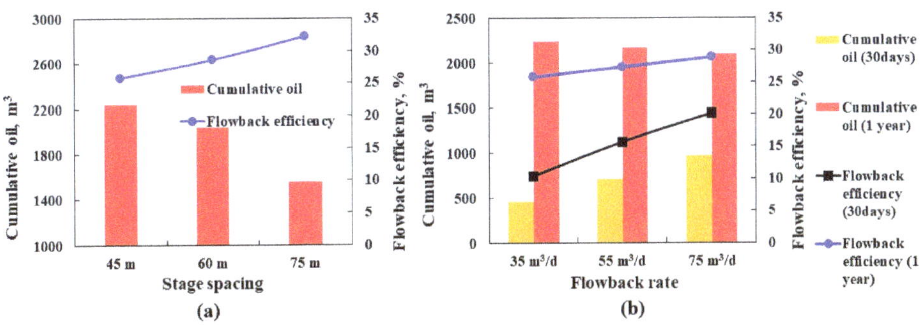

Figure 9. Cumulative oil and flowback efficiency profiles: (**a**) stage spacing. (**b**) flowback rate.

As shown in Figure 10a, the diffusion area of pressure drop in case 3 is the smallest, while those in the other two cases are relatively close after one year of production. It follows that for multi-stage fracturing treatment, reducing the stage spacing can improve productivity to some extent. However, if the total length of horizontal interval remains unchanged, it also means an increase in the number of fracturing stages, resulting in a rising fracturing operation cost and tool cost. Therefore, the design of stage spacing is not the smaller the better, and needs to be combined with the actual reservoir conditions.

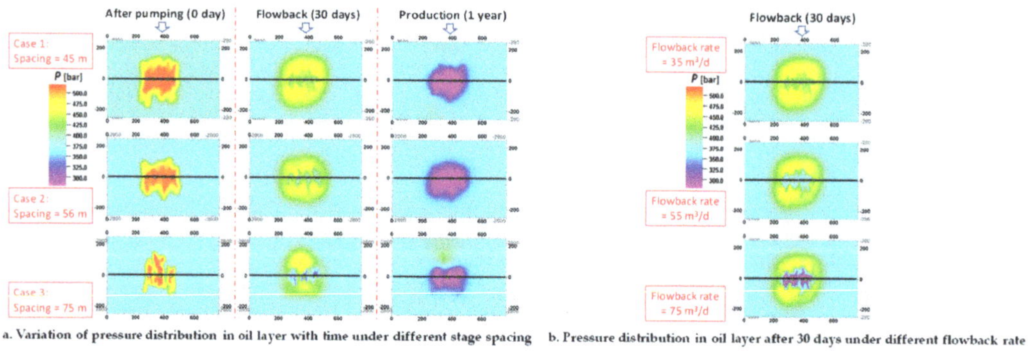

Figure 10. Variation of pressure distribution in oil layer: (**a**). stage spacing. (**b**). flowback rate.

3.5. Flowback Rate

The fracturing fluid recovery is controlled by the adjustment of choke size. Slow-back with a small choke size may delay the payback point, while fast-back with a large choke size may damage the fracture conductivity or matrix permeability [34]. To demonstrate the impact of choke size on well performance, three cases with different flowback rates of 35, 55 and 75 m^3/d were simulated.

Figure 9b shows that: (1) After 1 month of production, both the water recovery and cumulative oil production increase almost linearly with the flowback rate. (2) After 1 year of production, the water recovery is still positively correlated, while the oil production is negatively correlated with the flowback rate. This reversal is due to the fact that the fast-back will force fracture closure to accelerate, and the flowback fluid is mainly water, which has not yet entered the deep matrix. Thus, the pressure depletion will be excessively consumed in water recovery, which is not conducive to later oil drainage. Additionally, according to Figure 10b, the pressure depletion near the well bore increases with the flowback rate, resulting in gradually significant damage to fracture conductivity and matrix permeability. Therefore, it is suggested to adopt a slow-back mode with a small choke size when designing the cleanup schedule of shale oil fractured wells.

4. Conclusions

Based on the typical data of a shale oil fractured well, a series of numerical models coupling complex fracture propagation and flowback simulation were established. The history match of the basic model verifies the adaptability of the proposed workflow. Accordingly, the distributions of pore pressure and fracturing fluid underground during the pumping, cleanup and production periods are accurately described, which provides a reliable basis for the study of oil-water production performance.

Pumping rate, slick water ratio and cluster spacing all affect the flowback efficiency of shale fractured wells, which is mainly reflected in the complex fracture configuration differences formed under these engineering factors, thus changing the distribution law of fracturing fluid in the formation. With the increase in pumping rate, water recovery decreases and oil production increases. The flowback efficiency increases with the increase of the slick water ratio, but from the perspective of productivity there is an optimal ratio for hybrid fracturing fluid. Significant fracture interference may occur if the perforation clusters are too dense, so the cluster spacing is not the smaller the better.

With the decrease of stage spacing along the horizontal well, the fracture merging between adjacent fracturing stages becomes more significant. Thus, the fracture complexity increases and the water recovery decreases accordingly. Single well productivity increases with the decrease of stage spacing, but the increase gradually decreases. When the total length of the horizontal well remains the same, considering the cost of fracturing operations and tool use, the interval design is not the smaller the better, but needs to be combined with the actual reservoir conditions.

For the schedule of flowback in shale fractured wells, the slow-back mode should be adopted. This is because fast-back will cause the supplementary formation pressure during fracturing to be excessively consumed in the subsequent water recovery. It may also cause significant pressure-sensitive damage in the near-well area, which is not conducive to long-term production.

Author Contributions: Methodology, K.L.; validation, K.L., J.Z. and X.S.; writing—original draft preparation, K.L. and J.Z.; writing—review and editing, S.Z.; language check, G.R. All authors have read and agreed to the published version of the manuscript.

Funding: This research was funded by Research Foundation of China University of Petroleum-Beijing at Karamay (No. XQZX20220003); Natural Science Foundation of Xinjiang Uygur Autonomous Region (No. 2022D01B79).

Institutional Review Board Statement: Not applicable.

Informed Consent Statement: Not applicable.

Data Availability Statement: Not applicable.

Acknowledgments: This work was jointly supported by the Research Foundation of China University of Petroleum-Beijing at Karamay and the Natural Science Foundation of Xinjiang Uygur Autonomous Region.

Conflicts of Interest: The authors declare no conflict of interest.

References

1. Zhou, L.; Zhao, X.; Chai, G.; Jiang, W.; Pu, X.; Wang, X.; Han, W.; Guan, Q.; Feng, J.; Liu, X. Key exploration & development technologies and engineering practice of continental shale oil: A case study of Member 2 of Paleogene Kongdian Formation in Cangdong Sag, Bohai Bay Basin, East China. *Pet. Explor. Dev.* **2020**, *47*, 1059–1066.
2. Pu, C.; Zheng, H.; Yang, Z.; Gao, Z. Research status and development trend of the formation mechanism of complex fractures by staged volume fracturing in horizontal wells. *Acta Pet. Sin.* **2020**, *41*, 1734–1743.
3. Palisch, T.T.; Vincent, M.C.; Handren, P.J. Slickwater Fracturing: Food for Thought. *SPE Prod. Oper.* **2010**, *25*, 324–344. [CrossRef]
4. Gao, Z.; Qu, X.; Huang, T.; Xue, T.; Cao, P. Stress sensitivity analysis and optimization of horizontal well flowback system for shale oil reservoir in Ordos Basin. *Nat. Gas Geosci.* **2021**, *32*, 1867–1873.
5. Dutta, R.; Lee, C.H.; Odumabo, S.; Ye, P. Experimental Investigation of Fracturing-Fluid Migration Caused by Spontaneous Imbibition in Fractured Low-Permeability Sands. *SPE Reserv. Eval. Eng.* **2014**, *17*, 74–81. [CrossRef]
6. Hun, L.; Shicheng, Z.; Fei, W.; Ziqing, P.; Jianye, M.; Tong, Z.; Zongxiao, R. Experimental Investigation on Imbibition-front Progression in Shale Based on Nuclear Magnetic Resonance. *Energy Fuels* **2016**, *30*, 9097–9105. [CrossRef]
7. Xu, G.; Shi, Y.; Jiang, Y.; Jia, C.; Gao, Y.; Han, X.; Zeng, X. Characteristics and Influencing Factors for Forced Imbibition in Tight Sandstone Based on Low-Field Nuclear Magnetic Resonance Measurements. *Energy Fuels* **2018**, *32*, 8230–8240. [CrossRef]
8. Xu, G.; Jiang, Y.; Shi, Y.; Han, Y.; Wang, M.; Zeng, X. Experimental Investigation of Fracturing Fluid Flowback and Retention under Forced Imbibition in Fossil Hydrogen Energy Development of Tight Oil Based on Nuclear Magnetic Resonance. *Int. J. Hydrogen Energy* **2020**, *45*, 13256–13271. [CrossRef]
9. McClure, M. The potential effect of network complexity on recovery of injected fluid following hydraulic fracturing. In Proceedings of the SPE Unconventional Resources Conference, The Woodlands, TX, USA, 1–3 April 2014.
10. Ezulike, D.O.; Dehghanpour, H.; Virues, C.J.; Hawkes, R.V.; Jones, R.S. Flowback Fracture Closure: A Key Factor for Estimating Effectuve Pore Volume. *SPE Reserv. Eval. Eng.* **2016**, *19*, 567–582. [CrossRef]
11. Fu, Y.; Dehghanpour, H.; Ezulike, D.O.; Jones, R.S.J. Estimating effective fracture pore volume from flowback data and evaluating its relationship to design parameters of multistage-fracture completion. *SPE Prod. Oper.* **2017**, *32*, 423–439. [CrossRef]

12. Alamdari, B.B.; Kiani, M.; Kazemi, H. *Experimental and Numerical Simulation of Surfactant-Assisted Oil Recovery in Tight Fractured Carbonate Reservoir Cores*; SPE Improved Oil Recovery Symposium: Tulsa, OK, USA, 2012.
13. Jing, W.; Huiqing, L.; Genbao, Q.; Yongcan, P.; Yang, G. Investigations on spontaneous imbibition and the influencing factors in tight oil reservoirs. *Fuel* **2019**, *236*, 755–768. [CrossRef]
14. Wang, C.; Cui, W.; Zhang, H.; Qiu, X.; Liu, Y. High efficient imbibition fracturing for tight oil reservoir. In Proceedings of the SPE Trinidad and Tobago Section Energy Resources Conference, Port of Spain, Trinidad and Tobago, 25–26 June 2018.
15. Dehghanpour, H.; Lan, Q.; Saeed, Y.; Fei, H.; Qi, Z. Spontaneous Imbibition of Brine and Oil in Gas Shales: Effect of Water Adsorption and Resulting Microfractures. *Energy Fuels* **2013**, *27*, 3039–3049. [CrossRef]
16. Bennion, D.B.; Thomas, F.B.; Bietz, R.F.; Bennion, D.W. Water and hydrocarbon phase trapping in porous media-diagnosis, prevention and treatment. *J. Can. Pet. Technol.* **1996**, *35*, 29–36. [CrossRef]
17. Wang, M.; Leung, J.Y. Numerical investigation of fluid-loss mechanisms during hydraulic fracturing flow-back operations in tight reservoirs. *J. Pet. Sci. Eng.* **2015**, *133*, 85–102. [CrossRef]
18. Khormali, A.; Petrakov, D.G.; Farmanzade, A.R. Prediction and Inhibition of Inorganic Salt Formation under Static and Dynamic Conditions–Effect of Pressure, Temperature, and Mixing Ratio. *Int. J. Technol.* **2016**, *7*, 943–951. [CrossRef]
19. Khormali, A.; Bahlakeh, G.; Struchkov, I.; Kazemzadeh, Y. Increasing inhibition performance of simultaneous precipitation of calcium and strontium sulfate scales using a new inhibitor—Laboratory and field application. *J. Pet. Sci. Eng.* **2021**, *202*, 108589. [CrossRef]
20. Zhang, T.; Li, X.; Yang, L. Effects of shut-in timing on flowback rate and productivity of shale gas wells. *Nat. Gas Ind.* **2017**, *37*, 48–60.
21. Zhang, Z.; Clarkson, C.; Williams-Kovacs, J.D.; Yuan, B.; Ghanizadeh, A. Rigorous Estimation of the Initial Conditions of Flowback Using a Coupled Hydraulic-Fracture/Dynamic-Drainage-Area Leakoff Model Constrained by Laboratory Geomechanical Data. *SPE J.* **2020**, *25*, 3051–3078. [CrossRef]
22. Zhang, T.; Li, X.; Li, J.; Feng, D.; Li, P.; Zhang, Z.; Chen, Y.; Wang, S. Numerical investigation of the well shut-in and fracture uncertainty on fluid-loss and production performance in gas-shale reservoirs. *J. Nat. Gas Sci. Eng.* **2017**, *46*, 421–435. [CrossRef]
23. Liao, K.; Zhang, S.; Ma, X.; Zou, Y. Numerical Investigation of Fracture Compressibility and Uncertainty on Water-Loss and Production Performance in Tight Oil Reservoirs. *Energies* **2019**, *12*, 1189. [CrossRef]
24. Li, L.; Jiang, H.; Li, J.; Wu, K.; Meng, F.; Chen, Z. Modeling tracer flowback in tight oil reservoirs with complex fracture networks. *J. Pet. Sci. Eng.* **2017**, *157*, 1007–1020. [CrossRef]
25. Chen, X.; Liao, K.; Lv, Z.; Zhu, J.; Wang, J.; Li, Y.; Wang, F. Numerical Simulation Study on Optimal Shut-in Time in Jimsar Shale Oil Reservoir. *Front. Energy Res.* **2022**, *10*, 849064. [CrossRef]
26. Weng, X.; Kresse, O.; Chuprakov, D.; Cohen, C.E.; Prioul, R.; Ganguly, U. Applying complex fracture model and integrated workflow in unconventional reservoirs. *J. Pet. Sci. Eng.* **2014**, *124*, 468–483. [CrossRef]
27. Yan, X.; Mou, J.; Tang, C.; Xin, H.; Zhang, S.; Ma, X.; Duan, G. Numerical Investigation of Major Impact Factors Influencing Fracture-Driven Interactions in Tight Oil Reservoirs: A Case Study of Mahu Sug, Xinjiang, China. *Energies* **2021**, *14*, 4881. [CrossRef]
28. Baocheng, W.U.; Jianmin, L.I.; Yuanyue, W.U.; Le, H.; Tingfeng, Z.; Yushi, Z. Development practices of geology-engineering integration on upper sweet spots of Lucaogou Formation shale oil in Jimsar sag, Junggar Basin. *China Pet. Explor.* **2019**, *24*, 679–690.
29. Cipolla, C.L.; Fitzpatrick, T.; Williams, M.J.; Ganguly, U.K. Seismic-to-simulation for unconventional reservoir development. In Proceedings of the SPE Reservoir Characterisation and Simulation Conference and Exhibition, Abu Dhabi, United Arab Emirates, 9–11 October 2011.
30. Jurus, W.J.; Whitson, C.H.; Golan, M. Modeling water flow in hydraulically-fractured shale wells. In Proceedings of the SPE Annual Technical Conference and Exhibition, New Orleans, LA, USA, 30 September–2 October 2013.
31. Ehlig-Economides, C.A.; Ahmed, I.; Apiwathanasorn, S.; Lightner, J.; Song, B.; Vera, F.; Xue, H.; Zhang, Y. Stimulated shale volume characterization: Multiwell case Study from the Horn River shale: II. Flow perspective. In Proceedings of the SPE Annual Technical Conference and Exhibition, San Antonio, TX, USA, 8–10 October 2012.
32. Zhu, W.; Ma, D.; Zhu, H.; An, L.; Li, B. Stress sensitivity of shale gas reservoir and influence on productivity. *Nat. Gas Geosci.* **2016**, *27*, 892–897.
33. Wang, M.; Leung, J.Y. Numerical investigation of coupling multiphase flow and geomechanical effects on water loss during hydraulic-fracturing flowback operation. *SPE Reserv. Eval. Eng.* **2016**, *19*, 520–537. [CrossRef]
34. Fu, Y.; Dehghanpour, H.; Motealleh, S.; Lopez, C.M.; Hawkes, R. Evaluating fracture volume loss during flowback and its relationship to choke size: Fastback vs. slowback. *SPE Prod. Oper.* **2019**, *34*, 615–624. [CrossRef]

processes

MDPI

Article

Fracture Characteristics and Distribution in Slant Core from Conglomerate Hydraulic Fracturing Test Site (CHFTS) in Junggar Basin, Northwest China

Shanzhi Shi [1], Renyan Zhuo [2,*], Leiming Cheng [1], Yuankai Xiang [1], Xinfang Ma [2] and Tao Wang [1]

1 Engineering Technology Research Institute, PetroChina Xinjiang Oilfield Company, Karamay 834000, China
2 State Key Laboratory of Petroleum Resources and Prospecting, China University of Petroleum-Beijing, Beijing 102249, China
* Correspondence: zhuorenyancucbm@163.com

Abstract: Hydraulic fracture networks, especially fracture geometry, height growth, and proppant transport within the networks, present a critical influence on productivity evaluation and optimization of fracturing parameters. However, information about hydraulic fracture networks in post-fractured formations is seldom available. In this study, the characteristics (density and orientation) of hydraulic fractures were obtained from field observations of cores taken from conglomerate hydraulic fracturing test site (CHFTS). A large number of fractures were observed in the cores, and systematic fracture description was carried out. The fracture analysis data obtained includes fracture density, fracture depth, fracture orientation, morphology, fracture surface features, apertures, fill, fracture mechanical origin (type), etc. Our results show that 228 hydraulic fractures were intersected in a span of 293.71 m of slant core and composed of irregularly spaced single fractures and fracture swarms. One of the potential sources of the observed fracture swarms is near-wellbore tortuosity. Moreover, for regions far away from the wellbore, reservoir heterogeneity can promote complex hydraulic fracture trajectories. The hydraulic fractures were mainly cross-gravel and high-angle fractures and align with maximum horizontal stress (S_{Hmax}) $\pm$ 15°. The fracture density, orientations, and types obtained from the core fracture description provided valuable information regarding fracture growth behavior. For the near-wellbore area with a transverse distance of less than 25 m from the hydraulically-fractured wellbore, tensile fractures were dominant. While for the area far away from the wellbore, shear fractures were dominant. Our results provide improved understanding of the spatial hydraulic fracture dimensions, proppant distribution, and mechanism of hydraulic fracture formation. The dataset acquired can also be used to calibrate numerical models and characterize hydraulic fracture geometry and proppant distribution.

Keywords: CHFTS; slant coring well; hydraulic fracture characteristics; fracture swarms; hydraulic fracture formation mechanism

Citation: Shi, S.; Zhuo, R.; Cheng, L.; Xiang, Y.; Ma, X.; Wang, T. Fracture Characteristics and Distribution in Slant Core from Conglomerate Hydraulic Fracturing Test Site (CHFTS) in Junggar Basin, Northwest China. *Processes* **2022**, *10*, 1646. https://doi.org/10.3390/pr10081646

Academic Editors: Linhua Pan, Yushi Zou, Jie Wang, Minghui Li, Wei Feng and Lufeng Zhang

Received: 22 July 2022
Accepted: 16 August 2022
Published: 19 August 2022

1. Introduction

Tight conglomerate reservoirs are unable to obtain natural productivity due to their poor physical properties and hydraulic fracturing is thus strongly demanded. Conglomerate reservoirs are characterized by strong heterogeneity and large horizontal stress difference, which brings great challenges to the process of fracturing [1]. The primary challenge is to define artificial fracture morphology in conglomerate formations.

The artificial fracture propagation mode during the rock failure process in conglomerate reservoirs mainly comprises three types: penetrating through gravel, bypassing gravel, and embedding in gravel, but the mechanism of artificial fracture propagation is deeply understood. Former studies regarding the mechanical characteristics and fracture propagation mechanism of conglomerate reservoirs were carried out through physical experiments on surface rocks and finite element simulation based on different materials.

While these techniques provided many useful insights, the verification of results through direct the observation of hydraulic fractures is needed, and the best method of achieving this verification is to examine core intervals that have been hydraulically fractured [2].

The knowledge of the extent and density of hydraulic fracture networks in conglomerate reservoirs is generally limited. The hydraulic fracture distribution in multi-stage and multi-well operations is irregular and can be difficult to predict even with robust subsurface constraints [3]. During the stimulation of stages with multiple perforation clusters, complex interactions between fractures take place, resulting in different numbers of propagated fractures, which were usually considered greater than one but less than the number of the perforation clusters [4]. Whereas recent subsurface data from Hydraulic Fracturing Test Site-1 (HFTS1) in the Midland Basin [2,5] and Hydraulic Fracturing Test Site-2 (HFTS2) in the Delaware Basin [6,7] indicated that hydraulic fractures were not evenly distributed through the slant cores, they tended to occur in clusters, and the number of fractures was generally greater than the number of perforations [8]. Previous work in HFTS1 and HFTS2 has shown the value of cores, which were recovered from a stimulated volume, in providing information on the geometry and extent of hydraulic fractures [9,10]. The core fracture information is basic data regarding reservoir-scale simulations of hydraulic fracturing and production. Field data from the Hydraulic Fracturing Field Test (HFTS) provided upscaled parameters for calibrating the reservoir scale hydraulic fracturing model, which could accurately capture both the average length and the total aperture of the fractures in the fracture swarms [11,12]. Based on the hydraulic fracturing model calibrated to the HFTS, a fully coupled hydraulic fracturing, reservoir, and geomechanic simulator was used to perform an economic optimization of design parameters, including well spacing, landing depth, and completion design parameters [13].

The CHFTS is in the Mahu oilfield of the Junggar basin, the slant core well is located near hydraulically fractured wells. A slant core through the stimulated volume was acquired above and below the adjoining stimulated wells, which provided direct information about hydraulic fractures. In this paper, we characterized and measured hydraulic fractures, drilling-induced fractures, and core-cutting-induced fractures in the slant core, and further classified the hydraulic fractures according to their characteristics and mechanical origin. Moreover, the hydraulic fracture density and spatial distribution were quantified based on the acquired fracture dataset. The results provide a basis for understanding hydraulic fracture characteristics and the mechanical mechanism of the conglomerate reservoir. Findings can help verify indirect diagnostic results, such as microseismic monitoring and tracer monitoring.

2. CHFTS Project Overview

2.1. Test Site

CHFTS is a field-based hydraulic fracturing research experiment performed in the Junggar basin. Figure 1 is a 3D view of the CHFTS wells. A total of eleven horizontal wells were drilled in T_1b_3 and T_1b_2 formations, in which seven wells are T_1b_3 with a well spacing of 100 m and five wells are T_1b_2 with a well spacing of 150 m. The horizontal section length of the horizontal wells are 1800 m, and the measured depths are 4597–5040 m. The horizontal wells were drilled from north to south in a three-dimensional staggered arrangement, which was approximately perpendicular to the predicted direction of maximum horizontal stress. The thickness of the T_1b_2 formation is 6–8 m, and the thickness of the T_1b_3 formation is 14–16 m. The T_1b_2 and T_1b_3 wells are separated vertically with an interlayer of approximately 13–20 m thickness. The reservoir lithology is dominated by conglomerates (gravel diameter 5–70 mm). The gravel composition is mainly pyrolith, followed by metamorphic rock. The inter-gravel is mainly filled with sand, mud, or fine gravel and the overall reservoir is highly heterogeneous.

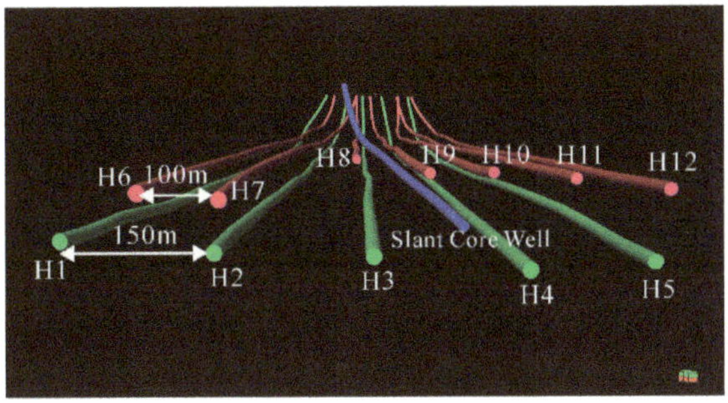

Figure 1. Three-dimensional view of CHFTS wells. Wells in red are T_1b_3 and those in green are T_1b_2. The slant core well is in blue, and it passes through the fracture networks of wells H8/H9 in the T_1b_3 formation and well H4 in the T_1b_2 formation.

2.2. Completion Overview

The 12 horizontal wells were divided into three factory operating well pads. The main body adopted a single segment with three clusters. The cluster spacing was 20 m. Meanwhile, three types of tests were conducted: cluster spacing (10 m/20 m/30 m), proppant concentrations (1.0–1.8 m^3/m), and limited entry and temporary plugging fracturing technology.

Fracture growth is a point of concern during the completion of the adjacent horizontal wells. Thus, comprehensive monitoring data were collected, including advanced diagnostics, such as microseismic tracer data [14]. Figure 2 shows the relative position of the horizontal wells, test wells, and slant core well. A microseismic was deployed in eleven wells to help identify hydraulic fracture dimensions (H4 was microseismic monitoring well), and fluid tracers (water-based and oil-based) were placed in two wells to evaluate horizontal well profile heterogeneity.

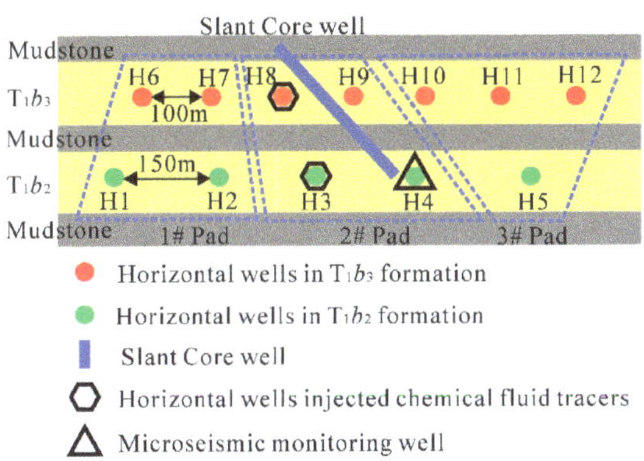

Figure 2. Elevation view of CHFTS wells, showing the relative position of the horizontal wells, test wells, and slant core well.

2.3. Coring through the Stimulated Reservoir Volume (SRV) and the Results

The prominent task of CHFTS is to acquire a four-inch diameter whole core in close proximity to both the T_1b_3 and T_1b_2 formation wells. The well that accomplished this task was the slant core well positioned within the 2# operating well pad (horizontal wells are H8/H9/H3/H4), with an azimuth of 175.0° and an inclination of 80.3°. Figure 3 shows a perspective view of the slant core well trajectory with respect to the adjacent horizontal wells. The core well is shown as a blue line in Figure 3 and will hereafter be referred to as S1. In the T_1b_3 formation, S1 is located between two horizontal wells as H8 and H9, near the middle of the horizontal section. The cores were drilled from the east side of H8 with a lateral distance of 18.6 m from the nearest completion stage, and sloped down to the bottom of H9 with a vertical distance of 14.6 m. In the T_1b_2, S1 is close to H4 on the west side, the lateral distance between the cores and completion stage is in the scope of 20.3–51.8 m.

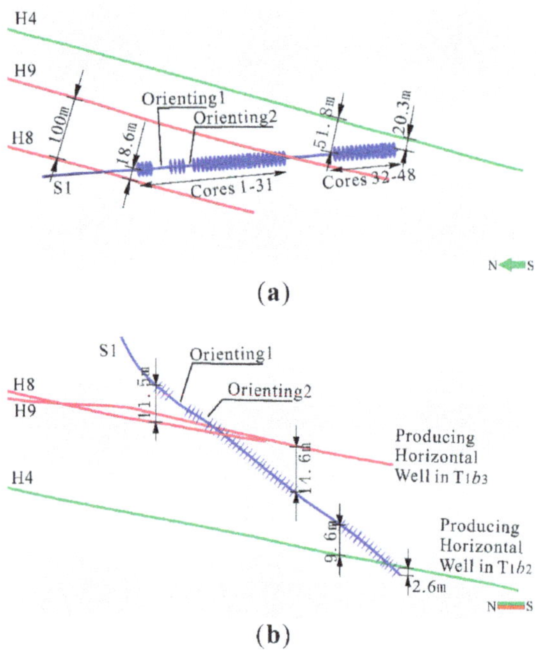

Figure 3. The slant core well trajectory with respect to the adjacent horizontal wells. (**a**) Vertical view; (**b**) side view.

The S1 trajectory enabled the collection of cores with varying lateral distances from the adjacent producing wells, providing insights into both the vertical and horizontal fracture geometries [15].

This slant core well recovered approximately 323.13 m of core, and 293.71 m of those were applied for study; a total of 48 cores, each with a length of about 6.5 m. Cores 1 to 31 are located in the T_1b_3 formation, sized approximately 194.95 m, and cores 32 to 48 are located in the T_1b_2 formation, sized approximately 98.76 m.

3. Slant Core Methods: Handling and Process for Core Description

A complete set of slant core methods was presented, as shown in Figure 4. Each of the 4-inch-diameter cores was contained within an aluminum tube. To conduct the CT scans of the core, the core was cut into segments of about 1 m length without removal from the aluminum tube, which assists in maintaining the integrity of the core. The core CT scans were utilized to compare the core description dataset to differentiate in situ fractures from fractures created while removing the core from the aluminum tube. Then, the core was removed from the aluminum tube. The methodology for extracting cores was required to maintain the condition of the core and fractures as close to their original state. A clam-shell core barrel extraction method was utilized, referred to as clam-shell methodology, via cutting the core barrel along its length on both sides to then exposing the core for study [15].

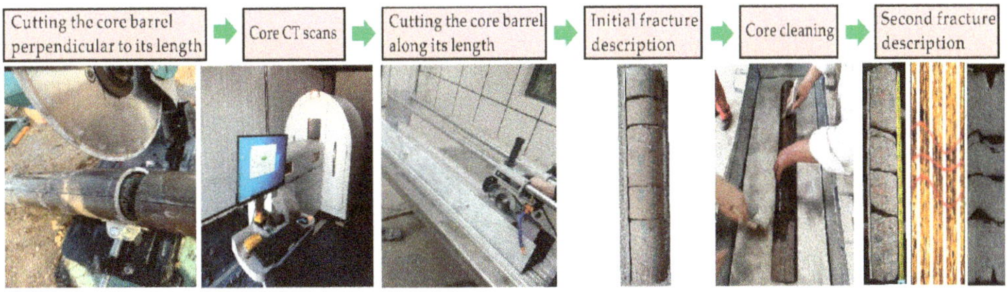

Figure 4. Workflow chart of the slant core methods: handling and process for core description.

The recovered core provides a unique opportunity to obtain a high-quality research dataset of hydraulic fracture networks in post-fractured formations. The systematic core description was conducted as follows:

The recovered core captured hundreds of fractures, which were numbered from top to bottom, denoted as "# core-segment-fracture". The fracture depth was determined by measuring the length of the core segment and the length from the fracture to the top of the core segment.

Initial fracture description was performed prior to core cleaning. Comprehensive fracture description data include fracture number, fracture depth, fracture orientation, morphology/fracture surface features, apertures, fill, fracture mechanical origin (type), etc.

Since proppant was pumped during the fracturing slurry in the adjacent horizontal wells, it was anticipated that proppant would be discovered in the collected whole core within hydraulic fractures. Following initial fracture description, sludge residue from coring operations on all fracture faces, the exterior core surface, and within the core sleeves, including drilling mud, rock cuttings, proppant, and aluminum shavings from the clamshell process, was recovered for detecting proppant [15].

A second fracture description was performed after core cleaning. Removing sludge residue on all core surfaces provided a clear view of the lithologic interface, gravel size, and gravel morphology on the fracture surface, etc. The repositioning of the core fractures to the in situ position was executed via combining the characteristics of lithologic interfaces with fractures through interpretation by FIM image logging. Considering that the core barrel had been slightly offset and twirled in the slant core wellbore during the process of coring, the core fractures repositioning assisted in correcting the fracture depth, fracture orientation, and dip angle. The fractures interpreted by FIM image logging were compared with the core fractures at the corresponding depth to identify hydraulic fractures.

Few natural fractures were observed, both filled and unfilled. Criteria was developed for distinguishing between hydraulic (tensile and shear), drilling-induced, and core cutting-induced fractures by examining the features of all fractures, combining a CT scan of the core and FIM image logging.

4. Fracture Characteristics

4.1. Hydraulic Fracture

A total of 228 hydraulic fractures were observed in the slant core with an average fracture density of 0.78 fractures per meter. According to the characteristics of fractures formed under different mechanical conditions, hydraulic fractures were further subdivided into tensile fractures and shear fractures, of which 52 are tensile fractures and 176 are shear fractures. The tensile fractures are the principal fractures, and the shear fractures swarms are adjacent to the principal fractures.

The primary evidence used to identify the hydraulic fractures was FIM image logging, presenting broad dark bands. Moreover, CT scans of the core were utilized to eliminate core cutting-induced fractures. The fracture morphology features of the CHFTS slant core mainly included straight, microwave, and crushed zones, as shown in Figure 5. The fracture edges are incomplete and mostly scattered with gravel. The weak plane with low cementation strength is prone to forming a crushed zone.

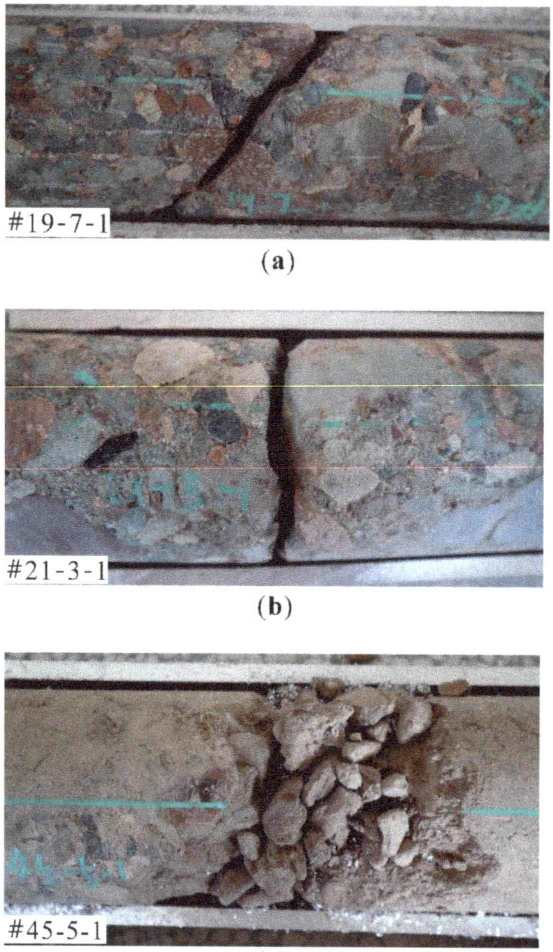

Figure 5. Morphology features of hydraulic fractures. (**a**) Straight hydraulic fracture, oriented 107°, deviates slightly from the east–west direction. (**b**) Microwave hydraulic fracture. The fracture edge is incomplete. (**c**) Crushed zone with scattered gravel.

The hydraulic fracture surfaces are all rough and uneven, many have through-penetrating gravel surfaces and some have bypassing gravel surfaces. In some other cases, both through-penetrating gravel and bypassing gravel existed in a single surface, as shown in Figure 6.

(a)

(b)

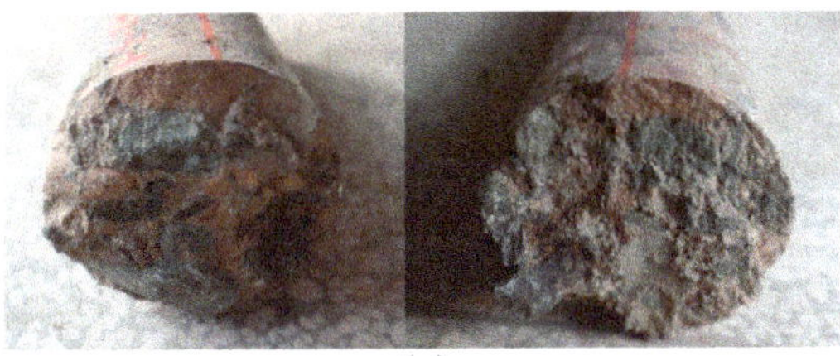

(c)

Figure 6. Surface character of hydraulic fractures. (**a**) Through-penetrating gravel surfaces, planar surface with split gravel. (**b**) Bypassing gravel surfaces, uneven surface with raised unbroken gravel. (**c**) Bypassing gravel and through-penetrating gravel occur in a single surface.

Figure 7 shows the filling materials in hydraulic fractures. Hydraulic fractures in the core are completely open with large apertures and the fracture surfaces are filled with materials, including mud, mud sediment, and proppant.

(a)

Figure 7. *Cont.*

(b)

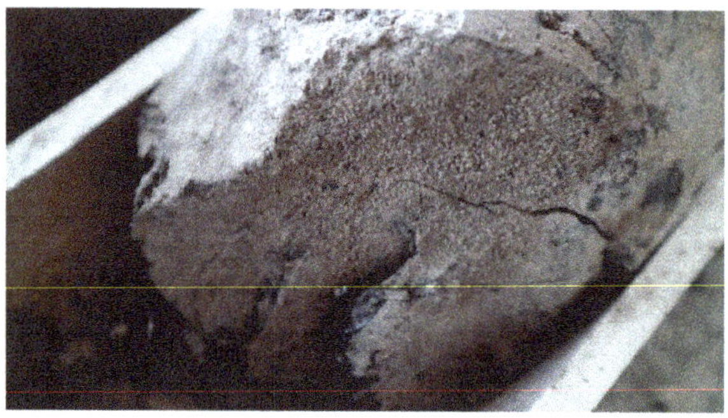

(c)

Figure 7. Filling materials in hydraulic fracture. (**a**) A film of dried mud on the surface. (**b**) Thick drilling mud sediment in the fracture. (**c**) Proppant attached to the fracture surface.

All fractures are high-angle, largely in the east–west direction, oriented at 70°–110°, and align with the maximum horizontal stress (±15°). Difference in the morphology of the created fractures was commonly observed in cores, and the orientations of certain through-penetrating gravel fractures deviate slightly from the main direction (Figure 5a). Changes of those kind are evidence for the formation of complex fracture networks.

Hydraulic fractures are unevenly distributed along the coring wellbore, and they occurred in both single and clusters [6]. Doublets and fracture swarms are quite common. For example, doublets (Figure 8a), triplets (Figure 8b), a 5-fracture swarm (Figure 8c), and even a 9-fracture swarm (Figure 8d) were extant.

(a)

Figure 8. *Cont.*

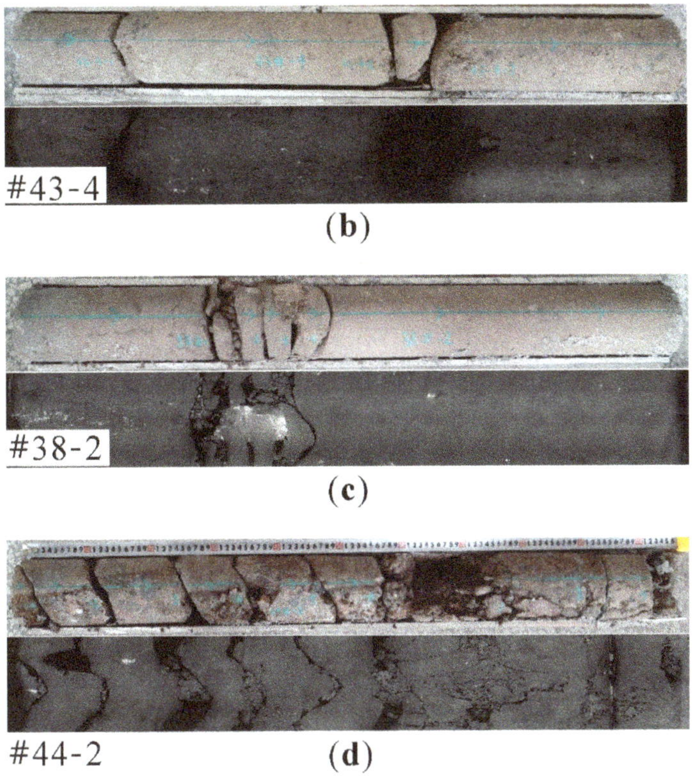

Figure 8. Fracture swarms in the slant core. (**a**) Doublets; (**b**) triplets; (**c**) 5-fracture swarm; (**d**) 9-fracture swarm.

The characteristics of hydraulic fractures are controlled by the dynamic geostress field. In the near-wellbore area, under the condition of low stress difference and high pore pressure, tensile failure mainly occurs, and the tensile fractures extend along the weak-cemented planes or gravel edge with large effective apertures. Proppant enters the wedge-shaped fractures and forms the propped fractures. In the area far from the wellbore, under the conditions of high dynamic stress difference at the fracture tip, shear failure mainly occurs, and the shear fracture extends along the through-penetrating gravel fracture surface or forms a crushed zone. The shear displacement leads to a certain volume expansion of the fracture and forms a self-supported fracture.

According to fracture characteristics under different mechanical states, hydraulic fractures were subdivided into tensile fractures and shear fractures. Tensile fracture damage along the weak cementation surfaces of gravel and matrix under tensile stress forms gravel-edge fractures. Thus, tensile fractures are mostly microwave in morphology, with a large aperture; they have uneven surfaces, mainly have bypassing gravel surfaces, shown as Figure 9.

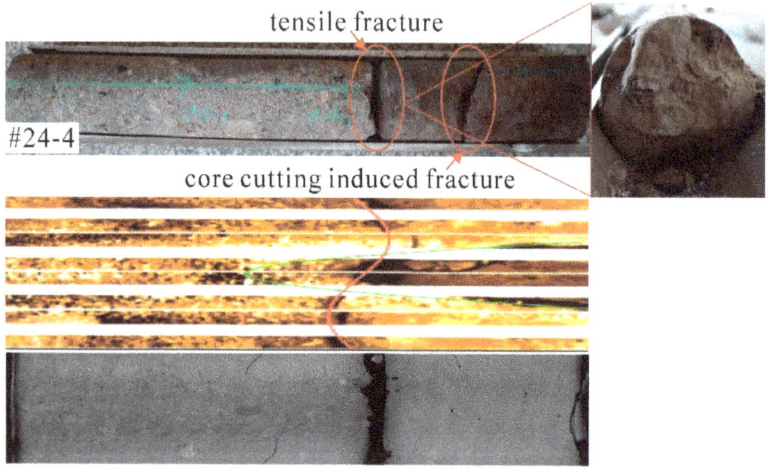

Figure 9. Tensile fracture is microwave in morphology and has a bypassing gravel surface.

Shear fractures are created by an induced stress field with the growth of the principal hydraulic fractures along natural weak planes. These shear fractures can also be formed by shear failure or the bifurcation and splitting of principal fractures during propagation. Thus, shear fractures are mainly straight and crushed zones in morphology. These fractures mainly have small apertures and through-penetrating gravel surfaces. Where shear fractures occur, fracture orientations change slightly and fracture swarms and crushed zones are common. Figure 10 shows conjugate shear fracture and crushed zone in the slant core. Conjugate shear fractures were observed in cores, forming scissor-shaped fracture planes. Moreover, crushed zones are prone to forming when shear slip occurs at the weak-cemented planes; some of these occur close to fracture swarms.

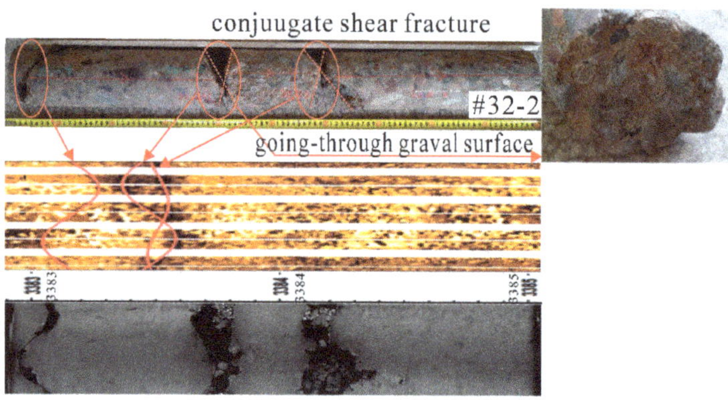

(a)

Figure 10. *Cont.*

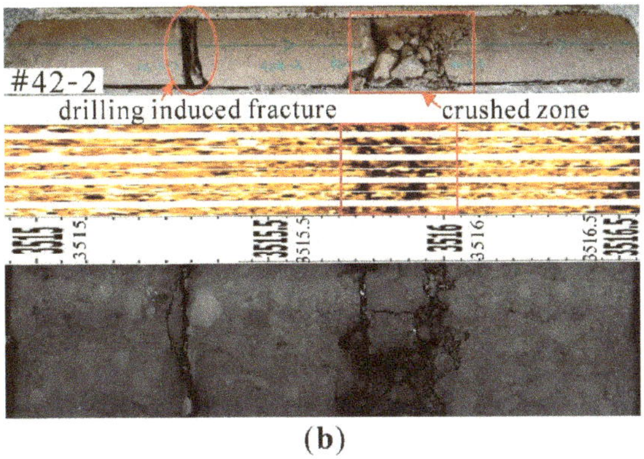

Figure 10. Conjugate shear fracture and crushed zone in a slant core. (**a**) Scissor-shaped conjugate shear fracture is straight in morphology and has through-penetrating gravel surface. (**b**) Crushed zone, which forms when shear slip occurs at the weak-cemented planes.

4.2. Drilling-Induced Fractures and Core Cutting-Induced Fractures

Both drilling-induced and core cutting-induced fractures were present, with characteristics different from hydraulic fractures. Drilling-induced fractures have complete morphology, many are twisted and serrated, and have bypassing gravel surfaces with uneven drilling mud attached (Figure 11).

Figure 11. Drilling-induced fracture in slant core with irregular twisted fracture morphology. Fracture surface comprises bypassing gravel with drilling mud attached.

Core cutting-induced fractures were mostly formed near the two ends of the core segment; they are highly closed and serrated and display a fresh surface with no drilling mud (Figure 12).

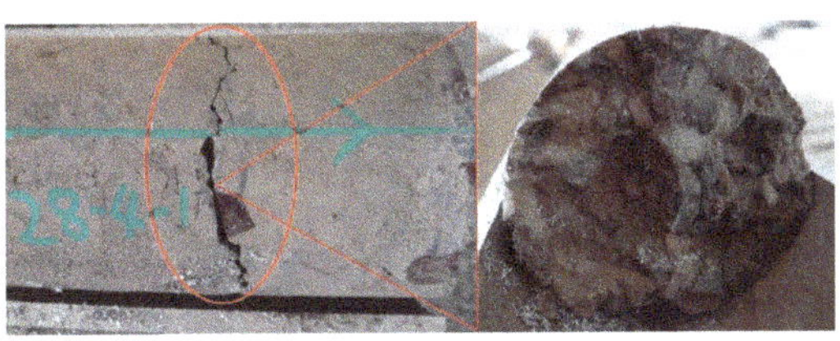

Figure 12. Core cutting-induced fracture in a slant core. Fracture morphology is highly closed and serrated. Fracture surface is fresh with no drilling mud.

4.3. Proppant Observation

Obvious quartz particles were observed in two fractures, which was consistent with the sand used in fracturing in terms of particle size, roundness, and uniformity. The quartz particle was identified as proppant, as shown in Figure 13. The two fractures lie in the T_1b_3 formation. A thin layer of proppant appeared as patches on the fracture surface with mud attached, indicating that drilling mud flowed into the hydraulic fracture during coring. In Figure 13, the proppant is the light-grey signal in the CT scan. The bright signal in the fractures and matrix is likely pyrite [2].

Figure 13. Proppant attached to the hydraulic fracture surface. In the CT scan, proppant is the light-grey signal, and the bright signal is likely pyrite. The characteristic was also observed in Ref. [2].

Moreover, sandy mud deposits were observed in multiple fractures. The drilling mud, mud deposits, and proppant particles were collected from parted fractures for further analyses, regarding whether it contained proppant. However, a large level of sand pack was not found in the core hydraulic fractures. The possible reason for this is that the fracture surfaces were separated during the coring process and drilling mud flowed into the hydraulic fracture. The proppant could not adhere to the fracture surface and was washed away by drilling mud. Therefore, drilling mud and cutting samples were collected during the coring operation to detect and quantify the spatial distribution of proppant along the cored interval.

5. Core Fracture Visualization and Analysis

A total of 371 fractures were recorded in the slant core, and these fractures were systematically described and classified. A total of 228 hydraulic fractures were identified, including 52 tensile fractures and 176 shear fractures. One of the potential sources of the observed fracture swarms is near-wellbore tortuosity [16]. Moreover, for regions far away from the wellbore, reservoir heterogeneity can promote complex hydraulic fracture trajectories and form fracture swarms. The variations of rock mechanical properties and in situ stress may lead to variable fracture-front speeds and potential fracture splitting and segmentation [17,18].

The data (type, orientation, and measured depth of fractures) from the core description were used to visualize fracture orientations/types along the core wellbores and the perforation clusters of the infield-scale in three-dimensional space, especially the relative locations of fractures to the fracturing wells/stages/clusters [8]. After filtering other types of fractures (fractures induced by drilling and core cutting), only the hydraulic fractures were visualized as lines with their orientations at their corresponding measured depths. The tensile fractures are in blue, and the shear fractures are in red. To better understand the spatial location of fractures relative to their initiation points (perforation clusters), the adjacent perforation clusters were also visualized by disks with different colors for each treatment stage.

Figure 14 shows a perspective view of hydraulic fractures along the core wellbore and the adjacent perforation clusters. The completion interval length is 60 m. Wells H8, H9, and H4 have two clusters, three clusters, and six clusters in each interval, respectively. The cores 1~31 are located in the T_1b_3 formation, in which the cores 1~8 are closest to the stages 2 to 4 of the well H8, and the cores 9~31 are closest to the stages 3 to 5 of the well H9. The cores 32~48 are located in the T_1b_2 formation, closest to stage 4 and stage 5 of well H4.

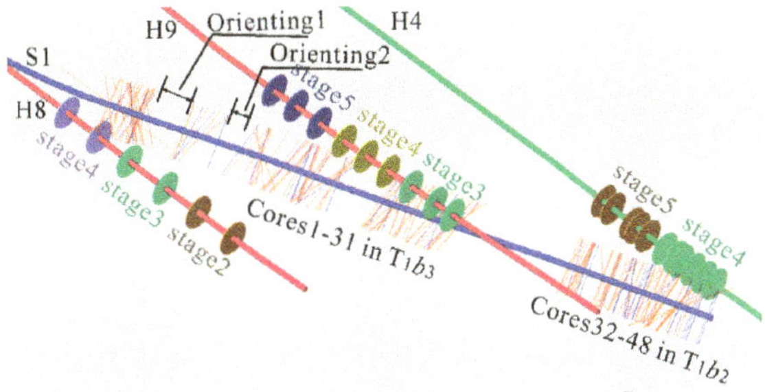

Figure 14. Perspective view of hydraulic fractures along the core wellbore and perforation clusters in the infield-scale in three-dimensional space.

Hydraulic fractures were observed in cores within the 100 m space between wells H8 and H9, indicating that fractures extended laterally over a distance of 50 m. Vertically, the distribution range of hydraulic fractures is 20.2 m above the adjacent well and 9.8 m below it. It can be inferred that the fractures extended over 30 m in a vertical direction, realizing the full coverage of hydraulic fracture network in the reservoir.

5.1. Presentation of Fractures in Cores 1-31

Cores 1-31 are located between H8 and H9 in the T_1b_3 formation. Figure 15 shows the perspective view of the hydraulic fractures in cores 1-31. Cores 1-8 are close to the two perforating clusters from stage 3 and the heel-side perforating clusters from stage 2 of well H8. Fractures observed in cores 1-8 are mainly from these clusters, and the fractures and clusters have a lateral distance of 21.4–45.1 m and a vertical distance in the range of 20.2 m above to 9.8 m below the clusters. The length of cores 1-8 is 50.0 m. The number of hydraulic fractures is 57, including 8 tensile fractures and 49 shear fractures.

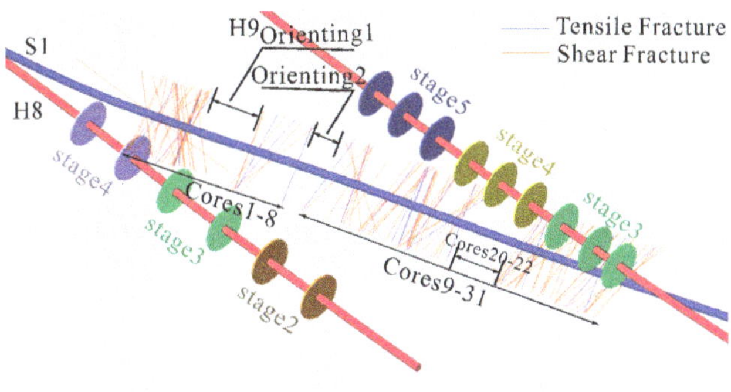

Figure 15. Perspective view of fractures in cores 1-31 from the southwest.

Cores 9-31 are close to the two toe-side perforating clusters from stage 5, three perforating clusters from stage 4, and the heel-side perforating clusters from stage 3 of the well H9. Fractures observed in cores 9-31 are mainly from these clusters, and the fractures and clusters have a lateral distance of 3.5–44.7 m, and the vertical distance is in the range of 0.8 m above to 14.5 m below the clusters. The length of the cores 9-31 is 144.95 m. The number of hydraulic fractures is 79, including 15 tensile fractures and 64 shear fractures.

No hydraulic fractures were observed in cores 20–22, in which the core lithology is reddish-brown silty mudstone. This indicates that argillaceous rocks are not conducive to fracture propagation. In addition, cores 20–22 are close to the middle perforation cluster from stage 4 of well H9. The lack of fractures in the middle perforation cluster seems to indicate that the growth of the middle fractures, lagging behind the two outside fractures, is suppressed. A significant proportion of the literature attributes this uneven distribution mainly to the stress shadow phenomenon [4].

5.2. Presentation of Fractures in Cores 32–48

Cores 32–48 are located on the west side of well H4 in the T_1b_2 formation, close to the six perforating clusters from stage 4 and the three toe-side perforating clusters from stage 5 of well H4. Figure 16 shows them in a side view from the west. The lateral distance between the cores and the clusters is 19.8–51.5 m, and the vertical distance is in the range of 8.8 m above to 2.3 m below the clusters. The length of cores 32–48 is 98.76 m, and the number of hydraulic fractures is 92, including 24 tensile fractures and 68 shear fractures.

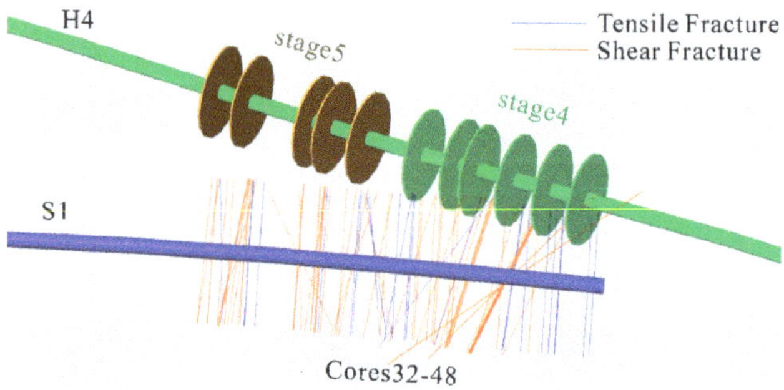

Figure 16. Side view of fractures in cores 32–48. The viewpoint is from the west.

As in previous cores, the large number of fractures observed in cores seems to indicate that tensile fractures are formed in the near-wellbore area and the extension of tensile fractures induces a large number of shear fractures in the area far from the wellbore.

Based on hydraulic fracture distribution data along the coring wellbore, a statistical analysis was conducted to better understand the spatial distribution of hydraulic fractures. Hydraulic fractures were considered to be derived from adjacent wells with the closest horizontal distance. Thus the spatial corresponding relationship between the hydraulic fractures distribution in cores and adjacent wells was established.

Figure 17 presents the relationship between hydraulic fracture density in the cores and the lateral distance from the horizontal well. Figure 17a shows that the hydraulic fracture density decreases with the increase of distance in the area far from the wellbore for well H8. The coring trajectory starts about 19 m away from H8, suggesting that cores in near-wellbore area were not drilled. In Figure 17b,c, the hydraulic fracture density is small in the near-wellbore area, where the lateral distances from wells H9 and H4 are in the range of <10 m and <25 m, respectively. In particular, for well H9, the hydraulic fracture density decreases with the increase of distance in the area far from the wellbore with a lateral distance range of >10 m.

As a general trend, the fracture density of the near-wellbore area is lower than that of the far-wellbore area. For the near-wellbore area, one of the potential sources of the dense fractures is near-wellbore tortuosity, which can propagate in parallel even in the presence of strong stress shadowing [16]. Meanwhile, for regions far away from the wellbore, reservoir heterogeneity can promote complex hydraulic fracture trajectories [17,18]. During fracture propagation, the local net pressure of fracture front decreases gradually, which results in decreasing fracture density in the area far from the wellbore. The near-wellbore area can be considered to have a lateral distance of less than 25 m from the horizontal wellbore.

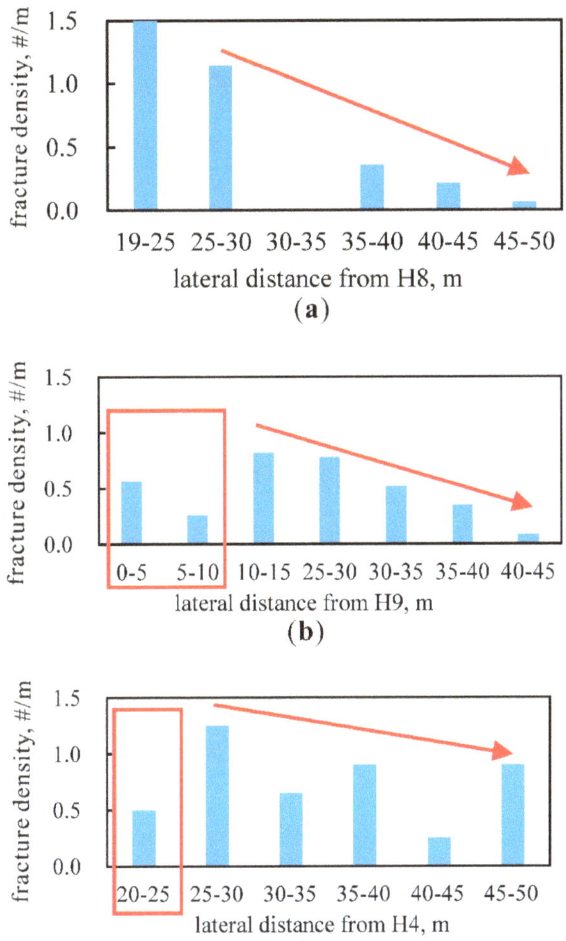

Figure 17. The relationship between the hydraulic fracture density in cores and lateral distance from the horizontal well. (**a**) H8; (**b**) H9; (**c**) H4.

6. Conclusions

(1) Hydraulic fracturing formed planar fracture swarms. The conglomerate hydraulic fractures have varied morphology; mainly straight with through-penetrating gravel surfaces. Fracture swarms are quite common. Fracture density along the coring wellbore is 0.78 fractures per meter, and hydraulic fractures are largely in the east–west direction, oriented at 70°–110°. Obvious proppant particles were observed in two fractures.

(2) For regions far away from the wellbore, a large number of additional fractures were created by the growth of principal fractures. These additional fractures may be shear failures along the natural weak plane created by the induced stress field or the bifurcation and splitting of principal fractures during propagation.

(3) In the near-wellbore area, tensile fractures are mainly formed, and the fracture density of the near-wellbore area is lower than that of the far-wellbore area. In the far-wellbore area, shear fractures are mainly formed, and the hydraulic fracture density decreases with the increase of distance. The near-wellbore area can be considered to have a lateral distance of less than 25 m from the horizontal wellbore.

Author Contributions: Conceptualization, S.S. and X.M.; methodology, R.Z.; software, R.Z.; validation, L.C., Y.X. and T.W.; formal analysis, R.Z.; investigation, R.Z.; resources, S.S.; data curation, Y.X.; writing—original draft preparation, R.Z.; writing—review and editing, X.M.; visualization, R.Z.; supervision, L.C.; project administration, Y.X. and T.W. All authors have read and agreed to the published version of the manuscript.

Funding: This research was funded by Xinjiang Oilfield Improvement of Shale/Tight Oil and Gas Volume Reconstruction Technology and Field Test Project (2021CGCGZ006).

Institutional Review Board Statement: Not applicable.

Informed Consent Statement: Not applicable.

Data Availability Statement: The raw data required to reproduce these findings cannot be shared at this time as they also form part of an ongoing study.

Acknowledgments: This work was supported by the Xinjiang Oilfield Improvement of Shale/Tight Oil and Gas Volume Reconstruction Technology and Field Test Project.

Conflicts of Interest: The authors declare no conflict of interest.

References

1. Li, W.; Song, L.; Xi, C.; Xie, S.; Rong, K. Drilling and Completion Technology for Horizontal Well with Ultra-Long Horizontal Section in Tight Conglomerate Reservoir of Xinjiang Mahu Oilfield. *XINJIANG OIL & GAS* **2021**, *17*, 86–91.
2. Gale, J.F.W.; Elliott, S.J.; Laubach, S.E. Hydraulic Fractures in Core from Stimulated Reservoirs: Core Fracture Description of HFTS Slant Core, Midland Basin, West Texas. In Proceedings of the Unconventional Resources Technology Conference, Houston, TX, USA, 23–25 July 2018.
3. Busetti, S. Hydraulic Fracture Swarms and Irregular Stress Shadows. In Proceedings of the 53rd US Rock Mechanics/Geomechanics Symposium, New York, NY, USA, 23–26 June 2019.
4. Damjanac, B.; Maxwell, S.; Pirayehgar, A.; Torres, M. Numerical Study of Stress Shadowing Effect on Fracture Initiation and Interaction Between Perforation Clusters. In Proceedings of the Unconventional Resources Technology Conference, Houston, TX, USA, 23–25 July 2018.
5. Courtier, J.; Chandler, K.; Gray, D.; Martin, S.; Thomas, R.; Wicker, J.; Petroleum, L.; Ciezobka, J. Best Practices in Designing and Executing a Comprehensive Hydraulic Fracturing Test Site in the Permian Basin. In Proceedings of the Unconventional Resources Technology Conference, Austin, TX, USA, 24–26 July 2017.
6. Gale, J.F.W.; Elliott, S.J.; Rysak, B.G.; Ginn, C.L.; Zhang, N.; Myers, R.D.; Laubach, S.E. Fracture Description of the HFTS-2 Slant Core, Delaware Basin, West Texas. In Proceedings of the Unconventional Resources Technology Conference, Houston, TX, USA, 26–28 July 2021.
7. Zhao, Y.; Bessa, F.; Sahni, V.; Pudugramam, S.; Liu, S. Key Learnings from Hydraulic Fracturing Test Site-2 (HFTS-2), Delaware Basin. In Proceedings of the Unconventional Resources Technology Conference, Houston, TX, USA, 26–28 July 2021.
8. Shrivastava, K.; Hwang, J.; Sharma, M. Formation of Complex Fracture Networks in the Wolfcamp Shale: Calibrating Model Predictions with Core Measurements from the Hydraulic Fracturing Test Site. In Proceedings of the 2018 SPE Annual Technical Conference and Exhibition, Dallas, TX, USA, 24–26 September 2018.
9. Craig, D.P.; Hoang, T.; Li, H.; Magness, J.; Ginn, C.; Auzias, V. Defining Hydraulic Fracture Geometry Using Image Logs Recorded in the Laterals of Horizontal Infill Wells. In Proceedings of the Unconventional Resources Technology Conference, Houston, TX, USA, 26–28 July 2021.
10. Maity, D.; Ciezobka, J. A Systematic Interpretation of Subsurface Proppant Concentration from Drilling Mud Returns: Case Study from Hydraulic Fracturing Test Site (HFTS-2) in Delaware Basin. In Proceedings of the Unconventional Resources Technology Conference, Houston, TX, USA, 26–28 July 2021.
11. Birkholzer, J.T.; Morris, J.; Bargar, J.R.; Brondolo, F.; Cihan, A.; Crandall, D.; Deng, H.; Fan, W.; Fu, W.; Fu, P.; et al. A New Modeling Framework for Multi-Scale Simulation of Hydraulic Fracturing and Production from Unconventional Reservoirs. *Energies* **2021**, *14*, 641. [CrossRef]
12. Fu, W.; Morris, J.P.; Sherman, C.S.; Fu, P.; Huang, J. Controlling Hydraulic Fracture Growth Through Precise Vertical Placement of Lateral Wells: Insights from HFTS Experiment and Numerical Validation. *Rock Mech. Rock Eng.* **2022**, *55*, 5453–5466. [CrossRef]
13. Pudugramam, S.; Irvin, R.J.; McClure, M.; Fowler, G.; Bessa, F.; Zhao, Y.; Han, J.; Li, H.; Kohli, A.; Zoback, M.D. Optimizing Well Spacing and Completion Design Using Simulation Models Calibrated to the Hydraulic Fracture Test Site 2 (HFTS-2) Dataset. In Proceedings of the Unconventional Resources Technology Conference, Houston, TX, USA, 20–22 June 2022.
14. Ciezobka, J.; Reeves, S. Overview of Hydraulic Fracturing Test Sites (HFTS) in the Permian Basin and Summary of Selected Results (HFTS-I in Midland and HFTS-II in Delaware). In Proceedings of the Latin America Unconventional Resources Technology Conference, Online, 16–18 November 2020.
15. Ciezobka, J.; Courtier, J.; Wicker, J. Hydraulic Fracturing Test Site (HFTS)—Project Overview and Summary of Results. In Proceedings of the Unconventional Resources Technology Conference, Houston, TX, USA, 23–25 July 2018.

16. Joseph, M.; Fu, P.; Sherman, C.; Wu, H.; Huang, J.; Fu, W. An Investigation of Candidate Mechanisms for Hydraulic Fracture Swarming through High-Fidelity Numerical Modeling. In Proceedings of the SPE Hydraulic Fracturing Technology Conference and Exhibition, The Woodlands, TX, USA, 1–3 February 2022.
17. Manchanda, R.; Shrivastava, K.; Zheng, S.; Sharma, M. A New Mechanism for the Formation of Hydraulic Fracture Swarms. In Proceedings of the SPE Hydraulic Fracturing Technology Conference and Exhibition, The Woodlands, TX, USA, 4–6 February 2020.
18. Fu, W.; Morris, J.P.; Fu, P.; Huang, J.; Sherman, C.S.; Settgast, R.R.; Ryerson, F.J. Developing Upscaling Approach for Swarming Hydraulic Fractures Observed at Hydraulic Fracturing Test Site through Multiscale Simulations. *SPE J.* **2021**, *26*, 2670–2684. [CrossRef]

processes

Article

Optimization of Water Injection Strategy before Re-Stimulation Considering Fractures Propagation

Guangcong Ren [1], Xinfang Ma [1,*], Shicheng Zhang [1], Yushi Zou [1], Guifu Duan [2] and Qiyong Xiong [3]

1 State Key Laboratory of Petroleum Resource and Prospecting, China University of Petroleum (Beijing), Beijing 102249, China
2 Research Institute of Petroleum Exploration and Development, China National Petroleum Corporation, Beijing 100083, China
3 Engineering Technology Research Institute, Xinjiang Oilfield Company, PetroChina, Karamay 834000, China
* Correspondence: maxinfang@cup.edu.cn; Tel.: +86-13366505131

Abstract: Water injection before re-stimulation has a positive effect to mitigate the "frac hit" and increase oil production in tight reservoirs. However, the study of water injection strategy optimization has not been thoroughly investigated. Some conclusions can be found in the existing literature, but the pressure and stress distribution, fractures morphology and oil production were not considered as a whole workflow during the study. In addition, the different reservoir deficit was not considered. Although technical experience and economic benefit have been obtained in some field tests, failed cases still exist. To fill this gap, a series of numerical models are established based on a tight reservoir located in northwest China. Under the different re-stimulation timing, the pressure distribution, stress distribution, and fractures morphology after water injection of different injection/production ratios are calculated, respectively. The oil and water production are predicted. The results show that, after a short period of production with a small deficit, the degree of "frac hit" is slight. Injecting water has an obvious effect on increasing oil production for both parent and infill well. After a long period of production with a large deficit, the problem of "frac hit" is very severe. Injecting water has an obvious effect on increasing oil production only for the parent well. The production of infill well is influenced by the fractures' interference and pressure increasing comprehensively. For the well group, measured by the final cumulative oil production, the optimal injection/production ratio is different, but the water injection volume is similar, which is about 15,000 m^3.

Keywords: water injection; frac hit; re-stimulation; parent well; infill well

Citation: Ren, G.; Ma, X.; Zhang, S.; Zou, Y.; Duan, G.; Xiong, Q. Optimization of Water Injection Strategy before Re-Stimulation Considering Fractures Propagation. *Processes* **2022**, *10*, 1538. https://doi.org/10.3390/pr10081538

Academic Editors: Linhua Pan, Jie Wang, Minghui Li, Wei Feng and Lufeng Zhang

Received: 11 July 2022
Accepted: 4 August 2022
Published: 5 August 2022

1. Introduction

The hydraulic fracturing of horizontal wells is a necessary method in unconventional reservoirs to create production capacity, which faces the problems of rapid production decline and low recovery. Drilling infill well is becoming a standard operational practice to increase production in unconventional reservoirs including tight oil and shale. Estimations from literature suggest that 50% of the remaining potential area in major US basins could be stimulated by drilling infill wells [1]. Similarly, as the largest conglomerate reservoir in the world, Mahu Oilfield in northwest China needs to drill infill wells to increase oil production and enhance oil recovery for the large well spacing between parent wells put into operation earlier.

However, according to the research and practice results, new fractures of infill well tend to propagate into the low-stress zones caused by pressure depletion after a long productive time, which is also known as "frac hit" [2,3]. There are some other descriptions [4–7] including frac bushing, well interference, and fracture-driven interaction. Frac hit can be detrimental to both parent wells and infill wells. In earlier times, Fisher reported a case in which five vertical wells were killed because of the hydraulic fracturing of nearby wells in Barnett [8,9]. In recent years, frac hit has been widely reported and analyzed under the context of more drilled infill wells.

Courtier [10] reported an infill well in Wolfcamp which was monitored through microseismic recorders during hydraulic fracturing and the results indicate that fractures are asymmetric along the horizontal wellbore. The half-length close to the parent well is obviously larger than another. Cipolla [11] reported a horizontal well group including a parent well and two infill wells in Bakken, North Dakota. Asymmetric fractures of infill well were monitored, and a numerical simulation method was used to study the influence of pressure depletion to well interference and group production. Chittenden [12] reported a case of Delaware Basin in Texas, which included a parent well and two infill wells. After the parent well has been put into production, 900 ft away, an infill well was fractured and preferential growth of fractures to depleted zones happened. For another infill well, which is farther away, the parent well has symmetrical fractures along horizontal wellbore. In addition to these field reports, a numerical study of infill well hydraulic fractures propagation rules can be retrieved in the literature [13].

The cases mentioned above focus on the phenomenon of well interference, and further discussion of its damage [14,15] to production is well investigated and understood in recent years. Rainbolt [16] reported a horizontal well in Wolfcamp losing the production rate during the infill well fracturing. Joslin [17] reported a case in Montney for which production of an infill well is only a half of the parent well when the cluster number is obviously greater. It seems like infill well drilling is a risky operation, and the interaction relationship between parent and infill wells needs further study. The degree of influence depends on the characteristic parameters of hydraulic fractures and the matrix [18]. The greater the fracture conductivity, the greater the number of fractures and the smaller the matrix permeability, and the greater the damage to the parent wells. King [4] suggested that SRV reduction and flow path obstruction may be the reasons for reducing the production rate. Ratcliff [19] found that the conductivity damage parameters can replicate the productivity degradation. Fowler [20] described three damage processes including conductivity damage, skin damage, and water block damage. Wang [21] suggested that the existence of natural fractures can lead to frac hit. Many possible reasons including physical plug, mechanical, and chemical problems are also mentioned.

Therefore, many researchers propose to carry out repeated fracturing or water injection for parent wells, and the energy increasing effect of liquid on the formation will lead to the increase of formation pressure and horizontal principal stress in depleted zones, which can avoid or mitigate the problem of frac hit. In Montana and North Dakota, production of five parent wells has been improved by water injection [22], which indicates that well interference was mitigated effectively. Bommer [23] reported that frac hit was mitigated successfully by injecting water to parent wells in Bakken, North Dakota. Whitfield [24] reported a successful case of mitigating frac hit by injecting water. They thought that the injected water volumes can vary from the small pre-loads of 500–1000 bbls to the large rate of more than 18,000 bbls. Gala [25] reported another case, and the effects of different injecting fluids including water and gas are compared. Due to compressibility of gas, gas injection needs more time than water. As for the influence of water injection to the stress reorientation, Singh [26] simulated the stress distribution around the injectors. The results showed that the maximum stress trajectories tend to align along injection wells. Safari [27] suggested that water injection does not change the stress significantly. Guo [28] suggested that the appropriate cumulative water injection depends on the reservoir volume production. The results of numerical simulation of stress distribution and fractures propagation based on the field data in Eagle Ford showed that, when the injection/production ratio is greater than 76.9%, fractures of infill well are more symmetrical. The cumulative production corresponding to different fracture morphology is not considered. Li [29] studied the key factors on successful water injection. They mentioned that the injection pressure should be equal to the initial reservoir pressure. Lower injection pressure could not mitigate the frac hit effectively. Furthermore, the integrated models [30] considering fracture propagation and production prediction were established to optimize the type of energy increasing medium and the cumulative injection. In this study, fracture growth of the infill well was simulated; meanwhile, no fracture growth or widening are expected in the parent well. Other measures which can increase pressure and stress in the depleted zones have the similar effects, for instance, re-fracturing [17,31] and surface shut-in [32]. In recent years, some new types of fluid medium [33–37] are introduced. The geomechanical controls [38] and increased offset spacing [39] are also mentioned. For the conglomerate reservoir in Mahu Oilfield, which is our main concern, it is necessary not only to drill infill wells, but to carry out re-fracturing of the parent well. The horizontal wells of Mahu conglomerate reservoir have high output in the initial stage after primary fracturing, but daily production decreases rapidly. It is necessary not only to ensure that the infill wells are unaffected, but also to reactivate the productivity of the parent well. It is worth noting that, before re-fracturing the parent wells, some protective measures should be conducted. After a long period of production, stress distribution around the parent wells is uneven. The directions of new fractures created by re-refracturing are not vertical to the horizontal wellbore, which will have a negative effect on increasing production. Injecting water to parent wells before re-fracturing is proposed to increase reservoir energy, control the new fractures direction, and reduce the negative fractures as well as well interference. At the same time, the water injected into the formation can replace the crude oil if the formation has strong water absorption. Some oil companies have conducted a field test of injecting water to the parent wells before re-fracturing. There were some successful cases in tight oil reservoirs located in northwest China [40,41]. The strategy of water injection before re-fracturing including total water injection, water injection rate, soaking time, and injection timing has been optimized [42], but fracture propagation was not considered.

In this paper, we propose a new re-stimulation measure of water injection before re-fracturing parent as well as drilling and fracturing infill well in the Mahu conglomerate reservoir. Based on an unconventional fractures model (UFM), the fractures morphology of parent well and infill well is simulated. The stress distribution before re-stimulation was calculated by single way coupling. The cumulative water injection volume was optimized based on the predicted cumulative oil production under different conditions.

2. Methods

The new re-stimulation measure consists of water injection before re-fracturing parent well as well as drilling and fracturing infill well. As shown in Figure 1, there are two wells at a certain distance in the model, and an ideal parent well will be put into production for a period and form a depleted zone around the parent well. The water will be injected into the parent well, increasing formation pressure and horizontal principal stress. Re-stimulation measures of re-fracturing parent well and fracturing infill well will be conducted simultaneously to avoid fractures interference. It should be noted that new fractures of the parent well are created during re-fracturing, and perforations are located in the center between existing primary fractures. Parent well and infill well will be put into production at the same time. Based on the cumulative oil production after re-stimulation, the cumulative water injection volume will be optimized under different conditions.

2.1. Model Description

A series of numerical models are established. To avoid boundary effects, a model with a size of 5 km × 3 km is established based on parameters of the Mahu conglomerate reservoir. The whole reservoir model consists of five layers with the total thickness of 200 m. The oil-bearing bed is in the center, with the thickness of 40 m, and the parameters are as shown in Table 1. The perforation is in the oil layer. The horizontal section length of ideal parent well is 1000 m. The primary fracturing of the parent well consists of eight stages with three clusters per stage. As mentioned earlier, new perforations are located in the center between primary fractures; therefore, the number of re-fracture clusters are the same as the primary. Hydraulic fractures of infill well and re-fractures of parent well are distributed as zipper-type.

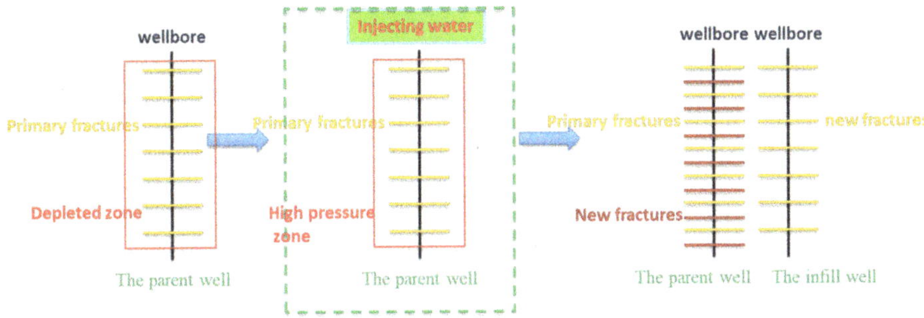

Figure 1. Illustration of the re-stimulation process.

Table 1. Field data for oil layer of the base case.

Parameters	Values	Parameters	Values
Matrix permeability, mD	0.01	Minimum horizontal principal stress, MPa	45
Matrix porosity	0.1	Maximum horizontal principal stress, MPa	55
Initial reservoir pressure, MPa	39	Overburden stress, MPa	60
Initial oil saturation	0.5	Young's modulus, GPa	35
Reservoir thickness, m	200	Poisson's ratio	0.3
Bubble-point pressure, MPa	15	Tensile strength, MPa	5
Total compressibility, kPa^{-1}	2×10^{-6}	Reservoir temperature, °C	85

Fractures propagation is simulated by the Unconventional Fractures Model [43]. It proposed a fully coupled solution of fracture propagation, rock deformation, and fluid flow simulation. UFM has a similar assumption and governing equations to a conventional pseudo-3D model but solves the equations for the complex fractures. The complex fractures are divided into a lot of connected elements with different heights which are constructed in a similar way to those in the pseudo-3D fracture model. Natural fractures are not very developed in the Mahu reservoir, and they are not considered in this study.

Considering the flow in fractures as laminar flow in a flat plate, the Poiseuille law is used to express the flow of the power law fluid [44]:

$$\frac{\partial p}{\partial s} = -\alpha_0 \frac{1}{\overline{w}^{2n'}} \frac{q}{H_{fl}} \left| \frac{q}{H_{fl}} \right|^{n'-1} \tag{1}$$

$$\alpha_0 = \frac{2k'}{\varnothing(n')^{n'}} \left(\frac{4n'+2}{n'} \right)^{n'} \tag{2}$$

$$\varnothing(n') = \frac{1}{H_{fl}} \int \left(\frac{w(z)}{\overline{w}} \right)^{\frac{2n'+1}{n'}} dz \tag{3}$$

where p is fluid pressure, Pa; s is the distance along the fracture, m; $\overline{w}$ is the average fracture width, m; n' is fluid power-law index; q is the local flow rate in the fracture, m^3/s; H_{fl} is the height of the fluid in the fracture, m; k' is the fluid consistency index; $w(z)$ is fracture width as a function of depth z, m; z is the depth, m.

The material balance condition is given by the continuity equation:

$$\frac{\partial p}{\partial s} + \frac{\partial\left(H_{fl}\overline{w}\right)}{\partial t} + \frac{2h_L C_L}{\sqrt{t-\tau_0(s)}} = 0, t > \tau_0(s) \tag{4}$$

The following condition should be met:

$$\int_0^t Q(t)dt = \int_0^{L(t)} H\overline{w}ds + \int_0^{L(t)}\int_0^t \frac{2h_L C_L}{\sqrt{t-\tau_0(s)}}dtds \tag{5}$$

The sum of flow rate into all open perforations should be equal to injection rate:

$$\sum q_i(t) = Q(t),\ i = 1, \ldots,\ N_{perf} \tag{6}$$

where t is injecting time, s; h_L is the leakoff zone height, m; C_L is the total leakoff coefficient, $m/s^{0.5}$; $\tau_0(s)$ is the time when each fracture element is first exposed to fluid, s; $Q(t)$ is the total injection rate, m^3/s; $L(t)$ is the total fracture length at time t, m; H is the fracture height, m; q_i is the injection rate into each cluster, m^3/s; N_{perf} is the number of clusters.

As mentioned earlier, the fractures are divided into a series of connected elements with different heights which are constructed in a similar way to the pseudo-3D fracture model. The 2D plane–strain solution for fracture width in pseudo-3D models is adopted. In a vertical fracture, the width profile can be determined by following equations:

$$w(z) = \frac{4}{E}\left[p_{cp} - \sigma_n + \rho_f g\left(h_{cp} - \frac{H}{4}\right)\right] + \sqrt{\frac{2}{\pi h}}\sum_{i=1}^{n-1}(\sigma_{i+1} - \sigma_i)\left[\frac{H}{2}\arccos\left(\frac{H-2H_i}{H}\right) + \sqrt{z(H-z)}\right] \tag{7}$$

where E is the Young's Modules, Pa; p_{cp} is the pressure at perforation depth h_{cp}, Pa; σ_n and σ_i are the minimum horizontal principal stress at the section and ith layer, Pa; ρ_f is the fluid density, kg/m^3; H_i is the fracture height of i_{th} layer, m.

Based on the fractures morphology, the unstructured grid was generated automatically to establish a numerical model. To avoid the problem of differences between fracture width and matrix grid size, a single-porosity medium is used. Mesh refinement was applicated around the hydraulic fractures. The water-oil two-phase model was used since the bottom flow pressure is always greater than bubble-point. Relative permeability curves of matrix and fracture are shown in Figure 2.

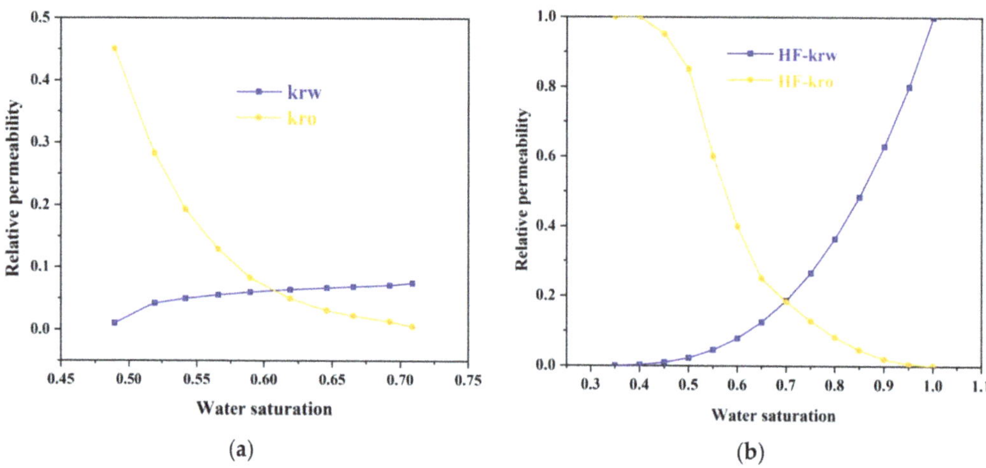

Figure 2. Relative permeability of: (**a**) matrix; (**b**) hydraulic fractures.

After a period of production, a one-way coupled method is used to calculate the stress distribution. Generally, there are three coupling methods [45] including a fully coupled method [46,47], iterative coupled method [48,49], and a one-way coupled method [50–52]. In a one-way coupled model, the fluid flow equations are generally solved using a reservoir simulator, the results of pressure, temperature, and saturation are passed to the geomechanical simulator, but no results from the geomechanical simulator are used in the reservoir simulator. A one-way coupled method has the advantage of high computational efficiency, but the disadvantage of low accuracy relatively. However, our main concern is the water injection, and the commercial one-way coupled simulator (Visage [45]) is used in this study.

2.2. History Match

In order to confirm the reliability of the models we established, a production history match has been completed. The production history data are from a horizontal well in Mahu Oilfield. The well was drilled in 2013, and the primary fracturing was conducted then with 12 stages. It is a flowing well with an oil production rate of 35 m^3/d during the initial stage. With production, formation energy and bottom flow pressure decreased, and the well was transformed into pumping after about 1000 days. Daily oil production and wellhead pressure were recorded precisely by a ground monitor. Wellhead pressure should be converted into bottom flow pressure by calculating liquid column pressure. Since the calculation of bottom flow pressure after the well started pumped involves more factors, only the production history of flowing period has been matched.

The control mode of bottom flow pressure was used in conjunction with field historical data to form well constraints. As shown in Figure 3, a simulated oil production rate is essentially in agreement with the data monitored, which confirms the reliability of the model. Based on the model after history match, the next work is to build the ideal wells as mentioned before.

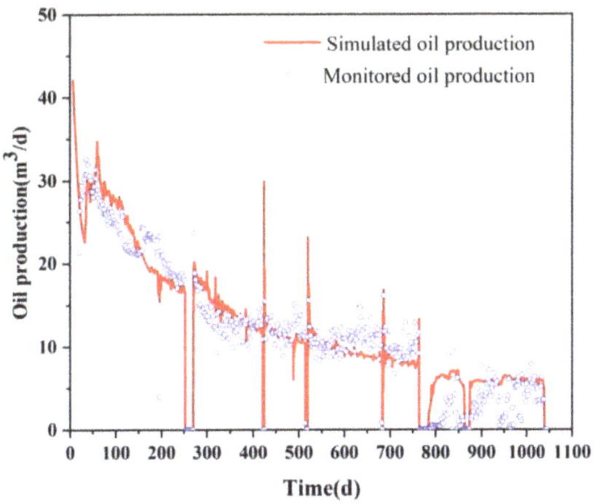

Figure 3. Oil production history match.

3. Results and Discussion

Well A1 is a parent well fractured primarily earlier, then put into production with the daily oil production rate of 50 m^3/d and the bottom hole pressure of 15 MPa. Well A2 is an infill well fractured primarily after that well A1 has been re-stimulated by injecting water and re-fracturing. The parent-infill well spacing is 250 m. To study the influence of re-stimulation timing on the optimized cumulative water injection volume, cumulative oil production of parent well and infill well are predicted separately after the parent well has been opening for 1 year and 4 years.

3.1. 1st Year

As shown in Figure 4, after the parent well has been put into production for one year with the final cumulative oil production of 18,068 m^3 and water production of 6826 m^3, reservoir pressure decreases to 25.8 MPa and minimal horizontal principal stress decreases to 41.2 MPa near fractures. In the middle zone between two fractures, pressure also decreases, but only slightly to 37.5 MPa. Accordingly, the decline range of minimal horizontal principal stress is also small. There is a deficit of 24,894 m^3 in the total. The ratios of water injection to production are set to 0.2, 0.4, 0.6, 0.8, 1.0, and 1.2. The corresponding cumulative water injections are 4978.8 m^3, 9957.6 m^3, 14,936.4 m^3, 19,915.2 m^3, 24,894 m^3, and 29,872.8 m^3, respectively. According to earlier finds [42], long soaking time after water injection has a positive effect on increasing production. In this study, the soaking time is two months to ensure that pressure spreads more uniformly.

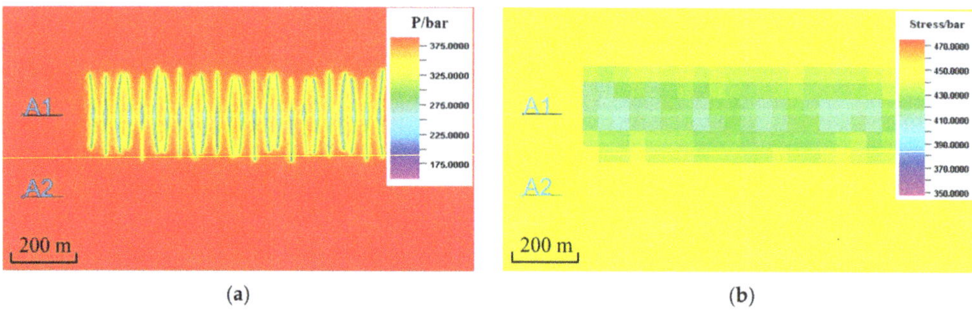

Figure 4. Distribution of (**a**) pressure; (**b**) stress.

Pressure distribution corresponding to different injection/production ratios is shown in Supplementary Materials (Figure S1). With the increasing of injection/production ratio, reservoir pressure near the fractures increases gradually. The high-pressure areas are concentrated near fractures. Like the situation before water injection shown in Figure 4, in the middle zone between two primary fractures, reservoir pressure changes sightly compared to the initial pressure. It should be noted that reservoir pressure near the wellbore of A2 is unchanged.

Based on the pressure distribution, the stress distribution is calculated by a one-way coupled model. Minimal horizontal principal stress distribution corresponding to different injection/production ratios is shown in Supplementary Materials (Figure S2). With the increasing of the injection/production ratio, a minimal horizontal principal stress around wellbore increases gradually. However, the difference of stress distribution among these situations is slight. The minimum horizontal principal stress around wellbore is about 42.4 MPa while the injection/production is 0.2. The stress value in a few areas is slightly greater than the initial value if the ratio is greater than 1.0. As the short production time and small reservoir deficit, the problem of "frac hit" caused by reservoir depletion is slight.

The fractures propagation of parent well and infill well after 1-year production is simulated utilizing the UFM model. Fracture morphology when no-water is injected is shown in Figure 5. Compared to primary fractures of the parent well, re-fractures are affected by ununiform stress distribution. Within one stage, the fracture length of each cluster varies greatly. As the direction of maximum horizontal principal stress changed from the initial state, some re-fractures divert and connect to the primary fractures uncontrollably, causing the fractures interference. As for the infill well, some fractures propagate into the depleted zone and connect to the factures of the parent well; in other words, frac hit happens. There are four clusters of infill well among twenty-four clusters in total communicating with the fractures of the parent well, which means the degree is slight.

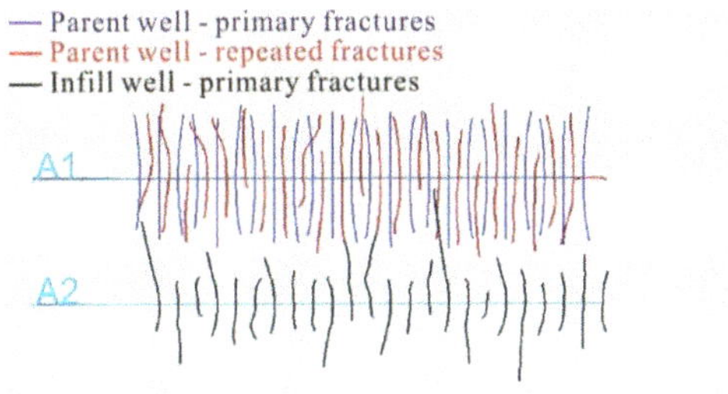

Figure 5. Fractures morphology of parent well and infill well after 1-year production (no water injected).

Fractures morphology of parent well and infill well corresponding to different injection/production ratios is shown in Supplementary Materials (Figure S3). In general, the problem of "frac hit" is not serious because the production time is short and cumulative liquid production is not large enough. In each situation of different injection/production ratios, fracture interference and well interference happen. The number of infill well fractures which are communicating with fractures of parent well is small, generally less than five. The cluster spacing of parent wells becomes smaller because of the new perforations, and fractures interference usually happens among primary and repeated fractures of the parent well. Compared to the situation of no-water injection, the length of parent well re-fractures is larger.

Based on fractures morphology, the numerical models with unstructured grids are generated. The cumulative liquid production of four years after re-stimulation corresponding to different injection/production ratios is predicted. The same constraint condition with primary production is used under different situations, which include the initial oil production of 50 m^3/day and the given minimal bottom well pressure of 15 MPa. The predicted cumulative oil production is shown in Figure 6. It should be noted that, in the case of no water injection, the parent well has been soaked for a few days before being put into production, which is in accordance with the filed operation. Injecting water has an obvious effect on increasing oil production for both parent well and infill well, especially the parent well. Compared to oil production, under the condition of no water injection, different injection/production ratios work equally well for increasing oil production. There is no strict correlation between final oil production and injection/production ratio because of the difference of stress and pressure distribution and the resulting different degrees of fracture and well interference. The larger the ratio is, the higher the average formation pressure is, which favors oil production. However, the degree of stress and pressure heterogeneity corresponding to each injection/production ratio is different. If one fracture interferes with others severely, the final production could be influenced obviously, to a certain extent.

To describe the effect of injecting water on the liquid production of the parent well, the cumulative liquid production corresponding to different injection/production ratios is shown in Figure 7. It is obvious that, with the increase of the injection/production ratio from 0 to 1.2, the liquid production including water and oil keeps the increasing trend, which shows that more water injection will increase reservoir energy and benefit the producing of more liquid. As shown in Figure 8, with the injection/production ratio increases from 0.2 to 1.2, the water cut of the parent well is increasing with a positive correlation. However, water cut corresponding to the situation of no-injection is higher than those corresponding to the ratios of 0.2 to 0.6, and almost the same as the 0.8 and 1.0. As mentioned before, in the simulation case of no-injection, the parent well has been soaked for a period before the production to meet the field operation of re-stimulation. Therefore, water cut of the parent well during the initial production stage is 0. After a period of production, water cut is restored to a high level that is close to the state after 1-year production. In addition, as the capillary pressure is considered, water cut corresponding to the situation of no-injection is higher than water-injection.

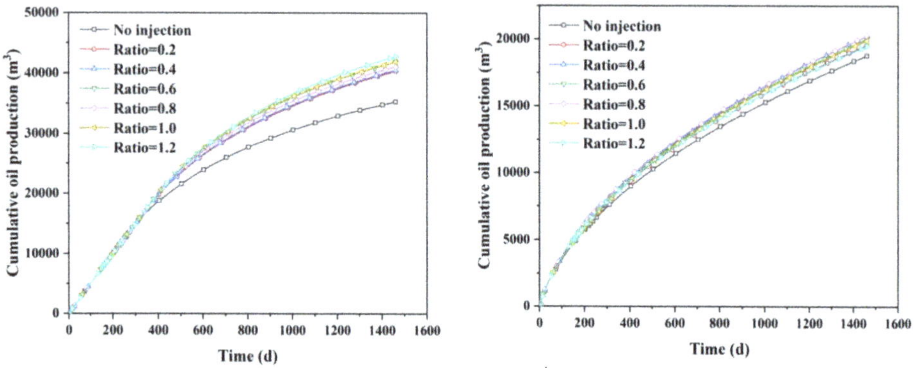

(**a**) (**b**)

Figure 6. Cumulative oil production of different injection/production ratio, (**a**) parent well; (**b**) infill well.

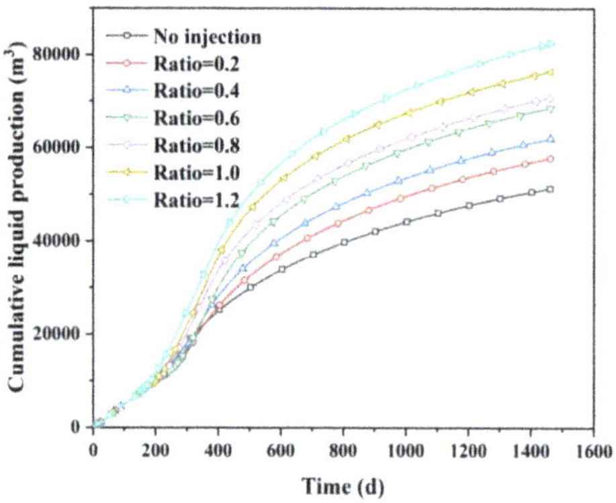

Figure 7. Cumulative liquid production of parent well.

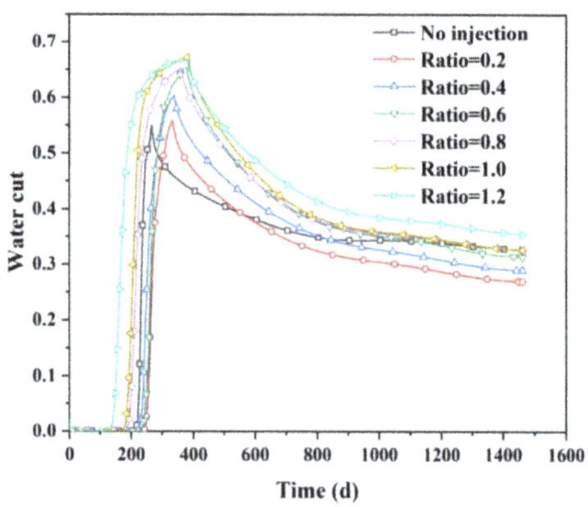

Figure 8. Water cut of parent well.

As for the infill well, the cumulative liquid production is shown in Figure 9. There is no strong correlation between injection/production ratio with cumulative liquid production. The cumulative liquid production corresponding to the situation of no water injecting is the lowest. Basically, as shown in Figures 9 and 10, the cumulative liquid production and water cut corresponding to each situation are close because the degree of well interference is slight and universal.

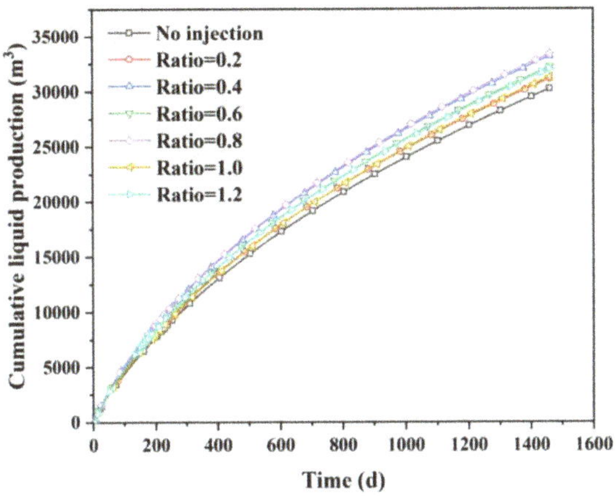

Figure 9. Cumulative liquid production of infill well.

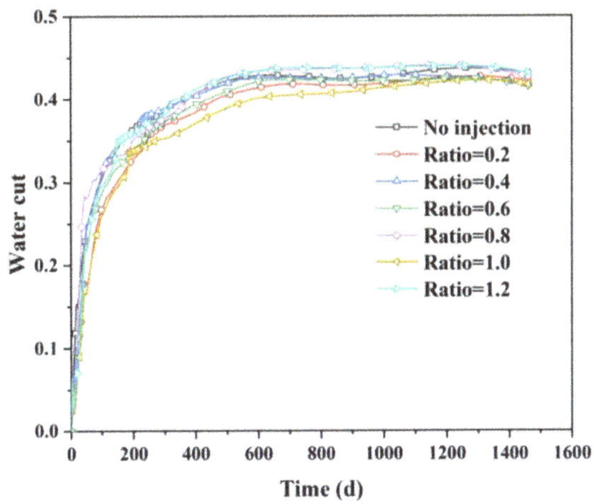

Figure 10. Water cut of infill well.

Finally, the final cumulative oil production of the well group including parent well and infill well is used as a standard for determining the optimal injection/production ratio. As Figure 11 shows, when the injection/production ratio is 1.2, the maximum final cumulative oil production of the well group is 62,331 m^3 with the final water cut of 0.455. However, the second highest oil production is 61,909 m^3 with the final water cut of 0.386 when the injection/production ratio is 0.6. It means that producing 422 m^3 of more oil brings 13,336 m^3 of more water. In this case, the injection/production ratio of 0.6 is the optimal, and the injection volume is 14,396 m^3.

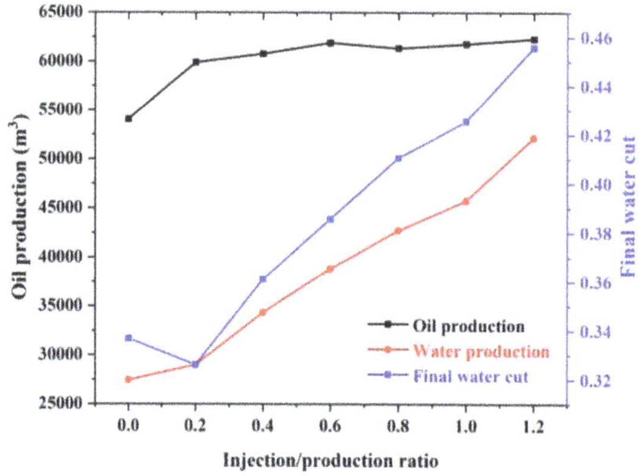

Figure 11. Liquid production and water cut of the well group.

3.2. 4th Year

As shown in Figure 12, after the parent well has been put into production for four years with the final cumulative oil production of 39,908 m^3 and water production of 36,219 m^3, the reservoir pressure decreases to 22.9 MPa, and the minimal horizontal principal stress decreases to 40.4 MPa near fractures. Unlike the situation after 1-year production, in the middle zone between two fractures, pressure also decreases significantly to 25.8 MPa. Accordingly, minimal horizontal principal stress decreases greatly too. There is a deficit of 76,127 m^3 in the total. The ratios of water injection to production are set to 0.2, 0.4, 0.6, 0.8, 1.0, and 1.2. The corresponding cumulative water injection volumes are 15,225.4 m^3, 30,450.8 m^3, 45,676.2 m^3, 60,901.6 m^3, 76,127 m^3, and 91,352.4 m^3, respectively. The soaking time is two months to ensure that pressure spreads more uniformly.

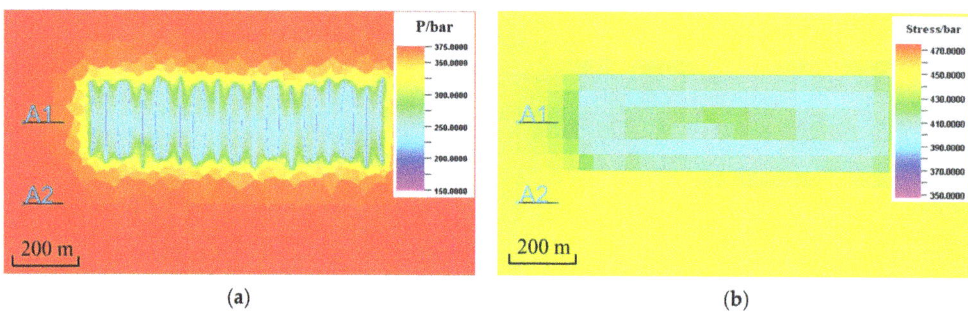

Figure 12. Distribution of: (**a**) pressure; (**b**) stress.

Pressure distribution corresponding to different injection/production ratios after a 4-year production is shown in Supplementary Materials (Figure S4). Compared to the pressure distribution after 1-year production, there are more significant differences among various situations of different injection/production ratios. Unlike the phenomenon of high-pressure areas only being concentrated near fractures after one year, with the increasing of the injection/production ratio, reservoir pressure in the whole depleted zone increases gradually. When the injection/production ratio reaches 0.8, the high-pressure zone is gradually formed, in which the pressure is higher than the initial value. The reason for this is that the reservoir deficit of liquid is huge, which leads to a great amount of water injection in the short run when the injection/production ratio is larger. The pressure propagation distance is limited. However, there is a transitional zone between a high-pressure zone after injecting water and the initial undeveloped reservoir.

Minimal horizontal principal stress distribution corresponding to different injection/production ratios after four years is shown in Supplementary Materials (Figure S5). With the increasing of the injection/production ratio, minimal horizontal principal stress around wellbore increases gradually. Compared to the situation after one year, the difference in stress distribution among different ratios is more obvious. The minimum horizontal principal stress around wellbore is about 42.7 MPa, while the injection/production is 0.2, and 45.8 MPa while the injection/production is 1.2. A few elements start with the slightly greater value than the initial state when the ratio is greater than 0.8. From the point of view of the stress distribution, when the injection/production ratio is greater than 0.8, the "frac hit" could be mitigated as the existence of high-stress areas.

Fractures morphology of parent well and infill well after a 4-year production as well as no-water injection is shown in Figure 13. The repeated fractures of parent well propagate in different directions, which is not perpendicular to the wellbore, especially in the middle stages. The fractures direction of the third stage is even parallel to the wellbore. After a long period of production, the interference of primary fractures with the repeat of the parent well is serious. As for the fractures of the infill well, the half-fractures close to the parent well have an obviously longer length than the other side, which means "frac hit" happens. It should be noted that the fracture length is longer, compared to those of the situation after 1-year production. Almost all fractures propagate to the depleted zones. Accordingly, the fracture width is smaller.

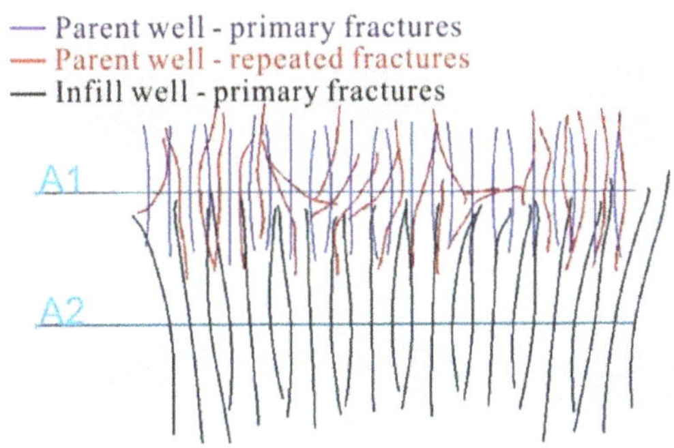

Figure 13. Fractures morphology of parent well and infill well after 4-year production (no water injected).

Fractures morphology of parent well and infill well corresponding to different injection/production ratios is shown in Supplementary Materials (Figure S6). In general, the problem of "frac hit" is very serious because the production time is long and the cumulative liquid production is very large. In each situation of different injection/production ratios, fractures interference and well interference happen. For the repeated fractures of parent wells, the length is getting larger with the increase of the injection/production ratio. When the injection/production ratio is 0.2, there are still some re-fractures that divert from the initial propagation direction. When it increases to 0.4 and 0.6, re-fractures morphology becomes similar to the primaries. However, as the ratio continues to increase, the re-fractures divert again. As mentioned before, some elements start with the slightly greater value than the initial state when the ratio is greater than 0.8. It shows that the injection/production of 0.8 is too large considering the stress distribution and fractures propagation. For fractures of the infill well, even if the injection/production ratio increases to 1.2, the fractures of infill well are still communicating with the primary fractures of the parent well, and a few propagate to the parent wellbore. When the injection/production ratio is greater than 0.6, although the fractures interference with the parent well happens, the fractures of infill well also propagate fully to the other side. In other words, when the injection/production ratio is greater than 0.6, on one side of the infill well, fractures interference happens, which is harmful to the production. On the other side, fractures propagate excessively, which is a benefit to the production. It shows that mitigating "frac hit" needs not only increases formation energy but also decreases the volume of re-fracturing fluid.

The cumulative oil production corresponding to different injection/production ratios is also predicted. The same constraint condition with primary production and re-frac production after one year is used, which includes the initial oil production of 50 m^3/day and the given minimal bottom well pressure of 15 MPa. The predicted cumulative oil production is shown in Figure 14. The parent well has also been soaked for a few days before being put into production, which is same as the situation after one year.

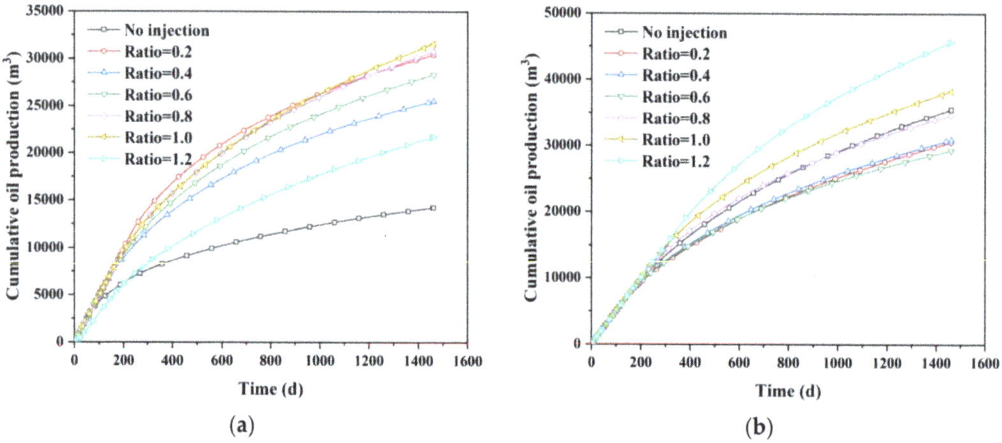

Figure 14. Cumulative oil production of different injection/production ratio, (**a**) parent well; (**b**) infill well.

When the infill well is stimulated after 1-year production of the parent well, injecting water has an obvious effect on increasing oil production for both parent well and infill well, especially the parent well. However, if the re-stimulation timing is the 4th year, injecting water may have a negative effect. For the parent well, when there is no water injected, cumulative oil production is the lowest, and the next is the ratio of 1.2. When the injection/production ratios are 0.2, 0.8, and 1.0, the parent well has the adjacent high cumulative oil production. For the infill well, the opposite has occurred. When the injection/production ratio is 1.2, cumulative oil production is the highest. Except for the ratios of 1.0 and 1.2, the cumulative oil productions of other situations are lower than the production of no-injecting.

Similarly, the cumulative liquid production corresponding to different injection/production ratios is shown in Figure 15. It is also obvious that, with the increasing of the injection/production ratio from 0 to 1.2, the liquid production of the parent well including water and oil keeps the increasing trend. From the slope of the cumulative liquid production curve, when the injection/production ratio is greater than 0.2, the oil production rate during the initial stage becomes higher quickly. However, as Figure 16 shows, different from the situation after 1-year production, the water cut keeps increasing as well. When there is no water injected into the parent well, the water cut remains at a low level. When the injection/production ratio is 0.2, the water cut firstly increases within the first 210 days, and then begins to decrease. When the injection/production ratio is greater than 0.2, the water cut remains at a high level. Comprehensively considering the water cut and the cumulative oil production of the parent well, the injection/production ratio of 0.2 is the best option for the parent well.

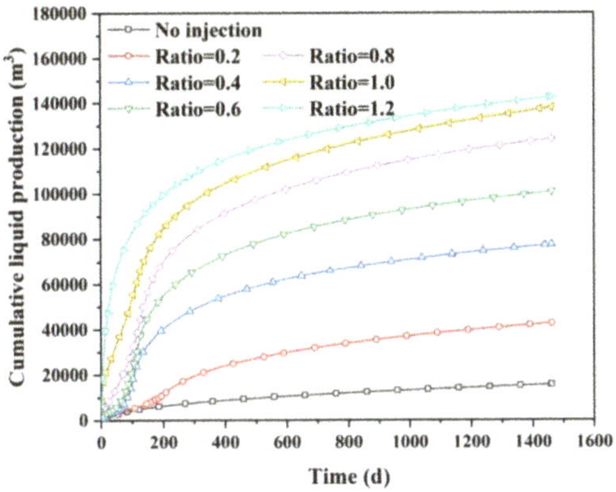

Figure 15. Cumulative liquid production of parent well.

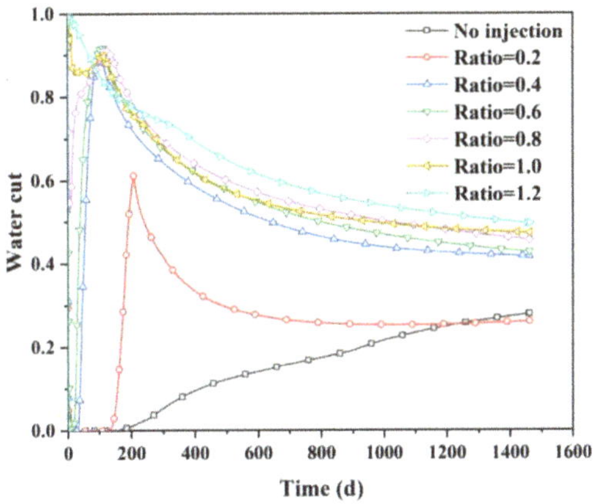

Figure 16. Water cut of parent well.

As for the infill well, the cumulative liquid production is shown in Figure 17. There is no strong correlation between the injection/production ratio with cumulative liquid production. The cumulative liquid production corresponding to the situation of no water injecting is higher than the ratios of 0.2 to 0.6. Considering the fracture morphology of parent well and infill well corresponding to different injection/production ratios, the fractures interference does have a negative effect on liquid production. At the same time, excessive propagation has a positive effect, as the stimulated reservoir area and volume are getting, larger. For the infill well, as shown in Figure 18, the water cut corresponding to the different injection/production ratio is numerically close because the "frac hit" is ubiquitous. Compared to the parent well, the average water cut of infill well is smaller.

Similarly, the final cumulative oil production of the well group including parent well and infill well is used as a standard for determining the optimal injection/production ratio. In this case, the injection/production ratio corresponding to the maximum final group production is 1.0, with the final cumulative oil production of 69,827 m^3, and the final water cut is 0.656. However, as Figure 19 shows, the water production and water cut increase rapidly when the injection/production ratio is larger than 0.2. The final cumulative oil production of well group is 60,925 m^3, and the final water cut is 0.342, when the injection/production ratio is 0.2. It means that producing 8902 m^3 of more oil will bring 101,838 m^3 of more water. In addition, the situations of ratios of 0.8 and 1.2 are similar.

Therefore, as the re-stimulation timing is the 4th year, the optimal injection/production ratio is 0.2, and the injection volume is 15,224 m^3.

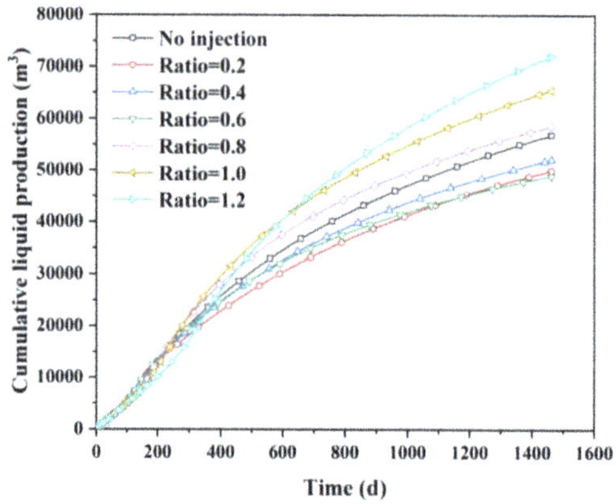

Figure 17. Cumulative liquid production of infill well.

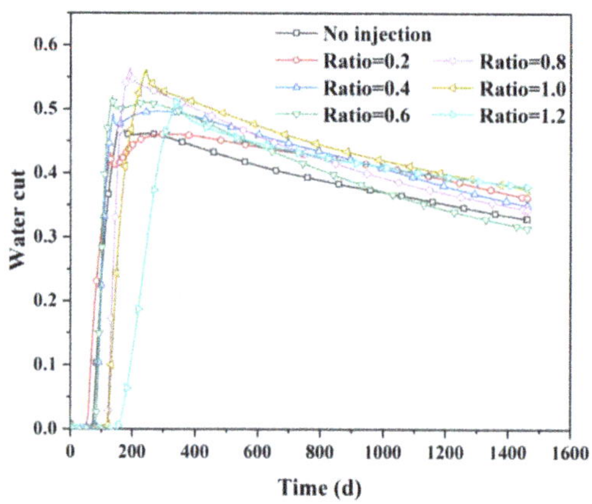

Figure 18. Water cut of infill well.

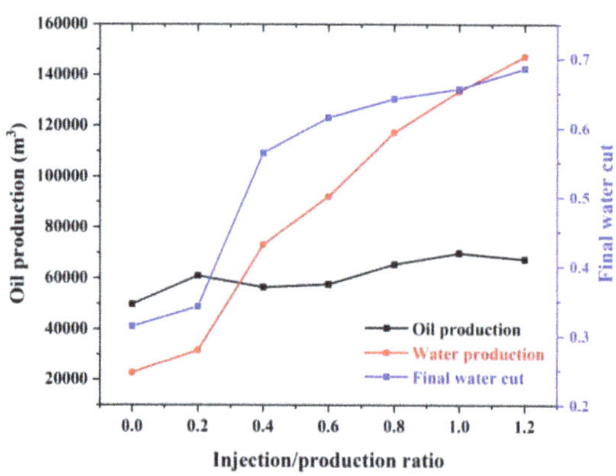

Figure 19. Liquid production and water cut of the well group.

We can see that the optimal injection/production ratio for different re-stimulation timing is different, but the corresponding injection volume is almost the same. It suggests that the maximum water injection volume, which can be effective for increasing oil production, is finite for a reservoir with certain physical properties. More water injection means more water production, which is mainly produced from the parent well after re-stimulation. Although the reservoir deficit volume is very large after a long period of production, the water injection volume is not suitable to be too large. From the pressure distribution mentioned before, it shows that the pressure increases in all of the depletion zones. However, the water saturation does not. The seepage distance of injected water is far less than the propagation distance of pressure. Therefore, the water centralized distribution near the fractures will be produced first. That is why the water cut of the parent well increases rapidly during the initial stage and remains at a high level, as the injection/production ratio is large. In addition, if the capillary pressure is greater than the field data we used, the optimal injection/production ratio and injection volume will be larger.

In addition, we think that the optimal water injection volume has more to do with the reservoir conditions than the liquid production or reservoir deficit. Whether the optimal water injection volume should be differentiated depends on the reservoir parameters. Therefore, the optimal injection/production ratio for a well block is not the best choice for another.

Furthermore, to mitigate the frac hit, the influence of changed stress distribution to fractures length of infill well must be considered. Increasing reservoir energy by injecting water is not enough; reducing the volume of fracturing fluid will play an important role in mitigating well interference.

4. Conclusions

When the re-stimulation timing is the 1st year and reservoir deficit is small, the high-pressure areas are concentrated near primary fractures of parent well after water injection. With the increasing of the injection/production ratio, minimal horizontal principal stress around wellbore increases gradually. In general, the problem of "frac hit" happens universally but is slight. Injecting water has an obvious effect on increasing oil production for both parent well and infill well, especially the parent well. For both parent well and infill well, the final cumulative oil productions corresponding to different injection/production ratios after re-stimulation are close. For the parent well, there are strong positive correlations between the final cumulative liquid production and the injection/production ratio. For the infill well, the cumulative liquid production corresponding to different situation is close.

When the re-stimulation timing is the 4th year and reservoir deficit is large, the reservoir pressure in the whole depleted zone increases apparently. With the increasing of the injection/production ratio, minimal horizontal principal stress around wellbore increases gradually. There are more obvious differences among different situations. For the parent well, the interference of primary fractures with the repeated is serious. The fractures length is longer than the situation after 1-year production. For the infill well, the half-fractures close to the parent well have an obviously longer length than the other side, which means "frac hit" happens in a severe way. Only for the parent well does injecting water have an effect on increasing oil production. There are also strong positive correlations between the final cumulative liquid production and the injection/production ratio for the parent well. For the infill well, under the comprehensive influence of excessive propagation and fractures interference, the cumulative liquid production corresponding to the situation of no water injecting is higher than the ratios of 0.2 to 0.6.

The optimal injection/production ratio for different re-stimulation timing is different, but the corresponding injection volume is almost the same, which is about 15,000 m^3. When the water injection volume is large, more water injection means more water production, which is mainly from the parent well after re-stimulation. The water centralized distribution near the fractures will be produced first. The water cut of the parent well increases rapidly during the initial stage and remains at a high level as the injection/production ratio is large.

Supplementary Materials: The following supporting information can be downloaded at: https://www.mdpi.com/article/10.3390/pr10081538/s1, Figure S1: Pressure distribution corresponding to different injection/production ratios after 1-year production; Figure S2: Minimal horizontal principal stress distribution corresponding to different injection/production ratios after 1-year production; Figure S3: Fractures morphology of parent well and infill well corresponding to different injection/production ratios after 1-year production; Figure S4: Pressure distribution corresponding to different injection/production ratios after 4-years production; Figure S5: Minimal horizontal principal stress distribution corresponding to different injection/production ratios after 4-years production; Figure S6: Fractures morphology of parent well and infill well corresponding to different injection/production ratios after 4-years production.

Author Contributions: Conceptualization, X.M.; methodology, S.Z. and G.R.; software, G.D.; validation, X.M., S.Z. and Y.Z.; formal analysis, G.R.; investigation, G.R.; resources, G.R.; data curation, Y.Z.; writing—original draft preparation, G.R. and Q.X.; writing—review and editing, G.R. All authors have read and agreed to the published version of the manuscript.

Funding: This research was funded by the Research Foundation of the CNPC Strategic Cooperation Science and Technology Project (Key Technologies of Mahu Conglomerate Reservoir, ZLZX 2020-01-04-01).

Conflicts of Interest: The authors declare no conflict of interest.

References

1. Miller, G.; Lindsay, G.; Baihly, H.; Xu, T. Parent well refracturing: Economic safety nets in an uneconomic market. In Proceedings of the SPE Low Perm Symposium, Denver, CO, USA, 5–6 May 2016. [CrossRef]
2. Daneshy, A. Analysis of horizontal well fracture interactions, and completion steps for reducing the resulting production interference. In Proceedings of the SPE Annual Technical Conference and Exhibition, Dallas, TX, USA, 24–26 September 2018. [CrossRef]
3. Abivin, P.; Vidma, K.; Xu, T.; Boumessouer, W.; Bailhy, J.; Ejofodomi, E.; Sharma, A.; Menasria, S.; Sergey, M. Data analytics approach to frac hit characterization in unconventional plays: Application to Williston Basin. In Proceedings of the International Petroleum Technology Conference, Dhahran, Kingdom of Saudi Arabia, 13–15 January 2020. [CrossRef]
4. King, G.; Rainbolt, M.; Swanson, C. Frac hit induced production losses: Evaluating root causes, damage location, possible prevention methods and success of remedial treatments. In Proceedings of the SPE Annual Technical Conference and Exhibition, San Antonio, TX, USA, 9–11 October 2017. [CrossRef]
5. Ajani, A.; Kelkar, M. Interference study in shale plays. In Proceedings of the SPE Hydraulic Fracturing Technology Conference, The Woodlands, TX, USA, 6–8 February 2012. [CrossRef]
6. Rassenfoss, S. Rethinking fracturing: The problems with bigger fracs in tighter spaces. *J. Pet. Technol.* **2017**, *69*, 28–34. [CrossRef]
7. Daneshy, A.; King, G. Horizontal well frac-driven interactions: Types, consequences, and damage mitigation. *J. Pet. Technol.* **2019**, *71*, 45–47. [CrossRef]
8. Fisher, M.; Wright, C.; Davidson, B.; Goodwin, A.; Fielder, E.; Buckler, W.; Steinsberger, N. Integrating fracture mapping technologies to optimize stimulations in the Barnett Shale. In Proceedings of the SPE Annual Technical Conference and Exhibition, San Antonio, TX, USA, 29 September–2 October 2002. [CrossRef]
9. Fisher, M.; Heinze, J.; Harris, C.; Davidson, B.; Wright, C.; Dunn, K. Optimizing horizontal completion techniques in the Barnett Shale using microseismic fracture mapping. In Proceedings of the SPE Annual Technical Conference and Exhibition, Houston, TX, USA 26–29 September 2004. [CrossRef]
10. Courtier, J.; Gray, D.; Smith, M.; Stegent, N.; Carmichael, J.; Hassan, M.; Ciezobka, J. Legacy well protection refrac mitigates offset well completion communications in joint industry project. In Proceedings of the SPE Liquids-Rich Basins Conference–North America, Midland, TX, USA, 21–22 September 2016. [CrossRef]
11. Cipolla, C.; Motiee, M.; Aicha, K. Integrating microseismic, geomechanics, hydraulic fracture modeling, and reservoir simulation to characterize parent well depletion and infill well performance in the Bakken. In Proceedings of the SPE/AAPG/SEG Unconventional Resources Technology Conference, Houston, TX, USA, 23–25 July 2018. [CrossRef]
12. Chittenden, H.; Cannon, D.; Jeziorski, K.; Bowman-Young, S.; Lindsay, S. Understanding the role of well sequencing in managing reservoir stress response in the permian: Implications for child-well completions using high-resolution microseismic analysis. In Proceedings of the SPE Hydraulic Fracturing Technology Conference and Exhibition, The Woodlands, TX, USA, 31 January–2 February 2020. [CrossRef]
13. Ajisafe, F.; Solovyeva, I.; Morales, A.; Ejofodomi, E.; Matteo, M. Impact of well spacing and interference on production performance in unconventional reservoirs, Permian Basin. In Proceedings of the SPE/AAPG/SEG Unconventional Resources Technology Conference, Austin, TX, USA, 20–22 July 2017. [CrossRef]
14. Jacobs, T. What is really happening when parent and child wells interact? *J. Pet. Technol.* **2021**, *73*, 28–31. [CrossRef]

15. Shahri, M.; Tucker, A.; Rice, C.; Lathrop, Z.; Ratcliff, D.; McClure, M.; Fowler, G. High fidelity fibre-optic observations and resultant fracture modeling in support of planarity. In Proceedings of the SPE Hydraulic Fracturing Technology Conference and Exhibition, Virtual, 4–6 May 2021. [CrossRef]
16. Rainbolt, M.; Jacey, E. Frac hit induced production losses: Evaluating root causes, damage location, possible prevention methods and success of remediation treatments, Part II. In Proceedings of the SPE Hydraulic Fracturing Technology Conference and Exhibition, The Woodlands, TX, USA, 23–25 January 2018. [CrossRef]
17. Joslin, K.; Ranjbar, E.; Shahamat, S.; Kiran, S. Pressure sink mitigation: Effect of preloading parent wells to better control infill hydraulic fracture propagation. In Proceedings of the SPE Canada Unconventional Resources Conference, Virtual, 28 September–2 October 2020. [CrossRef]
18. Yu, W.; Wu, K.; Zuo, L.; Tan, X.; Ruud, W. Physical models for inter-well interference in shale reservoirs: Relative impacts of fracture hits and matrix permeability. In Proceedings of the SPE/AAPG/SEG Unconventional Resources Technology Conference, San Antonio, TX, USA, 1–3 August 2016. [CrossRef]
19. Ratcliff, D.; McClure, M.; Fowler, G.; Elliot, B.; Austin, Q. Modelling of parent child well interactions. In Proceedings of the SPE Hydraulic Fracturing Technology Conference and Exhibition, The Woodlands, TX, USA, 31 January–2 February 2022. [CrossRef]
20. Fowler, G.; Ratcliff, D.; Mark, M. Modeling frac hits: Mechanisms for damage versus uplift. In Proceedings of the International Petroleum Technology Conference, Riyadh, Saudi Arabia, 21–23 February 2022. [CrossRef]
21. Wang, L.; Du, X.; Qiu, K.; Wu, S.; Zhuang, X.; Bai, X.; Wang, L.; Pan, Y. Frac hit in complex tight oil reservoir in ordos basin: The challenges, the root causes and the cure. In Proceedings of the SPE Annual Technical Conference and Exhibition, Virtual, 21–22 October 2020. [CrossRef]
22. Vincent, M. Restimulation of unconventional reservoirs: When are refracs beneficial? *J. Can. Pet. Technol.* **2011**, *50*, 36–52. [CrossRef]
23. Bommer, P.; Bayne, M.A. Active well defense in the Bakken: Case study of a ten-well frac defense project, McKenzie County, ND. In Proceedings of the SPE Hydraulic Fracturing Technology Conference and Exhibition, The Woodlands, TX, USA, 23–25 January 2018. [CrossRef]
24. Whitfield, T.; Watkins, M.; Dickinson, L.J. Pre-loads: Successful mitigation of damaging frac hits in the Eagle Ford. In Proceedings of the SPE Annual Technical Conference and Exhibition, Dallas, TX, USA, 24–26 September 2018. [CrossRef]
25. Gala, D.P.; Manchanda, R.; Sharma, M. Modeling of fluid injection in depleted parent wells to minimize damage due to frac-hits. In Proceedings of the SPE/AAPG/SEG Unconventional Resources Technology Conference, Houston, TX, USA, 23–25 July 2018. [CrossRef]
26. Singh, V.; Roussel, N.P.; Sharma, M. Stress reorientation around horizontal wells. In Proceedings of the SPE Annual Technical Conference and Exhibition, Denver, CO, USA, 21–24 September 2008. [CrossRef]
27. Safari, R.; Lewis, R.; Ma, X.; Mutlu, U.; Ghassemi, A. Infill-well fracturing optimization in tightly spaced horizontal wells. *SPE J.* **2017**, *22*, 582–595. [CrossRef]
28. Guo, X.; Wu, K.; Cheng, A.; Tang, J.; Killough, J. Numerical investigation of effects of subsequent parent-well injection on interwell fracturing interference using reservoir-geomechanics-fracturing modeling. *SPE J.* **2019**, *24*, 1884–1902. [CrossRef]
29. Li, N.; Wu, K.; Killough, J. Numerical investigation of key factors on successful subsequent parent well water injection to mitigate parent-infill well interference. In Proceedings of the SPE/AAPG/SEG Unconventional Resources Technology Conference, Denver, CO, USA, 22–24 July 2019. [CrossRef]
30. Zheng, S.; Manchanda, R.; Gala, D.; Mukul, S. Preloading depleted parent wells to avoid fracture hits: Some important design considerations. *SPE Drill Completion* **2021**, *36*, 170–187. [CrossRef]
31. Garza, M.; Baumbach, J.; Prosser, J.; Pettigrew, S.; Elvig, K. An Eagle Ford case study: Improving an infill well completion through optimized refracturing treatment of the offset parent wells. In Proceedings of the SPE Hydraulic Fracturing Technology Conference and Exhibition, The Woodlands, TX, USA, February 2019. [CrossRef]
32. Gupta, J.; Zielonka, M.; Albert, R.; El-Rabaa, W.; Burnham, H.; Nancy, H. Integrated methodology for optimizing development of unconventional gas resources. In Proceedings of the SPE Hydraulic Fracturing Technology Conference, The Woodlands, TX, USA, 10 February 2012. [CrossRef]
33. Xu, L.; Ogle, J.; Collier, T. Fracture hit mitigation through surfactant-based treatment fluids in parent wells. In Proceedings of the SPE Liquids-Rich Basins Conference—North America, Odessa, TX, USA, 7–8 November 2019. [CrossRef]
34. Telmadarreie, A.; Li, S.; Bryant, S. Effective pressure maintenance and fluid leak-off management using nanoparticle-based foam. In Proceedings of the SPE Canadian Energy Technology Conference, Calgary, AL, Canada, 16–17 March 2022. [CrossRef]
35. Cedeno, M. Unloading frac hits in gas wells: How does the nitrogen injection rate and pressure affect the unloading process? In Proceedings of the SPE Trinidad and Tobago Section Energy Resources Conference, Virtual, 28–30 June 2021. [CrossRef]
36. Rasheed, M.; Shihab, S.; Sabah, O. An investigation of the structural, electrical and optical properties of graphene-oxide thin films using different solvents. *J. Phys. Conf. Ser.* **2021**, *1795*, 012052. [CrossRef]
37. Abbas, M.; Rasheed, M. Solid state reaction synthesis and characterization of Cu doped TiO2 nanomaterials. *J. Phys. Conf. Ser.* **2021**, *1795*, 012059. [CrossRef]
38. Kumar, D.; Ghassemi, A. Geomechanical controls on frac-hits. In Proceedings of the SPE International Hydraulic Fracturing Technology Conference & Exhibition, Muscat, Oman, 11–13 January 2022. [CrossRef]

39. Haghighat, A.; Ewert, J. Child/Parent Well interactions; study the solutions to prevent frac-hits. In Proceedings of the SPE Canadian Energy Technology Conference, Calgary, AB, Canada, 15–16 March 2022. [CrossRef]
40. Xiang, H. Refracturing practice of tight oil reservoirs in Ma 56 Block, the Santanghu Basin. *Spec. Oil Gas Reserv.* **2017**, *24*, 157–160. [CrossRef]
41. Sui, Y.; Liu, D.; Liu, J.; Jiang, M.; Liu, J.; Zhang, N. A new low-cost refracturing method of horizontal well suitable for tight oil reservoirs: A case study on Ma 56 Block in Tuha Oilfield. *Oil Drill. Prod. Technol.* **2018**, *40*, 369–374. [CrossRef]
42. Ren, G.; Ma, X.; Zhang, S.; Zou, Y. Optimization of water injection strategy before re-fracturing. In Proceedings of the ARMA/DGS/SEG 2nd International Geomechanics Symposium, Virtual, 1–4 November 2021.
43. Weng, X.; Kresse, O.; Cohen, C.; Wu, R.; Gu, H. Modeling of hydraulic-fracture-network propagation in a naturally fractured formation. *SPE Prod. Oper.* **2011**, *26*, 368–380. [CrossRef]
44. Nolte, K. Fracturing-pressure analysis for nonideal behavior. *J. Pet. Technol.* **2011**, *43*, 210–218. [CrossRef]
45. Zhu, H.; Song, Y.; Tang, X. Research progress on 4-dimensional stress evolution and complex fracture propagation of infill wells in shale gas reservoirs. *Pet. Sci. Bull.* **2021**, *6*, 396–416. [CrossRef]
46. Lewis, R.W.; Sukirman, Y. Finite element modelling for simulating the surface subsidence above a compacting hydrocarbon reservoir. *Int. J. Numer. Anal. Methods Geomech.* **1994**, *18*, 619–639. [CrossRef]
47. Gutierrez, M. Fully coupled analysis of reservoir compaction and subsidence. In Proceedings of the European Petroleum Conference, London, UK, 25–27 October 1994. [CrossRef]
48. Settari, A.; Walters, D.; Behie, G. Reservoir geomechanics: New approach to reservoir engineering analysis. In Proceedings of the Technical Meeting/Petroleum Conference of The South Saskatchewan Section, Regina, 15–17 October 1999. [CrossRef]
49. Chin, L.; Thomas, L.; Sylte, J.; Pierson, R. Iterative coupled analysis of geomechanics and fluid flow for rock compaction in reservoir simulation. *Oil Gas Sci. Technol.* **2002**, *57*, 485–497. [CrossRef]
50. Fung, L.; Buchanan, L.; Wan, R. Couplled geomechanical-thermal simulation for deforming heavy-oil reservoirs. *J. Can. Pet. Technol.* **1994**, *33*. [CrossRef]
51. Koutsabeloulis, N.; Hope, S. "Coupled" stress/fluid/thermal multi-phase reservoir simulation studies incorporating rock mechanics. In Proceedings of the SPE/ISRM Rock Mechanics in Petroleum Engineering, Trondheim, Norway, 8–10 July 1998. [CrossRef]
52. Minkoff, S.; Stone, C.; Arguello, J.; Bryant, S.; Eaton, J. Staggered in time coupling of reservoir flow simulation and geomechanical deformation: Step 1-one-way coupling. In Proceedings of the Annual Simulation Symposium, Houston, TX, USA, 3–6 October 1999. [CrossRef]

processes

MDPI

Article

Study and Mechanism Analysis on Dynamic Shrinkage of Bottom Sediments in Salt Cavern Gas Storage

Baocheng Wu [1], Mengchuan Zhang [2], Weibing Deng [1], Junren Que [1], Wei Liu [1], Fujian Zhou [2], Qing Wang [2], Yuan Li [2] and Tianbo Liang [2,*]

[1] CNPC Engineering Technology Research Institute of Xinjiang Oilfield Company, Karamay 834000, China; wubc@petrochina.com.cn (B.W.); dengweibing@petrochina.com.cn (W.D.); quejunren@petrochina.com.cn (J.Q.); liuwei2006@petrochina.com.cn (W.L.)

[2] State Key Laboratory of Oil and Gas Resources and Prospecting, China University of Petroleum at Beijing, Beijing 102249, China; mcmengchuan@163.com (M.Z.); zhoufj@cup.edu.cn (F.Z.); 18811731154@163.com (Q.W.); 2017310210@student.cup.edu.cn (Y.L.)

* Correspondence: liangtianboo@163.com

Abstract: Underground salt cavern gas storage is the best choice for the production peak adjustment and storage of natural gas, and is a basic means to ensure the safe supply of natural gas. However, in the process of these caverns dissolving due to water injection, argillaceous insoluble sediments in the salt layer will fall to the bottom of the cavity and expand, occupying a large amount of the storage capacity and resulting in the reduction of the actual gas storage space. Effectively reducing the volume of sediments at the bottom of the cavity is a potential way to expand the storage capacity of the cavity. In this study, a method to reduce the volume of argillaceous insoluble sediments with particle sizes ranging from 10 mesh to 140 mesh, via a chemical shrinkage agent, has been proposed. Firstly, the inorganic polymer shrinkage agent PAC30 was synthesized, and then a set of dynamic shrinkage evaluation methods was established to evaluate the influence of temperature, particle size, concentration, and other factors on the shrinkage performance. Finally, by means of a scanning electron microscope (SEM), the Zeta potential, and static adsorption experiments, the mechanism of the interaction between PAC30 and cavity-bottom sediments was described and verified in detail. The experimental results show that the optimal concentration of PAC30 for dynamic shrinkage is 20 ppm. The shrinkage performance of PAC30 decreases with an increase in temperature, and the smaller the particle size of the insoluble sediments, the worse the shrinkage performance. According to the adsorption experiment and Zeta potential, PAC30 can be effectively adsorbed on the surface of insoluble sediments, and the SEM images show that, after adding PAC30, the particles are tightly packed, and the volume of insoluble sediments is significantly reduced. In the large-scale model experiment, the expansion rate of PAC30 reached 20%, which proves that the shrinkage agent is a potential method to expand the gas storage volume.

Keywords: shrinkage agent; underground salt cavern gas storage; insoluble sediments; dynamic shrinkage experiment

Citation: Wu, B.; Zhang, M.; Deng, W.; Que, J.; Liu, W.; Zhou, F.; Wang, Q.; Li, Y.; Liang, T. Study and Mechanism Analysis on Dynamic Shrinkage of Bottom Sediments in Salt Cavern Gas Storage. *Processes* **2022**, *10*, 1511. https://doi.org/10.3390/pr10081511

Academic Editors: Linhua Pan, Minghui Li, Lufeng Zhang, Jie Wang, Wei Feng and Yushi Zou

Received: 11 July 2022
Accepted: 28 July 2022
Published: 1 August 2022

1. Introduction

Underground energy storage (e.g., natural gas, oil, H_2) is one of the most important topics for the sustainable development of the national economy [1]. Underground salt cavern gas storage is the best choice for the peak shaving of natural gas production and stocking up of natural gas resources and is the most basic means to ensure the safe supply of natural gas [2]. Gas storage refers to underground gas banks. Although extensive in the construction period, salt cavern gas storage is relatively stable in performance and allows the fast injection and quick recovery of gas. Traditional gas storage built with depleted sandstone gas reservoirs can only realize one cycle of gas injection in summer and gas recovery in winter every year. Traditional gas storage is similar to growing rice in one season per year, while salt cavern gas storage is similar to growing rice in more than one season a year, which can meet the demand of peak adjustment in more diversified ways [3].

Salt layers in China have the characteristics of a dense structure, low porosity, low permeability, and self-healing ability, making them ideal media for oil and gas storage [4]. However, there are still some problems in building gas storage in salt layers. Salt cavern gas storage in China is characterized by multiple interlayers and high contents of insoluble sediments. For example, Jintan and Pingdingshan salt cavern gas storage facilities have insoluble sediments occupying 33–66.2% (on average 46.8%) of the total volume of the salt caverns, leaving limited cavity space for gas storage [5]. This is a problem worthy of attention [6].

Many scholars have studied insoluble sediments in gas storage. Li [7] studied the distribution characteristics of insoluble sediments in gas storage and proved that particle size was in a linear relationship with the volume of insoluble sediments. Sun [8] analyzed the factors affecting the sedimentation rate of insoluble residues. The results showed that the settling velocity of insoluble sediments was mainly affected by its own particle size and brine Bome, whereby the settling velocity of particles increased with the increase in particle size and decreased with the increase in brine Bome. Gan [9] investigated the expansion rates of insoluble sediments under different calcium sulfate mass fractions, temperatures, and particle sizes. They found that the expansion rate of insoluble sediments was negatively correlated with the mass fraction of calcium sulfate, and positively correlated with the temperature and particle size of insoluble sediments. Ren [10] measured the void volume of insoluble sediments using the gas expansion method and carried out an experimental study and analysis of the air injection and halogen discharge of insoluble sediments based on a physical modeling device. Chen [11] explored the influences of the shape, size, arrangement, and particle composition of broken rocks on the expansion coefficient using ideal experiments and logical analysis. They concluded that the reasons the volume of insoluble sediments at the bottom was larger than that in the interlayer were that salt rock became larger in volume when crushed and clay minerals absorbed water and expanded.

However, due to the late start of the study on the volume expansion of insoluble sediments in gas storage, there are few research and application examples in the oil field [12]. Moreover, most of the current studies focus on the sedimentation rate and accumulation morphology of insoluble sediments in gas storage [13–16], and there are few reports on reducing the volume of insoluble sediments. Investigating how to reduce the volume of insoluble sediments via the chemical method according to the existing experiences and rules remains to be studied.

In this study, in view of the expansion of insoluble sediments in salt cavern gas storage, a type of shrinkage agent suitable for insoluble sediment particles of 10–140 mesh at the bottom of gas storage was synthesized. Through a series of evaluation experiments, the shrinkage effect of the shrinkage agent was explored, and the interaction mechanisms between the shrinkage agent and insoluble sediments were discussed. The experimental results show that the shrinkage agent has the best shrinkage effect at the concentration of 20 ppm and can expand the effective volume of gas storage.

2. Materials and Methods

2.1. Materials

Sodium chloride (NaCl, 99%), aluminum chloride ($AlCl_3$, 99%), and potassium hydroxide (KOH, 99%) were purchased from China National Pharmaceutical Group. The insoluble sediments were taken from gas storage in Chuzhou, China.

2.2. Preparation of Shrinkage Agent PAC30

The shrinkage agent prehydrolyzed polyaluminum is the most widely used shrinkage agent and has been extensively studied. Compared with traditional $AlCl_3$, the polyaluminum shrinkage agent has a higher charge neutralization ability and faster aggregation speed [4]. It is generally believed that Al_{13}, with its strong structural stability, high charge neutralization capacity, and nanoscale molecular diameter, is the core component of the polyaluminum shrinkage agent. However, in order to treat the insoluble sediments in gas storage better, its charge neutralization ability and molecular diameter need to be further optimized.

$Al_{30}([Al_{30}O_8(OH)_{56}(H_2O)_{24}]^{18+})$ polymer (PAC30) is a polycation with a Keggin structure in a hydrolyzed polyaluminum solution [17]. Al_{30} is composed of two δ-Al_{13} connected by four Al monomers. The two-tetrahedral coordinated Al in Al_{30} produces wide ^{27}Al NMR signals at d = 70 ppm. Al_{30} has a higher resistance to high temperatures and stronger charge neutralization ability than Al_{13}, and it has 18 positive charges and a unique nanomolecular size, making it another promising shrinkage agent besides Al_{13} [18].

PAC30 was synthesized according to the method of Chen (2006). Specifically, 0.7 mol/L KOH was slowly added to 100 mL of a 1 mol/L $AlCl_3$ aqueous solution and the mixture was stirred at 60 °C until the alkalinity reached 2.4. Sufficient shear force was needed in the mixing process to prevent the formation of amorphous products. Then the solution was heated at 90 °C for 24 h under the conditions of stirring and condensation reflux. Then it was aged for five days. The final product was PAC30.

2.3. XRD Analysis of Insoluble Sediments

The mineral analysis was carried out by a whole-rock X-ray diffractometer (XRD) in three steps: The first step was to crush the rock powder to 400 mesh; the second step was to put the sample into the sample chamber and press the fixed-spring plate; the third step was to turn on the instrument for testing and analysis.

2.4. Establish a Shrinkage Agent Evaluation Method

At present, there is no standard test method for the evaluation of the shrinkage agent. According to the field construction process, this paper designed a set of evaluation methods for the shrinkage agent:

(1) Dynamic dissolution shrinking experiment.

Salt cavern gas storage in China is built by layered salt with a high content of insoluble sediments (15~40%), a thin salt layer (approximately 150 m), and a thick interlayer (3~11 m). After completion, a large amount of argillaceous insoluble matter will have accumulated in the cavity. In order to verify the shrinkage effect of the synthesized PAC30 and explore the influence of different factors on the shrinkage effect, the dynamic corrosion shrinkage and expansion experiment was established, and the experimental device was set up according to Figure 1. The experimental steps were designed as follows:

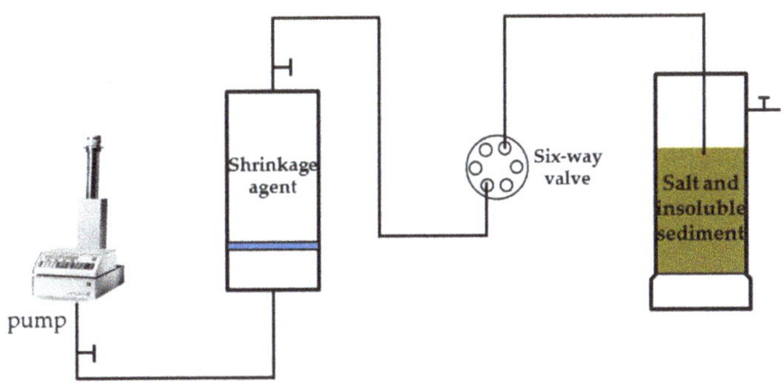

Figure 1. Schematic diagram of dynamic shrinkage experiment device.

Firstly, the insoluble sediments from the gas storage were ground into powder of corresponding meshes, washed with a dichloromethane solution to remove oil and salt for 24 h, and dried in an oven at 105 °C for 24 h until a constant weight was reached.

Secondly, a total of 200 g of insoluble sediment and NaCl were taken. The insoluble sediment: NaCl was evenly mixed at a mass ratio of 30:70 and then poured into a three-mouth flask. We then dropped different shrinkage agents or distilled water into the three-port flask at a constant flow rate of 0.2 mL/min and recorded the volume of the remaining solid every time for 168 h. In the range where the difference between the results was less than 5%, all the results were taken as the arithmetic average of the two groups. The calculation formula of the shrinkage rate is [19]:

$$K_1 = \frac{V_1 - V_2}{V_1} \times 100\% \tag{1}$$

where K_1 is the shrinking rate (%), V_2 is the expansion volume of insoluble sediments in the shrinkage agent solution (mL), and V_1 is the expansion volume of insoluble sediments in water (mL).

(2) Large-scale Dynamic Salt Solution Simulation.

The experimental device is shown in Figure 2. Figure 2a,b shows the self-built device, which is a cuboid container of 24 cm × 28 cm × 40 cm for storing insoluble sediments. Figure 2c shows the large-scale dynamic salt solution simulation device consisting of a shrinkage agent, a circulating pump, and a simulation of the salt layer. By using the circulating pump, the liquid is injected into the rock powder to dissolve the rock powder in the interior. After the dissolution is basically completed, the brine is discharged by the circulating pump to realize the complete cavity construction process. The whole simulation process roughly restores the field cavity construction process, and the experimental results are closer to reality.

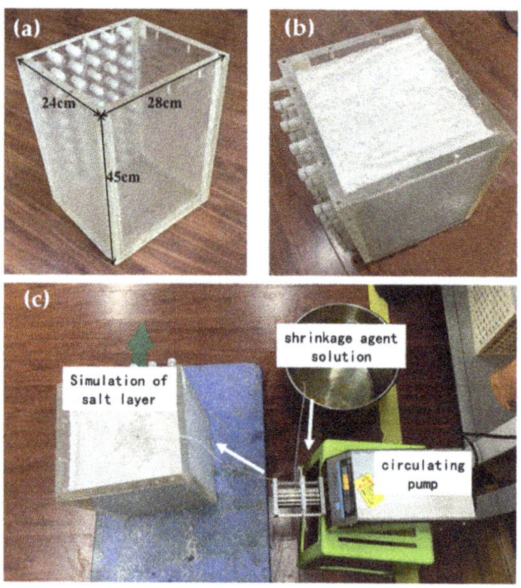

Figure 2. Large-scale dynamic salt solution simulation device. (**a**) A container of insoluble sediments, (**b**) a container filled with insoluble sediments, and (**c**) a large-scale dynamic salt solution simulation device.

The experimental steps are as follows: Firstly, 35 kg of rock powder was prepared according to the mass ratio of stratum insoluble matter and salt of 30:70 and then mixed evenly and poured into the experimental container in Figure 2. Then, at a constant flow rate of 27.6 mL/min, the shrinkage agent or distilled water was injected into the container where the rock powder was released and mixed evenly. The dissolution rate of the rock powder could be judged by discharging the liquid density, and the volume V of residual solid was finally recorded. When the difference between the measured results was less than 5%, all the results were taken as the average values of the two groups. The final shrinkage is calculated by Formula (1).

2.5. Mechanism Analysis of Shrinkage Agent

The mechanism of the shrinkage effect of PAC30 on insoluble sediments was analyzed by an adsorption experiment, the Zeta potential, and SEM.

(1) Static adsorption of the shrinkage agent.

We added 10 mL of the 200 ppm shrinkage agent to 1.5 g of 70/100 mesh insoluble sediments [20] and let it stand at room temperature for a period of time. The experimental time was set to 20 min, 40 min, 1 h, 3 h, 6 h, and 24 h, respectively. Then, the mixture solutions were centrifuged for 15 min at 4000 RPM, and the supernatant fluids were taken for spectrophotometric measurements. From the change in absorbance over time, the loss of the shrinkage agent absorbed by insoluble sediments can be quantified.

(2) Scanning electron microscope (SEM) analysis of insoluble sediments.

The insoluble sediments after the experiment were first put into the electrical thermostat drying box and dried at 210 °C for 12 h. Then, the dried insoluble sediments were stuck onto an electron microscope metal sheet covered with conductive adhesive. To increase the electrical conductivity of the rock powder, the metal sheet with the bonded rock powder was put into a sputtered gold injector to be coated with gold. Afterward, the metal sheet coated with gold was pushed into the electron microscope cavity.

(3) Zeta potential.

The Zetasizer Nanolaser analyzer was used to measure the potential of the insoluble sediments from the gas storage and shrinkage agents. We added 140 mesh rock powers to water, stirred well, and let stand for 4 h. The supernatant was taken to test the zeta potential. This value is considered to be the potential value of insoluble sediments.

3. Results and Discussion

In this part, the application potential of PAC30 in salt cavern gas storage will be analyzed from the aspects of XRD analysis of insoluble sediments, the shrinkage effect, and the shrinkage mechanism of the shrinkage agent PAC30.

3.1. XRD Analysis of Insoluble Matter in Salt Cavern Gas Storage

Different from most of the rock salts in other countries, the extensively distributed rock salts in eastern China are all lacustrine thinly bedded types, characterized by low thickness, a high proportion of impurities, and numerous intersecting non-salt interlayers [21]. These interlayers have a wide range of thicknesses, and the content of argillaceous insoluble sediment in the interlayer ranges between 15% and 35% [22], generally accumulating a large amount of insoluble residue in the cavity after construction.

Figure 3 shows the multi-interlayer salt rock derived from well A in the Chuzhou block. It can be seen from the figure that there are obvious mudstone interlayers distributed in the salt rock. During the salt cavern leaching process, the salt in the rock is dissolved and the insoluble mudstone falls to the bottom and expands in volume. The insoluble sediments were analyzed by rock mineral composition (XRD). It can be seen from Table 1 that the formations of the insoluble matter are mainly plagioclase, quartz, clay minerals, and calcite. Table 2 is the experimental results of the clay mineral composition analysis. The main clay minerals include montmorillonite, illite, kaolinite, an illite–montmorillonite mixed layer, and a green montmorillonite mixed layer.

Figure 3. Multi-interlayer salt rock derived from well A.

Table 1. Mineral composition analysis of insoluble sediments in well A.

Sample Number	Plagioclase	Quartz	Calcite	TCCM
Content (wt%)	35	31	22	12

Table 2. Clay composition analysis of clay minerals in insoluble sediments from well A.

Sample Number	S	It	Kao	I/S	C/S
Content (wt%)	48	7	4	10	31

Note: S is montmorillonite, it is illite, Kao is kaolinite, I/S is montmorillonite mixed layer, C/S is green montmorillonite mixed layer.

3.2. *Characterization of Shrinkage Agent PAC30*

In recent years, ^{27}Al NMR technology has been used for the determination of aluminum morphology in many fields [23–27]. Compared with other methods used for the determination of aluminum morphology, the outstanding advantage of NMR technology is that it can go deep into the substance without destroying the sample, and it will not change the chemical morphology of Al in the measured solution. What is directly obtained on the NMR spectrum of Al in the solution is a single weighted average peak for a certain morphology [28]. Therefore, ^{27}Al NMR is a clear morphological identification technology for hydroxylated polyaluminum.

The ^{27}Al NMR spectra of PAC30 are shown in Figure 4. δ = 80 ppm represents sodium aluminate used as the internal standard [29]. The δ values of 63 ppm and 70 ppm represent the resonance peaks of the aluminum oxygen tetrahedron in Al_{13} and Al_{30}, respectively. As the polymerization form of Al_{13} is an aggregate of 13 aluminum atoms with a Keggin structure formed by 12 hexadecylated Al octahedrons surrounding a tetrahedron with a tetrahedron through hydroxyl and oxygen bridging, only the tetrahedron with a tetrahedron at the symmetry center can produce a resonance peak at δ 63 ppm [30]. Similarly, the Al_{30} polymerization form is a 30-Al aggregate formed by two δ-Al_{13} polymerization forms linked by four six-coordinated Al/O octahedrons, among which only the two tetracoordinated al atoms in the Cc symmetry can produce resonance peaks at δ 70 ppm [31].

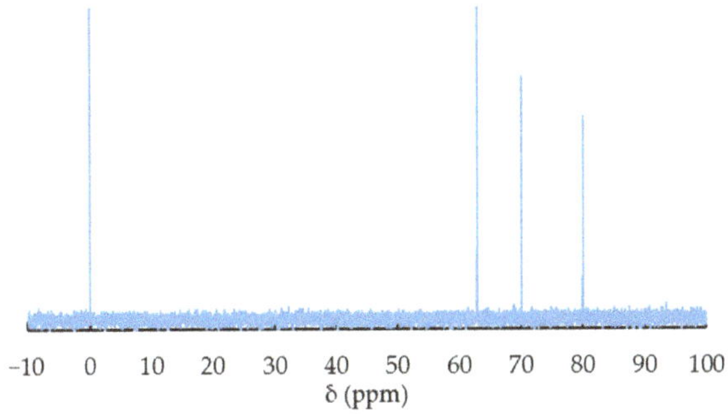

Figure 4. ^{27}Al NMR spectra of PAC30.

3.3. *Analysis of Factors Influencing the Effect of Shrinkage Agent*

In the field application, the complex application environment affects the effect of the shrinkage agent. Through the shrinkage evaluation method established above, the influence of different factors on the shrinkage effect of PAC30 was studied.

3.3.1. Effect of Concentrations of PAC30 on Dynamic Shrinkage

The amount of shrinkage agent is the main factor affecting the shrinkage rate. Five different concentrations of PAC30 were selected to analyze the shrinkage effect. The five concentrations were 5 ppm, 10 ppm, 20 ppm, 200 ppm, and 500 ppm. The temperature was set at 25 °C. The number of insoluble sediment mesh used in the experiment was 80–140 mesh. The experimental results are shown in Figure 5. It can be seen from Figure 5 that when the dosage of PAC30 was 20 ppm, the shrinkage rate reached 30.02%.

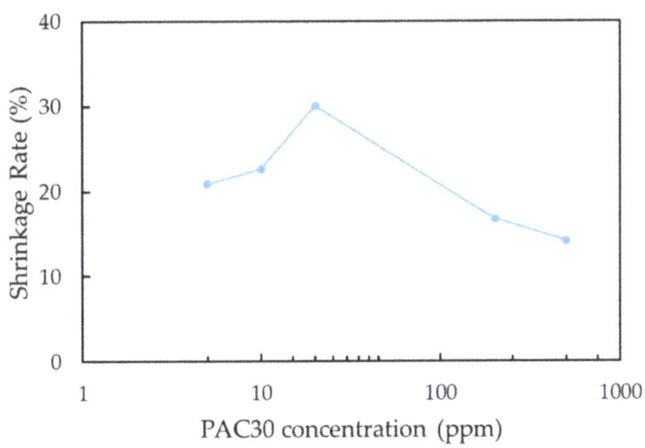

Figure 5. Curve of PAC30 concentration and dynamic shrinkage rate.

When the PAC30 dosage was less than 20 ppm, the shrinkage rate increased with the increase in the PAC30 concentration. When PAC30 dosage was higher than 20 ppm, the shrinkage rate decreased with the increase in the PAC30 concentration. When PAC30 is insufficient, the content of Al_{30} is insufficient, so on the one hand, the charge on the surface of insoluble matter cannot be completely shielded; on the other hand, only small particle aggregates are formed, and the small aggregate particles cannot gather into large ones. Therefore, the optimum dosage of PAC30 is 20 ppm.

The addition of the shrinkage agent mainly affects the zeta potential of insoluble sediments. Due to the effect of electrostatic force, the shrinkage agent will adsorb on the surface of rock powder particles to neutralize the negative charges on the surface and reduce the expansion of insoluble sediments. At the same time, the particles can be connected by intermolecular forces such as chain and hydrogen bonds to further compress the pore space between particles and enhance the shrinkage effect. However, when the concentration of the shrinkage agent further increases, excess positive charges will bring positive charges to the surface of the insoluble sediments, causing the electrostatic repulsion to increase and the effect of the shrinkage agent to decrease. This explains why, after reaching the peak, the shrinkage effect decreases [32].

In the process of the experiment, continuous injection of the PAC30 solution will result in more of the shrinkage agent in the insoluble sediments, so less shrinkage agent is needed to achieve the optimal volumetric shrinkage. The PAC30 solution with a 0.002% concentration was used for the dynamic shrinking test in subsequent experiments.

3.3.2. Effect of Temperature on Shrinkage Effect of PAC30

Under formation conditions, the temperature is one of the important factors affecting liquid performance. Figure 6 compares the shrinkage effects of PAC30 at 25 °C and 60 °C. It can be seen from Figure 6 that the shrinkage rate gradually increased with the concentration at two temperatures. The expansion rate at 60 °C is significantly lower than that at 25 °C. This is because when the temperature increases, the Brownian motion of PAC30 can accelerate, which speeds up the interaction between the PAC30 molecules and the upper rock particles. So, the shrinkage rate was accelerated at the beginning of 60 °C. However, when the volume of upper rock particles was compacted (as shown in Figure 7), it further hindered the diffusion of PAC30 at this depth, which resulted in the reduction of shrinkage. Moreover, when the temperature is too high, the hydrolysis speed of PAC30 is too fast, leading to the aging, shrinkage, and fracture of flocculant molecules, and finally, to the continuous decline of the flocculation capacity of PAC30 [33].

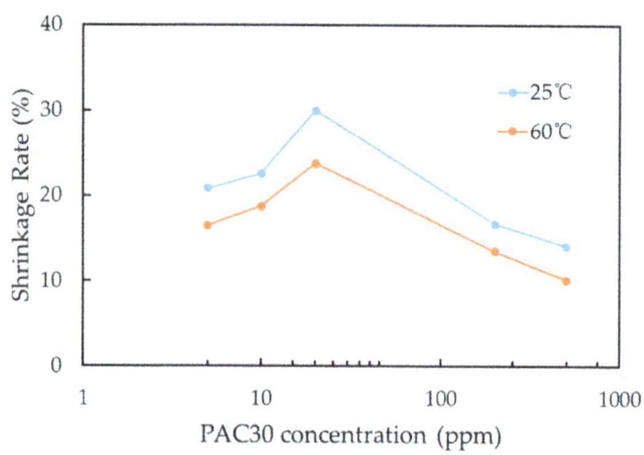

Figure 6. Curve of expansion rate of PAC30 with temperature and concentration.

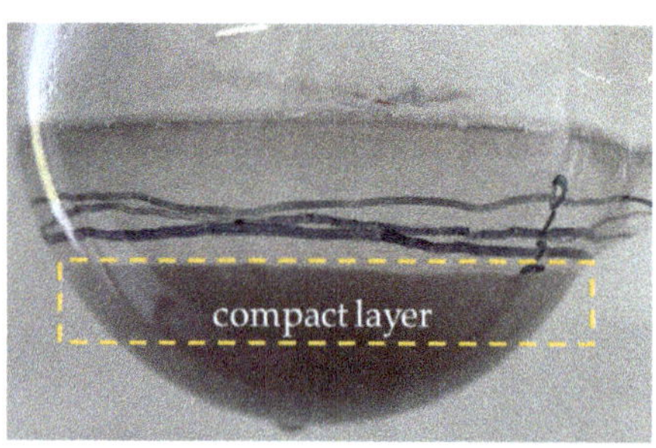

Figure 7. Compact layer formed after injection of PAC30 solution.

3.3.3. Effect of Particle Size of Insoluble Sediments on Shrinkage Agent PAC30

The mesh size of insoluble sediments used to evaluate the dynamic shrinkage effect of PAC30 is 80–140 mesh, and the shrinkage rate is 30.02%. However, the particles of insoluble sediments from the gas storage have a wide size range. In order to evaluate the corresponding relationship between the shrinkage agent and the particle size of insoluble sediments, the particle size composition of two wells was simulated based on field sampling results [3,21].

Table 3 shows the particle size distribution data of this experiment. Particle size combination 1 designed in the experiment was significantly smaller than that of particle size combination 2. The results of the dynamic dissolution experiment are shown in Figure 8. It is concluded that the smaller the particle size of the rock powder is, the larger the shrinkage rate is and the better the shrinkage effect is. It is proposed that there are two main reasons: First, the smaller the particle size, the larger the specific surface, and the larger the area between the particle and the shrinking agent. It resulted in stronger adsorption and compaction. Secondly, it can be seen in Figure 9 that there are many small particles suspended on the liquid surface. They can be taken away with the flow of brine, which is consistent with the literature results. Sun found that the particles with a size less than 0.1 mm do not settle in brine and are discharged with the transport of brine in the cavity [33–35]. With the discharge of the parts of small particles, the space occupied by insoluble sediments at the bottom of the salt cavern is reduced to a certain extent, so that the space of the gas storage is further enlarged.

Table 3. The particle size distribution combination used in the experiment.

Particle Size (mesh)	10~14	14~24	24~32	32~60	60~80	80~140
particle size combination 1 (%)	70.0	16.0	4.0	4.0	4.0	2.0
particle size combination 2 (%)	88.0	3.4	4.0	1.6	2.0	1.0

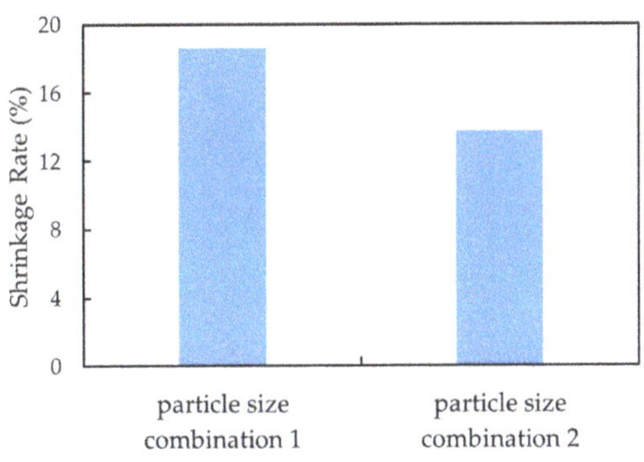

Figure 8. Relationship between shrinkage rate and particle size combination of PAC30.

Figure 9. Experimental diagram of particle size dynamic shrinkage. (**a**) The experimental diagram of particle size combination 1, (**b**) the experimental diagram of particle size combination 2.

3.3.4. Physics Simulation Experiment with Large Size Device

To get closer to the cavern construction process on site, a large experimental device was developed, as shown in Figure 2.

First, 35 kg of rock powder was prepared according to the mass ratio of 30:70 of insoluble sediments and stone salt. Then it was mixed, stirred evenly, and poured into the experimental container. Again, at a constant flow rate of 27.6 mL/min, the shrinkage agent solution or distilled water was injected into the container. Whether salt rock was completely dissolved could be judged by the density of the discharged liquid, and the final volume of the remaining solid was recorded. The whole dissolution time of each experiment was approximately 10 days. The volume of mixed rock powders in this experiment was 26,880 cm^3. The residual volume after the injection of tap water was 10,214.4 cm^3 and accounts for 38% of the total volume. After the injection of the 20 ppm PAC30 solution, the volume of the accumulation was reduced to 8064 cm^3, and the volume was reduced by 21.05% compared with that of tap water. If the volume of insoluble sediments accounted for 40% of the total volume of the gas storage, it can be calculated that the storage volume increased by 8% after the injection of the 20 ppm PAC30 solution. The results of this experiment directly proved that the shrinkage effect of PAC30 is excellent. PAC30 can effectively reduce the volume of insoluble sediments and increase the storage volume of a bedded salt cavern.

3.4. Study on Mechanism of Shrinkage

Due to the negative charge on the surface of the rock powder particles after hydration, the surrounding ions with opposite charges are attracted, and these opposite charges are distributed in a diffusion state at the two-phase interface to form a diffusion double layer. The surface of the rock powder is closely connected to water molecules connected by hydrogen bonds and some cations with hydration shells, which together constitute the adsorption solvation layer, which is macroscopically reflected as volume expansion. The main function of the shrinkage agent is to adsorb the surface of the rock powder, neutralize the charge, aggregate the rock powder, and reduce the space occupied by the rock powder. The adsorption of PAC30 on different mineral surfaces was studied in this paper.

3.4.1. Adsorption of PAC30 on Mineral Surface

According to the experimental parameters given in Section 2.4, the total injection amount of PAC30 was 40.32 mg and the ratio of PAC30 to insoluble sediments was 0.672 mg/g (PAC30/insoluble sediments) in 168 h. The adsorption experiment should ensure that the content of PAC30 in the solution is higher than the adsorption amount on the insoluble surface. The amount of insoluble sediments used in the adsorption experiment is 1.5 g, and at least 1mg of PAC30 is required. Therefore, 200 ppm was selected as the concentration used in the adsorption experiment.

The static adsorption capacities of rock powder at different times in the adsorption process were measured. The experimental results are shown in Figure 10, and the adsorption capacities of different minerals to PAC30 are shown in Table 4.

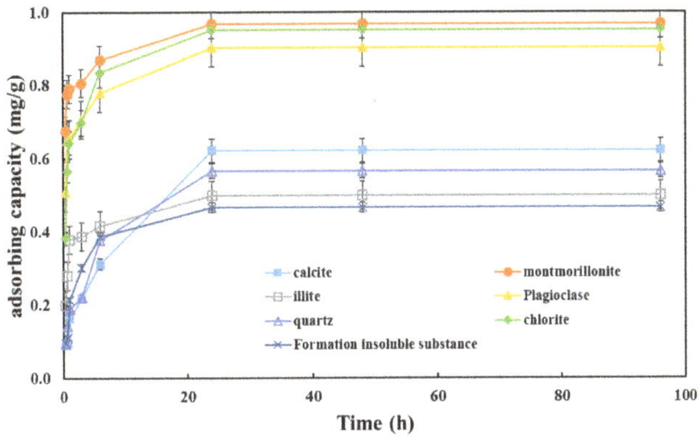

Figure 10. Adsorption curves of PAC30 in different minerals.

Table 4. Adsorption capacity of different minerals on PAC30 and charge of mineral surface.

Rock Powder Minerals	Montmorillonite	Quartz	Plagioclase	Illite	Calcite	Chlorite	Insoluble Sediments
Zeta potential (mV)	−37	−4.43	−17.1	−8.48	−1.96	−18.8	−18.1
adsorption (mg/g)	0.967	0.566	0.902	0.499	0.622	0.951	0.466

The adsorption capacity is a measure of the strength of mutual repulsion or attraction between shrinkage agents and minerals. A large adsorption capacity indicates a strong interaction between the shrinkage agent and the mineral, and a small adsorption capacity indicates a weak interaction between the shrinkage agent and the mineral [36]. Only when the force between the shrinkage agent and the mineral is strong can the effect of shrinkage be achieved.

It can be seen from Table 3 that the PAC30 has a larger adsorption capacity on the surface of clay minerals than on the surfaces of other minerals. The adsorption capacity of PAC30 is proportional to the surface charge of the mineral. For example, montmorillonite is easily hydrated and dispersed in water, with a surface charge of −37 mV. PAC30, with positive charges on the surface, can form a strong electrostatic force with montmorillonite, and has an adsorption capacity of 0.967 mg/g on the surface of montmorillonite. Its adsorption capacity on the surface of the insoluble sediments was 0.466 mg/g, indicating that PAC30 can fully interact with the insoluble sediments. Moreover, the adsorption of PAC30 on the mineral surface became stable 24 h into the experiments; with the further extension of the adsorption time, its adsorption capacity did not change significantly.

3.4.2. Zeta Potential

The Zeta potential is an important characteristic parameter of the diffusion layer, and its value can reflect the thickness of the diffusion layer [37–39]. The lower the absolute value of the Zeta potential is, the fewer adsorbed cations in the diffusion layer, and the thinner the diffusion layer is. Conversely, the higher the absolute value of the Zeta potential, the more adsorbed cations there are and the thicker the diffusion layer is.

In order to study the Zeta potential changes after the interaction between different concentrations of PAC30 and insoluble sediments, according to the adsorption capacity of PAC30 on the insoluble sediment's surface, PAC30 with a concentration of 100 ppm to 500 ppm was mixed with 15 g of insoluble sediments, and its Zeta potential was tested.

It can be seen from Table 5 that with the increase in the PAC30 concentration, Zeta potential gradually turns from negative to positive, indicating that when the concentration of PAC30 added is lower than 200 ppm, the thickness of the diffusion double layer decreases, the interparticle repulsion decreases, and the insoluble sediment particles aggregate. When the concentration of PAC 30 added exceeds 200 ppm, the concentration of positive charges between particles increases and particles begin to disperse again. In the process of decline of potential absolute values, PAC30 can compress the electric double layer by electric neutralization to cause the water film combined with clay particles to become thinner and the rock powder particles aggregate with each other, achieving the purpose of shrinkage of insoluble sediments. However, when the added cations exceed a certain concentration, the surfaces of clay particles are positively charged, which, in turn, attract negatively charged anions and form new diffusion double layers, so the repulsion between clay particles begins to increase again, resulting in an increase in the diffusion layer thickness.

Table 5. Zeta potential values of insoluble sediments after adding different concentrations of PAC30 solution.

Insoluble Sediments Amount (g)	PAC30 Concentration (ppm)	Zeta Potential (mV)
	0	−18.10
	100	−11.40
15	200	0.36
	300	4.73
	500	14.90

3.4.3. SEM Analysis

In order to further explore the shrinkage mechanisms of the shrinkage agent, the microstructure of the underlying insoluble sediments was observed by scanning electron microscopy to study the insoluble sediment treated with the shrinkage agent.

Figure 11a is an SEM image of the insoluble sediments. It can be seen from this figure that the particles were in loose accumulation, with no aggregation. Figure 11b shows the SEM image of the insoluble sediments after being treated by PAC30. The PAC30 with a molecular size of more than 5 nm can adsorb at multiple points onto rock particles, causing particles of expanded insoluble sediments to agglomerate into a larger whole, thereby reducing the spacing between the particles. The smaller the floc size, the higher the shrinkage efficiency is.

Figure 11. SEM images of insoluble sediments. (**a**) The insoluble sediments; (**b**) the insoluble sediments treated with PAC30.

The basic principle of the shrinkage agent is to compress the double electric layer via bridge multipoint adsorption to achieve volume compression. Inorganic polymer shrinkage agents, with large molecular sizes, can absorb on the surface of insoluble sediments to neutralize negative charges, and cause insoluble sediments particles to become more compacted through bridging and multi-point adsorption, as shown Figure 12. PAC30 is one such material, with a Keggin structure and a molecular weight between 2000 and 4000 g/mol, which can form multi-point adsorption and bridges between rock powder particles.

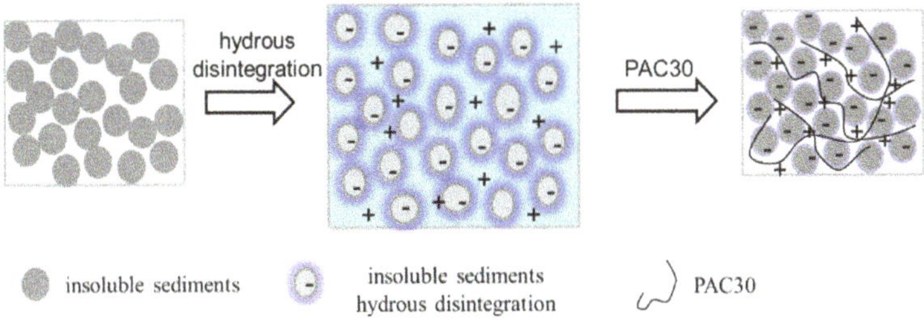

Figure 12. Schematic diagram of action mechanism of shrinkage agent.

4. Conclusions

In order to reduce the volumetric expansion of the insoluble sediments at the bottom of the salt cavern gas storage, a chemical shrinkage agent, PAC30, was designed and synthesized. A dynamic shrinkage experimental apparatus was established to study the effects of temperature, particle size, and concentration on the shrinkage efficiency of PAC30. Furthermore, its shrinkage mechanism was clarified by SEM and Zeta potential tests. The main conclusions are as follows:

Firstly, the self-synthesized PAC30 is an effective chemical shrinkage agent, especially when the concentration is 20 ppm. The result of the NMR experiment shows that it has a Keggin structure and forms a 30-Al polymer consisting of two δ-Al13 structures. Secondly, the shrinkage efficiency is mainly controlled by adsorption and electrostatic interactions. When the Zeta potential of the insoluble sediments after PAC30 treatment was close to 0 mv, the volume of the insoluble sediments was the smallest. Through SEM observations, insoluble sediments accumulated more closely after adding PAC30. The main shrinkage mechanism is that PAC30, with a molecular weight between 2000 and 4000 g/mol, can absorb on the surface of insoluble sediments to neutralize negative charges and cause insoluble sediments particles to become more compacted through bridging and multi-point adsorption, which is conducive to reducing the expansion volume. Finally, considering the field conditions, such as temperature and particle size, the shrinkage rate of PAC30 was intensively evaluated. A large dynamic shrinkage device was built, and the maximum dynamic shrinkage rate of PAC30 was up to 21.05%. With the increase in insoluble sediments' particle size, the shrinkage efficiency of PAC30 decreased rapidly. This is because large particles have a smaller specific surface area. A high temperature accelerated the appearance of the compact layer and reduced the shrinkage rate. The adsorption capacity of PAC30 on the insoluble sediments' surface was 0.466 mg/g at the optimal dosage. The experimental results of this paper provide a method to expand the capacity of salt cavern gas storage, and understanding the shrinkage mechanism will guide the design of new chemical shrinkage agents.

Author Contributions: Conceptualization, F.Z.; methodology, B.W. and M.Z.; validation, W.D., Q.W. and J.Q.; formal analysis, W.L.; investigation, T.L.; resources, F.Z.; data curation, Y.L.; writing—original draft preparation, Y.L.; writing—review and editing, Y.L. and T.L.; visualization, B.W.; supervision, F.Z. and J.Q. All authors have read and agreed to the published version of the manuscript.

Funding: This work was financially supported by the Strategic Cooperation Technology Projects of CNPC and CUPB (ZLZX2020-01).

Conflicts of Interest: The authors declare no conflict of interest.

References

1. Furuoka, F. Natural gas consumption and economic development in China and Japan: An empirical examination of the Asian context. *Renew. Sustain. Energy Rev.* **2016**, *56*, 100–115. [CrossRef]
2. Mac Kinnon, M.A.; Brouwer, J.; Samuelsen, S. The role of natural gas and its infrastructure in mitigating greenhouse gas emissions, improving regional air quality, and renewable resource integration. *Prog. Energy Combust. Sci.* **2018**, *64*, 62–92. [CrossRef]
3. Song, S.; Li, T.; Liu, P.; Li, Z. The transition pathway of energy supply systems towards carbon neutrality based on a multi-regional energy infrastructure planning approach: A case study of China. *Energy* **2021**, *238*, 122037. [CrossRef]
4. Tang, H.; Xiao, F.; Wang, D. Speciation, stability, and coagulation mechanisms of hydroxyl aluminum clusters formed by pacl and alum: A critical review. *Adv. Colloid Interface Sci.* **2015**, *226*, 78–85. [CrossRef] [PubMed]
5. Chen, M.; Wu, W. Preliminary advice for constructing the natural gas storage in Guangdong province. *Chem. Eng. Oil Gas* **2010**, *39*, 193–195.
6. Wanyan, Q.; An, G.Y. Current status and development direction of salt cavern gas storage technology. *Pet. Drill. Prod. Technol.* **2020**, *42*, 5.
7. Chen, J.; Lu, D.; Liu, W.; Fan, J.; Jiang, D.; Yi, L.; Kang, Y. Stability study and optimization design of small-spacing two-well (sstw) salt caverns for natural gas storages. *J. Energy Storage* **2020**, *27*, 101131.1–101131.11. [CrossRef]
8. Zheng, Y.L.; Qiu, X.S.; Ding, G.S. Experimental research on using residue void space in salt cavern gas storage. *J. Salt Sci. Chem. Ind.* **2019**, *48*, 14–19.
9. Ren, Z.X.; Li, J.J.; Wang, H.M. Particle size distribution characteristics of salt stone reservoir cavity bottom accumulation based on fractal theory. *Oil Gas Storage Transp.* **2017**, *36*, 5.
10. Sun, C.L.; Hou, J.R.; Ding, G.S. Study on the settlement pattern of insoluble residue particles in brine from salt cavern gas storage reservoirs. *Salt Sci. Chem.* **2017**, *46*, 3.
11. Gan, S.T.; Xiao, Y.M.; Tang, N. Experiment on the expansion coefficient of insoluble material in salt rock. *Oil Gas Storage Transp.* **2018**, *37*, 5.
12. Ren, Z.X.; Ba, J.H.; Ren, Z.X. Experimental study on gas injection and brine discharge in salt rock reservoir piles. *Oil Gas Field Ground Eng.* **2018**, *37*, 4.
13. Wang, X.; Liu, Y.; Hou, J.; Li, S.; Kang, Q.; Sun, S.; Ji, L.; Sun, J.; Ma, R. The relationship between synsedimentary fault activity and reservoir quality—A case study of the Ek1 formation in the Wang Guantun area, China. *Interpretation* **2020**, *8*, SM15–SM24. [CrossRef]
14. Wang, X.; Zhou, X.; Li, S.; Zhang, N.; Ji, L.; Lu, H. Mechanism Study of Hydrocarbon Differential Distribution Controlled by the Activity of Growing Faults in Faulted Basins: Case Study of Paleogene in the Wang Guantun Area, Bohai Bay Basin, China. *Lithospher* **2022**, *2021*, 7115985. [CrossRef]
15. Wang, X.; Zhang, F.; Li, S.; Dou, L.; Liu, Y.; Ren, X.; Chen, D.; Zhao, W. The Architectural Surfaces Characteristics of Sandy Braided River Reservoirs, Case Study in Gudong Oil Field, China. *Geofluids* **2021**, *2021*, 8821711. [CrossRef]
16. Chen, X.; Zhang, L.; Li, Y.; Ma, H.; Ji, G. Experimental investigation on bulking-expansion coefficient of sediment of storage in bedded salt. *Min. Res. Dev.* **2013**, *33*, 34–37.
17. Chen, Z.; Fan, B.; Peng, X.; Zhang, Z.; Fan, J.; Luan, Z. Evaluation of Al30 polynuclear species in polyaluminum solutions as coagulant for water treatment. *Chemosphere* **2006**, *64*, 912–918. [CrossRef]
18. Chen, Z.; Liu, C.; Luan, Z.; Zhang, Z.; Li, Y.; Jia, Z. Effect of total aluminum concentration on the formation and transformation of nanosized Al13 and Al30 in hydrolytic polymeric aluminum aqueous solutions. *Chin. Sci. Bull.* **2005**, *50*, 2010–2015. [CrossRef]
19. Yao, E.; Xu, H.; Zhang, K.; Liu, S.; Sheng, L.; Li, B.; Zhou, F. Experimental and Mechanistic study of mudstone volumetric swelling at the bottom of salt cavern gas storage. *Arabian J. Chem.* **2022**, *15*, 104082. [CrossRef]
20. Liu, S.; Yao, E.; Luo, Y.; Li, Y.; Zhou, F.; Ran, L.; Yuan, C. Study on Factors Influencing the Shrinkage of Swelled Deposits at the Bottom of Salt Cavern Gas Storage. Presented at the 55th U.S. Rock Mechanics/Geomechanics Symposium, Virtual, 18–25 June 2021.
21. Liu, W.; Zhang, Z.; Chen, J.; Fan, J.; Jiang, D.; Daemen, J.J.K. Physical simulation of construction and control of two butted-well horizontal cavern energy storage using large molded rock salt specimens. *Energy* **2019**, *185*, 682–694. [CrossRef]
22. Ban, F.S. Status and development trend of solution mining technologies used for salt-cavern gas storage. *Oil Gas Storage Transp.* **2017**, *36*, 754–758.
23. Li, Y.; Zhou, F.; Li, B.; Cheng, T.; Zhang, M.; Wang, Q.; Liang, T. Optimization of Fracturing Fluid and Retarded Acid for Stimulating Tight Naturally Fractured Bedrock Reservoirs. *ACS Omega* **2022**, *7*, 25122–25131. [CrossRef]
24. Karlik, S.J.; Elgavish, G.A.; Eichhorn, G.L. Multinuclear NMR studies on aluminum (III) complexes of ATP and related compounds. *J. Am. Chem. Soc.* **1983**, *105*, 602–609. [CrossRef]
25. Beck, L.W.; Haw, J.F. Multinuclear NMR studies reveal a complex acid function for zeolite beta. *J. Phys. Chem.* **1995**, *99*, 1076–1079. [CrossRef]
26. Templin, M.; Wiesner, U.; Spiesss, H.W. Multinuclear solid-state-NMR studies of hybrid organic-inorganic materials. *Adv. Mater.* **1997**, *9*, 814–817. [CrossRef]
27. Li, Y.; Zhou, F.; Wang, J.; Li, B.; Xu, H.; Yao, E.; Zhao, L. Influence of Nanoemulsion Droplet Size of Removing Water Blocking Damage in Tight Gas Reservoir. *Energies* **2022**, *15*, 5283. [CrossRef]
28. Titova, Y.Y.; Schmidt, F.K. What ^{27}Al NMR Spectroscopy Can Offer to Study of Multicomponent Catalytic Hydrogenation Systems? *J. Organomet. Chem.* **2022**, *975*, 122410. [CrossRef]

29. Chen, Z.Y.; Luan, Z.K.; Fan, B.; Zhang, Z.G.; Li, Y.Z.; Jia, Z.P. Quantitative studies on Keggin polycation Al-13 and Al-30 in hydrolytic polymeric aluminum aqueous solution by Al-27 nuclear magnetic resonance. *Chin. J. Anal. Chem.* **2006**, *34*, 38–42.
30. Chen, F.; Ye, L.; Ma, H.; Shi, X.; Liu, H.; Yang, C. Subsidence above gas storage in salt caverns predicted with viscoelastic theory. *J. Nat. Gas Sci. Eng.* **2022**, *103*, 104620. [CrossRef]
31. Li, P.; Li, Y.; Shi, X.; Zhao, K.; Liang, X.; Ma, H.; Liu, K. Compaction and restraining effects of insoluble sediments in underground energy storage salt caverns. *Energy* **2022**, *249*, 123752. [CrossRef]
32. Ran, L.; Liu, S.; Wanyan, Q.; Yao, E.; Bai, S. Optimization and Evaluation of Chemical Shrinking Agent for Deposits in Salt Cavern Gas Storage. In Proceedings of the ASME 2021 40th International Conference on Ocean, Offshore and Arctic Engineering, Virtual, 21–30 June 2021; American Society of Mechanical Engineers: New York, NY, USA, 2021; Volume 85208, p. V010T11A057.
33. Li, J.L.; Shi, X.L.; Wang, T.T.; Yang, C.H.; Li, Y.P.; Ma, H.L.; Ma, X.Q.; Shi, H. A prediction model of the accumulation shape of insoluble sediments during the leaching of salt cavern for gas storage. *J. Nat. Gas Sci. Eng.* **2016**, *33*, 792–802. [CrossRef]
34. Yang, Y.; Li, L.; Yu, W.; Zhou, Y.; Zhu, K.; Yuan, B. The Application of Breakthrough Pressure in the Evaluation of the Sealing Ability of Cement–Casing Interface and Cement Matrix in Underground Gas-Storage Wells. *Processes* **2022**, *10*, 620. [CrossRef]
35. Molíková, A.; Vítězová, M.; Vítěz, T.; Buriánková, I.; Huber, H.; Dengler, L.; Urbanová, I. Underground gas storage as a promising natural methane bioreactor and reservoir? *J. Energy Storage* **2021**, *47*, 103631. [CrossRef]
36. Tiwari, S.; Namsani, S.; Singh, J.K. Effect of salt on the adsorption of ionic surfactants at the air-water interface. *J. Mol. Liquids* **2022**, *360*, 119498. [CrossRef]
37. Sultana, H.; Bokhari, T.H.; Usman, M. Adsorptive micellar flocculation (surfactant-based phase separation technique): Theory and applications. *J. Mol. Liquids* **2021**, *323*, 115001. [CrossRef]
38. Bakhtiari, M.; Shad, S.; Zivar, D.; Razaghi, N. Coupled hydro-mechanical analysis of underground gas storage at Sarajeh field, Qom formation, Iran. *J. Nat. Gas Sci. Eng.* **2021**, *92*, 103996. [CrossRef]
39. Farhadian, A.; Naeiji, P.; Varfolomeev, M.A.; Peyvandi, K.; Kiiamov, A.G. Reconsideration of the micellization theory: Promotion or inhibition of gas hydrate formation for gas storage and flow assurance applications. *Chem. Eng. J.* **2022**, *427*, 131852. [CrossRef]

processes

MDPI

Article

Effect of Shear Flow on Drag Reducer Performance and Its Microscopic Working Mechanism

Zhiyu Liu [1], Zaifu Tian [2], Haoren Yuan [2], Yuan Li [1], Hongkui Ge [1,*] and Fujian Zhou [1,*]

[1] State Key Laboratory of Petroleum Resources and Engineering, China University of Petroleum, Beijing 102249, China; zhiyuliu@cup.edu.cn (Z.L.); 2017310210@student.cup.eud.cn (Y.L.)
[2] Oil Production Technology Research Institute, Shengli Oilfield Luming, China Petrochemical Corporation (Sinopec Group), Dongying 257000, China; giraffe8868@163.com (Z.T.); yuanhr1981@126.com (H.Y.)
* Correspondence: lzy8863766@163.com (H.G.); zhoufj@cup.edu.cn (F.Z.)

Abstract: As the development of unconventional oil and gas resources goes deeper, the stimulation of reservoirs goes deeper year by year. Flow in longer wellbores poses a challenge to the stability of drag-reduction performance of fracturing fluid. However, at present we have limited understanding of the mechanism of drag-reduction damage caused by shear flow, especially the microscopic mechanism. Therefore, in this work, the variation pattern of drag reducer solution performance with shear rate has been analyzed by using a high precision loop flow drag test system. The test results show that there is a critical shear rate for the performance damage of the drag reducer solution, and high strength shear flow and cumulative shear flow time are the main factors leading to the performance degeneration of the drag reducer. Based on the nanometer granularity distributions, rheological properties and microscopic structures observed with a transmission electron microscope of drag reducer solutions subjected to shear flows of different velocities, it is confirmed that the damage to the microscopic structure of the solution is the main reason leading to its performance degeneration. The destruction of the microscopic structure causes the drag reducer solution to degrade in non-Newtonian characteristics, so it becomes poorer in its capability of reducing turbulent dissipation and drops in drag-reduction capability. This research can provide a reference for improving and optimizing drag-reduction capability of fracturing fluid.

Keywords: drag reducer for slick water fracturing fluid; shear failure; grain size distribution; rheological properties; micro-mechanism of shear failure

Citation: Liu, Z.; Tian, Z.; Yuan, H.; Li, Y.; Ge, H.; Zhou, F. Effect of Shear Flow on Drag Reducer Performance and Its Microscopic Working Mechanism. *Processes* **2022**, *10*, 1485. https://doi.org/10.3390/pr10081485

Academic Editors: Linhua Pan, Minghui Li, Lufeng Zhang, Jie Wang, Wei Feng and Yushi Zou

Received: 13 July 2022
Accepted: 24 July 2022
Published: 28 July 2022

1. Introduction

Drag reducers are a kind of key additive in fracturing fluid for large scale hydraulic fracturing [1], which can reduce the friction of fracturing fluid flowing from ground surfaces to reservoirs effectively. As reservoirs developed become deeper year by year, the performance degeneration of drag reducers caused by flow shear degradation during flow has been becoming worse [2]. Currently, drag reducers commonly used are polymer type [3], so examining the variation patterns of polymer drag reducer performance caused by shear flow, the working mechanism and control mechanism of shear flow degradation is of great significance for improving the stability of drag reducer and fracturing effect.

The drag-reduction phenomenon is also called TOMs; that is, the addition of a small amount of polymer into water can reduce the flow resistance of water significantly [4,5]. Drag-reduction performance is affected by a number of factors. The degeneration of drag-reduction performance caused by mechanical degradation has always been a focus of study [6]. Earlier studies are the basis of subsequent ones. A lot of studies have shown that in the course of mechanical degradation of polymer, there might be break in the polymer chain structure. Among them, Merril and Horn proved that turbulent flow could cause the polymer drag-reduction capability to degenerate through straight pipe flow experiments, and concluded that structure breaking happened mostly at the midpoints of large molecules [7,8]. On this basis, Odell found through experiments that when subjected to shear higher than a critical rate, isolated large molecules would drop to half of their original molecular weights [9–11]. Vanapalli et al. verified the theory that the break of large molecules at midpoints caused mechanical degradation of polymers by chromatographic analysis [12]. After a large number of physical experiments, Brostow et al. proposed a theoretical model of molecule degradation in turbulent flow, which fitted well with experimental data [13]. Kim et al. investigated mechanical degradation of polymer with a novel rotating disc device and found that the degradation degree of a solution was affected by the solubility of the solvent [14–16]. Shanshool et al. focused their study on the concentration of drag reducer and found that high concentration solutions were not susceptible to mechanical degradation [17,18]. Research that focused on empirical equations of mechanical degradation is represented by the studies of Hénautet (2012) and Pereira (2012) [19,20]. Studies on how to prevent mechanical degradation are represented by the one by Zadrazil et al., 2012 [21]. All these studies promoted the study of mechanical degradation of drag reducer solutions. In recent years, on the basis of previous research results, Edson J examined the variation patterns of drag reducer performance due to mechanical degradation with concentration, molecular weight, temperature, and Reynolds number [22,23]. By introducing the abrupt contractions method into their study, Ivanor M. et al. found the relationship between drag-reduction rate and molecular weight and proposed the corresponding drag-reduction model [24]. Mohamm M. et al. explored the variation patterns of viscoelastic properties of drag reducer solution when affected by mechanical degradation, and concluded that drag-reduction performance was not affected by the small amplitude of oscillatory shear, but was related to continuous tensile failure [25]. The above studies have provided references for and promoted the study of mechanical degradation of drag reducer solution.

In the process of reservoir stimulation fracturing, fracturing fluid with the drag reducer additive is often injected into the target formation to create the hydraulic fractures by fracturing pumping trucks with a high flow rate and high shear rate, which can significantly affect the drag-reduction efficiency performance. However, the effect of flow shear rate on drag-reduction efficiency of the drag reducer has rarely been investigated due to the lack of a fluid pipe flow device with high shear rates. The novelty of this work is to build a novel pipe flow device with a high shear rate to investigate the relationships between the drag-reduction rate of the drag reducer solution and the shear rate under different flow conditions. Moreover, the effects of shear rate on grain size of agglomerates and rheological properties of drag reducer solution were also analyzed. In addition, the SEM was conducted to investigate the effects of shear rate on the microstructure of the drag reducer.

In this work, the relationships between the drag-reduction rate of the drag reducer solution and the shear rate under different flow conditions were researched by pipe flow experiments. Our results show that shear degradation of the drag reducer occurs at the high flow shear rate stage. The SEM results demonstrated the relationship between the degeneration of drag-reduction performance and mechanical degradation of the polymer drag reducer. Concerning the mechanical element of drag-reduction degeneration, it is concluded that the destruction of net-like structure makes the polymer solution weaker in non-Newtonian fluid features (a drop in viscoelasticity), increases the turbulent dissipation, and causes deterioration in drag-reduction capability. This result will provide a useful reference for the field application of the drag reducer when it is injected during hydraulic fracturing. The results will provide construction guidance for pumping the temporary plugging agent and its fracturing fluid in the field, helping to avoid the risk of drag reducer degradation at high shear rates.

2. Experiment Apparatus and Materials

2.1. Experiment Apparatus

In the experiments, a Zetasizer Nano ZS was used to analyze the distributions of grain sizes of polymer agglomerates in the drag reducer solutions, a Hakke MARSIII rheometer was used to test the rheological properties of the drag reducer solutions, and a JEM-100 transmission electron microscope made by JEOL was used to observe the micro-structures of the drag reducer solutions. The drag-reduction tests of drag reducer solutions were carried out in a high precision loop flow drag test system. With this system, the drag-reduction performance of a solution under the effect of shear flow can be tested under the circulation of fluid. The loop flow drag test system, modified from the friction tester, consists of a fluid supply system, an experimental pipe system, a pressure test system, and a data collection system. The structure of the loop flow drag test system is shown in Figure 1.

Figure 1. Loop flow drag test system.

To improve the accuracy of drag-reduction rate test, the fluid supply system and power delivery unit of the friction test equipment were modified. The fluid supply system after modification is composed of 3 tanks, which are used to store the solution to be tested, the back-flow solution, and fresh water, separately, to eliminate the interference between tested fluids. The power delivery unit was modified into a screw pump with a maximum pumping rate of 2.5 m^3/h to effectively eliminate the shear failure in the polymer solution caused by the turbine pump. The flow drag test system can measure the drags of fresh water and drag reducer solution under the same flow conditions accurately, thus enhancing the precision of experimental measurements. The pipe system mainly consists of three stainless steel pipelines of 6 mm, 8 mm, and 10 mm in diameter, respectively, and 3 m in length each. To minimize the additional effect from the corners connected to each pipeline, pressure measuring points are set at 0.25 m away from both ends, so each pipeline has an effective test length of 2.5 m. Pressure drop in the pipeline is measured with the differential pressure transducer connected between two pressure test points. As the pipelines of different diameters may vary widely in flow drag pressure drop, two differential pressure transducers of different ranges, 5 MPa (0.5–5.0 MPa, $\pm$0.1 kPa) and 0.5 MPa (0–0.5 MPa, $\pm$0.02%), are installed. In actual tests, the differential pressure transducer that meets the measurement accuracy requirements can be switched according to the pressure drop in flow in the pipeline.

2.2. Experimental Materials

The drag reducer used in the experiments was a kind of anion polymer emulsion, DR-800, made by Shengli Petrochemical Company Ltd. (Zibo, China). The effective component of the drag reducer is polyacrylamide copolymer synthesized from a mixture of acrylamide (AM), acrylic acid (AA), 2-acrylamido-2-methylpropanesulfonic acid (AMPS), and butyl acrylate (BA) in aqueous solution by polymerization process. The drag reducer solution used in the experiments was 0.05% in volume concentration.

3. Experimental Principle and Method

3.1. Experimental Principle

Drag reducer solution can make the flow resistance of fluid drop significantly when entering the flow pipe. This is shown as an increase in pumping rate and decrease in the friction pressure drop in the fluid. When the fluid in the pipe is pumped at a constant pressure, the effect of drag reduction is shown as increase in the fluid flow rate; when the fluid is pumped at a constant flow rate, the effect of drag reduction is shown as a decrease in friction pressure drop.

To better characterize this physical phenomenon, drag-reduction rate was introduced to evaluate the drag-reduction performance of the drag reducer solution. The drag-reduction rate (*DR*) is the most significant physical parameter in the present study; it is the physical parameter used to estimate the drag-reduction degree in the different shear rates of flow. As the drop in flow resistance of the fluid in the pipe is the result of the decrease in the flow friction coefficient in essence, the drag-reduction rate is defined as the reduction rate of the friction coefficient:

$$DR\% = \frac{\lambda_0 - \lambda_{DR}}{\lambda_0} \times 100\% \tag{1}$$

where:

$DR\%$—Drag-reduction rate;

λ_0—Friction coefficient of fresh water with no drag reducer added;

λ_{DR}—Friction coefficient of drag reducer solution.

In the experiments, it is difficult to measure the friction coefficient directly, so concerning the general formula of flow resistance in the straight pipe, the Fanning equation is adopted:

$$\lambda = \frac{2D}{\rho u^2 L} \Delta P \tag{2}$$

where:

D—Inner diameter of the pipe, m;

P—Fluid density, kg/m^3;

u—Flow velocity of the fluid, m/s;

L—Pipe length, m;

ΔP—Friction pressure drop in the fluid, Pa.

For experiments in the same pipe, both D and L are constant values. Meanwhile, as the amount of drag reducer added is very small, the density variation of the fluid after addition of the drag reducer is negligible. Therefore, under the same flow parameter (average flow velocity) of the fluid, the relation between the drag-reduction rate and the friction pressure drop in the pipe can be obtained by substituting Equation (2) into Equation (1):

$$DR\% = \frac{\Delta P_0 - \Delta P_{DR}}{\Delta P_0} \times 100\% \tag{3}$$

Hence, based on Equation (3), the friction pressure drops of the fluids with and without the drag reducer at two ends of the experiment pipe at the same flow velocity can be measured by differential pressure gauge, and then the drag-reduction rate of the drag reducer solution can be worked out by using the measured friction pressure drops.

In the experiments, mass flowmeter was used to measure the average flow velocity of fluid in the pipe. As the experiments aimed to find out the relationship between drag-reduction rate and shear rate of fluid flow, the relation between flow velocity and shear rate of fluid flow under a certain pumping rate was needed to calculate the corresponding shear rate of fluid flow. The formula for calculating shear rate of fluid flow was:

$$\gamma = \frac{4Q}{\pi R^3} = \frac{8u}{D} \tag{4}$$

where γ is the shear rate near the pipe wall, s^{-1}. As the shear failure of the fluid flow mainly concentrates near the pipe wall, this value is used to characterize the shear rate of the fluid flow. Based on the above principle, with the flow velocity of fluid regulated by the mass flowmeter, the friction pressure drops in fresh water and drag reducer solution and the flow velocity were measured, and then the relationship between drag-reduction rate and shear rate of fluid flow was worked out.

3.2. *Experimental Method*

3.2.1. Effect of Flow Shear on Drag-Reduction Performance of Drag Reducer

The loop flow drag test system was used to test the variation pattern of drag-reduction performance of the drag reducer solution with shear rate of fluid flow. A 30 L amount of the drag reducer solution at 0.05% concentration was prepared in the tank and blended evenly. By changing the pumping rate of the screw pump, pressure drops of the fluid in the pipe at different shear rates of fluid flow were measured. Then, the drag-reduction rates at different shear rates were calculated by using the pressure drop data to find out the variation pattern of drag-reduction rate with shear rate of fluid flow. Specifically, in the experiments, a frequency changer was used to change the pumping rate to make the shear rate of fluid flow in the pipe rise at a specific step length to the set value and then fall at the same step length, and this flow process was repeated several times. The effects of shear rate of fluid flow and cumulative shear time on drag-reduction performance of the drag reducer solution were analyzed based on the data obtained from the above experiments. To further analyze the effect of shear rate of fluid flow on drag-reduction performance, the drag reducer solution of the same concentration was prepared anew, with the maximum shear rate of fluid flow set at 2900 s^{-1} and 3500 s^{-1}, respectively; variation patterns of drag-reduction rate with shear rates of fluid flow were tested in different ranges of shear rates.

3.2.2. Effects of Flow Shear on Grain Size and Rheological Properties of Drag Reducer Solution

To find out the effect of fluid shear rate on agglomerate grain size and rheological properties of drag reducer solution, samples of drag reducer solution not subjected to shearing, drag reducer solution subjected to repeated low-rate shearing, and drag reducer solution subjected to repeated high-rate shearing were taken and coded as A, B, and C, respectively. Then, Zetasizer Nano ZS was used to obtain the grain size distributions of agglomerates in these samples. Finally, Hakke MARSIII rheometer was used to analyze the rheological characteristic curves of the three samples to find the relationship between agglomerate grain size and rheological behavior.

3.2.3. Effect of Flow Shear on Micro-Structure of the Drag Reducer Solution

The three samples were observed with a JEM-100CX transmission electron microscope made by JEOL. In the observations, the amplifications were kept the same as far as possible to compare the micro-structures of the polymers subjected to flow shear of different rates.

4. Results and Discussion

4.1. Effect of Flow Shear on Drag-Reduction Performance

The drag-reduction rates at different shear rates of flow were tested. The curve of the drag-reduction rate with shear rate (from 1000 s^{-1} to 6000 s^{-1}) of the 0.05% volume concentration drag reducer solution is shown in Figure 2.

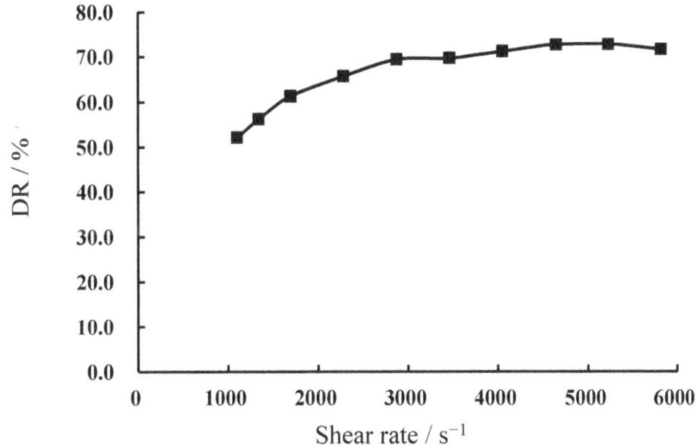

Figure 2. Relationship curve of drag-reduction rate with shear rate.

It can be seen from Figure 2 that with the gradual rise of shear rate the drag-reduction rate goes up gradually at first, but when the shear rate exceeds a critical value, the drag-reduction rate increases slowly and then even decreases. From this curve, an optimum drag-reduction rate can be found in the process of shear rate variations.

To verify this regularity, the drag-reduction rates of the drag reducer solution when the shear rate dropped from 6000 s^{-1} to 1000 s^{-1} (at the same variation interval when the shear rate rose) were tested. The tested results are plotted in the same chart of Figure 2 to enable a comparison, as shown in Figure 3.

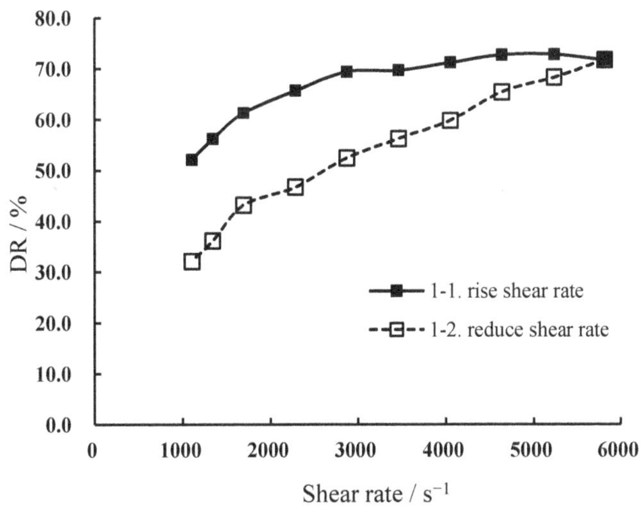

Figure 3. Curve of drag-reduction rate with rise and fall (at the same variation interval) of shear rate.

It can be seen from Figure 3 that the drag-reduction rate dropped significantly with flow shear, but the drag-reduction rate dropped different degrees at different shear rates. The curve shows large values in the left and smaller values in the right, indicating that the significant drop in drag-reduction rate appeared in the flow stages of the middle and left parts of the curve, while the drop in drag-reduction rate in the right side flow stage was small. This is because the long duration and the high rate of shear have a larger impact on the performance of the drag reducer solution and can lead to a considerable decrease in drag-reduction rate. However, when the cumulative shear flow time is short, the failure is not noticeable either. That is why the drag-reduction rate only dropped 6.4% at the shear rate of 5224 s^{-1}, while that at the shear rate of 1103 s^{-1} it dropped 35.3%; in other words, at the shear rate of 1103 s^{-1}, the drag reducer solution had experienced the whole shear process fortwo times longer than the process rise to the shear rate of 5224 s^{-1}, so the drag-reduction rate at this point dropped more significantly.

To verify this finding further, we repeated the flow experiment another three times. The shear rate in the four experiments varied in the same pattern and step length. The drag-reduction rates measured in the four experiments were plotted in the same chart (Figure 4).

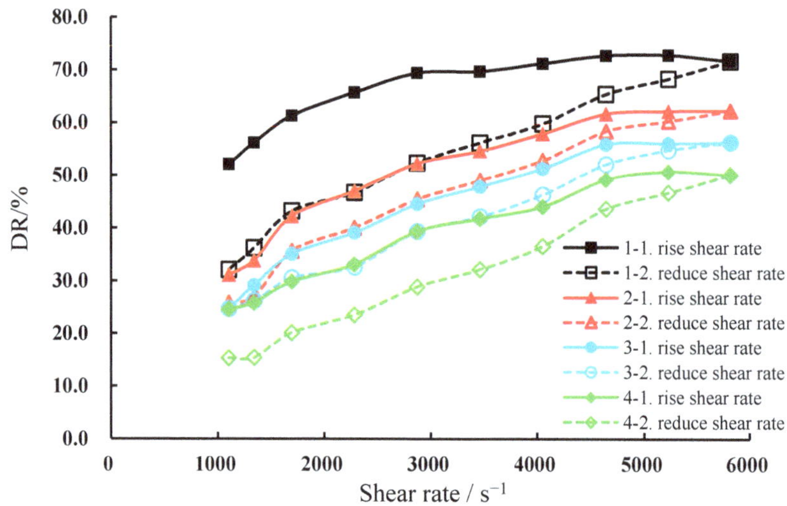

Figure 4. Variations of drag-reduction rate with shear rate from repeated experiments.

In Figure 4, curves of the same color represent the shear flow processes of rate rise and fall, respectively, in the same experiment; solid curves represent the process with shear rate rising gradually, and dotted lines represent the process with shear rate falling gradually. The curves from top to bottom show the drag-reduction rates measured in the first to fourth experiments, respectively. Comparison of the four groups of curves shows that each group of curves shows a variation pattern in drag-reduction rate similar to that mentioned above. Meanwhile, at high shear rate (5224 s^{-1}), all of them show a drop in drag-reduction rate of different degrees (3–10%). In particular, it is noted that the drag-reduction rate curve in the first shear rate decrease process almost coincides with that in the second shear rate increase process (the black dotted line and the red solid line sections at the shear rates between 1000 s^{-1} and 2900 s^{-1}). This means the performance of the drag reducer solution did not change much in this flow stage; in other words, the drag-reduction rate did not change much. Moreover, in the other repeated experiments, the relationship curves between shear rate and drag-reduction rate show similar features in the same flow stage. The drag reducer solution in this flow stage had experienced the nth time of high-rate shear, but did not experience the n + 1th time of high-rate shear. The coincidence of the two drag-reduction rate curves indicates that flow at low shear rates (less than 2900 s^{-1}) has little effect on the performance of the drag reducer. To sum up, high-rate shear may be the major factor leading to the degeneration of drag reducer performance; moreover, there is a critical value for the flow shear rate; only when the flow shear rate exceeds this critical value will the performance of the drag reducer be affected significantly. At flow shear rates over this critical value, longer cumulative shear time will make the performance of the drag reducer deteriorate further. To validate this finding, flow experiments at low flow shear rate (less than 2900 s^{-1}) and high flow shear rate (less than 3500 s^{-1} at maximum) were carried out. The experiments took the same shear rate variation pattern in the above experiments and were each repeated four times. To make the comparison of the experimental results easy, the drag-reduction curves were plotted with the method same as that of Figure 4.

4.1.1. Effect of Low-Rate Shear Flow on the Performance of Drag Reducer Solution

The variation pattern of the drag-reduction rate at low shear rate (less than 2900 s^{-1}) was further examined. The drag reducer solution at the volume concentration of 0.05% was prepared anew, and the same experiment method as that mentioned above was used. However, in this experiment, the maximum shear rate of flow was limited below 2869 s^{-1}, and the drag-reduction rate curves after four iterations of repeated shear flow are shown in Figure 5.

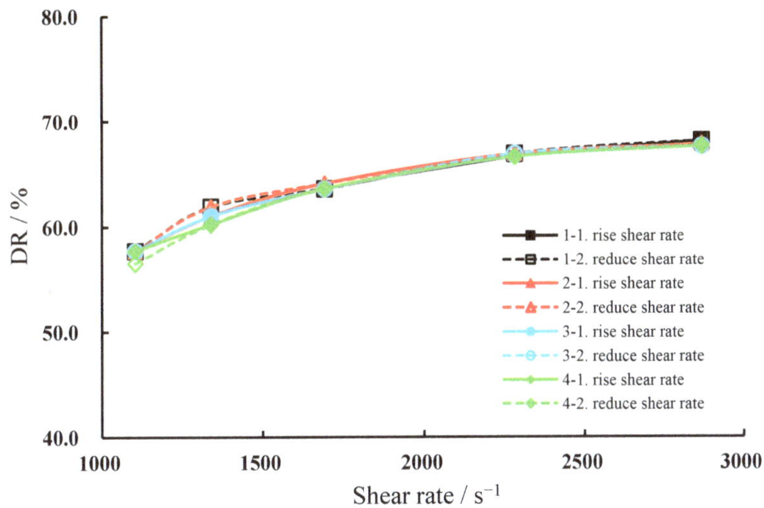

Figure 5. Variations of drag-reduction rate with low shear rate (controlled below 2900 s^{-1}).

Figure 5 shows shear rates below 2869 s^{-1}, the drag-reduction rate curves of the four repeated experiments basically coincide, and the drag-reduction rate variation at the same shear rate is only 1.5% at most. This means that with the maximum shear rate limited below 2900 s^{-1}, the drag reducer solution after four iterations of shear flow in the pipe caused little deterioration in drag-reduction performance.

4.1.2. Effect of High Shear Rate Flow on Performance of the Drag Reducer Solution

In this section, the variation pattern of drag-reduction rate at high shear rate of flow (3500 s^{-1} at maximum) was analyzed. Similarly, the drag reducer solution of 0.05% volume concentration was prepared anew, and the same experiment method as mentioned above was used. However, in this experiment, the maximum shear rate of flow was limited to 3458 s^{-1}, and the drag-reduction rate curves of four iterations of repeated shear flow are shown in Figure 6.

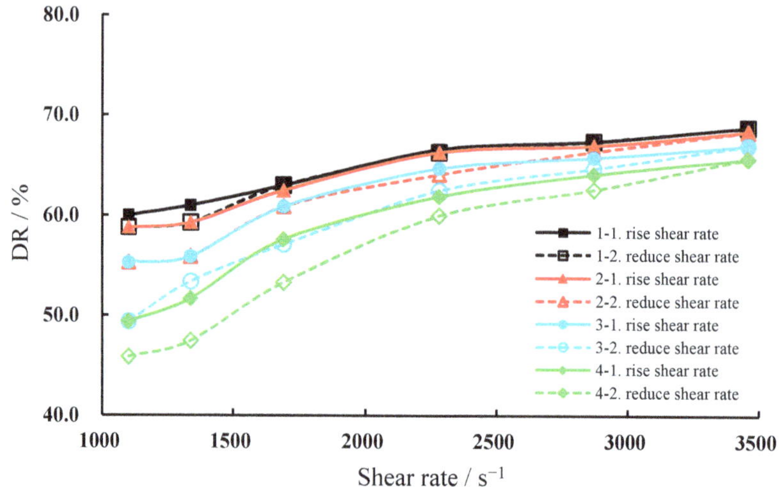

Figure 6. Curves of drag-reduction rate with shear rate (controlled below 3500 s^{-1}).

Figure 6 shows that with maximum shear rate was limited to 3458 s^{-1}; the drag-reduction rate varies in its pattern similar to that in the first flow experiment. The drag-reduction rates tested after four iterations of shear flow dropped significantly in comparison with thefirst test. In particular, at the shear rate of 1103 s^{-1} (the black solid line and green dotted line in this figure), the drag-reduction rate dropped by 23.5%. That is to say the drag reducer solution caused significant degeneration in drag-reduction performance after being subjected to high-rate shear for certain iterations cumulatively.

According to the understanding of the drag-reduction mechanism from mainstream studies, it is believed that the specific structure of polymers in drag reducer solutions leads to the non-Newtonian behavior of the solutions, and in turn to drag reduction. Looking back at the flow experiments in this work, it is found that the stable polymer structure in the drag reducer solution is damaged after high shear rate flow, and only when the cumulative damage reaches a certain degree will the drag reducer solution become worse concerning drag-reduction performance. Moreover, the shear rate of flow has a critical value; only when the shear intensity exceeds this critical value will the structure of the drag reducer be damaged, leading to degeneration in the drag-reduction rate. If the shear rate does not reach the critical value, no matter how long the cumulative flow time, the drag reducer rate will not change much. To validate this conclusion, rheological experiments and micro-structure analysis were carried out to explore the matter more deeply.

4.2. Variation Patterns of Grain Sizes and Rheological Behavior of Drag Reducer Solution under Shear Flow

The laser nano-granularity analysis results of samples A, B, and C are shown in Figure 7.

Through analysis it can be seen that the sample A not subjected to flow shear had agglomerate sizes concentrated between 200 nm and 500 nm. With the increase in shear rate of flow, the peak sizes of grains in the solution gradually shifted toward the left of the axis; that is to say the sizes of the agglomerates in the solution decreased with the increase in shear rate. In comparison, the sample subjected to low-rate shear flow had a wider span of peak values of grain sizes, indicating that it was affected by shear flow, part of the agglomerates were broken into smaller agglomerates, but due to different degrees of effect, the agglomerates in the solution ranged between 70 nm and 250 nm in size; in other words, they were smaller in value but wider in span in peak size. The sample subjected to high-rate shear flow had smaller agglomerates in the range between 30 nm and 80 nm; that is to say, the agglomerates in the solution had smaller peak size values in a smaller range.

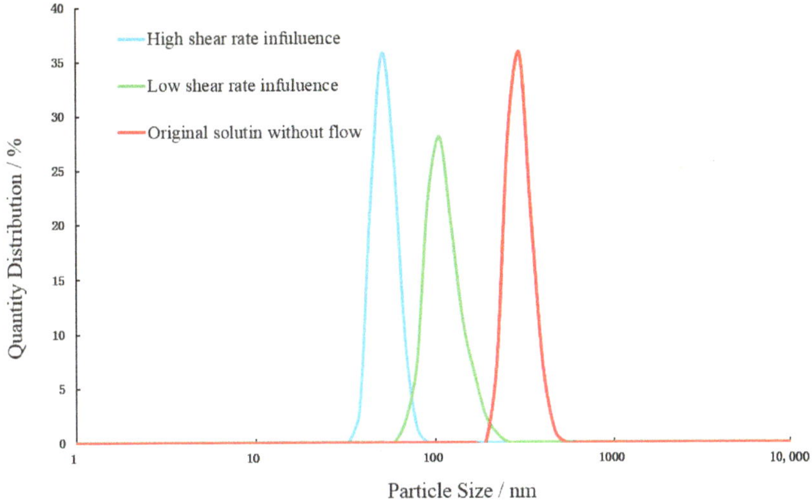

Figure 7. Distributions of grain sizes of the drag reducer solutions subjected to shears of different rates.

The rheological test results of the three samples are shown in Figure 8.

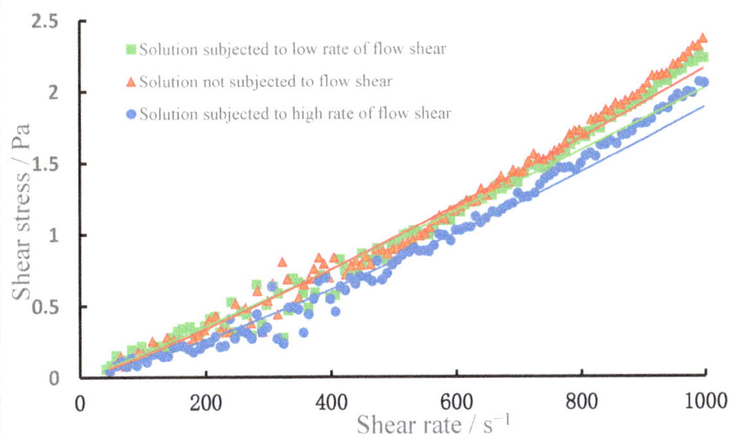

Figure 8. Rheological characteristic curves of the drag reducer solutions with shear rate.

By comparing the three curves of shear stress with shear rate, it is found that the three samples were similar in rheological behavior, all showing rheological features of power-law fluid (typical non-Newtonian fluid). When affected by shear flow, the drag reducer solution would turn weaker in power-law behavior and tend to be more like Newtonian fluid (fresh water). The more serious the shear failure of the solution, the weaker the non-Newtonian feature of the solution would be. Correspondingly, the solution decreased in viscosity and became weaker in viscoelasticity.

4.3. Variation Pattern of Micro-Structure of the Drag Reducer Solution

The micro-structure of the sample A not subjected to flow shear observed under electron microscope is shown in Figure 9.

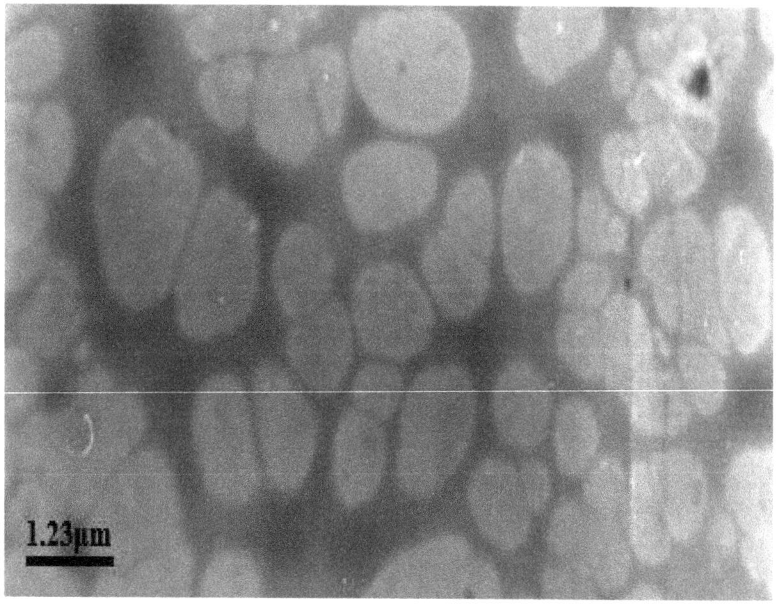

Figure 9. Micro-structure of the drag reducer not subjected to flow shear.

The darker parts in Figure 9 are polymer agglomerates observed. It can be seen from Figure 9 that most of the agglomerates were connected in a net-like structure, with obvious nodes observed. Some of the agglomerates were connected and extended into the mesh wall with a certain thickness, and the net-like structure was spread evenly in the solution. The stable agglomerate structure can improve the viscoelasticity of the drag reducer solution and reduce the turbulent dissipation during fluid flow effectively, thus lowering the flow friction.

The micro-structure of the sample B subjected to low-rate shear is shown in Figure 10.

It can be seen from Figure 10 that parts of the meshes were damaged, while the remaining parts of the meshes remained intact. The damaged meshes remained in the original distribution state but failed to form a complete wall structure and only the basic mesh shape remained. The drag reducer net-like structure in the solution changed in micro-structure after experiencing low shear rate flow. Although the main net-like structure remained, the wall of part of the meshes became thinner. Affected by shear stress of flow, the net-like structure was stretched, but as the shear rate of the flow was low, the shear stress did not destroy the net wall completely. Therefore, the drag reducer solution did not drop significantly concerning drag-reduction performance.

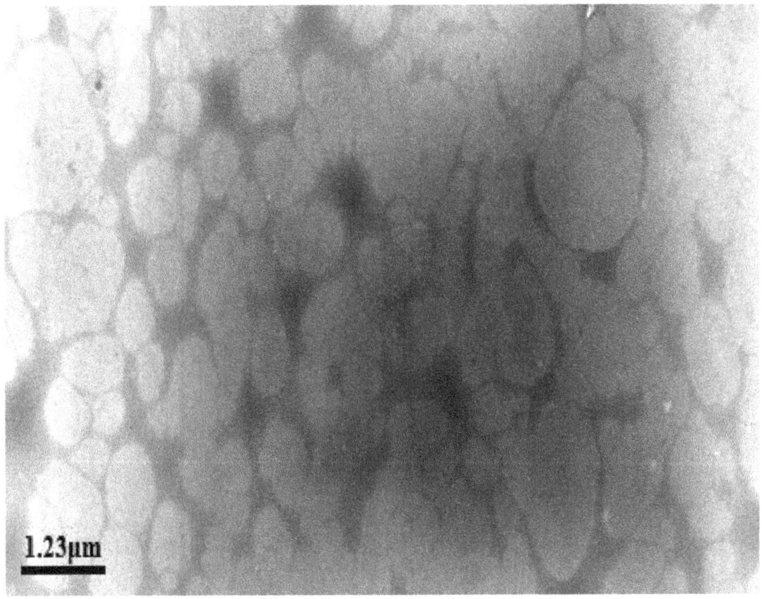

Figure 10. Micro-structure of the drag reducer after being subjected to low rate of flow shear.

The micro-structure of sample C experiencing high shear rate flow is shown in Figure 11, which is quite different from that of samples A and B.

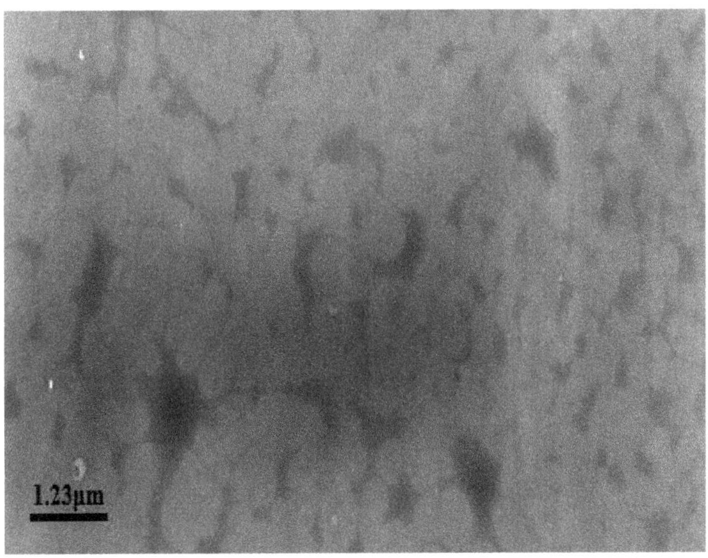

Figure 11. Micro-structure of the drag reducer after being subjected to high rate of flow shear.

It can be seen from Figure 11 that most of the net-like structure was destroyed, there were hardly any complete meshes, and most drag reducer agglomerates appeared in an isolated state. Moreover, the agglomerates were quite different in size and were distributed unevenly, and only a small number of nodes had an unstable structure connected by a small amount of stretched net-like structure.

In a word, from the above experiments of effect of flow shear rate on drag reducer micro-structure, it is concluded that the degeneration of drag-reduction performance of drag reducer solution is closely related to the destruction of the net-like structure of the drag reducer. The damage degree of the net-like structure of the drag reducer determines the degeneration degree of drag-reduction performance of the drag reducer solution, and the more severe the micro-structure damage, the more significant the drop in drag-reduction rate. However, after the damage reaches a certain degree (the net-like structure is completely destroyed), the drag-reduction rate tends to be stable again. The case in the low shear rate flow stage is different from that in the high shear rate flow stage. In the low shear rate flow stage, although the net-like structure is damaged to some degree, the enhancement of drag-reduction performance outweighs the negative effect of net-like structure stretch with the increase in flow velocity of the drag reducer solution. Moreover, part of the energy during structure stretching is stored in the drag reducer agglomerates, reducing the energy consumption, which is good for decreasing energy dissipation caused by turbulent flow. Therefore, in the initial stage, with the rise of flow velocity, the drag-reduction rate does not drop noticeably. However, even at low shear rate, a long time shear would result in damage the same as that of high shear rate flow, and finally the drag-reduction rate would become stable (that is, the drag-reduction rate after complete destruction of drag reducer structure).

5. Conclusions

In this work, the variation pattern of drag-reduction rate with shear rate was analyzed with a high precision loop drag test system. The reason why shear flow can influence the performance of the drag reducer was also explored. The distributions of grain sizes, rheological properties, and micro-structures of drag reducer solution samples were analyzed. On this basis, the mechanism of the high strength shear flow causing micro-structure damage, changes in rheological properties and degeneration of drag-reduction performance of the drag reducer solution were figured out. Through this research, the following conclusions have been reached:

1. With the increase in shear rate, the drag-reduction rate of the drag reducer solution will first rise to a peak value and then fall gradually. Then, with the increase in cumulative shear flow time, the drag-reduction rate will drop more significantly. Meanwhile, the shear rate causing drag-reduction drop has a critical value; only when the flow shear rate reaches this value will the performance of the drag reducer start to be damaged.
2. When subjected to flow shear higher than the critical rate, polymer agglomerates in the drag reducer solution will decrease in size. The higher the flow shear rate, the smaller the peak values of grain sizes and the higher the peak intensity of agglomerates in the solution.
3. The drag reducer solution shows typical characteristics of non-Newtonian fluid, which can be weakened by flow shear. The stronger the flow shear, the weaker the non-Newtonian fluid characteristics and the poorer the rheological behavior of the solution.
4. The damage of the net-like structure of the drag reducer is the main cause of degeneration of drag-reduction performance. Micro-mechanism research shows that the destruction of net-like structure caused by shear failure makes the polymer solution weaker in non-Newtonian fluid features (drop in viscoelasticity), leads to an increase in turbulent dissipation, and eventually results in the deterioration of drag-reduction capability.

Author Contributions: Conceptualization, H.G. and F.Z.; investigation, Z.T. and H.Y.; methodology, Z.L. and Y.L.; project administration, Z.L.; writing—original draft, Z.L.; writing—review and editing, Y.L. All authors have read and agreed to the published version of the manuscript.

Funding: This study is funding supported by Shengli Oil Field Luming Co., Ltd., "Research and application of temporary plugging technology in water fracture displacement of beach- bar sand reservoir water fracture displacement" (LMKJ2021-05), Xinjiang Deli Zhongcheng Petroleum Engineering Co., Ltd., "Research on new drag-reduction materials suitable for different fluid flow conditions" (ZC-20201209).

Data Availability Statement: The data presented in this study are available on request from the corresponding author.

Conflicts of Interest: The authors declare no conflict of interest.

References

1. Yang, B.; Zhao, J.; Mao, J.; Tan, H.; Zhang, Y.; Song, Z. Review of friction reducers used in slickwater fracturing fluids for shale gas reservoirs. *J. Nat. Gas Sci. Eng.* **2019**, *62*, 302–313. [CrossRef]
2. Muntasheri, A.; Critical, G.A. Review of Hydraulic Fracturing Fluids over the Last Decade. In Proceedings of the SPE Western North American and Rocky Mountain Joint Meeting, Denver, CO, USA, 17–18 April 2014.
3. Barati, R.; Liang, J. A Review of Fracturing Fluid Systems Used for Hydraulic Fracturing of Oil and Gas Wells. *J. Appl. Polym. Sci.* **2014**, *131*, 40735.
4. Toms, B.A. Some observations on the flow of linear polymer solutions through straight pipelines at large reynolds numbers. In Proceedings of the 1st International Congress on Rheology, Amsterdam, The Netherlands, 21–24 September 1948; Volume 2, pp. 135–141.
5. Toms, A. On the early experiments on drag reduction by polymers. *Phys. Fluids* **1977**, *3*, 20. [CrossRef]
6. Pereira, A.S.; Mompean, G.; Thompson, R.L.; Soares, E.J. Elliptical, parabolic, and hyperbolic exchanges of energy in drag reducing plane couette flows. *Phys. Fluids* **2017**, *29*, 115106. [CrossRef]
7. Horn, A.F.; Merrill, E.W. Midpoint scission of macromolecules in dilute solution in turbulent flow. *Nature* **1984**, *312*, 140–141. [CrossRef]
8. Merrill, E.W.; Horn, A.F. Scission of macromolecules in dilute solution: Extensional and turbulent flows. *Polym. Commun.* **1984**, *25*, 144–146.
9. Odell, J.A.; Muller, A.J.; Narh, K.A.; Keller, A. Degradation of polymer-solutions in extensional flows. *Macromolecules* **1990**, *2*, 23. [CrossRef]
10. Muller, A.J.; Odell, J.A.; Carrington, S. Degradation of semidilute polymer solutions in elongational flows. *Polymer* **1992**, *33*, 12. [CrossRef]
11. Nguyen, T.Q.; Kausch, H.H. Effects of solvent viscosity on polystyrene degradation in transient elongational flow. *Macromolecules* **1990**, *23*, 5137–5145. [CrossRef]
12. Vanapalli, S.A.; Ceccio, S.L.; Solomon, M.J. Universal scaling for polymer chain scission in turbulence. *Proc. Natl. Acad. Sci. USA* **2006**, *103*, 16660–16665. [CrossRef] [PubMed]
13. Brostow, W.; Ertepinar, H.; Singh, R.P. Flow of dilute polymer solutions chain conformations and degradation of drag reducers. *Macromolecule* **1990**, *23*, 5109–5118. [CrossRef]
14. Kim, C.A.; Kim, J.T.; Lee, K.; Choia, H.J.; Jhon, M.S. Mechanical degradation of dilute polymer solutions under turbulent flow. *Polymer* **2000**, *41*, 7611–7615. [CrossRef]
15. Kim, K.; Islam, M.T.; Shen, X.; Sirviente, A.I.; Solomon, M.J. Effect of macromolecular polymer structures on drag reduction in a turbulent channel flow. *Phys. Fluids* **2004**, *16*, 4150–4162. [CrossRef]
16. Kim, K.; Sirviente, A.I. Turbulence structure of polymer turbulent channel flow with and without macromolecular polymer structures. *Exp. Fluids* **2005**, *38*, 739–749. [CrossRef]
17. Shanshool, J.; Al-Qamaje, H.M.T. Effect of molecular weight on turbulent drag reduction with polyisobutylene. *NUCEJ Spatial.* **2008**, *11*, 52–59.
18. Shanshool, J.; Marwa, F.; Jabbar, A.; Sulaiman, I.N. The influence of mechanical effects on degradation of polyisobutylenes as drag reducing agents. *Pet. Coal* **2011**, *53*, 218–222.
19. Hénaut, I.; Glénat, P.; Cassar, C.; Gainville, M.; Hamdi, K.; Pagnier, P. Mechanical Degradation Kinetics of Polymeric DRAs. In Proceedings of the 8th North American Conference on Multiphase Technology, Banff, AB, Canada, 20–22 June 2012.
20. Pereira, A.S.; Soares, E.J. Polymer degradation of dilute solutions in turbulent drag reducing flows in a cylindrical double gap rheometer device. *J. Non-Newton. Fluid Mech.* **2012**, *9*, 179–180. [CrossRef]
21. Zadrazil, I.; Bismarck, A.; Hewitt, G.F.; Markides, C.N. Shear layers in the turbulent pipe flow of drag reducing polymer solutions. *Chem. Eng. Sci.* **2012**, *72*, 142–152. [CrossRef]
22. Edson, J.S. Review of mechanical degradation and de-aggregation of drag reducing polymers in turbulent flows. *J. Non-Newton. Fluid Mech.* **2020**, *276*, 104225.

23. Vincenzi, D.; Watanabe, T.; Ray, S.; Picardo, J. Polymer scission in turbulent flows. *J. Fluid Mech.* **2021**, *912*, A18. [CrossRef]
24. Martins, I.; Soares, E.J.; Siqueira, R.N. Mechanical scission of a flexible polymer (polyethylene oxide) under highly turbulent flows through abrupt contractions. *J. Non-Newton. Fluid Mech.* **2022**, *301*, 104740. [CrossRef]
25. Mohammadtabar, M.; Sanders, R.S.; Ghaemi, S. Viscoelastic properties of flexible and rigid polymers for turbulent drag reduction. *J. Non-Newton. Fluid Mech.* **2020**, *283*, 104347. [CrossRef]

MDPI
St. Alban-Anlage 66
4052 Basel
Switzerland
www.mdpi.com

Processes Editorial Office
E-mail: processes@mdpi.com
www.mdpi.com/journal/processes